The Minerals, Metals & Materials Series

Eric A. Ott · Joel Andersson · Chantal Sudbrack ·
Zhongnan Bi · Kevin Bockenstedt · Ian Dempster ·
Michael Fahrmann · Paul Jablonski ·
Michael Kirka · Xingbo Liu · Daisuke Nagahama ·
Tim Smith · Martin Stockinger · Andrew Wessman
Editors

Proceedings of the 10th International Symposium on Superalloy 718 and Derivatives

TMS

Springer

Editors
Eric A. Ott
General Electric
Cincinnati, OH, USA

Joel Andersson
University West
Trollhättan, Sweden

Chantal Sudbrack
National Energy Technology Laboratory
Albany, OR, USA

Zhongnan Bi
Central Iron and Steel Research Institute
Beijing, China

Kevin Bockenstedt
ATI Specialty Materials
Monroe, NC, USA

Ian Dempster
Wyman Gordon/PPC
Houston, TX, USA

Michael Fahrmann
Haynes International
Kokomo, IN, USA

Paul Jablonski
National Energy Technology Laboratory
Albany, OR, USA

Michael Kirka
Oak Ridge National Laboratory
Oak Ridge, TN, USA

Xingbo Liu
West Virginia University
Morgantown, WV, USA

Daisuke Nagahama
Honda R&D Co., Ltd.
Wakō, Saitama, Japan

Tim Smith
NASA Glenn Research Center
Cleveland, OH, USA

Martin Stockinger
Montanuniversität Leoben
Leoben, Austria

Andrew Wessman
The University of Arizona
Tucson, AZ, USA

ISSN 2367-1181 ISSN 2367-1696 (electronic)
The Minerals, Metals & Materials Series
ISBN 978-3-031-27446-6 ISBN 978-3-031-27447-3 (eBook)
https://doi.org/10.1007/978-3-031-27447-3

This Springer imprint is published by the registered company Springer Nature Switzerland AG
The registered company address is: Gewerbestrasse 11, 6330 Cham, Switzerland

Preface

Following with tradition since 1989, this edition of the International Symposium on Superalloy 718 and Derivatives continues the legacy of journal-quality, peer-reviewed, technical papers published concurrently with oral presentations at the in-person conference in Pittsburgh, Pennsylvania, USA. As the editorial team and conference organizers, we are proud to continue to offer a forum where the most relevant research, development, and application of superalloy materials of the family including alloy 718 can be shared with the technical community and can be archived for perpetuity in the form of this published electronic proceedings. The 2023 session is the 10th symposium in the series, and this proceedings volume provides 50 new papers representing the most recent advances of technical efforts in the field of superalloys adding to the more than 650 published papers in this overall series covering over 30 years.

While our experiences as a community over the last several years have led to strengthened electronic communication ties and channels, we look forward to this symposium's in-person gathering offering the opportunity to bring researchers and users together in one forum to discuss technical learnings, define new relationships, and renew and build peer networks. With the pace of progress in the technical field moving more and more quickly, we have chosen to expand access to the latest work by soliciting "late news abstracts" as part of the in-person conference. Although we are not able to publish detailed papers tied to these added topics due to timing, we expect that these poster presentations along with the core group of papers in these proceedings will help to add to and promote more collaboration and spirited technical debate in the field on high temperature superalloy materials. As a celebration of the long legacy created by our predecessors in the field, we also have added a special historical look at the progress of technology in the field in this proceeding's introduction from the perspective of one of our Keynote speakers from ATI. A look back through time helps us to appreciate where we've come from technically and helps to plot the future trajectory of the technology.

This proceedings volume represents contributions from all corners of the world. Author affiliations include participation from eight countries across three continents, and the demographics further underline the breadth of the perspectives that have historically made up these proceedings with academia, laboratories, and companies representing 40%, 15%, and 45% of the papers, respectively. We would like to thank our editorial and organizing committees for their investment in continuing this symposium series and would like to extend a special thanks to all of the paper, presentation, and poster authors for sharing their experiences and technical findings with us and the community. Finally to our symposium attendees, we want to extend our deepest gratitude for making the 718 Conference series a success for the tenth time.

Eric A. Ott
Lead Editor

Joel Andersson
Organizer

Historical Introduction: Meeting the Challenges of the Future by Understanding Our Past

Introduction

Understanding a material and its future performance in a potential application often begins by studying its history. If known, a detailed review of the processing it has undergone is most helpful, and if not available, an examination of its microstructure. While the correlation of structure, properties, and performance is not quite as connected in the evolution of an industry, understanding the past often provides significant insights to the potential of the future. It's important to review this history through a lens of understanding of what has been done, what was the outcome, and most importantly, why was it done. The "why" is critical to understanding the past limitations in the drive for continued advancement. This reflection brings an appreciation of how far the superalloys' history has come along with a humble realization of how far there is yet to go.

ATI's history is intertwined in the industrial advancement of superalloys to the ubiquitous applications powerhouse they are today. ATI was formed by merging Teledyne and Allegheny Ludlum in 1996. Teledyne owned two of the business units still part of ATI: Specialty Alloys and Components, then Wah Chang, and Specialty Materials, then Allvac. Teledyne also owned a slew of other companies, most of which were spun-off or sold.

Allegheny Ludlum was created in a 1938 merger between Allegheny Steel and Ludlum Steel. Both companies were pioneers in the steel industry with vestiges going back to the production of cannon balls for both the British and Continental Army in the Revolutionary War. Allegheny Steel primarily made flat products, was the first to use an electric arc furnace and supplied stainless steel for the Chrysler Building. Ludlum Steel produced steels primarily in bar, rod, and wire form and supplied for the construction of the Empire State building. The merged Allegheny Ludlum company initially focused primarily on specialty stainless steel products. However, they were active in the growing aerospace industry and had engineers focused on developing heat resistant alloys for jet engines, and methods to make them in production quantities. They ventured into numerous other alloys, including titanium (a joint venture with National Lead), Uranium (to support the US government) and Nickel. In 1950 they produced the first large (10 ton) heat of an aluminium and titanium strengthened superalloy in Watervliet, NY. This was followed three years later with the first production scale vacuum arc remelting. Allegheny Ludlum's remaining facilities operate under the name ATI Specialty Rolled Products.

Allvac was formed in 1957 as an "All-Vacuum" melting company by Jim Nesbit, a former General Electric engineer who believed that vacuum melting was critical to the continued advancement of nickel melting. The very first heat in tank 1 was Waspaloy for Pratt and Whitney blade bar. From the very beginning, Allvac's focus was clearly on clean melting for the aerospace and defense markets. By 1962 they had installed what was then the world's largest vacuum induction melting (VIM) tank at 12,000 lbs. capacity, a bet that the future of superalloys would require the cleanliness vacuum melting delivered. Allvac was sold to Vasco in 1965. Vasco Metals Corporation was acquired by Teledyne in 1966 and this unit now operates as ATI Specialty Materials.

The fourth current business unit of ATI, Forged Products, was an acquisition of the Ladish Company in 2010. Ladish began in 1905 when a supervisor for American Malting Company, Herman Ladish, started looking around for someone to make more reliable machine components. He partnered with John Obenberger, who ran a local forge shop, purchased a steam powered hammer forge, and began making axles and other forgings. While initially focused on small parts, by the 1930's they were making larger parts for industrial and farming applications as well as serving the burgeoning aerospace industry. They were making parts for jet engines by the mid-1940s.

Melting

The initial large melts of alloy 718 were air melted, including at Allegheny Ludlum. With this process melting is done with Argon Oxygen Decarburization (AOD) and may be followed by Electro Slag Remelting (ESR). The air melting process of the era lacked the chemistry control to handle the higher levels of titanium and aluminium additions. While there are much tighter controls in air melt today, the use is largely restricted to applications outside of the aerospace and aeroderivative industries. The introduction of vacuum melting provided both chemistry control and an increased cleanliness in the ingot to drive improved mechanical properties. Initial Vacuum Induction Melting (VIM) tanks did not have the ability to make additions during the melt process, Fig. 1. Ingots were static cast without remelting. As a result, the chemistry was far less controllable overall, and the chemical segregation was substantial. The shrink pipes were enormous (potentially dominating the length of the ingot), Fig. 2.

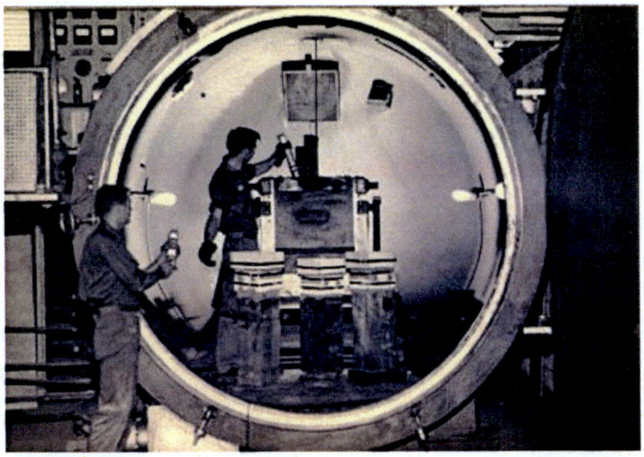

Fig. 1 Initial charging of a VIM tank in 1958

Fig. 2 Shrink pipe from a statically cast VIM electrode

Later VIM furnace designs included the ability to make chemistry additions. During melting, Nickel alloys are best thought of as a soup versus a cake. It's important to "taste" via in-process dip sampling to test the chemistry and make the appropriate modifications. This allows the chemistry to be dialed in prior to casting the liquid into a mold. Further improvements include using a hot topping procedure to insulate the upper portion of the electrode and slow down solidification, resulting in a smaller pipe cavity, thus a more solid electrode for further processing.

Remelting is critical whether via ESR, Vacuum Arc Remelting (VAR) or both. A schematic of the remelting processes can be seen in Fig. 3. An ESR remelt offers additional cleaning of the melt as the molten droplets go through a slag. During the exposure to the slag impurities can be removed via reaction, floatation, or dissolution.

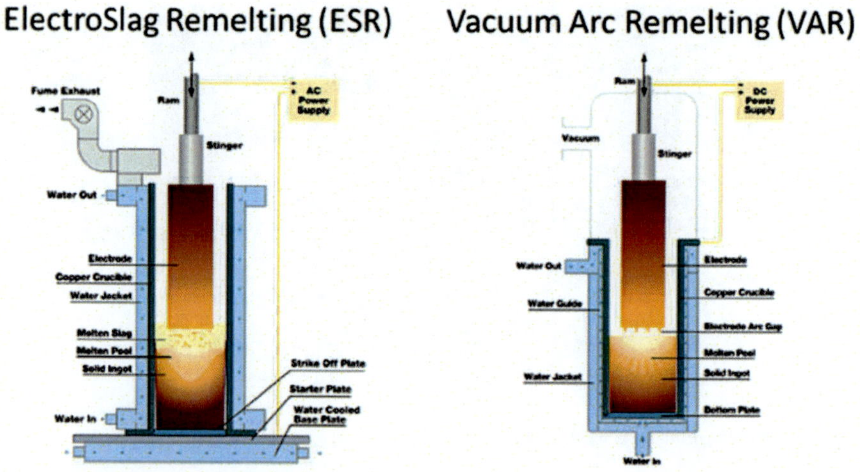

Fig. 3 Schematic of electro slag and vacuum arc remelting processes

Table 1 Comparison of electro slag and vacuum arc remelting processes

Electro Slag Remelting	Vacuum Arc Remelting
Removes sulfur	Removes gases, nitrogen retention is difficult
Oxidizing process, results in loss of reactive elements	Removes volatile elements, including Mn and Cu at higher concentrations
Can greatly improve cleanliness	Moderate improvement in cleanliness
Deeper melt pool due to slag skin	Shallower melt pool due to Helium cooling, if used
Less sensitive to white spot formation	More sensitive to white spot formation
More sensitive to freckle formation	Less sensitive to freckle formation
Higher melt rate	Lower melt rate

However, ESR has a slower solidification rate resulting from the slag insulating the molten pool, which can increase the propensity for some deleterious segregation. The VAR process, being under a vacuum, is considered best for removing gases and other volatile elements and has a faster solidification rate than ESR. A comparison of the advantages of both remelting options can be seen in Table 1.

Regardless of the melting process, the ability to control the process parameters has improved enormously. Remelting process control began as straight voltage control, with a crude aim point, to deliver constant current throughout the melt while manually adjusting the electrode height above the molten pool. Today's remelt furnaces typically run by controlling drip shorts, that is controlling the voltage and ingot gap and/or melt rate control which adjusts the voltage based on the weight of the hanging electrode. These control schemes allow engineers to reduce process variability by

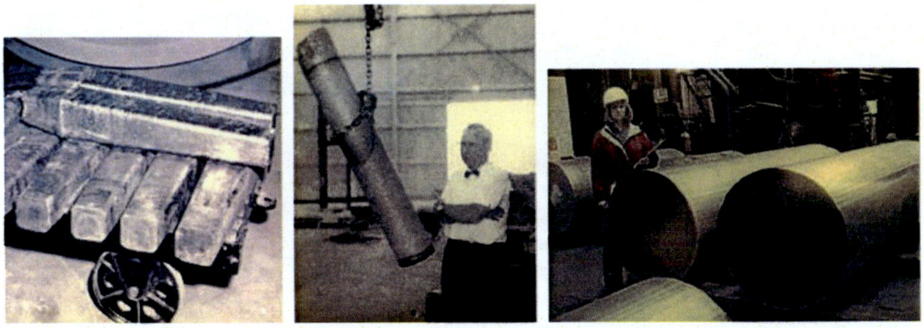

Fig. 4 Left to right: first VIM ingots, first VAR ingot, large 718 ingots for power generation

maximizing the time with consistent equilibrium solidification conditions. Furthermore, these controls have allowed the diameter of the ingots produced to increase substantially, making for a more cost-effective manufacturing process. Initially VIM ingots were small, 9" diameter, 42" long while VAR ingots grew to roughly 12" in diameter. Today 718 ingots can be made up to 36" inches diameter for power generation applications, Fig. 4. Melt anomalies nearly always relate back to an area of non-equilibrium solidification. As the ingot diameter grows, the requirements for stronger process control increase significantly, both for maintaining chemical homogeneity during the melt and in ensuring the integrity of the solidified ingot for downstream processing.

By the 1960s, in addition to the cast-wrought alloys and processes being developed there were also advances in powder-based superalloys. Initially driven by Pratt and Whitney, powder-based alloys have the advantage of less segregation due to high solidification rates achieved in small droplets that form the powder. While more costly, these alloys can include more complicated chemistries and thermomechanical processing paths. They also extend the temperature range for superalloys in jet engines, which have gotten progressively hotter during operation.

The criticality of controlling the melt process cannot be overstated. All future processing relies on the material behaving consistently in process and in application, and industry has been humbled when any defect has made it to an in-service product, including events with the tragic loss of human life.

Thermomechanical Processing

Much of the initial thermomechanical processing was developed to obtain a certain geometry for later stage processing. The mills available were limited in size, tonnage, and ability to reheat material so there was little ability to optimize the process. ATI had a simple reversing mill to convert ingots down to the standard rectangles and round-cornered squares required for forgings. There was no intermediate cutting, and a simple barrel furnace would warm the surface between passes. Once the material

Fig. 5 Left to right: initial blooming mill, early press, and radial forges installed at Allvac

went into the mill, it was committed. Furthermore, the mills weren't designed for the tonnages required for superalloy processing. As an example, all the rolls broke on one of the first mills Allvac acquired for blooming. The process was full of variability—and the engineers claim they could identify the mill operator by looking at the billet structure. All the operators were compliant with the process instructions, but it was clear there was significant impact from uncontrolled nuances within the processes. This drove an intense focus on process control and understanding, still a focus at ATI today. By the 1960s engineers were tracking every step in the process by hand, mining for differences—a time consuming and manual predecessor to today's drive to fully digitize. What they found drove tighter controls and tolerances on reduction sequences, transfer times and temperatures in all areas. By the 1970s, as the material got tougher with the introduction of new alloys and melt technology that allowed for production of larger ingots, mills with higher tonnage became necessary to process superalloys into useful forms. The company began using a press and radial forge as the forging industry was forced to shift to accepting round vs round-cornered square billet. Photos of the initial blooming mill, press forge and radial forge are shown in Fig. 5.

The shift to round billet was more than a geometry change. The use of new equipment required new recipes and a renewed focus on process variability to obtain the refined, homogenous structures needed for use.

Powder alloys were initially converted into parts from the as hot isostatically pressed (HIP) condition. This proved to be unsatisfactory for the fatigue strength needed for many part applications. As a result, a change was made for as-HIP'd cans to undergo billet conversion such as extrusion or press forging prior to the closed die part forging.

For making parts, Ladish had been forging Astrolloy and Waspaloy since the 1930s, primarily with conventional presses, hammers, and ring rolling equipment. As more complex and powder-based superalloys were developed, controlling the microstructure was the goal. Parts were made by a combination of isothermal forging (where the die and billet are at the same temperature), hot die forging (where the die is heated but not quite as hot as the material) and conventionally forging (colder die) to get the necessary ultrasonic transparency. At the same time parts were growing substantially. While initially billets were 4 inches in diameter, newer engines required

6- and 10-inch billets. With a billet that large, or larger, the chemical homogeneity and fine microstructure play a significant role. The structure is critical to obtain something that is inspectable. In the 80s the typical ultrasonic requirement was a #5 (5/64") flat bottomed hole. Today powder billets are inspected to ~10% of a #1 (1/64") flat bottomed hole. ATI's vertically integrated structure offers advantages for material inspectability as the process history from melt or powder to billet, and to part is readily shared. This has resulted in significant leveraging of learnings across the full ATI value chain.

In addition to parts getting larger, the forged discs got more complicated. The industry moved from a "there's a part in there somewhere" approach to one that cared deeply about material yield and the control of the condition and shape of the part for forging, heat treatment and ultrasonic inspection. Near-net shape forging with minimal material waste and consistent processes became the key focus. This requires expert die design and know-how. Initially there were no computers to help with forging design. Engineers used wood and plasticine models to watch die fill and lap formation. With the growth of computing resources, engineers could focus on temperature rise, strain, tonnage, last areas to fill as well as lapping and folding, but this did not provide a fast solution. A model could run for days for a single axisymmetric disc evaluation. Today's models and advanced computer resources are much faster.

As the aeroengine industry has largely moved their design focus from microstructural control to designs which minimize the defect flaw size the superalloy forgings used in discs are primarily hot die or isothermally forged. Some, but far fewer, are still conventionally forged 718 parts. Just as a forging billet requires homogeneity and fine structure, the requirements for final part structures have become more specific. The industry has improved its ability to design and control the forging and heat-treating process. Designs moved to be more damage tolerant by requiring a coarser grain size to improve crack propagation. This requires heat treating above the solvus and a tight control over cooling. The SuperCooler shown in Fig. 6 was designed in the late 90s to meet this need. The device allows for tailored grain sizes and reduces residual stress in a given location. The equipment controls the cooling rate over a broader range than liquid quenching or ducted air nozzles. This precise, computer-controlled cooling profile results in improved mechanical properties, reduced residual stresses and higher yield since the as-forged shape requires less additional machining.

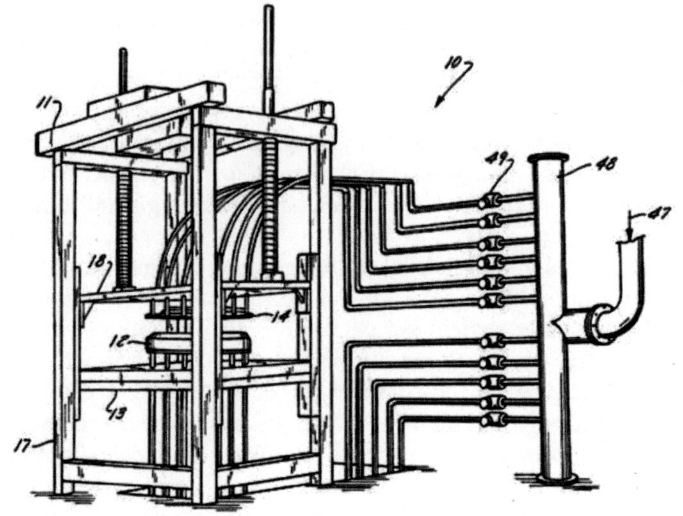

Fig. 6 Schematic of the supercooler

Computational Process Simulation and Modeling

Modeling has been used since the early stages of metallurgical working and the growth of computing power and characterization tools has only strengthened its role. Physics-based models can be more accurate, but they are time consuming, expensive, and challenging to produce. Advances in machine learning have allowed for complementary empirical models that may accelerate physics-based models by narrowing their focus to targeted areas. The expanding development and use of new characterization tools such as electron backscattered diffraction (EBSD) help reveal the impact of processing far beyond an average grain size. Tool advances help drive understanding as the debits in properties are usually related back to a local anomaly, a larger grain, area of segregation, etc. The ability to see and measure these rarities drives the ability to reduce, and ultimately eliminate them.

Compositional modeling has grown with the ability to leverage combinatorial methods to predict properties. High-throughput automated methods allow for rapid screening to confirm results. These methods were initially used primarily for chemical composition to strengthen Calculation in Phase Diagrams (CALPHAD) thermodynamics databases. Today there are automated methods to look at microstructures, suggest process conditions and test mechanical properties of materials. Much of this remains in research laboratories, but industry has implemented automated testing and analysis (mechanical, or microstructural) of materials. ATI uses Monte Carlo simulations to understand process sensitivity around chemical composition, for example. This allows an understanding of how tight the processing window needs to be and what is controllable to meet the specification requirements.

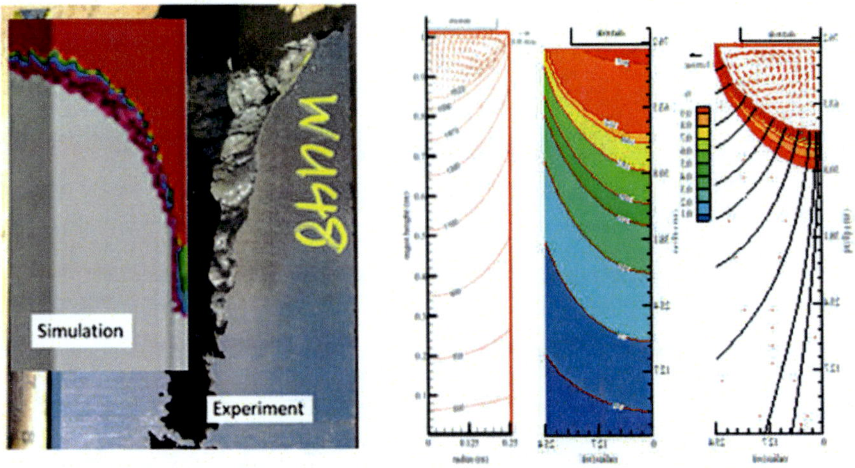

Fig. 7 Left to right: simulation and physical trial of a VIM melt and simulation of a VAR melt

The job of the melter is to be clean and maximize throughput by going fast in large diameters. However, advanced alloys push melting processes slower and in smaller diameters to maintain cleanliness and chemical homogeneity. Two dimensional, and some three-dimensional models exist. Fig. 7, left, shows the modeling results and physical trial side by side to predict the shrink pipe of a VIM electrode. On the right side, Fig. 7 shows the temperature and melt pool shape of a VAR ingot.

The models can get quite sophisticated balancing the complicated dynamics in a melting process and can be used to develop melt profiles or aid in a defect investigation. There continues to be room for advancement in melt modeling. It is known that equilibrium solidification is key to a clean melt but there remain portions of the melt process and melt pool that are invisible today. The ability to link a process controller directly to the melt pool would be powerful.

Process modeling is routinely used to evaluate the conversion of ingots to fine uniform grain-sized billet. The thermomechanical history as well as grain size evolution can be predicted by this means. Figure 8 shows an example of temperature monitoring during open-die modeling. Billet conversion modeling requires the stringing together of a long sequence of heating and deformation simulation steps. While software tools exist that have made this up-front process definition simple, there continues to be room for improvement. This includes a realistic definition of the sequence of operations (for example matching elapsed times and die speeds), as well as limitations on the size of finite element mesh that will yield reasonable run times.

At ATI every new part forging is modeled. Whether at a high level to quote an opportunity or for a detailed process design to target specific properties in the final product. Advances in computer time now allow dozens, or even hundreds of models to be completed prior to a single push. This saves time and money. Furthermore, the use of automated recipe-based production systems ensure that the engineer's

Fig. 8 Open die modeling of ingot conversion processes: upsetting, cogging, and radial forging

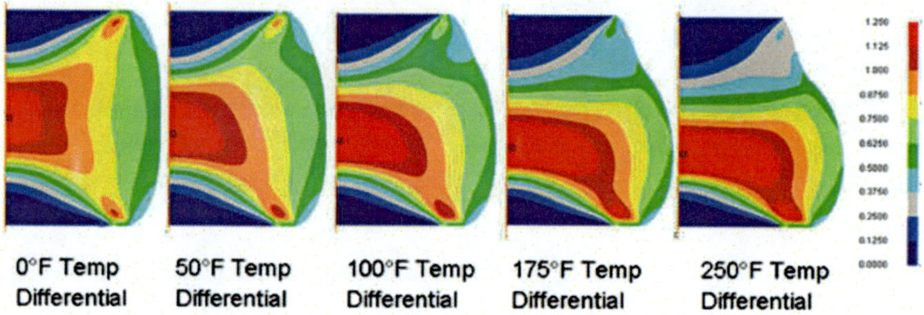

Fig. 9 Impact of temperature non-uniformity in a forged part

intended process is what the part sees, reducing variability. Our goal at ATI is to get to the point where we can accurately model everything we do. When combined with doing everything in a consistently repeatable way at production scale it has the potential to allow the creation of the ultimate digital twin containing all the process information from the first melt to the final machine operations. Figure 9 shows the impact of temperature non-uniformity in a forging process.

Advances in modeling have also shown the benefit of concurrent engineering. Bringing together the system designer and the part fabricator allows for an efficient process to produce a manufacturable part in the most cost-efficient manner. Rather than designing in silos, working together can allow for part reduction, modest adjustments for a big benefit to both parties and a better understanding of intent and need. Materials suppliers, including ATI, often work closely with jet engine manufacturers years before their new engines are introduced to ensure manufacturability of the parts and components.

Additive manufacturing continues to bring growth in alloy 718 and its derivatives. Opening the doors to integrate and model new shapes and designs that were not possible before, along with a whole new process to understand. The potential to reduce weight and handle, or change, loading conditions not just on one part but potentially on an entire system is huge. Industrial uses of new processes require a pay-off, but we are seeing clearly today that may come less from cheaper parts, although possible, and more from better performance at the part or system level.

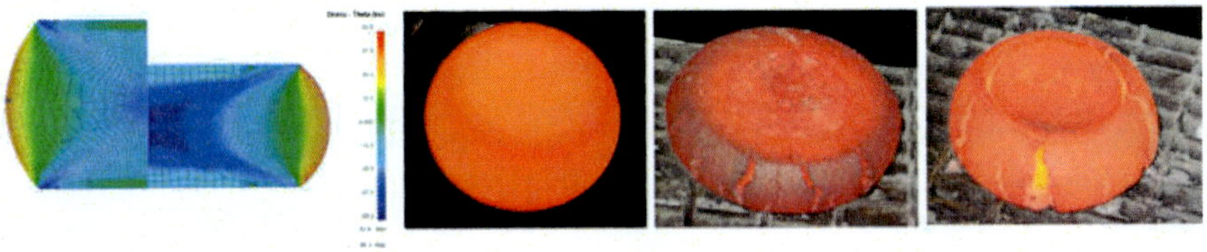

Fig. 10 Finite element modeling compared with physical trial results

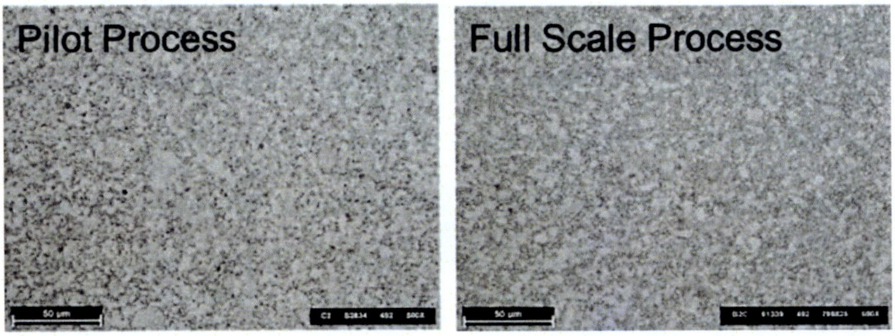

Fig. 11 Microstructures taken from a subscale powder billet (left) processed on pilot scale equipment designed to simulate full size billets processes on production equipment (right)

Finally, this paper would be incomplete without stressing the importance of physical experiments performed at the pilot scale. While modeling is fast, it requires physical validation. ATI's experience shows that the proper design of pilot scale experiments can rapidly, and cost effectively, advance materials to production sizes and specification requirements. Finite element modeling can accurately predict the shape, thermal profile, adiabatic heating, and strain. However, these models are not yet able to fully capture the probability of cracking related to differences in reduction and die dwell as shown in Fig. 10.

Pilot scale trials are most valuable when they can both combine with modeling and represent an accurate simulation of the production process. Microstructural results are shown in Fig. 11 comparing the results of a subscale powder billet processed on ATI's pilot scale equipment compared to the results obtained from a full-size billet on ATI's production equipment. The microstructures showed a difference of only 0.5 ASTM grain size (12.5 at subscale versus 13 at full scale). The value of a strong pilot scale process laboratory cannot be overstated.

Conclusions

A rigorous focus on consistent, repeatable processing at scale is important to ATI, and other materials suppliers. Knowledge and understanding of materials processing is compiled into models and pilot scale work to advance even the most challenging derivatives. The improvements seen in the sixty-plus years of alloy 718 and its derivatives are directly tied to improvements in process control and understanding. It is worthwhile to revisit the history of these developments periodically to remove artificial constraints. Advances in understanding, inspectability, measurability, and controllability allow a deeper understanding and highlight more aspects to evaluate.

Melissa Martinez
ATI
Dallas, TX, USA

Acknowledgments This paper would not have been possible without the many conversations and documents provided by ATI engineers and historians, in particular: Tal Harris, Dan Kahlke, Henry Lippard, Erin McDevitt, Vikas Saraf, Ramesh Minisandram, and David Bergstrom. While they provided significant historical references, photos, and images, any errors or omissions are the author's alone.

References

1. Donachie, M. and Donachie, S., *Superalloys, A Technical Guide*, ASM International Second Edition, 2002.
2. Kracke, A., Superalloys, the Most Successful Alloy System of Modern Times-Past, Present, and Future, 7th International Symposium on Superalloy 718 and Derivatives, 13-50, (2010), https://doi.org/10.1002/9781118495223.ch2.
3. Boyce, B.L., Uchic, M.D. Progress toward autonomous experimental systems for alloy development. *MRS Bulletin* **44**, 273–280 (2019). https://doi.org/10.1557/mrs.2019.75.
4. McDevitt E (2010) Effect of temperature and strain during forging on subsequent delta phase precipitation during solution annealing in ATI 718plus® alloy. In: Paper presented at the 7th international symposium on superalloy 718 and derivatives, pp 307–319. DOI:10.1002/9781118495223.ch23.
5. Hardy, M.C., Detrois, M., McDevitt, E.T. *et al.* Solving Recent Challenges for Wrought Ni-Base Superalloys. Metall Mater Trans A 51, 2626–2650 (2020). https://doi.org/10.1007/s11661-020-05773-6.
6. Patel AD, Minisandram RS, "Pool Profiles During Low and High Current Vacuum Arc Melting," in TMS Annual Meeting (2004) https://www.onemine.org/document/abstract.cfm?docid=230733.
7. Patel AD, Minisandram RS, Evans D, "Modeling of Vacuum Arc Remelting of Alloy 718 Ingots," Proceedings in Superalloys 2004 (2004) https://doi.org/10.7449/2004/superalloys_2004_917_924.
8. McDevitt, E., Minisandram, R., Garcia-Avila, M. (2018). The Case for Physical Experiments in a Digital Age. In: Proceedings of the 9th International Symposium on Superalloy 718 & Derivatives: Energy, Aerospace, and Industrial Applications. The Minerals, Metals & Materials Series. Springer, Cham. https://doi.org/10.1007/978-3-319-89480-5_63.

9. Bockenstedt et al (2016). Conventionally Forged RR1000 Billet for Forged Turbine Components In: Proceedings of the 13th Intenational Symposium of Superalloys, http://doi.org/10.1002/9781119075646.ch51.
10. SuperCooler Heat Treatment Technology for Aerospace Components Materials World, Vol. 11, No. 5 pp. 10-12, May 2003.
11. Bunge (2001) *Method and apparatus of cooling heat-treated work pieces,* US6394793B1
12. Lemsky. (2005) Assessment of NASA Dual Microstructure Heat Treatment Method Utilizing Ladish SuperCooler™ Cooling Technology. NASA/CR—2005-213574, https://ntrs.nasa.gov/api/citations/20050161950/downloads/20050161950.pdf.
13. Gurda, John. Forging Ahead: A Centennial History of Ladish Co. Milwaukee: Ladish Co., Inc., 2005.
14. International Directory of Company Histories, Vol. 8. St. James Press, 1994. (history of Allegheny Ludlum)
15. International Directory of Company Histories, Vol. 62. St. James Press, 2004. (history of Allvac)
16. History (atimaterials.com), accessed Jan 8, 2023

Contents

Part V High Temperature Fe-, Ni, and, Co-based Alloys

Part VI Additive: Powder and Processing

Conference Organizing Committee

General Chair

Joel Andersson, University West

Co-chair

Chantal Sudbrack, National Energy Technology Laboratory

Committee Members

Eric A. Ott, General Electric
Zhongnan Bi, Central Iron and Steel Research Institute
Ted Asare, Pratt & Whitney
Kevin Bockenstedt, ATI Specialty Materials
Ian Dempster, Wyman Gordon/PPC
Michael Fahrmann, Haynes International
Paul Jablonski, National Energy Technology Laboratory
Michael Kirka, Oak Ridge National Laboratory
Xingbo Liu, West Virginia University
Daisuke Nagahama, Honda R&D Co., Ltd.
Tim Smith, NASA Glenn Research Center
Martin Stockinger, Montanuniversität Leoben
Andrew Wessman, The University of Arizona

Part I
Melting, Forging, and Wrought Processes

The Formation of Downward Freckles in Nb-Containing Superalloy Remelt Ingots

A. Mitchell and S. Hans

Abstract A revised mechanism for the formation of downward freckles is proposed. The mechanism accounts for the observations of freckle formation in industrial ingot production. In particular, the association of freckling with process instability is emphasized and is part of the proposed mechanism. Additionally, we suggest that the shrinkage which takes place as the final eutectic solidifies can explain the formation of freckling when the basic gravitational force produced by negative buoyancy in the solidifying liquid appears to be insufficient for the initiation of interdendritic fluid flow. We conclude that the key factor in preventing freckle formation in industrial ingot remelting is the maintenance of process stability.

Keywords Freckle · Superalloy · Remelting

Introduction

The segregation feature known as "freckle" is unfortunately familiar to producers of superalloy ingots and has been frequently reported in publications [1–4]. Although the industry has been able to develop melting parameters which produce alloy ingots of most superalloys not containing freckle, the very occasional occurrence of the defect is of concern. In Nb-containing superalloys, the freckle takes a form different from that found in many other superalloys. The freckle appears to have originated from liquid flow downwards in the structure as opposed to the more general case of upward flow. This difference lies in the segregation mechanism which produces a remaining liquid on solidification which is denser than the bulk liquid rather than less so as in the other superalloy classes. This feature is due principally to the segregation tendency of niobium and molybdenum as well as the inverse segregation of chromium and iron [5]. The downward freckle mechanism has been studied and the presently

A. Mitchell (✉)
Department of Materials Engineering, University of British Columbia, Vancouver, BC, Canada
e-mail: alec.mitchell@ubc.ca

S. Hans
Aubert Duval Co, Les Ancizes, France

E. A. Ott et al. (eds.), *Proceedings of the 10th International Symposium on Superalloy 718 and Derivatives*, The Minerals, Metals & Materials Series,
https://doi.org/10.1007/978-3-031-27447-3_1

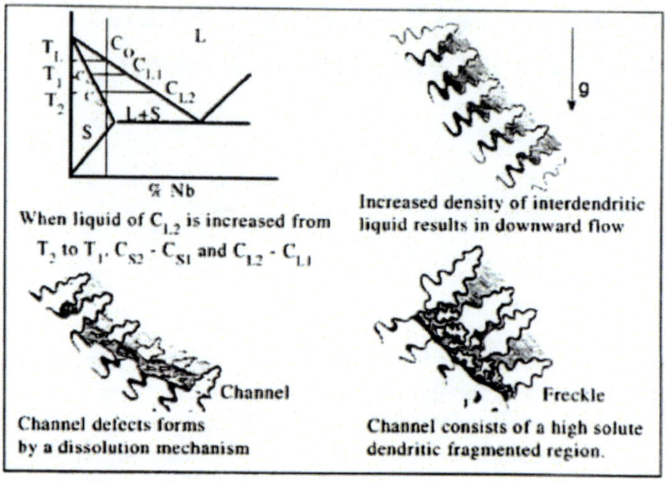

Fig. 1 Proposed downward freckle formation mechanism [6]

accepted model is that proposed by Brooks [6], as shown in Fig. 1. The heavy liquid is assumed to flow downwards in the columnar-dendritic structure at an angle that is slightly more horizontal than the ingot isotherms. In doing so, the liquid enters a higher temperature region and is able to dissolve secondary dendrites and increase the local permeability. The process terminates as the ingot progressively freezes. The model has been studied using CFD approaches [7] from which it has been proposed that the process can be represented by the dimensionless Rayleigh Number (Ra). Following Valdes [8], freckling in the Nb-containing superalloys is predicted for a range of Ra values (0.4–0.8) depending on the individual alloy content, which concept has been used to design melting parameters for both ESR and VAR. There are two notable features of the reports of Ra computations. First, it appears that there is no unique universal value of Ra at which freckles should form. Second, it appears that the Ra values found for freckle formation are alloy dependent but it is unlikely that the alloy dependence arises solely from uncertainties in the alloy properties.

The Freckle Process Model

The existing models for computing the Ra values relevant to freckle formation contain the assumption derived from the Blake-Korzey equation [9] that the permeability coefficient (k) governing the freckle flow is approximately $10^{-10}m^2$ given the general conditions of solidification rate (in the region of 0.1 K/s) in the freckle-prone regions of a remelt ingot. However, experimental and theoretical studies [9, 10] on the permeability of a columnar-dendritic network similar to that found in the mid-radius region of a remelt ingot have shown that the permeability coefficient in the "normal direction" (across the primary growth direction) is $10^{-12}m^2$ at a fraction solid (Fs) corresponding to the composition of the computed maximum remaining liquid density in these alloys (Fs approximately 0.6–0.7) [1]. Reports have suggested that the average freckle composition in IN718 is 9.3%Nb [5], 9.5%Nb [11], and 11.5%Nb [12]. It is to be noted that the "average" composition of the freckle is difficult to determine by a scanning EDAX analysis of the area due to the inhomogeneous nature of the freckle. However, careful adjustment of successive analysis windows [13] indicates a value of between 12.5 and 13%Nb for the lower part of the freckle which is assumed to represent the liquid initially flowing before substantial dilution by dendrite solution has taken place. The lower Nb content values reported represent the average content of the area covered by the disturbed structure between the points at which the regular dendritic structure can be observed. The value of 11–13%Nb corresponds to the liquid composition at a fraction solid of 0.6–0.7% in IN718 [14]. This latter value has a permeability coefficient which then is too small to permit significant flow under the calculated gravity-driven negative buoyancy. In consequence, it is clear that a freckle can form once flow starts and dissolves a small section of the dendrite network, but in a regular dendrite array such as the ones typically found in remelt ingots, it is unlikely for substantial flow to spontaneously start under the negative buoyancy force in the solidification structure present at $0.6 < Fs < 0.7$.

The negative buoyancy arising from the gravitational effect of the density gradient in the interdendritic liquid varies considerably from alloy to alloy. Figure 2 illustrates computed values for the liquid density in several Nb-containing alloys. It can be seen that the largest values are those for IN718, which (as found in practice) is hence very freckle-sensitive. IN625 is less so, but IN706 presents almost zero density change during solidification and should not freckle. However, IN706 has been found to produce freckled ingots (Fig. 3) [15] in both ESR and VAR ingots when melted under conditions which were not substantially different from normal freckle-free production values.

It is observed in practice that ingots produced from ESR furnaces are more likely to contain freckles than those from VAR. The generally accepted reason is that the thermal balance differs between the processes and hence the interdendritic spacings at any given point in the ingot are smaller in VAR than in ESR. However, computations of the thermal gradients in typical ingots of the two processes, when carried out at the same melting rate and same ingot diameter, do not show a very substantial difference in dendritic structure [16, 17]. Both processes should exhibit the same tendency of freckle formation since the permeability of the dendrite network is similar.

Observations of freckling in industrial ingots show two features that must also be explained by the model. First, freckling exhibits a degree of randomness with occasional ingots showing a small number of freckles distributed at various points in the ingot's vertical growth. Second, the freckle trail is often seen to follow the isotherm pattern rather than being consistently at a lesser angle to the growth direction as required by the model.

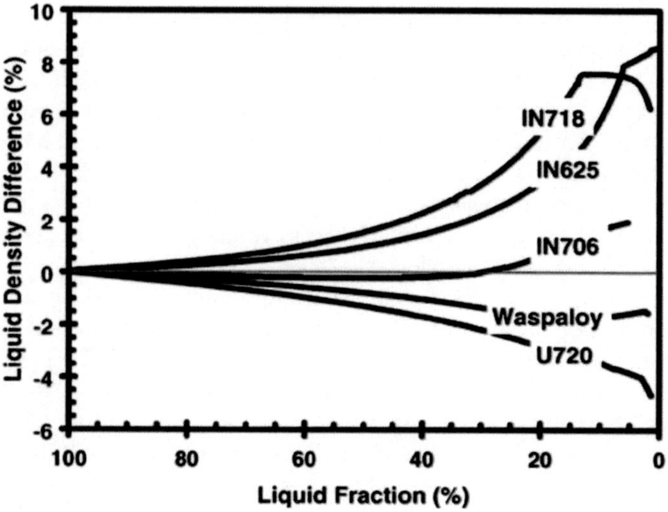

Fig. 2 Density variations in Nb-containing superalloys [12]

Fig. 3 IN706 freckles in a part forged from a 900 mm diameter ESR ingot

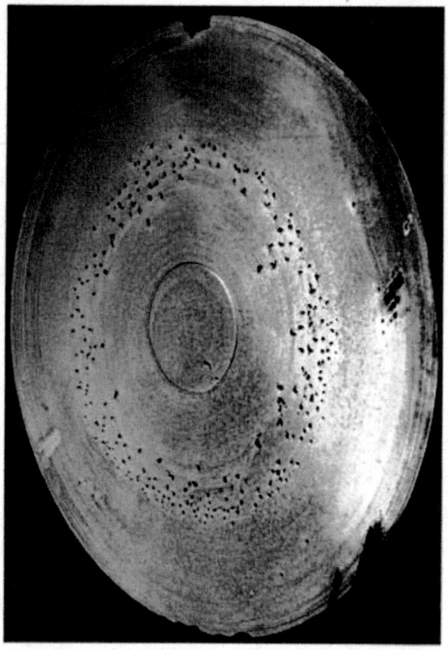

It appears from the above critique of the existing model for downward freckle formation that the initial proposal for simple downward gravity-driven flow does not fully explain the observed experimental facts of the feature's occurrence.

A Proposed Revised Model for the Freckling Process

The example of IN706 (Fig. 3) suggests that a force in addition to that produced by the negative buoyancy is necessary to explain the flow producing the freckles. It is probable that this additional force arises in the shrinkage taking place as the final eutectic liquid freezes. There are many reports of micro-porosity in these alloys [18], produced by eutectic shrinkage. One report links the shrinkage to freckle formation [19] in single crystal castings although not in a Nb-containing superalloy. All of these latter alloys form a final eutectic, including IN706, and this additional force would apply to all alloys of this class. Given the extent of eutectic formation, the magnitude of the force would fall in the sequence IN718 > IN626 > IN706, but is evidently still large enough to provoke freckling in the case of the smallest buoyancy force. The existence of this additional force could explain the possibility of freckle-forming flow when the permeability coefficient of the dendrite network is smaller than that initially assumed from the Blake-Korzeny equation [9]. It is also interesting to speculate that this force could also account for the difference in freckling tendency between ESR and VAR. Assuming that the shrinkage does not contain gas, the resulting hydrostatic force on the segregating liquid in VAR would be the pressure exerted by the liquid head, including the liquid metal pool. In ESR, the same force would be the pressure exerted not only by the metal liquid head in the ingot but also by the weight of the slag and the pressure of the atmosphere above the slag. An estimation of the effect suggests that the difference between the two cases would almost double the downward force in ESR as opposed to VAR. The cumulative effect would be that segregating liquid in VAR experiences a substantially lower gravitational pressure than that in ESR, hence accounting for the difference in freckling tendency.

The use of the Ra value in predicting freckle tendency is also sensitive to the above issues. Modeling studies of remelting conditions permit the calculation of Ra distribution in the ingot and can hence be used to identify potential freckle formation on that basis. However, from the industrial results, it is clear that ingots can be made under conditions not nominally subject to freckling (using the Valdes criteria) but which produce freckles as development experience with 500 mm ESR and VAR ingots of IN718 has demonstrated in practice. Ingots can also be made which should produce freckles but which are freckle-free. An example of the latter is shown in Fig. 4 for IN706 [17, 20] indicating from the Ra values [8] that there is no region subject to freckling (using as input parameters the melting conditions generally used to produce the alloy) although freckling has been observed in practice ([15], Fig. 3). It appears that the use of the Ra signature has a general application but cannot be used as a unique predictor of freckling possibility in remelt ingots. Equally, although alloy

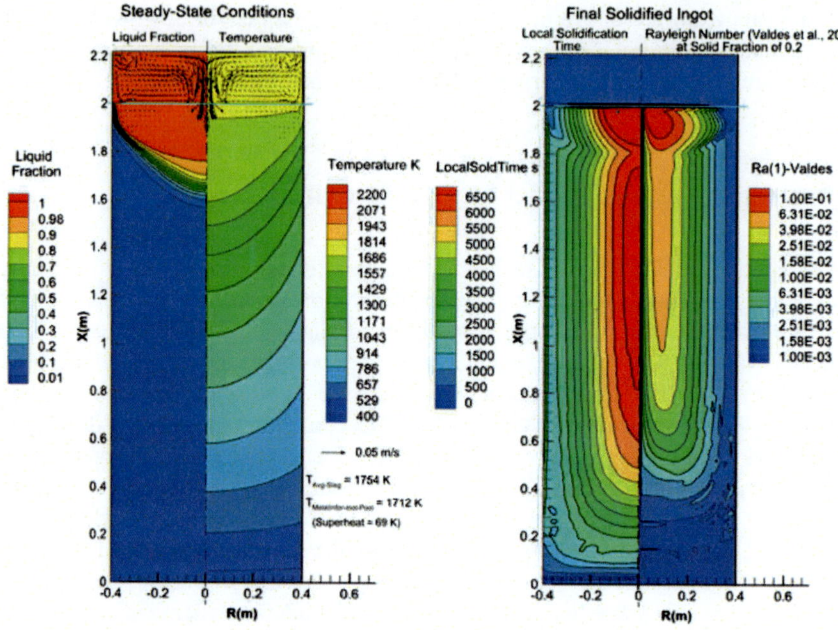

Fig. 4 Computed thermal conditions in IN706 880 mm diameter ESR at 480 kg/h melt-rate

composition modification may assist in reducing freckle tendency (as computed by the Ra technique), it cannot alone be used as the countermeasure.

Although the above proposals could account for the forces necessary for freckling, the question of initiation remains together with the requirement to account for the occasional random occurrence of freckle in practice and its association with process instability. It has been shown through experiment and modeling that direct transfer of pool disturbance into the interdendritic liquid deep in the structure is not possible [21] which leads to the need for an alternate explanation of the freckle-instability linkage. The necessary local random increase in permeability required for initiation when the regular general permeability is represented by a permeability coefficient value of $10^{-12}m^2$ is most probably created by imperfections in the dendrite network. The dendrite structure in an industrial ingot is seldom the regular array which is portrayed in, for example, CAFE models of solidification. Ingots may display random incidence of small equiax grains (Fig. 5), irregular primary growth direction (Fig. 6), and intersecting primary dendrites (Fig. 7). Reports of studies of real-time solidification using synchrotron X-radiation [22] illustrate that breakage of the primary dendrite tips seen in Fig. 8 can result in the type of structural irregularities shown in Fig. 5. (The effect shown in Figs. 5 and 8 is a formation mechanism for the columnar/equiax transition (CET) in the ingot structure [18]). Combinations of these effects have been identified as the cause of freckling in superalloy single crystal castings [23]. It has been noted that it is essential to use melting parameters that do not cause the liquidus isotherm to enter into the thermal region which has been experimentally identified to provoke the CET in remelted ingots since it brings unacceptable general segregation. However, the present indication is that it is also hazardous to approach this condition on account of potential freckle formation. The consequence of these irregularities is a local increase in the permeability providing a potential opportunity for freckle initiation. The irregularities do not necessarily all lead to freckles but can also produce small "patches" of segregation from collecting liquid which does not expand to establish the flow forming a freckle, but solidifies to produce larger-than-normal primary precipitates. In the case of the IN718, the precipitates are primary carbonitrides, but the local solute concentration is not sufficiently high in Nb content that it produces Laves phase after heat treatment. Such randomly distributed patches of large primary precipitates are frequently also found in tool and bearing steels (which can form downward freckles) where the larger primary carbide clusters are detrimental to properties. It is possible that the structure irregularities are also the initiation factor in starting the flow required for the upward freckle feature in alloys experiencing the positive buoyancy force produced by light element segregation. For example, in single crystal castings, freckles readily form at the casting surface and at sharp changes in the casting section, both likely to have formed the required irregularities at those points and so increasing the local permeability.

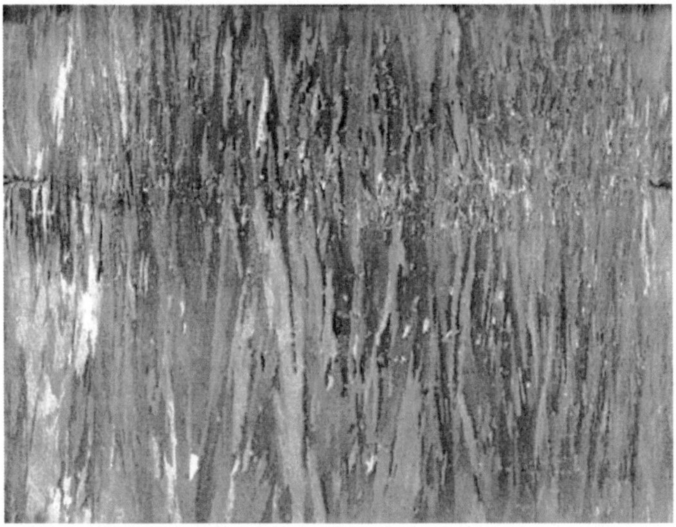

Fig. 5 ESR ingot section 450 mm diameter, IN718

Fig. 6 Irregular primary dendrite growth directions, ESR ingot section 450 mm diameter

Fig. 7 Intersecting dendrites leading to residual porosity ESR ingot section, 450 mm diameter

Practical Implications

Although the industry has, over the years, developed melting procedures which can be relied upon to produce ingots which very seldom contain freckles, the consequent restriction of ingot sizes and melting rates carries a significant economic cost. As a result, there is pressure to push the process parameters to the maximum acceptable limits with the accompanying hazards of segregation and freckle that occasionally threaten quality. The effects and mechanisms described above contain one crucial factor in relation to this maximization. In addition to the generic ones of alloy composition and process geometry as used in the Ra prediction methods, we must also account for the danger of making a dendritic structure that contains irregularities leading to the defects described. The irregularities are the direct product of process

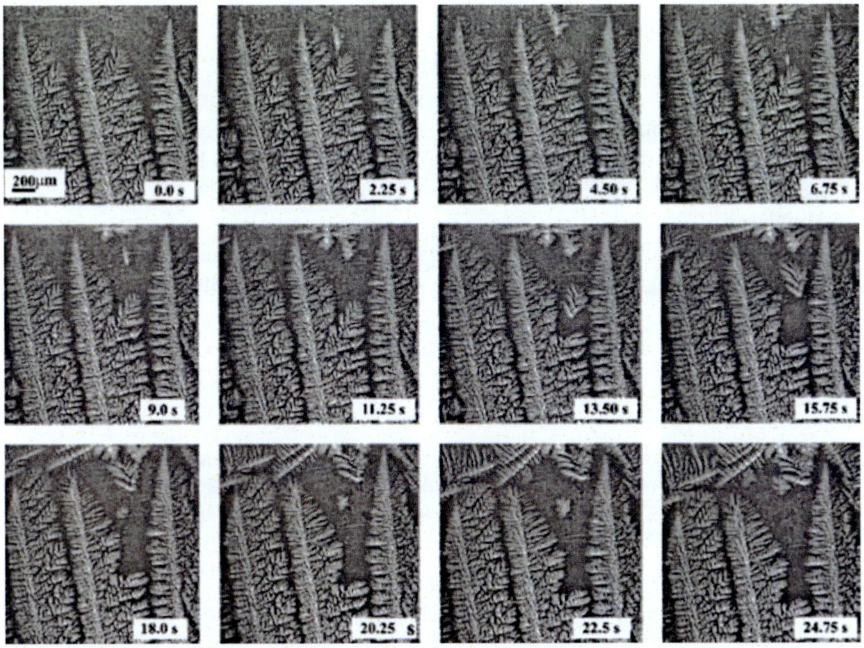

Fig. 8 Real-time X-ray transmission micrographs showing the dendrite tip-breaking mechanism in an Al + Cu alloy [22]

instability. Examples of instability may be found as arc control deficiencies and erratic helium injection in VAR; uncontrolled mold current and immersion depth in ESR; electrode quality in both processes. The corollary is that for a freckle-free ingot we must have full control of the process to produce the maximum regularity of columnar-dendritic growth even when working within a nominally freckle-free set of process parameters.

Definition, measurement, and quantification of control stability are all significant challenges. Even greater is the corresponding requirement that process stability be monitored in such a way that a forensic quality audit can identify any potential deviations from the required values, so making it possible to confirm that a formal process specification agreement with the end-user has been fulfilled. It appears from the above discussion that although the Ra computational technique can give a general indication of freckle tendency, the degree of process stability can override its indications. Instability is responsible for the rare but finite incidence of the defect in an otherwise correctly designed process. Since the freckle defect is difficult to identify in many alloys by conventional NDT routines (and in some cases even by macroetching), it is essential that the melting operation be carried out under conditions where process stability is maintained and can be verified.

Conclusions

1. Freckle initiation is the result of process instability; freckle tendency is the result of alloy composition.
2. Once initiated, the freckle behaviour can be described by the Ra number computation but must be modified by the existence of eutectic solidification shrinkage.
3. Alloy modification to prevent freckling must take into account eutectic formation.
4. The key to freckle-free operation in practice lies primarily in stable process operation.
5. Attention must be addressed to the question of stable process control and its measurement.

Acknowledgements The authors would like to acknowledge the assistance of Dr. K. Kelkar in providing the data shown in Fig. 4 and for permission from Dr. R. Mathiesen to reproduce the data shown in Fig. 8. We are also grateful for the assistance of many colleagues in the industry for comments on freckle observations.

References

1. W Yang, W Chen, K-M Chang, S Mannan and J deBarbadillo, (2001) "Freckle criteria for Upward Directional Solidification of Alloys", Met Trans A, 32A, 397–406.
2. M C Schneider, J P Gu, C Beckermann, W J Boettinger and U R Kattner; "Modeling of Micro- and Macro-Segregation and Freckle Formation in Single Crystal Nickel-Base Superalloy", Met Trans A, 1997, 28A, pp. 1517–1531.
3. M A Dexin, Wu Qiang, A Buhrig-Polaczek; "Some New Observations on Freckle Formation in Directionally-Solidified Superalloy Components", Met Trans B, 2012, 43B, pp. 344–352.
4. S Tin and T M Pollock; "Prediction of Freckle Formation in Single Crystal Nickel-base Superalloys", J Material Science, 2004, 39, pp. 7199–7205.
5. P B L Auburtin; "Determination of the Influence of Growth Front Angle on Freckle Formation in Superalloys" PhD Thesis University of British Columbia, 1998.
6. J A Brooks, J S Krafcik, J A Schneider, J A VanDenAvyle and F Spadafora, (1999) Fe Segregation in Ti-10V-2Fe-3Al 30 Inch Ingot Beta-Fleck Formation, Proc LMPC 1999, AVS New York, eds A Mitchell. 130–144.
7. C Beckermann, J P Gu and W J Boettinger, "Development of a Freckle Predictor via the Rayleigh Number Method for Single Crystal Nickel-base Superalloy Castings", Met Trans A, 2000, 31A, pp. 2545–2557.
8. J Valdes, P King and X Liu, (2010) On the Formulation of a Freckling Criterion for Nickel-base Superalloys in VAR, Met Trans A, 2010, 41A, 2408–2416.
8. J. Madison et al., (2012) Metall. Mater. Trans. A, 43 369.
10. C Y Liu, K Murakami and T Okamoto, "Permeability of Dendrite Network of Cubic Alloys", J Materials Science and Technology, 1989, 5, pp. 1148–1152.
11. P Auburtin, T Wang, S L Cockcroft and A Mitchell, "Freckle Formation and Freckle Criteria in Superalloy Castings", Met Trans B, 2000, 31B, pp 802–811.
12. F Zanner, R Williamson and R Erdmann, "On the Origin of Defects in VAR Ingots", Proc LMPC 2005, publ ASM Metals Park OH, ed. P D Lee, pp. 13–27.

13. S, Hans, J-M Ruppert and A Mitchell, "Spot Segregation in Alloy 718", Met Trans, to be published.
14. Zhengdong Longa, Xingbo Liua, Wanhong Yang, Keh-Minn Chang and Ever Barbero, "Thermodynamic assessment of liquid composition change during solidification and its effect on freckle formation in superalloys" Materials Science and Engineering, 2004, A 386 pp. 254–261.
15. A D Helms, C B Adasczik and L A Jackman; "Extending the Limits of Cast/Wrought Superalloy Ingots", "Superalloys 1996", eds R D Kissinger, D J Deye, D L Anton, A D Cetel, M V Nathal, T M Pollock and W A Woodford, TMS, Warrendale, 1996, pp. 427–433.
16. T Morita, T. Suzuki, T. Taketsuru, D. G. Evans, and W. Yang "The Tendency for Freckle Formation in Alloy 718", Proc. Superalloys 718 625, 706 and Various Derivatives, ed. E Loria, TMS, Warrendale, 2001, pp. 149–160.
17. K M Kelkar, S V Patankar, S K Srivatsa, R S Minisandram, D G Evans, J J deBarbadillo, R H Smith, R C Helmink, A Mitchell, H A Sizek; "Computational Modeling of ESR", Proc. Liquid Metal and Casting Conference 2013, eds. M J M Krane and A Jardy, TMS, Warrendale, 2013, pp. 11–21.
18. J LeComte-Beckers, "Study of Microporosity Formation in Ni-Base Superalloys", Met Trans A, 1988, 19A, pp. 2341–2348.
19. D Han, W Jiang, J Xiao, K Li, Y Lu, and L Lou, "Influence of Geometric Structure and Feeding Behavior of Casting on Freckle Formation during Directional Solidification of a Ni-Based Single-Crystal Superalloy", Crystal Research and Technology Journal, 2021, https://doi.org/10.1002/crat.20200019.
20. MeltFlow-ESR Technical Background and Users Manual, Innovative Research, LLC, Plymouth, MN 55446.
21. K Sasaki, Y Sugitani, S Kobayashi and S Ishimura, 'Effect of Fluid Flow on the Formation of the Negative Segregation Zone in Steel Ingots", ISIJ, 1979, 65, pp. 60–69.
22. R H Mathiesen, L Arnberg, P Beuet and A Somogyi, "Crystal Fragmentation and Columnar-Equiax Transition in Al-Cu Studied by Synchrotron X-Ray Video Microscopy", Met Trans A, 2006, 37A, pp. 2515–2524.
23. T M Pollock and W.H. Murphy, "The Breakdown of Single-Crystal Solidification in High Refractory Nickel-Base Alloys", Met Trans A, 1996, 27A, pp. 1081–1094.
24. A Mitchell "Comment on the Columnar/Equiax transition in Remelt Ingots" Proc. LMPC 2022, ed M Krane, publ TMS, New York, 2022.
25. James A. Van Den Avyle, John A. Brooks, and Adam C. Powell, "Reducing Defects in Remelting Processes for High-Performance Alloys", JOM, 1998, pp. 22–26.

Manufacturing Large Superalloy Pipe Bends

J. J. de Barbadillo and B. A. Baker

Abstract The U.S. Department of Energy, through its Offices of Fossil Energy and Energy Efficiency and Renewable Energy, has funded programs to develop materials technology for high-efficiency energy systems that utilize supercritical steam or carbon dioxide as working fluids. These plants would operate in the 700–800 °C temperature range and require the use of nickel-base alloys to meet design creep-rupture life requirements for welded structures. Specific components include large-diameter pipes, induction bends, and forged fittings. INCONEL® alloy 740H® (UNS N07740) is a γ'-strengthened nickel-base superalloy that was developed for this application and down-selected for a manufacturing demonstration program designated AUSC ComTest. Work was recently completed on the Phase 2 program under DOE contract DE-FE0025064 that had the goal of demonstrating the ability of US industrial supply chain to manufacture full-scale components. In another DOE contract, DE-EE0008367, that concluded in the Fall of 2022, the production of large-diameter longitudinally seam-welded pipe was demonstrated. Alloy 740H contains 16–20% γ' and is sensitive to thermal stress cracking and auto-aging with implications for each stage of the manufacturing operation. Previous publications have described the production of ingot, pipe, and welded pipe made for these demonstration projects. This paper describes induction bending and subsequent heat treatment of these pipes. The results demonstrate the capability of US manufacturers to make these bends with satisfactory dimensional control and properties. The investigation and mitigation of process cracking encountered in this work will be discussed. The test articles are now stored at Oak Ridge National Laboratory awaiting future programs for more detailed material characterization.

Keywords Extruded pipe · Induction bending · Mechanical properties · Strain-age cracking

® Registered trademarks of Special Metals a PCC Company.

J. J. de Barbadillo (✉) · B. A. Baker
Special Metals, 3200 Riverside Drive, Huntington, WV 25705, USA
e-mail: jdebarba@specialmetals.com

E. A. Ott et al. (eds.), *Proceedings of the 10th International Symposium on Superalloy 718 and Derivatives*, The Minerals, Metals & Materials Series,
https://doi.org/10.1007/978-3-031-27447-3_2

Introduction

Projects to develop and demonstrate high-efficiency advanced energy systems that could operate continuously at 700 °C and above have been underway in the United States and many other countries for more than 20 years. The primary goal of these projects has been to reduce the quantity of carbon dioxide generated per unit of delivered electric power. One project is the US Department of Energy/Industry collaborative project known as the Advanced Ultra-Supercritical (AUSC) Steam Boiler initiative. This project was administered by the National Energy Technology Lab (NETL) with significant technical input from the Electric Power Research Institute (EPRI) and Oak Ridge National Lab (ORNL). The goal of this project was to develop technology to enable the development of a coal-fired Rankine cycle steam power plant that would operate at 760 °C. The initial phase of the project included an analysis and definition of material requirements for critical components of the boiler and steam turbine, a broad survey of properties of candidate materials, and a down select of materials for more extensive characterization. Alloy 740H was selected for further evaluation for boiler tubing and steam transfer piping [1]. The second phase of the project was to produce full-scale components including pipes and bends, wye and turbine rotor forgings, a nozzle carrier casting, and fabricated prototype structures and valves. This work has been completed and recently reported [2].

In another project, the DOE Solar Energy Technology Office has developed process designs and materials capability for Generation 3 Concentrating Solar Power. This concept is similar to the current solar power tower plants that are operating in several countries; but works at a higher temperature and utilizes supercritical CO_2 as the working fluid to drive the turbine/generator power unit. In an effort to reduce manufacturing costs for alloy 740H pipe, this project supported the demonstration of the capability to make seam-welded pipe [3]. This product is made from a plate that is rolled into a tube shape and then continuously welded and subsequently heat treated. The improved product yield translates to a 30–40% reduction in delivered pipe cost. In common with steam power plant piping, the design of a solar thermal plant requires return bends and elbows that are most economically made by induction bending. The purpose of the present paper is to describe the results of these experimental bends.

Alloy 740H is a γ' strengthened nickel-base alloy that was developed specifically for use as a tube and pipe in the 700–800 °C temperature range [4]. Characterization of the alloy by NETL, ORNL, and various consortium members supported a data package that resulted in the first ASME Boiler & Pressure Vessel Code Case for an age-hardened nickel-base alloy for use in welded construction for service in the time-dependent property temperature range [5]. Many technical papers describing the microstructure and properties of the alloy have been published and summarized in a review of AUSC technology by Di Gianfrancesco [6]. While this alloy has a relatively low volume fraction of γ' (16–20%), it auto-ages readily and is subject to strain-age cracking under certain conditions. This behavior complicates the manufacturing of large-diameter thin and heavy wall components. Previous work had shown that alloy

740H pipes with OD of up to 254 mm could be successfully bent and heat treated [7]; however, it was anticipated that larger diameter pipes would be much more difficult. Nomura, et al. reported that they successfully induction bent 510 mm OD nickel-base alloy HR6W pipe; however, that alloy has a much more sluggish age-hardening mechanism than alloy 740H [8]. To the authors' knowledge, induction bending of seam-welded age-hardened nickel alloy pipes had never been previously reported.

Due to the very high forces required to bend large heavy-wall pipes, an induction heated bending process is generally applied. Shaw Clearfield, LLC in Clearfield, Utah (formerly McDermott/CB&I) was engaged for this project. Shaw had prior experience in bending alloy 740H pipes up to 168 mm OD for previous demonstration projects [7]. Induction bending is a type of incremental metal forming process where deformation forces are minimized by confining the strain to a narrow band created by heating with an external circumferential induction coil. In the heated zone, the temperature is controlled within a range where the material strength is greatly reduced. In the case of alloy 740H, the required forming temperature range is 1037–1148 °C. A bird's eye drawing of a typical induction bending machine operating in tension mode is shown in Fig. 1. The front end of the pipe is gripped by a hydraulic clamp mounted on a track that describes the radius of the bend to be formed. The back end of the pipe is pushed through a series of guide rollers and the induction coil as illustrated in the drawing. A water-cooling ring is often mounted behind the induction coil; however, for this work water was not used due to cracking concerns. As can be seen in Fig. 2, the actual hot zone is quite short, with the bending occurring within an approximately 50 mm wide zone. One consequence of tension bending is thinning of the pipe wall at the extrados and thickening at the intrados. The degree of thinning varies with the OD/W ratio; but is on the order of 10–15% for large pipes. This wall thickness reduction must be taken into consideration for the piping design. The wall thinning can be reduced or eliminated by bending in compression mode; however, this is a very complex machine design that is not available in the US for large pipes.

Manufacturing Process

Seamless Pipes

The seamless precursor pipes were made from 914 mm diameter VIM/VAR ingots HT9039JY and HT9067JY with the composition shown in Table 1. Details of melting, homogenization, and conditioning of the ingots were reported previously [2, 10]. The nominally 11,340 kg ingots were cut in half and extruded to produce four pipes of two different dimensions at Wyman-Gordon in Houston, TX. The extrusion details have been previously reported [2, 11]. The ingot bottom piece of heat HT9067JY was used to make a nominal 560 mm OD × 100 mm W "header" pipe for the bending trials. This pipe was 3.16 m long and weighed 3658 kg. The ingot bottom piece of Heat

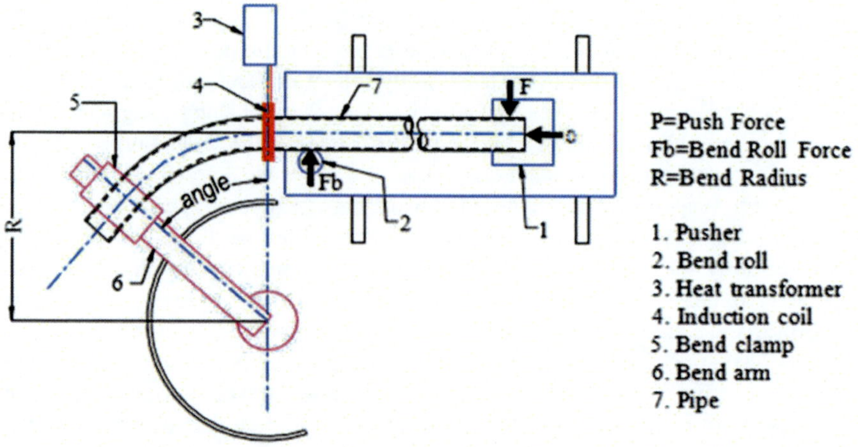

Fig. 1 Schematic of induction bending machine from A.K. Dev, *Introduction to Induction Bending* [9]

Fig. 2 Two views of induction bending alloy 740H pipe at Shaw Clearfield, LLC. The discolored surface distinguishes the section of pipe that has already been bent

HT9039JY was used to make a nominal 710 mm OD × 38 mm W "reheater" pipe for bending. This pipe was 5.09 m long and weighed 3835 kg. These pipe dimensions were selected as they posed different challenges for the extrusion process and, as will be discussed later, different challenges for bending. Both pipes were solution annealed at the mill prior to being released for bending. Non-destructive pipe testing included dimensional verification, hydro pressure testing, and full ultrasonic testing

Table 1 Composition of alloy 740H heats. Sample from head, mid-radius

Heat number		C	Mn	Fe	Si	S	Ni	Co	Cr	Mo	Al	Ti	Nb
HT9039JY		0.03	0.2	0.6	0.14	0.0006	49.3	20.1	24.5	0.48	1.49	1.42	1.5
HT9067JY		0.03	0.3	0.4	0.15	0.0003	49.4	20.1	24.6	0.48	1.49	1.42	1.5
HT6309JK		0.03	0.3	0.2	0.15	0.0005	49.7	20.0	24.6	0.49	1.40	1.48	1.5
UNS N07740	Max	0.08	1.0	3.0	1.0	0.03	Bal	22.0	25.5	2.0	2.0	2.5	2.5
	Min	0.005						15.0	23.5		0.2	0.5	0.5

Fig. 3 Pipes prior to bending. Header pipe on left, reheater pipe on right

to ASTM E213 with a 5% of wall ID/OD reference notches. Photographs of the conditioned pipes are shown in Fig. 3.

Welded Pipe

The seam-welded pipe was made from 19 mm thick plate rolled from VIM/ESR slab ingot HT6309JK. The composition of this heat is shown in Table 1. The plate was formed and welded at Swepco Tube, LLC in Clifton, NJ into a 356 mm OD × 6.1 m long pipe. The pipe wall is slightly thickened in the forming process. The pipe was welded in ten passes using a GTAW process. The details of pipe fabrication have been previously reported [3]. After welding the entire pipe was solution annealed and cut in half. One half weighing 420 kg was used for bending while the other half was aged and used for mechanical property testing. Non-destructive testing included 100% radiography of the weld and ultrasonic testing using a 10% wall thickness notch. Figure 4 shows the pipe on the welding stand.

Fig. 4 Welded pipe on the welding stand at Swepco Tube

Bending

Shaw Clearfield was tasked with making 90° bends with a 3D radius on each of the three pipes and then solution annealing and aging them. While the pipes were initially in the solution annealed condition, it was known that substantial auto-aging was likely to have occurred. These bends were expected to be difficult since microstructure changes were expected to occur sequentially as the bending process progressed. Potential issues included insufficient force to bend, cracking due to low temperature at the ID, strain-age cracking on reheating for the solution anneal, and distortion and buckling. In addition, there was the possibility of intergranular cavitation at the bend extrados due to the slow tensile deformation process. If present, cavitation would degrade the material properties.

Shaw used their internal bending procedure that is compliant with ASME B&PV Code 16.49 and FMAI TPA-1BS-98. The bends were conducted on their Cojafex PB1600 bending machine which is the largest that is manufactured. Due to the relatively short length of the 740H pipes, similar diameter steel extenders were welded to the pusher end of the pipe. The welded pipe was oriented so that the weld was on the neutral axis (recommended commercial practice). The nominal bending temperature was 1093 °C and a quench ring was not used. Specific process parameters such as travel speed and coil power were proprietary to Shaw Clearfield. Despite the relatively high strength of alloy 740H, the actual bending was accomplished with no difficulty. The tensile yield strength of alloy 740H had previously been determined to be 32 MPa at 1093 °C. Photographs of the completed pipe bends are shown in Fig. 5.

Fig. 5 Completed pipe bends. **a** Header pipe, **b** Reheater pipe, and **c** Welded pipe

Results

Dimensions and NDT

Two aspects of pipe bending are of primary concern: cracking, and dimensions and in particular wall thinning on the extrados. For these pipe bends, the vendor conducted limited contact sonic testing to verify extrados wall uniformity. In addition, the header pipe and the welded pipe were sectioned at the apex to collect detailed wall thickness measurements around the circumference. Plots showing the wall thickness at 10° increments are shown in Fig. 6a and b. The measured wall thinning of −10.2 and −13.9% is within the range of 10–15% recommended for 3D bends of large-diameter pipe [12, 13]. The extrados wall thinning for the reheater pipe, as measured by UST, varied from −7.5 to −17.5%. However, this range may be influenced by the more variable wall thickness and eccentricity of the original extruded pipe. Ovality within the bend area of all three pipes ranged from 0.8 to 1.3% which is well within the commercial standard of 3%. All the pipes were liquid penetrant tested to the Shaw internal standard. An external crack that was found on the extrados of the reheater pipe is discussed below. Extensive radiography was conducted by an external vendor after the pipes were returned to Special Metals. These findings are discussed below.

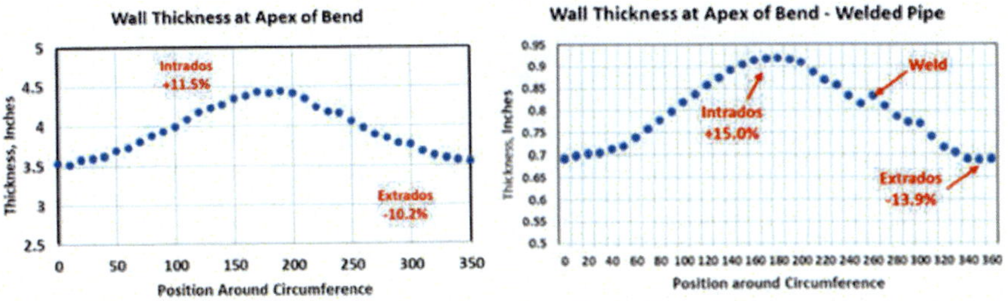

Fig. 6 Wall thickness around circumference at apex of bend. Left 22-in OD Pipe, Right 14-in welded pipe

Mechanical Properties

Mechanical property testing was limited in this investigation to verification of compliance with ASME Code Case 2702 minimum properties and general verification that no microstructural damage such as cavitation had occurred. The bending of the reheater pipe occurred too late in the project to allow for cutting and property determination. The uncut pipe bend is now stored at Oak Ridge National Lab awaiting testing under a future project. Room temperature tensile properties of the header pipe are contained in Table 2. All tests met the ASME requirements; however, the yield strength at all locations was lower than that of the unbent pipe. This difference is likely due to the difference in effective aging time between the shop-aged pipe bend and the lab-aged unbent pipe. There was no significant yield or tensile strength difference between the intrados, extrados, or neutral axis despite the obvious difference in strain applied (nominally 13% tensile strain in the outermost fiber of the extrados). As expected, the circumferential and through wall ductility was somewhat lower than the longitudinal ductility. This is thought to be primarily due to the orientation of carbide inclusion stringers rather than grain shape or texture.

The room temperature tensile properties of the seam-welded pipe are shown in Table 3. Again, all tensile properties met the ASME minima and the yield strength of the bent pipe was lower than the unbent pipe. In this case, both pipes were mill aged, the unbent pipe at Swepco and the bent pipe at Shaw. Through wall properties could not be determined on this relatively thin wall pipe. A more detailed study of high-temperature properties of this pipe has been conducted and will be reported

Table 2 Room temperature tensile properties of the header pipe

Location	Test	0.2% YS, MPa*	UTS, MPa	Elong. (%)	RA (%)
Neutral axis	Longitudinal	651	1138	29	30
	Circumferential	661	1119	24	21
	Through wall	628	1058	21	24
Exrados	Longitudinal	665	1134	32	33
	Circumferential	682	1111	29	29
	Through Wall	672	1064	22	21
Intrados	Longitudinal	678	1133	32	34
	Circumferential	658	1121	28	27
	Through wall	670	1062	21	20
Unbent pipe	Longitudinal	751	1138	38.4	41
	Circumferential	727	1079	30.2	26.5
	Through wall	689	1069	24.4	22.1
ASME 2702 Min		621	1034	20	

*Average of 2

Solution anneal 1107 °C/WQ + Age 8 h 800 °C

Table 3 Room temperature tensile properties of seam-welded pipe

Item	Location	Orientation	0.2% Offset YS, MPa	UTS, MPa	Elong, %	RoA, %
Welded pipe	Base metal	Circumferential	697	1114	40.4	40.2
"	"	Longitudinal	707	1117	39.6	39.5
"	Weld	All weld metal	743	1129	30.0	30.9
"	"	Cross weld	702	1094	30.9	33.3
Bent pipe	Extrados	Longitudinal	670	1104	40.2	42.8
"	"	Circumferential	659	1097	38.1	39.6
"	Intrados	Longitudinal	679	1102	39.8	38.3
"	"	Circumferential	685	1109	39.9	36.6
"	Neutral/weld	All weld metal	722	1141	31.6	30.2
"	"	Cross weld	686	1094	31.6	33.4
ASME: YS 621 MPa, UTS 1034 MPa, Elong. 20%						
Heat Treatment: Solution Anneal 1107 °C, Age Harden 4 h at 800 °C						

in future publications. Creep-rupture properties were found to correspond with the scatter band for previously generated data used to compute design stress allowables.

Cracking Investigations

Header pipe: ID cracking in a band at the extrados side in the bend start area was visually noticed and subsequently radiographed at Shaw (Fig. 7a). This area was marked for cutting a 140 mm slice for further examination. The macroscopic appearance of the ID is shown in Fig. 7b. The cracks were circumferential in an arc of 60° and approximately 6 mm deep. Figure 7c shows a vertical section through the cracked area. Micrographs of the cracked area (Fig. 8) show that the cracks are predominantly intergranular with an oxidized surface. Profuse unconnected intergranular voiding was present near the ID surface in the cracked region. It was noted in the operations log that an excessive dwell time was experienced at the start of the bending operation that was caused by slow penetration of the heat from the induction coil in this very thick pipe. This may have resulted in an embrittled zone where the temperature never reached the γ' and carbide solvus. There is no direct evidence of an embrittled structure because the bends were solution annealed later. Since this cracked area was within the bend thinning zone, this damage could not be removed by spot grounding and consequently the pipe would have been rejected. One solution to this problem would be to begin the bend after the pipe passed through the affected area. Another possibility is to internally heat the pipe with a gas torch in the starting area. It was

Fig. 7 Images of ID cracks in header pipe. **a** Radiograph, **b** ID surface view, and **c** Cross-section

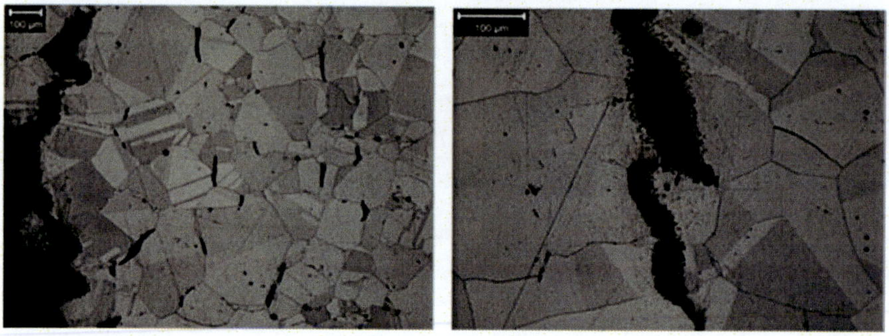

Fig. 8 Microstructure in ID cracked region of header pipe

also noted that the cracks tended to initiate at visible grinding marks suggesting that a smoother final grinding pass would have been beneficial.

Reheater pipe: Shaw Clearfield reported that LP testing showed a single isolated 50 mm long OD crack at the apex of the bend. UST examination after the bent pipe was returned to Special Metals indicated that the crack was about 12.5 mm deep. This crack was not excised for examination; however, it is possible that it initiated at a pre-existing, undetected flaw in the extruded pipe. This crack did not correspond to an indication on the previous pipe UST scan, but based on the 5% of wall rejection standard, a 5 mm deep flaw would not have been rejected.

Visual inspection of the pipe ID showed a network of cracks in the region of the bend stop. Mistras Group, Inc. conducted a full radiographic investigation of the area. Representative radiographs (Fig. 9) show a network of fine branching cracks in a band about 300 mm wide that extended around the entire circumference of the pipe. Due to time constraints, this cracking could not be further investigated; however, it strongly resembles the pattern observed in the welded pipe which was attributed to strain-age cracking.

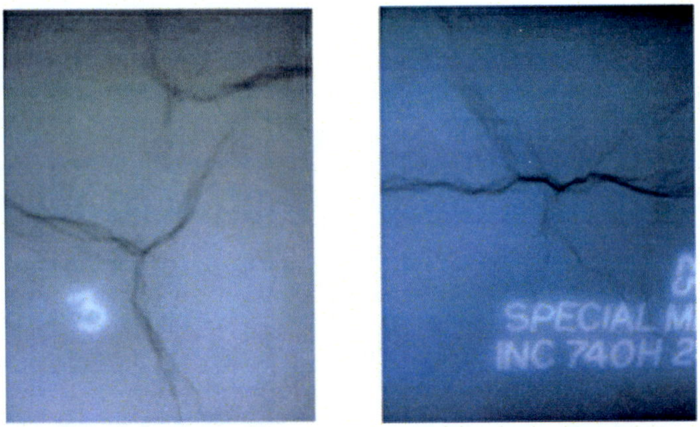

Fig. 9 Radiographs showing ID cracks in reheater pipe

Seam-welded pipe: A branching crack was detected on the ID of the pipe at the bend stop area. This cracked area was roughly 125 mm in the circumferential direction and 75 mm longitudinal direction as shown in Fig. 10a and b. Destructive metallographic examination conducted by EPRI showed that the crack penetrated about half of the pipe wall at maximum depth. It appears to have initiated in the base metal and stopped in the weld metal. Since the entire weld had previously been radiographed and showed no indications, a pre-existing weld flaw could not be the cause of the cracking. The main crack and side branches were predominantly intergranular. The most plausible explanation is that this was a strain-age crack that occurred during reheating for the post-bending solution anneal. The observation of oxide within the crack is consistent with this idea. Ding et al. have reported that an area on the ID about 5° from the neutral axis in the stop zone of an induction bent pipe is a region of maximum biaxial residual stress [14]. While the actual amount of γ' present in the crack region at that point is unknown due to the very complex thermal conditions in this transient, it is likely that the residual stress in this area combined with a relatively slow heating rate combined to generate strain-age cracking. A modification of the clamping procedure at the bending stop stage may also be beneficial in reducing residual stress.

(a) (b) (c)

Fig. 10 ID crack in seam-welded pipe. **a** Radiograph, **b** ID view, and **c** Cross-section

Discussion

Alloy 740H has been successfully made in all mill product forms in a wide range of product sizes. The mechanical properties have been shown to be remarkably consistent. However, many fabricators in the energy supply chain are unfamiliar with γ′ strengthened alloys and the precautions required to avoid surface checking and strain-age cracking during forming and subsequent heat treatment. This is particularly true of large components required for power and chemical plants. Strain-age or stress relief cracking has been extensively studied in welded structures of alloy 740H [13]. These studies and other ongoing work by EPRI under DOE sponsorship will develop practical solutions to concerns for shop and field welding and heat treatment of welds in complex structures.

The present project illustrates the need to apply this knowledge to unique metal forming operations such as induction bending where robust process models are not currently available. Thick and thin wall pipe bends have different issues that cannot be readily solved by trial and error due to the high cost of the material. Figures 11 and 12 illustrate the issues involved with 740H and other γ′ strengthened alloys. Figure 11a is a phase diagram for the nominal 740H composition simulated with JMatPro. The γ′ solvus is predicted to be 973 °C. The η solvus is 1040 °C, so for a bending temperature of 1093 °C, the strengthening phases in the material in the hot zone should be fully dissolved. However, for heavy-wall pipe as noted previously the pipe ID temperature may be substantially lower. Tensile ductility for 740H drops significantly as the temperature drops below the solvus. The simulated continuous cooling diagram in Fig. 11b shows that γ′ forms rapidly on cooling. This means that in a heavy section component the interior is significantly auto-age hardened. This is illustrated in Fig. 12a which is a hardness traverse across the wall of the mill solution annealed header pipe. The maximum Rb reading of 105 at the center of the pipe wall equates to approximately Rc 29 while the ID surface is Rb 88. After aging, a uniform pipe wall hardness of Rc 30–32 was achieved. This hardness disparity must be managed when making a process transition such as a start or stop of a forming operation.

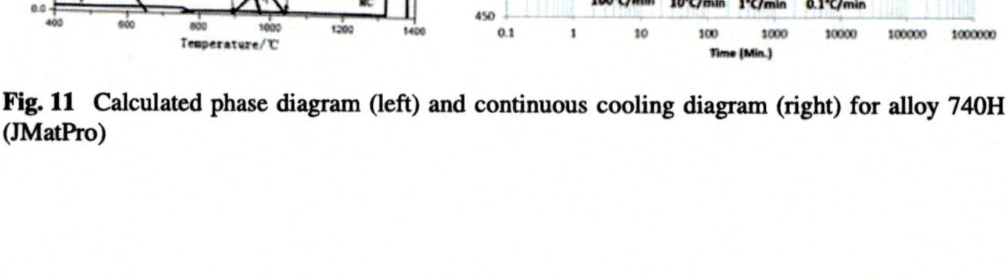

Fig. 11 Calculated phase diagram (left) and continuous cooling diagram (right) for alloy 740H (JMatPro)

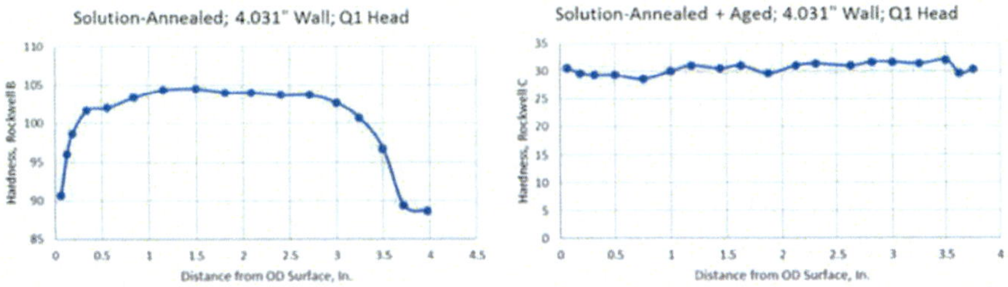

Fig. 12 Hardness through wall of header pipe. Left as solution annealed and water quenched, right after aging at 800 °C

There is a continuing effort underway to understand the complexities of induction pipe bending of age-hardened alloys. One activity involves the development of a bending model at Argonne National Laboratory under the Department of Energy's HPC4Mfg program. The initial version of the model is strictly thermo-mechanical and at the time of this writing is undergoing beta testing. Future enhancements may incorporate microstructure evolution. In a separate program, an unbent 710 mm OD pipe from the AUSC ComTest program is slated for an instrumented bending trial to provide partial validation of the model.

Conclusions

1. Induction bending of relatively heavy-wall alloy 740H pipe less than 250 mm OD has been done successfully by multiple commercial tube benders. Induction pipe bends are now in service in two sCO2 applications.
2. This work has shown that it is physically possible to bend alloy 740H pipes of up to 710 mm OD and to obtain acceptable mechanical properties and dimensional control in 560 mm OD heavy-wall pipe.
3. These "best effort" trials showed that the alloy is sensitive to cracking from pre-existing surface quality as well as specific operational practices interacting with a strain-aging mechanism. Continuing work will focus on developing robust production procedures through the use of process models and instrumented tests.

Acknowledgements The manufacture of seamless extruded pipe and bends was sponsored by the US Department of Energy, Office of Fossil Energy, award DE-FE0025064 administered by the National Energy Technology Lab., Prime contractor Energy Industries of Ohio. The manufacture of seam-welded pipe was sponsored by the US Department of Energy, Office of Energy Efficiency and Renewable Energy, Solar Energy Technology Office award DE-EE0008367, Prime Contractor Electric Power Research Institute. The authors acknowledge the critical role of Firdosh Kavarana Shaw Clearfield UT (now at Tulsa Tube Bending Co.) in planning and executing the induction bends and many technical conversations with John Shingledecker, EPRI.

References

1. Purgert, R, et al., "Boiler Materials for Ultra-Supercritical Coal Power Plants", Final Technical Report to US Department of Energy for Cooperative Agreement: DE-FG26-01NT41175, December 2015, https://www.osti.gov/scitech/biblio/1346714.
2. Purgert, R, et al., "Materials for Advanced Ultra-Supercritical Steam Turbines (AUSC) – A-USC Component Demonstration", Final Technical Report, July 2022, OSTI report 1875111, https://doi.org/10.2172/1875111.
3. de Barbadillo, J, et al., "Recent Developments in Manufacturing and Fabrication of Alloy 740H Components for sCO2 Applications", Proc. 7th Int. Supercritical CO_2 Power Cycles Symposium, Feb 21–24, 2022, San Antonio, TX, http://sCO2symposium.com.
4. Smith, R, Baker, B, and Smith, G, "Nickel Alloys for the Next Generation of Steam Boilers", VGB Materials and Quality Assurance Conference, Dortmund, Germany, March 2004.
5. ASME Case 2702, "Seamless Ni-25Cr-20Co Material" Section 1, Cases of the ASME boiler and pressure vessel code, BPV-Supp. 7, American Society of Mechanical Engineers, 2011.
6. Di Gianfrancesco, A, "Materials for Ultra-Supercritical and Advanced Ultra-Supercritical Power Plants", Woodhead Publishing Series in Energy: Number 104, Elsevier, Duxford, UK, 2017.
7. de Barbadillo, J, et al., "INCONEL Alloy 740H – Development of Fittings Capability for AUSC Applications", Proc. 8th Int. Conf.: Advances in Materials Technology for Fossil Power Plants, ASM Int., Materials Park, OH, 2016, 101–112.
8. Nomura, K, et al., "Establishing Induction Bending Technique for Nickel-Based Alloy HR6W Large Piping", Joint EPRI-123 HiMat International Conference on Advances in High Temperature Materials, ASM Int., 2019, 869–879.
9. A.K Dev, "Introduction to Induction Bending", http://WhatisPiping.com.
10. de Barbadillo, J, Baker, B, and O'Connell, C, "Manufacturing Large Superalloy Ingots for Advanced Energy Systems", 2022 Liquid Metal Processing and Casting Conference, Philadelphia, PA.
11. de Barbadillo, J, Baker, B, and O'Connell, C, "Manufacture of Large Superalloy Ingots and Extruded Pipes" 4th International Ingot Casting, Rolling and Forging Conference, to be published by AIST, Warrendale PA, 2022.
12. Recommended Standards for Induction Bending of Pipe and Tube, TPA-IBS-98, Tube & Pipe Association, International, Rockford, IL, 1998.
13. Pipe Bending Methods, Tolerances, Process and Material Requirements, PFI Standard ES-24, Pipe Fabrication Institute, New York, NY, 2015.
14. Ding, Y, et al., "Residual Stress Modeling of Induction-Bent Pipes", Proc. of the ASME 2012 Pressure Vessels and Piping Division Conference, July 15–19, 2012, (PVP2012-78153).
15. Shingledecker, J., et al., "Investigation of Weldment Cracking During Fabrication of a 700°C Fired sCO_2 Heater", 7th International Supercritical CO_2 Power Cycles Symposium, February 21–24, 2022, San Antonio, TX, http://sCO2symposium.com.

The Effect of Microstructure on the Strength of VDM Alloy 780

M. C. Hardy, M. Hafez Haghighat, C. Argyrakis, R. C. Buckingham, A. La Monaca, and B. Gehrmann

Abstract Nickel-base alloys have been developed, which offer higher temperature capability from ingot metallurgy than Alloy 718. However, these alloys have higher volume fractions of gamma prime (γ') precipitates, require further processing steps and consequently, have higher material processing costs, show a greater propensity for freckle and are not readily electron beam (EB) welded. There is an appetite for an alloy that shows improved forgeability, that is EB weldable and can be used at temperatures of up to 700 °C. This study examines VDM Alloy 780, first using laboratory compression tests and heat treatment experiments to determine suitable thermo-mechanical processing (TMP) conditions. Subsequently, pancakes that were 21–22 mm in height and 130–133 mm in diameter were forged from 70 mm in diameter × 76 mm high bars, which were extracted from mid-radius locations of 8-inch diameter billet. From these, test piece blanks were extracted, and heat treated for tensile tests at temperatures of 20, 650 and 750 °C. Results from these tests were compared with microstructure and data from 30 mm thick pancake forgings that received different TMP conditions from work at VDM Metals International. The combined experiences provide an insight into the effect of forging and heat treatment conditions on the microstructure and tensile test properties of VDM Alloy 780. They show that strength levels for fine grain Alloy 720Li can be achieved if specific sizes of γ grains and γ' precipitates can be produced.

Keywords Disk rotors · VDM Alloy 780 · Forging · Heat treatment · Microstructure · Tensile properties

M. C. Hardy (✉) · C. Argyrakis · R. C. Buckingham · A. La Monaca
Rolls-Royce plc, PO Box 31, Derby DE24 8BJ, UK
e-mail: mark.hardy@rolls-royce.com

M. Hafez Haghighat · B. Gehrmann
VDM Metals International GmbH, Kleffstrasse 23, 58762 Altena, Germany

E. A. Ott et al. (eds.), *Proceedings of the 10th International Symposium on Superalloy 718 and Derivatives*, The Minerals, Metals & Materials Series,
https://doi.org/10.1007/978-3-031-27447-3_3

Introduction

Fine grain Alloy 718 is used widely for disk rotor applications in aircraft engines. It has many appealing attributes, such as high strength levels that minimise component size and weight, robust and low-cost material and component manufacture, including the ability to join disk rotors together by electron beam (EB) welding, but with low levels of melt anomalies and excellent fatigue performance, particularly when surfaces are shot peened. Unfortunately, the temperature capability of the alloy is limited to circa 600–650 °C by coarsening and dissolution of small gamma double prime (γ'') and gamma prime (γ') precipitates [1, 2], with the possibility of precipitating α-Cr in the vicinity of delta (δ) from longer and higher temperature excursions [3] and most significantly, very high rates of time dependent crack growth in air due to oxygen embrittlement, accentuated by the fine grain size and the presence of Nb at grain boundaries [4, 5].

Nickel-base alloys [6–13] have been developed, which offer higher temperature capability from ingot metallurgy than 718. Unlike 718, however, alloys such as 720Li, René 65, AD730 and M647 are rich in Al and Ti, rather than Nb and are exclusively precipitation strengthened by gamma prime (γ') rather than gamma double prime (γ'') and γ'. They have (i) higher volume fractions of strengthening precipitates (Table 1), (ii) require further processing steps [14, 15] and, consequently, have higher material processing costs, (iii) show a greater propensity for carbo-nitride stringers [7, 16], freckle [17, 18], and partially or unrecrystallized grains in billet [19–21] and (iv) are not readily electron beam (EB) welded [22]. This paper has examined an alternative composition, VDM Alloy 780 [10, 11] which shows similar Nb levels to 718 and ATI 718Plus [12] but a lower fraction of γ' precipitates than 720Li, René 65, AD730 and M647. Like A718Plus [13], it can form stacked delta (δ) and eta (η) precipitates during forging and heat treatment [23]. ATI 718Plus precipitates η/δ in billet as it is soaked at sub-η/δ solvus forging temperatures [24]. The precipitates then develop into blocky plates during forging with a preferred orientation in the direction of material flow [24, 25]. Whilst lamellar η also forms in recrystallized grains during heat treatment [24, 25], it is the blocky η/δ plates that enable reduced time dependent crack growth (TDCG) rates for material that is subject to radial and tangential stresses. Unfortunately, much higher rates of time dependent crack growth occur for material loaded in the axial direction [26].

Resistance to TDCG is a critical material property for disk rotors as it can limit the design life of the component or the interval between inspections. This is more relevant in today's engines as high climb rates are increasingly required by commercial airlines to move aircraft more quickly to altitude to reduce fuel burn [27]. As optimized resistance to TDCG is produced with grain sizes that are larger than ASTM 9–8.5 (16–19 μm), acceptable strength, which determines the size and weight of disks, must be achieved with effective precipitation strengthening, i.e., a distribution of fine γ' particles (<30 nm in size) that show high anti-phase boundary energy [28] and tight control of grain size in closed die forgings. In this case, the aim is to produce strength levels that approach or are equivalent to those shown by fine grain 720Li.

Table 1 Composition of nickel-based superalloys, produced by ingot metallurgy for disk rotors, in atomic %. Approximate (calculated) fractions of strengthening precipitates are given in furthest right column. B and C levels for VDM 780 are similar to those for 718

Alloy	Ni	Cr	Co	Fe	Mo	W	Nb	Al	Ti	% γ'/γ''
718	51.4	21.2	0.0	19.7	1.8	0.0	3.3	1.3	1.1	23
720Li	54.9	17.3	14.3	0.0	1.8	0.4	0.0	5.2	5.9	44
René 65	54.7	17.9	13.1	0.9	2.4	1.3	0.4	4.5	4.5	38
AD730	57.9	17.3	8.3	4.1	1.9	0.8	0.7	4.8	4.1	38
M647	49.4	17.6	19.4	0.0	1.8	0.8	1.7	7.0	2.2	43
A718 +	51.5	20.1	8.9	9.9	1.6	0.3	3.4	3.2	0.9	30
VDM 780	45.0	20.0	24.5	0.6	1.8	0.0	3.4	4.3	0.4	32

This is particularly challenging or perhaps even impossible with ingot metallurgy and compositions that show relatively low fractions of γ' precipitates.

Sharma et al. [29] have reported that VDM Alloy 780 shows slow dynamic recrystallization (DRX) kinetics, compared to 718 but that full recrystallization can be achieved from fast post-DRX. Note that static recrystallization (SRX) and metadynamic recrystallization (MDRX) are two different recrystallization mechanisms, which can contribute to post-DRX [30]. It was found that material forged to a strain of 1.3, from a super-η/δ-solvus upset at 1050 °C and a strain rate of 0.1 per s, was only 50% recrystallized. However, a 5-min soak after deformation at 1050 °C achieved full recrystallization and an average grain size of 26 μm. The starting billet microstructure was 38 μm, with no η/δ particles.

Disk forgings are often complex in shape due to drive arms and, in the case of high pressure (HP) turbine disc forging, have a large volume of material in the hub. As such, a forging practice using a conventional hydraulic press will also have to contend with adiabatic heating and a variation in temperature across the forging.

There is considerable experimental evidence in the literature regarding the nucleation and coarsening of precipitates in 718 [1, 2] and ATI 718Plus [31], which can inform the development of an ageing heat treatment for VDM 780. At the current time, similar data is not available in the literature. It is understood in ATI 718Plus [31] that precipitation of γ' can be suppressed completely if fast rates of cooling (>3 °C/s) from solution heat treatment, such as from oil quenching, are used, which means that the entire volume fraction of γ' can be precipitated from ageing heat treatment. This offers the opportunity to develop a narrow distribution of very fine γ' precipitates (10–30 nm). However, there is the possibility that such a distribution of γ' precipitates will create a lack of polydispersity, typically found from secondary and tertiary γ' precipitates, and produce a significant loss of ductility [32]. It was proposed that this occurs from localised strain concentrations that arise as dislocations pile up on γ/γ' interfaces rather than shearing γ' precipitates [32].

This paper describes experimental work to understand the influence of grain size and γ' precipitate size on tensile strength and ductility of VDM Alloy 780.

Starting Microstructure

The starting material for the study was produced by triple melting, i.e., vacuum induction melting (VIM), electro-slag remelting (ESR) and vacuum arc remelting (VAR) and hot deformed into billet with a diameter of 203 mm (8 inch). Test pieces and bars for compression tests and pancake forgings were extracted from the mid-radius of the billet, such that the compression direction was parallel with the long axis of the billet axis. Two billets were used, the first, for compression tests and pancake forgings, showed an average grain size of ASTM 6.5 (38 μm). The second billet, for complementary work to produce pancake forgings at VDM Metals International, showed a slightly coarser average grain size of ASTM 5 (65 μm) (Fig. 1). Whilst images of etched samples from optical microscopy indicate 100% recrystallization, electron backscattered diffraction was not undertaken to confirm this. However, Sharma et al. [29] reported that similar as-received VDM Alloy 780 billet material was partially recrystallized with fine γ′ precipitates. In this and the Sharma study, no η/δ particles were detected in the as-received billet.

The γ′ solvus temperature (T_{solvus}) of the material is 990–993 °C. This was determined from thermal expansion, specific heat capacity measurements and differential scanning calorimetry, conducted at the Fraunhofer Institut IKTS in Dresden, Germany. The η/δ T_{solvus} was determined from laboratory heat treatments on samples from billet, which were polished and examined in a scanning electron microscope (SEM) at VDM Metals International. From these SEM images, the highest T_{solvus} of the η/δ phases was found to be 1040 °C. Small amounts of η/δ have been detected at grain boundaries in material that has been exposed for 1 h above 1000 °C.

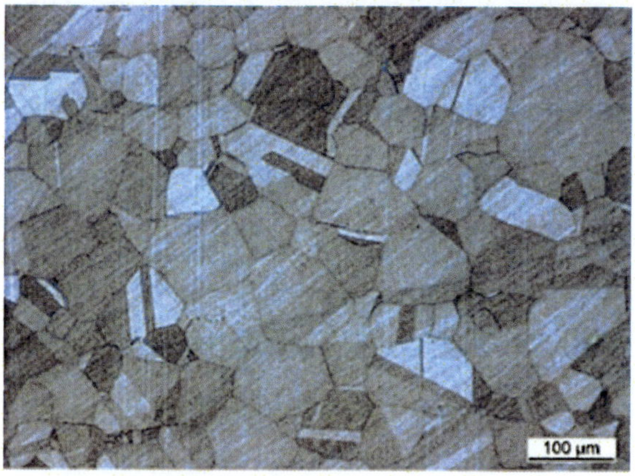

Fig. 1 Optical microscopy image of an etched sample from the mid-radius of 203 mm (8 inch) diameter billet, which was used for forging at VDM Metals International. The average grain size was 65 μm (ASTM 5). The sample was etched with 60 ml HCl (37%) and 10 ml HNO_3 (65%)

Experimental Work

Initial experimental work involved compression tests on right circular cylinder (RCC) test pieces that were nominally 16 mm in diameter and 24 mm in height. Test pieces were soaked at the specified forging temperatures of 985 °C (1805 °F), 1015 °C (1859 °F) and 1050 °C (1922 °F) for 8 h prior to testing. The soak time is a worst-case exposure for billet that is intended for large diameter forgings. Compression tests were conducted at the Illinois Institute of Technology in Chicago, USA at a constant strain rate of 0.1 per s using a servo-hydraulic frame and loading bars fitted with Si_3N_4 anvils. Temperature control was provided by K-type thermocouples, which were attached to test pieces. Boron nitride coating was applied to both test pieces and anvils. After forging down to 12 mm in height (50% upset), test pieces were transferred within 30 s to a laboratory furnace and re-exposed to the forging temperature for at least 5 min then furnace cooled at rates of about 100°F/h (0.015 °C/s). This was intended to promote post-DRX. One semi-circular pie slice from each test piece was left in the as forged and soaked condition. The other half was available for solution heat treatment (SHT). A semi-circular pie slice from the test piece forged at 985 °C was solution heat treated at 1000 °C for 1 h and static air cooled. Pie slices were ground, polished and etched using Kalling's reagent. Optical microscopy images were taken in the centre and mid-radius locations at the axial mid-height position.

Forging simulations using DEFORM were undertaken. Figure 2 indicates that the mid-radius, MR, and centre, C, positions for optical microscope images were subject to effective strains of 0.87 and 1.175 respectively.

The results of the compression tests and heat treatment trials established conditions for subsequent forging of bars that were machined from the mid-radius of the starting billet material (with an average grain size of 38 μm). These were nominally 70 mm (2.76 inch) in diameter and 76 mm (3 inch) high. In preparation for forging, bars were coated with hexagonal boron nitride and soaked for 8 h at the forging temperature. Bars were forged at ATI Materials using a pilot scale hydraulic press, fitted with nickel superalloy plates that were exposed to the forging temperature. The total elapsed time from installing the plates and work piece was about 20 s. Bars were forged down to an actual cold height of 0.83–0.86 inch (21–22 mm) and a diameter of 130–133 mm (5.1–5.2 inch) using a forging temperature of 985 °C (1805 °F). The nominal ram speed was 0.1 inch/inch per s. The forged pancakes were returned to the furnace for post-DRX at 985 °C after 60 s from the end of upset. They were soaked at 985 °C for 5 min and furnace cooled at a nominal cooling rate of 100°F/h (0.015 °C/s) until they reached 600 °C (1112 °F), at which point, they were air cooled. In Table 2, these forging and heat treatment details are presented as Condition 1. It was recognised that adiabatic heating occurred during the short forging duration (about 120 s). A process simulation, using a heat capacity correction, suggested that for the 985 °C test, the temperature in the centre of the pancake may have reached 1007 °C.

Test piece blanks were extracted from the pancake forgings according to the schematic plan in Fig. 3 (left). Blanks for tensile tests were 10 mm in diameter ×

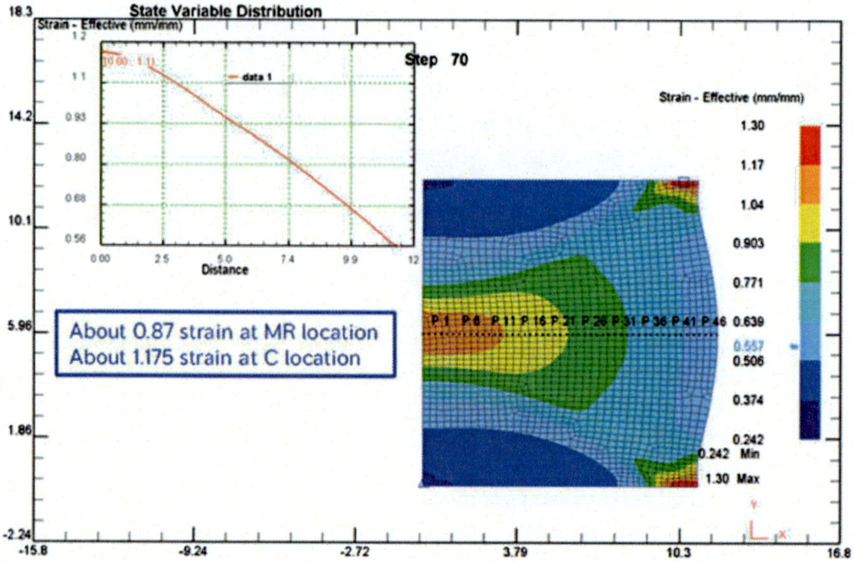

Fig. 2 Results from a DEFORM forging simulation on a RCC test piece. Images shows one half of a RCC pancake, 22.6 mm in diameter and 12 mm in height

Table 2 Forging conditions for pancake forgings and heat treatment conditions for test piece blanks. T = temperature, t = time, SR = strain rate, SHT = solution heat treatment, PSHT = post solution heat treatment

Condition	Forge T (°C)	Soak t (h)	SR (/s)	SHT T (°C)	SHT t (h)	PSHT 1	PSHT 2
1	985	8	0.1	1000	1	843 °C for 2 h	788 °C for 8 h
2	950	1	0.05	955	1	800 °C for 8 h	750 °C for 7 h
3	1000	1	0.05	955	1	800 °C for 8 h	750 °C for 7 h
4	1050	1	0.05	955	1	800 °C for 8 h	750 °C for 7 h

52 mm long. The red blank which spans the full diameter of the pancake had a square section and was used to confirm grain size prior to machining test pieces. Blanks were solution heat treated at 1000 °C (1832 °F) for 1 h and fast cooled to room temperature at rates of nominally 3 °C/s from 1000 °C (1832 °F) to 871 °C (1600 °F). A 2-stage post-solution heat treatment (PSHT) or age was then applied. This consisted of 2 h at 843 °C (1550 °F) and 8 h at 788 °C (1450 °F) and was defined based on information from VDM Metals International, which indicated that the precipitation kinetics of VDM Alloy 780 were slower than those reported by Srinivasan et al. [31] for ATI

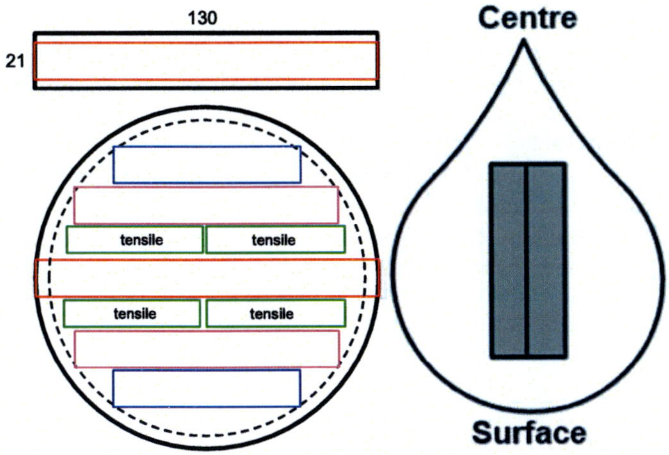

Fig. 3 Schematic plans showing extraction of test piece blanks from 130 mm diameter × 21 mm height pancakes forgings (left) and forged triangular section pieces (right), which were about 150 mm from surface to centre × 30 mm in height. Surface and centre refer to the location of the as-received material, extracted from billet

718Plus. The aim was to nucleate the maximum fraction of γ' precipitates during the first stage of the PSHT then continue to coarsen the precipitates during the second, lower temperature exposure. Blanks were furnace cooled at a rate of about 100 °F/h (0.015 °C/s) in between the 2 ageing temperatures. After completing the second stage age, blanks were static air cooled.

Complementary experimental work was conducted at VDM Metals International in Altena, Germany. Triangular cross section bars, 95 mm (3.74 inch) in height were extracted from 203 mm (8 inch) diameter billet. Bars were soaked for 1 h at the forging temperatures given in Table 2 for Conditions 2, 3 and 4, before being forged using an industrial scale hydraulic press, with high temperature insulation mats between forging dies and the bar to minimise temperature reduction during deformation. The total elapsed time from a bar leaving the furnace to forging was about 12 s. Bars were forged down to an actual cold height of 30 mm (1.2 inch) using a nominal ram speed was 0.05 mm/mm per s. Adiabatic heating in the middle of the section was predicted to 30–40 °C. Square section blanks, 20 × 20 × 80 mm were cut from the middle of the pancakes, as indicated in Fig. 3 (right), from mid-height locations. They were then solution heat treated at 955 °C (1751 °F) for 1 h and air cooled. Whilst γ' T_{solvus} is 990–993 °C, γ' precipitates have not been detected, due to very slow precipitation kinetics, until after a partial SHT at 955 °C. A process simulation was conducted to understand the cooling rates in a circular section test piece blank. Simple radiation was assumed, with emissivity of 0.8 and natural convection (0.02 mW/mm^2). The cooling rates between 955 and 700 °C, in the mid-length of the test piece, were predicted to be greater than 2.5 °C/s. Blanks were aged at 800 °C (1472 °F) for 8 h, cooled at 50 °C/h (0.014 °C/s) to 750 °C (1382 °F) and soaked at 750 °C for 7 h.

After completing the second stage age, blanks were static air cooled. Although the PSHT for Conditions 2, 3 and 4 is different to that for Condition 1, the aims of the 2-stages are the same, as described earlier, albeit with finer γ′ precipitates expected from 7 h at 750 °C, compared to 8 h at 788 °C. Forging and heat treatment details are summarised in Table 2.

Test pieces that conform to 5D designs were machined from heat treated blanks. Tensile tests were conducted at Element Plzen, Czech Republic in accordance with ASTM E8 and E21 at room temperature, 650 and 750 °C. Strain rates were initially 0.005 per minute for proof stress measurement then 0.05 per minute to rupture.

Results

True stress, true strain data from compression tests on RCC test pieces are presented in Fig. 4. For the applied strain rate of 0.1 per s, the press loads during forging at 985 °C are 40–60% higher than those at 1050 °C and show a variation in peak true stress of about 60 MPa in 2 test pieces. There is very little variation in true stress, true strain data from compression tests at 1015 and 1050 °C.

Optical microscope images of microstructure after forging and post forge DRX soak are presented in Figs. 5 and 6. It is evident from Fig. 5 that sub-solvus forging at 985 °C produces only partial recrystallization, with relatively small differences in strain between the centre (1.175) and mid-radius (0.87) locations giving rise to significant differences in % recrystallization. Much of the mid-radius location shows a deformed billet grain size, with very little recrystallization, at least from the optical

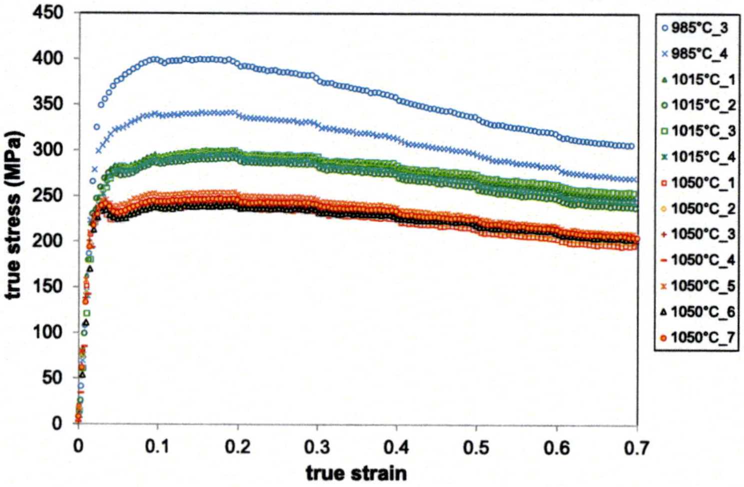

Fig. 4 True stress, true strain data from compression tests on RCC test pieces at a strain rate of 0.1 per s. Test pieces were soaked for 8 h

microscope image. Figure 6 shows images of the mid-radius locations from the RCC test pieces forged above the γ' T_{solvus} at temperatures of 1015 and 1050 °C. The average grain size of RCC test pieces forged at 1015 °C was rated as ASTM 6–5.5 (45–53 μm), using the comparison method in ASTM E112, with grains as large as (ALA) ASTM 3–2.5 (127–151 μm). The latter were determined from ASTM E930. Similarly, the average grain size of RCC test pieces forged at 1050 °C was rated as ASTM 5–4 (64–90 μm), with ALA grains of ASTM 2.5–1.5 (151–214 μm). Whilst it is not possible to comment on the degree of recrystallization in these images, a larger grain size has been produced at 1050 °C, which suggests grain growth rather than recrystallization. This finding is unexpected given the reported work by Sharma et al. [29] and will be discussed later. As noted earlier, no or very small quantities of η/δ phases were detected in billet or forged material. Grain growth at temperatures above 1000 °C is therefore not attributed to dissolution of grain boundary η/δ needles.

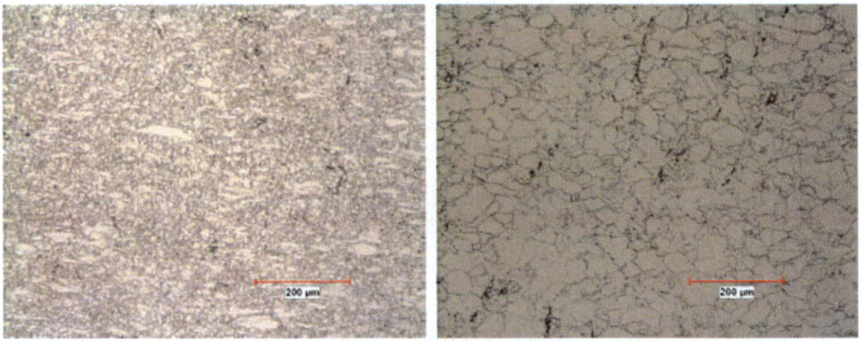

Fig. 5 Optical microscope images of microstructure from mid-height/centre (left) and mid-height/mid-radius (right) of RCC test piece forged at 985 °C. Etched with Kalling's reagent. Predicted strains were 1.175 and 0.87 at centre and mid-radius location respectively

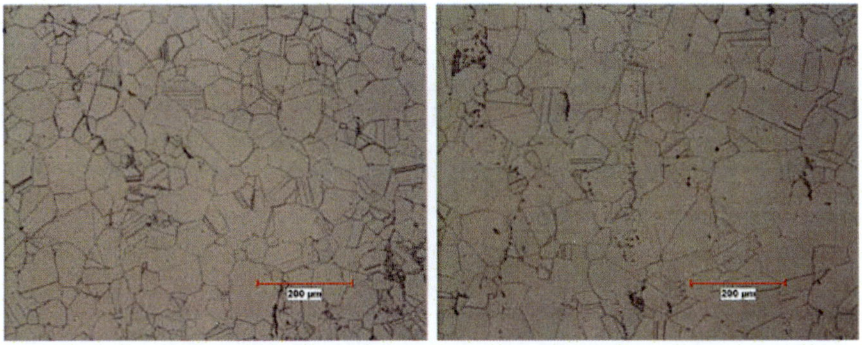

Fig. 6 Optical microscope images of microstructure from mid-height/mid-radius of RCC test piece forged at 1015 °C (left) and 1050 °C (right). Etched with Kalling's reagent. The predicted strain was 0.87 at the mid-radius location

The images in Figs. 5 and 6 also show particles, presumably primary carbide (NbC) stringers that are broadly orientated in a direction, which is parallel to the forging axis.

After heat treatment, the grain size of edge, mid-radius and centre locations for the red blank in Fig. 3 was examined. The average grain size was rated, using the comparison method, to be ASTM 8–7 (22–32 μm), with occasional grains ALA ASTM 4.5–4 (75–90 μm). These grain size values are for Condition 1 in Table 2. The corresponding grain sizes (Table 3) for Conditions 2, 3 and 4 in Table 2 were determined using the linear intercept method (in ASTM E112). They indicate the degree of recrystallization. Whilst a reasonably uniform grain structure is produced after SHT (Fig. 7), remnant un-recrystallized grains remain as the SHT temperature was below γ' T_{solvus}.

The results of tensile tests at 20, 650 and 750 °C on VDM Alloy 780 in Conditions 1–4 are compared with 0.2% proof stress and tensile strength data for alloy 720Li in Fig. 8. The data for 720Li are typical for forgings in the fine grain condition, solution heat treated at 1095 °C, oil quenched and aged at 760 °C for 16 h. In this condition, the average grain size is between ASTM 12–7 (6–32 μm), with occasional grains ALA ASTM 1 (254 μm). The strength values for VDM Alloy 780 in Condition 2 are very similar to those for 720Li, i.e., within 1–2% either side of the 720Li data. The 0.2% proof stress values for VDM Alloy 780 in Condition 3 are also very similar to those for 720Li but the tensile strength values are 5–7% lower. However, the strength values for VDM Alloy 780 in Condition 1 and 4 are generally lower than those for 720Li, notably at 650 °C, showing reductions of 15 and 19% in 0.2% proof stress

Table 3 Average grain size values after heat treatment for Condition 2, 3 and 4 (see Table 2)

Condition	Ave. fine grain size (μm)	ASTM No	%	Ave. coarse grain size (μm)	ASTM No	%
2	10	10.5	85	119	3.5	15
3	15	9.5	90	109	3.5	10
4	19	8.5	95	152	2.5	5

Fig. 7 Optical microscope images showing the grain structure for VDM 780 after Condition 2 (left), 3 (middle) and 4 (right). The forging direction is vertical. Samples were etched with 60 ml HCl (37%) and 10 ml HNO_3 (65%)

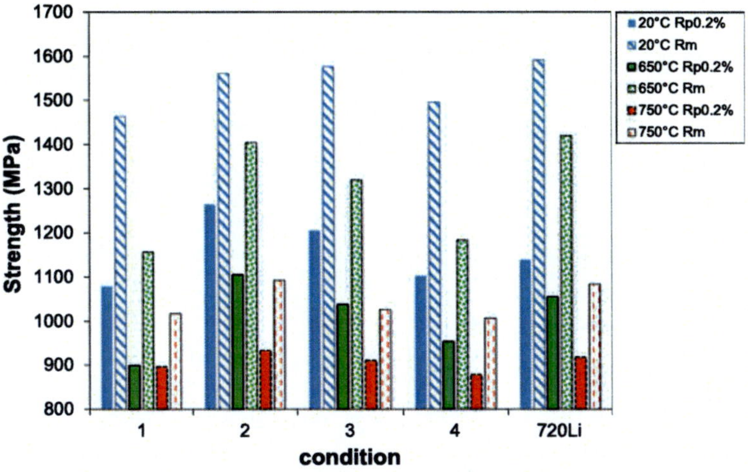

Fig. 8 0.2% Proof stress (Rp0.2%) and tensile strength (Rm) data for VDM Alloy 780 in Conditions 1–4 and for alloy 720Li at 20, 650 and 750 °C

and tensile strength respectively for Condition 1 and reductions of 10 and 17% in 0.2% proof stress and tensile strength respectively for Condition 4.

Figure 9 presents ductility data for VDM Alloy 780 in Conditions 1–4. The alloy exhibits good ductility at room temperature in all the Conditions examined. However, elongation is no greater than 7% at 650 and 750 °C, with reduction of area values of less than 2% for Conditions 3 and 4 at 650 °C. In comparison, typical ductility values for alloy 720Li are greater 18% at these temperatures.

Discussion

From the results of experimental work, it is evident that strength of VDM Alloy 780 is highly sensitive to changes in the size of γ grains and γ' precipitates. The least appealing microstructure in terms of strength was produced from Condition 1. A screw thread at one end of the broken test piece from the 20 °C test was polished and examined in a SEM. A backscattered electron image in Fig. 10 (left) shows the grain structure in this test piece, which is consistent with the rated average grain size of ASTM 8–7 (22–32 µm) for the red square section blank (Fig. 3, left).

The polished section from the screw thread end was subsequently electrolytically etched with 10% sodium metabisulfite ($Na_2S_2O_5$) solution for 1–2 s using 3 V. Whilst the periphery of the sample was over-etched, a sufficient area was etched correctly to produce images such as the in-lens image in Fig. 10 (right), which shows γ' precipitates. As expected, these are fine but perhaps more rounded cuboidal rather than spherical. Two X 200,000 magnification images were analysed using Image J

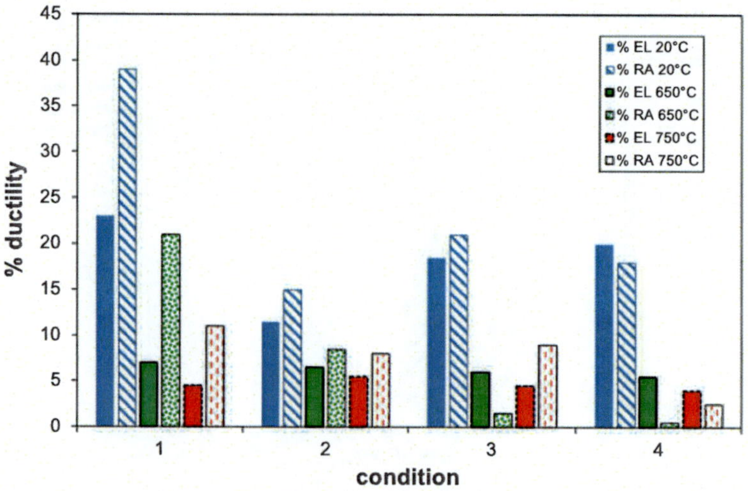

Fig. 9 % Elongation (EL) and reduction in area (RA) data for VDM Alloy 780 in Conditions 1–4 at 20, 650 and 750 °C

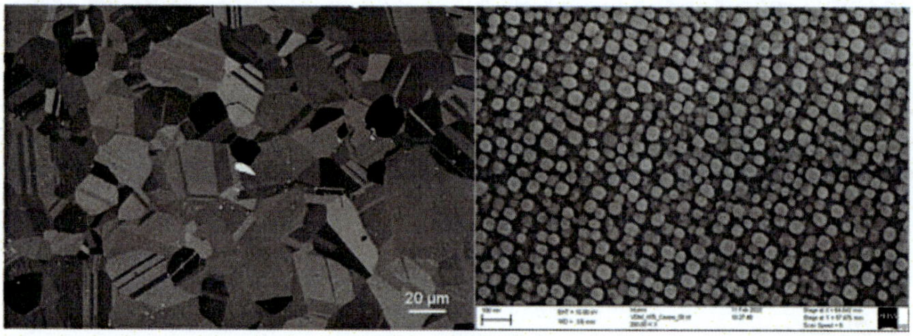

Fig. 10 Backscattered electron image (left) and in-lens image (right) showing γ grains and γ′ precipitates from VDM Alloy 780 in Condition 1 after heat treatment

software to determine the equivalent circular diameter of γ′ precipitates from area measurements. Over 900 precipitates were analysed for each condition. The results of image analyses are presented in Fig. 11. Given the use of an etched sample, and the necessity for high magnifications, it was not possible to reliably determine the area fraction of γ′ precipitates nor sizes of γ′ precipitates that are smaller than 10 nm.

Similar image analyses are required to understand the size of γ′ precipitates for Conditions 2, 3 and 4. This is beyond the scope of the current study. However, remaining blanks for tensile test pieces (Fig. 3, left) in Condition 1 were subsequently re-solution heat treated at 1000 °C for 20 min, water quenched and aged at 843 °C (1550 °F) for 20 min, static air cooled to room temperature then aged at 720 °C (1328 °F) for 8 h and static air cooled. This Condition was called Condition 1+ .

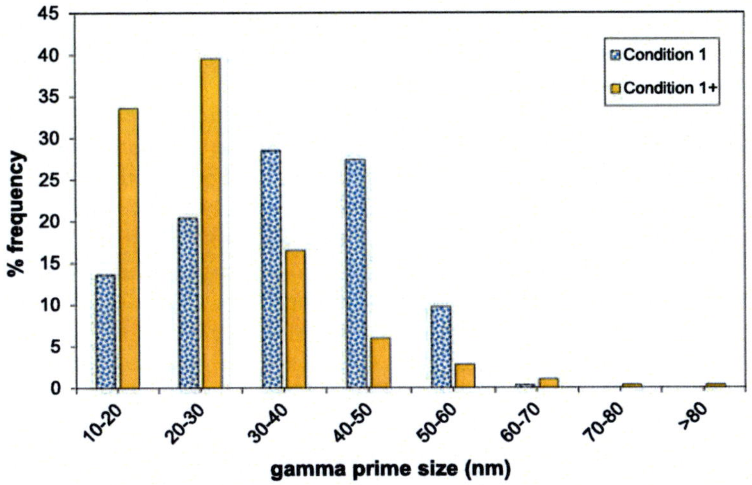

Fig. 11 Histogram showing the size distribution of γ′ precipitates (equivalent circular diameter) for VDM Alloy 780 in Condition 1 and 1+ after heat treatment. The average γ′ size is about 35 nm for Condition 1 and 26 nm for Condition 1+

Tensile tests were conducted to identical conditions as the earlier tests. It was found that the 0.2% proof stress value for Condition 1+ at 20 °C was within 0.5% of the Condition 1 but showed 3% improvement at 650 °C. A 5% reduction in 0.2% proof stress was recorded at 750 °C in Condition 1+ compared to Condition 1. In terms of tensile strength, Condition 1+ showed a very similar value, within 0.5% of Condition 1 but reduced values at 650 °C and 750 °C, of 6 and 10% respectively. Reduction in areas values for Condition 1+ were at least 14% at all temperatures although the % elongation values were 8.5 and 3% at 650 and 750 °C respectively. The size distribution of γ′ precipitates for Condition 1+ are also plotted in Fig. 11. It is evident that the PSHT for Condition 1+, compared to that for Condition 1, has reduced the average size of γ′ precipitates from 35 to 26 nm. It is unclear whether this is due to the reduced time at 843 °C or reduced temperature used for the second, lower temperature stage or a combination of the 2.

Despite the reduced size of γ′ precipitates, an improvement in yield stress or tensile strength was not achieved in Condition 1+, compared to Condition 1. This highlights the importance of grain size. Whilst improved strength levels may be possible with an ASTM 8–7 (22–32 μm) grain size, from smaller γ′ precipitates, such changes are likely to reduce ductility, compared to alloy 720Li, as indicated in Conditions 2, 3 and 4. Optimising the size of γ′ precipitates from ageing heat treatment is discussed further in the following paragraphs. However, to conclude the discussion regarding the effect of grain size of VDM Alloy 780 on strength, it appears that strength levels of 720Li could be achieved if γ grains are no greater than ASTM 9 (16 μm), as shown from tests on VDM Alloy 780 material in Conditions 2 and 3. Such a grain size may be difficult to produce in ingot metallurgy alloys, particularly in large forgings with complex shapes and varying strain levels. Finer grains are also likely to have less appealing time dependent crack growth behaviour, which may present a significant limitation.

One issue that remains unresolved in this study is the observed grain growth in RCC test pieces from compression tests at 1050 °C and a strain rate of 0.1 per s. Using very similar or identical billet, Sharma et al. [29] produced an average grain size of 26 μm, from identical forging conditions. Similarly, the average grain size in material produced to Condition 4 (Table 3) was found to be 19 μm. One observed difference is the soak time for billet material prior to forging. The RCC test pieces were soaked for 8 h at 1050 °C, whereas the other billet was soaked for 30 min in the work reported by Sharma et al. and 1 h for Condition 4. It is likely then that the extended time above the γ′ T_{solvus} has grown the grains considerably in the billet so that the starting material prior to forging is not equivalent. This is confirmed in Fig. 12, from subsequent heat treatment trials on billet material at 1050 °C. The fitted curve indicates that an 8 h soak at 1050 °C is expected to produce an average grain size of 108 μm, which is significantly larger than that for the as-received billet material (38 μm).

The SHT trials at 1050 °C also provided an opportunity to show that γ′ precipitation is suppressed from fast cooling rates. The samples were 1–4 mm in thickness and were static air cooled from 1050 °C. The cooling rates for such a sample are expected to be higher than 3 °C/s. Gamma prime precipitates could not be resolved at a magnification of X 200,000 from a polished sample that was electrolytically etched using the procedure described earlier.

Without further experimental evidence to confirm the size of γ′ precipitates after heat treatment, the coarsening kinetics of ATI 718Plus have been considered using

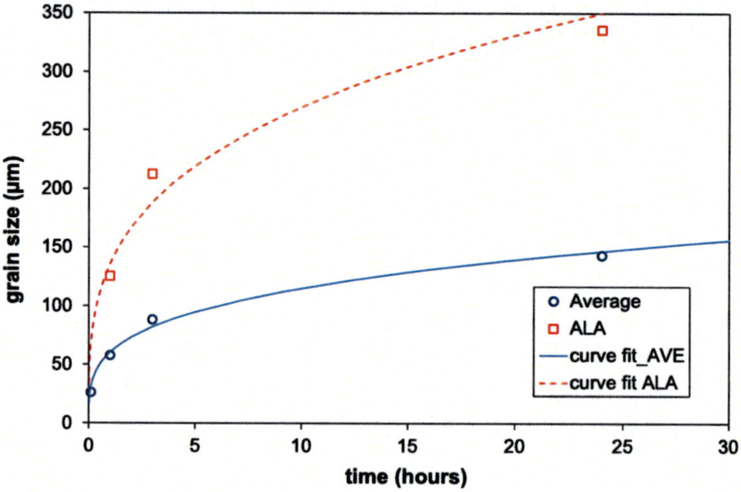

Fig. 12 Grain size values for billet material soaked for 1, 3 and 24 h at 1050 °C

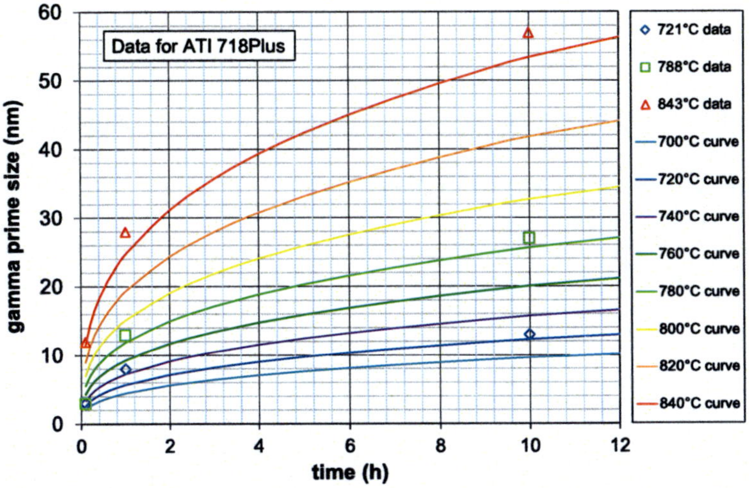

Fig. 13 Curves based on the Oswald ripening model and data from Srinivasan et al. [31] to characterise the coarsening of γ′ precipitates in ATI 718Plus

the data from Srinivasan et al. [31]. An Oswald ripening model, i.e., Eqs. 1 and 2 below, was used to fit curves to experimental data [31], where r is particle radius at time t (r_t) or t of 0 (r_0) and θ is temperature in °C. In Fig. 13, curves were produced using C and n values of −23.265 and 0.037 respectively.

$$\bar{r}_t^3 - \bar{r}_0^3 = k.t \tag{1}$$

$$k = \exp(n\theta + C) \tag{2}$$

Using these C and n parameters, the ageing heat treatments for Conditions 1, 2–4 and 1+ would be expected to produce γ′ precipitates that have an average size of 59, 46 and 29 nm in ATI 718Plus. As smaller γ′ precipitates were measured in VDM Alloy 780 material from Conditions 1 and 1+ , the precipitation and/or coarsening kinetics of VDM Alloy 780 are confirmed to be slower than those for ATI 718Plus. Further heat treatment experiments are recommended on VDM Alloy 780 to establish an Oswald ripening model.

All the conditions of VDM Alloy 780 examined in this study showed low ductility values at 750 °C and a small difference between yield stress and tensile strength. This indicates that a working temperature of 750 °C is likely to be too high for any future disk rotors made from VDM Alloy 780. A peak temperature of 700 °C is probably more realistic.

Summary and Conclusions

- From this study on VDM Alloy 780, it was found that forging temperature has a significant effect on the fraction of recrystallized grains after forging at a strain rate of 0.1 per s.
- Forging temperatures below the gamma prime (γ') temperature T_{solvus} require relatively high press loads (>300 MPa) and produce low fractions of recrystallized grains, unless forging strains are greater than 1.2. However, a uniform grain structure can be produced after solution heat treatment (SHT). The resultant average grain size depends on the solution heat treatment (SHT) temperature. A short duration above γ' T_{solvus} can produce an average grain size of ASTM 8–7. SHT below γ' T_{solvus} can produce an average grain size finer than ASTM 9 but with some remnant unrecrystallized grains.
- No or very small quantities of η/δ phases were detected in billet or forged material. Grain growth at temperatures above 1000 °C is not attributed to dissolution of grain boundary η/δ needles.
- A forging temperature of 1050 °C, above γ' T_{solvus}, requires lower press loads (circa 250 MPa) and produces a high fraction of recrystallized grains. An average grain size of ASTM 8.5 to 7 can be produced after forging from dynamic recrystallization (DRX) and post-DRX. However, the soak time for billet at 1050 °C, prior to forging, should be minimized to prevent grain growth.
- Grain size and the size of γ' precipitates determine tensile behaviour of VDM Alloy 780.
- Strength levels of 720Li can be achieved in VDM Alloy 780 if γ grains are no greater than ASTM 9 (16 μm) and provided an appropriate size distribution of γ' precipitates is produced.
- Like alloy ATI 718Plus, precipitation of γ' in VDM Alloy 780 can be suppressed if fast rates of cooling (>3 °C/s) from SHT are used, so that the entire volume fraction of γ' is precipitated from ageing heat treatment. Whilst γ' T_{solvus} is 990–993 °C, γ' precipitates have not been detected, due to very slow precipitation kinetics, until after a partial SHT at 955 °C.
- Further heat treatment experiments are recommended on VDM Alloy 780 to establish an Oswald ripening model. This is required to determine an optimized ageing heat treatment. 2 h at 843 °C + 8 h at 788 °C produced an average γ' size of 35 nm. 20 min at 843 °C + 8 h at 720 °C produced an average γ' size of 26 nm.
- Narrow size distributions of fine γ' precipitates have produced low ductility values in VDM Alloy 780, compared to those shown by alloy 720Li. % elongation in VDM Alloy 780 was no greater than 7% at 650 and 750 °C, with reduction of area values of less than 2% at 650 °C for some forging and heat treatment conditions examined.
- Tensile testing at 750 °C produced low ductility and a small difference between yield stress and tensile strength. Forging and heat treatment conditions had little influence on tensile properties at 750 °C.

Acknowledgements The authors would like to thank Rolls-Royce and VDM Metals International for permission to publish this work. Rolls-Royce would like to thank the following individuals for their contributions to this work: Bradley Hostetler, Ramesh Minisandram from ATI Materials for pancake forging and blank heat treatment, Kathy Ho, Sammy Tin from Illinois Institute of Technology for compression testing, and Stuart White from Intertek Derby for SEM work.

References

1. Han Y-F, Deb P, Chaturvedi MC (1982) Coarsening behavior of γ''- and γ'- particles in Inconel alloy 718. Met. Sci. 16 (12): 555–561.
2. Devaux A, Nazé L, Molins R, Pineau A, Organista A, Guédou, JY, Uginet JF, Héritier (2008) Gamma double prime precipitation kinetics in Alloy 718. Mater. Sci. Eng. A (486): 117–122. https://doi.org/10.1016/j.msea.2007.08.046.
3. Xie X, Wang G, Dong J, Wu C, Radavich R, Shen G, Lindsley BA (2001) Alpha chromium formation in alloy 718 and its effect on creep crack propagation. In: Loria, EA (ed) Superalloys 718, 625, 706 and Various Derivatives. The Minerals, Metals & Materials Society, Warrendale, PA, USA, p. 399–410.
4. Gao M, Dwyer DJ, Wei RP (1994) Chemical and microstructural aspects of creep crack growth in Inconel 718 alloy. In: Loria, EA (ed) Superalloys 718, 625, 706 and Various Derivatives. The Minerals, Metals & Materials Society, Warrendale, PA, USA, p. 581–592.
5. Molins R, Hochstetter G, Chassaigne JC Andrieu E (1997) Oxidation effects on the fatigue crack growth behaviour of alloy 718 at high temperature. Acta Mater. 45(2): 663–674.
6. Keefe PW, Mancuso SO, Maurer GE (1992) Effects of heat treatment and chemistry on the long-term phase stability of a high strength nickel-base superalloy (1992). In: Antolovich, SD et al. (eds) Superalloys 1992. The Minerals, Metals & Materials Society, Warrendale, PA, USA, p. 487–496.
7. Heaney JA, Lasonde ML, Powell AM, Bond BJ, O'Brien CM (2014) Development of a new cast and wrought alloy (René 65) for high temperature disk applications. In: Ott, E et al. (eds) 8th International Symposium on Superalloy 718 and Derivatives. The Minerals, Metals & Materials Society, Warrendale, PA, USA, p. 67–77.
8. Devaux A, Picqué B, Gervais MF, Georges E, Poulain T, Héritier P (2012) AD730™ – a new nickel-based superalloy for high temperature engine rotative parts. In: Huron, ES et al, (eds) Superalloys 2012: 12th International Symposium on Superalloys. The Minerals, Metals & Materials Society, Warrendale, PA, USA, p. 911–919.
9. Kanno N, Higashi M, Takai R, Ishikawa S, Sasaki K, Sugiyama K Sumi Y (2020) Development and application of new cast and wrought Ni-base superalloy M647 for turbine disk. In: Tin, S et al. (eds) Superalloys 2020. The Minerals, Metals & Materials Society, Springer, p. 82–90. https://doi.org/10.1007/978-3-030-51834-9.
10. Fedorova T, Rösler J, Gehrmann B, Klöwer J (2014) Invention of a new 718-type Ni-Co superalloy family for high temperature applications at 750°C. In: Ott, E et al. (eds) 8th International Symposium on Superalloy 718 and Derivatives. The Minerals, Metals & Materials Society, Warrendale, PA, USA, p. 587–599.
11. Rösler J, Hentrich T, Gehrmann B (2019) On the development concept for a new 718-type superalloy with improved temperature capability. Metals 9, 1130 https://doi.org/10.3390/met9101130.
12. Kennedy, RL (2005) Allvac® 718Plus™, superalloy for the next forty years. In: Loria, EA (ed) Superalloys 718, 625, 706 and Derivatives 2005. The Minerals, Metals & Materials Society, Warrendale, PA, USA, p. 1–14.
13. Pickering EJ, Mathur H, Bhowmik A, Messé OMDM, Barnard JS, Hardy MC, Krakow R, Loehnert K, Stone HJ, Rae CMF (2012) Grain-boundary precipitation in Allvac 718Plus. Acta Mater. 60: 2757–2769. https://doi.org/10.1016/j.actamat.2012.01.042.

14. Sczerzenie FE, Maurer GE (1984) Development of Udimet 720 for High Strength Disk Applications. In: Gell, M et al (eds) Superalloys 1984: Proceedings of the Fifth International Symposium on Superalloys. Metallurgical Society of AIME, Warrendale, PA, USA, p. 573–580
15. Fahrmann M, Suzuki A (2008) Effect of Cooling Rate on Gleeble Hot Ductility of Udimet Alloy 720 Billet. In: Reed RC et al (eds) Superalloys 2008. The Minerals, Metals & Materials Society, Warrendale, PA, USA, p. 311–316.
16. Bond BJ, O'Brien CM, Russell JL, Heaney JA, Lasonde ML (2014) René65 Billett Material for Forged Turbine Components. In: Ott, E et al. (eds) 8th International Symposium on Superalloy 718 and Derivatives. The Minerals, Metals & Materials Society, Warrendale, PA, USA, p. 107–17.
17. Genereux PD, Borg CA (2000) Characterization of Freckles in a High Strength Wrought Nickel Superalloy. In: Pollock TM et al (eds) Superalloys 2000. The Minerals, Metals & Materials Society, Warrendale, PA, USA, p. 19–27.
18. Wang X, Ward RM, Jacobs MH, Barratt MD (2008) Effect of Variation in Process Parameters on the Formation of Freckle in INCONEL 718 by Vacuum Arc Remelting. Metall. Mater. Trans. A (39): 2981–2989.
19. Minisandram RS, Jackman LA, Russell JL, Lasonde ML, Heaney JA, Powell AM (2014) Recrystallization Response During Thermo-Mechanical Processing of Alloy René 65 Billet. In: Ott, E et al. (eds) 8th International Symposium on Superalloy 718 and Derivatives. The Minerals, Metals & Materials Society, Warrendale, PA, USA, p. 95–105.
20. Crozet C, Devaux A, Forestier R, Charmond S, Hueller, Helm D, Buchmann W (2016) Effect of Ingot Size on Microstructure and Properties of the New Advanced AD730TM Superalloy. In: Hardy MC et al (eds) Superalloys 2016: Proceedings of the 13th International Symposium on Superalloys. The Minerals, Metals & Materials Society, Warrendale, PA, USA, p. 437–446.
21. Coyne-Grell A, Blaizot J, Rahimi S, Violatos I, Nouveau S, Dumont C, Nicoläy, Bozzolo N (2022) Recrystallization Mechanisms and Associated Microstructure Evolution During Billet Conversion of a Gamma-Gamma' Nickel Based Superalloy. J. Alloys Compd. (916) https://doi.org/10.1016/j.jallcom.2022.165465.
22. Kalinowski JM (1994) Weldability of a Nickel-Based Superalloy. NASA Contractor Report 195376, Lewis Research Center https://ntrs.nasa.gov/api/citations/19950004780/downloads/19950004780.pdf. Accessed 21 December 2022.
23. Ghica C, Solís, Munke J, Stark A, Gehrmann B, Bergner M, Rösler, Gilles R (2020) HRTEM analysis of the high-temperature phases of the newly developed high-temperature Ni-base superalloy VDM 780 Premium. J. Alloys Compd 814: https://doi.org/10.1016/j.jallcom.2019.152157.
24. Casanova A, Hardy M, Rae C (2014) Morphology and kinetics of grain boundary precipitation in alloy ATI 718Plus®. In: Ott, E et al. (eds) 8th International Symposium on Superalloy 718 and Derivatives. The Minerals, Metals & Materials Society, Warrendale, PA, USA, p. 573–586.
25. Casanova A, Loehnert K, Huenert D, Hardy M, Rae C (2016) Systematic evaluation of microstructural and thermo-mechanical effects on the as-forged condition of alloy ATI 718Plus®. In: Hardy, M et al. (eds) Superalloys 2016: Proceedings of the 13th International Symposium on Superalloys. The Minerals, Metals & Materials Society, Warrendale, PA, USA, p. 427–436.
26. Huenert D, Proebstle M, Casanova A, Schluetter R, Krakow R, Buescher M, Randelzhofer P, Evans A, Loehnert K, Witulski T, Neumeier S, Rae C (2016) ATI718Plus® - new nickel based disc alloy and its capability. In: Hardy, M et al. (eds) Superalloys 2016: Proceedings of the 13th International Symposium on Superalloys. The Minerals, Metals & Materials Society, Warrendale, PA, USA, p. 783–792.
27. https://www.icao.int/Meetings/EnvironmentalWorkshops/Documents/ICAO-TransportCanada-2006/Anderson_ops.pdf.
28. Crudden DJ, Mottura A, Warnken N, Raeisinia B, Reed RC (2014) Modelling of the influence of alloy composition on flow stress of high-strength nickel-based superalloys. Acta Mater. 75: 356–370.

29. Sharma J, Hafez Haghighat, Gehrmann B, Moussa C, Bozzolo N (2020) Dynamic and post-dynamic recrystallization during supersolvus forging of the new nickel-based superalloy-VDM Alloy 780. In: Tin, S et al. (eds) Superalloys 2020. The Minerals, Metals & Materials Society, Springer, p. 450–460.
30. Nicoläy A, Fiorucci G, Franchet JM, Cormier J, Bozzolo N (2019) Influence of Strain Rate on Subsolvus Dynamic and Post-Dynamic Recrystallization Kinetics of Inconel 718. Acta Mater. 174: 406–417.
31. Srinivasan D, Underwood Lawless L, Ott EA (2012) Experimental determination of TTT diagram for alloy 718Plus®. In: Huron, ES et al, (eds) Superalloys 2012: 12th International Symposium on Superalloys. The Minerals, Metals & Materials Society, Warrendale, PA, USA, p. 759–768.
32. Collins DM, Stone HJ (2014) A modelling approach to yield strength optimisation in a nickel-base superalloy. Int. J. Plast. 54: 96–112.

Local Assessment of Mechanical Properties in Forged Alloy 718 Components Based on the Simulation of the Microstructure Evolution During Production

Christian Gruber, Peter Raninger, Aleksandar Stanojevic, Flora Godor, Hans-Peter Gänser, Stefan Marsoner, and Martin Stockinger

Abstract The characterization of local and global fracture mechanical properties is carried out with destructive testing methods and is increasingly required in the specifications of forgings. Especially in the case of alloy 718 aircraft parts, the numerical estimation of local material properties is essential for lightweight design, geometry optimization, and a significant reduction of development and experimental characterization costs. This leads to a demand for numerical models to capture initial microstructural inhomogeneities, describe the forming history, and reflect the local microstructure and properties of the final product. Therefore a digital twin for the complex forging process was developed in order to reproduce and evaluate the resulting local microstructure across the complete process chain. Since the microstructure determines the mechanical properties like yield stress and fracture toughness, a dedicated model was implemented to describe the local evolution of the relevant microstructural features.

Keywords Forging · Alloy 718 · FE simulation · Microstructure modeling · Multi-class grain size distribution · Fracture toughness · Precipitation · Delta phase

C. Gruber (✉) · A. Stanojevic · F. Godor
Voestalpine BÖHLER Aerospace GmbH & Co KG, Kapfenberg, Austria
e-mail: christian.gruber2@voestalpine.com

P. Raninger · H.-P. Gänser · S. Marsoner
Materials Center Leoben Forschung GmbH, Leoben, Austria

M. Stockinger
Department for Product Engineering, Montanuniversität Leoben, Leoben, Austria

E. A. Ott et al. (eds.), *Proceedings of the 10th International Symposium on Superalloy 718 and Derivatives*, The Minerals, Metals & Materials Series,
https://doi.org/10.1007/978-3-031-27447-3_4

Introduction

In order to estimate the microstructure and the mechanical properties of forged aircraft components, the microstructure of the pre-material and the complete thermomechanical history of the component are key parameters. The initial microstructure varies locally throughout the volume of a billet and has a major influence on the final structure and the local distribution of mechanical properties. Therefore, a coupled thermo-mechanical and microstructural simulation tool of the production process takes these pre-material inhomogeneities into account. In order to characterize the local grain size, grain size distribution, precipitations and delta phase as well as the mechanical properties such as yield/ultimate strength and fracture toughness, the billet material was thoroughly analyzed and digitized [1, 2]. The analysis of several feedstock samples showed a pronounced variation of the microstructure due to different production routes involving processes like vacuum arc re-melting, homogenization, upsetting, and cogging. A digital twin [3, 4] of the forging process takes these variations of the billet material as initial conditions depending on the analyzed position into account and provides quantities like local temperature, strain, and strain rate. Based on the microstructural data and the thermomechanical history, a newly developed multi-class grain size model [5, 6], parametrized by a series of specially designed experiments, reconstructs the complete microstructure evolution during the different forming processes. In this simulation recrystallization, grain growth, and precipitation processes are taken into account. A special focus is set on the recrystallization behavior, which can be subdivided into dynamic, meta-dynamic, and static recrystallization [7–9]. The locally different occurrence and interaction of the previously mentioned phenomena significantly impact the resulting microstructure in terms of the grain size distribution. As the fracture toughness of alloy 718 results from the specific combination of microstructural features such as the grain size distribution, size, and amount of precipitates (carbides, carbonitrides and nitrides) and the delta phase fraction, the local fracture toughness can be estimated in dependence of the local microstructure, the load direction and the assumed crack orientation [1, 10]. These relationships are considered in the model and used to estimate the local and global yield strength [11] and fracture mechanical properties. For validation of the digital twin with its fracture property model, a series of tensile and fracture tests was performed. These tests were carried out not only on the final part but also on the billet material in order to determine the inhomogeneities and initial property values at different positions. The investigated positions are then tracked through the FE simulation and local thermomechanical data are used as input for the multi-class grain size model, which finally yields the resulting microstructure at the tracked material points. This modeling approach establishes the link between a) the initial microstructure of the billet, b) the forging process, and c) the final distribution of the fracture toughness in the component.

Material

The used and tested materials are cast and wrought alloy 718 in different conditions in terms of their processing routes. Three different billet materials in two dimensions as well as subsequent forged final aircraft parts from these billets are characterized and analyzed. The investigated billet materials consist of 1) a round 8″ double melt (DM) material, 2) a 10″ double melt material, and 3) a 10″ triple melt (TM) material from the same manufacturer, summarized in Table 1. To understand and model the connection between the microstructures and properties of the billet and the final part, two specific aircraft parts—a turbine disk and a structural part (engine mount)—are forged from the same billets and tested for their mechanical and microstructural properties.

To emphasize the importance of the local assessment of each tested material, Fig. 1 represents the distribution and variation of the microstructure at the investigated positions of the pre-material. The scanning electron microscopy (SEM) image using a backscatter detector (BSE) clearly shows the variation in grain size, grain size distribution, and delta phase fraction.

In addition, the number, size, and distribution of other precipitates (niobium- and titanium-rich precipitates) than delta phase were investigated on a large scale for all billet materials and stored in a database in order to reflect the possible initial conditions for the forged parts and track the local evolution of all these microstructural features. The microstructural features grain size as well as delta phase fraction and size strongly vary throughout the thermomechanical process, which can be calculated with the multi-class grain size model. The coarse carbon and nitrogen-containing

Table 1 Material list

Nr.	Material	Aircraft part
1	8″ DM	Structural part
2	10″ DM	Structural part
3	10″ TM	Turbine disk

Fig. 1 SEM 10″ TM billet; left: center; right: edge; white: delta phase

precipitates only shift with the local position due to the geometry and material flow during the closed die forging.

Experimental

To quantify mechanical properties like the yield strength and the fracture mechanical properties, a series of tests were performed with samples representing different microstructural states, wherein the biggest difference in microstructural features was detected from the center to the edge on the 10″ TM billet material. The tensile tests and fracture toughness tests were performed according to ASTM E8a [12] and ASTM E399 [13]. Within this test series, the following microstructural features were analyzed as they have the highest impact on the fracture mechanical properties and can be quantified through testing and simulations: grain size, delta phase area fraction, particle density in terms of precipitates per area. These features vary in the three different materials (8″ DM, 10″ DM, and 10″ TM); therefore three specimens for each material and each of the three tested orientations were extracted from two different positions. This scheme was subsequently applied to the associated demonstrator components. The number of three specimens is used for statistical evaluation. The three orientations serve to vary the direction of crack propagation (R-C, R-L, and L-R, Fig. 2). With this sampling strategy, all crack propagation directions correlated with the material flow in the pre-materials are covered. With the investigation of the two different positions in the billet material, the microstructural impact is evaluated, since there is a strong gradient of grain size from the inside to the outside. The exact sampling plan for the billets is shown in Fig. 2, where the fracture toughness specimens (Compact Tension (CT) specimen, width 20 mm) are shown. The tensile specimens (6 mm diameter) were taken directly adjacent to the CT specimens, with the tensile direction perpendicular to the resulting fracture surface plane. In order to make the fracture toughness values comparable, the strength levels of all samples from the billet materials were brought to that of the demonstrator components by means of a heat treatment (solution annealing (SA) and aging for structural parts; direct aging (DA) for turbine disk). This heat treatment was also necessary to compare the billet materials with each other, since hardness mappings of each in "as delivered" condition (hardness according to ISO 6507 [14]) showed a strong variation and indicated a difference in yield strength, which would also affect the fracture toughness and bias the relation between the microstructure and the mechanical properties.

For fatigue crack growth determination, additional 8-point bending specimens (100 $\times$ 20 $\times$ 6 mm) from the 10″ DM material were extracted and used for the evaluation of the fatigue threshold value according to ISO 12108 [15] and for comparison with results from specimens extracted from demonstrator parts. These correlations and modeling results, including a detailed analysis of fracture surfaces, are given in [16].

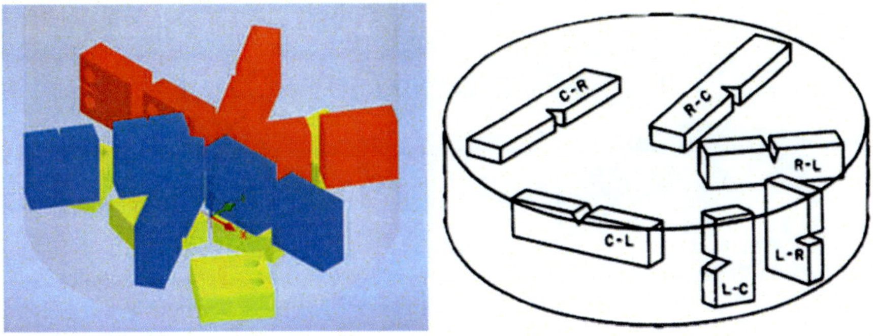

Fig. 2 Compact tension (CT) specimen positions in the billet and crack orientation directions according to ASTM E399 [13]

Modeling and Simulation

For the model development, a hybrid statistical model is defined based on data from historical fracture toughness tests and tensile tests of various components by different test labs. The model is parameterized by the influencing features described before. For this purpose, the statistical model serves as a first approach and shall be augmented with physical knowledge based on further tests and findings. In principle, this approach is applicable for all mechanical parameters, but the main emphasis in this study is set on the fracture toughness, in particular the calculation of K_{IC}. The estimation of the fracture toughness is of great importance for structural components in aviation due to the loading characteristics, and the possibility of a numerical calculation of local K_{IC} would be advantageous for the design and optimization of new parts. In case for determining the local yield strength the applied approximation is described in [11] and further provides a relevant input parameter for the digital twin by representing the local and global fracture toughness properties.

Historical test reports for all relevant structural components from voestalpine BÖHLER Aerospace GmbH & Co KG from alloy 718 were used as a data basis for model parametrization and the resulting model constants were validated within the experimental test series. The mechanical parameters (K_{IC}/K_Q, yield strength) were supplemented with the microstructural features (ASTM grain size, δ-phase fraction, number of precipitates/particle density). The crack propagation orientation was defined by means of angular functions between the crack propagation direction and the forging direction (Fig. 3), as well as the position of the crack plane with respect to the main forging direction (Fig. 4). The fracture toughness results from the obtained database were used for parameterization with multilinear regression.

The local arrangement of the particles relative to the crack propagation direction is a key factor (Fig. 4). The particle arrangement is directly dependent on the local material flow within the component, which is influenced during forging. The material flow direction does not necessarily coincide with one of the principal axes of the global coordinate system. This is especially true for (a) multi-step forming where

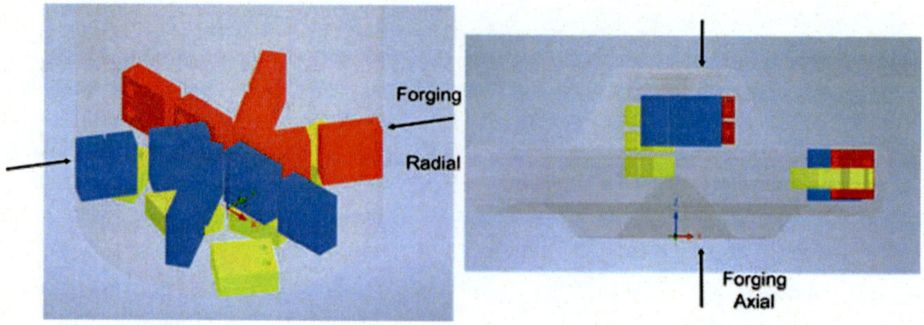

Fig. 3 Forging directions of the tested materials, left: radial forging of billet and right: die forging of the final geometry

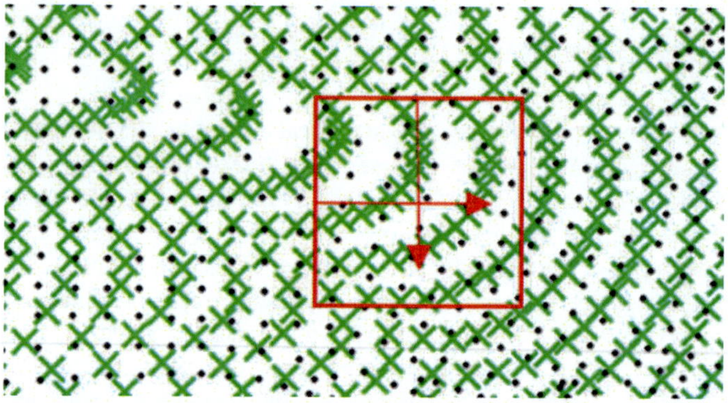

Fig. 4 Material flow in a turbine disk; green cross: shift of particles; red square marks the CT specimen and the arrows the crack propagation direction

the part is turned between steps and (b) for complex shaped structural components. Therefore, a spatial variable must be introduced that describes the local orientation dependence of fracture toughness and can be used for a modeling approach. Accordingly, it is of great importance to know the material flow. Figure 3 shows the difference in forging direction for two different operations relative to the orientation of an axisymmetric component. In case of the pre-material, the forging direction is radial and the primary material flow is axial. This behavior is more complex in the case of closed die forging. In the case of a turbine disk, the forging direction is mainly axial, but the material flow is strongly characterized by the geometry of the die.

The relationship between the forging direction and the material flow or better the crack propagation direction to the material flow direction can therefore be best characterized by angular functions between the two directions and serves as an input variable for the model. Furthermore, the crack propagation direction relative to the material flow direction was identified as one of the most influential parameters. Also here angular functions between these two quantities were introduced. A point tracking, which is part of the digital twin [1], can be used as a tool to determine the angle between the two directions since the displacement of a point over time can be considered as a vector between two-time increments and its orientation describes the material flow. Accordingly, this is an input for the fracture toughness model since the forging direction in the simulation is only vertical at all times. The influence of the crack propagation direction relative to material flow is illustrated in Fig. 4 for the 2D simulation of the forged turbine disk of the 10″ TM material. The initially homogeneous distribution of precipitates (green crosses) is displaced during deformation and changes their position according to the material flow. In the final configuration, the precipitation-plane with the highest density of green crosses corresponds to the crack propagation direction yielding the lowest fracture toughness in the model.

As mentioned, the statistical model was fitted using multilinear regression according to the Ordinary Least Square (OLS) [17] method. Here, the input parameters are combined in a linear function, whereby the parameterization is carried out on the basis of minimum of the distance squares to the measuring points. This fit was performed using the library statsmodels [18] in Python, whereby the delta phase, the grain size in ASTM, the yield strength, the precipitation density, the angle between the crack plane and the crack propagation direction relative to the forging direction were used as model input variables. This approach was chosen since all these variables can also be simulated or calculated within the digital twin from the multi-class grain size model. According to the OLS method, the modeling approach results in Eq. 1:

$$\begin{aligned} K_{Q(calculated)} = & c_0 + c_1(\delta - Phase[\%]) + c_2(Grain\ size\ [ASTM]) \\ & + c_3(Yield\ strength\ [MPa]) + c_4(Precipitation\ density) \\ & + c_5 \sin(crack\ plane\ normal - forging\ direction) \\ & + c_6 \cos(crack\ propagation\ direction - forging\ direction) \end{aligned} \quad (1)$$

The parameters c_0–c_6 result in the smallest deviations and a coefficient of determination (R-value) of 0.747, which is a reasonable value regarding the rather small database consisting of historical data only.

Results

The complete results matrix with all associated parameters (sample ID, heat treatment condition, grain size and direction/orientation) is listed in Table 2 for all materials and for all associated demonstrators (average of three specimens). In the following subsections the influencing factors on the mechanical properties (strength, elongation at final fracture and fracture toughness) will be discussed.

Table 2 Test series results matrix; average of three measured samples

Sample ID	Heat treatment condition	Grain size	Direction K_{IC}	Direction tensile test	Y0,2	TS	A	Z	K_Q
					[MPa]	[MPa]	[%]	[%]	[MPa$\sqrt{m}$]
10″ DM	Aged	Coarse	R-C	C	1233.3	1423.3	13.3	17.7	109.5
10″ DM	Aged	Fine	R-C	C	1260.0	1443.3	16.0	22.3	102.2
10″ DM	Aged	Coarse	R-L	R	1233.3	1420.0	15.8	18.7	110.9
10″ DM	Aged	Fine	R-L	R	1273.3	1466.7	17.0	19.3	100.4
10″ DM	Aged	Coarse	L-R	L	1243.3	1453.3	23.2	39.0	127.7
10″ DM	Aged	Fine	L-R	L	1290.0	1486.7	22.4	39.3	127.9
10″ TM	Aged	Coarse	R-C	C	1233.3	1423.3	15.7	23.7	118.5
10″ TM	Aged	Fine	R-C	C	1236.7	1436.7	19.4	25.7	108.2
10″ TM	Aged	Coarse	R-L	R	1236.7	1426.7	18.8	24.7	118.8
10″ TM	Aged	Fine	R-L	R	1253.3	1453.3	19.2	25.0	104.5
10″ TM	Aged	Coarse	L-R	L	1236.7	1436.7	27.0	42.7	138.6
10″ TM	Aged	Fine	L-R	L	1246.7	1463.3	23.6	39.3	131.8
Turbine disk 10″ TM	As Forged	Coarse	R-C	C	941.7	1306.7	30.0	34.0	111.8
Turbine disk 10″ TM	As Forged	Fine	R-C	C	971.7	1256.7	31.3	42.7	116.6
Turbine disk 10″ TM	As Forged	Coarse	R-L	R	946.7	1303.3	22.3	26.0	118.6
Turbine disk 10″ TM	As Forged	Fine	R-L	R	975.0	1240.0	34.4	47.0	110.5
Turbine disk 10″ TM	As Forged	Coarse	L-R	L	968.3	1280.0	18.9	20.7	107.8
Turbine disk 10″ TM	As Forged	Fine	L-R	L	973.3	1220.0	26.8	34.3	109.0
Turbine disk 10″ TM	DA	Fine	R-C	C	1390.0	1540.0	17.4	33.7	117.7
Turbine disk 10″ TM	DA	Fine	R-L	R	1383.3	1523.3	19.2	28.7	121.7

(continued)

Table 2 (continued)

Sample ID	Heat treatment condition	Grain size	Direction K_{IC}	Direction tensile test	Y0,2	TS	A	Z	K_Q
Turbine disk 10″ TM	DA	Fine	L-R	L	1396.7	1526.7	16.1	32.0	106.7
Structural Part 8″ DM	SA + aged	Coarse	L-S	L	1166.7	1443.3	23.0	33.0	111.9
Structural Part 8″ DM	SA + aged	Fine	L-S	L	1203.3	1480.0	22.8	39.0	115.5
Structural Part 8″ DM	SA + aged	Coarse	S-L	S	1186.7	1463.3	11.9	13.3	87.4
Structural Part 8″ DM	SA + aged	Fine	S-L	S	1173.3	1476.7	23.5	39.3	93.2
Structural Part 8″ DM	SA + aged	Coarse	L-T	T	1190.0	1453.3	19.8	22.3	111.8
Structural Part 8″ DM	SA + aged	Fine	L-T	T	1146.7	1450.0	16.9	18.3	115.8
Structural Part 10″ DM	SA + aged	Coarse	L-S	L	1176.7	1466.7	21.5	29.3	108.5
Structural Part 10″ DM	SA + aged	Fine	L-S	L	1166.7	1456.7	21.9	30.3	108.3
Structural Part 10″ DM	SA + aged	Coarse	S-L	S	1143.3	1456.7	15.4	16.0	97.5
Structural Part 10″ DM	SA + aged	Fine	S-L	S	1170.0	1463.3	16.5	16.0	102.4
Structural Part 10″ DM	SA + aged	Coarse	L-T	T	1170.0	1446.7	24.8	39.3	112.7
Structural Part 10″ DM	SA + aged	Fine	L-T	T	1166.7	1456.7	24.6	41.0	118.4
Structural Part 10″ DM	SA + aged	Fine	L-T	L	1160.0	1456.7	24.1	38.3	111.6
Structural Part 10″ DM	SA + aged	Coarse	T-L	S	1156.7	1453.3	16.1	18.7	104.7
Structural Part 10″ DM	SA + aged	Fine	L-S	T	1133.3	1450.0	21.8	32.7	117.2

- Yield strength

The strength properties were tested on each of the three tensile specimens for the previously defined positions and orientations according to ASTM E8 with a specimen diameter of 6 mm. The individual strength contributions are correlated with the hypotheses of the physically based modeling approach in [19] and corresponding results from the digital twin. Furthermore, the test series results are compared with each other in groups and the changes in strength are assigned to the respective varied microstructural features. The samples of the turbine disk in "as forged" condition were additionally used as a supplement to identify the contribution of the nano-precipitates (γ' and γ'' [20]) from the heat treatment to the total strength.

With regard to the strength values, the highest difference can be seen for the variation of the heat treatment conditions. The "as forged" condition delivers 400 MPa lower yield strengths than the same specimen positions and orientations in the direct aged condition. Accordingly, a 40% increase in strength can be attributed to the γ' and γ'' precipitates.

In contrast, the solution-annealed and aged specimens (structural part 8″ DM and 10″ DM) show lower yield strengths than the aged pre-materials, but the tensile strength has increased slightly. This behavior can also largely be attributed to the nano-precipitates. However, the solution annealing process leads to a higher niobium content in solution, because of a reduced phase fraction of the δ-phase and results in a coarser grain size and accordingly the changes in the strength cannot be assigned to a single microstructural feature of alloy 718.

After heat treatment the strength levels of the double and the triple melt billets are the same, which is the prerequisite for the comparability of fracture toughness experiments and the obtained results. Similar strength values have been confirmed from all investigated billet samples (10″ DM + 10″ TM).

The orientation of the tensile specimens only affects the elongation at final fracture and necking. While the strength within a tested group appears almost independent of grain size and orientation, the elongation and necking show a preferential orientation for improved values. While the pre-material has the highest plastification in axial orientation, the comparable rotationally symmetric demonstrator (turbine disk) shows a maximum in the radial direction. The samples from the structural components exhibit a similar behavior regarding the influence of the orientation. The change of the sample orientation with the best test results from axial in the billets to radial in the components is due to the orientation of the specimen in relation to the material flow in the respective component. This can be attributed to the formation of a specific arrangement of microstructural components especially the coarse carbide and carbonitride precipitates, which move with the material flow during forging. This hypothesis is corroborated by the analysis of the fracture toughness results since it includes the influence of the crack propagation direction and has already been identified as an important criterion from a previous analysis.

- Fracture toughness

A similar but more detailed approach for the yield strength is applied for the fracture toughness, as the statistical modeling approach will be deployed and validated based on that series of experiments. The same sampling scheme was used to determine the K_{IC} values according to ASTM E399 on CTW40 specimens with chevron notches. The tested materials showed consistently the same strength level, which is essential in order to compare the fracture toughness results with each other and correlate them with the corresponding microstructural feature. The comparison of the microstructure, the direction of crack propagation, the used pre-material, and the occurrence of precipitates are now precisely documented in this analysis. All these characteristic values were quantified in the pre-material as part of the test series and applied as parameters for the derivation of the fracture mechanic model.

The difference in the impact of the crack propagation direction for the pre-material compared to the demonstrator part of the turbine disk is striking. As with the fracture necking, the orientation dependence cannot be directly assigned to the crack propagation direction, but is dependent on the material flow during forging. Similarly, the comparisons from the 10″ DM billet material to the associated demonstrator component show a change in maxima and minima depending on the crack propagation direction. Therefore, the consideration of the orientation of the crack propagation relative to the material flow direction in the model can be confirmed as relevant input. The investigations further show that specimens with coarse microstructures and mostly combined with lower delta phase contents tend to have a higher fracture toughness of about 2 $MPa\sqrt{m}$ for each percent less delta phase.

However, the carbide and carbonitride precipitates in terms of number and size (density) deliver by far the highest impact on the fracture toughness. These incoherent precipitates remain unchanged and are only displaced over the entire thermomechanical process after casting. Their characteristics influence to a large extent the level of fracture toughness, which can be concluded from fractography because they are always exposed at all fracture surfaces and accordingly the shortest path from particle to particle describes the path of the fracture surface. The local rearrangement of the precipitate with the material flow thus determines the crack propagation direction where a maximum or minimum of the fracture toughness occurs. This behavior is shown in detail in Fig. 5 since here the specimens were taken at the same position in the pre-material billet and only the direction of crack propagation was changed. Thereby, the orientation of the precipitates in rows is clearly visible at the fracture surface. While the R-C orientation reflects a horizontal alignment, the R-L sample has a vertical alignment of the precipitates. This orientation again corresponds to the main material flow of the pre-material during cogging in the process chain of billet production. A detailed image of these precipitations rows (mainly niobium carbides) is shown in Fig. 6, where the exposed rows of precipitates clearly show the correspondence with a minimum toughness in this direction.

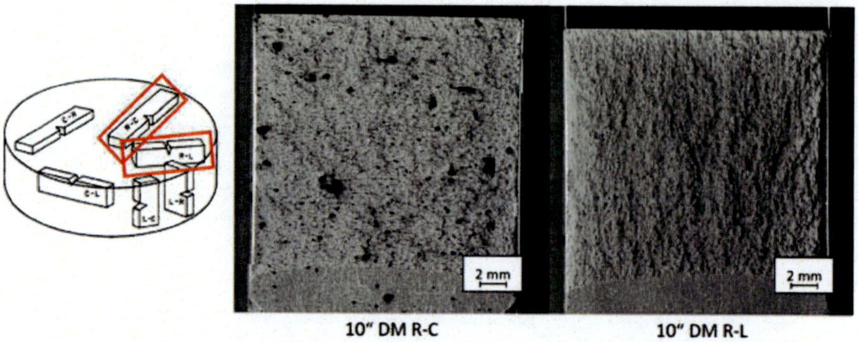

Fig. 5 Fractography of different crack propagation directions (10″ DM R-C vs. 10″ DM R-L)

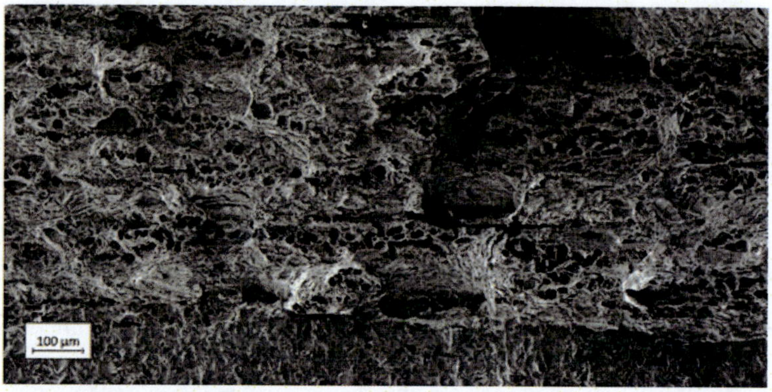

Fig. 6 Crack front of 10″ DM billet, edge position, R-C orientation

The application of the model in comparison with the data from the specific test series shows good accuracy. The comparison of the measured to the calculated fracture toughness values is shown in Fig. 7. All investigated billets and demonstrator parts from the test series lie within a range of ± 10 MPa$\sqrt{m}$. However, the impact of the sample preparation and the process of crack initiation within the limits of the specification remains as uncertainties and cannot be traced or quantified, neither from the test laboratory nor in the simulation.

Conclusion

The approach of a statistical hybrid model as part of the digital twin serves as a good starting point for understanding influencing features of the microstructure on the mechanical properties, especially on the fracture toughness. Using multilinear

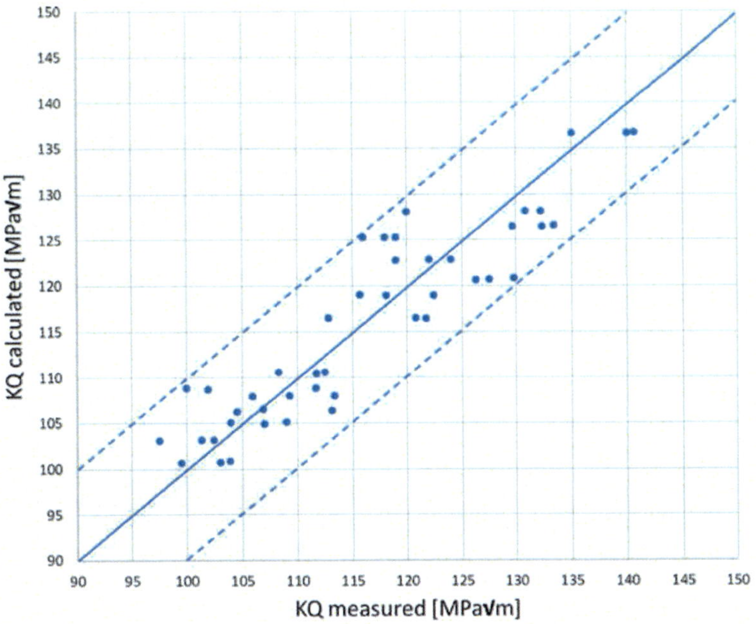

Fig. 7 Fracture toughness: calculated versus measured results

regression, the selected input parameters (delta phase, grain size, yield strength, precipitation density, the angle between the crack plane and the crack propagation direction relative to the forging direction) already provides characteristic results on local assessment within a range of ±10 MPa√m. This accuracy is sufficient for the application to predict local and global fracture toughness values. Additional influencing factors such as specimen geometry and preparation have similar effects on the spread of the result. To extent the model to a physically driven one, significantly more microstructural states would have to be analyzed in terms of heat treatment conditions, variations in chemistry, and in the number, size, and distribution of precipitates. However, this additional effort would incur disproportionate costs due to the strictly limited billet materials and controls within the aerospace industry. Only an increased understanding of the density of precipitates and how they act during the fracture process in determining fracture toughness will be beneficial. Furthermore, this generated knowledge would also help to better understand the crack propagation process. The results of the tensile tests corroborate the findings from previously applied models and hypotheses. In both model approaches (yield strength and fracture toughness), the mean grain size provides already adequate calculated mechanical properties. On the other hand, the multi-class representation of the microstructure is of high importance by determining the fatigue threshold value and a mandatory part of the digital twin for representing complex microstructures. The combination of all developed models builds a simulation tool for designing new aircraft parts of alloy 718 and will support the optimization of die geometries and process parameters to fulfil the highest customer demands, achieving the best quality by simultaneous reduction of local variations in microstructural features and mechanical properties.

Acknowledgements The research leading to these results has received funding from the TakeOff program. TakeOff is a Research, Technology, and Innovation Funding Program of the Austrian Federal Ministries for Climate Action, Environment, Energy, Mobility, Innovation, and Technology (BMK) (867403). The Austrian Research Promotion Agency (FFG) has been authorized for the Program Management.

References

1. C. Gruber, Digitalization of forged alloy 718 parts with implemented microstructure and fracture mechanic model, PhD thesis, Montanuniversität Leoben, 2022.
2. D. Winkler, Entwicklung einer Methode zur verbesserten Analyse von großflächigen Makroscheiben, Bachelor thesis, Montanuniversität Leoben, 2022.
3. A. Schwarz-Gsaxner and M. Stockinger, ENTWICKLUNG EINER ECAP ANLAGE MIT INTEGRIERTEM DIGITALEN SCHATTEN, Tagungsband 11, Ranshofener Leichtmetalltage: Leichtmetalle, Werkstoff- und Prozesstechnologien im Kontext von Dekarbonisierung und Digitalisierung, 2020, 80–87.
4. M. Lindthaler, Kurzschlusserkennung in der Kupferelektrolyse durch neuronale Netzwerke, Master thesis, Montanuniversität Leoben, 2020.
5. C. Gruber, P. Raninger, M. Stockinger and C. Bucher, Multi-class grain size model for forged alloy 718 aircraft parts, Material Science Forum, Volume 1016, 2021, 499–508.
6. C. Gruber, P. Raninger, A. Stanojevic, F. Godor, M. Rath, E. Kozeschnik and M. Stockinger, Simulation of Dynamic and Meta-Dynamic Recrystallization Behavior of Forged Alloy 718 Parts Using a Multi-Class Grain Size Model, Materials 14(1), 111, 2020.
7. M. Stockinger, Microstructural Simulation of Closed Die Forging of Nickel Base Alloys. Ph.D. Thesis, University of Technology Graz, 2003.
8. M. Stockinger and J. Tockner, Optimizing the forging of critical aircraft parts by the use of finite element coupled microstructure modelling, Superalloys 718, 2005, 625–706.
9. D. Huber, C. Stotter, C. Sommitsch, S. Mitsche, P. Poelt, B. Buchmayr and M. Stockinger, Microstructure modeling of the dynamic recrystallization kinetics during turbine disc forging of the nickel based superalloy Allvac 718Plus, Superalloys 2008, 855–861.
10. S. Atschreiter, Untersuchung der Variation der Rissbruchzähigkeit bei Inconel 718, Master Thesis, Montanuniversität Leoben, Leoben, Austria, 2018.
11. B. Oberwinkler, A. Fischersworring-Bunk, M. Hüller and M. Stockinger, Integrated process modeling for the mechanical properties optimization of direct aged alloy 718 engine disks, Proc. Int. Symp. Superalloys, 2016, 513–521.
12. ASTM E8/E8M-13a, Standard Test Methods for Tension Testing of Metallic Materials, American Society for Testing and Materials, 2013.
13. ASTM E399–20a, Standard Test Method for Linear-Elastic Plane-Strain Fracture Toughness of Metallic Materials, American Society for Testing and Materials, 2020.
14. ISO 6507-1:2018: Metallische Werkstoffe - Härteprüfung nach Vickers, International Organization for Standardization, 2018.
15. ISO 12108:2018: Metallic materials – Fatigue testing – Fatigue crack growth method. International Organization for Standardization, 2018.
16. C. Gruber, P. Raninger, J. Maierhofer, H.-P. Gänser, A. Stanojevic, A. Hohenwarter and R. Pippan, Microstructural Impact on Fatigue Crack Growth Behavior of Alloy 718, Metals, 2022, 12(5): 710.

17. The Method of Least Squares, https://www.jmp.com/en_in/statistics-knowledge-portal/what-is-regression/the-method-of-least-squares.html [05.03.2022].
18. Statsmodels, https://www.statsmodels.org/dev/index.html [18.12.2021].
19. V. V. Riellia, F. Godor, C. Gruber, A. Stanojevic, B. Oberwinkler and S. Primig, Effects of processing heterogeneities on the micro- to nanostructure strengthening mechanisms of an alloy 718 turbine disk, Materials & Design 212, 2021.
20. F. Theska, A. Stanojevic, B. Oberwinkler and S. Primig, Microstructure-property relationships in directly aged Alloy 718 turbine disks, Mater. Sci. Eng. A. 776, 2020.

Towards Enhancing Hot Tooling to Form High-γ' Superalloys

Arthi Vaasudevan, Fernando D. León-Cázares, Enjuscha Fischer, Thomas Witulski, Catherine Rae, and Enrique Galindo-Nava

Abstract Ni-superalloys are well-established for use in high temperature applications in aerospace, power generation, and automotive sectors, yet, are seldom considered as materials for hot tooling. The operational conditions of hot forming dies potentially exceed those experienced by aircraft turbine discs. Fortunately, new disc alloys have pronounced elevated temperature capabilities and the current study focuses on implementing two advanced alloys, VDM 780 and Haynes 282 (H282) as hot tool materials. There is, however, inadequate evidence of their life-limiting properties and mechanisms in the in-service temperature regime of 700–900 °C. Thus, realistic operating conditions were replicated by combining interrupted short and long-term thermal-mechanical tests. Initially, isothermal ageing in the furnace was used to compare the extent of γ' coarsening between the alloys, and subsequent in-situ ageing and compression testing measured the accompanying loss in strength. Compression creep testing at stresses near the yield points (250–750 MPa)

A. Vaasudevan (✉) · C. Rae (✉) · E. Galindo-Nava (✉)
Department of Materials Science and Metallurgy, University of Cambridge, 27 Charles Babbage Rd., Cambridge CB3 0FS, UK
e-mail: av511@cam.ac.uk

C. Rae
e-mail: cr18@cam.ac.uk

E. Galindo-Nava
e-mail: e.galindo-nava@ucl.ac.uk

E. Galindo-Nava
Department of Mechanical Engineering, University College London, 510a Roberts Engineering Building Torrington Place, London WC1E 7JE, UK

E. Fischer · T. Witulski
Materials Science, Otto Fuchs KG, Derschlager Str. 26, 58540 Meinerzhagen, Germany
e-mail: enjuscha.fischer@otto-fuchs.com

T. Witulski
e-mail: thomas.witulski@otto-fuchs.com

F. D. León-Cázares
Sandia National Laboratories, 7011 East Avenue, Livermore, CA 94550, USA
e-mail: fleonca@sandia.gov

E. A. Ott et al. (eds.), *Proceedings of the 10th International Symposium on Superalloy 718 and Derivatives*, The Minerals, Metals & Materials Series,
https://doi.org/10.1007/978-3-031-27447-3_5

revealed accelerated creep rates at high temperatures. The results indicated that even as exposure duration, temperature, and applied stress all influence microstructural evolution, the exposure temperature was pivotal in determining the effective life of these γ' strengthened alloys. Dissolution kinetics of γ' around near-solvus temperatures was crucial and was governed by elemental additions. As a result, the research paves the way for a better understanding and design of superalloys with improved thermal integrity for hot tooling.

Keywords VDM 780 · H282 · Superalloys · Hot tooling · Gamma prime · Thermal stability · Creep testing · Microstructure · Delta · Carbides · Turbine discs · Hot forging

Introduction

There is an increasing emphasis on developing high temperature materials to cater to the transportation and power requirements [1]. The largest users of high temperature materials are the gas turbine engines in modern jet aircrafts and the land-based turbine power generators. Other areas of application include turbocharger rotors, pressure vessels, heat exchanger tubing, etc. Nickel superalloys satisfy the requirements with their ability to withstand appreciable loading at operating temperatures close to the melting point ($T_o/T_m > 0.6$) for extended periods of time and resist mechanical and chemical degradation, without undergoing failure. The implementation of these materials results in improved fuel economy and reduced carbon emissions [1, 2].

The superior performance of Ni superalloys, especially the ones employed as aircraft engine components, stems from having a high γ' volume fraction (V_f), and demand the forming of these parts at temperatures around 1000–1100 °C [3]. This consequentially, exerts severe expectations on the hot tooling, imposing tool temperatures in the range of 700–900 °C. Popular steel dies are inadequate to handle such increased thermal–mechanical stresses. The current work takes a novel approach to exploring the application of low γ' superalloys as hot forming materials.

Over the last 50 years, alloy Inconel 718 (IN718) has been a widely used material among low γ' alloys owing to its combination of significant strength, workability, corrosion resistance, affordability, and excellent weldability [4]. Despite these appealing attributes, IN718 suffers from the thermal instability of γ'' phase. At 650 °C, the metastable phase overages and rapidly transforms into the thermodynamically stable, but plate-like and brittle δ. Hence, the alloy's application for prolonged time is restricted to temperature lower than 650 °C to retain strength and creep resistance properties. ATI Specialty Materials developed alloy 718 Plus with a 50 °C advantage over IN718 (up to 704 °C) [5], via promoting the more stable $L1_2$ phase γ'. More recently, the development of the new alloy VDM 780, with the prospect of retaining good properties until 750 °C has gained attention for its use at even higher temperatures [6].

VDM 780, developed by VDM Metals, is an improved derivative of IN718 for higher service temperatures. The γ′ solvus is ~995 °C and solvus of δ is ~1020 °C [7]. Like 718 Plus, there has been an ambiguity concerning the nature of the high temperature phase, either δ and/or η [8, 9]. The high temperature phase was proved by C. Ghica, et al. [10] using HRTEM to be a layered structure consisting of alternating η and δ phases, with a majority of η in the precipitates [10]. For the sake of unambiguous discussion in this work, the high temperature phase will be referred to as δ. Nevertheless, no trace of γ″ has been observed. The presence of a bimodal distribution of γ′ was revealed, with the Vickers hardness values increasing in heat treated specimens holding increased V_f of smaller precipitate sizes [9]. Apart from the potential of having high strength capabilities up to 750 °C, there is no data published in open literature about the elevated temperature strength, thermal stability, and creep resistance of VDM 780.

The current study concentrates on investigating the thermal stability and creep properties of VDM 780 at temperatures between 700 and 900 °C, and importantly draws a comparison between VDM 780 and an established, low γ′ alloy, Haynes 282 (H282), with a view to potentially implementing them as materials for hot forming tools. H282 is an appealing γ′-strengthened alternative to IN718, with adequate manufacturability. Its modest elevated temperature strength is compensated by excellent microstructure stability over long exposure times [11]. The alloy is said to demonstrate resistance to γ′ coarsening and to forming deleterious TCP phases even with 1000 h of exposure at temperatures between 760 and 870 °C [11], and hence, was chosen for comparison with VDM 780. In the current study, VDM 780 and H282 were subject to isothermal ageing, in-situ strength testing, and compression creep testing, in the temperature range of 700–900 °C, followed by thorough microstructural examination to understand the effects of stress, temperature, and ageing time on the microstructural evolution and stability.

Experimental Methods and Procedures

Alloys VDM 780 and H282 were made available by Otto Fuchs KG in the form of forging stocks. The nominal compositions of the alloys are presented in Table 1. The alloys were forged to produce billets and were subsequently heat treated as follows: **VDM 780**: Solution treatment (ST) 1020 °C/1H/AC, Precipitation anneal (PA) 720 °C/8H/FC → 620 °C/8H/AC; and **H282**: ST 1050 °C/1H/AC, PA 1010 °C/2H/AC + 788 °C/8H/AC.

Table 1 Nominal compositions of VDM 780 and H282 in weight percent

Wt.%	Cr	Co	Al	Ti	Nb	Mo	C	B
VDM 780	18	25	2.1	0.2	5.4	3	–	–
H282	20	10	1.5	2.1	–	8.5	0.06	0.005

Metallographic Preparation and Microstructural Examination

Samples in as-received (AR), as-aged, and crept conditions were cut in the direction transverse to the longitudinal axis of the cylinders using Struers-Secutom precision cutting machine. The cut pieces were mounted in conductive Bakelite and were ground progressively with fine SiC papers starting from 1200 to 4000 grits, then polished with 1 μm diamond solution followed by a final 0.06 μm colloidal silica polish. Samples were subsequently electrochemically etched using the γ etchant, 10% orthophosphoric acid in distilled water. The microstructures were examined under the secondary electron and backscatter electron modes available on a scanning electron microscope (SEM) Zeiss Gemini 300.

Isothermal Ageing

Cylinders were sealed in quartz tubes filled with Ar gas and subsequently were isothermally aged in an induction furnace. Following aging, the samples were rapidly quenched in ice water to freeze and retain the as-aged microstructures.

Mechanical Testing

All compression tests in this work were carried out using TA Instrument's DIL 805 A/D, a differential dilatometer amenable to quenching (A) and deformation (D) modes as it ensured good temperature and strain control. Miniaturized specimens were machined out of the heat-treated billets to yield solid cylinders of diameter 5 mm and length 10 mm, appropriate for dilatometry. Prior to testing, the cylindrical specimens were thoroughly rinsed in acetone and then in ethanol in an ultrasonic bath. They were ground with 1200-grit emery sheet to remove the peripheral oxide layer to expose the bare alloy surface to spot welding, given a final rinse in industrial methyl sulphate and blow dried. Accurate measurements of initial length and diameter of all prepared dilatometer samples were recorded with a vernier caliper having an accuracy of 0.01 mm to be input into the program. Maximum care was taken to maintain parallelism between the ends of the cylinders and not to grind away excess material. Spot welding of S-type 0.2 thermocouple at the center of the cylinder was carried out under Ar atmosphere and two Mo-alloy discs (of diameter 10 mm and thickness 0.5 mm, each) were spot welded onto the ends of the cylinders to provide lubrication and prevent easy cooling of samples from the tools. The ensemble of Mo-discs, sample and thermocouples were carefully placed inside the dilatometer, ensuring that the platens were strictly parallel.

Hot compression tests were conducted using strain-control at test temperatures from 700 to 900 °C to derive proof stresses and flow curves behaviour. Final true strain for deformation was set to 0.15 for all conditions. Held under vacuum, the cylinders were ramped to test temperatures at 30 °C/s, homogenised for 10 min (to attain thermal equilibrium), and deformed until the set strain was achieved or the applied force exceeded 20 kN (maximum permissible force). Time, displacement, force, true strain, and true stress were recorded at 3000 deformation steps. A strain rate of 0.01/s (or deformation rate 0.1 mm/s) was selected as it closely replicated the industrial deformation rate. The samples were then cooled to room temperature at the rate of 10 °C/s under pure argon gas and were stabilised for a minute before venting the chamber.

Compression creep testing was conducted under load-control at temperatures from 700 to 900 °C, and at constant stresses between 250 and 750 MPa, input as equivalent force in Newtons. Test specimens were again held under vacuum, ramped to test temperatures at 30 °C/s, soaked for 10 min, and deformed for 3 or 7 h, depending on the test. Time, true strain, applied stress, change in length measured with silica pushrods and change in diameter measured using a monochromatic laser beam were recorded for 6000 deformation steps. Following the deformation, the samples were cooled to room temperature at the rate of 10 °C/s under pure argon gas and were stabilised for a minute before venting the dilatometer chamber.

All test data collected were processed and analysed using OriginPro [12], a proprietary computer software for scientific graphing and data analysis.

Image Analysis

SEM micrographs were processed using the image analysis software ImageJ to estimate γ′ area % and mean size. A normal distribution of the particle size was generated with automatic binning, along with the summary of mean particle size and standard deviation. The data was imported to OriginPro, wherein the measurements of areas of observed precipitates were assumed to be perfectly circular and were processed to yield their resultant diameter.

Results

Initial Microstructure

VDM 780 and H282 are both low γ′ alloys, i.e., <35% V_f, yet they are fundamentally different in terms of their composition; standard heat treatments; grain size; the presence, morphologies, and distributions of secondary phases. Starting microstructures of VDM 780 and H282 were examined to understand these differences. Micrographs imaged at lower and higher magnifications are shown in Fig. 1. (VDM 780) and Fig. 2. (H282).

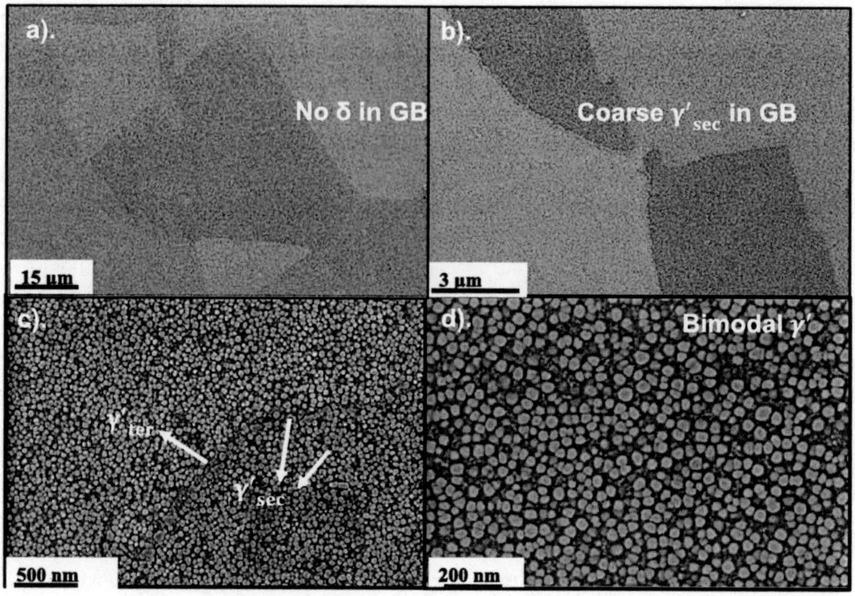

Fig. 1 Initial microstructure of VDM 780 in the as-heat treated state (**a**) secondary electron (SE) image showing grain boundaries devoid of δ phase, (**b**) In-lens SE (IL) image at higher magnification showing coarse γ'_{sec} at grain boundaries, (**c**) IL image showing γ'_{sec} and γ'_{ter} at the boundaries, and (**d**) IL image displaying bimodal distribution of intragranular precipitates in γ matrix

VDM 780 was composed of nearly equiaxed γ grains with a weighted average grain diameter of 93.8 ± 2.1 μm and a fair number of annealing twins. The grain sizes were measured from EBSD data using Aztec crystal software. High angle grain boundaries were designated as boundaries having misorientation above 15°; twin boundaries were excluded from the grain size calculation. The grain boundaries in VDM 780 were δ-free but were populated by coarse γ′. This was expected as the solution heat treatment temperature was around the δ solvus. This attribute can potentially cause grain coarsening at elevated temperatures due to the absence of an effective grain boundary pinning phase. When viewed at higher magnification, the grain interiors showed a bimodal distribution of γ′, containing secondary precipitates $\left(\gamma'_{sec}\right)$ of mean diameter 35 ± 5.24 nm, formed on cooling from solution treatment and tertiaries $\left(\gamma'_{ter}\right)$ of diameters ranging between 10 and 15 nm, nucleated during cooling from solution treatment and developed further during aging treatment. The average area % of γ'_{sec} was measured to be 26.8%.

The γ′ particle size distribution in VDM 780 is shown in Fig. 3a.

H282 revealed a combination of fine and coarse grains with grain diameters ranging between 30 and 273 μm. The mean grain diameter was greater than that of VDM 780 and was measured to be 171.2 ± 13.9 μm. The sub-grain microstructure at a finer scale contained monomodal γ'_{sec} distribution with an average particle diameter of 37 ± 4.6 nm. The average area % was calculated to be 22.2%. The γ′ particle size

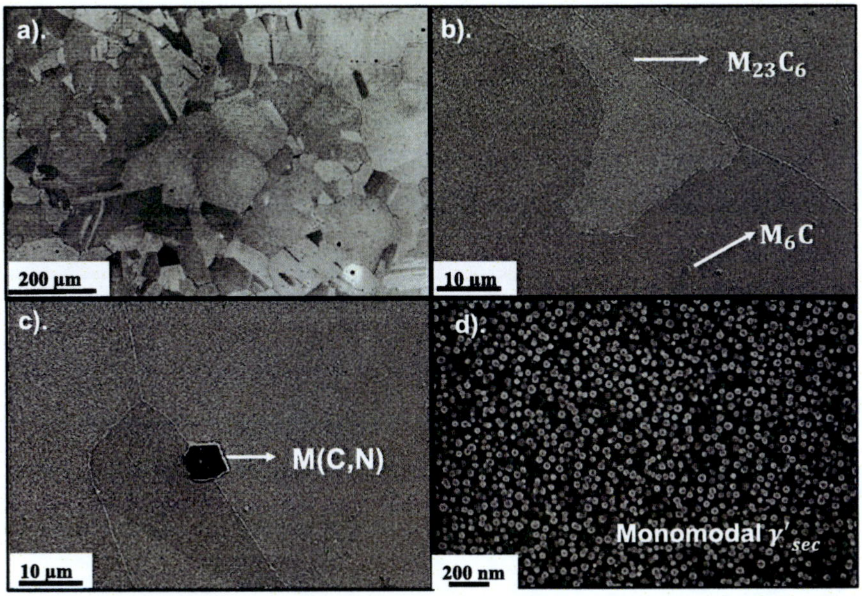

Fig. 2 Initial microstructure of H282 in the as-heat treated state (**a**) SE image displaying a mix of small and coarse grains, (**b**) SE image depicting the presence of $M_{23}C_6$ on boundaries and M_6C intragranularly, (**c**) SE image indicating the presence of a carbo-nitride, and (**d**) IL image of monomodal distribution of γ′ in γ matrix

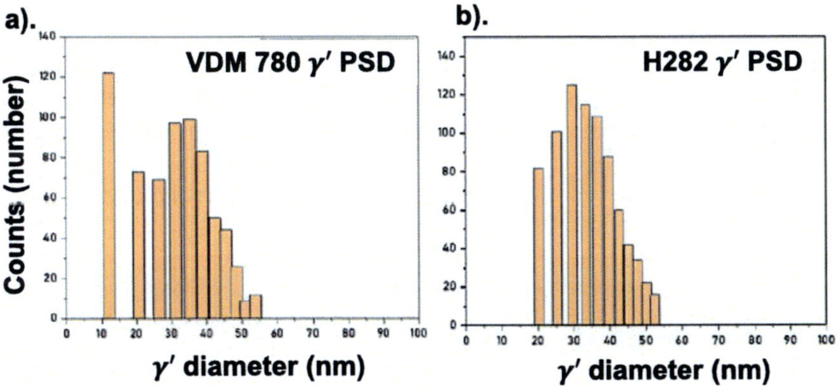

Fig. 3 γ′ particle size distribution in the alloys, (**a**) shows bimodal distribution exhibited by VDM 780, and (**b**) shows monomodal distribution exhibited by H282

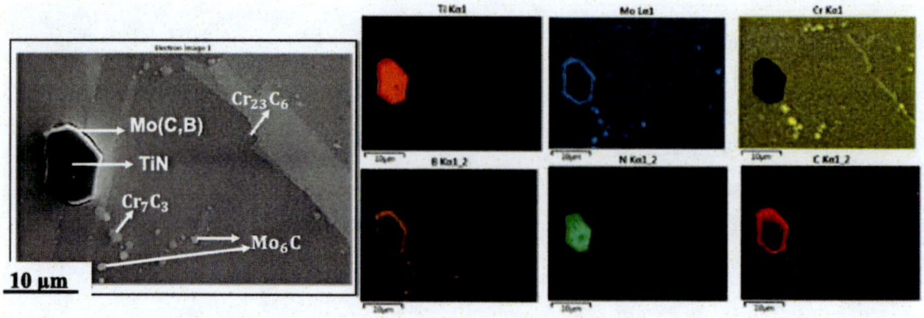

Fig. 4 EDX mapping of different types of carbides present in H282

distribution in H282 is shown in Fig. 3b. The grain boundaries were decorated with discontinuous films of $M_{23}C_6$ and grain interiors showed a mix of other carbides such as MC, M_6C, along with borides, nitrides, carbo-nitrides, or carbo-borides. To understand the nature of carbides and other secondary phases present apart from γ', and their morphology and distribution in the AR material; energy dispersive X-ray spectroscopy (EDX) was performed, and the results are presented in Fig. 4.

From the images, a Cr-rich discontinuous carbide network was observed in the grain boundaries, and Mo-rich globular carbides were observed intragranularly and were taken to be $M_{23}C_6$ and M_6C, respectively, similar to observations made in [13, 14]. Apart from these, Ti–rich nitrides were also located inside the grains and owing to high density of Ti, they appeared dark under the SEM. It was reported by Hanning, Fabian, et al. [15], that Mo-rich borides were absent in H282 following TEM analysis. But EDX images presented in Fig. 4 suggest a possibility of Mo-rich boride precipitation surrounding TiN precipitates. It is indeed ambiguous whether Mo borides envelope Ti Carbo-nitrides or Mo carbo-borides envelope Ti-rich nitrides. A thorough TEM examination is necessary to draw conclusions on the identity of these secondary phases.

Nevertheless, the preliminary differences in the starting microstructures of VDM 780 and H282 were understood.

Elevated Temperature Strength

There is lack of information on yield strengths and flow stress behaviours of these alloys, especially VDM 780, at the temperatures between 700 and 900 °C. Hot forming dies, and other tools are mostly under the influence of compressive stresses, and H282 and VDM 780 were subjected to uniaxial compression testing. The knowledge of flow curves and strengths at this temperature range is very useful to identify the possibility of plastic deformation during forming cycles when compared against the simulation predicted results of Von Mises stresses on the tools, wherever applicable.

The flow curves are presented in Fig. 5a. (VDM 780), and b (H282), along with proof stress information in Fig. 6. At 700, 800, and 850 °C strain hardening was observed in both alloys while at 900 °C strain softening and dynamic recovery occurred. Proof stresses (as measured to be true stresses corresponding to 0.2% true strain) are higher for VDM 780 at lower temperatures but are nearly equal at 900 °C. High strength of VDM 780 could be attributed to high γ′ V_f and finer precipitate size than that in H282.

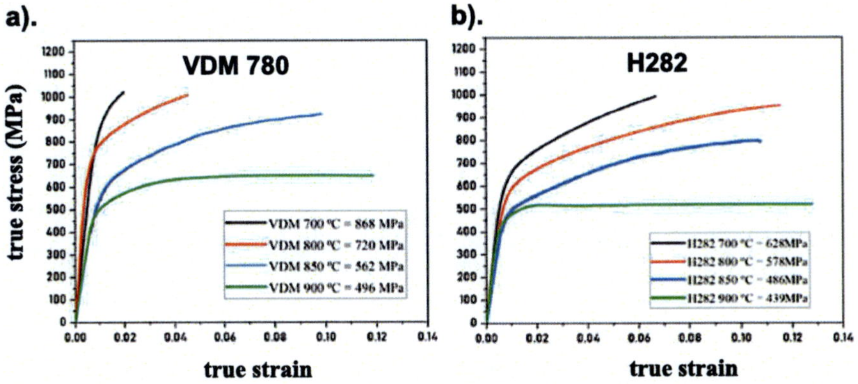

Fig. 5 Flow curves displaying the variation of true stress vs true strain (**a**) in VDM 780 and (**b**) in H282

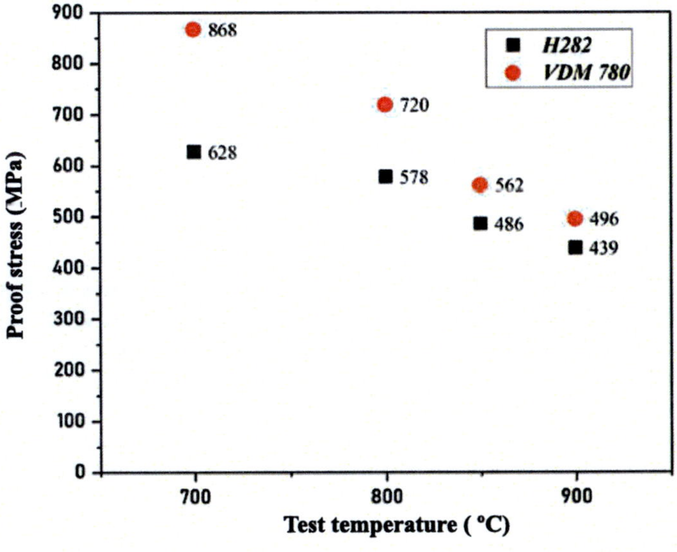

Fig. 6 A plot of variation of proof stresses as a function of test temperatures for VDM 780 (red) and H282 (black)

Effect of Thermal Exposure on Microstructure and Mechanical Properties

Hot forming tools and dies are expected to remain in service for a long time with good structural integrity for which the thermal stability of microstructure is essential. In γ' strengthened alloys, the microstructural instability is reflected as a loss of phase %, undesirable phase transformation or in-situ precipitation of deleterious phases and leads to a loss of strength and ductility [11].

During hot forming, workpieces are generally maintained between 1000 and 1200 °C [16, 17] and the temperature of hot work tool steels are raised to a high temperature (~550 °C) to accommodate severe thermal gradients between the workpiece and the tool; this is detrimental to the tool life [3]. To prevent this, in the current study with superalloys, the background temperature of the tool is to be increased further. Based on simulation, the expectation is that the temperature of the tools will vary between 700 and 900 °C. The surface temperature might raise closer to ~900 °C. Thus, temperature regime of 700–900 °C is considered. Specimens of both alloys were isothermally aged for 40 h at 700 °C, 18 h at 800 °C, to mimic long-term in-service conditions, and at 900 °C for 3 h to mimic the short-term spikes in tool surface temperatures during operation.

To gain a deeper understanding of time-driven microstructural changes, a shorter exposure duration of 7 h was also chosen at both 700 and 800 °C to be compared to the 40 h and 18 h microstructures, respectively.

Microstructures of both alloys were examined after isothermal aging in the furnace. The aged microstructures of VDM 780 are presented in Fig. 7 and that of H282 are presented in Fig. 8. Average secondary γ' particle diameters along with the standard deviations, and area % for each aged condition were measured using ImageJ.

VDM 780

In VDM 780, at 700 °C after 7 h exposure little change in precipitate size or V_f was observed in comparison with the AR microstructure, while after 40 h of exposure, even though the γ'_{ter} was intact, a modest increase in the size of both secondaries and tertiaries was observed. At 800 °C distinct changes in the microstructure were noticed. Soaking for 7 h led to the disappearance of γ'_{ter} and the sizes of γ'_{sec} were higher than the AR specimen, but comparable to those present in 700 °C, 40 h sample. Increasing the soaking period to 18 h at 800 °C drove a substantially increased coarsening rate of the secondary precipitates. Additionally, δ phase was observed after 7 and 18 h.

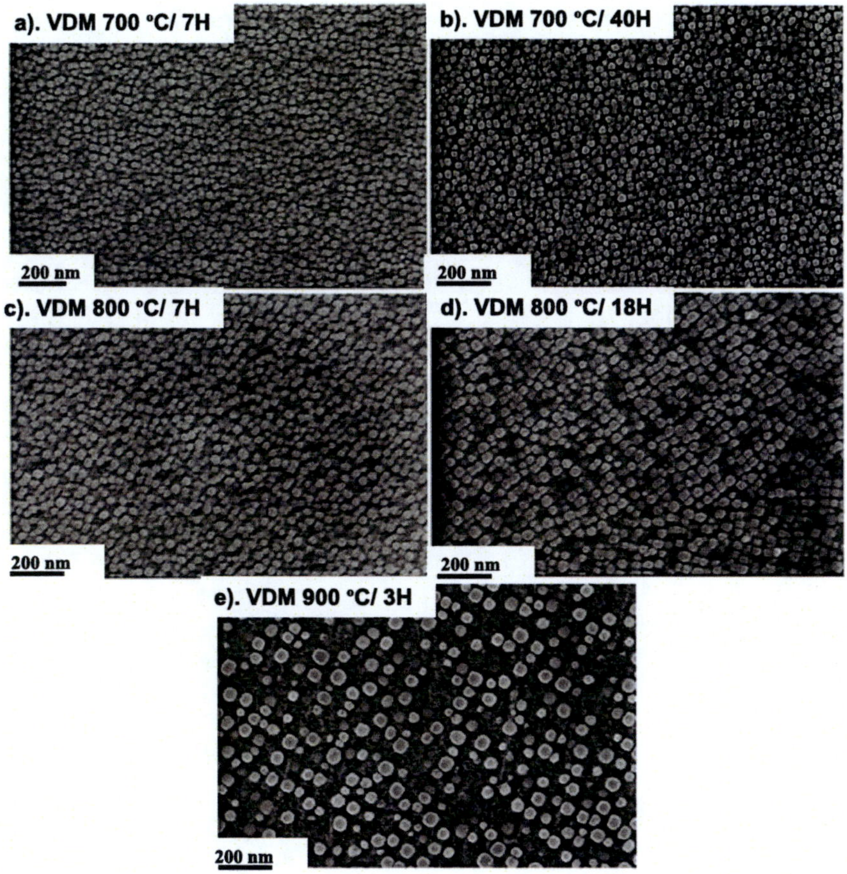

Fig. 7 IL images displaying as-aged microstructures of VDM 780 after furnace ageing for conditions (**a**) 700 °C/ 7H; (**b**) 700 °C/ 40H; (**c**) 800 °C/ 7H; (**d**) 800 °C/ 18H; and (**e**) 900 °C/ 3H

Figure 9 explains the relevant features noted after 800 °C treatment. Incoherent δ precipitates were sporadically observed to be precipitated from the grain boundaries and propagated intragranularly into the grains, serving no purpose apart from consuming the γ′ forming elements, Al and Nb and thus, leading to the decline in the γ′ V_f. At 900 °C even a small exposure period of 3 h caused a severe decline in secondary V_f along with much larger precipitate sizes.

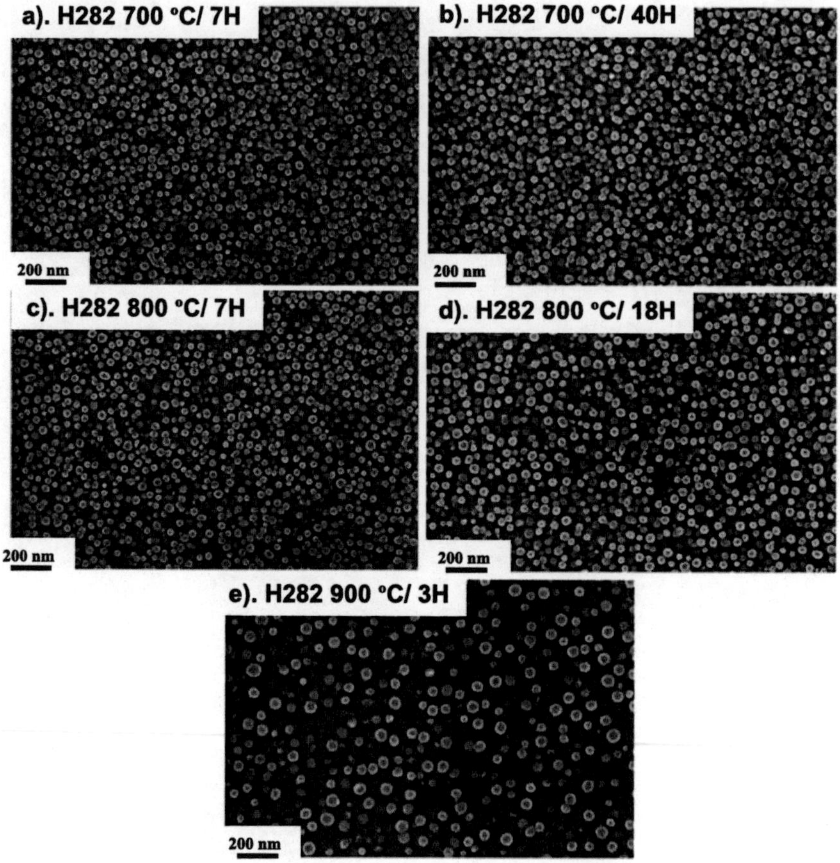

Fig. 8 IL images displaying as-aged microstructures of H282 after furnace ageing for conditions (**a**) 700 °C/ 7H; (**b**) 700 °C/ 40H; (**c**) 800 °C/ 7H; (**d**) 800 °C/ 18H; and (**e**) 900 °C/ 3H

H282

In H282, there was no visible change in γ' size and V_f between the AR specimen and those aged for 7 h at 700 and 800 °C. However, the sample aged at 700 °C for 40 h and 800 °C for 18 h displayed a slight increase in size compared to the initial microstructure and to samples aged at shorter times at the respective temperatures. At 900 °C with 3 h of exposure, there was a substantial decline in the secondary γ' V_f with pronounced coarsening.

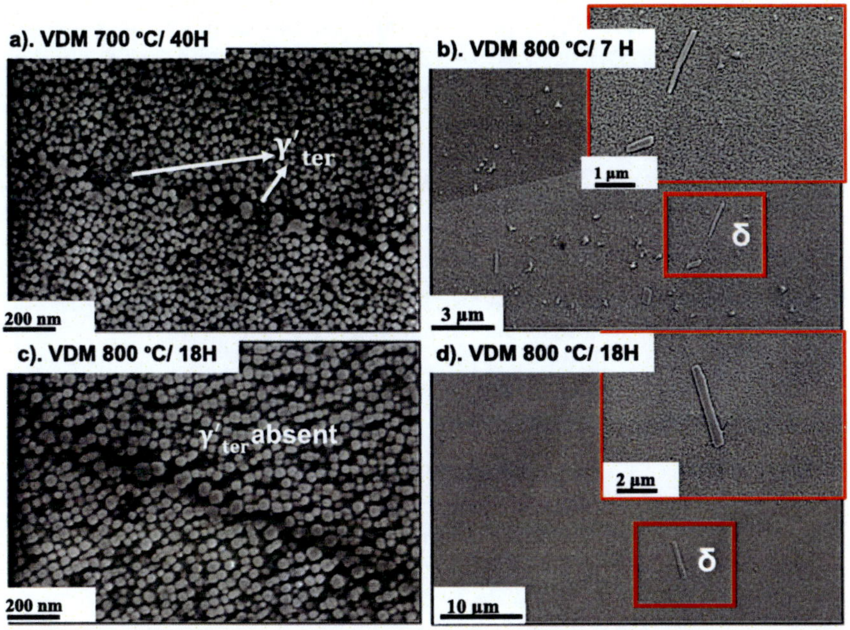

Fig. 9 Explains the differences observed in the microstructures of VDM 780 after iso-ageing, (**a**) 700 °C/ 40H: IL image showing γ'_{sec} and γ'_{ter}, and no δ; (**b**) 800 °C/ 7H: showing the presence of δ; (**c**) 800 °C/ 18H: revealing the absence of γ'_{ter} precipitates; and (**d**) 800 °C/ 18H: showing the presence of δ phase

Figure 10 displays the evolution of grain boundary $Cr_{23}C_6$ carbides with temperature. The density of the carbides decreased from an almost continuous network along grain boundaries at 700 °C, 40 h to discontinuous precipitation at 800 °C, 18 h to a much-depleted discrete precipitation at 900 °C, 3 h.

Effect of Exposure Time and Temperature on γ′

Figures 11 and 12 present details on the effect of ageing temperature and times on γ′ coarsening for VDM 780 and H282, respectively. It is evident that VDM 780 is unaffected by exposure at 700 °C. The particle size and area % are unchanged between AR and 40 h at 700 °C; and there is an overlap between sizes of 7 and 40 h at 700 °C, along with no additional phase precipitation or transformation, implying that the microstructure is very stable with temperature and time at 700 °C. A substantial increase in size and decline in area % from AR is witnessed at 800 °C/18H; so is the increase in particle sizes from 800 °C/7H to 800 °C/18H. The temperature is favourable for δ nucleation and growth, and it can be postulated that this accelerates

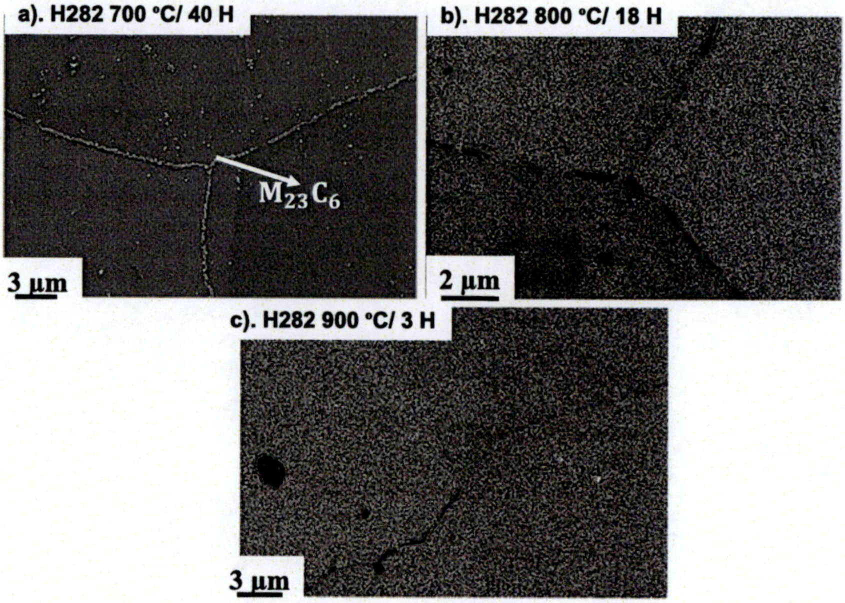

Fig. 10 IL images revealing the decreasing $Cr_{23}C_6$ density with ageing temperature in H282 for ageing conditions (**a**) 700 °C/ 40H; (**b**) 800 °C/ 18H; and (**c**) 900 °C/ 3H

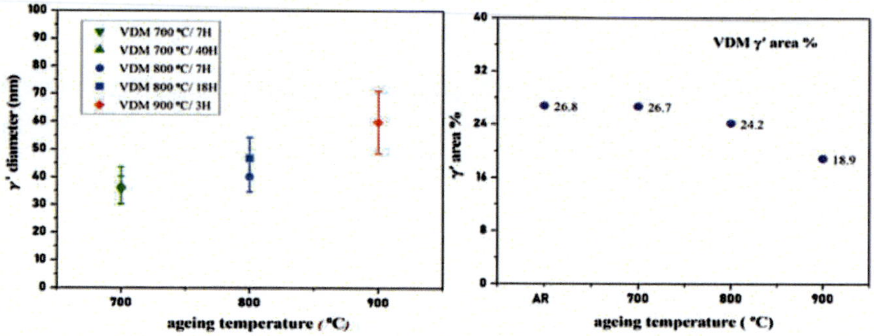

Fig. 11 Examining the effect of ageing temperature and time on the γ' size and V_f in VDM 780

the coarsening and dissolution of γ'. At 900 °C, a serious 8% decline in area % occurs along with a 25 nm increase in the size.

In H282, the microstructure is stable with respect to ageing time and temperature at 700 and 800 °C. Despite increased size between AR and 700 °C/40H, there is no change in area % among 700 °C/7H, 700 °C/40H, 800 °C/7H, and 800 °C/18H. This again validates the hypothesis that H282 exhibits γ' stability at elevated temperatures. However, at 900 °C, like VDM 780, H282 also succumbs to γ' coarsening and

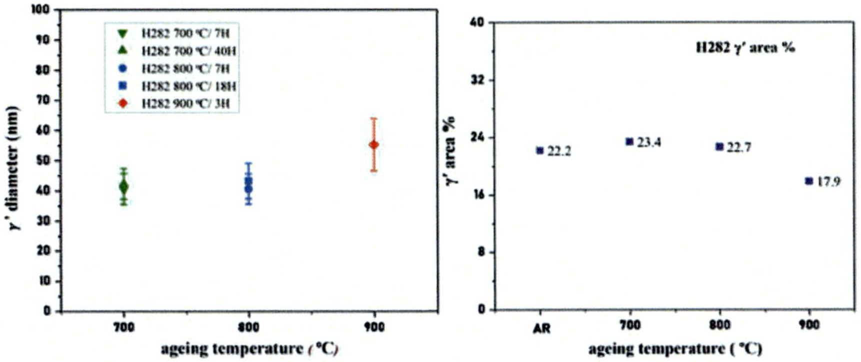

Fig. 12 Explaining the effect of ageing temperature and time on the γ′ size and V_f in H282

dissolution as the temperature is near the phase solvus. It is noteworthy that even as the γ′ content and sizes at 900 °C are comparable between the alloys, the rate of the decrease is more pronounced in VDM 780 than in H282.

Impact of Precipitate Evolution on Yield Strength

To learn the effects of thermal ageing on mechanical strength, in-situ ageing and uniaxial testing were done at soaking temperatures and proof stresses were determined. In-situ ageing and uniaxial tensile testing was done by Otto Fuchs KG on both VDM 780 and H282. In-situ ageing and compression testing on VDM 780 was carried out in the dilatometer for conditions 700 °C/40H, 800 °C/18H, and 900 °C/3H. Where duplicated, the tests were closely repeatable giving confidence in the accuracy of the technique and the results. During in-situ ageing, holding times were matched with corresponding ageing times in the furnace, to mimic furnace ageing and avoid any changes in the microstructure arising from quenching and re-heating.

Figure 13 shows the flow stress curves of VDM 780 before and after ageing at the test temperatures. Figure 13a shows the difference of proof stresses between aged and unaged samples as a function of test temperatures. The slight increase in strength after 40 h ageing at 700 °C can be attributed to the growth in size of tertiaries that may fall under the category leading to transition from weak to strongly coupled pairs of dislocations. The decrease in strength observed at 800 °C after 18 h can be attributed to the dissolution of tertiaries, coarsening of secondaries and the onset of δ precipitation. However, additional tensile tests presented in Fig. 14a (VDM 780), b (H282), exhibit near-equal strengths for VDM 780 at 850 °C and above. As expected, near-equal proof stresses in H282 were observed from the tensile tests for different test temperatures and times. Therefore, we can conclude that the relative variation in strength for the conditions tested is only small.

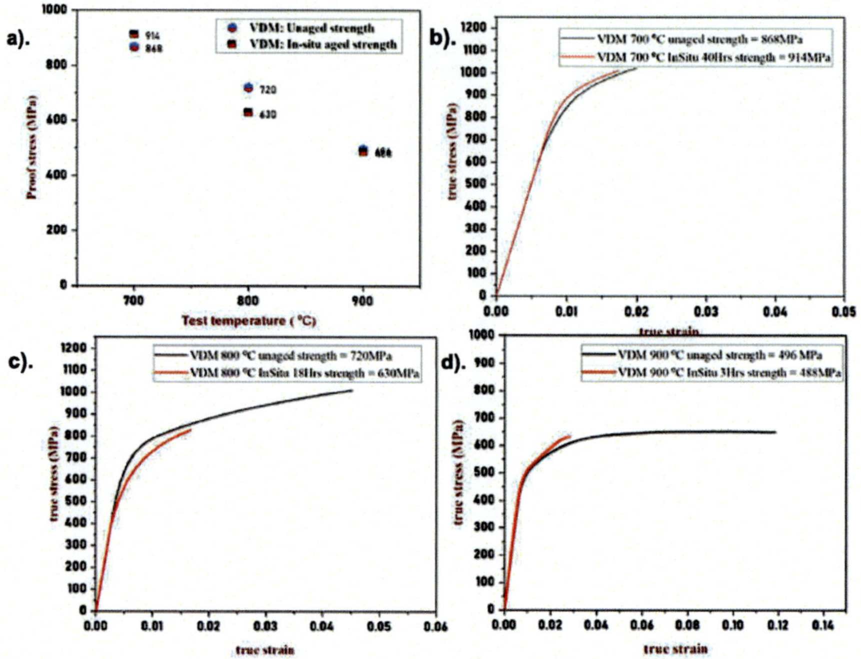

Fig. 13 (**a**) Variation of compressive proof stresses of VDM 780 with test temperature prior to and after ageing. (**b**) Flow curves of VDM 780 tested at 700 °C in unaged and 40 H aged condition. (**c**) Flow curves of VDM 780 tested at 800 °C in unaged and 18 H aged condition. (**d**) Flow curves of VDM 780 tested at 900 °C in unaged and 3 H aged condition

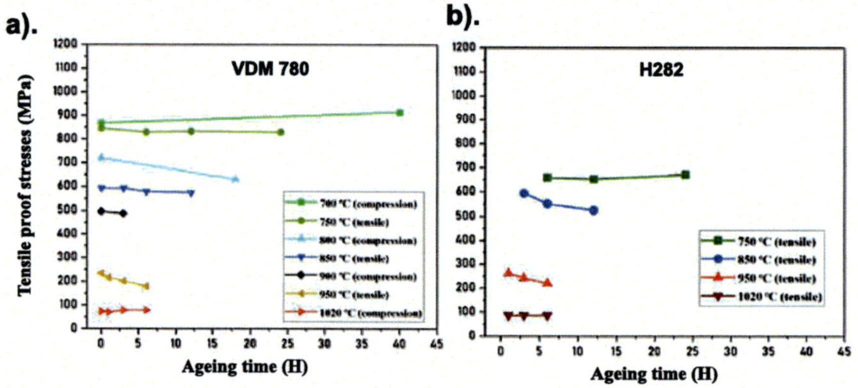

Fig. 14 Presenting the variation of tensile proof stresses with exposure times in hours for different temperatures in (**a**) VDM 780 and (**b**) H282

Creep Testing and Microstructural Evolution

The macroscopic creep deformation response of VDM 780 and H282 are presented as true strain vs time curves in Fig. 15. Tables 2 (H282) and 3 (VDM 780) display the creep strain values along with strength information for different test conditions. The influence of both temperature and stress on creep could be deduced as specimens were crept at a combination of high and low stresses at temperatures between 700 and 900 °C. Creep was interrupted at low strains and exposure times were curtailed to subject alloys to limited deformation. For each of the conditions tested, observed minimum creep rates were computed and are plotted as a function of creep temperature as shown in Fig. 15c. As expected, the creep rates increased with increasing temperatures. Overall, for similar testing conditions, VDM 780 displayed negligible creep strains at lower temperatures compared to H282. In contrast, at the highest test temperature the creep strain of VDM 780 is almost twice than that experienced by H282.

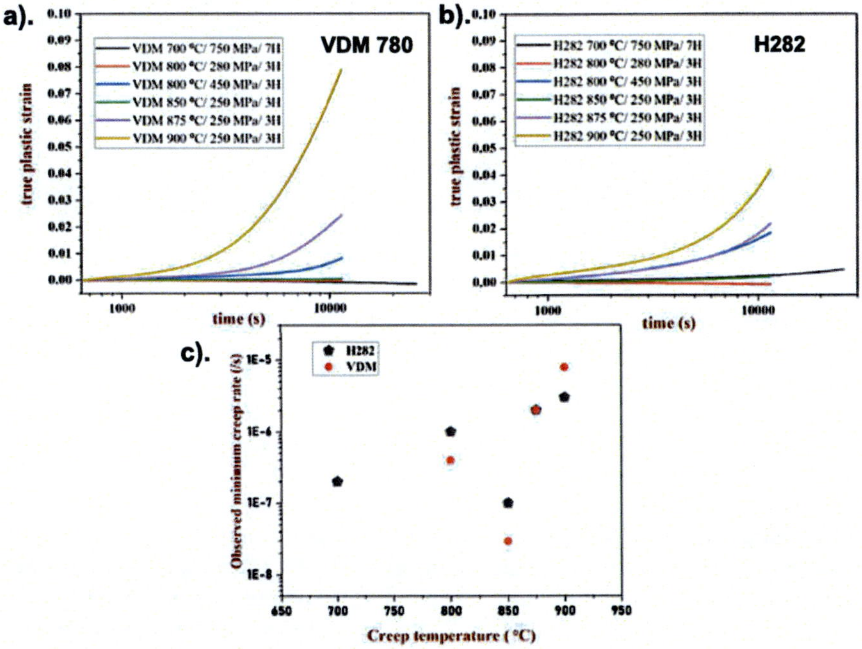

Fig. 15 Plots displaying creep plastic strain verses time for (**a**) VDM 780 and (**b**) H282; (**c**) presenting the variation of observed minimum creep rate as function of creep temperature for VDM 780 (red) and H282 (black)

Table 2 H282 Creep strain

T (°C)	Stress (MPa)	t (hours)	Strain H282	Strength H282
700	750	7	0.5%	628
800	280	3	Nil	578
800	450	3	1.8%	578
850	250	3	0.2%	486
875	250	3	2.2%	–
900	250	3	4.2%	439

Table 3 VDM 780 Creep strain

T (°C)	Stress (MPa)	t (hours)	Strain VDM 780	Strength VDM 780
700	750	7	Nil	868
800	280	3	Nil	720
800	450	3	0.8%	720
850	250	3	0.05%	562
875	250	3	2.4%	–
900	250	3	7.8%	496

Creep Response of H282

At 700 °C, with relative stress level, $\sigma_{applied}/\sigma_{yield}$ (σ_a/σ_y) of 119.4%, the alloy sustained principally primary creep with limited secondary creep. At 800 °C, for creep condition of σ_a/σ_y around 50%, no creep was observed. The applied stress was further increased to 450 MPa at 800 °C to result in σ_a/σ_y of 77.85%. This led to a noticeable 1.8% creep strain. Hence, it can be noted that at lower temperatures, applied stress dictated creep strain. However, at higher temperatures of 875 and 900 °C, the resulting creep strain was significant. At 50% relative stresses, there was a two-fold increase in strains experienced between 875 and 900 °C.

Microstructures of the samples were examined at selected conditions 700 °C/750 MPa/7H, 800 °C/450 MPa/3H, and 900 °C/250 MPa/3H; to reveal the effects of temperature and stresses. From the micrographs (in Fig. 16), no new stress-induced microstructural changes were witnessed apart from expected coarsening of γ'_{sec} with concomitant decline in γ'_{sec} and $M_{23}C_6$ V_f, quite consistent with the microstructures observed after isothermal ageing.

Creep Response of VDM 780

Creep behaviour of VDM 780 was distinct from that observed in H282. Between 700 and 850 °C, low creep strains were experienced with none of the tests fully reaching steady state creep, notwithstanding, a few tests showing nil creep strains. 700 °C

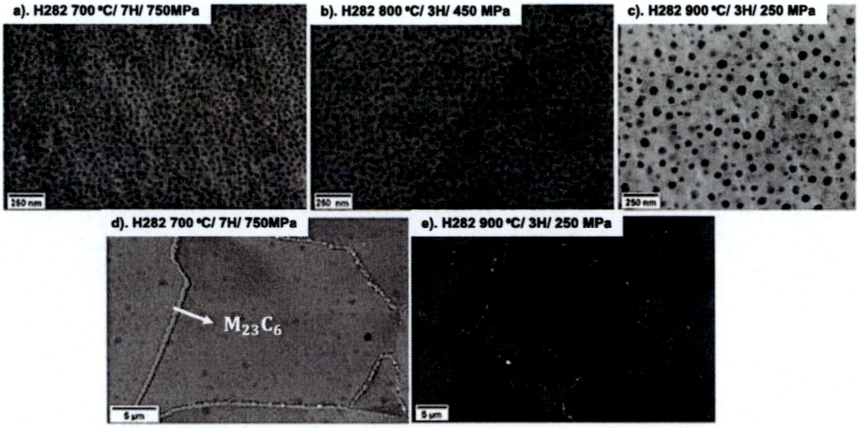

Fig. 16 Backscatter SEM (BSE) micrographs of un-etched H282 specimens after creep at (**a**) 700 °C/7H/750 MPa, (**b**) 800 °C/3H/450 MPa, and (**c**) 900 °C/3H/250 MPa; all displaying γ'_{sec}. BSE micrographs of H282 (**d**) and (**e**) displaying continuous and discrete precipitation of $Cr_{23}C_6$ in the boundaries in 700 °C/7H/750 MPa and 900 °C/3H/250 MPa conditions, respectively

testing with σ_a/σ_y of 86.4%, 800 °C/280 MPa/3 H with σ_a/σ_y of 38.9% and 850 °C/250 MPa/3 H with σ_a/σ_y of 44.5% all displayed no plastic strain. An increase in the applied stress to 450 MPa (σ_a/σ_y of 62.5%) at 800 °C/3H resulted only in 0.8% creep strain, which indicated that VDM 780 was less sensitive to creep stress at lower test temperatures. At temperatures above 850 °C, there was a non-linear increase in the creep plastic strains observed. At 875 °C with σ_a/σ_y of approximately around 50%, a substantial plastic strain was observed equal to that of H282 at the same condition, with more secondary creep regime. At 900 °C, maximum creep plastic strain of 8% was experienced. This implies that VDM 780 is more temperature sensitive and employing it above 850 °C could be precarious over long duration.

Any observations of negative creep were considered non-physical and were attributed to perturbations arising from Mo-discs undergoing deformation or any inaccuracies while recording the initial dimensions of the samples as the DIL 805 is very sensitive. For compression deformation with the DIL 805 A/D, the change in length is continuously decreasing and is reflected as positive creep strain. A negative creep strain means that the sample expanded against the applied compressive load. Such instances of negative creep are well documented for tensile creep in single crystal superalloys undergoing directional coarsening of γ', with precipitate rafts aligning perpendicular to the stress axis [18], which is not a possibility in these alloys. Nonetheless, microstructure examination under TEM would be necessary to conclusively negate the presence of any creep strain.

Figure 17 displays the SEM micrographs of the crept VDM 780 samples, sequentially in various test conditions. It was observed that at 700 °C, the microstructure was comparable to the AR one with retained tertiaries, secondaries, and coarsened secondaries at the grain boundaries. At 800 °C testing with lower stress, the microstructure

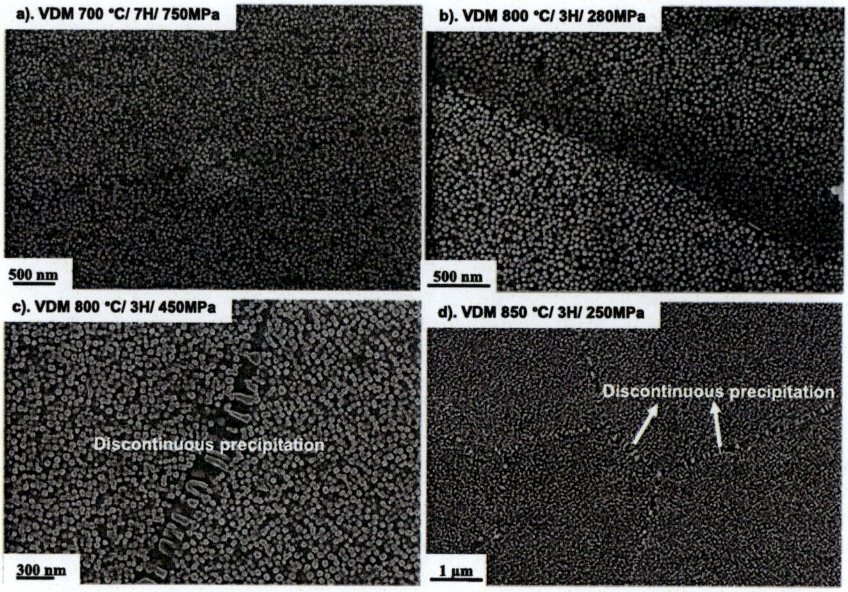

Fig. 17 IL micrographs of VDM 780 crept specimens, (**a**) 700 °C/ 7H/ 750 MPa showing both γ'_{sec} and γ'_{ter}; (**b**) 800 °C/ 3H/ 280 MPa displaying γ'_{sec}; (**c**) 800 °C/ 3H/ 450 MPa; and (**d**) 850 °C/ 3H/ 250 MPa revealing discontinuous grain boundary precipitation with intragranular γ'_{sec}

was intact, but at higher stresses a peculiar stress-induced effect of grain boundary migration followed by discontinuous grain boundary γ' precipitation was observed. The discontinuous precipitation was seen in all samples tested at 850 °C and above. Discontinuous precipitation constitutes "the formation of a two-phase lamellar structure behind moving grain-boundaries" and is well explained in Williams, David B. et al. [19]. The precipitation reaction is of interest as it leads to a detriment in mechanical properties.

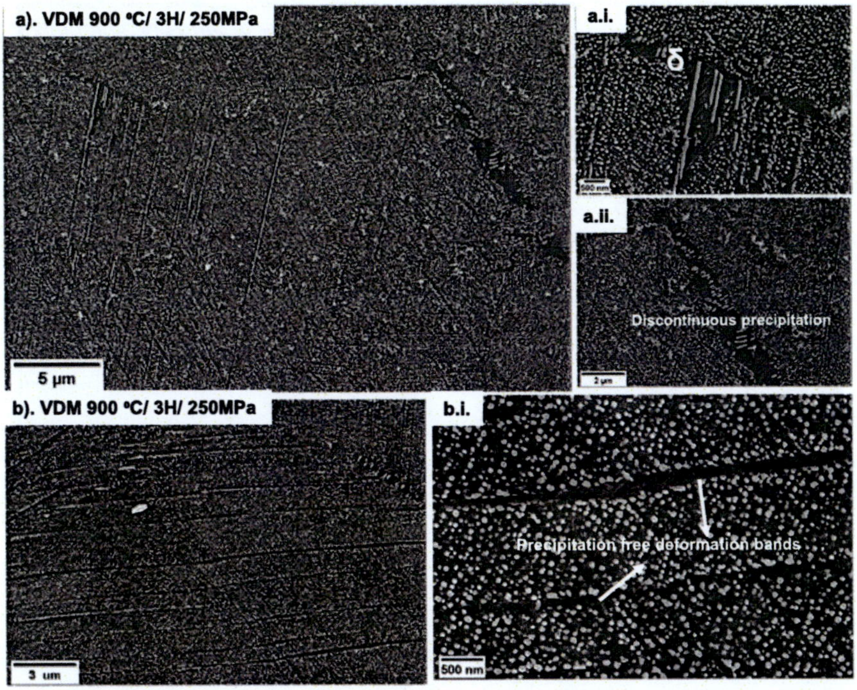

Fig. 18 VDM 780 creep at 900 °C/ 3H/ 250 MPa: IL micrographs showing (**a.i**) thick rod-like δ and (**a.ii**) discontinuous precipitation of lamellar γ′ behind migrated grain boundaries; (**b.i**) BSE micrographs showing γ′-free deformation bands

At 900 °C, presented in Fig. 18, thick rods of δ precipitation growing inside the grains were observed along with discontinuous γ′ precipitation. Also, at 900 °C, there were γ′-free deformation bands. Figure 19 presents the rough schematic for the δ phase TTT curve for alloy VDM 780. With phase solvus at 1020 °C, the nose of δ precipitation (shown as green curve in the figure) could be somewhere between 850 and 900 °C for time between 1 and 3 h. At 800 °C, δ starts to form between 3 and 7 h.

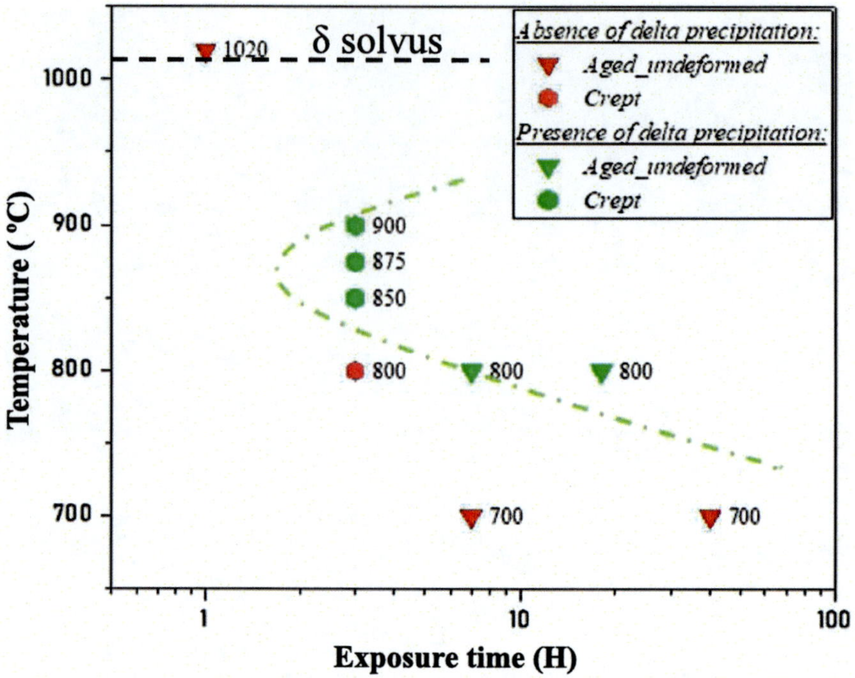

Fig. 19 Time temperature transformation curve of δ phase in VDM 780, showing the presence or absence of δ for different exposure temperatures and times

Discussion

A comparison is drawn between the alloys VDM 780 and H282 for hot tooling applications. However, the larger aim of this paper is also to identify and design optimised nickel superalloys for this application. Usually in hot forming, the workpiece is maintained at a higher temperature than the tools and due to this thermal gradient, there is a superficial increase in the temperature of the tools. In the current study with superalloy candidates, the aim is to increase the background or core temperature capability of the tools. The tools are expected to experience temperatures between 700 and 800 °C at most locations, for major parts of their life, with the surface that encounters the workpiece, to experience temperatures above 800 °C up to 900 °C. From the experimental evidence, it was observed that both VDM 780 and H282 performed satisfactorily for this requirement and were better than the conventional nickel superalloy, IN718. But between the alloys, VDM 780 was better at lower temperatures while H282 was more reliable at higher temperatures.

The elevated temperature strength of VDM 780 is about 200 MPa greater than H282 at lower temperature and the difference diminishes as temperature increases. With no practical variation in grain size due to temperature, the grain boundary strengthening can be calculated using Hall–Petch relationship using $\sigma_D = \frac{K_Y}{\sqrt{D}}$, where D is the mean grain size and K_Y is the Hall–Petch constant. The value used for K_Y in calculations for yield strength models proposed by Galindo-Nava et al. [20] is 710 MPa $\mu m^{1/2}$. Using the value for K_Y, the relationship yields a difference of about 20 MPa between the alloys due to differences in grain size. This implies that the effect of grain size towards strengthening is minor. Similarly, the effect of solid solution strengtheners is also likely to be small. In these alloys, the primary strength arises from γ'; size, distribution and V_f, all dictate the penetrability of dislocations. In the case of fine sized precipitates with low fraction, the leading and trailing dislocations do not occupy γ' simultaneously and constitute weak pair coupling, while in case of large precipitate sizes, they give rise to strongly coupled dislocation pairs wherein both dislocations reside in a particle at the same time. Critical resolved shear stresses (CRSS) quantify the stresses required to drive these dislocations into the precipitates and generally, a maximum is observed when there is a transition between weak to strong coupling. The bimodal distribution of γ' is advantageous in VDM 780 at lower temperatures for providing high strength. At 700 °C, with increasing ageing time, the size of tertiaries increases. This will increase the strength when the growth in size of γ'_{ter} is sufficient to result in strong coupling. However, at higher temperatures of 800 and 900 °C, due to dissolution of γ'_{ter} and coarsening of γ'_{sec}, the particle shearing stress will be absent and the strength of the alloy will diminish.

From the iso-aged microstructures of H282, it can be observed that, apart from a decreased density of $M_{23}C_6$ carbides, no noticeable difference in γ' area percent or particle diameter occurs between 700 and 800 °C ageing treatments. For all practical purposes, equal measurements could be assumed across 700 °C/7H, 700 °C/40 H, 800 °C/7H, and 800 °C/18 H treatments. With ageing time, the retention in strength and resistance of γ' coarsening in H282 is induced by high Ti/Al ratio and high Mo additions. The higher the Ti/Al ratio, the lower is the γ-γ' misfit and the rate of coarsening of γ' is reduced. The slow diffusivity of Mo retards the coarsening of γ'. Additionally, Ti also increases the antiphase boundary energy (APBE) and confers resistance to precipitate shearing. This plays a positive role in time dependent creep deformation. Thus, high Ti/Al ratio and high Mo additions in H282 impart thermal stability [21, 22].

In VDM 780, the high Al and Nb content (2 and 5.4 Wt.% as opposed to 0.2 and 0% in H282, respectively) confers higher γ' V_f and higher elevated temperature yield strength [6]. However, higher Al/Ti ratio makes the γ' susceptible to coarsening due to increased γ-γ' misfit. The lack of Ti in VDM 780 is compensated by Nb additions. Nb segregates to γ' and its low mobility helps retain finer precipitates [22]. Thus, in VDM 780, at low temperatures, γ' kinetics is controlled by Nb. But as the temperature increases closer to γ' solvus and δ forming temperatures, Nb is depleted by the formation of δ (Ni_3Nb, DO_a orthorhombic phase). Also the presence of 25% Co stabilises δ phase in VDM 780 once it is formed [23]. Hence, above 850 °C, there are no elemental contributions precluding γ' coarsening and dissolution, thus VDM 780 is more susceptible to loss of strength and creep resistance when compared to H282.

In creep, no new or unexpected microstructural changes arise in H282 in addition to the trend observed after iso-ageing. The microstructure is stable with no undesirable incoherent phases or TCP phases, similar to reports by Pike, L et al. [11]. However, in VDM 780, as temperature increases, lamellar γ' and δ phases appear. When δ precipitates form, they are enveloped by a γ'-depleted zone, which is inherently weaker. Likewise, the occurrence of coarser lamellar γ' will contribute little towards strengthening. Additionally, at 900 °C, γ'-free deformation bands are observed to exist in majority of grains. Sundararaman et al. [24], has reported that these bands originate due to shearing of precipitates on <111> planes, possibly accompanied by cross slipping to <100> planes, all taking place within the formed band. Once cut, the halves get redistributed and dissolved, thus leaving an empty band. The applied stress alone cannot be solely attributed to the phenomenon. It is agreed that the fragmentation occurs owing to the mechanical influence, however, the dissolution of the redistributed fragments, each of which is below the critical radius, is temperature driven. This occurrence could be one of the reasons for the detrimental response of this alloy to the applied stress at 900 °C. Thus, VDM 780 is less able to withstand the high surface temperatures of the tools.

Some of the damages encountered by the tool surfaces include adhesive and abrasive wear, erosion, plastic deformation, thermal–mechanical fatigue, and surface or gross cracking [25]. In this study, the susceptibility to plastic deformation in candidate alloys is studied through the means of short-term strength testing and long-term creep testing. Also, the extent of deterioration of properties with time is examined using isothermal ageing. It can be concluded that VDM 780 is more prone to die chilling effect and can be used for operations having lower demanding surface temperatures, while H282 can be employed for more demanding surface temperature applications.

In future, to implement γ' superalloys as tool materials, the following factors must be considered: 1. γ' coherency with γ matrix and APBE required by dislocations to disorder γ' precipitates, 2. γ' V_f and variation with increasing temperature, and 3. γ' size, stability, and extent of coarsening. These parameters rely largely on the alloy chemistry but can be modified by applying the appropriate processing conditions.

Conclusions

A study is presented wherein two low γ' strengthened derivatives of IN718, a newly developed alloy VDM 780 and another commercially popular alloy H282, are compared for the purpose of application as hot tooling materials. These alloys have gained popularity for their appreciable fabricability and weldability, however, they remain largely unexplored with respect to their mechanical properties in the temperature regime of 700–900 °C, the range of interest to hot tooling. The tools are expected to withstand continuous loading cycles at high temperatures for prolonged times. Therefore, elevated temperature strength, thermal stability through furnace ageing, and time-dependent creep deformation are evaluated. The mechanisms driving deformation are studied by examining the evolution of γ′ and other secondary phases with applied temperature, time, and stress. An insightful correlation between process parameters, mechanical outcomes, and microstructure is established.

Between 700 and 800 °C, VDM 780 is observed to be much stronger and creep resistant, potentially due to the presence of bimodal distribution of fine tertiaries and higher V_f of secondaries. The γ′ (monomodal) V_f is smaller in H282, comparatively. Nevertheless, both the alloys display microstructural stability with no new undesirable precipitation that might weaken the microstructure.

Above 800 °C and below 900 °C, wherein the tests mimic short-term spikes in tool temperatures during operation, H282 is observed to possess better creep resistance with reasonable strength due to a more stable microstructure. In VDM 780, the microstructure is weakened by (a) occurrences of stress-induced discontinuous grain boundary precipitation of lamellar γ′ and (b) incoherent δ precipitation emanating from the boundaries and travelling into the grains, enveloping γ′-depleted zones around them. Additionally, coarsening and dissolution of intragranular γ′ is more pronounced in VDM 780 than in H282, when compared with the as-received microstructures.

In summary, both alloys display better mechanical performance at different conditions. VDM 780 is superior at lower temperatures (<850 °C) with high operating stresses while H282 is more tolerant of higher operating temperatures at lower stresses. Overall, between temperature, time, and stress; temperature has the overpowering control. The differences arising in γ′ precipitate size scale, distribution, and V_f, evolution with temperature is revealed to have the most impact on the macroscopic response of these alloys. The current study provides an impactful comparison of these two alloys and paves way into designing new nickel superalloys for hot tooling applications.

References

1. Kracke, A. Superalloys, the most successful alloy system of modern times-Past, present, and future. In Superalloy 718 and Derivatives (2012). https://doi.org/10.1002/9781118495223.ch2.
2. Roger C. Reed, The superalloys: Fundamentals and Applications. Cambridge: Cambridge University Press, 2002.
3. Lu, B.S. & Wang, L.G. & Huang, Y: Effect of deformation rate on interfacial heat transfer coefficient in the superalloy GH4169 hot forging process (2016). Applied Thermal Engineering. 108. https://doi.org/10.1016/j.applthermaleng.2016.07.167.
4. Kennedy, R.L.(2005). Allvac 718Plus, Superalloy for the Next Forty Years. Proceedings of the International Symposium on Superalloys and Various Derivatives. 1–14. https://doi.org/10.7449/2005/Superalloys_2005_1_14.
5. Xie, Xishan & Xu, Chunmei & Wang, Gailian & Dong, Jianxin & Cao, Wei-Di & Kennedy, Richard. (2005). TTT diagram of a newly developed Nickel-base superalloy - Allvac 718Plus. Proceedings of the International Symposium on Superalloys and Various Derivatives. https://doi.org/10.7449/2005/Superalloys_2005_193_202

6. Fedorova T, Rösler J, Klöwer J, Gehrmann B (2014) Development of a new 718-type Ni-Co superalloy family for high temperature applications at 750 °C. In: MATEC web of conferences, p 14 (2014).
7. Solís, C., Munke, J., Bergner, M. et al. Effect of heat treatment on microstructure and mechanical properties of VDM Alloy 780 Premium (2018) Metall and Mat Trans A 49: 4373. https://doi.org/10.1007/s11661-018-4761-6.
8. Casanova, A., Hardy, M. and Rae, C. (2014). Morphology and kinetics of grain boundary precipitation in alloy ATI 718Plus®. In 8th International Symposium on Superalloy 718 and Derivatives. https://doi.org/10.1002/9781119016854.ch45.
9. Solís, C., Munke, J., Bergner, M., Kriele, A., Mühlbauer, M., Cheptiakov, D., Gehrmann, B., Rösler, J. and Gilles, R. (2018). In situ characterization at elevated temperatures of a new Ni-based superalloy VDM-780 Premium. Metallurgical and Materials Transactions A, 49(9), pp. 4373–4381.
10. C. Ghica, C. Solís, J. Munke, A. Stark, B. Gehrmann, M. Bergner, J. Rösler, R. Gilles, HRTEM analysis of the high-temperature phases of the newly developed high-temperature Ni-base superalloy VDM 780 Premium, Journal of Alloys and Compounds, Volume 814, 2020, 152157, ISSN 0925-8388. https://doi.org/10.1016/j.jallcom.2019.152157.
11. Pike, L. (2010). Long Term Thermal Exposure of HAYNES 282 Alloy. 645–660. https://doi.org/10.7449/2010/Superalloys_2010_645_660
12. "Origin: Data Analysis and Graphing Software". Originlab.Com, 2022, https://www.originlab.com/origin. Accessed 27 Aug 2022.
13. Polkowska, A., Polkowski, W., Warmuzek, M. et al. Microstructure and Hardness Evolution in Haynes 282 Nickel-Based Superalloy During Multi-Variant Aging Heat Treatment. J. of Materi Eng and Perform 28, 3844–3851 (2019). https://doi.org/10.1007/s11665-019-3886-0.
14. Joseph, C. (2015). Microstructural characterization of Haynes 282 after heat treatment and forging. CHALMERS UNIVERSITY OF TECHNOLOGY.
15. Hanning, Fabian, et al. "Investigation of the Effect of Short Exposure in the Temperature Range of 750–950 degrees C on the Ductility of Haynes (R) 282 (R) by Advanced Microstructural Characterization." *Metals* 9.12 (2019).
16. J.J. deBarbadillo,14 - INCONEL alloy 740H, Editor(s): Augusto Di Gian Francesco, Materials for Ultra-Supercritical and Advanced Ultra-Supercritical Power Plants, Woodhead Publishing, 2017, Pages 469–510, ISBN 9780081005521. https://doi.org/10.1016/B978-0-08-100552-1.00014-2.
17. B.-A. Behrens, A. Bouguecha, I. LÃken, A. Klassen, D. Odening, Near-Net and Net Shape Forging, Comprehensive Materials Processing, Elsevier, 2014, Pages 427–446, ISBN 9780080965338. https://doi.org/10.1016/B978-0-08-096532-1.00323-X.
18. Louchet, F. (1995). A model of negative creep in nickel-based superalloys. *Scripta metallurgica et materialia, 33*(6).
19. Williams, David B. and E. P. Butler. "Grain boundary discontinuous precipitation reactions." International Materials Reviews 26 (1981): 153–183.
20. E. I. Galindo-Nava, L. D. Connor and C. M. F. Rae, "On the Prediction of Yield Stress of Unimodal and Multimodal Gamma Prime Nickel-Base Superalloys," Acta Materialia, vol. 98, pp. 377–390, 2015.
21. Hawk, J. A., Cheng, T. L., Sears, J. S., Jablonski, P. D., & Wen, Y. H. (2015). Gamma prime stability in Haynes 282: theoretical and experimental considerations. Journal of Materials Engineering and Performance, 24(11), 4171–4181.
22. M. Donachie and S. Donachie, Superalloys: A technical guide. Materials Park: ASM International, 2002.
23. McDevitt, E. (2012). Effect of temperature and strain during forging on subsequent delta phase precipitation during solution annealing in ATI 718Plus Alloy. In Superalloy 718 and Derivatives. https://doi.org/10.1002/9781118495223.ch23.
24. Sundararaman, M., Chen, W., Singh, V., & Wahi, R. P. (1990). TEM investigation of γ' free bands in nimonic PE16 under LCF loading at room temperature. *Acta metallurgica et materialia, 38*(10), 1813–1822.

25. Lavtar, L., Muhič, T., Kugler, G., & Terčelj, M. (2011). Analysis of the main types of damage on a pair of industrial dies for hot forging car steering mechanisms. Engineering Failure Analysis, 18(4), 1143–1152. https://doi.org/10.1016/j.engfailanal.2010.11.002.

In-Situ HT-EBSD Measurements and Calibration of Multi-class Model for Grain Growth and δ-phase Dissolution Kinetics of Alloy 718

P. Raninger, C. Gruber, W. Costin, A. Stanojevic, E. Kozeschnik, and M. Stockinger

Abstract In the aerospace industry the microstructure evolution of alloy 718 during forging and heat treatment and the resulting mechanical properties are decisive in view of the high-quality requirements of aircraft components. During thermomechanical processing the temperature control and adiabatic heating lead to grain growth and, if δ-solvus temperature is exceeded, to the dissolution of the δ-phase, which further results in accelerated grain growth. To describe the history of the microstructure in terms of grain size during and after forging or heat treatment an existing multi-class microstructure model is optimized with a focus on grain growth kinetics and parameterized by experimental results. The multi-class model describes the microstructure and the coarsening during processing more precisely in terms of the grain size distribution than previously used single-class models. A topic for other but related work is the prediction of mechanical properties such as tensile strength, fracture toughness and creep resistance, which requires a precise prediction of grain size characteristics in terms of sizes and size distributions as provided by the model presented in this work. The experimental data basis stems from in-situ high-temperature electron backscatter diffraction (HT-EBSD) investigations and supporting experiments. The experimental setup and the results are discussed in detail.

Keywords Alloy 718 · δ-phase dissolution · Grain growth kinetics · In-situ HT-EBSD · SEM analysis · Multi-class modeling

---

P. Raninger (✉) · W. Costin
Materials Center Leoben Forschung GmbH, Leoben, Austria
e-mail: peter.raninger@mcl.at

C. Gruber · A. Stanojevic
Voestalpine BÖHLER Aerospace GmbH & Co KG, Kapfenberg, Austria

E. Kozeschnik
TU Wien, Institute of Materials Science and Technology, Vienna, Austria

M. Stockinger
Department for Product Engineering, Montanuniversität Leoben, Leoben, Austria

E. A. Ott et al. (eds.), *Proceedings of the 10th International Symposium on Superalloy 718 and Derivatives*, The Minerals, Metals & Materials Series,
https://doi.org/10.1007/978-3-031-27447-3_6

Introduction

Forged components in the aerospace industry are subject to increasing demands in terms of mechanical properties and geometry, while development times are becoming shorter and shorter. This development is based on the requirements for lower component weights, longer service lives, and accurate prediction of properties [1]. This should ensure more efficient use of the components while at the same time reducing emissions as required for meeting sustainable development goals [2]. Furthermore, the costs for development and production must be kept as low as possible for the forging manufacturing plant. This applies especially to the most highly stressed components in the aero engine.

In order to meet these requirements, finite element (FE) based computer simulations including microstructural models are used to predict the desired properties already in the design phase in the sense of integrated computational materials engineering (ICME) [3]. Accordingly, the relevant material characteristics defining the final properties and depending on the manufacturing process are to be incorporated into models in order to have a predictive tool for aircraft parts available. In addition, all influencing factors such as inhomogeneities from the pre-material production, as well as the varying forming characteristics in terms of energy input from the forging aggregates as well as the final heat treatment must be taken into account finally leading to a digital twin for alloy 718 aircraft parts. An overview of the manufacturing process and relevant process parameters is given in [1, 4].

The present work aims the investigation of mechanisms taking place during the final heat treatment as well as a suitable numerical implementation in a model [5–7]. The modeling approach uses grain size classes, which allows the calculation of the growth of fractions in favor of thermodynamically disadvantaged classes and to simultaneously monitors the evolution of various grain fractions. The model captures dynamic and meta-dynamic recrystallization processes in dependence on forging conditions. It is based on semi-empirical models of the type described by Sellars [8] that have been further modified for alloy 718 but need to be expanded to capture grain growth within the multi-class framework.

A substantial experimental basis for the first tests of a new model implementation is created with a focus on grain growth kinetics with and without the presence of the δ-phase fractions. In-situ HT-SEM and HT-EBSD measurements are carried out, which allows observation of changes in the microstructure during isothermal heating as described e.g. in [9] for the case of a precipitation hardenable steel. For the current study on alloy 718, two in-situ experiments with different initial microstructures (with and without δ-phase) are performed with a focus on isothermal holding at 1045 °C for 90 min. These experiments are supplemented with a series of ex-situ annealing treatments where temperatures and holding times are varied and results are documented with light microscopy. All image files from the different experiments are evaluated regarding grain size distribution and δ-phase area fraction by the use of proper image analysis techniques. Furthermore, the δ-phase dissolution kinetics is evaluated with an in-situ SEM video and the influence of the δ-phase on grain growth inhibition is visualized. After complete δ-dissolution the re-building of Ni_3Nb precipitates is investigated with a focus on kinetics and resulting morphology.

Material and Experimental Methods

The material for the investigation of growth kinetics is an aerospace-certified triple melt material from a 10 inch billet of alloy 718 in as-forged condition after the cogging process. This material was specifically chosen since it shows a variation of grain size and δ-phase content in its cross-section thus providing optimal conditions for a systematic investigation of grain growth in dependence of the δ-phase content at various grain sizes. The variation in grain size over its radius is shown in Fig. 1, wherein the EBSD IPF colored grain maps represent the initial conditions for the conducted in-situ HT-EBSD measurements. At the local position of the relative radius (R) of 0.05, the average grain size was 10.0 ASTM with nearly no grain growth inhibiting δ-phase (1.2% area fraction, measured at 7 positions) compared to the edge sample at R = 0.98 with a grain size of 11.5 ASTM and 5.2% δ-phase.

Furthermore, 72 (36 edge, 36 center) samples from the same local positions with nearly equal grain sizes and δ-phase fractions were extracted for ex-situ heat treatments in a lab furnace. The samples were annealed at 5 temperature levels (980, 1010, 1040, 1070, and 1100 °C) for 5–240 min and analyzed for the resulting average grain sizes with light microscopy [10]. The results are a substantial basis for the calibration of the multi-class grain growth model capturing varying heat treatment conditions and initial states in terms of grain size and δ-phase fraction.

In contrast to the classic ex-situ heat treatments and subsequent microscopy, the in-situ HT-EBSD experiments provide several interpolation points in a single run. To ensure the highest recording quality after grinding and polishing, an additional flat milling (Hitachi IM4000) [11] was performed for 5 min at 6 kV. The operation

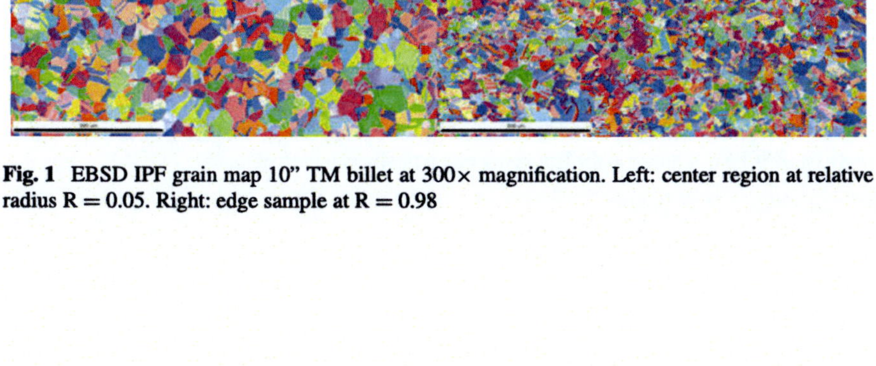

Fig. 1 EBSD IPF grain map 10" TM billet at 300× magnification. Left: center region at relative radius R = 0.05. Right: edge sample at R = 0.98

is illustrated in Fig. 2a). The specimen for preparation had a dimension of 5 × 7 × 1 mm, with a thermocouple on the upper surface in the vicinity of the measured area as shown in Fig. 2b). The heating was carried out with a second-generation high-temperature heating stage HSEA-1000 from TSL Solutions KK [12]. The position of thermocouple close to the region of interest and not on the reverse side is essential for precise temperature control during the experiments since the heating stage technically supplies the heat to the reverse side of the sample and there might be a temperature gradient over the specimen thickness.

The test setup in the SEM (Cross Beam 340) is shown in Fig. 3a). Before an in-situ experiment was started, a representative SEM image (BSE) has been taken (Fig. 3b)), which on the one hand documents the initial condition of the specimen and on the other hand provides the basis for the selection of the measuring parameters in terms of magnification, electron beam voltage and the frame size of the EBSD detector (Oxford Instruments—CMOS Symmetry). The starting microstructure in Fig. 3b) shows the importance of the pre-material selection since its high concentration of δ-phase will dissolve above δ-solvus and lead to a change of grain growth kinetics.

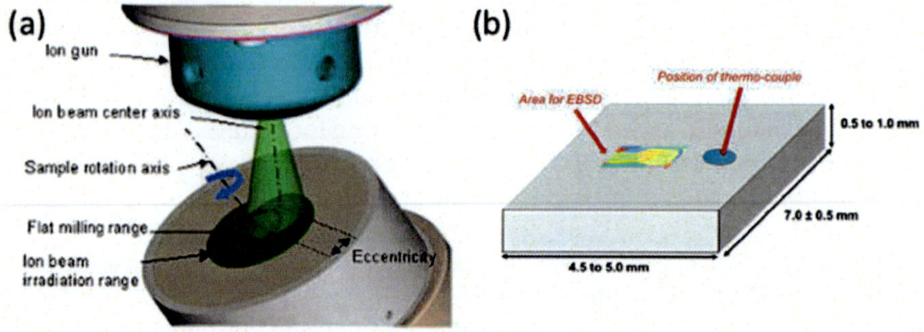

Fig. 2 **a** Illustration of the work principle of flat milling [Hitachi], **b** Illustration of the position of the thermocouple and the region of interest for EBSD

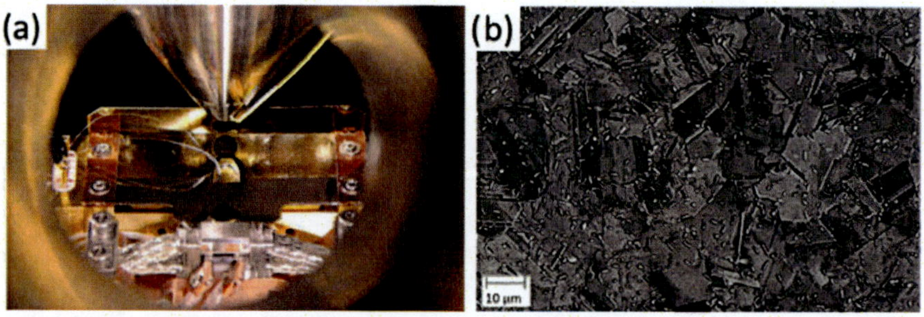

Fig. 3 **a** Measurement setup in the SEM including the heating stage, **b** representative back-scatter electron (BSE) SEM image at the beginning of the in-situ experiment

Because the mentioned dissolution kinetics is of high importance, the in-situ HT-EBSD experiments were conducted at 1045 °C, which corresponds to the specific maximum of the used heating module and lies above the δ-solvus temperature of about 1025 °C. The thermal history is depicted in Fig. 4 starting with pre-heating to 200 °C and degassing for 30 min before the first EBSD image is taken representing the starting condition of the sample. Heating continues with a maximum heating rate of 250 °C/min up to 980 °C (below δ-solvus) where the temperature is kept constant for 3 min in order to refocus on the previously chosen measurement area, which is necessary due to the thermal drift during ramp up of the specimen. However, the impact of the short holding time at 980 °C on the microstructure can be considered negligible. After the final ramp up to 1045 °C a SEM video is started providing in-situ imaging of the dissolution process of the δ-phase and the grain growth of each grain size fraction (small grains grow into coarser ones). The δ-solution process finishes in under 5 min, which marks the end of the SEM video recording, instantly followed by the sequential recording of EBSD images. One EBSD is recorded every 3 min over the course of the predefined holding time of 1.5 h. To be able to characterize the precipitation of the δ-phase too, an additional 1 h holding time at 900 °C was attached with EBSD imaging every 15 min.

The systematic and continuous recording of SEM and EBSD data within in a single experiment is a significant advantage for the purpose of this work. No uncertainties arise from the sampling positions in case of an alternative multi-specimen approach. The influence of local variations of the microstructure in the pre-material in case of individual sampling is eliminated since only a single sample is required and a specific area can be consistently traced throughout the applied thermal history. Additionally, every single image would require an ex-situ heat treatment and individual characterization, which would lead to higher expenses in terms of time and costs compared to the in-situ method in the SEM.

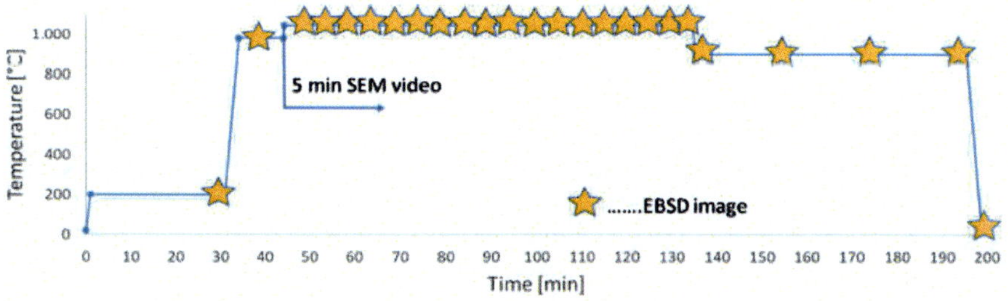

Fig. 4 Temperature evolution for the two in-situ EBSD experiments. The timing of the EBSD measurements is indicated

Modeling

The modeling of grain growth in the framework of grain classes is supposed to extend the capabilities of a simulation tool that already captures the evolution of grain sizes during forging where recrystallization is the dominant mechanism. The basis for the description of grain size evolution are semi-empirical approaches as outlined in [5]. Since the multi-class model uses area-based grain sizes, single-class growth models are not directly compatible with the multi-class framework due to the following mathematical problems. The class-dependent grain growth based on a standard single class model for each size fraction would realistically reflect the faster growth kinetics of small grains, but they would only grow within their grain fractions towards the next defined grain class. The approach would lead to expected coarsening, but it happens parallel to and independently of the growth of other fractions leading to a violation of the assumption of a constant volume. Correspondingly, the model would not represent the process of grain growth in a physically sound way.

A correct numerical representation could be built on physically, computationally intensive models, which however would take too long to be applicable in a digital twin. Therefore, a semi-empirical approach is applied, where the distribution of the grain classes is fixed during coarsening and the mean value follows single-class growth kinetics. Since the calculation is area-based, the conversion into ASTM classes (which follow a logarithmic scale of the grain sizes) can be well established and should mitigate the deviation to physical models. An insight into obtainable results based on these assumptions with fixed distributions is shown in Fig. 5, wherein the sum of the area to be reproduced as well as the density function of the grain distribution does not change, and only the average mean value of the grain size grows (represented by the peak value in Fig. 11, which will be discussed further below).

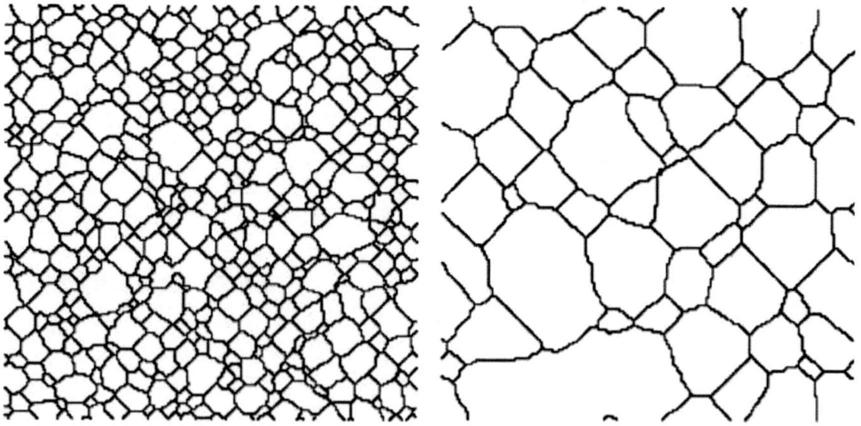

Fig. 5 Illustration of grain growth modeling based on increase of mean size and fixed size distribution [13]

For the growth of the mean grain size, which, however, is still size-dependent, the approaches of Buken and Kozeschnik [14, 15] are chosen and parameterized based on the results of the test series. The resulting grain diameter D in Eq. 1 is the product of the grain boundary mobility M and the size-dependent driving force for grain growth $P_{d,i}$, where the index i refers to the respective size class. The scaling factor η_H, material constant k_d, and the surface energy γ_{HA} from Eq. 2 can be combined into a single temperature and grain size dependent constant.

$$\dot{D} = \frac{dD}{dt} = MP_{D,i} \tag{1}$$

$$P_{D,i} = 2\eta_H k_d \gamma_{HA} (\frac{1}{D_i} - \frac{1}{D_{mean}}) \tag{2}$$

$$\frac{1}{M} = \frac{1}{M_{Precipitates}} + \frac{1}{M_{Solute\ drag}} \tag{3}$$

The grain boundary mobility M in the case of samples containing δ-phase is clearly different from those with hardly any δ-phase as will be illustrated in the results section based on model results. At the current development stage, the model has been calibrated and tested with data from the literature and part of the presented experimental results. The full range of datasets obtained from the in-situ as well as ex-situ experiments will be used for a consistent calibration of model parameters for the tested parameter range and for validation purposes.

Results and Discussion

The results of the two separate test series, i.e. in-situ HT-EBSD and ex-situ heat treatment and optical light microscopy (OLM), are individually discussed and afterward compared in the context of grain growth behavior. The in-situ data, which is available as a sequence of EBSD images every 3 min, are intended to show the progression of grain growth visualized in the 20 ASTM grain classes that are also used in the model. Since in-situ data was only recorded for two initial microstructures and one dissolution temperature, the ex-situ tests are a valuable supplement to the data set. They were evaluated for the evolution of the mean value of the grain size with time to quantify the influence of varying temperatures on the δ-phase dissolution and coupled grain growth.

HT-EBSD

As with all metallic materials, the large grain fractions grow in favor of the finer fractions. This behavior is illustrated in Fig. 6 with three representative EBSD grain maps from the same position of the R = 0.98 sample. Especially the comparison between 30 and 90 min holding time shows how the largest grains grow at the expense of the surrounding smaller grains, while their orientation and the position of the twins remain approximately constant.

Additionally, it was found that the δ-phase in sample R = 0.98 was already completely dissolved after 5 min at 1045 °C. This δ-phase dissolution kinetics is representative for this high-temperature experiment and both are confirmed by the SEM video at the beginning of the holding time and the second recorded EBSD image. After the complete dissolution of the δ-phase, the kinetics of the grain growth was faster and comparable to results of the sample at R = 0.05 with a low δ-phase fraction right from the beginning.

It should be noted that in the case of EBSD data, as opposed to OLM data, the grain size evaluation can be easily carried out with or without consideration of the twin boundaries, depending on the goal of the analysis. Physically more meaningful is the removal of the twin boundaries but grain size data from OLM typically includes them and therefore the EBSD evaluation for the model calibration was chosen accordingly. This way the data are compatible with the OLM data from the ex-situ experiments and historic and future OLM data from standard testing where the removal of twins is not possible. Modeling results will thus predict the commonly determined grain sizes, that include twin boundaries. The difference is illustrated in Fig. 7 for the pre-material at R = 0.98. The comparison shows that the evaluation convention leads to finer size distributions and this must be kept in mind for consistent interpretation of experimental and simulation results.

Figure 8 shows the total evolution of the grain size distribution, including twin boundaries, in the corresponding classes. It can be seen that especially the finest fractions (smaller than ASTM 12.0) completely disappear after only 5 min at these temperatures. After that, a wide-range grain size distribution can be seen, which increases with an approximate maximum in the middle of the distribution. After

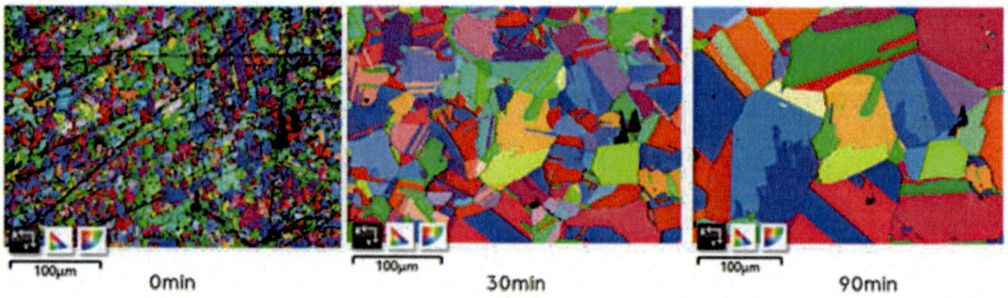

Fig. 6 Representative EBSD images of a specific position of the R = 0.98 sample at the beginning of the holding time and after 30 and 90 min

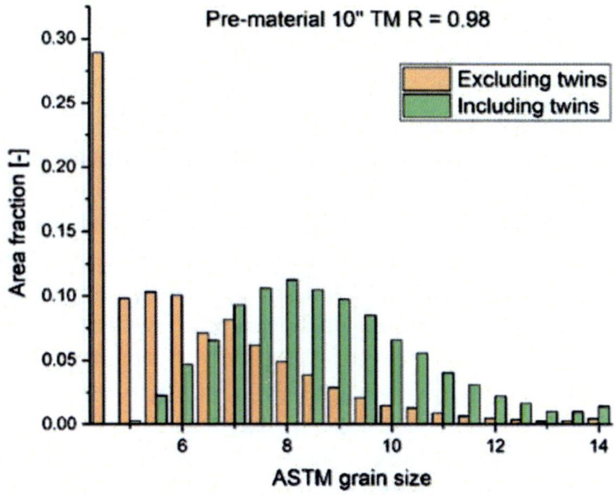

Fig. 7 Illustrative comparison of grain size evaluation from EBSD data with and without consideration of twin boundaries for 10" TM pre-material

about 30 min a maximum at ASTM 4.5 forms, due to some extremely coarsened grains. However, this maximum is also due to the non-linear slope of the ASTM grain classes and the fact that no coarser class than ASTM 4.5 is considered for the evaluation in this study. For this reason, all larger grains accumulate in this class and do not move on to a coarser fraction.

Apart from the very beginning of the holding time the results of the R = 0.05 sample show the same evolution of grain growth kinetics as R = 0.98. The difference is the starting condition, with coarser grains and no δ-phase. For R = 0.98, a comparable state was reached after 5 min, and from this point on results match.

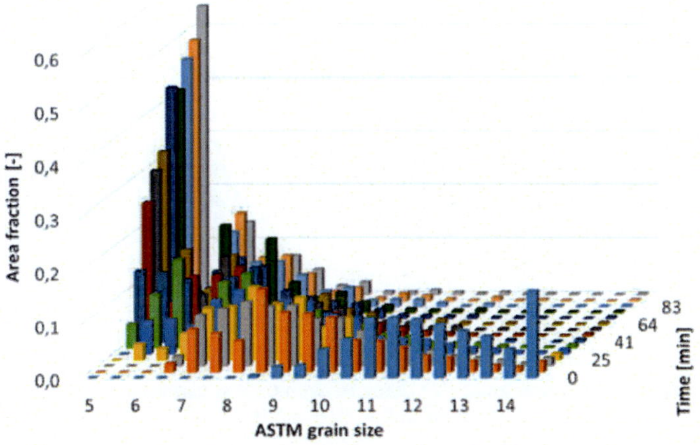

Fig. 8 Grain size distribution evolution; HT-EBSD R = 0.98; T = 1045 °C

Ex-Situ Heat Treatments

In order to supplement the in-situ experiments, the samples from ex-situ annealing were evaluated using OLM data using the algorithm described in [4]. The focus was laid on the grain growth kinetics below and above the δ-solvus temperature. For the ex-situ samples, the evolution of the mean grain diameters was determined at 980, 1010, 1040, 1070, and 1100 °C and is shown in Fig. 9. Please note that the edge samples were extracted at R = 0.95 and not at R = 0.98 as for the in-situ experiments. However, the positions on the radius are sufficiently close to exclude microstructural differences. As expected, the grain size increases with time and is more pronounced at higher temperatures. The distinction between the edge samples (R = 0.95) and the center samples (R = 0.05) with a lower proportion of δ-phase is clearly visible since these already exhibit increased growth kinetics below δ-solvus. This phenomenon changes above the δ-solvus temperature since the edge samples usually have a higher dislocation density due to the radial forging, which accelerates the growth kinetics accordingly. Related effects are discussed in [16, 17].

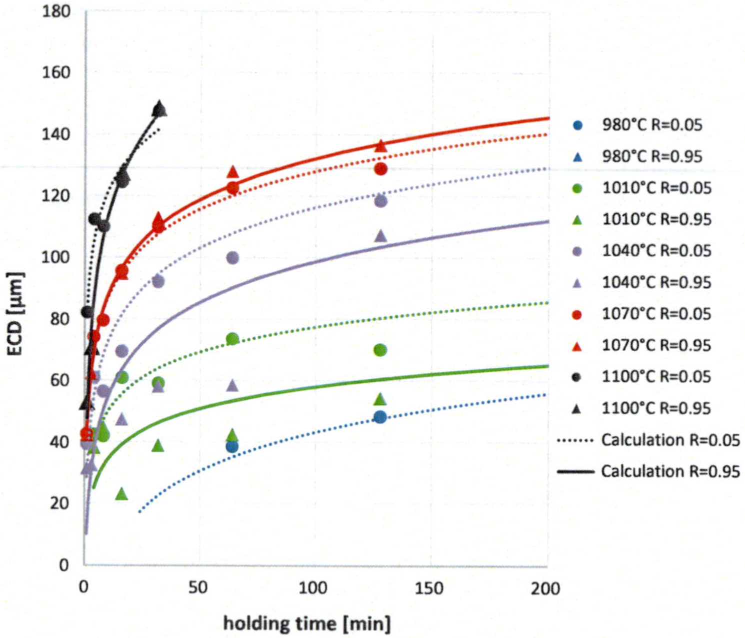

Fig. 9 Evolution of the mean grain size diameter with time at various temperatures and for two different initial conditions corresponding to R = 0.05 and R = 0.95. Symbols refer to experimental results, lines to model predictions

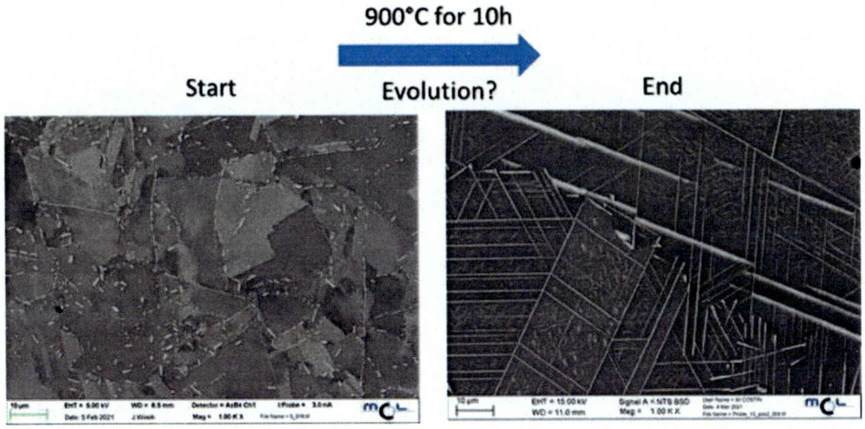

Fig. 10 Delta dumping – change in phase fraction and morphology of precipitating δ-phase at 900 °C after 10 h holding time

Precipitation of δ-phase

Due to the design of the in-situ experiments also the precipitation of δ-phase could be investigated, which is kinetically significantly slower than dissolution. According to the literature, the fastest precipitation kinetics occur around 900 °C [18]. In addition to the 1 h holding time at this temperature in the in-situ experiments, an ex-situ experiment was carried out to characterize the extreme condition after 10 h at 900 °C, which is depicted in Fig. 10. The δ-phase forms a needle-like morphology and grows partly through grains as reported in a similar study of Xie et al. [18]. In technical applications this condition is avoided due to the significant degradation of the mechanical properties. In general, the consideration of δ-phase precipitation in the model is of lower relevance compared to δ-phase dissolution, because the cooling through the relevant temperature range in the process chain of forging and heat treatment is always sufficiently fast and does not affect the fraction or the morphology of the δ-phase.

Model Results

The logarithmic trend lines fitted to the experimental data in Fig. 9 represent the first parameter identification for the implemented modeling approach. The predictions are in good accordance with the measurement results even without detailed parametrization. The full parametrization of the model at all temperature levels has still to be carried out including the coupling with δ-phase kinetics as shown for the single-class approach [7], in order to capture the change of grain boundary mobility

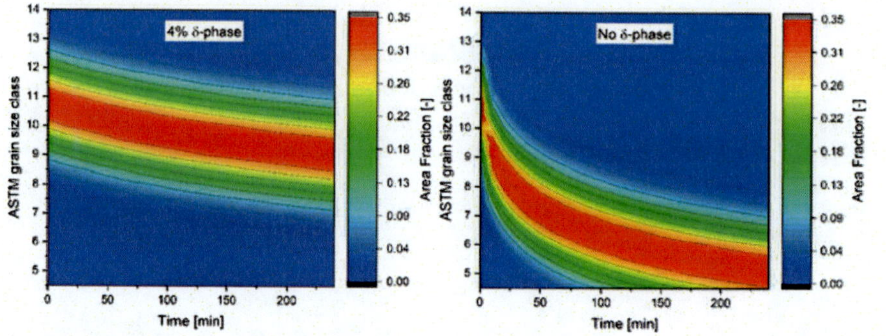

Fig. 11 Visualization of results from the multi-class grain growth model for grain growth at 1000 °C for a microstructure with 4% δ-phase (left) and no δ-phase (right)

and grain growth within one consistent model. Nevertheless, the model can already be used to illustrate the effect of the strongly varying grain boundary mobility in microstructures with and without δ-phase. Figure 11 shows the growth kinetics for both cases at a temperature below δ-solvus and starting with the same initial grain size distribution. The growth kinetics without δ-phase is significantly faster than in a corresponding simulation with 4% δ-phase. At 1000 °C with 4 h holding time the initial average grain size of 10.5 ASTM (average of all measured ex-situ samples) with its characteristic distribution function grows to 9.0 ASTM in the case of the sample containing 4% δ-phase and to 5.0 in the case of the sample free of δ-phase. The resulting area-based fractions of each grain class can be determined using the known statistical distribution with respect to the individual grain classes, as mentioned in the modeling section.

However, there are limitations of the modeling approach for grain growth, since the evolution of more complex grain size distributions cannot be described. A single mean value and standard deviation cannot describe duplex distributions like "as large as" (ALA) grains or bimodal distributions. This capability is by definition an essential advantage of the multi-class grain size framework, which still applies to the recrystallization part of the model. Regarding grain growth the chosen approach poses a limitation of this advantage, especially when simulated size distributions are subsequently used for predicting mechanical properties. A possible solution is the use of two separate distribution functions for the small and the larger grain fractions that are simulated separately and merged after every calculated time increment. In practice, bimodal distributions are avoided by the optimization of the forging process, which can be done based on the recrystallization part of the model and for the common continuous grain size distributions as depicted in Fig. 8, no such limitations arise for the grain growth part.

Conclusion

The combined experimental approach with the in-situ and ex-situ characterization of δ-phase and grain growth kinetics for physical understanding and model development provides several advantages compared to classical methods using ex-situ experiments only. Besides the time and cost efficiency of the presented in-situ investigations using a heating stage inside an SEM for continuous recording of SEM and EBSD images, the results can be interpreted without uncertainties arising from the common inhomogeneities within the pre-material and their unknown effect in approaches using multiple samples. Due to their easier accessibility, ex-situ experiments can effectively supplement the in-situ results for additional temperatures and specifically for long holding times.

The first model results confirm that the simple approach for the extension of the existing multi-class grain size model for recrystallization effectively describes the evolution of grain growth obeying the law of volume constancy when size fractions grow. In the implementation the average grain size resulting from all size classes grows according to a semi-empirical model and is converted back to individual grain classes based on the initial statistical size distribution. However, care must be taken when bimodal or duplex distributions are the initial state for grain growth because limitations arise from the chosen model approach in this case.

With the extensive and consistent experimental data the first auxiliary parametrization of driving force and grain boundary mobility in the model based on literature data and parts of the presented results can now be extended and optimized for all relevant temperature ranges and microstructural initial states providing a model covering recrystallization during forging and δ-phase dependent grain growth during subsequent heat treatment.

Acknowledgements The research leading to these results has received funding from the TakeOff program. TakeOff is a Research, Technology, and Innovation Funding Program of the Austrian Federal Ministries for Climate Action, Environment, Energy, Mobility, Innovation, and Technology (BMK) (867403). The Austrian Research Promotion Agency (FFG) has been authorized for the Program Management.

References

1. Stockinger, M.; Stanojevic, A.; Wieser, V. & Raninger, P., Development of an automated property simulation tool for direct aged alloy 718 engine disk forgings, Minerals, Metals and Materials Series, 2018, 2018-June, 353–367
2. https://sdgs.un.org/goals
3. Horstemeyer, M., Integrated Computational Materials Engineering (ICME) for Metals: Using Multiscale Modeling to Invigorate Engineering Design with Science, Integrated Computational Materials Engineering (ICME) for Metals: Using Multiscale Modeling to Invigorate Engineering Design with Science, 2012
4. C. Gruber, Digitalization of forged alloy 718 parts with implemented microstructure and fracture mechanic model, PhD thesis, Montanuniversität Leoben, 2022

5. Gruber, C.; Raninger, P.; Stanojevic, A.; Godor, F.; Rath, M.; Kozeschnik, E. & Stockinger, M., Simulation of dynamic and meta-dynamic recrystallization behavior of forged Alloy 718 parts using a multi-class grain size model, Materials, 2021, 14, 1–15
6. Gruber, C.; Raninger, P.; Stockinger, M. & Bucher, C., Multi-class grain size model for forged alloy 718 aircraft parts, Materials Science Forum, 2021, 1016 MSF, 499–508
7. M. Stockinger and J. Tockner, Optimizing the forging of critical aircraft parts by the use of finite element coupled microstructure modelling, Superalloys 718, 2005, 625–706
8. Sellars, C. & Whiteman, J., Recrystallization and grain growth in hot rolling, Metal Science, 1979, 13, 187–194
9. Brandl, D.; Lukas, M.; Stockinger, M.; Ploberger, S. & Ressel, G., Evidence of austenite memory in PH 15-5 and assessment of its formation mechanism, Materials and Design, 2019, 176
10. D. Winkler, Entwicklung einer Methode zur verbesserten Analyse von großflächigen Makroscheiben, Bachelor thesis, Montanuniversität Leoben, 2022
11. https://www.hitachi-hightech.com/us/product_detail/?pn=em-im4000plus
12. https://www.tsljapan.com/products/in-situ/ebsd-kanetsu
13. Wright, S. A.; Plimpton, S. J. & Swiler, T. P., Potts-model grain growth simulations: Parallel algorithms and applications, 1997
14. Buken, H. & Kozeschnik, E., Modeling Static Recrystallization in Al-Mg Alloys, Metallurgical and Materials Transactions A: Physical Metallurgy and Materials Science, 2021, 52, 544–552
15. Kozeschnik, E., Mean-Field Microstructure Kinetics Modeling, Encyclopedia of Materials: Metals and Alloys, 2021, 521–526
16. Aoki, C.; Ueno, T.; Ohno, T. & Oikawa, K., Influence of hot-working conditions on grain growth of superalloy 718, Journal of Materials Processing Technology, 2019, 267, 26–33
17. Stanojevic, A.; Bucher, C.; Gruber, M.; Oberwinkler, B. & Stockinger, M., Optimization of the forging process window in respect of AGG, IGG and direct age effect in alloy 718 engine disks, Minerals, Metals and Materials Series, 2018, 2018-June, 691–709
18. Xie, X.; Xu, C.; Wang, G.; Dong, J.; Cao, W.-D. & Kennedy, R., TTT diagram of a newly developed nickel-base superalloy- Allvac® 718Plus™, Proceedings of the International Symposium on Superalloys and Various Derivatives, 2005, 193–202

Abnormal Grain Growth Maps of Wrought Ni-Base Superalloys

M. G. Fahrmann and D. A. Metzler

Abstract The presence of abnormally large, overgrown grains upon sub-solvus annealing of certain hot-worked structures has been reported for several cast and wrought and powder metallurgy superalloys. These overgrown grains typically feature a high density of annealing twins and a precipitate distribution similar to the one in the adjacent fine-grained matrix. Predicting the propensity of a given thermo-mechanical processing path to trigger said abnormal grain growth (AGG) has proven difficult since key aspects such as the distribution of stored energy in the as-hot-worked structure are generally not known a priori. Instead, alloy-specific AGG maps are proposed that provide a potential risk assessment for AGG to occur. Such maps were generated exemplarily for two very different Ni-base superalloys, HAYNES® 244® alloy and HAYNES® 233™ alloy, built on a sizeable number of different hot-worked and sub-solvus annealed product forms. The usefulness and limitations of such maps are discussed. In addition, some insight into possible mechanisms of AGG upon sub-solvus annealing of the studied alloys is provided.

Keywords Polycrystalline superalloys · Abnormal grain growth maps · Overgrown grains · Zener pinning · Stored energy

Introduction

Abnormally large grains forming upon annealing of certain hot-worked structures have been reported for a number of commercial cast and wrought and powder metallurgy superalloys [1–12]. This phenomenon has been referred to as Abnormal Grain Growth (AGG), Critical Grain Growth (CGG), or Inhomogenous Grain Growth (IGG), and has typically been associated with certain thermo-mechanical processing paths. Of particular interest is the occurrence of abnormally large grains during sub-solvus annealing of components since sub-solvus annealing is specifically designed to control grain size by virtue of the pinning force of grain boundary precipitates.

M. G. Fahrmann (✉) · D. A. Metzler
Haynes International, Inc, 1020 W. Park Avenue, Kokomo, IN 46904-9013, USA
e-mail: mfahrmann@haynesintl.com

E. A. Ott et al. (eds.), *Proceedings of the 10th International Symposium on Superalloy 718 and Derivatives*, The Minerals, Metals & Materials Series,
https://doi.org/10.1007/978-3-031-27447-3_7

Using the term AGG throughout this paper, AGG was encountered in laboratory sub-solvus annealing studies of several hot-worked HAYNES alloys and superalloys. While those annealing conditions are not of immediate commercial relevance for the studied alloys, they offer unique insight into this important phenomenon in anticipation of future product needs. Initially, an attempt was made [12] to rationalize the observed AGG on the basis of a framework recently proposed by Charpagne et al. [10]. However, it quickly became obvious that the two key inputs in the framework, i.e., the distribution of stored energy in the as-hot-worked structure and the distribution of grain boundary pinning precipitates, are generally not known a priori, thus limiting the predictive capability of such maps.

Taking a different approach, we propose an engineering AGG map that reflects several key aspects that can be reasonably well controlled in a manufacturing environment: the local dynamically recrystallized grain size in the as-hot-worked condition, annealing temperature relative to the solvus temperature of the main grain boundary pinning phase, and annealing time. Such maps were illustratively constructed for two alloys featuring very different grain boundary pinning precipitates by incorporating a sizeable number of different hot-worked product forms. These empirical maps were supplemented by detailed microstructure analyses.

Material and Experimental Methods

The two alloys selected are HAYNES 244 alloy and HAYNES 233 alloy. The nominal compositions of these alloys are shown in Table 1. The predominant grain boundary pinning phase at the relevant hot working and annealing temperatures in the former is μ phase [13], in contrast to secondary carbides in the latter [14].

The annealing response of a sizeable number of product forms, particularly in 233 alloy, was studied by optical metallography and scanning electron microscopy (SEM), supplemented by energy-dispersive spectroscopy (EDS) and electron backscatter diffraction (EBSD). Metallographic preparation methods and imaging/analyses conditions were reported elsewhere [12].

Table 1 Nominal compositions (in wt.%) of the two studied cast and wrought HAYNES superalloys

Alloy	Ni	Co	Fe	Cr	Mo	W	Mn	Al	Ti	C	B
244 alloy	bal	1[a]	2[a]	8	22.5	6	0.8[a]	0.5[a]	–	0.03[a]	0.006[a]
233 alloy[1]	bal	19	1.5[a]	19	7.5	0.3[a]	0.4[a]	3.3	0.5	0.1	0.004

[a] Maximum [1] also contains 0.5% Ta and 0.03% Zr

Results and Discussion

AGG Map of 244 Alloy

Initially, AGG was observed upon sub-solvus annealing of a 9.5 mm (3/8″) diameter hot-rolled bar that was rolled from 1204 °C (2200 °F) across multiple roll stands to a total of 80% reduction. Sub-solvus annealing of bar samples at 1066 °C (1950 °F) for 30 min resulted in typical features shown in Fig. 1: large overgrown, heavily twinned grains featuring intra-granular precipitates whose distribution appears to be similar to that in the adjacent fine-grained matrix of ASTM 10–12 grain size. The bright specks in this backscattered electron (BSE) image are Mo- and W-rich precipitates, believed to be μ phase. Note that these precipitates largely reside in the grain boundaries in the fine-grained matrix.

Note also that the solvus temperature of these precipitates in 244 alloy is approximately 1121 °C (2050 °F). While significant grain growth ensued at the higher, commercially relevant, annealing temperatures of 2050 °F and above, no abnormally large grains were observed upon super-solvus annealing of this particular material, as well as many other products.

The microstructure shown in Fig. 1 is characteristic of this type of AGG and was also encountered in other product forms of this alloy at similar higher levels of work. However, AGG was not encountered when annealing material that was warm-worked or only partially recrystallized as a result of hot working. The contrasting as-hot-worked microstructures of these two conditions are shown in Fig. 2.

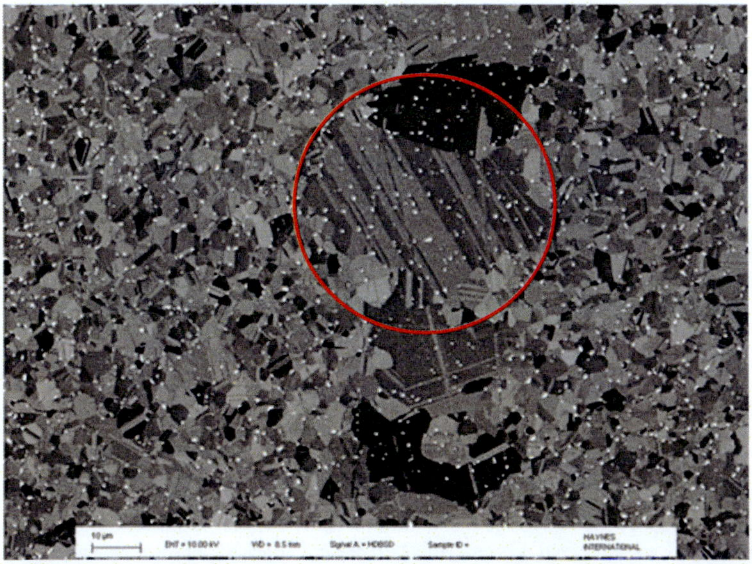

Fig. 1 Overgrown grains (red circle) in a hot rolled and subsequently sub-solvus annealed bar of HAYNES 244 alloy. The bright specks are Mo- and W-rich precipitates. SEM BSE micrograph

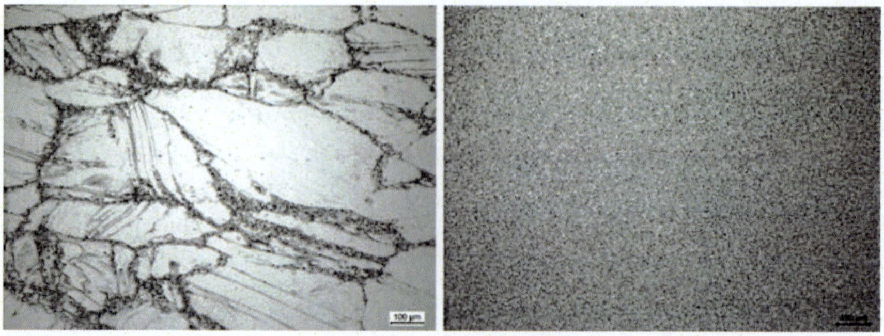

Fig. 2 Micrographs of as-hot-rolled microstructures of 244 alloy bar products after 40% reduction (left, no AGG upon sub-solvus annealing) and 80% reduction (right, same condition as Fig. 1 where AGG was encountered during sub-solvus annealing). Both micrographs have 100× magnification

These findings led us to suspect that the (meta) dynamically recrystallized grain size (DRX GS) might be a key factor for triggering AGG. Additionally, the annealing temperature relative to the solvus temperature of μ phase was surmised to be another key factor in this alloy. Accordingly, the AGG map depicted in Fig. 3 was constructed based on the experimental evidence.

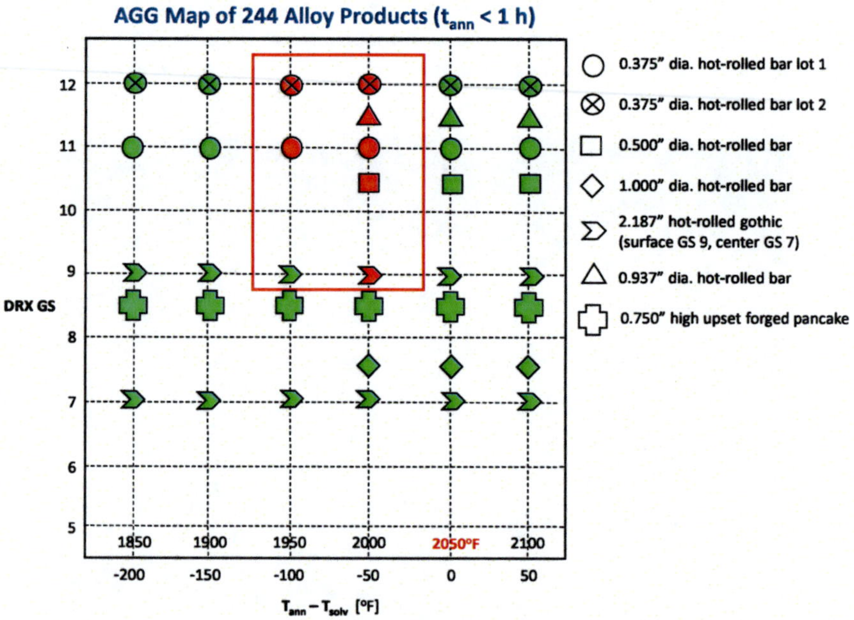

Fig. 3 AGG map of hot-worked products of 244 alloy. The red symbols represent conditions for which AGG was observed. Green symbols represent either a locked-in structure (sub-solvus annealed below 2050 °F) or normal grain growth when super-solvus annealed (above 2050 °F)

The conditions that triggered AGG upon annealing are delineated by the red box. Inspection of the microstructures of all the other sub-solvus annealed conditions represented by the green symbols suggests that Zener pinning of the grain boundaries was effective in preventing AGG. One important takeaway from this map is that only sufficiently fine-grained and largely (if not fully) dynamically recrystallized hot-worked structures seem prone to this type of AGG. Note that this map is relevant for annealing times of less than one hour, commensurate with the section thicknesses of the products studied. In fact, excessively long sub-solvus annealing times did eventually trigger AGG in some of these products. While those annealing times are not commercially relevant, this finding points at annealing time to be a third dimension in the parameter space of AGG with implications for the kinetics of AGG.

AGG Map of 233 Alloy

A similar map was generated for 233 alloy (Fig. 4), following the same methodology, albeit including a significantly greater number of products. The relevant secondary, grain boundary pinning phase in this alloy are Cr- and Mo-rich secondary carbides that go into the solution at around 2100 °F (1149 °C).

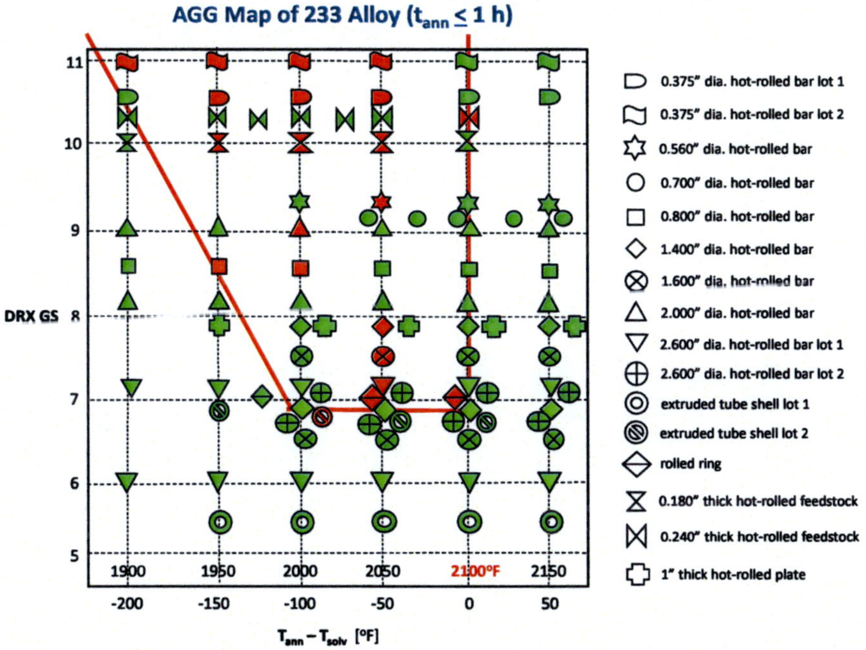

Fig. 4 AGG map of 233 alloy. The red symbols represent conditions for which AGG was observed. Green symbols reflect conditions under which AGG was not encountered

As was the case for 244 alloy, a sufficiently fine (meta) dynamically recrystallized grain size seems to be one of the prerequisites for AGG to occur. However, the delineation of the AGG-prone region in 233 alloy is less obvious, as indicated by the coexistence of normal grain growth and AGG within the red box. This finding led to a rigorous scrutiny of the corresponding microstructures. It was discovered that several of the green symbols (corresponding to normal grain growth) within the AGG-prone region represent microstructures featuring largely clean grain boundaries that had apparently broken away from the precipitates during annealing, as shown illustratively in Fig. 5. This type of microstructure was not encountered in any of the studied (albeit more limited in number) 244 alloy products. At present, it is not clear what might have triggered this distinctly different annealing response.

However, some of the microstructures (e.g., the 0.240″ thick hot-rolled feedstock in Fig. 4) did feature fine grains pinned by secondary carbides and, yet appeared to be stable, at least within one hour of annealing. Upon closer inspection of the 0.240″ feedstock, numerous features were identified that might act as nuclei for large overgrown grains, featuring twin boundaries and intra-granular precipitates as shown in Fig. 6. Their characteristics were distinctly different from those of the surrounding matrix where grain boundaries were still pinned by precipitates.

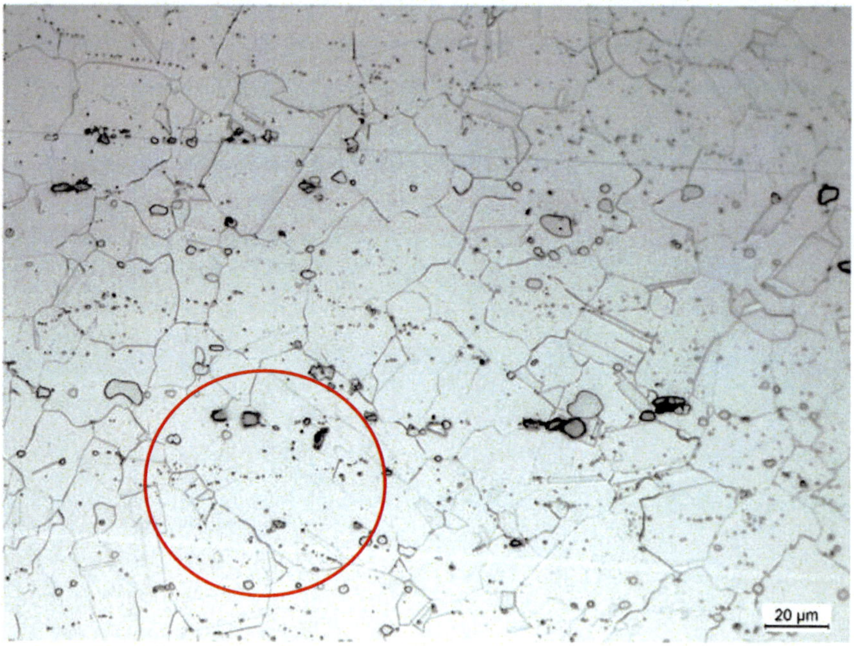

Fig. 5 Microstructure of a 2″ diameter 233 alloy bar sample annealed at 2000°F (1093 °C) for 30 min in the AGG-prone region. AGG was not observed. Note that the grain boundaries are not pinned by secondary carbides—the latter delineate prior grain boundaries as shown in the red circle

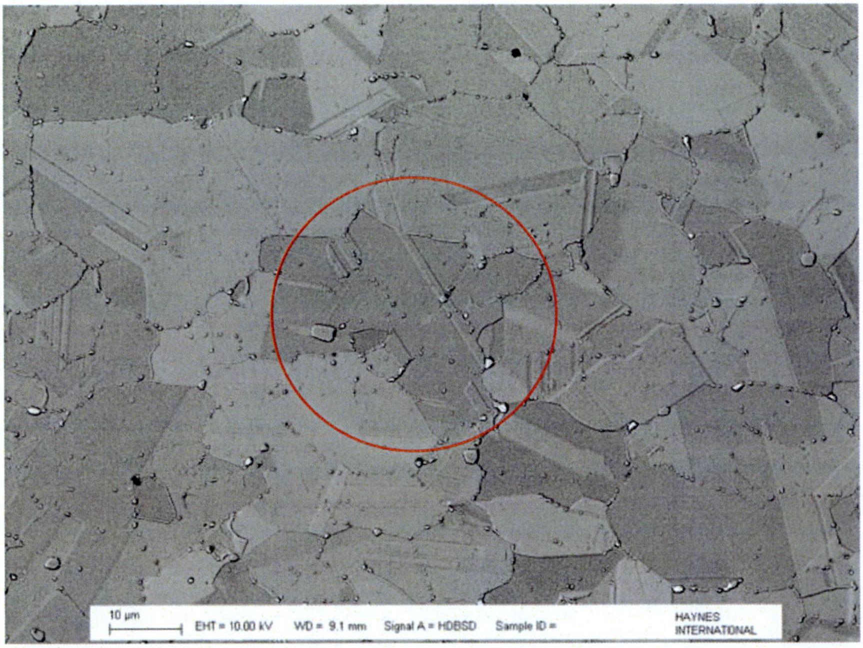

Fig. 6 Microstructure of 0.240″ thick hot-rolled 233 alloy feedstock that was annealed at 2025 °F (1107 °C) for 1 h in the AGG-prone region. No AGG was encountered. However, numerous features that might represent AGG nuclei (one is encircled) were observed

After three hours of annealing, samples from the same 0.240″ feedstock exhibited massive AGG as shown in Fig. 7. However, not all of these potential nuclei resulted in abnormally large grains since their density appears to be orders of magnitude higher than that of the actual abnormally large grains found after three hours of sub-solvus annealing.

These observations suggest that grain boundary mobility, allowing for sustained unabated grain growth, might be an important factor. To gain further insight, samples from the 0.180″ thick hot-rolled feedstock were subjected to sub-solvus annealing at various temperatures and staggered annealing times; in 15 min increments. The data in Table 2 provides strong evidence for a temperature-dependent, thermally-activated incubation period.

Note that experimental evidence for the potential AGG nuclei shown in Fig. 6 could be established at all studied annealing temperatures within the first 15 min of annealing, suggesting that it is the formation of certain high-mobility grain boundaries that triggers AGG.

While our studies have provided additional insights into this type of AGG in the presence of a grain boundary pinning phase, identifying the particular processing paths that trigger AGG upon sub-solvus annealing remains challenging. Still, it is expected that AGG maps such as those presented in Figs. 3 and 4 have merit since they

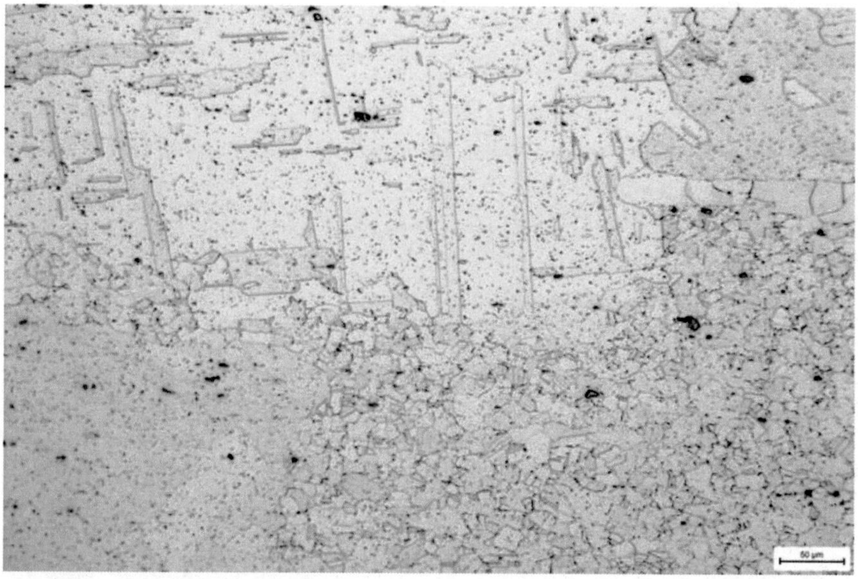

Fig. 7 Microstructure of 0.240″ thick hot-rolled 233 alloy feedstock that was annealed at 2025°F (1107 °C) for 3 h in the AGG-prone region. Massive AGG was encountered

Table 2 Incubation time for massive AGG encountered upon sub-solvus annealing of 0.180″ thick hot-rolled feedstock of 233 alloy

Annealing temperature [°F (°C)]	Incubation time [min]
1950 (1066)	135
2000 (1093)	60
2025 (1107)	30

allow for an engineering risk assessment of the impacts on quality, yield, and cost. State-of-the-art computational tools such as DEFORM® coupled with microstructure evolution models can provide an assessment of the location-specific as-hot-worked (forged, rolled, extruded …) structure—a key input for the proposed AGG maps.

The main limitations of this method are: (a) the maps appear to be alloy-specific, and (b) the map is empirical, thus requiring a significant amount of experimental data to generate sufficiently accurate regime boundaries.

Summary and Conclusions

1. The occurrence of abnormal grain growth (AGG) upon sub-solvus annealing of certain hot-worked structures was studied in two very different cast and wrought Ni-base superalloys.

2. The defining features of AGG in these alloys were very large overgrown, heavily twinned grains.
3. Compelling experimental evidence for a thermally-activated incubation period for AGG is provided, the conjecture being that AGG is triggered by the formation of certain high-mobility grain boundaries in the course of the incubation period.
4. For engineering purposes, the findings were presented in the form of alloy-specific AGG maps, (meta) dynamically recrystallized grain size and sub-solvus annealing temperature being the key inputs. A third dimension, annealing time, needs to be considered in the context of the temperature-dependent incubation period.

Acknowledgements The authors are grateful to the Haynes International Research & Technology staff. HAYNES and 244 are registered trademarks of Haynes International, and 233 is a trademark of Haynes International, Inc. DEFORM is a trademark of Scientific Forming Technologies Corp., Columbus, OH.

References

1. Uginet JF, Pieraggi B (1997) Study of secondary grain growth on 718 alloy. Proc. Intl. Symp. on Superalloys 718, 625, 706 and Various Derivatives, ed. E. A. Loria, (TMS/1997), p. 343–352.
2. Huron H, Srivatsa S, Raymond E (2000), Control of grain size via forging strain rate limits for R'88DT. Proc. 9th Intl. Symp. Superalloys, eds. T. M. Pollock et al., (TMS/2000), p. 49–58.
3. Cho YK, Yoon DY, Henry MF (2001), The Effects of Deformation and Pre-Heat-Treatment on Abnormal Grain Growth in RENE 88 Superalloy. Met. Mat. Trans. 32A: 3077–3090.
4. Whitis DD (2004), Recovery and Recrystallization after critical strain in the nickel-based superalloy RENE 88DT. Proc. 10th Intl. Symp. Superalloys, eds. K. A. Green et al. (TMS/2004), p. 391–400.
5. Bozzolo N, Agnoli NA, Souai N, Bernacki M, Loge R (2013) Strain induced abnormal grain growth in nickel base superalloys. 5th International Conference on Recrystallization and Grain Growth, Sydney, 2013. https://doi.org/10.4028/www.scientific.net/MSF.753.321.
6. Magnoli A, Bernacki M, Loge R, Franchet JM, Laigo J, Bozzolo N (2015) Selective Growth of Low Stored Energy Grains during δ Sub-Solvus Annealing in the Inconel 718 Nickel-Based Superalloy. Met. Mat. Trans. A 46: 4405–4421.
7. Aoki C, Ueno T, Ohno T (2016) Influence of hot working conditions on grain growth behavior of Alloy 718. Proc. 13th Intl. Symp. Superalloys, eds. M. Hardy et al., (TMS/2016), p. 609–617.
8. Miller VM, Johnson AE, Torbet CJ, Pollock TM (2016) Recrystalliztion and the Development of Abnormally Large Grains after Small Strain Deformation in a Polycrystalline Nickel-Based Superalloy. Met. Mat. Trans. 47A: 1566–1574.
9. Parr IMD, Jackson TJ, Hardy MC, Child DJ, Argyrakis C, Severs K, Saraf V, Stumpf JM (2016) Inhomogeneous grain coarsening behavior in supersolvus heat treated nickel-based superalloy RR1000. Proc. 13th Intl. Symp. Superalloys, eds. M. Hardy et al., (TMS/2016), p. 447–456.
10. Charpagne M-A, Franchet J-M, Bozzolo N (2018) Overgrown grains appearing during sub-solvus heat treatment in a polycrystalline γ-γ ′ Nickel-base superalloy. Materials and Design 144: 353–360.
11. Wang X, Huang Z, Cai B, Zhou N, Magdysyuk O, Gao Y, Srivatsa S, Tan L, Jiang J (2019) Formation mechanism of abnormally large grains in a polycrystalline nickel-based superalloy during heat treatment processing. Acta Mat. 168: 287–298.

12. Fahrmann MG and Metzler DA, (2020) Abnormal Grain Growth in the Presence of Grain Boundary Pinning Precipitates. Proc. 14th Intl. Symp. Superalloys, eds. S. Tin et al., (TMS/2020), p. 519–532.
13. Fahrmann MG, Srivastava SK, Pike LM (2012) Development of a new 760 °C (1400 °F) capable low thermal expansion alloy. Proc. 12th Intl. Symp. Superalloys, eds. E. S. Huron et al., (TMS/2012) p. 769–777.
14. Pike L, Srivastava K, Foresythe A (2019) A New Alumina Forming Ni-Co-Cr Base Alloy for Service in Gas Turbine Engine Combustors and Other High Temperature Applications, https://doi.org/10.1007/s11665-019-03950-2.

Alloy Design and Development of a Novel Ni-Co-Based Superalloy GH4251

Hongyao Yu, Hailong Qin, Xizhen Chen, Guangbao Sun, Bin Gan, Yu Gu, Teng An, Jinglong Qu, Jinghui Du, and Zhongnan Bi

Abstract The need to develop new high-temperature materials has increased significantly in the last decade owing to the demand of higher engine operating temperature. This demand has motivated the development of a new Ni-Co-based superalloy GH4251 with service temperature up to 700–800 °C. Based on disk alloy U720Li, the GH4251 alloy is designed by adjusting the content of Co, Cr, Ti, Nb, and other elements. On the one hand, by increasing the Co content to 25 wt. %, the stacking fault energy is effectively reduced, which makes it easy to form nano-twins and other substructures that strengthening the alloy together with γ' precipitates. The yield strengths of the newly designed alloy can achieve 1100 MPa at 750 °C and the creep-rupture life is more than 500 h at 750 °C under 530 MPa with fine grain size (ASTM 8), which is superior to U720Li. On the other hand, a certain Nb element is added to replace Ti element, which can reduce the γ'solvus temperature and its precipitating dynamics, along with a changed thermal deformation behavior caused by lower stacking fault energy, leading to a significant better hot work ability and weld ability compared with U720Li. Besides being used as disk or ring forgings, this novel GH4251 alloy can also be well processed by additive manufacturing due to its low cracking tendency.

Keywords Ni-Co based superalloy · Alloy design · Disk forging · Additive manufacturing

H. Yu · H. Qin · X. Chen · G. Sun · B. Gan · J. Du · Z. Bi (✉)
Beijing Key Laboratory of Advanced High Temperature Materials, Beijing 100081, China
e-mail: bizhongnan21@aliyun.com

Y. Gu · T. An · J. Qu · J. Du · Z. Bi
Gaona Aero Material Co. Ltd., Beijing 100081, China

E. A. Ott et al. (eds.), *Proceedings of the 10th International Symposium on Superalloy 718 and Derivatives*, The Minerals, Metals & Materials Series,
https://doi.org/10.1007/978-3-031-27447-3_8

Introduction

Ni-based superalloys have been developed for wide use in critical components of aircraft engines, such as turbine disks and blades. Currently, higher requirements are put forward on the mechanical properties, temperature-bearing capacity, and processing properties of Ni-based superalloys. In general, strengthening methods, such as grain boundary hardening [1], solid-solution hardening [2], and precipitation hardening [3], are conventionally employed to improve the mechanical properties and temperature-bearing capacity of Ni-based superalloys. However, these strengthening methods inevitably sacrifice the processing properties due to the increased strengthening effects in the temperature range of processing. Therefore, additional strengthening mechanisms are required to improve the mechanical properties of Ni-based superalloys without compromising the processing properties.

Over the past few years, nanotwin strengthening has been proposed to improve the mechanical properties of materials due to their unique interaction with dislocations during plastic deformation [4, 5]. Moreover, the high thermal/mechanical stabilities of twin boundaries due to their intrinsic low-energy state enable the advantages for high-temperature applications [6]. For superalloys with low stacking fault energy (SFE), nanotwins are often observed when they are plastically deformed at intermediate temperature. Furthermore, the introduction of high-density nanotwins into grains can effectively enhance both yield strength and creep resistance at intermediate temperature without sacrificing other mechanical properties including low-cycle fatigue, crack growth, and hot workability [7, 8]. The results shed light on enhancing the mechanical properties and temperature-bearing capacity of Ni-based superalloys via nanotwinning rather than adding solid-solution atoms or increasing the volume fraction of γ' particles.

Based on these advantages of nanotwin substructure, a novel Ni-Co based superalloy, named GH4251 alloy, with favorable mechanical properties and processing properties used for turbine disks, turbine blades, and receivers has been designed. The chemical compositions of GH4251 and two typical superlloys U720Li and TMW-4 are shown in Table 1. GH4251 alloy contains 25 wt.% Co content and 8.0 wt.% Al + Ti + Nb + Ta content. In the newly designed GH4251 superalloy, increasing Co content can reduce SFE and thus promotes the twinning ability; meanwhile, adding Al + Ti content ensures moderate volume fraction of γ' particles; furthermore, adding Nb content reduces the precipitation rate of γ' particles which widens processing time.

Table 1 Nominal chemical compositions (wt.%) of GH4251, U720Li, and TMW-4

Alloy	C	Cr	Co	W	Mo	Al	Ti	Nb + Ta	Zr	B	Ni
U720Li	0.02	16.0	15.0	1.3	3.0	2.5	5.0	0	0.03	0.02	Bal.
TMW-4	0.02	13.5	25.0	1.2	2.8	2.3	6.2	0	0.03	0.02	Bal.
GH4251	0.01–0.03	12.0–14.0	24.0–26.0	1.0–1.5	2.6–3.0	2.2–2.6	4.3–4.8	0.8–2.0	0.01–0.05	0.01–0.03	Bal.

Materials and Experimental Procedures

Preparation of C&W GH4251

During a conventional casting and wroughting (C&W) process, the full-size disk, which is 500 mm in diameter and 65 mm in thickness, with an average grain size of about 15 μm was successfully fabricated using GH4251 alloy. Samples were taken in tangential direction from the center to 1/2 of the disk rim, and the specimens were solid solution treated at 1080 °C for 4 h, then air cooled to room temperature, followed by two-stage aging treatment at 650 °C/24 h/AC and 760 °C/16 h/AC.

Preparation of C&W U720Li and TMW-4M3

The pancakes of U720Li and TMW-4M3 were produced by a conventional C&W processing route including triple melting, billet making, and pancake forging. The specimens used were cut from the pancakes and then were heat treated in the standard heat-treated condition (1100 °C/4 h followed by oil quench(OQ) and then aging at 650 °C/24 h/OQ + 760 °C/16 h/OQ).

Preparation of SLM-Processed GH4251

SLM-processed specimens were manufactured by FS301M (Fasoon Technologies) laser-selective melting and forming equipment. The scanning power was 200–260 W, and the scanning speed was 1100–1400 mm/s. Hot isostatic pressing (HIP), including a heat preservation at 1200 °C/140 MPa for 4 h and then cooling in furnace, followed by solution treatment 1070 °C/4 h/AC and then two-step aging treatments at 760 °C/16 h/AC.

Mechanical Tests

The tensile test specimen with the size of $\Phi 5 \times 27$ mm was prepared by slow wire cutting and mechanical polishing. Then tensile tests were performed according to ASTM E8/E21 standard test methods. The stress rupture specimen with the geometry of $\Phi 5 \times 27$ mm was tested at various temperatures and stresses to rupture.

Microstructure Characterization

For Optical Microscope (OM), the mechanically polished specimens were etched in a solution consisting of CuCl2 (5 g) + HCl (25 ml) + ethanol (25 ml) at room temperature for 2–3 min. For electron backscatter diffraction (EBSD) analysis, the specimens were sliced, mechanically polished, and then electro-polished in a solution with 80% methanol and 20% H_2SO_4 at 20 V for 5–8 s. The specimens for SEM were prepared by the electron-polishing and then electro-chemically etched in a solution (15 g Cr_2O_3 + 10 ml H_2SO_4 + 150 ml H_3PO_4) at 4.5 V for 4–6 s.

Results and Discussion

C&W GH4251

GH4251 alloy is provided with considerable hot workability and weldability due to its intrinsic component characteristics mentioned above. When compared to U720Li alloy, it is can be found that GH4251 exhibits much better hot plasticity in the temperature range of 1000–1200 °C, shown in Fig. 1. This is mainly attributed to two factors. On the one hand, replacing Ti with Nb can reduce the precipitation rate of γ' particles which widens processing time. On the other hand, increased Co content can promote the formation of twin boundaries, which provide nucleation sites for recrystallized grains and facilitate the dynamic recrystallization softening.

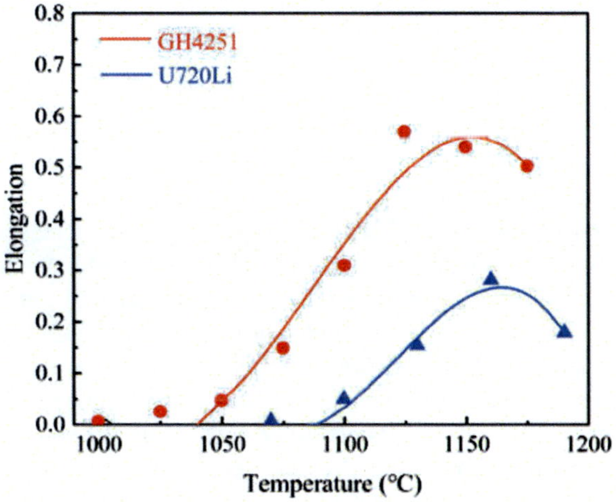

Fig. 1 Comparison of the elongation at various temperatures of GH4251 and U720Li alloys

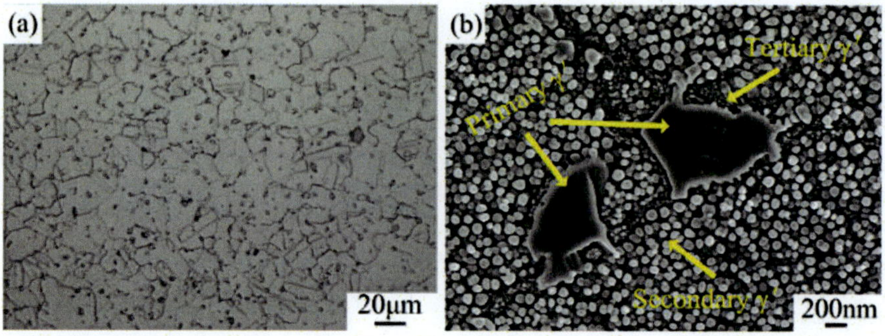

Fig. 2 Typical microstructures of specimen after solution and aging treatments, **a** OM image; **b** morphology of γ′ particles

Microstructures

Figure 2 shows the microstructure of the GH4251 alloy after solution and aging treatment. The microstructure consists of equiaxed grains with an average size of about 15 μm (Fig. 2a). The primary γ′ particles with a size of 200–500 μm are mainly distributed at the grain boundaries, while the secondary (an average size of 20–100 nm) and tertiary γ′ particles (an average size of 20 nm) are uniformly distributed in the γ matrix (Fig. 2b). The total volume fraction of primary, secondary, and tertiary γ′ particles is about 45%.

Tensile Properties

Figure 3 shows the tensile properties of GH4251, TMW, and U720Li alloy at different temperatures. Clearly, both the yield strength and ultimate strength of the three alloys decrease with the increase in temperature. The results of tensile tests (Fig. 3a) show that the yield strength of GH4251 is higher than that of both TMW-4M3 and U720Li in the range of 25–750 °C, and the difference increases as the temperature increases. The ultimate strength of GH4251 is comparable to that of TMW-4M3 in the range of 25–750 °C, but higher than that of U720Li by about 10% (Fig. 3b).

Figure 4 shows the TEM images of GH4251 after tensile tests at different temperatures. At 25 °C, planar glide of dislocations is the dominant deformation mechanism. A large number of dislocations are blocked at the γ/γ′ interfaces owing to the coherency strain strengthening induced by the mismatch between the γ matrix and γ′ particles (Fig. 4b). Some dislocations enter and entangle with each other in the primary γ′ particles (Fig. 4a). Meanwhile, isolated stacking faults are generated in the secondary γ′ particles (Fig. 4c) because stress concentration at the γ/γ′ interfaces induced by the motion of dislocations reaches the critical value of twinning, resulting in the dissociation of a/2 <110> dislocation into two a/6 <112> Shockley partials and

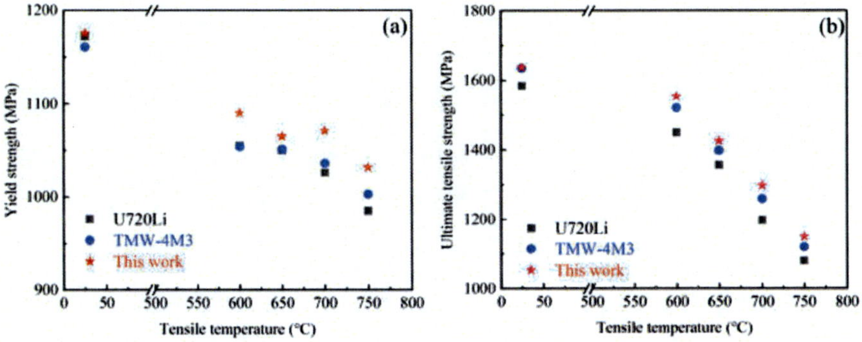

Fig. 3 Tensile properties of GH4251, TMW [9] and U720Li [10] at different temperatures **a** yield strength; **b** ultimate strength

thus the formation of stacking faults [11]. At 750 °C, stacking faults and microtwins shearing both the γ' particles and γ matrix is the dominant deformation mechanism. A large number of stacking faults in the primary γ' particles are observed, and the stacking faults in different slip systems interact with each other, forming V-shape structure (Fig. 4d). The length and thickness of the stacking faults increase when compared with the deformation case at 25 °C, which indicates that SFE decreases with the increasing temperature (Fig. 4e). As a consequence, the critical resolved shear stress of dissociated dislocations shearing the γ' particles decreases with the increasing temperatures [12]. In addition, microtwins (MTs) have also been observed (Fig. 4f). The formation of the MTs involves the dissociation of a/2 <110> dislocations at the γ matrix and the reordering of atoms on the adjacent {111} planes [13]. The reordering process is mediated by the exchange between the vacancies and the atoms. Therefore, the reordering process can be accelerated with the increase in temperature, promoting the formation of the MTs [14].

Stress Rupture Properties

Figure 5 shows the Larson-Miller comparison of stress rupture life for GH4251 alloy with U720Li [15] and TWM-4 [15] alloys. It can be seen that GH4251 alloy achieves a large temperature gain in both low and high stresses when compared with U720Li alloy. And the performance of GH4251 alloy is comparable with TMW-4 alloy at higher stress, while obtains a certain temperature gain at low-stress level.

High-density MTs are observed in the specimen after rupture at intermediate temperature, as shown in Fig. 6, which indicates that micro twinning becomes the dominant mechanism. The MTs can increase the resistance of cross-slip and decrease the average free path of the mobile dislocations and thus delay the primary and secondary creep stages [16, 17], resulting in the improvement of the creep properties.

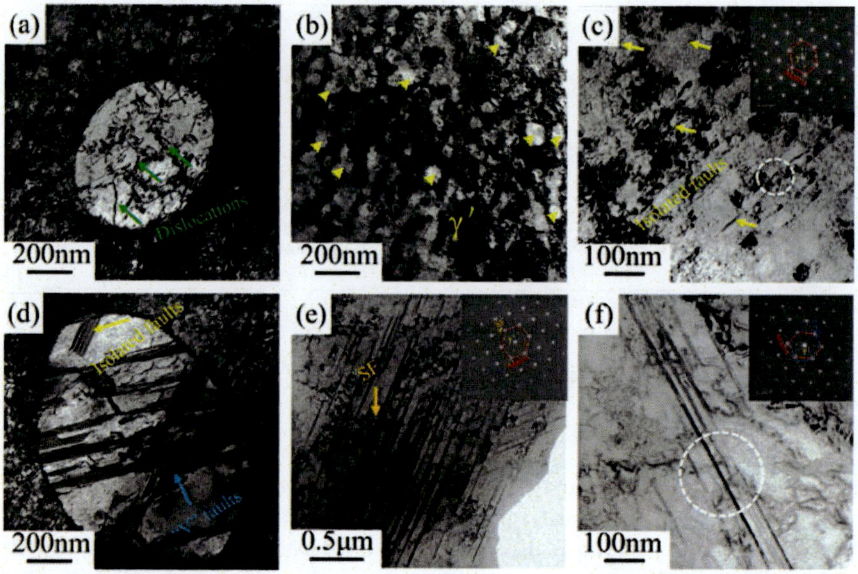

Fig. 4 TEM images of specimens tensiled at (a–d) 25 °Cand (e–f) 750 °C. **A** Dislocations in primary γ' particles; **b** Dislocations around secondary γ' particles; **c** Isolated stacking faults; **d** V-shape stacking faults in γ' particles; **e** Continuous stacking faults; **f** Stacking faults and microtwins

Fig. 5 Larson-Miller plot of stress rupture life of GH4251 alloy in comparison with reported U720Li and TMW-4 alloys

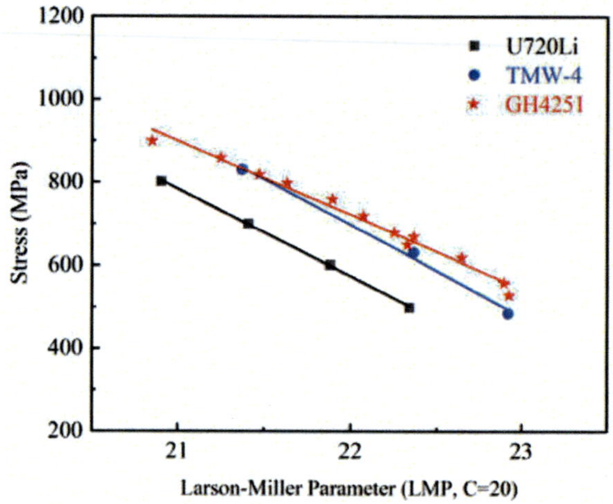

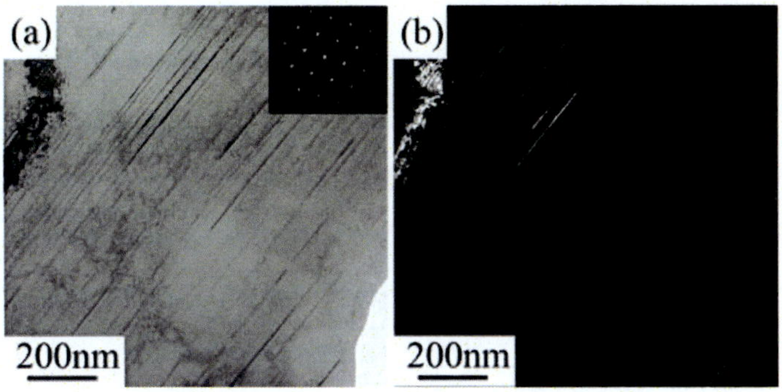

Fig. 6 TEM images of the specimen after stress ruptured at 750 °C/530 Mpa

SLM-Processed GH4251

Since the considerable processing properties of GH4251 alloy, such as hot workability and weldability, it can be judged that the alloy also has excellent additive process performance.

Printing Parameters

SLM-processed GH4251 were manufactured by FS301M (Fasoon Technologies) laser-selective melting and forming equipment. The scanning power and scanning speed are initially set as the range of 200–260 W and 1100–1400 mm/s, respectively. Figure 7 shows the ratio of defects (porosity, microcracks, and un-fusions) of the SLM-processed GH4251 specimens under different scan powers and speeds. The SLM-processed specimens maintain a low ratio of defect under various process parameters due to the wide processing window of GH4251 alloy. When the scan power is 200 W and scan speed is 1100 mm/s, the defect ratio reduces to a minimum of only 0.08%. Under the process parameter, microcracks, un-fusions, and porosity rarely exist in the specimen (Fig. 8). Therefore, the scan power of 200 W and scan speed of 1100 mm/s were determined to be the optimal printing parameters for GH4251 alloy.

Microstructures

Figure 9a, b shows the OM images of the cross and vertical sections of SLM-processed unpolished specimens after hot isostatic pressing (HIP) and solution treatments (ST). No obvious cracks or pores were observed on the surface of the specimen,

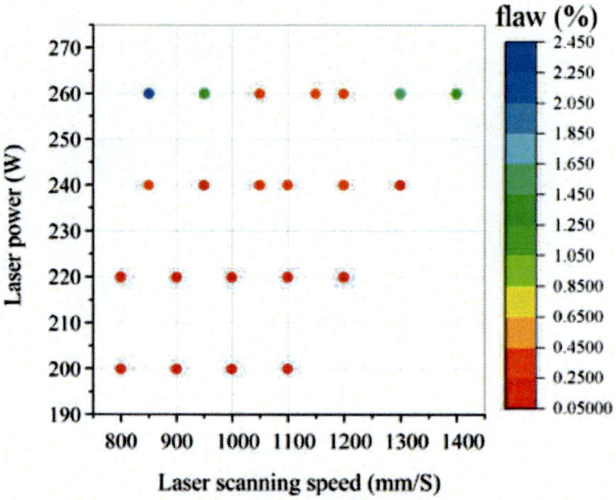

Fig. 7 Ratio of defects of the SLM-processed specimens under different scan powers and speeds

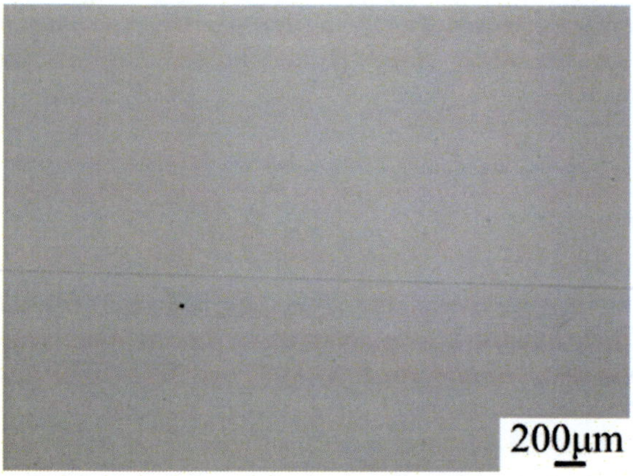

Fig. 8 Morphology of SLM-processed specimens at scanning power of 200 W and scanning speed of 1100 mm/s

indicating that the formation of pores and cracks is effectively suppressed. Figure 9c, d shows the OM images of the cross and vertical sections of SLM-processed polished specimens after HIP and ST. It can be seen that the microstructures consisted of abundant equiaxed grains, indicating that the specimens have been fully recrystallized and the anisotropy has been significantly eliminated.

Figure 10a, b shows the EBSD images of the cross and vertical sections of SLM-processed specimens after HIP and ST. The microstructures consisted of abundant randomly oriented equiaxed grains at either section. The average size of the cross and vertical sections is about 72 and 79 μm, respectively, resulting in comparable mechanical properties in two orientations. Figure 10c, d shows the γ' morphology of the SLM-processed specimens after HIP and ST. The uniformly distributed primary

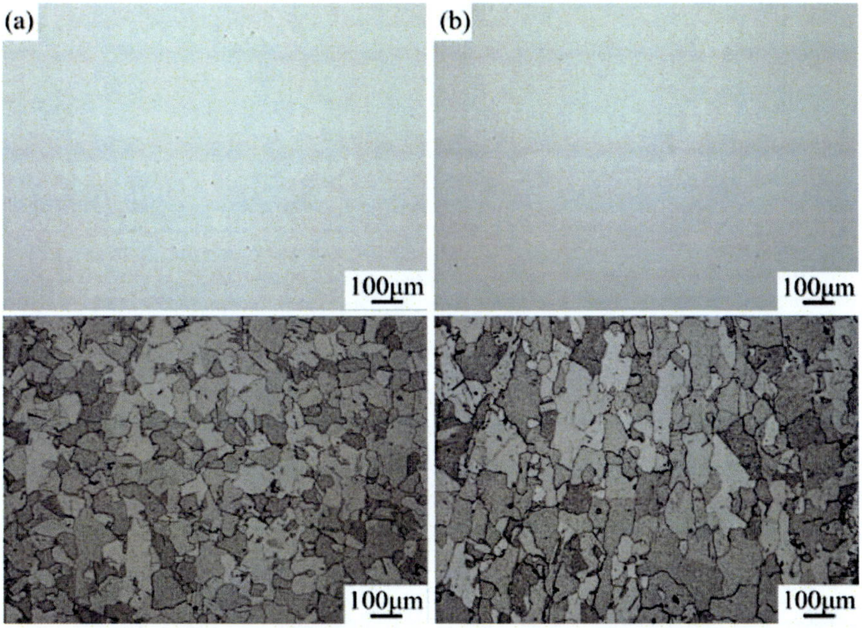

Fig. 9 Microstructures of specimens after HIP and ST treatments **a** and **c** OM image of cross-direction; **b** and **d** OM image of vertical-direction

and secondary γ' particles are observed at the grain boundaries and in the γ matrix. The primary γ' particles with an average size of 1.5 μm in the γ matrix are nearly spherical, and the one at the grain boundary is irregular in shape. The secondary and tertiary γ' particles are nearly spherical and the average size is about 54 and 15 nm.

Tensile Properties

Figure 11 shows the tensile properties in the range of 25–1000 °C of the specimens in vertical section after HIP and ST treatments. It can be seen that the specimens maintain high strength in the range of 25–1000 °C. When the temperature exceeds 900 °C, the strength of the specimens significantly decreases. Furthermore, the SLM-processed specimens exhibit comparable mechanical properties in both cross and vertical sections. And the yield and ultimate strengths of the SLM-processed specimens are comparable to that of the C&W specimens at 25 and 750 °C.

Figure 12 shows the fracture morphology of specimens after tensile tests in different directions at 800 °C. It can be seen that there is no obvious necking in the fracture (Fig. 12a). Abundant cleavage facets and some secondary cracks along

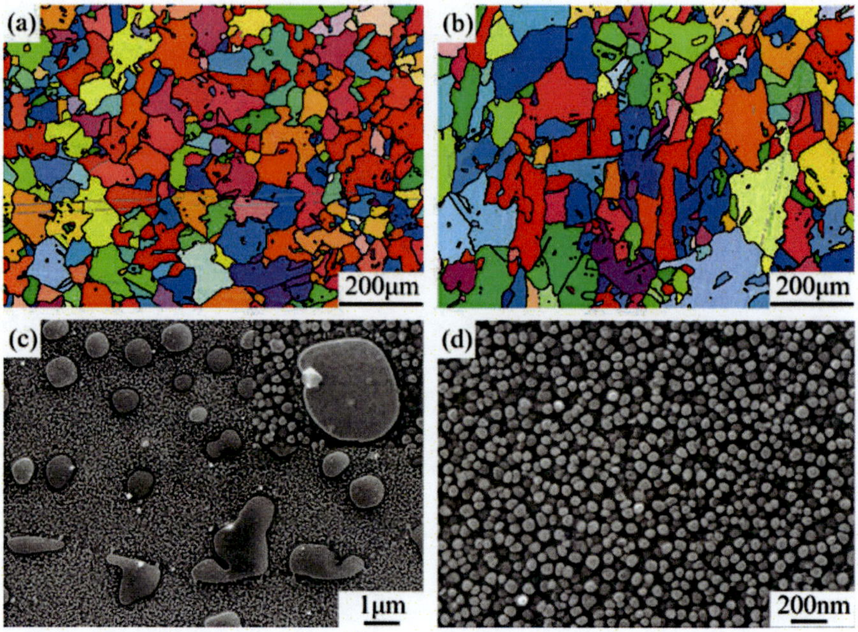

Fig. 10 Microstructures of specimens after HIP and ST treatments **a** EBSD image of cross-direction; **b** EBSD image of vertical-direction; **c**, **d** Morphology of γ' particles

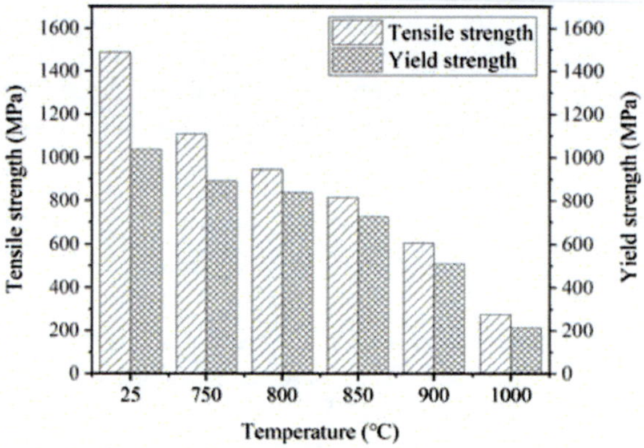

Fig. 11 Tensile properties of specimens in vertical section after HIP and ST treatments

the grain boundaries are on the fracture surface (Fig. 12b). Also, intergranular cracks are observed near the fracture of the specimen (Fig. 12c).

Fig. 12 SEM micrograph of the fracture surfaces of the specimen in vertical section tested at 800 °C

Stress Rupture Properties

Figure 13 shows the stress rupture properties of specimens after HIP and ST treatments at 700 °C/700 MPa in different directions. It illustrates that the stress rupture life of the specimen in the vertical section is better than that of the cross one.

Figure 14 shows the fracture morphology of GH4251 alloy after stress ruptured at 700 °C/700 MPa. Both transverse and vertical fractures have no obvious necking (Fig. 14a and d), and cleavage facets, small equiaxial dimples and small irregular dimples are observed on the fracture surface (Fig. 14b and e). Some secondary cracks are formed in the surface of cross-sectional specimen, while intergranular cracking is suppressed in the vertical-section specimen (Fig. 14c and f).

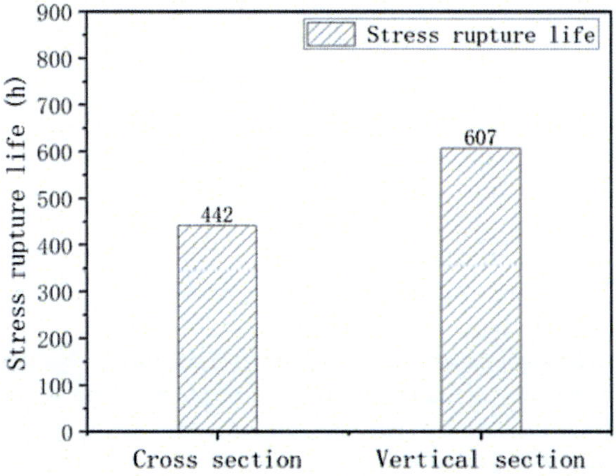

Fig. 13 Stress rupture properties of the specimens at 700 °C/700 MPa in cross and vertical sections

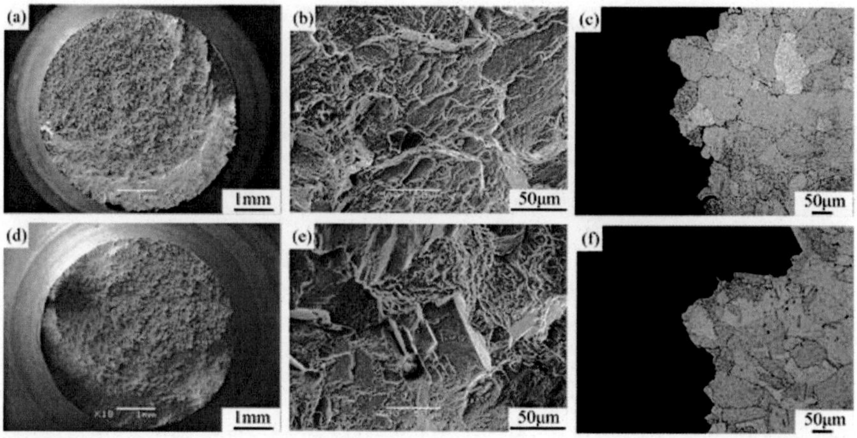

Fig. 14 SEM micrograph of the fracture surfaces of the specimens after stress ruptured 700 °C/700 MPa **a** cross section; **b** vertical section

Conclusion

A novel Ni-Co-based superalloy GH4251 with both favorable mechanical and processing properties has been designed for applications in turbine disks, turbine blades, and receivers. The newly designed superalloy was prepared by C&W and SLM processes, respectively. The tensile and stress rupture tests at various conditions were conducted. The evolution of microstructure is examined. The findings are summarized as follows.

(1) The new alloy can achieve a yield strength of 1100 MPa at 750 °C and its creep rupture life is more than 500 h at 750 °C under 530 MPa with fine grain size (ASTM 8), superior to U720Li.
(2) The prominent deformation mechanisms at 750 °C include isolated stacking faults shearing the primary γ' particles and abundant continuous stacking faults and microtwins cutting through both the secondary γ' particles and γ matrix.
(3) The feature of high twinning ability sheds light on the applications of high-strength sheet and fastener via the method of pre-set nanotwins structure.
(4) Besides being used as disk or ring forgings, this novel GH4251 alloy has a significant better hot work ability and weld ability compared with U720Li and can also be processed by additive manufacturing.
(5) The scan power 200 W of and scan speed 1100 mm/s of are determined as the optimal printing parameters for SLM-processed GH4251 alloy.
(6) The SLM-processed GH4251 alloy has excellent tensile properties in the range of 25–900 °C as well as good stress rupture properties at various conditions.

Acknowledgements This work was supported by the National Key Research and Development Program of China (Grant No. 2021YFB3702500) and the National Natural Science Foundation of China (Grant No. 12205055).

References

1. W.G. Li, J.Z. Ma, H.B. Kou, J.X. Shao, X.Y. Zhang, Y. Deng, Y. Tao, D.N. Fang, Modeling the effect of temperature on the yield strength of precipitation strengthening Ni-base superalloys, International Journal of Plasticity 116 (2019).
2. M.S. Chen, Z.H. Zou, Y.C. Lin, H.B. Li, G.Q. Wang, Y.Y. Ma, Microstructural evolution and grain refinement mechanisms of a Ni-based superalloy during a two-stage annealing treatment, Materials Characterization 151 (2019) 445–456.
3. A. Heckl, S. Neumeier, M. Goken, R.F. Singer, The effect of Re and Ru on γ/γ' microstructure, γ-solid solution strengthening and creep strength in nickel-base superalloys, Materials Science and Engineering A 528 (2011) 3435–3444.
4. L. Lu, X. Chen, X. Huang, K. Lu, Revealing the maximum strength in nanotwinned copper, Science 323 (2009) 607–610.
5. Y.J. Wei, Y.Q. Li, L.C. Zhu, Y. Liu, X.Q. Lei, G. Wang, Y.X. Wu, Z.L. Mi, J.B. Liu, H.T. Wang, H.J. Gao, Evading the strength–ductility trade-off dilemma in steel through gradient hierarchical nanotwins, Nature Communications 5 (2014) 3580.
6. K. Lu, Stabilizing nanostructures in metals using grain and twin boundary architectures, Nature Reviews Materials 1 (2016) 1–13.
7. Y. Yuan, Y.F. Gu, C.Y. Cui, T. Osada, T. Yokokawa, H. Harada, A novel strategy for the design of advanced engineering alloys—strengthening turbine disk superalloys via twinning structures, Advanced Engineering Materials 13 (2011) 296–300.
8. Y. Yuan, Y.F. Gu, T. Osada, Z.H. Zhong, T. Yokokawa, H. Harada, A new method to strengthen turbine disc superalloys at service temperatures, Scripta Materialia 66 (2012) 884–889.
9. Y.F. GU, T. Fukuda, C. Cui. Comparison of Mechanical Properties of TMW Alloys, New Generation of Cast-and-Wrought Superalloys for Disk Applications, Metallurgical and Materials Transactions A 40A(13): 3047–3050.
10. L. Kovarik, R.R. Unocic, Ju Li, P. Sarosi, C. Shen, Y. Wang, M.J. Mills, Microtwinning and other shearing mechanisms at intermediate temperatures in Ni-based superalloys, Progress in Materials Science 54 (2009) 839–873.
11. G. Viswanathan, P. Sarosi, M. Henry, D. Whitis, W. Milligan, M. Mills, Investigation of creep deformation mechanisms at intermediate temperatures in René 88 DT, Acta Materialia (2005) 3041–3057.
12. M. Kolbe, The high temperature decrease of the critical resolved shear stress in nickel-base superalloys, Materials Science and Engineering A 319–321 (2001) 383–387.
13. D.Q. Qi, B.D. Fu, K. Du, T.T. Yao, C.Y. Cui, J.X. Zhang, H.Q. Ye, Temperature effects on the transition from Lomer-Cottrell locks to deformation twinning in a Ni-Co-based superalloy, ScriptaMaterialia 125 (2016) 24–28.
14. Y.M. Eggeler, K.V. Vamsi, T.M. Pollock, Precipitate shearing, sault energies, and solute segregation to planar faults in Ni-, CoNi-, and Co-base superalloys, Annual Review of Materials Research 51 (2021) 209–240.
15. Y. GU, Z. Zhong, Y. Yuan, T. Osadala, C Cui, T. Yokokawa , H. Harada, An advanced cast-and-wrought Superalloy (TMW-4M3) for turbine disk applications beyond 700°C, Superalloys 2012: 12th International Symposium on Superalloys, TMS (The Minerals, Metals & Materials Society), 2012.
16. B. Décamps, S. Raujol, A. Coujou, F. Pettinari-Sturmel, N. Clément, D. Locq, P. Caron, On the shearing mechanism of γ' precipitates by a single (a/6)112 Shockley partial in Ni-based superalloys, Philosophical Magazine 84 (2014) 91–107.

17. D. Barba, E. Alabort, S. Pedrazzini, D.M. Collins, A.J. Wilkinson, P.A.J. Bagot, M.P. Moody, C. Atkinson, A. Jérusalem, R.C. Reed, On the microtwinning mechanism in a single crystal superalloy, Acta Materialia 135 (2017) 314–329.

Part II
Microstructure and Properties

Preferential γ' Precipitation on Coherent Annealing Twin Boundaries in Alloy 718

Semanti Mukhopadhyay, Fei Xue, Hariharan Sriram, Robert W. Hayes, Emmanuelle A. Marquis, Yunzhi Wang, and Michael J. Mills

Abstract Early strain localization, parallel and adjacent to annealing twin boundaries (ATBs), has been reported in several superalloys. While strain localization is generally attributed to local shear stresses developed near ATBs due to elastic anisotropy, the role of local microstructural features near ATBs is unclear. Precipitate free zones (PFZ) parallel to ATBs in a γ''–strengthened alloy 945X have been reported and were found to greatly influence strain localization in the alloy. However, it is unclear if such PFZs near ATBs occur in other superalloys, potentially influencing strain localization. The present work investigates local microstructures near ATBs in a Ni-based superalloy—718, which is strengthened by both γ' and γ'' phases. Based on characterization experiments, this paper reports that Alloy 718 shows a high density of herringbone-like $\gamma'-\gamma''$precipitates at ATBs. However, the ATBs exhibit a much higher fraction of γ' (25.3 $\pm$ 2.8%) than γ'' (18.4 $\pm$ 2.4%), ultimately causing the coprecipitate fraction at ATBs to be as high as 43%. A local HCP phase at the ATBs within γ' precipitates exhibiting Nb segregation and Al depletion is also reported. On the contrary, no appreciable change in the ATB composition was observed within γ''. Finally, a novel mode of heterogeneous precipitation at ATBs is proposed.

S. Mukhopadhyay (✉) · H. Sriram · Y. Wang · M. J. Mills (✉)
Department of Materials Science and Engineering, The Ohio State University, 177 Watts Hall, 2041 N. College Road, Columbus, OH 43210, USA
e-mail: mukhopadhyay.20@buckeyemail.osu.edu

M. J. Mills
e-mail: mills.108@osu.edu

S. Mukhopadhyay · M. J. Mills
Center for Electron Microscopy and Analysis, The Ohio State University, Columbus, OH 43212, USA

F. Xue · E. A. Marquis
Department of Materials Science and Engineering, University of Michigan, 2300 Hayward St, Ann Arbor, MI 48109, USA

R. W. Hayes
Metals Tech. Inc, 19801 Nordhoff St, Northridge, CA 91324, USA

E. A. Ott et al. (eds.), *Proceedings of the 10th International Symposium on Superalloy 718 and Derivatives*, The Minerals, Metals & Materials Series,
https://doi.org/10.1007/978-3-031-27447-3_9

Keywords Alloy 718 · Annealing twin boundaries (ATBs) · Precipitation

Introduction

Most Ni-based alloys, including Ni-base superalloys, exhibit a very high density of coherent annealing twin boundaries (ATBs) [1–4]. In fact, ATB densities as high as $0.2\mu m^{-1}$ have been reported for a Ni-base superalloy (Rene 88DT) with a grain size of 20μmADDINZOTERO_TEMP. Such a high density would cause every grain in the microstructure to exhibit at least one ATB. In addition, this high density of ATBs may now influence many material properties. Since Ni-base superalloys are utilized in a high temperature and stress environment that is simultaneously severely oxidizing and corrosive, the role of ATBs in such alloys has been extensively investigated.

ATBs have been reported to negatively influence strain localization under monotonic as well as fatigue-loading, fatigue crack initiation, and hydrogen-assisted cracking in Ni and Ni-base superalloys [3–22]. Conversely, ATBs are reported to improve mechanical strength and the alloy's response to stress corrosion cracking (SCC) [23]. In addition, ATBs are also reported to impede crack propagation during H-assisted tensile tests [24]. Furthermore, ATBs are shown to be highly resistant to oxidation and are reported to cause added strengthening by acting as strong barriers for dislocation motion [23, 25–27]. Thus, the ATBs exert a significant influence over the mechanical properties of Ni-base superalloys, especially on crack nucleation and propagation. This is particularly important because by causing strain localization, crack initiation, and aiding in the propagation of cracks, ATBs might ultimately be responsible for the premature failure of the superalloys during service.

Recently, Zhang et al. discovered that precipitate free zones (PFZs) adjacent and parallel to ATBs in a γ''-strengthened ($D0_{22}$ crystal structure, Ni_3Nb) Ni-base superalloy 945X caused ATB failure in this alloy [28]. Surprisingly, the alloy was reported to exhibit PFZs adjacent to a modified arrangement of γ'' precipitates at its ATBs. Further, such a weak PFZ parallel to a slip system was found to exacerbate intense strain localization and cracking [28].

Subsequently, Zhang et al. suggested that the γ'' precipitate arrangement at the ATBs was promoted by an energy benefit due to a "δ-like complexion" at the ATB within the γ'' precipitates [28]. The δ-phase ($D0_a$ crystal structure, Ni_3Nb) is a thermodynamically stable phase in many Ni-base superalloys within the 718 family, so the energy argument is potentially intuitive. However, it is not clear if the reported complexion has a different chemical composition than the surrounding γ''. This is important because an ATB in γ' with compositional variations would resemble "local phase transformations" (or LPT), similar to a recently proposed strengthening mechanism in superalloys [29–33]. On the contrary, in the absence of a difference in chemical composition, the complexion is merely caused by the atomic arrangements at the ATB. The ATB naturally causes a 3-layer local HCP stacking (ABA) in the FCC system (ABCABC stacking). Now, consider such an ATB within an ordered precipitate. Because the planes A, B, and C themselves have an ordered arrangement of atoms, the expected HCP stacking at the ATB inherits this ordered arrangement. Then, the atomic arrangement on this 3-layer ordered HCP stacking consists of alternate layers of A and B atoms (from the coherent A_3B-type ordered precipitate), which implies that such a 3-layer stacking at the ATBs is not out of the ordinary. In fact, this purported "δ-like complexion" is expected on the ATB in any ordered precipitate. Whether this complexion, without a chemical driving force, truly has an energetic advantage for every such system is questionable.

Since most Ni-, Fe-, and Co-based superalloys exhibit ordered precipitates, it is unknown whether all these systems exhibit precipitates and subsequent PFZs near their ATBs. Further, most of these alloys are strengthened by the coherent ordered precipitate γ' with an $L1_2$ crystal structure: (Ni, Co)$_3$(Al, Nb, Ti, W ...). Thus, it is presently unreported whether an LPT might promote a modified arrangement of γ'on ATBs.

Thus, this work characterized the microstructure of the important and widely utilized Ni-base superalloy (Alloy 718) which is strengthened by both γ' and γ'' precipitates. Strain localization in this system has been reported under monotonic and fatigue conditions [7, 12, 34], which makes the investigation of its ATB microstructure crucial. For this investigation, a detailed characterization of the ATB microstructure was carried out for a commercially processed and aged 718 using scanning transmission electron microscopy (STEM), electron dispersive X-ray spectroscopy (EDS), and atom probe tomography (APT).

Methods

Materials and Heat Treatments

Chemical composition (in wt.%) of alloy 718 is listed in Table 1. The commercially aged 718 sample was obtained from Metals Technology Inc., Northridge, CA. The commercial processing of alloy 718 usually follows the ASTM standard B637-18 wherein the forget bars are solution treated and then subjected to a two-step age—first at 720 °C for 8 h and then at 620 °C for a total aging time of 18 h [35].

STEM-based Imaging and Chemical Composition Mapping

To prepare samples for STEM investigations, first, the aged 718 sample was prepared as per standard metallographic techniques. Then, edge-on ATBs (that could be viewed along their common $[011]_{twin}$ zone axis) were selected in this sample using electron backscatter diffraction (EBSD) in a ThermoFischer scientific Apreo High resolution

Table 1 Chemical composition of Alloy 718

Element	Cr	Mo	Ti	Al	Fe	Nb	C	Ni
Wt%	19	3.05	0.9	0.5	18.9	5.13	0.02	52.5

SEM at 20 kV and 6.4 nA. The formulation below was used to accurately identify grains of interest and prepared electron-transparent samples using a Thermo Fischer Scientific Helios 600 Focused Ion Beam (FIB). For this purpose, a Pt cap was first deposited to protect the region of interest using a Ga ion beam at 30 kV and 0.26 nA. Then, a deep trench was created adjacent to the Pt cap at 30 kV and 21 nA. Following this, the trenched foil was undercut at 2.6 nA and attached to a Cu Omniprobe grid. The foil was then thinned at 30 kV using several ion beam currents and cleaned at 5 kV and 71 nA until the foil was electron transparent.

All STEM-based experiments, including energy dispersive X-ray spectroscopy (EDS) were carried out at 300 kV in a Thermo Fischer Scientific Titan^3 60–300 STEM with Super-X EDS detectors. A relatively high current of 1 nA was used for efficient EDS map acquisitions, and 60 pA was used for imaging.

Identifying Grains of Interest for STEM-based Evaluation

Because most STEM-based microscopes allow a limited range of tilt, it was crucial to prepare the FIB foil such that the ATBs could be investigated in an "edge-on" configuration and with favorable zone axes for high resolution imaging during STEM experiments. To accurately identify grains of interest for this purpose, this study considered the geometry of the sample as presented in Fig. 1. Here, it is assumed that the ATB plane is $(1\bar{1}1)_{twin}$.But the ATB does not have to be edge-on when imaged in the SEM or FIB. Then, because the STEM sample needs to be viewed along the common zone axis $[011]_{twin} = [0\bar{1}\bar{1}]_{parent:intwinframe}$, and the STEM sample plane is perpendicular to the SEM imaging plane (beam directions in Fig. 1a), it follows that $[011]_{twin}$ lies in the plane of the sample during SEM (or FIB) imaging.

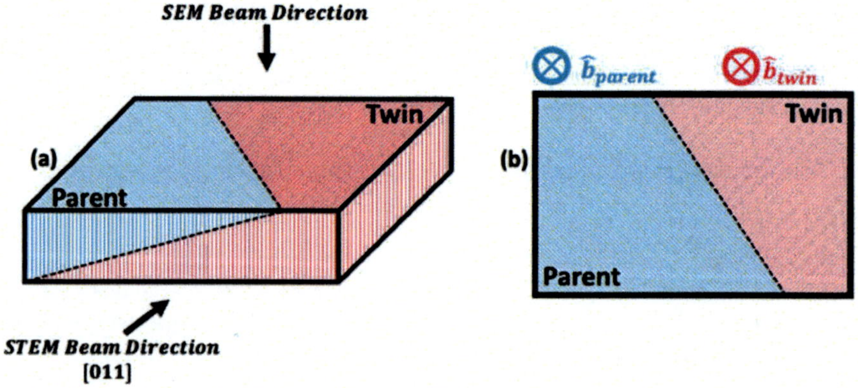

Fig. 1 Selecting regions of interest for STEM imaging: **a** A 3D section containing two twin-related domains, and **b** 2D section, as seen in an SEM from **a** containing two twin-related domains. Note that the twin boundary makes an arbitrary angle w.r.t. the sample frame of reference

Next, when imaging in the SEM, arbitrary miller indices are assigned to the plane normal of the twin and parent domains $[hkl]$ and $[h'\ k'\ l']$, respectively. Because $[011]_{twin}$ lies in the sample plane,

$$[hkl] \cdot [011] = 0 \Rightarrow l = -k$$

Further, coordinate transformations can be carried out to represent $[h'\ k'\ l']$ as per the twin domain's orthonormal bases so that both the twin-related domains may be presented with respect to the same orthonormal bases. Thus

$$\begin{bmatrix} h' \\ k' \\ l' \end{bmatrix} = \frac{1}{3}\begin{pmatrix} -1 & -2 & 2 \\ -2 & -1 & -2 \\ 2 & -2 & -1 \end{pmatrix} \times \begin{bmatrix} h \\ k \\ l \end{bmatrix} = \frac{1}{3}\begin{bmatrix} -h-2k+2l \\ -2h-k-2l \\ 2h-2k-l \end{bmatrix} = \frac{1}{3}\begin{bmatrix} -h-4k \\ -2h+k \\ 2h-k \end{bmatrix}$$

This relationship can now help us identify grains of interest such that a FIB foil lifted out would contain an edge-on ATB which can be imaged along the common <011> zone axes. For example, suppose the twin normal = $[hkl]$ = [100], according to this formulation, the parent normal (in twin frame) = $[h'\ k'\ l'] = [\bar{1}\bar{2}2]$. However, because of multiplicity, it follows that the parent normal (in the above case, where twin normal belongs to the <100> family of directions), has to merely belong to the family of directions <$\bar{1}22$>. Similarly, the above formulation may be used to identify several other twin-related domains which will allow in an edge-on ATB along a common <011> zone in the STEM sample.

Atom Probe Tomography

Site-specific APT specimens on twin boundaries were prepared using a standard lift-out process with Thermo Fisher Helios 650 Nanolab dual SEM/FIB instrument. APT acquisition was conducted using Cameca LEAP 5000 XR Atom Probe in laser mode at 50 K, a pulse energy of 50 pJ, a dynamic pulse rate from 200 to 313 kHz to keep a maximum mass to charge state ratio of 110 Da, and a dynamic detection rate from 0.5 to 1.5% with a constant areal evaporation rate. Datasets were analyzed using the Cameca AP Suite 6.1 software. Reconstruction was based on voltage evolution using the evaporation field of Ni (35 nm/V) with an image compression factor between 1.55 to 1.65 and a field factor k_f from 3.3 to 4.2 to optimize the reconstructed microstructure close to TEM observation, e.g. straight twin boundary, similar precipitate size and parallel {001} shared plane for γ'/γ'' compact precipitates on each set of {001} planes.

Results and Discussion

Fraction of the Phases at the ATB

Figure 2 presents the ATB microstructures obtained in aged 718. Figure 2a shows a STEM EDS map obtained from an edge-on ATB. Because the compositions of γ' and γ'' phases are quite different (γ' shows Al enrichment, while γ'' exhibits a much higher Nb concentration), an EDS map showing a mix of Al and Nb concentration can help identify or distinguish the three phases present (γ, γ', and γ''). Thus, in Fig. 2a, regions in red indicate γ'' phase, while the green regions are γ' precipitates.

Based on this EDS map, the precipitate arrangement at the ATB appears to be different from the bulk. If a line parallel to the ATB were constructed and moved across this EDS map, the line would exhibit the highest number of green pixels at the ATB. Ultimately, this qualitatively shows that the ATB is more densely populated with precipitates.

At a first glance, it appears that the γ' content at the ATB is much higher than γ''. However, because this map was obtained along a [011] zone axis, two variants of γ'' (out of three) are inclined and not clearly visible in this map. In addition, because STEM-based characterization samples all the precipitates through the thickness of a FIB lamella, determination of accurate fraction of ATB phases is challenging. To resolve this issue, this study investigated the ATB of the same alloy (aged 718) with APT.

Figure 2b–d presents APT data from a {111}-type twin plane in the sample. Note that precipitates from the bulk of the sample (or, non-ATB precipitates) were removed so only the γ' and γ'' precipitates that intersect the ATB plane are visible. The measured area fraction of γ'in the ATB plane, i.e., the area of the plane occupied by γ' precipitates, is $25.3 \pm 2.8\%$, and that of γ'' is $18.4 \pm 2.4\%$, with the overall coprecipitate area fraction around 44%. In contrast, the bu"k volume fractions of

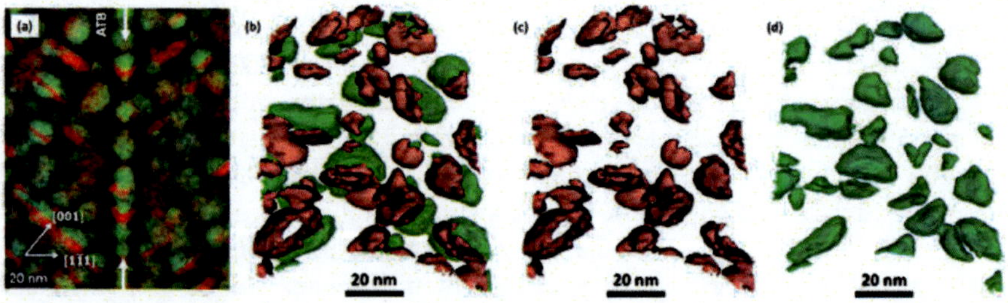

Fig. 2 ATB microstructures in aged alloy 718 investigated via **a** STEM EDS mapping, and **b–d** APT. Note that **b** only presents the precipitates on the ATB plane to assist in easy visualization. **a** Colors in STEM EDS mapping correspond as follows: Green indicates Al and Red indicates Nb. This map shows a high precipitate fraction at ATB. **b–d** Red shows Nb 8.5 at.% isosurface, and green presents Al 5.5 at.% isosurface on the ATB plane. No compositional differences were noted for the ATB precipitates w.r.t. their bulk counterparts

γ' and γ'' far away from the twin boundary were estimated to be $4.9 \pm 0.2\%$ and $10.4 \pm 0.2\%$, with a total precipitate fraction of 15% and γ''/γ' ratio of 2.1. For the bulk precipitate volume fraction estimation, three APT tips (with 74 M, 60 M, and 32 M ranged atoms for each) were prepared from three separate grains. Next, phase compositions for the three phases—γ, γ', γ'', and tip compositions were measured separately in each tip. Then, relative volume fractions were estimated for each tip by applying the level rule between precipitate compositions and tip compositions. The volume fractions and the γ''/γ' ratio estimated in the present work are in agreement with previous reports on wrought 718 subjected to similar heat treatment— γ' and γ'' volume fractions: 6.5 and 17.8%, respectively, with a γ''/γ' ratio of 2.74 [36]. Assuming that area and volume fractions are equivalent, per Underwood [37], 2.9 times higher precipitate volume fractions are observed at the ATBs with a 1.8 times increase in γ'', and 5.2 times increase in γ' fraction. Therefore, while the volume fraction of both the precipitates exhibits an increase at the ATBs, the γ' phase exhibits a significantly higher preference for ATB sites. In that sense, alloy 718 exhibits clear preferential precipitation of γ' phase at all ATBs.

This is a surprising result because the coherent $\{111\}$ $\Sigma = 3$ twin boundary is a low energy boundary [38]. Thus, the energy minimization argument, which is generally used to explain grain boundary precipitation, might not lead to a significant energy reduction. Furthermore, the energy minimization argument presumes a complete consumption of the grain boundary (or fault) by the precipitate [39]. In other words, heterogeneous nucleation of a grain boundary phase lowers the grain boundary energy by consuming a part of it. However, in case of the ATB precipitates in the present work, as well as in Zhang et al. [28], the ATB itself is not being consumed during the precipitation of γ' and γ''at the ATB. Since the ATBs form during the recovery/recrystallization/grain growth process of the heat treatment, ATB formation can be assumed to precede γ' or γ'' precipitation. In this case, the presence of ATBs within γ' and γ'' indicates that this ATB precipitation deviates from the classical theory of heterogeneous nucleation [39].

Composition and Segregation to ATB

Next, chemical segregation to the ATB was characterized using STEM EDS mapping, Fig. 3. For this purpose, a "through-thickness" γ' precipitate was sampled to eliminate any contribution to the acquired EDS data from an overlapping precipitate or γ matrix. Again, a mix of Al and Nb at.% maps was used to delineate the three phases—γ, γ', and γ''.

The composition profile across γ'ATB in Fig. 3b indicates a 0.5 at.% segregation of Nb, and a 0.4 at.% depletion of Al. Incidentally, no appreciable segregation or depletion was measured at the γ''ATB. Thus, while the ATB in γ' shows a compositional difference from its parent phase, the ATB in γ'' does not. In that case, the ATB in γ' satisfies the commonly accepted definition of a phase—"a phase is the portion

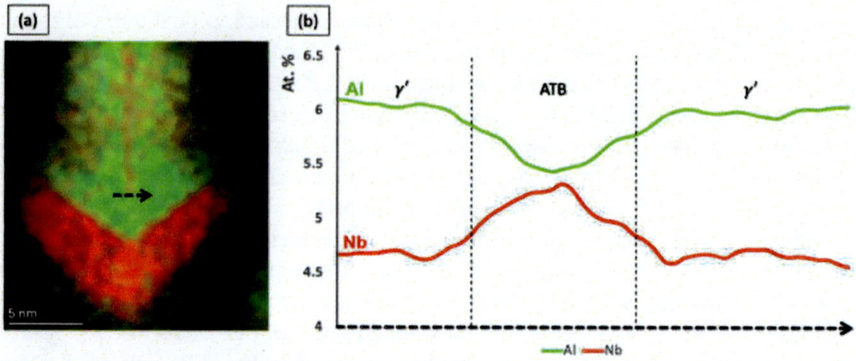

Fig. 3 STEM EDS data from a through-thickness γ' and γ'' precipitate at the ATB: **a** EDS map showing Nb in red and Al in green, and **b** line profile along the black dashed line in **a**. Line profile excludes other elements (Fe, Cr, Ni, Mo, and Ti) as no other peaks ($>3\sigma$) or valleys ($<3\sigma$) were observed

of a system whose properties and composition are homogeneous, and which is physically distinct from other parts of the system" [40]. Following this definition, the ATB within the γ' precipitate shows the compositional difference from the rest of γ' and is physically distinct from bulk γ' by virtue of its ABA stacking. Further, a bulk phase with multiple layers of this ABA stacking, along with Nb segregation and Al depletion would resemble a $D0_{19}$-type χ-phase. Because the properties of χ-phase are expected to be different from γ', it follows that the 3-layer ATB within the γ' may also exhibit properties distinct from γ'. Thus, the γ'ATB in Fig. 3 may now be regarded as a 3-layer "local" phase (or an LPT discussed previously). This finding is similar to the LPT reported on the deformation microtwins in γ'-strengthened alloys like RRHT-5 etc. [29, 30]. Note that at present, it is not possible to conclusively determine this 3-layer local phase at the γ'ATBs. This is because the compositional segregation and the ABA stacking at the ATBs are consistent with three different bulk phases—χ ($D0_{19}$, ABAB stacking), η ($D0_{24}$, ABACA stacking), and δ ($D0_a$, ABA stacking). More experiments are currently being carried out to resolve this issue and will be communicated subsequently. Thus, for now, this study will refer to the local 3-layer phase at the γ'ATBs as an 'HCP-phase' based on the HCP-like stacking ABA at the ATB. Based on these results indicating a 520% increase in γ' area fraction at the ATB as compared to the bulk, it is possible that the local 'HCP-phase' is responsible for this increase. In that case, this work suggests that LPTs might influence γ' precipitation behavior at the ATBs.

The segregation along the γ'ATB and increased γ' phase fraction at the boundary potentially suggests that these compositional variations might precede γ' nucleation. This is because once the local HCP phase forms at the ATB, it must reject Al onto the bulk. Since Al is a strong γ' strengthener and partitioner, Al segregation adjacent to the HCP phase might be expected to effectively cause some form of constitutional undercooling for the γ' phase. In addition, the HCP phase may provide its interface as sites for heterogeneous nucleation to decrease the HCPphase$|\gamma$ interfacial energy. Thus, this study hypothesizes that the Al segregation and ready availability of heterogeneous nucleation sites are responsible for the increased γ' precipitation at the ATB. This same local HCP phase may not increase γ'' precipitation primarily because of Nb segregation to this local phase. Because of Nb segregation to the 3-layer HCP phase, its immediate surrounding may be expected to be depleted in Nb. Such a decrease in Nb might not support γ'' phase precipitation adjacent to this HCP phase.

Therefore, based on the higher volume fraction of $\gamma'$and the ATB segregation, the following sequence of events are proposed. First, as compositional fluctuations at the ATB arise during the cooling step of the imposed heat treatment, some of these compositional fluctuations resembling the HCP phase composition may reject Al to their surroundings. Now, Al rejection as well as the local HCPphase$|\gamma$ interface might enable heterogeneous γ' nucleation. In addition, some heterogeneous γ'' nucleation might also be expected at this stage. As the phases grow, more γ'' is expected to nucleate at the $\gamma' \mid \gamma$ interface to reduce the interfacial energy of the system. This is sequence of events also supports the observation of finer γ' and γ'' precipitates at the ATB when compared with those in the bulk, as well as the overall herringbone-like shape of the coprecipitates at the ATB.

Finally, it is qualitatively observed that alloy 718 exhibits a region of low precipitate density parallel to its ATB (Fig. 2a). This decrease in precipitate density adjacent to ATBs is presently the subject of quantitative study and will be communicated subsequently. Such a region with lower precipitate density is likely caused by depletion of solutes because of the preferential precipitation at the ATB. Because this region is parallel to the ATB, and therefore, to a slip plane, it will allow easy activation of this particular slip system, contingent upon the Schmid factor. In addition, elastic anisotropy which introduces additional local shear stresses near the ATB can additionally favor slip parallel to the ATB [7, 41]. Thus, the strain localization in alloy 718 parallel to ATBs might be further intensified by local microstructural features and needs to be investigated in more detail.

Conclusions

The present work investigates the local ATB microstructures in aged alloy 718. The purpose of this work was to study the precipitates present at the ATBs in this alloy and to assess if a microstructural cause might exist for strain localization near ATBs.

1. Alloy 718 shows a high density of coprecipitates at its ATB in the aged condition. However, the γ''/γ' ratio is very different at the ATB than in bulk. ATBs exhibit a much higher fraction of γ' (25%) than γ'' (18%), ultimately causing the coprecipitate fraction at ATB to be around 43%.
2. The significant increase in γ' fraction at the ATB might be influenced by a local HCP-phase formation at its ATB with Nb segregation and Al depletion. On the contrary, no appreciable change in ATB composition was observed in γ''.

Thus, this work reveals a novel case of preferential precipitation at all the coherent ATBs in alloy 718. In addition, preliminary results suggest that denuded regions of lower precipitate density might form adjacent to the ATBs which could affect strain localization under external loading. However, the direct influence of preferential precipitation on the mechanical behavior of alloy 718 remains unclear and must be critically assessed.

Acknowledgements Electron microscopy was performed at the Center for Electron Microscopy and Analysis (CEMAS) at The Ohio State University. This material is based upon work supported by the NSF DMREF program DMR—1922239.

References

1. R. A. Varin and J. Kruszynska, "Control of annealing twins in type 316 austenitic stainless steel," *Acta Metall.*, vol. 35, no. 7, pp. 1767–1774, Jul. 1987, https://doi.org/10.1016/0001-6160(87)90122-2.
2. C. S. Pande, M. A. Imam, and B. B. Rath, "Study of annealing twins in fcc metals and alloys," *Metall. Trans. A*, vol. 21, no. 11, pp. 2891–2896, Nov. 1990, https://doi.org/10.1007/BF02647209.
3. W. C. Lenthe, M. P. Echlin, J. C. Stinville, M. De Graef, and T. M. Pollock, "Twin related domain networks in René 88DT," *Mater. Charact.*, vol. 165, p. 110365, Jul. 2020, https://doi.org/10.1016/j.matchar.2020.110365.
4. N. Bozzolo, N. Souaï, and R. E. Logé, "Evolution of microstructure and twin density during thermomechanical processing in a γ-γ' nickel-based superalloy," *Acta Mater.*, vol. 60, no. 13, pp. 5056–5066, Aug. 2012, https://doi.org/10.1016/j.actamat.2012.06.028.
5. J. Miao, T. M. Pollock, and J. Wayne Jones, "Microstructural extremes and the transition from fatigue crack initiation to small crack growth in a polycrystalline nickel-base superalloy," *Acta Mater.*, vol. 60, no. 6–7, pp. 2840–2854, Apr. 2012, https://doi.org/10.1016/j.actamat.2012.01.049.
6. J. Miao, T. M. Pollock, and J. Wayne Jones, "Crystallographic fatigue crack initiation in nickel-based superalloy René 88DT at elevated temperature," *Acta Mater.*, vol. 57, no. 20, pp. 5964–5974, Dec. 2009, https://doi.org/10.1016/j.actamat.2009.08.022.
7. J. C. Stinville, N. Vanderesse, F. Bridier, P. Bocher, and T. M. Pollock, "High resolution mapping of strain localization near twin boundaries in a nickel-based superalloy," *Acta Mater.*, vol. 98, pp. 29–42, 2015.
8. J. C. Stinville, W. C. Lenthe, J. Miao, and T. M. Pollock, "A combined grain scale elastic–plastic criterion for identification of fatigue crack initiation sites in a twin containing polycrystalline nickel-base superalloy," *Acta Mater.*, vol. 103, pp. 461–473, 2016.
9. J. C. Stinville, M. P. Echlin, D. Texier, F. Bridier, P. Bocher, and T. M. Pollock, "Sub-grain scale digital image correlation by electron microscopy for polycrystalline materials during elastic and plastic deformation," *Exp. Mech.*, vol. 56, no. 2, pp. 197–216, 2016.
10. J. C. Stinville et al., "Fatigue deformation in a polycrystalline nickel base superalloy at intermediate and high temperature: Competing failure modes," *Acta Mater.*, vol. 152, pp. 16–33, 2018.
11. J.-C. Stinville, W. C. Lenthe, M. P. Echlin, P. G. Callahan, D. Texier, and T. M. Pollock, "Microstructural statistics for fatigue crack initiation in polycrystalline nickel-base superalloys," *Int. J. Fract.*, vol. 208, no. 1, pp. 221–240, 2017.
12. J.-C. Stinville et al., "Measurement of strain localization resulting from monotonic and cyclic loading at 650o C in nickel base superalloys," *Exp. Mech.*, vol. 57, no. 8, pp. 1289–1309, 2017.

13. J. C. Stinville, P. G. Callahan, M. A. Charpagne, M. P. Echlin, V. Valle, and T. M. Pollock, "Direct measurements of slip irreversibility in a nickel-based superalloy using high resolution digital image correlation," *Acta Mater.*, vol. 186, pp. 172–189, 2020.
14. J.-C. Stinville, V. M. Miller, and T. M. Pollock, "Effect of non-metallic ceramic inclusions on strain localization during low cycle fatigue of a polycrystalline superalloy," in *Superalloys 2016: Proceedings of the 13th Intenational Symposium of Superalloys*, 2016, pp. 897–905.
15. A. Bagri et al., "Microstructure and Property-Based Statistically Equivalent Representative Volume Elements for Polycrystalline Ni-Based Superalloys Containing Annealing Twins," *Metall. Mater. Trans. A*, vol. 49, no. 11, pp. 5727–5744, Nov. 2018, https://doi.org/10.1007/s11661-018-4858-y.
16. M. A. Charpagne et al., "Tuning strain localization in polycrystalline nickel-based superalloys by thermomechanical processing," in *Superalloys 2020*, Springer, 2020, pp. 471–481.
17. X. Zhang, J.-C. Stinville, T. M. Pollock, and F. P. Dunne, "Crystallography and elastic anisotropy in fatigue crack nucleation at nickel alloy twin boundaries," *J. Mech. Phys. Solids*, vol. 155, p. 104538, 2021.
18. D. Texier et al., "Crack initiation sensitivity of wrought direct aged alloy 718 in the very high cycle fatigue regime: the role of non-metallic inclusions," *Mater. Sci. Eng. A*, vol. 678, pp. 122–136, 2016.
19. M. Pinz, G. Weber, J. C. Stinville, T. Pollock, and S. Ghosh, "Data-driven Bayesian model-based prediction of fatigue crack nucleation in Ni-based superalloys," *Npj Comput. Mater.*, vol. 8, no. 1, pp. 1–15, 2022.
20. W. C. Lenthe, J.-C. Stinville, M. P. Echlin, and T. M. Pollock, "Prediction of fatigue-initiating twin boundaries in polycrystalline nickel superalloys informed by TriBeam tomography," *Microsc. Microanal.*, vol. 22, no. S3, pp. 1732–1733, 2016.
21. Z. Alam, D. Eastman, M. Jo, and K. Hemker, "Development of a High-Temperature Tensile Tester for Micromechanical Characterization of Materials Supporting Meso-Scale ICME Models," *JOM*, vol. 68, no. 11, pp. 2754–2760, Nov. 2016, https://doi.org/10.1007/s11837-016-2100-1.
22. L. Li, Z. Zhang, P. Zhang, and Z. Zhang, "A review on the fatigue cracking of twin boundaries: Crystallographic orientation and stacking fault energy," *Prog. Mater. Sci.*, p. 101011, Sep. 2022, https://doi.org/10.1016/j.pmatsci.2022.101011.
23. V. Randle, "Twinning-related grain boundary engineering," *Acta Mater.*, vol. 52, no. 14, pp. 4067–4081, Aug. 2004, https://doi.org/10.1016/j.actamat.2004.05.031.
24. M. Seita, J. P. Hanson, S. Gradečak, and M. J. Demkowicz, "The dual role of coherent twin boundaries in hydrogen embrittlement," *Nat. Commun.*, vol. 6, no. 1, Art. no. 1, Feb. 2015, https://doi.org/10.1038/ncomms7164.
25. S. Yamaura, Y. Igarashi, S. Tsurekawa, and T. Watanabe, "Structure-dependent intergranular oxidation in Ni–Fe polycrystalline alloy," *Acta Mater.*, vol. 47, no. 4, pp. 1163–1174, Mar. 1999, https://doi.org/10.1016/S1359-6454(99)00007-5.
26. F. Sansoz, H. Huang, and D. H. Warner, "An atomistic perspective on twinning phenomena in nano-enhanced fcc metals," *JOM*, vol. 60, no. 9, pp. 79–84, Sep. 2008, https://doi.org/10.1007/s11837-008-0124-x.
27. L. C. Lim, "Slip-twin interactions in nickel at 573K at large strains," *Scr. Metall.*, vol. 18, no. 10, pp. 1139–1142, Oct. 1984, https://doi.org/10.1016/0036-9748(84)90194-7.
28. Z. Zhang, Z. Yang, S. Lu, A. Harte, R. Morana, and M. Preuss, "Strain localisation and failure at twin-boundary complexions in nickel-based superalloys," *Nat. Commun.*, vol. 11, no. 1, Art. no. 1, Sep. 2020, https://doi.org/10.1038/s41467-020-18641-z.
29. A. J. Egan et al., "Effect of Nb Alloying Addition on Local Phase Transformation at Microtwin Boundaries in Nickel-Based Superalloys," in Superalloys 2020, Cham, 2020, pp. 640–650, https://doi.org/10.1007/978-3-030-51834-9_62.
30. A. J. Egan et al., "Local Phase Transformation Strengthening at Microtwin Boundaries in Nickel-Based Superalloys," Acta Mater., vol. 238, p. 118206, Oct. 2022, https://doi.org/10.1016/j.actamat.2022.118206.

31. T. M. Smith et al., "Utilizing local phase transformation strengthening for nickel-base superalloys," *Commun. Mater.*, vol. 2, no. 1, Art. no. 1, Oct. 2021, https://doi.org/10.1038/s43246-021-00210-6.
32. T. M. Smith et al., "Phase transformation strengthening of high-temperature superalloys," *Nat. Commun.*, vol. 7, no. 1, Art. no. 1, Nov. 2016, https://doi.org/10.1038/ncomms13434.
33. L. Feng et al., "Localized phase transformation at stacking faults and mechanism-based alloy design," *Acta Mater.*, vol. 240, p. 118287, Nov. 2022, https://doi.org/10.1016/j.actamat.2022.118287.
34. M. A. Charpagne et al., "Automated and quantitative analysis of plastic strain localization via multi-modal data recombination," *Mater. Charact.*, vol. 163, p. 110245, 2020.
35. "Standard Specification for Precipitation-Hardening and Cold Worked Nickel Alloy Bars, Forgings, and Forging Stock for Moderate or High Temperature Service." https://www.astm.org/b0637-18.html (accessed Sep. 21, 2022).
36. N. C. Ferreri, S. C. Vogel, and M. Knezevic, "Determining volume fractions of γ, γ', γ'', δ, and MC-carbide phases in Inconel 718 as a function of its processing history using an advanced neutron diffraction procedure," *Mater. Sci. Eng. A*, vol. 781, p. 139228, Apr. 2020, https://doi.org/10.1016/j.msea.2020.139228.
37. E. E. Underwood, "Stereology, or the quantitative evaluation of microstructures," *J. Microsc.*, vol. 89, no. 2, pp. 161–180, 1969, https://doi.org/10.1111/j.1365-2818.1969.tb00663.x.
38. J. Wang, N. Li, and A. Misra, "Structure and stability of $\Sigma3$ grain boundaries in face centered cubic metals," *Philos. Mag.*, vol. 93, no. 4, pp. 315–327, Feb. 2013, https://doi.org/10.1080/14786435.2012.716908.
39. J. W. Cahn, "The kinetics of grain boundary nucleated reactions," *Acta Metall.*, vol. 4, no. 5, pp. 449–459, Sep. 1956, https://doi.org/10.1016/0001-6160(56)90041-4.
40. D. A. Porter, K. E. Easterling, and M. Y. Sherif, *Phase Transformations in Metals and Alloys*, 4th ed. Boca Raton: CRC Press, 2021. https://doi.org/10.1201/9781003011804.
41. P. Neumann and A. Tönnessen, "Crack Initiation at Grain Boundaries in F.C.C. Materials," in *Strength of Metals and Alloys (ICSMA 8)*, P. O. Kettunen, T. K. Lepistö, and M. E. Lehtonen, Eds. Oxford: Pergamon, 1989, pp. 743–748. https://doi.org/10.1016/B978-0-08-034804-9.50116-9.

Tailoring the γ-γ'-γ'' Dual Superlattice Microstructure of INCONEL® 725 by High Temperature Aging and Nb/Ta Additions for Superior Creep Properties

Stoichko Antonov, Chang-Yu Hung, Jeffrey A. Hawk, Paul D. Jablonski, and Martin Detrois

Abstract Next-generation energy systems require superior resistance to creep deformation due to the considerably prolonged exposure times at operating stress and temperature. To improve the elevated temperature properties of INCONEL® 725 (IN725), a corrosion-resistant alloy, several variants with different Ti/Al ratios and judicious amounts of Nb and Ta were made. Furthermore, a high temperature aging (HTA) heat treatment, designed to promote favorable precipitate phase formation, was explored. These adjustments allowed to tailor the amount and type of precipitate strengthening which led to significant increases in time to failure. The Ti/Al ratio was used to favor the formation of γ' or γ'' precipitates. Compact morphology precipitates, consisting of γ' precipitates surrounded by a γ'' shell, were formed in alloys with a low Ti/Al ratio. The HTA increased the creep life of various alloy formulations up to a maximum improvement of 371% as compared to the standard aging heat treatment. The Nb and Ta additions had a similar effect on increasing creep life by promoting and stabilizing γ'' precipitation. The positive effect of the additions was even more pronounced when coupled with the HTA. A phase stability study with up to 10,000 h exposure at 700 °C revealed that the compact morphology helped in slightly reducing coarsening of the γ' precipitates, although the effect on creep was not significant. The findings of this study enable design of dual superlattice alloys through microstructural engineering that yields superior performance and can be applied to a wide range of alloys in the IN718 and derivatives family.

Keywords Ni-based superalloy · Inconel 725 · Creep · Phase stability · Dual superlattice

S. Antonov (✉) · C.-Y. Hung · J. A. Hawk · P. D. Jablonski · M. Detrois
National Energy Technology Laboratory, 1450 Queen Ave. SW, Albany, OR 97321, USA
e-mail: Stoichko.Antonov@netl.doe.gov

C.-Y. Hung
NETL Support Contractor, 1450 Queen Ave. SW, Albany, OR 97321, USA

E. A. Ott et al. (eds.), *Proceedings of the 10th International Symposium on Superalloy 718 and Derivatives*, The Minerals, Metals & Materials Series,
https://doi.org/10.1007/978-3-031-27447-3_10

Introduction

INCONEL® 725, i.e. IN725, is a derivative of INCONEL® 625 through the addition of Ti with aim of promoting the formation of γ' precipitates [1]. This precipitation strengthened Ni-based superalloy has high strength and good corrosion resistance [2], making it a material of choice for a wide variety of intermediate-temperature applications, especially considering that it does not contain costly Co additions. Interestingly, although IN625 was designed to be a solid-solution strengthened alloy, subsequent investigations have conclusively shown the precipitation of the γ'' phase at temperatures <750 °C. As a result, the typical microstructure of IN725 also consists of a face-centered cubic (FCC) matrix with embedded γ'' precipitates, however, the Ti addition also promotes the formation of γ' precipitates; the fraction of the two secondary phases can vary depending on the employed heat treatment [3]. The γ' precipitates are $Ni_3(Al, Ti)$ with ordered FCC $L1_2$ crystal structure, while the γ'' phase is typically Ni_3Nb and has an ordered tetragonal $D0_{22}$ crystal structure. Both precipitates have solubility for elements such as Co, Cr, Nb, and Mo among others. Generally, the γ' phase is preferred for high temperature applications of Ni-based alloys due to its higher solvus temperature and superior stability compared to that of the γ'' precipitates. However, the latter offers more potent strengthening due to the high coherency stresses with the matrix, that can significantly extend the creep life in temperature regimes where the γ'' precipitates are stable. Of note, when co-precipitated in a compact morphology, as reported by Cozar et al. [4, 5], the γ'' shell is thought to be beneficial in preventing significant coarsening of the γ' precipitates and further improve the mechanical properties of the alloy.

In addition to the γ' and γ'' phases, IN725 also contains carbides such as MC (Ti/Nb rich), $M_{23}C_6$ (primarily Cr-rich), and M_6C (Mo-rich) which are located along the grain boundaries (GBs) [3]. Furthermore, η phase has been reported to form at temperatures between 750 and 900 °C for times ranging from 10 to 100 h [6]. The η phase has an ordered hexagonal $D0_{24}$ crystal structure and a composition consisting of Ni_3Ti. Similar to δ phase precipitates, which have a $D0_a$ crystal structure and a composition of Ni_3Nb, the effect of η phase on the mechanical properties of Ni-based superalloys has been the subject of numerous investigations. Small and stable precipitates have been shown to be beneficial to alloy 718Plus by slowing down fatigue crack growth rates [7], while unstable precipitates were found to be detrimental to the creep performance of alloy 263 [8]. Small δ platelets are formed in IN718 using a high temperature aging heat treatment [9]. Finally, η-Ni_6AlNb and the coexistence of δ and η phases have been reported in several alloys [10, 11].

Using IN725 in high-temperature applications would be very valuable to fully utilize its superior corrosion resistance and relatively low cost. Particularly, alloys capable of operating for extended times at temperature in demanding environments are needed for next-generation land-based energy systems [12]. However, to date, there have been very few high-temperature mechanical properties reported for IN725, with most assessments being up to 550 °C or 650 °C [13, 14]. This investigation explores the high-temperature creep properties of several variants of IN725 that were designed with several key parameters. The stability of the γ′ and γ″ phases was explored using changes from the nominal composition of IN725 according to the Ti/Al ratio. Additions of Ta and Nb were chosen to promote the formation of desirable phases and a high temperature aging heat treatment was designed to increase the precipitation of those features. Compared to the baseline alloy tested at 700 °C/483 MPa, condition most representative of IN725, the creep life of the variants increased by up to 256%. The relationships between alloying, microstructural formation and evolution, and creep properties are presented and discussed.

Methods

The alloys were developed starting from the composition of commercial IN725, as described in Ref. [2]. A computational approach was utilized to obtain the targeted compositions using Thermo-Calc with the TCNi8 database for phase formation and JMatPro for kinetics. The baseline alloy, referred to as STD, consists of near-nominal composition for IN725 with a Ti/Al ratio of 4.3 (calculated using at.% throughout this study). A slightly higher Ta-containing version was made and named A. Then, the Ti/Al ratio was systematically decreased from variants A to D with values equal to 5.4, 1.3, 1.1, 0.7 in alloys A, B, C, and D, respectively. Finally, the amount of Nb + Ta was increased to 2.7 at.% in variants with the high and low Ti/Al ratio, namely STD and D, to promote the formation of GB phases, such as δ phase precipitates, according to plots from Ref. [11] or γ″ precipitates within the matrix. Tantalum was added to alloys STD and D to form alloys E1 and E, respectively. Similarly, Nb was added to alloys STD and D to form alloys F2 and F, respectively.

High-purity, industry grade, stock materials were used to constitute the compositions of the alloys and form 8 kg cylindrical ingots measuring 75 mm in diameter using vacuum induction melting (VIM). The charges were melted to a 50 °C superheat temperature under 200 Torr Ar partial pressure. Following solidification, ~5 mm thick slices were cut from the top of each ingot for chemistry analysis. Major elements were determined using X-ray fluorescence (XRF) on a Rigaku ZSX Primus II while C was determined using combustion analysis on LECO systems. The ingots were homogenized in a vacuum heat treatment furnace under 50 Torr Ar partial pressure and using Ar forced gas fan cooling. The heat treatment schedule was optimized computationally for the alloys based on their chemistry and the ingot microstructure to reduce the predicted residual elemental inhomogeneity to below 1% [15]. The heat treatment cycle consisted of 1030 °C for 1 h, 1065 °C for 3 h, 1090 °C for 3 h, 1115 °C for 3 h, 1135 °C for 6 h, and 1150 °C for 72 h. Hot working consisted of steps of forging followed by steps of hot rolling with reheat between each pass to produce plates measuring 10 mm in thickness with an equiaxed grain structure. The last reheat step was used as a solution heat treatment.

Two separate aging heat treatments were investigated. The first consists of the standard aging heat treatment for IN725 and is referred to as standard aging (SA): 730 °C for 8 h followed by cooling to 620 °C at 1 °C/min and holding at 620 °C for

8 h. The alternate aging heat treatment was designed to promote the formation of precipitate phases at higher temperatures, and thereby, is referred to as high temperature aging (HTA), described as follows: 800 °C for 20 h followed by cooling to 750 °C at 1 °C/min and holding at 750 °C for 8 h. The aging heat treatments were performed in the vacuum heat treatment furnace under 50 Torr Ar partial pressure and Ar forced gas fan cooling was performed at the end of the cycles. Specimens characterized and mechanically tested received either the SA or HTA heat treatments.

Cylindrical creep test specimens were produced with 76 mm in overall length, 10 mm in overall diameter with ANSI 3/8 × 16 threaded ends, and a reduced gage area. The reduced gage section measured 32 mm in length and 6.3 mm in diameter. Creep screening was performed according to ASTM E139 standard using a dead weight creep frame [16]. Four conditions were selected for constant load creep testing: 700 °C/483 MPa, 727 °C/414 MPa, 750 °C/345 MPa, and 790 °C/207 MPa. Not every alloy was tested at each condition, but all alloys were tested at 700 °C/483 MPa.

After standard metallographic sample preparation, observation of the microstructures was performed on a FEI Inspect F scanning electron microscope (SEM) in backscatter electron mode and equipped with an Oxford Instrument X-Max energy dispersive spectrometer (EDS) for phase characterization. Transmission electron microscopy (TEM) samples were prepared by extracting ∅3 mm discs from as-aged material and from the grip sections of post-creep specimens. Following grinding on progressive grit papers to ~70 μm, the TEM blanks were thinned to electron transparency using twin-jet polishing in a solution of 20% perchloric acid and 80% ethanol at −16 °C and 18 V. A JEOL JEM 2100PLUS, operating at 200 kV, was used for selected area diffraction (SAD) and bright field (BF), dark-field (DF) and high resolution (HR) imaging.

Results

Alloy Chemistry

The measured chemistries of the alloys are listed in Table 1. The high Ti/Al ratio alloys are STD, E1, F2, and A. Note: the Ti/Al ratio was lower in alloy STD likely due to minor Al loss in the other alloys which occurred during melting. In this series, the Ta content was increased to 3.6 wt.% in E1 and the Nb content was raised to 4.6 at.% in F with a sum Nb + Ta between 2.4 and 2.7 at.%. The Ti/Al ratio was decreased from A to D with values equal to 5.4, 1.3, 1.1, and 0.7 in A, B, C, and D, respectively. Finally, the low Ti/Al ratio series included alloy D and high-Ta variant E with 3.5 wt.% Ta and high-Nb variant F with 4.5 wt.% Nb. The sum of Nb + Ta is between 2.5 and 2.8 at.% for both alloys. Thus, this alloy series allows for the investigation of the effect of the Ti/Al ratio and additions of Nb or Ta to alloys with the highest and lowest Ti/Al ratios.

Table 1 Composition of the alloys with Ni balanced from XRF analysis (major elements), combustion analysis (C), and addition to the melt (B) in wt.% except for the ratio Ti/Al and sum Nb + Ta in at.%

Alloy	Cr	Mo	Ti	Al	Nb	Ta	Fe	Mn	Si	C	B	Ti/Al	Nb + Ta
High Ti/Al Ratio													
STD	21.2	7.1	1.82	0.24	3.4	0.2	3.9	0.04	0.02	0.039	0.003	4.3	1.8
E1	20.5	7.0	1.87	0.20	3.4	**3.6**	3.8	0.05	<0.01	0.049	0.003	5.4	**2.7**
F2	20.6	7.1	1.88	0.19	**4.6**	0.4	3.9	0.05	<0.01	0.049	0.003	**5.6**	**2.4**
Varying Ti/Al Ratio													
A	21.2	7.1	1.84	0.19	3.4	0.4	3.9	0.05	0.01	0.040	0.003	**5.4**	1.8
B	21.2	7.1	1.41	0.60	3.4	0.4	3.9	0.04	0.02	0.039	0.003	**1.3**	1.9
C	21.2	7.1	1.30	0.68	3.4	0.4	3.9	0.05	0.01	0.040	0.003	**1.1**	1.9
Low Ti/Al Ratio													
D	21.2	7.1	1.10	0.85	3.4	0.4	3.9	0.06	0.03	0.044	0.003	**0.7**	1.9
E	21.2	7.0	1.15	0.89	3.4	**3.5**	3.9	0.04	0.02	0.038	0.003	0.7	**2.8**
F	21.3	7.1	1.13	0.84	**4.5**	0.4	3.9	0.04	0.01	0.040	0.003	0.8	**2.5**

Microstructure

Analysis of the microstructure of the alloys following hot working and solution heat treatment revealed equiaxed grain structures with similar grain sizes ranging from 42 to 47 μm. The grain size was calculated using the linear intersect technique [17] with four regions considered, each used to perform four measurements. For each alloy, half of the material received the SA aging heat treatment while the other half received the HTA heat treatment. The microstructures observed using the SEM and including GB regions are shown in Fig. 1. First, the microstructures following SA (commercial heat treatment for IN725) revealed very fine precipitates that could hardly be resolved with the SEM, as shown in the insets of Fig. 1. The GBs of alloy F, however, contained larger precipitates when compared to alloys STD, D, and E. Images for E1 and F2 are not available at the time of writing.

The microstructures of the alloys following the HTA heat treatment revealed coarser precipitates. Particularly, a dense distribution of γ′ and/or γ″ particles can be resolved within the grain interior. Typically, the γ′ phase appears darker in Z-contrast backscatter images due to the higher content of light-element Al, while γ″ appears bright from the Nb and other heavy refractory elements. Various GB phases also formed. Dark, blocky precipitates were found in all alloys, and EDS analysis performed on similar but overaged specimens (to enable coarsening and more accurate EDS measurements) associated them with Cr-rich $M_{23}C_6$ carbides. The small, blocky, and bright precipitates are likely M_6C due to high Mo concentrations and traces of Si [18]. This correlates well with other investigations of IN725

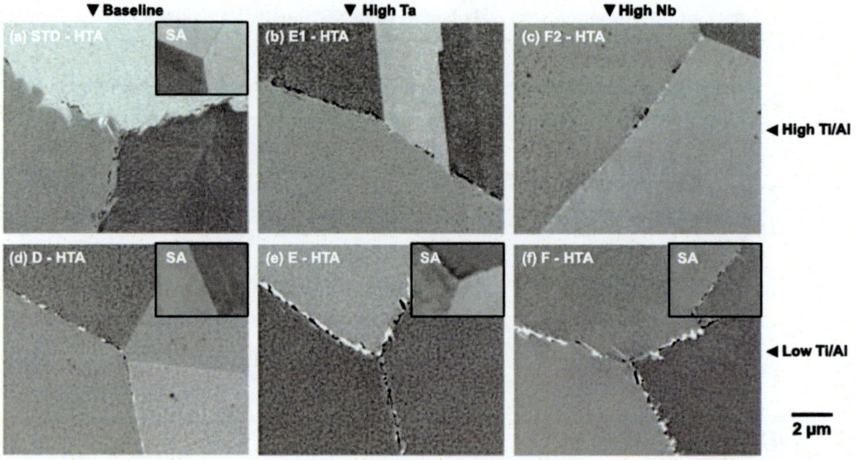

Fig. 1 SEM images of the microstructure of alloy **a** STD, **b** E1, **c** F2, **d** D, **e** E and **f** F including grain boundaries following the HTA heat treatment with insets for the SA heat treatment condition for alloys STD, D, E, and F

[3]. Overall, no significant differences are noticeable between the alloys, particularly STD, Fig. 1a, E1, b, F2, c and D, d. Alloys E, Fig. 1e and F, 1f, however, contained discrete needle/rod-shaped precipitates consistent with δ or η phase and a primarily $Ni_3(Nb, Ti)$ composition. Such precipitates were also found in STD and F2 to a lesser extent. Additional details on the EDS analysis can be found in Ref. [19].

Higher resolution images of the γ/γ' and/or γ'' were taken using TEM and the results for alloys STD, D, E, and F heat treated using either the SA or HTA are shown in Fig. 2. In general, the microstructural components and morphology of each alloy are the same between the two heat treatments, however, the size is much coarser following HTA. The STD alloy formed mostly $\gamma''/\gamma'/\gamma''$ sandwich-type structures, where the γ'' formed on only two faces of the γ' precipitates. Alloy D predominantly formed γ' precipitates. To better show the absence or presence of γ'' in this alloy, DF images using the shared γ'/γ'' (100) reflections are shown in Fig. 2b and f. Interestingly, a small fraction of γ'' precipitates were observed for the SA condition, while γ'' was not observed at all post-HTA. Alloys E and F show similar microstructures to each other following SA, Fig. 2c and d, respectively, and after HTA, Fig. 2g and h, respectively, where the γ'/γ'' precipitates exhibit compact morphologies in both aging conditions. The major microstructural difference between the two alloys being that alloy F has a slightly thicker and higher fraction of γ'' precipitates, and that the precipitates are significantly coarser after HTA compared to SA.

The microstructures of alloys E1 and F2 after HTA (these alloys were not subjected to the SA) were also observed by TEM and representative microstructures are shown in Fig. 3. The addition of Ta and Nb to E1 and F2, respectively, stabilized the γ'' precipitates, such that a γ/γ'' microstructure (with the absence of any γ' precipitates) formed. The two alloys exhibit similar γ/γ'' microstructure with three variants of

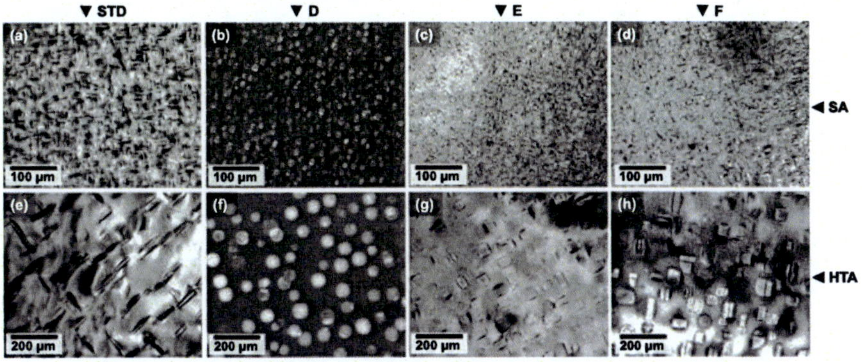

Fig. 2 TEM images of alloys **a, e** STD, **b, f** D, **c, g** E and **d, h** F for the γ′ and/or γ″ precipitates following the SA (**a–d**) or HTA (**e–h**) heat treatment. *Note* Images **a, c, d, e, g,** and **h** are BF, while **b** and **f** are DF

rather coarse γ″ precipitates. Figure 3c and d shows DF images of one of the γ″ variants, obtained using the shared γ′/γ″ (100) reflection. As can be seen, γ′ precipitates were not present, and the fraction of γ″ precipitates tends to be higher in alloy F, Fig. 3d. Such observations were performed at other locations and with the other shared γ′/γ″ reflections as to more conclusively determine whether γ′ formed in these alloys; it did not.

To better understand the formation and distribution of the γ″ precipitates in some of the alloys, HRTEM was performed. As shown in Fig. 2b and f, a small fraction of γ″ precipitates were observed only after SA in alloy D. The diffraction patterns showed faint γ″ reflections, even when such precipitates were not observed using DF imaging. Figure 4a shows a HRTEM image of a γ′ precipitate in alloy D after SA aging. Interestingly, 2–3 atomic layer thick γ″ structure was observed on the faces of the γ′ precipitate, and such a structure was observed on most of the γ′ precipitates in this alloy. Figure 4b shows an HRTEM image of alloy F after HTA, where the γ′/γ″ compact structure was readily apparent in Fig. 2h. Indeed, the faces of the γ′ precipitates are entirely covered (edge-to-edge) with γ″ that is tens of atomic layers thick. Finally, an HRTEM image from a γ″ precipitate in alloy E1 after HTA such as the ones in Fig. 3a and c is shown in Fig. 4c. The γ″ precipitate was not connected to any other type of precipitate, and γ′ structure was not observed.

Creep Properties

The results from all performed creep tests are shown in the form of a Larson-Miller parameter (LMP) plot in Fig. 5. The alloys were all tested at 700 °C and 483 MPa in either SA or HTA heat treatment condition, at the exception of alloys E1 and F2 only tested after HTA. Additional testing was performed at lower stresses for

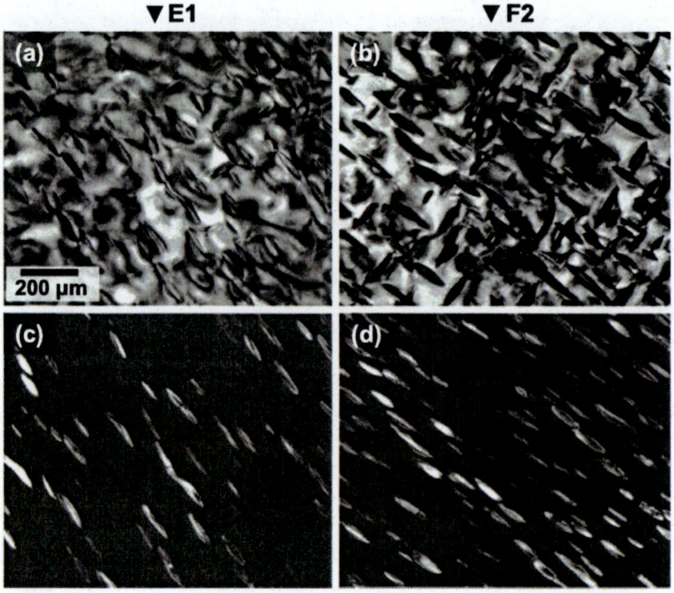

Fig. 3 TEM images of alloys **a**, **c** E1, and **b**, **d** F2, for the γ'' precipitates following HTA heat treatment, where **a**, **b** show BF images, and **b**, **d** show DF images at the same location

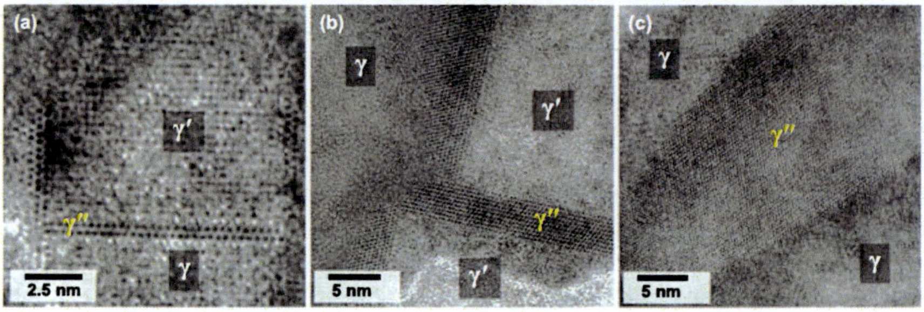

Fig. 4 HRTEM images of the $\gamma/\gamma'/\gamma''$ interfaces in alloys **a** D with the SA heat treatment, **b** F with the HTA heat treatment and **c** E1 with the HTA heat treatment

alloy STD to form a curve for the baseline creep properties. The testing temperature was raised for testing time constraints. From creep at 700 °C and 483 MPa, it is shown that the LMP decreases from 21.69 to 21.31 with decreasing Ti/Al ratio from alloy STD to D. However, using the HTA instead of the SA heat treatment increased the LMP values to above those for alloy STD with SA. The best-performing alloys were those with added Nb and Ta, namely E, F, E1 and F2, and with the HTA heat treatment. Particularly, alloy E1 HTA outperformed all other alloys on the LMP plot with additional testing performed to draw a fitting curve for comparison to alloy

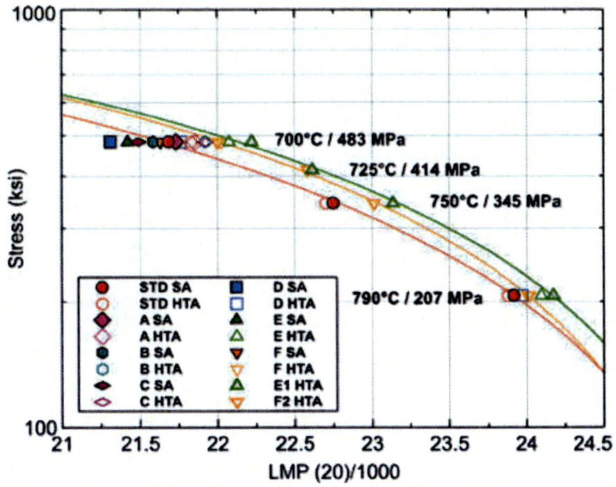

Fig. 5 Larson-Miller plot for testing of the alloys under four conditions and after the SA or HTA heat treatment

STD SA and F2 HTA. The LMP increased by 0.54 at 700 °C/483 MPa and 0.25 at 790 °C/207 MPa from alloy STD SA to alloy E1 HTA.

Discussion

A series of alloys based on IN725 was designed with several aspects that led to improvements in creep performance [20, 21]. The Ti/Al ratio was varied, additions of Nb and/or Ta were considered, and a high temperature aging (HTA) heat treatment was compared to the standard aging (SA) for IN725. From Fig. 5, the HTA combined with Nb or Ta additions led to the best properties. The effect of each design attribute is described in the following sections.

Effect of the Ti/Al Ratio

The Ti/Al ratio was decreased from alloys STD/A to D which led to changes in the microstructural features of each variant. Minimal differences were found along the GBs when comparing high Ti/Al ratio alloy STD and low Ti/Al ratio alloy D following the HTA heat treatment, Fig. 1a and d, respectively. More of the elongated δ/η platelets were found in alloy STD. A more significant microstructural difference was observed within the grains. As shown in Figs. 2 and 3, the high Ti/Al ratio alloys STD, E1, and F2 primarily contained γ″ precipitates while the low Ti/Al ratio alloys D, E, and F mostly contained γ′ precipitates surrounded by a varying fraction of γ″ precipitates. Only alloy D after HTA contained just γ′ precipitates. The decrease in the Ti/Al ratio consisted of a ~40% decrease in the Ti content and a ~4.5× increase

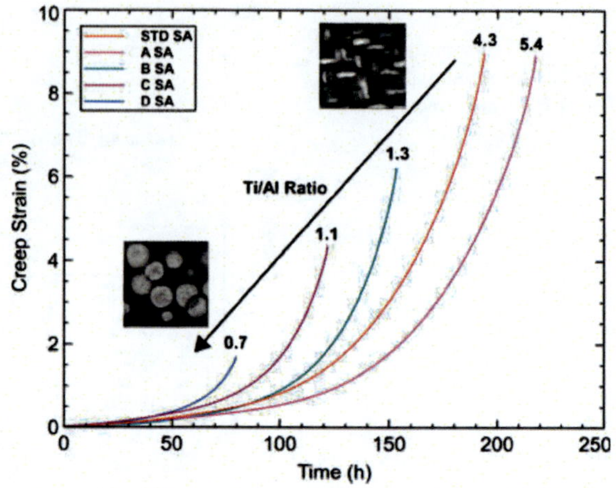

Fig. 6 Creep strain as a function of time for testing at 700 °C and 483 MPa for the alloys with decreasing Ti/Al ratio and following the SA heat treatment. TEM images of the precipitates in alloy STD and D in the aged condition are provided

in the Al content. Therefore, the decrease in the Ti/Al ratio stabilized the γ'-Ni_3Al precipitates over the γ''-Ni_3Nb particles.

The creep curves from testing of the alloys in the SA condition at 700 °C and 483 MPa are shown in Fig. 6 with the alloy Ti/Al ratio denoted for each curve. The decrease in Ti/Al ratio correlated well with a decrease in creep life and ductility. Thus, a 63% decrease in creep life and 81% decrease in elongation to failure were measured from alloys A to D. Testing of the alloys in the HTA condition revealed a similar trend with respect to creep ductility, as highlighted in Fig. 7 which shows the elongation to failure as a function of time to failure. However, the creep lives of B HTA and C HTA were greater than those for STD HTA, A HTA, and D HTA. Therefore, the switch from predominant γ'' to γ' precipitate strengthening was associated with a loss in creep ductility and creep life, the latter being more evident for the SA heat treatment. The γ'' phase is known to provide more potent strengthening due to the high coherency stresses with the γ matrix. The effect of the HTA heat treatment is further discussed in the next section.

Effect of the Aging Temperature

The HTA heat treatment had a pronounced effect on the microstructure of the alloys when compared to the SA heat treatment. First, coarser precipitates were found along the GBs which is shown in Fig. 1a, d, e, and f. The precipitation of secondary phases, such as δ and/or η, was promoted as well. Second, the γ' and/or γ'' precipitates were also coarser following HTA compared to SA, by comparing Fig. 2a-d to Fig. 2e-h. That is, the higher temperature of the HTA heat treatment accelerated the kinetics of phase formation. Furthermore, the first step of the HTA was not only higher than SA (800 °C compared to 730 °C) but also longer (20 h compared to 8 h) which enabled

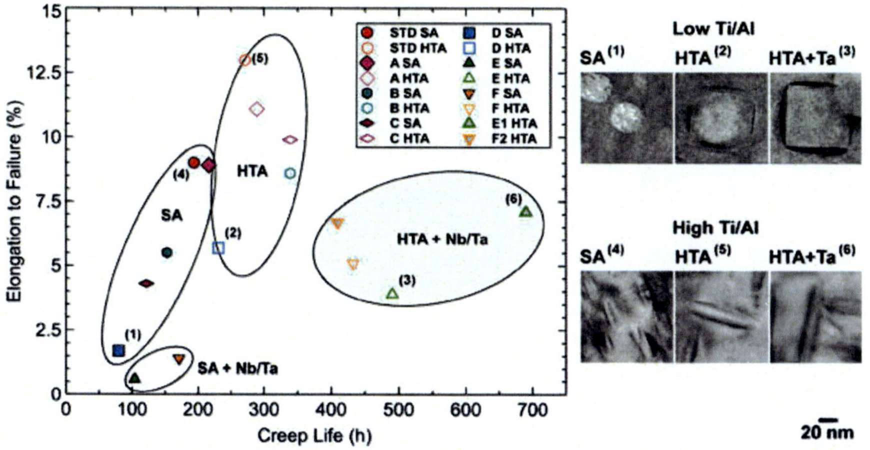

Fig. 7 Elongation to failure as a function of creep life for testing at 700 °C and 483 MPa for the alloys following the SA and HTA heat treatments with TEM images of the γ′/γ″ precipitates taken in the gage section

the coarsening of the various precipitate phases. Note: the solvus temperature of the γ″ phase was predicted from JMatPro to be between 820 °C and 860 °C depending on the alloy formulation, therefore, the first step of the HTA was below the solvus for γ″ and mechanical testing was also performed at temperatures below the γ″ solvus temperature. Secondary precipitation along the GBs has been the subject of numerous studies on IN718, particularly regarding δ phase precipitates. Typically, an additional step is used above the γ″ solvus temperature to promote the formation of δ precipitates [22]. With the highest temperature in the HTA heat treatment being below the γ″ solvus temperature, it is not surprising to see little to no precipitation of δ phase (depending on the variant) in the aged condition while fault-free γ″ precipitates are present in the microstructure. However, these γ″ precipitates are expected to transform (at least partially) to δ phase with fault accumulation during prolonged creep exposure or aging.

The effect of the HTA on the creep performance of the alloys was considerable. An overview is presented in Fig. 7 in which the alloys in the SA condition are compared to those in the HTA condition. Furthermore, the alloys with the Nb/Ta additions are also compared in the SA condition (SA + Nb/Ta) and HTA condition (HTA + Nb/Ta). Significant increases in both creep life and elongation to failure were measured between the SA and HTA heat treatments. Table 2 summarizes the percent increase in creep life and creep ductility when using the HTA instead of the SA heat treatment. The greatest improvements were measured for the alloys with the Nb and Ta additions, particularly alloy E which showed a 371% increase in creep life and 550% increase in creep ductility. Alloy STD SA, most representative of IN725, showed a 40% increase in creep life and 53% increase in creep ductility if the HTA heat treatment is used. With the HTA promoting precipitation strengthening

Table 2 Percent increase in creep life and elongation to failure from the SA to the HTA heat treatment for testing at 700 °C and 483 MPa

Alloy	Ta (at.%)	Nb (at.%)	Ti/Al	% increase in elongation to failure	% increase in creep life
STD	0.04	1.77	4.3	53	40
A	0.10	1.74	5.4	25	33
B	0.10	1.81	1.3	56	122
C	0.11	1.81	1.1	130	178
D	0.10	1.80	0.7	235	189
E	0.95	1.82	0.7	550	371
F	0.10	2.39	0.8	264	153

as well as precipitation at the GBs, additional obstacles to dislocation motion were formed which benefited the creep performance of the alloys. Furthermore, the right-hand side of the figure shows TEM micrographs of the post-creep microstructures. As can be seen, the γ′/γ″ precipitates remained relatively constant and only slight coarsening occurred compared to the aged conditions. Interestingly, alloy D HTA which did not form γ″ precipitates after aging, Fig. 2f, shows thin γ″ precipitates at the γ′ cube faces. These likely formed during the lower temperature creep exposure. The precipitate stability in terms of size is deemed beneficial for long-term creep exposures where precipitate coarsening typically leads to deterioration of the creep resistance.

The Role of Nb and Ta Additions

Additions of Nb or Ta were made to high Ti/Al ratio alloy STD to form alloys E1 (Ta-rich variant) and F2 (Nb-rich variant) and to low Ti/Al ratio alloy D to form alloys E (Ta-rich) and F (Nb-rich). The sum of Nb + Ta in alloys E1, F2, E, and F was kept nominally constant at 2.7 at.%. Adding Ta or Nb to STD, Fig. 2e, resulted in a greater density of γ″ precipitates, as shown in alloy E1, Fig. 3a, and F2, Fig. 3b. The addition of Nb was found more beneficial in promoting the formation of γ″ which was expected considering the composition of the γ″ phase, heavily comprised of Nb. In the low Ti/Al ratio alloys, the additions also led to the precipitation of γ″; however, since these alloys contained a dense repartition of γ′ precipitates, the γ″ phase formed at the γ/γ′ interfaces to form compact precipitates, Fig. 4b. It should be noted that the effect of the additions was mostly concentrated within the grains for the high Ti/Al ratio alloys which contained predominantly γ″ precipitates while the effect was more pronounced towards the GBs for the low Ti/Al ratio alloys, as shown in Fig. 1e and f, where δ phase precipitates were present. With more γ′ present in the low Ti/Al alloys, the Nb and Ta likely partitioned to the GB phases as opposed to γ″ in the high Ti/Al alloys. Nevertheless, avoiding δ phase formation at the GBs in a γ″

containing alloy may be preferable according to recent findings from Nicolay et al. [23] on IN718. The authors revealed improved mechanical properties by avoiding the first step of the three-step heat treatment, thereby reducing δ formation at the GBs and increasing the amount of γ″ precipitates within the grains, since those phases have similar compositions.

The Nb and Ta additions, as expected considering their impact on the microstructure of the alloys, were associated with significant improvements in the creep life of the alloys particularly when combined with the HTA heat treatment, as shown in Fig. 7. As such, the creep life of those alloys was between 409 and 690 h compared to 194 h for STD (and 271 h in the HTA condition) and 80 h for D (231 h in the HTA condition). The greater amount of precipitate strengthening was responsible for decreases in the minimum creep rate, as described in Ref. [19].

Phase Stability

A phase stability study was performed on the alloys following the HTA heat treatment. Specimens were exposed to 700 °C for times ranging from 500 to 10,000 h. Figure 8 provides an overview of the results after 5,000 h exposure. The GB phases, particularly δ-Ni_3Nb (identified using TEM-EDS, not shown) which appear as bright elongated platelets, coarsened when compared to the as-aged conditions of Fig. 1. Interestingly, alloys STD, E1, F2, and D were found to evolve in a similar manner. Alloy D did not contain as much δ precipitates compared to the other alloys. Alloys E and F, however, revealed extensive formation and growth of the δ phase as well as the formation of intragranular α-Cr precipitates, which appear black in the SEM images of Fig. 8e and f. Such α-Cr precipitate formation has been associated with the formation of intragranular δ-Ni_3Nb in IN718, as the δ phase rejects Cr at longer exposures and results in a local enrichment at the γ/δ interface [24]. Interestingly, in such cases, α-Cr is observed at temperatures <650 °C, while σ phase is observed at higher temperatures [25]. The stability range of the α-Cr precipitates in IN725 remains to be explored further (work underway) as IN718 contains a significantly higher level of Fe which would typically stabilize the σ phase. The lower Ti/Al ratio in alloys E and F predominantly stabilized the γ′ as opposed to γ″. In the high Ti/Al ratio alloys E1 and F2, the γ″ phase ties up elements prone to the formation of δ phase, such as Nb. Therefore, formation of δ phase was retarded in those alloys. Eventually, δ forms as shown in the inset of Fig. 8c after 10,000 h exposure at 700 °C. It is important to note that γ″ is metastable and eventually transforms into its stable δ form [26].

Higher resolution images of the γ/γ′/γ″ microstructures of D, E, and F in the HTA condition and after 10,000 h of exposure at 700 °C were taken using TEM and the results are shown in Fig. 9. The $[001]_\gamma$ diffraction patterns of each condition show distinct γ″ reflections at differing intensities (D being the weakest, while F shows strong intensity reflections). The shared (100) γ′/γ″ reflections (in red circle) were used for DF imaging and these images show that all three microstructures contain γ′ and γ″ precipitates, where the γ″ fraction increases in order of F > E > D. To confirm

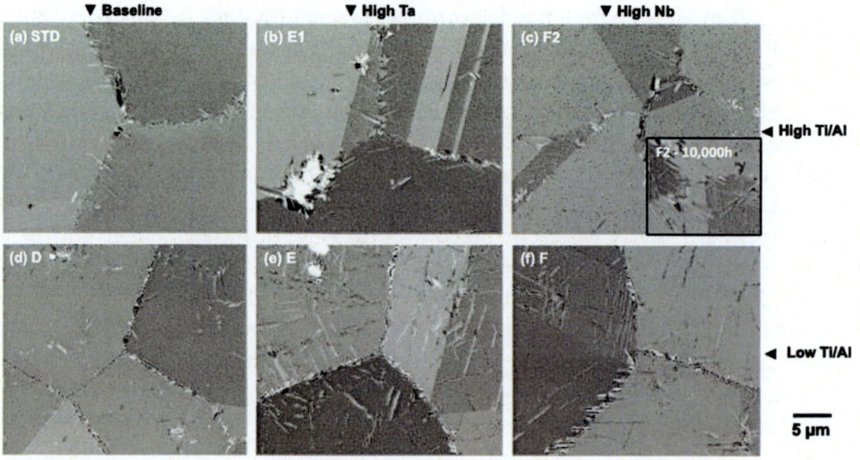

Fig. 8 SEM images of long-term exposure of alloy **a** STD, **b** E1, **c** F2, **d** D, **e** E, and **f** F at 700 °C for 5,000 h following HTA with inset for alloy F2 following 10,000 h exposure

the presence of the γ'' precipitates, intensity line profiles across the shown reflections were measured, and the results are shown in Fig. 9e. Notice that although weak, there is still a γ'' signal for alloy D, whereas this phase was absent after HTA, as shown in Fig. 2f, but present after creep at 700 °C and 483 MPa, Fig. 7. This suggests that the γ'' precipitates form during the 700 °C exposure at certain γ' precipitate faces (and quite often a γ'' precipitate is sandwiched between two γ' precipitates—the formation of this configuration remains to be explored further). As part of the analysis, the mean diameter of at least 40 γ' precipitates after HTA and 10,000 h exposure was manually measured, and the results are shown in Fig. 9d. The analysis shows that statistically, the starting γ' sizes are similar at ~45 nm (E having a slightly smaller average size of ~40 nm) and after 10,000 h at 700 °C, the precipitates coarsen to similar levels of ~60 nm (D coarsening slightly more up to ~68 nm). This represents a 50% increase in size after 10,000 h, which shows that these precipitates do not coarsen substantially (in accordance also to the post-creep microstructures), especially considering that 60 nm is still a very fine precipitate size. The alloys with added Ta (E) and Nb (F) and therefore a stronger compact structure (viz. thicker γ'' precipitates) showed lower γ' coarsening compared to alloy D.

Conclusions

Different alloy variants based on IN725 were designed with varying Ti/Al ratio, a high temperature aging (HTA) heat treatment and additions of Ta or Nb to improve the creep performance of the commercial alloy IN725. Compared to the baseline alloy tested at 700 °C/483 MPa, condition most representative of IN725, the HTA

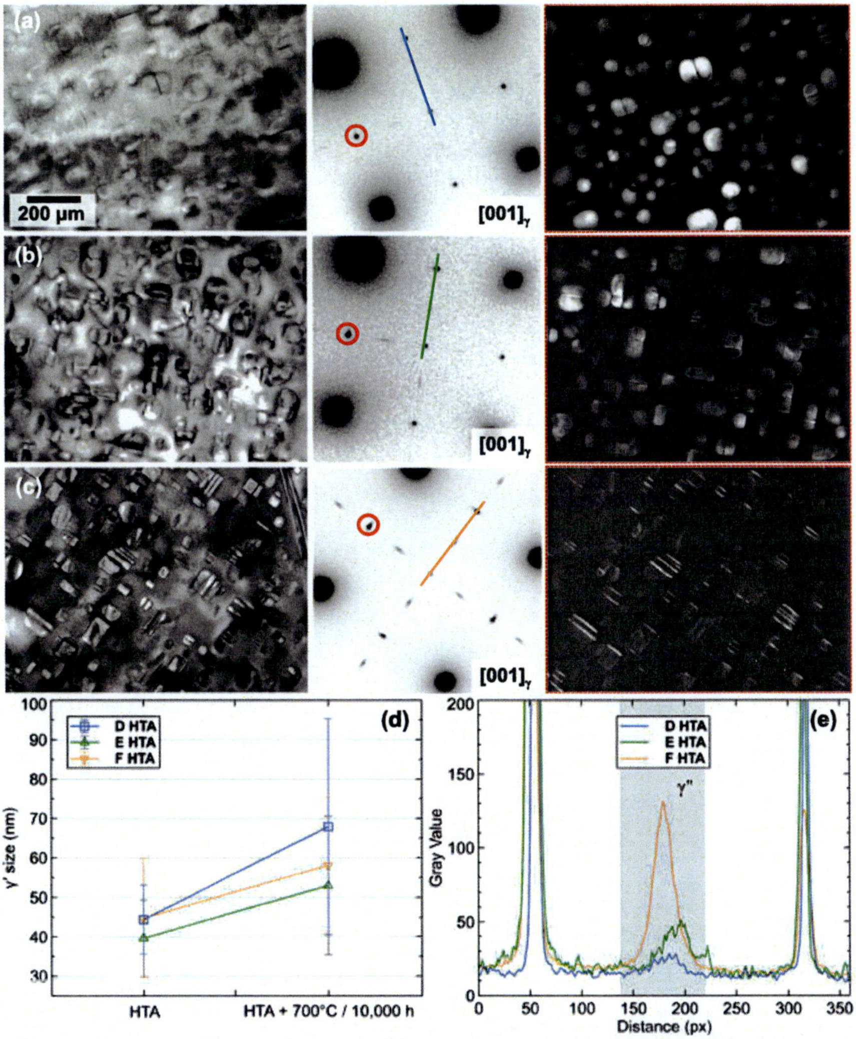

Fig. 9 **a–c** γ/γ′/γ″ microstructures and **d** coarsening behavior of the γ′ precipitates in alloys **a** D, **b** E and **c** F following 700 °C exposure for 10,000 h after HTA heat treatment with corresponding TEM BF, SAD and DF images. The DF images were produced with the reflections in red circles in the SAD patterns, while the intensity profiles along the given lines are plotted in **e**

and Nb addition resulted in 111–123% improvement in creep life (depending on the Ti/Al ratio), the HTA and Ta addition resulted in 138–256% improvement in creep life. The following can be concluded:

1. The Ti/Al ratio controlled the stability of γ′ vs. γ″ precipitates. While both phases formed in each variant, a high Ti/Al ratio ~5 favored formation of predominantly

γ'' precipitates while a low Ti/Al ~1 favored formation of predominantly γ' precipitates. The increase in Al was greater than the decrease in Ti. The creep life and ductility were reduced with decreasing Ti/Al ratio in the SA condition.
2. The HTA promoted coarsening of the γ'/γ'' precipitates as well as the formation and growth of GB phases. The HTA led to significant increases in creep life and ductility from the creation of additional obstacles for dislocation motion during creep.
3. Compact precipitates were formed in the low Ti/Al ratio alloys consisting of γ' precipitates surrounded by a γ'' shell. The faces of the γ' precipitates were entirely covered (edge-to-edge) with γ'' that was a few to tens of atomic layers thick in some alloys. The thickness of the γ'' increased with the additions of Nb or Ta.
4. The compact precipitates did not coarsen substantially after 10,000 h exposure at 700 °C, or after creep at 700 °C/483 MPa, and remained below 70 nm in equivalent diameter. The additions of Ta and Nb were associated with slightly reduced γ' coarsening likely due to a stronger compact structure (due to the thicker γ'' precipitates).
5. The long-term phase stability was impaired by the additions of Ta and Nb with significant growth of δ precipitates, particularly in the low Ti/Al ratio alloys. In the high Ti/Al ratio alloys, the growth of δ from the GBs was slowed down since the Ta and Nb additions mostly contributed to the increase in γ'' density within the grains.

Acknowledgements This work was performed in support of the US Department of Energy's Fossil Energy Crosscutting Technology Research Program. The authors would like to thank E.R. Argetsinger and J.A. Mendenhall for assistance in melting, C.D. Powell for mechanical testing, R.E. Chinn and C. McKaig for chemistry analysis, and M.B. Fortner for metallographic preparation.

References

1. L.E. Shoemaker, Alloys 625 and 725: Trends in properties and applications, 6th Int. Symp. Superalloys 718, 625, 706 Deriv. (2005) 409–418. https://doi.org/10.7449/2005/Superalloys_2005_409_418.

2. Special Metals Corporation, INCONEL alloy 725, 2005. SMC-066.
3. S. Mannan, F. Veltry, Time-Temperature-Transformation Diagram of Alloy 725, in: E.A. Loria (Ed.), Superalloys 718, 625, 706 Var. Deriv., TMS (The Minerals, Metals & Materials Society), 2001: pp. 345–356.
4. R. Cozar, A. Pineau, Influence of the Co/Ni Ratio on the y'and y" Precipitation in Fe-Ni-Co-Ta Alloys, Metall. Trans. 5 (1974) 2471–2472.
5. R. Cozar, A. Pineau, Morphology of γ′ and γ″ precipitates and thermal stability of inconel 718 type alloys, Metall. Trans. 4 (1973) 47–59. https://doi.org/10.1007/BF02649604.
6. S.K. Mannan, E.L. Hibner, B.C. Puckett, Physical Metallurgy of Alloys 718, 725, 725HS, and 925 for Service in Aggressive Corrosion Environments, Corrosion. (2003) 3125.
7. L. Viskari, Y. Cao, M. Norell, G. Sjöberg, K. Stiller, Grain boundary microstructure and fatigue crack growth in Allvac 718Plus superalloy, Mater. Sci. Eng. A. 528 (2011) 2570–2580. https://doi.org/10.1016/j.msea.2010.11.080.
8. M. Detrois, P.D. Jablonski, J.A. Hawk, The effect of η phase precipitates on the creep behavior of alloy 263 and variants, Mater. Sci. Eng. A. 799 (2021) 140337. https://doi.org/10.1016/j.msea.2020.140337.
9. A. Agnoli, M. Bernacki, R. Logé, J.M. Franchet, J. Laigo, N. Bozzolo, Understanding and Modeling of Grain Boundary Pinning in Inconel 718, in: E.S. Huron, R.C. Reed, M.C. Hardy, M.J. Mills, R.E. Montero, P.D. Portella, J. Telesman (Eds.), Superalloys 2012 12th Int. Symp. Superalloys, TMS (The Minerals, Metals & Materials Society), 2012: pp. 73–82. https://doi.org/10.1002/9781118516430.ch8.
10. E.J. Pickering, H. Mathur, A. Bhowmik, O.M.D.M. Messé, J.S. Barnard, M.C. Hardy, R. Krakow, K. Loehnert, H.J. Stone, C.M.F. Rae, Grain-boundary precipitation in Allvac 718Plus, Acta Mater. 60 (2012) 2757–2769. https://doi.org/10.1016/j.actamat.2012.01.042.
11. S. Antonov, M. Detrois, R.C. Helmink, S. Tin, Precipitate phase stability and compositional dependence on alloying additions in γ-γ′-δ-η Ni-base superalloys, J. Alloys Compd. 626 (2015) 76–86. https://doi.org/10.1016/j.jallcom.2014.11.155.
12. M.C. Hardy, M. Detrois, E.T. McDevitt, C. Argyrakis, V. Saraf, P.D. Jablonski, J.A. Hawk, R.C. Buckingham, H.S. Kitaguchi, S. Tin, Solving Recent Challenges for Wrought Ni-Base Superalloys, Metall. Mater. Trans. A. 51 (2020) 2626–2650. https://doi.org/10.1007/s11661-020-05773-6.
13. E.L. Hibner, H.W. Sizek, S.K. Mannan, Elevated temperature tensile and creep rupture properties of Inconel alloy 725, Superalloys 718, 625, 716 Deriv. (1997) 491–501.
14. M. Detrois, K.A. Rozman, P.D. Jablonski, J.A. Hawk, Compositional Design and Mechanical Properties of INCONEL® Alloy 725 Variants, in: E. Ott, X. Liu, J. Andersson, Z. Bi, K. Bockenstedt, I. Dempster, J. Groh, K. Heck, P.D. Jablonski, M. Kaplan, D. Nagahama, C. Sudbrack (Eds.), Proc. 9th Int. Symp. Superalloy 718 Deriv. Energy, Aerospace, Ind. Appl., Springer International Publishing, Pittsburgh, PA, 2018: pp. 421–437. https://doi.org/10.1007/978-3-319-89480-5_26.
15. P.D. Jablonski, J.A. Hawk, Homogenizing Advanced Alloys: Thermodynamic and Kinetic Simulations Followed by Experimental Results, J. Mater. Eng. Perform. 26 (2017) 4–13. https://doi.org/10.1007/s11665-016-2451-3.
16. ASTM International, ASTM E139-11(2018), Standard Test Methods for Conducting Creep, Creep-Rupture, and Stress-Rupture Tests of Metallic Materials, West Conshohocken, PA, 2018. https://doi.org/10.1520/E0139-11R18.
17. ASTM International, ASTM E112-13, Standard Test Methods for Determining Average Grain Size, West Conshohocken, PA, 2013. https://doi.org/10.1520/E0112-13.
18. Z. Xu, L. Jiang, J. Dong, Z. Li, X. Zhou, The effect of silicon on precipitation and decomposition behaviors of M6C carbide in a Ni–Mo–Cr superalloy, J. Alloys Compd. 620 (2015) 197–203. https://doi.org/10.1016/j.jallcom.2014.09.112.
19. M. Detrois, S. Antonov, K.A. Rozman, J.A. Hawk, P.D. Jablonski, Improved Creep and Tensile Properties of a Corrosion Resistant Ni-Based Superalloy Using High Temperature Aging and Nb/Ta Additions, Metall. Mater. Trans. A. 53 (2022) 2600–2613. https://doi.org/10.1007/s11661-022-06690-6.

20. M. Detrois, P.D. Jablonski, J.A. Hawk, NiCrMoNb Age Hardenable Alloy For Creep Resistant High Temperature Applications, And Methods of Making, US20220186343A1, 2022.
21. M. Detrois, P.D. Jablonski, J.A. Hawk, NiCrMoNb age hardenable alloy for creep-resistant high temperature applications, and methods of making, WO2022132928A1, 2022.
22. H.-Y. Bor, C.-N. Wei, H.T. Nguyen, A.-C. Yeh, C.-M. Kuo, Aging effects on the γ' and γ'' precipitates of Inconel 718 Superalloy, in: E.A. Ott, J.R. Groh, A. Banik, I. Dempster, T.P. Gabb, R. Helmink, X. Liu, A. Mitchell, G.P. Sjöberg, A. Wusatowska-Srnek (Eds.), 7th Int. Symp. Superalloy 718 Deriv., TMS (The Minerals, Metals & Materials Society), 2010: pp. 679–688.
23. A. Nicolaÿ, J. Franchet, N. Bozzolo, J. Cormier, Metallurgical Analysis of Direct Aging Effect on Tensile and Creep Properties in Inconel 718 Forgings, in: Tin S. et al. (Ed.), Superalloys 2020, Springer, Cham, 2020: pp. 559–569.
24. Z. Bi, J. Dong, M. Zhang, L. Zheng, X. Xie, Mechanism of α-Cr precipitation and crystallographic relationships between α-Cr and δ phases in Inconel 718 alloy after long-time thermal exposure, Int. J. Miner. Metall. Mater. 17 (2010) 312–317. https://doi.org/10.1007/s12613-010-0310-z.
25. J.F. Radavich, Effect of Alpha Chromium on Long Time Behavior of Alloy 718, in: Superalloys 718, 625, 706 Var. Deriv., TMS, 1997: pp. 409–415. https://doi.org/10.7449/1997/Superalloys_1997_409_415.
26. L.M. Suave, J. Cormier, P. Villechaise, A. Soula, Z. Hervier, D. Bertheau, J. Laigo, Microstructural evolutions during thermal aging of alloy 625: Impact of temperature and forming process, Metall. Mater. Trans. A 45 (2014) 2963–2982. https://doi.org/10.1007/s11661-014-2256-7.

Investigating Deformation Mechanisms in a Creep-Deformed 718-Variant Superalloy

Semanti Mukhopadhyay, Hariharan Sriram, Rich DiDomizio, Andrew J. Detor, Robert W. Hayes, Yunzhi Wang, and Michael J. Mills

Abstract Improving the efficiency of a land-based industrial gas turbine ultimately relies on novel alloy development for the turbine wheel. However, this alloy development task is challenging because it necessitates higher temperature capabilities along with phase stability during low cooling rate processing of a full-scale wheel. These challenges make 718-based variant alloys an attractive choice because of their superior thermal stability. However, to develop these novel alloys, their deformation behavior must also be accounted for. Thus, in the present work, we investigate the microstructure and creep deformation behavior of a novel 718-variant alloy. Detailed microstructural characterization reveals that the phase fraction of the γ'' phase in the variant alloy is much lower than 718. In addition, the presence of Mo causes detrimental grain boundary precipitation which leads to final failure during tensile creep deformation. The variant alloy accumulates creep strain faster than 718, ultimately fracturing at 0.6% strain. Finally, a detailed characterization of the deformed variant alloy reveals extensive microtwinning.

Keywords 718-variant superalloy · Creep deformation behavior · Microstructural characterization

S. Mukhopadhyay (✉) · H. Sriram · Y. Wang · M. J. Mills (✉)
Department of Materials Science and Engineering, The Ohio State University, Columbus, OH, USA
e-mail: mukhopadhyay.20@buckeyemail.osu.edu

M. J. Mills
e-mail: mills.108@osu.edu

S. Mukhopadhyay · M. J. Mills
Center for Electron Microscopy & Analysis (CEMAS), The Ohio State University, Columbus, OH, USA

R. DiDomizio · A. J. Detor
GE Global Research Center, Niskayuna, NY, USA

R. W. Hayes
Metals Technology Inc. (MTI) Northridge, Northridge, CA, USA

E. A. Ott et al. (eds.), *Proceedings of the 10th International Symposium on Superalloy 718 and Derivatives*, The Minerals, Metals & Materials Series,
https://doi.org/10.1007/978-3-031-27447-3_11

Introduction

The demand for higher efficiency of power generation is one of the major challenges facing the world today. This is because higher efficiency translates into reduced fuel consumption, cheaper power generation, and lower greenhouse gas emissions. Thus, unsurprisingly, this has resulted in a worldwide push to improve the efficiency of all the components involved in the power generation cycle.

A component of most power generation cycles is the industrial land-based gas turbine. Improving the efficiency of such a land-based gas turbine necessitates novel materials which can withstand higher operating temperatures. However, this alloy design problem is further complicated by the size of the turbine wheels. Due to their large sizes and variable thicknesses, full-scale turbine wheel forgings experience variable slow cooling rates, typically between 6 and 28 °C/min. Thus, the alloy development of the novel turbine wheel materials must ensure phase stability during slow cooling rates apart from addressing the challenge of higher operating temperatures. Recent work has suggested that, potentially, alloy 718 could be modified to meet both these requirements [1, 2].

Alloy 718 is a Ni-based superalloy strengthened by two precipitate phases $-\gamma'$ and γ'', in a Ni-solid solution matrix (γ) [3]. The γ' precipitates are based on the intermetallic phase Ni_3Al, exhibiting an $L1_2$ crystal structure. Next, the γ'' phase, which is the major strengthening phase in the alloy, exhibits a $D0_{22}$ crystal structure with a structural formula–Ni_3Nb. The γ' and γ'' precipitates often exist in a combined morphology, or as "coprecipitates", to minimize the overall interfacial energies of the $\gamma|\gamma'|\gamma''$ system [4, 5].

Many articles, old and new, have suggested that one such combined morphology, known as the "compact" morphology, exhibits very high thermal stability [2, 5–10]. Interestingly, the compact morphology is also reported to be highly stable in a $\gamma|\gamma'|\gamma''$ Al-Cu-Nb-Zr system [11]. In the compact morphology, γ'' discs coat all the {001} faces of the γ' cuboids so that γ'' prevents coarsening of the γ' core [6]. In addition, the $\gamma'' \rightarrow \delta$ transformation also becomes more sluggish, resulting in superior phase and morphological stabilities at high temperatures [10, 12]. Thus, their superior thermal stability makes these 718-variant alloys an interesting choice for next-generation higher temperature wheels.

However, slow-cooled 718-variant alloys exhibit poor creep behavior when compared to traditionally processed 718 [1, 12]. This outcome is nonintuitive because the superior thermal resistance of compact coprecipitates in the variant alloys is expected to improve their creep performance [13]. Instead, the variant alloys deform with a faster rate of creep deformation and decreased rupture lives [1]. Note that because the traditional processing of 718 involves oil quenching, the comparison between traditionally processed 718 and the slow-cooled 718-variant alloys might not be appropriate. This is because the much slower cooling rates imposed on the variant alloys are expected to influence the microstructure, and therefore the creep behavior of the alloy. However, currently, the creep behavior of slow-cooled 718 is not known. Further, it remains to be seen if the poor creep performance of the variant alloys is a result of a difference in the active deformation mechanisms, or fully attributable to slow-cooling which will be shown to induce grain boundary precipitates.

Thus, we investigated the microstructure and tensile creep deformation of a novel 718-based variant alloy. To increase γ'' phase fraction in the compact coprecipitate-strengthened alloy, we designed its composition with much higher Nb and 0% Ti, while also maintaining the (Al + Ti)/Nb ratio close to 1. Then, we compared the creep behavior and deformation mechanisms of the developed variant alloy with traditionally processed 718. For the current study, we subjected a slow-cooled variant alloy and a traditionally processed alloy 718 to uniaxial tensile creep tests at 649 °C (1200 °-F) and 689 MPa (100 ksi). Then, we characterized the deformed microstructures of the deformed variant alloy using various characterization techniques including diffraction-contrast scanning transmission electron microscopy (DC-STEM).

Methods

Materials and Heat Treatments

The nominal compositions of alloy 718 and the variant alloy (718–50) are provided in Table 1. Ingots of the variant alloy were vacuum induction melted and forged at GE Global Research Center, NY. Samples of traditionally processed 718 were obtained from Metals Technology Inc.

The "traditional" processing route for 718 involved a solution treatment above the δ-solvus temperature followed by oil quenching. Following the solution treatment, the sample was double-aged—first at 720 °C for 8 h, then cooled at 1.1 °C/min to 649 °C and aged for 8 h. The variant alloy was solution treated at 982 °C and cooled at 6 °C/min to room temperature. After this solution treatment, the samples also received a double-aging treatment—first at 720 °C for 8 h, slow cooled (1.1 °C/min) to 649 °C, and aged for 14 h.

Table 1 Nominal composition of alloy 718 and the variant alloy used in this study

Alloy name	Composition (Wt. %)							
	Ni	Cr	Fe	Al	Ti	Nb	Mo	C
718	52.5	19	18.9	0.5	0.9	5.13	3.05	0.02
Variant alloy	54.4	17.3	17.5	1.9	–	6.2	2.8	0.02

Creep Testing

Blanks for tensile creep testing at 649 °C (1200°F) and 689 MPa (100 ksi) were machined from both the aged samples listed in Table 1. Detailed sample preparation and test setup for creep testing are provided in Mukhopadhyay et al. [1].

Volume Fraction Measurement

Samples for microstructural analyses were prepared via standard metallographic techniques and swab etched with a dilute γ' etchant. This dilute etchant was prepared by mixing 50% (by volume) of glycerin with 50% of etching solution which comprised of 60% (by volume) lactic acid, 37% nitric acid, and 3% hydrofluoric acid.

Then, orientation mapping was carried out in a Thermo Fischer Scientific Apreo Scanning Electron Microscope (SEM) using electron backscatter diffraction (EBSD). For this purpose, a voltage of 20 kV and a current of 6.4 nA were utilized. Orientation mapping allowed us to select appropriately oriented grains. From {001} oriented grains, backscatter electron images were obtained at several magnifications. These electron micrographs were used for overall coprecipitate volume fraction measurements using the MIPAR™ software [14]. Note that this method only provides an estimate of the overall $\gamma'|\gamma''$ coprecipitate volume fraction because the etchant removes both γ' and γ'' from the sample surface.

Scanning Transmission Electron Microscopy (STEM)

Regions of interest for detailed analyses were selected in both aged and creep-deformed samples using EBSD and prepared using a ThermoFischer Scientific Helios 600 Focused Ion Beam (FIB). First, a Pt cap was deposited to protect the region of interest using Ga ion beam at 30 kV and 0.26 nA. Then, a deep trench was created at 30 kV and 21 nA. Following this, the trenched foil was undercut at 2.6 nA and attached to a Cu omniprobe grid. The foil was then thinned at 30 kV using several ion beam currents and cleaned at 5 kV and 71 nA until electron transparent.

All STEM-based experiments, including energy dispersive X-ray spectroscopy (EDS) were carried out at 300 kV in a ThermoFischer Scientific Titan^3 60–300 STEM with super-X EDS detectors. While a current of 1 nA was used to ensure sufficient counts during EDS experiments, 60 pA was used for imaging and diffraction-contrast STEM (or, DC-STEM) experiments. Detailed descriptions of the DC-STEM technique are provided by Miao et al. [15] and Phillips et al. [16].

Results and Discussion

Microstructure of the Developed Variant Alloy

Figure 1 shows an SEM micrograph and a STEM-EDS map (inset) of the developed "variant" alloy. Based on image analysis carried out on several SEM micrographs, the overall coprecipitate volume fraction was estimated to be about 28%. In addition, the average diameter of the coprecipitates was estimated to be ~200 nm. These measurements agree with further estimates carried out on STEM-EDS maps (for example: in inset, Fig. 1). Because Al and Nb preferentially partition to γ' and γ'' phases, respectively, STEM EDS mapping can highlight two phases by showing regions of high Al (γ') and high Nb (γ''). However, the mapping needs to be carried out in accordance with orientation relationships between $\gamma|\gamma'|\gamma''$. Thus, both the EDS map and the SEM micrograph in Fig. 1 were obtained along a <001>-type zone axis. Further, note that the coprecipitate morphology contains γ'' discs on all of its {001}-type faces. However, a <001>-type zone axis contains only two of the γ'' variants in an edge-on condition. The third γ'' variant with disc axis pointing out (or into) the plane of paper/screen, is expected to not be present in the FIB foil. This is because the coprecipitate size is larger than the FIB foil thickness, implying that the third variant must have been milled away during the thinning/cleaning steps involved in a FIB foil preparation. This is further demonstrated by the lack of a high Nb region almost in the middle of the γ' face in Fig. 1(inset). An absence of a high Nb region in the middle of the γ' face demonstrates that the third variant of γ'' does not appear in this FIB foil because although the third γ'' variant would not be viewed edge-on, a faint high Nb region may still be expected in the EDS map.

As pointed out above, this variant alloy contains a coprecipitate volume fraction of 28%. Thus, on a first glance, the developed variant alloy appears to exhibit potential

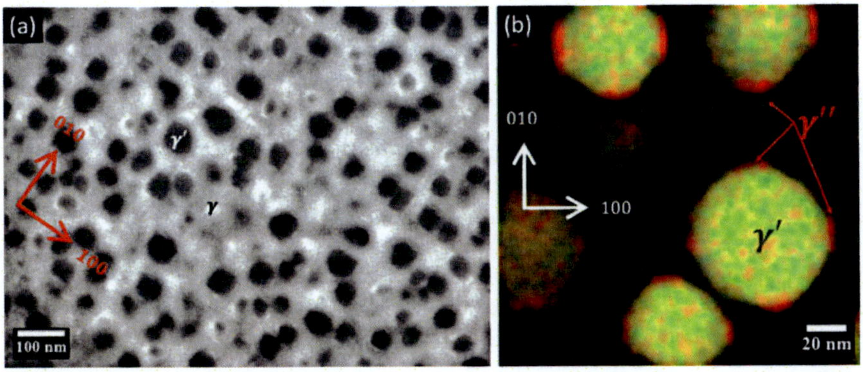

Fig. 1 Microstructure of the developed alloy analyzed via **a** SEM and **b** STEM-EDS mapping. In figure (**b**) the green color shows regions of Al segregation (γ' phase) and the red color shows the regions of Nb segregation (γ'' phase)

as a next-generation gas-turbine wheel material. However, a closer look shows that even though the volume fraction of coprecipitates is comparable to 718, the γ'' phase fraction is very low. Note that the estimation of the volume fraction of γ'' is difficult from either of the micrographs. First, in the SEM micrograph, the etching process removes the γ'' phase either partially or entirely, making any phase-fraction calculations impossible. Second, the EDS map samples a very small region from one grain of the alloy, and the map shows a projection through the thickness of the foil. This makes such an estimation cumbersome and unreliable. However, because the γ' and γ'' precipitates exist as compact coprecipitates, the phase fraction estimations can be carried out easily based on the geometry of this morphology.

Based on the STEM EDS map in Fig. 1(inset), the γ' precipitate may be considered a sphere with average diameter $d_{\gamma'}$. The γ'' precipitates, on the other hand, are discs with diameter $d_{\gamma''}$ and thickness $t_{\gamma''}$. Thus, by definition, average coprecipitate size is related to γ' diameter and γ'' thickness as $d_{coprecipitate} = d_{\gamma'} + 2\times\, t_{\gamma''}$.

Now, consider a volume element such that 28% of its volume is occupied by an average compact coprecipitate from this alloy. Because of the relatively large inter-particle spacing in the alloy, such a volume element without any other partial or full coprecipitate is realistic. In that case, average fraction $\frac{\gamma''}{\gamma'}$ may be calculated as:

$$\frac{Volume\ Fraction_{\gamma''}}{Volume\ Fraction_{\gamma'}} = \frac{6 \times Average\ volume\ of\ one\ \gamma''\ disc}{Average\ volume\ of\ one\ \gamma'\ sphere}$$

$$= \frac{6 \times \pi \times (\frac{d_{\gamma''}}{2})^2 \times t_{\gamma''}}{\frac{4}{3} \times \pi \times (\frac{d_{\gamma'}}{2})^3} = \frac{9 \times (d_{\gamma''})^2 \times t_{\gamma''}}{(d_{\gamma'})^3}$$

In the variant alloy, $d_{\gamma'} = 130 \pm 20\, nm$, $d_{\gamma''} = 50 \pm 10\, nm$, and $t_{\gamma''} = 20 \pm 40\, nm$. Thus, $\frac{\gamma''}{\gamma'} = 0.2$. Now, $Volume\ Fraction_{\gamma''} + Volume\ Fraction_{\gamma'} = Total\ Volume\ fraction$

Thus, $Volume\ Fraction_{\gamma'} = \frac{Total\ Volume\ fraction}{1.2} = 23.3\%$, and $Volume\ Fraction_{\gamma''} = \frac{Total\ Volume\ fraction}{6} = 4.7\%$

Thus, although the volume fraction of coprecipitates in the variant alloy is high, it comprises less than 5% of γ'' phase. In contrast, 718 contains between 12 and 18% γ'' with a $\frac{\gamma''}{\gamma'}$ the ratio of ~3 [5].

Thus, the microstructural characterization reveals three major things about the variant alloy. First, average compact coprecipitates are larger in size than monolithic, duplex or triplex coprecipitates in 718 [5], and second, the volume fraction of the major strengthening phase (γ'') is atleast 3 times less than in 718. Finally, based on the size of compact coprecipitates and the volume fractions measured, variant alloy is expected to exhibit large inter-particle spacing. All three factors are expected to significantly influence the deformation behavior of the developed alloy.

Creep Deformation in the Variant Alloy and Traditionally Processed 718

Figure 2 presents the creep data obtained during the tensile creep tests carried out at 649 °C (1200°F) and 689 MPa (100 ksi) in traditionally processed 718 and the newly developed variant alloy. Clearly, the variant alloy exhibits a faster rate of creep deformation ($1.6 \times 10^{-5} s^{-1}$). The variant alloy accumulated 0.6% strain in 10 h, after which the alloy fractured. In contrast, the 718-sample reached 0.5% strain in 45 h at an average strain rate $5.0 \times 10^{-6} s^{-1}$. In addition, the traditionally processed 718 samples did not fracture at this stage.

Thus, the creep data obtained for the variant alloy reveals two major differences between the variant alloy and the traditionally processed 718. First, at similar accumulated strains, the variant alloy fractured while the traditional 718 samples did not. This result is supported by our previous study on another slow-cooled 718-variant alloy (718–27) [1], in which we reported extensive intergranular failure during tensile creep deformation under similar testing conditions. Next, the second difference between the two alloys pertains to their strain rates. Such a difference in strain rate suggests the possibility of different deformation mechanisms in the two alloy systems.

In the creep-deformed variant alloy, we observed intergranular failure. Further, the deformed alloy samples revealed cavitation adjacent to σ phase precipitates (Fig. 3). This is consistent with our previously reported observation on the role of grain boundary σ phase on extensive intergranular failure during tensile creep experiments [1]. The σ precipitates are highly detrimental to the ductility and fracture toughness of any high-temperature superalloy because the incoherent $\sigma|\gamma$ interface acts as a

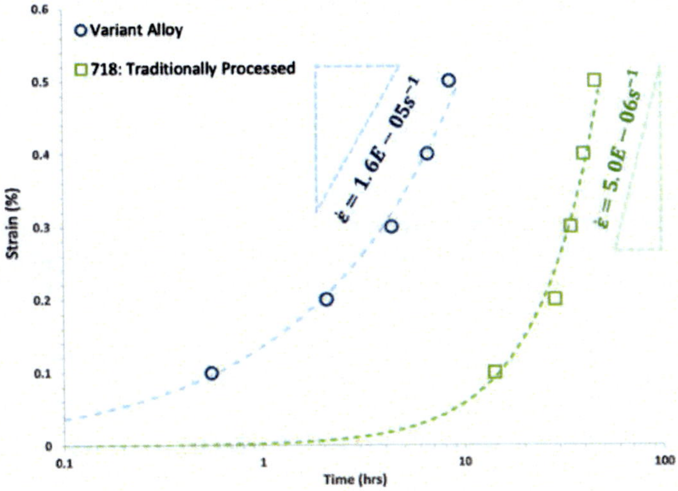

Fig. 2 Creep data for the newly developed variant alloy and traditionally processed 718 subjected to tensile creep deformation at 649 °C (1200 °F) and 689 MPa (100 ksi)

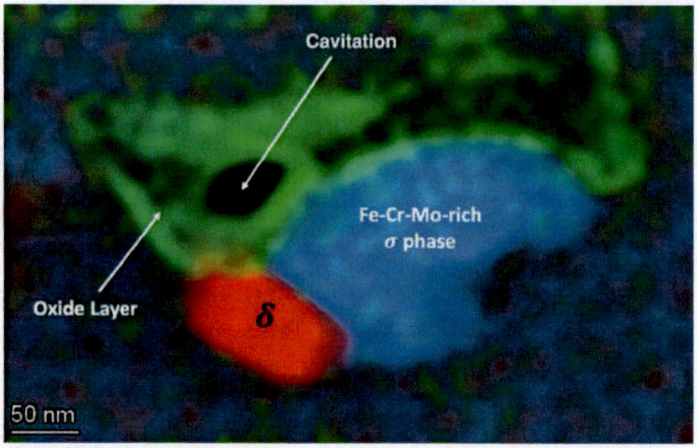

Fig. 3 EDS map showing grain boundary cavitation in creep-deformed variant alloy. In this EDS map, red shows Nb enrichment, green shows O enrichment, and blue shows Fe+ Cr+ Mo enrichment

site of high stress concentration leading to cavitation. Under further external loading, such cavitation then progresses to crack nucleation and final failure.

Our results are in agreement with a previous study carried out on differences between slow-cooled 718 and 706 reported that the slow-cooled 718 samples exhibited Mo-rich C14-type Laves phase precipitates which also caused extensive intergranular failure during tensile creep tests [17]. In contrast, a similar heat treatment with a slow-cooling rate did not induce detrimental grain boundary precipitation in 706 [17]. This argument is supported also by other coarsening studies carried out on 718 where Mo-rich precipitates were also observed at the high-angle grain boundaries [18]. Moreover, the role of Mo in causing such detrimental grain boundaries is further elucidated by our CALPHAD-based thermodynamic modeling results (Fig. 4). Figure 4 compares the equilibrium phase fraction of the stable phases as a function of temperature in our variant alloy, in 718, and in a Mo-free variant alloy. Note that to design the Mo-free variant alloy, the Mo content in our variant alloy was replaced by Ni. Unsurprisingly, the impact of the absence of Mo is insignificant on γ, γ' precipitates, and the thermodynamically stable δ phase. On the contrary, the Mo-free alloy does not exhibit σ phase, both 718 and our variant alloys do.

Thus, the thermodynamic calculations suggest that under slow cooling rates, which may mimic quasi-static equilibrium-like conditions, any Mo containing alloy (like our variant alloy, 718–27, or 718) will inadvertently cause σ phase precipitation. This result is consistent with studies on 706, which does not contain any Mo. Because 706 does not contain Mo, there is no report of brittle σ (or C14 Laves) phase formation even after very slow cooling rates. For example, 706 is regularly subjected to a ~0.9 °C/min cooling rate when it is cooled after its solution treatment to a "stabilization" hold [12, 19]. However, this heat treatment is apparently beneficial because it leads to the formation of a grain boundary η phase. Moreover, the authors are not

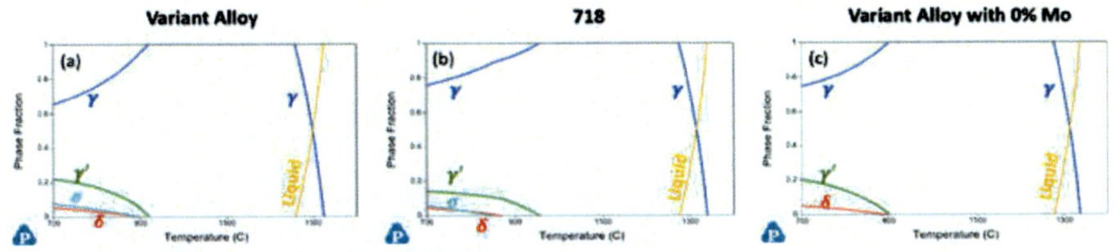

Fig. 4 CALPHAD-based thermodynamic phase calculations in **a** the newly developed variant alloy (Alloy-50), **b** 718, and **c** variant alloy with 0% Mo showing that σ phase stability is controlled by Mo content of the alloy

aware of any articles suggesting detrimental phase formation at the grain boundaries in 706 after such slow cooling rates.

Thus, when our variant alloy was subjected to the slow cooling rates described in the methods section, it must have inadvertently caused σ phase precipitates at the high-angle grain boundaries. These precipitates ultimately caused final failure in the developed variant alloy at 0.6% strain. On the contrary, because the 718 sample received its usual faster cooling rates (during oil quench from the solution temperature), σ phase precipitation did not occur owing to kinetic factors. In that sense, it might not be wise to compare our variant alloy to a traditionally processed 718 because of the kinetics of phase transformations occurring in the systems during the two different processing routes. At this stage, it is not known how a similarly processed 718 sample would deform under similar tensile creep conditions.

However, at the "intermediate" creep test temperature (649 °C or 1200 °F), grain boundary slipping/sliding is not expected to be the dominant deformation mechanism. Usually, grain boundary sliding is the dominant deformation mechanism at higher temperatures. Thus, while we expect some of the faster creep rate to be caused by the localized deformation and creep cavitation near σ phase precipitates, some of it might also be controlled by the intragranular deformation modes. Therefore, the creep test data suggests that both the intergranular and intragranular deformation mechanisms are different in the variant alloy when compared with the traditionally processed 718 considered in this investigation.

Characterization of Microtwinning in Creep-Deformed Variant Alloy

Figure 5 shows a [101] zone axis HAADF-STEM micrograph showing a microtwin that extends across the entire grain parallel to its $\Sigma = 3$ coherent annealing twin boundary (CTB). In Fig. 5(a), we will refer to the region to the right of CTB as "parent" and the left side as "twin" to facilitate the discussion.

The beam direction for the parent grain is about 1° (16.9 mrad) away from [928] and the loading direction has a rotation of about 36° counterclockwise to the vertical

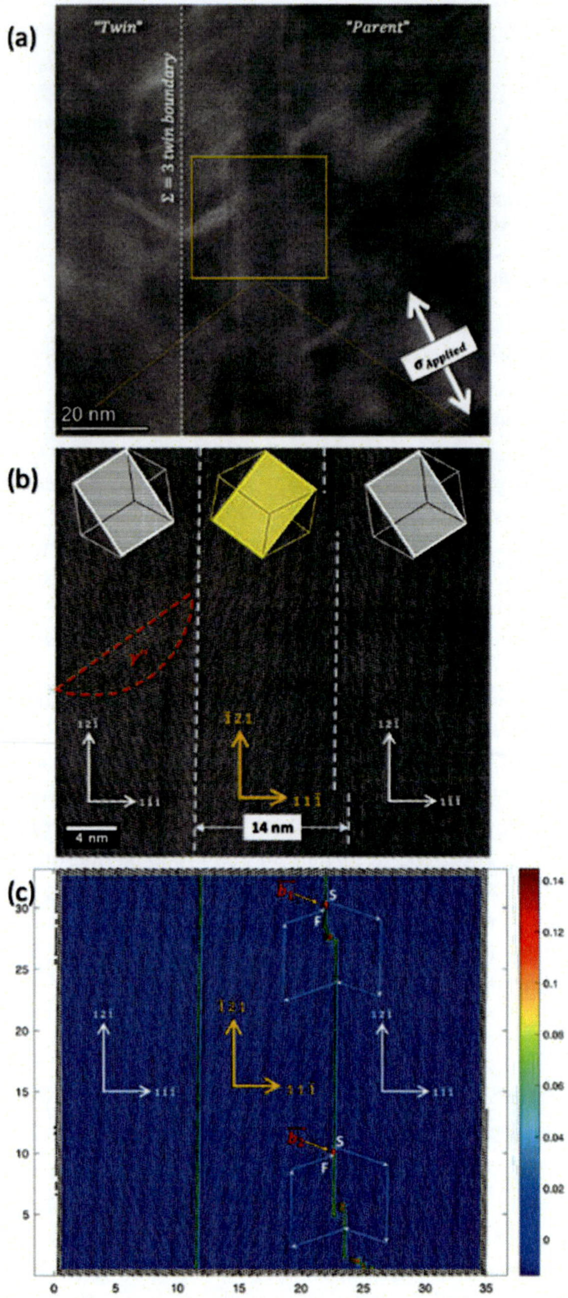

Fig. 5 Analysis creep deformed variant alloy exhibiting extensive microtwinning: **a** of HAADF-STEM micrograph showing microtwin observation parallel to an annealing twin boundary, **b** atomic resolution HAADF-STEM micrograph of the same microtwin showing a thickening process, and **c** COS map of figure (**b**) detailing the twin-thickening process via a ledge mechanism

axis of the image. Thus, using trace analysis, the annealing twin boundary plane was indexed as $(1\overline{1}\overline{1})$. Thus, the microtwin plane, which is parallel to the annealing twin boundary, is also $(1\overline{1}\overline{1})$. Interestingly, the Schmid factor for one of the possible slip systems ($(1\overline{1}\overline{1})[\overline{1}12]$) containing the twin boundary plane is 0.71. This indicates the likelihood of the $\frac{1}{6}[\overline{1}12]$ partial dislocation (or a $\frac{1}{3}[\overline{1}12]$ superpartial) being involved in the microtwinning process. As explained in Viswanathan et al. [20], microtwin formation may be attributed to consecutive partial dislocation activity on three adjacent {111}-type planes. Thus, the observed microtwin in the deformed variant alloy might have been created by the successive passage of the $\frac{1}{6}[\overline{1}12]$ Shockley partials on three adjacent $(1\overline{1}\overline{1})$ planes. Interestingly, the "twin" side of the annealing twin boundary (beam direction $[\overline{1710}]$) also exhibits a similar microtwin on its $(\overline{1}\overline{1}1)_{twin}$ plane. In this case, the slip system $(\overline{1}\overline{1}1)[2\overline{1}1]$ has a similarly high Schmid factor of 0.5.

Figure 5b and c elucidate the microtwin observed on the "parent" side of the annealing twin boundary. Figure 4c shows a Center of Symmetry (COS) map which highlights any deviation from the ABCABCABC stacking expected in an FCC system [21]. At the annealing twin and microtwin boundaries, the stacking locally resembles the HCP stacking of ABABAB. This deviation from the ABCABC stacking at the microtwin boundaries can be observed in the COS map in Fig. 4c. When the local atomic stacking near the microtwin boundary is considered, the stacking changes from ABCABC to the left of the boundary to ABA at the boundary and subsequently to ACBACB on the right side of the microtwin boundary. Thus, the map highlights only the boundary because both ABC and ACB stacking pertain to the quintessential FCC stacking.

As shown in Fig. 5b, the microtwin at the top is about 11.7 nm thick. On traversing downwards, the micrograph exhibits a several $(1\overline{1}\overline{1})$ layer-thick step, after which the microtwin thickens to about 12.11 nm. Further, a second step is observed closer to the bottom of the micrograph, which causes the microtwin to further thicken to 14 nm. As Fig. 5c shows, Frank's circuit analysis of both the steps indicates closure failure. For this Frank circuit analysis, we followed the Right-Hand Finish to Start (RHFS) rule, according to which we constructed a right-handed circuit and then evaluated the closure failure from the finish point of the circuit to its start point. According to this analysis, the Burgers vector for both the steps was found to be 6 units along $[12\overline{1}]$.

Because the microtwin continues thickening until it terminates at the grain boundary, the primary deformation mechanism of this grain (both twin and the parent) was found to be microtwinning. Incidentally, microtwinning is one of the most common mechanisms by which γ' strengthened disc alloys are reported to accommodate creep deformation under similar intermediate temperatures [20, 22–24]. On the other hand, creep deformation mechanism in Alloy 718 has been reported to be based on the glide or climb of dislocations [25–28]. Under slightly different creep test conditions, however, alloy 718 and another $\gamma'|\gamma''|\gamma$ superalloy (G4169) are also reported to exhibit deformation twinning [29, 30]. In contrast to our observation of microtwins spanning the entire grain, their post-mortem analyses revealed thinner deformation twins blocked by δ phase precipitates [29]. Overall, the observation of extensive microtwinning indicates a similarity between our variant alloy and many other γ' strengthened alloys instead of traditionally processed 718. This finding is consistent with our microstructural analyses in Sect. 3.1, because the γ' phase fraction in the variant alloy is five times higher than γ'' in the alloy (23.3%).

Conclusions

The promise of superior phase stability makes compact-coprecipitate strengthened 718-based alloys an exciting candidate for high-temperature gas turbine wheel applications. The present work was an attempt to investigate the differences in microstructure and the creep behavior between one such alloy and its traditionally processed parent alloy.

(1) When compared to 718, the microstructure of a newly developed alloy exhibited very low $\frac{\gamma''}{\gamma'}$ phase fraction and, consequently, a much lower γ'' volume fraction. Although the variant alloy was developed with much higher Nb and 0% Ti to promote higher γ'' content, the final microstructure comprised of less than 5% γ'' phase. In addition, the average precipitate size (120 nm) and large interparticle spacing were also in contrast to the much finer $\gamma'|\gamma''$ microstructure reported for 718. Such a drastically different microstructure was found to greatly influence creep behavior of the developed variant alloy.

(2) Next, the creep behavior of the alloy was found to differ from 718 in two ways.

   a. First, variant alloy fractured at about 0.6% strain, while 718 did not fail at similar strain levels. This was found to be the result of the slow cooling rates used in the processing of the variant alloy which σ phase precipitation at the grain boundaries. Because of its incoherent $\sigma|\gamma$ interface, the precipitates act as sites for stress concentration leading, ultimately, to cracking. Based on our thermodynamic calculations, precipitation of the σ phase is unavoidable in any Mo-containing alloy (including 718) when processed via slow cooling rates.

   b. Next, the creep deformed microstructure showed evidence for extensive microtwinning which spanned the entire grain. The microtwinning mode of deformation indicates a similarity between our developed variant alloy with many other γ'-strengthened alloys.

Ultimately, this work shows that although the current compositions and aging treatments of variant alloys are sub-optimum, the compact coprecipitate-strengthened variant alloys exhibit tremendous potential due to their higher thermal stability as well as an additional strengthening effect. This indicates that the compositional space of such variant alloys may enable the design of the next generation of high-temperature turbine wheel materials, and need to be further explored.

Acknowledgements Electron microscopy was performed at the Center for Electron Microscopy and Analysis (CEMAS) at The Ohio State University. This material is based upon work supported by the Department of Energy under Award No. DE-FE0031278. Additional support for S.M., Y.W., and M.J.M. was provided by the NSF DMREF program DMR—1922239.

References

1. S. Mukhopadhyay *et al.*, "Creep Behavior of Compact γ'-γ'' Coprecipitation Strengthened IN718-Variant Superalloy," *Metals*, vol. 11, no. 12, Art. no. 12, Dec. 2021, https://doi.org/10.3390/met11121897.
2. "Enabling Large Superalloy Parts Using Compact Coprecipitation of γ' and γ'' | SpringerLink." https://link-springer-com.proxy.lib.ohio-state.edu/article/. https://doi.org/10.1007/s11661-017-4356-7 (accessed Aug. 15, 2022).
3. M. J. Donachie and S. J. Donachie, *Superalloys: A Technical Guide, 2nd Edition.* ASM International, 2002.
4. J. M. Oblak, D. F. Paulonis, and D. S. Duvall, "Coherency Strengthening in Ni Base Alloys Hardened by D022 Precipitates," *Metall. Trans.*, p. 11.
5. R. Cozar and A. Pineau, "Morphology of y' and y" precipitates and thermal stability of inconel 718 type alloys," *Metall. Trans.*, vol. 4, no. 1, pp. 47–59, Jan. 1973, https://doi.org/10.1007/BF02649604.
6. R. Shi et al., "Growth behavior of γ'/γ'' coprecipitates in Ni-Base superalloys," *Acta Mater.*, vol. 164, pp. 220–236, Feb. 2019, https://doi.org/10.1016/j.actamat.2018.10.028.
7. Q. Guo et al., "A new type-γ'/γ'' coprecipitation behavior and its evolution mechanism in wrought Ni-based ATI 718Plus superalloy," *J. Mater. Sci. Technol.*, vol. 119, pp. 98–110, Aug. 2022, https://doi.org/10.1016/j.jmst.2021.12.033
8. J. He, G. Han, S. Fukuyama, and K. Yokogawa, "Interfaces in a modified Inconel 718 with compact precipitates," *Acta Mater.*, vol. 46, no. 1, pp. 215–223, Dec. 1998, https://doi.org/10.1016/S1359-6454(97)00221-8.
9. Z. Qiao, C. Li, H. Zhang, H. Liang, Y. Liu, and Y. Zhang, "Evaluation on elevated-temperature stability of modified 718-type alloys with varied phase configurations," *Int. J. Miner. Metall. Mater.*, vol. 27, no. 8, pp. 1123–1132, Aug. 2020, https://doi.org/10.1007/s12613-019-1949-8.
10. M. Dang *et al.*, "Formation Mechanism and Thermal Stability of Γ' and Γ." Rochester, NY, Jun. 27, 2022, https://doi.org/10.2139/ssrn.4147546.
11. S. Kumar Makineni *et al.*, "Enhancing elevated temperature strength of copper containing aluminium alloys by forming L12 Al3Zr precipitates and nucleating θ'' precipitates on them," *Sci. Rep.*, vol. 7, no. 1, Art. no. 1, Sep. 2017, https://doi.org/10.1038/s41598-017-11540-2.
12. T. Shibata, T. Takahashi, Y. Shudo, M. Kusuhashi, J. Taira, and T. Ishiguro, "Effect of Cooling Rate from Solution Treatment on Precipitation Behavior and Mechanical Properties of Alloy 706," in *Superalloys 718, 625, 706 and Various Derivatives (1997)*, 1997, pp. 379–388, https://doi.org/10.7449/1997/Superalloys_1997_379_388.
13. D. Hull and D. J. Bacon, *Introduction to Dislocations*. Butterworth-Heinemann, 2001.
14. J. M. Sosa, D. E. Huber, B. Welk, and H. L. Fraser, "Development and application of MIPARTM: a novel software package for two- and three-dimensional microstructural characterization," Integrating Mater. Manuf. Innov., vol. 3, no. 1, pp. 123–140, Dec. 2014, https://doi.org/10.1186/2193-9772-3-10.
15. J. Miao *et al.*, "STEM Characterization of the Deformation Substructure of a NiCoCr Equiatomic Solid Solution Alloy," Microsc. Microanal., vol. 23, no. S1, pp. 752–753, Jul. 2017, https://doi.org/10.1017/S1431927617004421.
16. P. J. Phillips, M. C. Brandes, M. J. Mills, and M. De Graef, "Diffraction contrast STEM of dislocations: Imaging and simulations," Ultramicroscopy, vol. 111, no. 9–10, pp. 1483–1487, Aug. 2011, https://doi.org/10.1016/j.ultramic.2011.07.001.

17. T. Takahashi, T. Ishiguro, K. Orita, J. Taira, T. Shibata, and S. Nakata, "Effects of Grain Boundary Precipitation on Creep Rupture Properties of Alloys 706 and 718 Turbine Disk Forgings," in *Superalloys 718, 625, 706 and Various Derivatives (1994)*, 1994, pp. 557–565, https://doi.org/10.7449/1994/Superalloys_1994_557_565.
18. D. Jianxin, X. Xishan, and Z. Shouhua, "Coarsening behavior of γ'' precipitates in modified inconel 718 superalloy," *Scr. Metall. Mater.*, vol. 33, no. 12, pp. 1933–1940, Dec. 1995, https://doi.org/10.1016/0956-716X(95)00446-3.
19. C. Kim, J. Park, H.-U. Hong, J. Gu, and Y. Song, "Role of stabilization heat treatment inducing γ'-γ'' co-precipitates and grain boundary η phase on tensile and creep behaviors of Inconel 706," *J. Alloys Compd.*, vol. 900, p. 163479, Apr. 2022, https://doi.org/10.1016/j.jallcom.2021.163479.
20. G. B. Viswanathan, S. Karthikeyan, P. M. Sarosi, R. R. Unocic, and M. J. Mills, "Microtwinning during intermediate temperature creep of polycrystalline Ni-based superalloys: mechanisms and modelling," *Philos. Mag.*, vol. 86, no. 29–31, pp. 4823–4840, Oct. 2006, https://doi.org/10.1080/14786430600767750.
21. V. A. Vorontsov, L. Kovarik, M. J. Mills, and C. M. F. Rae, "High-resolution electron microscopy of dislocation ribbons in a CMSX-4 superalloy single crystal," *Acta Mater.*, vol. 60, no. 12, pp. 4866–4878, Jul. 2012, https://doi.org/10.1016/j.actamat.2012.05.014.
22. R. W. Kozar, A. Suzuki, W. W. Milligan, J. J. Schirra, M. F. Savage, and T. M. Pollock, "Strengthening Mechanisms in Polycrystalline Multimodal Nickel-Base Superalloys," *Metall. Mater. Trans. A*, vol. 40, no. 7, pp. 1588–1603, Jul. 2009, https://doi.org/10.1007/s11661-009-9858-5.
23. D. Locq, P. Caron, S. Raujol, F. Pettinari-Sturmel, A. Coujou, and N. Clement, "On the Role of Tertiary γ';' Precipitates in the Creep Behaviour at 700 °C of a PM Disk Superalloy," in *Superalloys 2004 (Tenth International Symposium)*, 2004, pp. 179–187, https://doi.org/10.7449/2004/Superalloys_2004_179_187.
24. R. F. Decker, "The evolution of wrought age-hardenable superalloys," *JOM*, vol. 58, no. 9, pp. 32–36, Sep. 2006, https://doi.org/10.1007/s11837-006-0079-8.
25. M. C. Chaturvedi and Y. Han, "Strengthening mechanisms in Inconel 718 superalloy," *Met. Sci.*, vol. 17, no. 3, pp. 145–149, Mar. 1983, https://doi.org/10.1179/030634583790421032.
26. M. Chaturvedi and Y. Han, *Creep Deformation of Alloy 718*. 1989, p. 498, https://doi.org/10.7449/1989/Superalloys_1989_489_498.
27. T. Sugahara *et al.*, "Creep Behavior of the Inconel 718 Superalloy," *Defect Diffus. Forum*, vol. 326–328, pp. 509–514, 2012, https://doi.org/10.4028/www.scientific.net/DDF.326-328.509.
28. A. Drexler, A. Fischersworring-Bunk, B. Oberwinkler, W. Ecker, and H.-P. Gänser, "A microstructural based creep model applied to alloy 718," Int. J. Plast., vol. 105, pp. 62–73, Jun. 2018, https://doi.org/10.1016/j.ijplas.2017.11.003.
29. X. Lu, J. Du, Q. Deng, and J. Zhuang, "Stress rupture properties of GH4169 superalloy," *J. Mater. Res. Technol.*, vol. 3, no. 2, pp. 107–113, Apr. 2014, https://doi.org/10.1016/j.jmrt.2014.03.003.
30. Z. Xu *et al.*, "Creep property of Inconel 718 superalloy produced by selective laser melting compared to forging," *Mater. Sci. Eng. A*, vol. 794, p. 139947, Sep. 2020, https://doi.org/10.1016/j.msea.2020.139947.

Effect of Pre-straining on the Tensile and Stress-Rupture Properties of a Novel Ni-Co Based Superalloy

Bin Gan, Zhongnan Bi, Cheng Yang, Hongyao Yu, Rui Hu, and Jinhui Du

Abstract For Ni-Co-based superalloys, the higher level of Co content, the lower the stacking fault energy, which promotes the formation of deformation twins and affects the mechanical properties of the materials. Mechanical tests of a novel Ni-Co-based superalloy at temperatures between 25 and 760 °C revealed that its strength maintained a sufficiently high level when the testing temperature was up to 650 °C, above which the tensile strength decreased with test temperature, and there was a sharp drop at 760 °C. Pre-strain tests at room temperature indicated that the density of dislocation and stacking faults increased with strains and they became obstacles for dislocation motion at elevated temperatures, resulting in an increase in yield strength. In addition, the grain boundary of the pre-strained specimens has a higher dislocation density than the grain interior, and dislocation recovery occurred during the stress-rupture process, resulting in a lower stress-rupture life than the sample without a pre-strain.

Keywords Ni-Co based superalloys · Nanotwins · Pre-straining · Tensile tests · Stress-rupture

Introduction

Ni-based superalloys, possessing superior high-temperature strength and oxidation resistance, are widely used for manufacturing the hot-section components of aircraft engines [1–3]. To generate a higher thrust-to-weight ratio, a significant enhancement of their temperature-bearing capability is imperative [3], and this requirement is conventionally fulfilled by the meticulous design of composition to take advantage of solid solution hardening and precipitation strengthening [4]. However, an excessive

B. Gan (✉) · Z. Bi · C. Yang · H. Yu · J. Du
Beijing Key Laboratory of Advanced High Temperature Materials, Central Iron and Steel Research Institute, Beijing 100081, China
e-mail: ganbin@cisri-gaona.com.cn

C. Yang · R. Hu
State Key Laboratory of Solidification Processing, Northwestern Polytechnical University, Xi An 710072, China

E. A. Ott et al. (eds.), *Proceedings of the 10th International Symposium on Superalloy 718 and Derivatives*, The Minerals, Metals & Materials Series,
https://doi.org/10.1007/978-3-031-27447-3_12

amount of refractory elements may lead to the precipitation of TCP phases that are detrimental to mechanical properties [5–7]. The ultra-high strength superalloy is also difficult to forge. Therefore, alternative hardening mechanisms, such as deformation twinning, are worthy of investigation, so that a higher strength could be attained with a lower amount of strengthening phases [7].

As is well known, grain boundaries are effective sources of strengthening according to the Hall–Petch relationship. However, the further refinement of grain size to the nano-scale may lead to an undesirable loss of tensile ductility and a low thermal stability [5, 6]. Twin boundaries can improve the high-temperature mechanical properties of FCC materials such as pure Cu, high-entropy alloys, and Ni-based superalloys by lowering their intrinsic low energy state [6, 8–12]. Formation of nanotwins promotes a high work hardening rate by introducing extra abundant boundaries that act as barriers to dislocation motion. In addition, twin boundaries can also accommodate plastic deformation. Hence, nanotwin strengthening has the potential to increase the strength and ductility simultaneously [5–13].

In terms of the synergic strengthening of twin boundaries and intermetallic compounds, there are two classical routes. The first one is MP159 alloy, its ultra-high strength is realized with a high density of deformation twins in the matrix, and a high thermal stability is achieved by the formation of γ' particles around twin boundaries [14–20]. While for Ni-Co based superalloys with a low stacking fault energy, high strength is obtained with a medium level of secondary phases, complemented with deformation twinning. The low energy state of twin boundaries and the pinning effect of particles are beneficial for a high thermal stability [14–17]. It was reported that a high-density of twin boundaries can increase the strength and creep resistance of TMW-4M3 superalloy with a high Co content [21–25].

It is well-known that twin boundaries can be introduced by plastic deformation in Ni-Co-based superalloys. Therefore, it is promising to achieve the synergistic strengthening effect by mechanical pre-straining. However, for those alloys with a coarse grain, a longer stress-rupture life is accompanied with a lower yield strength, while for those fine grain alloys, a higher yield strength is coupled with a shorter stress-rupture life [2–6]. Hence, it is important to evaluate the feasibility of increasing the yield strength by pre-straining of a coarse grain superalloy, while maintaining a sufficient long stress-rupture life. In this work, a novel Ni-Co-based superalloy with low stacking fault energy was investigated. Different degrees of pre-straining were imparted to the samples, so as to assess the impact of pre-strained substructure on the tensile and stress-rupture properties. The microstructural changes at different states were systematically characterized and compared.

Experimental Procedures

Material and Heat Treatment Procedure

The chemical composition of the investigated alloy is shown in Table 1. Ingot was prepared by Vacuum Induction Melting (VIM), Electro Slag Remelting (ESR), and Vacuum Arc Remelting (VAR). The cast ingot was homogenized at 1180 °C for 60 h and then forged into a rod with a diameter of 40 mm at 1100 °C. The forged rods were then heat treated in the following sequences: a solution treatment at 1080 °C for 4 h with air cooling, a primary aging at 650 °C for 24 h with air cooling, and a secondary aging at 760 °C for 16 h with air cooling.

Pre-straining Treatment

Tensile rods were machined out and then strained at room temperature with a strain of 3% and 6%, denoted as PT3# and PT6#, respectively. A higher strain level was not imparted, considering that stress concentration at grain boundaries may lead to an early fracture during the tensile and stress-rupture tests at elevated temperatures.

Mechanical Tests

Dog-bone shape specimens with a size of Φ12 × 65 mm were machined out. Room temperature tensile tests were conducted with an electronic universal testing machine with a strain rate of 3×10^{-4} s^{-1}. The pre-strained tensile rods were machined into dog-bone shape tensile and stress-rupture tests samples with a size of Φ3 × 15 mm. The tensile tests were conducted at 760 °C with a strain rate of 3×10^{-4} s^{-1}. The stress–strain profiles were measured. Stress-rupture tests were carried out at 760 °C/480 MPa and the rupture time was recorded.

Table 1 Chemical composition of the Ni-Co based superalloy (wt.%)

Element	Co	Cr	W	Al	Ti	Nb	C	Ni
Content (wt.%)	20.0	16.5	5	2.5	2.5	2.5	0.02	Bal.

Microstructural Characterization

Scanning Electron Microscopy (SEM), Electron Backscattered Diffraction (EBSD), Transmission Electron Microscopy (TEM), and Selected Area Electron Diffraction (SAED) were used to characterize the deformed microstructure. For EBSD analysis, samples were sliced, mechanically polished, and then electro-polished with a voltage of 20 V in a solution containing 80% methanol and 20% H_2SO_4 for 5–7 s. EBSD analysis was performed with a JEM-7200F SEM equipped with channel 5 software. After the above electro-polishing, the SEM sample was electrolytically etched with the following etchant (15 g Cr_2O_3+ 10 ml H_2SO_4+ 150 ml H_3PO_4) at a voltage of 4.5 V for 4–6 s. SEM samples were observed with a JEM-7800F SEM. The TEM samples were mechanically thinned to 50 μm and then thinned by ion beam milling. TEM and SAED analyses were performed with an FEI Tecnai G2 F20 TEM equipped with Bruker EDS, with an operating voltage of 200 kV.

Results and Discussion

Initial Microstructure

Figure 1 shows the original microstructure of the Ni-Co superalloy with a standard heat treatment. Figure 1a shows the equiaxed grains, and the average grain size was 110.7 ± 9.6 μm. As shown by Fig. 1b of the bright field TEM image, there were only a few dislocations. The dark field TEM image (Fig. 1c) shows the presence of large and small γ' particles in the γ matrix. The average size of γ' phases was 64.4 ± 3.4 nm, while their volume fraction was 38.8%.

Figure 2 shows the EBSD images of the Ni-Co based superalloy after the standard heat treatment. As shown by Fig. 2(a_1-d_1), the microstructure was composed of randomly-oriented, equiaxed grains and the fraction of Σ3 boundaries varied from 30.9 to 35.2%, as shown in Fig. 2(a_2-d_2). γ' particles were uniformly distributed in γ matrix, as shown in Fig. 2(a_3–d_3). The average size of γ' particles varied from 54.2

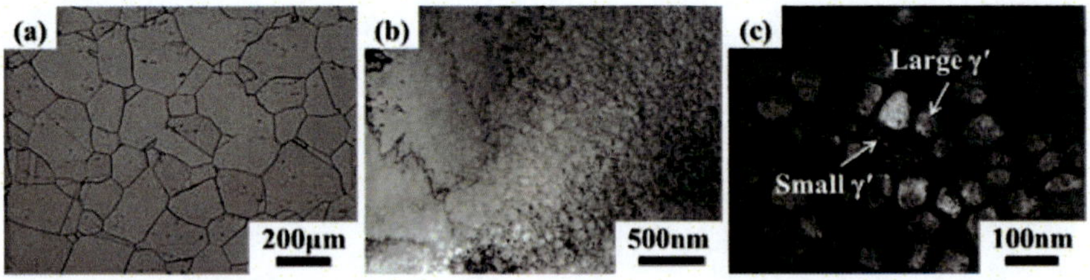

Fig. 1 Original microstructure: **a**n OM image of grain morophology; **b** bright field TEM image, and **c** center dark field TEM image of superalloy with a standard heat treatment

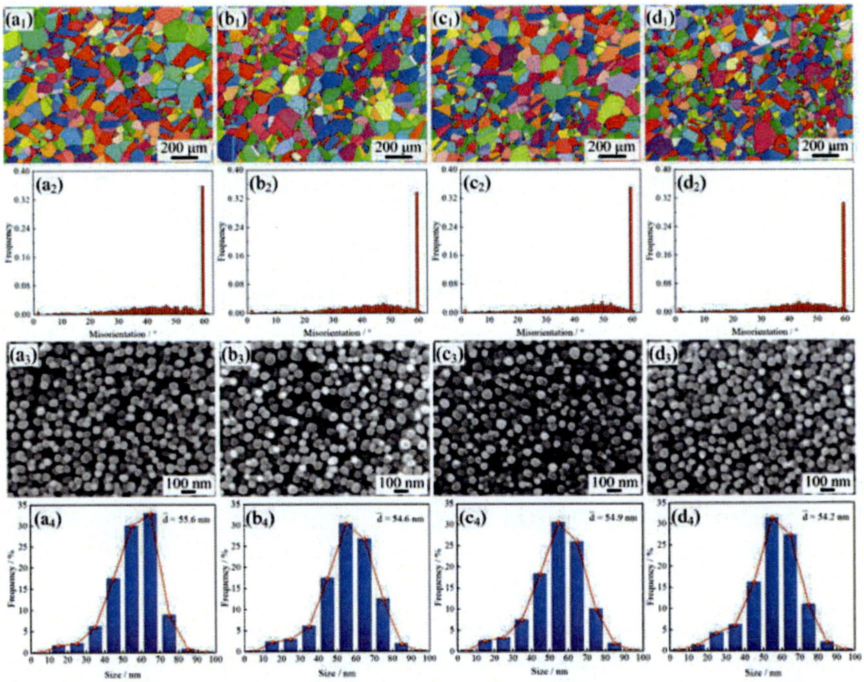

Fig. 2 Initial microstructure of the Ni-Co based superalloy before tensile tests: **a_1–a_4** 25 °C; **b_1–b_4** 650 °C; **c_1–c_4** 700 °C; **d_1–d_4** 760 °C

to 55.6 nm, as shown in Fig. 2(a_4–d_4). Their volume fraction varied from 36.9 to 37.7%, which was consistent with calculated values by JMatPro software.

Tensile Properties

Figure 3 shows the strength and elongation of the Ni-Co based superalloy with a standard heat treatment tested from 25 °C to 760 °C, which clearly confirmed that the strength maintained a sufficiently high level when the test temperature is lower than 650 °C. From 650 °C to 760 °C, the tensile strength decreased with the increment of test temperature, and there was a sharp drop at 760 °C. It is necessary to improve the mechanical properties of the Ni-Co based superalloy at 760 °C or even higher temperatures.

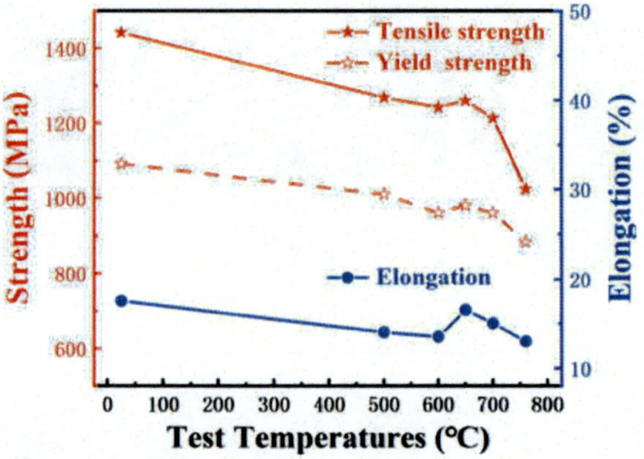

Fig. 3 Tensile properties of the Ni-Co based superalloy at different temperatures

Microstructure After Tensile Tests

Figure 4 shows the morphology and size distribution of γ' particles for the samples tested at different temperatures. Spherical γ' particles were uniformly distributed in the γ matrix, as shown in Fig. 4a_1–d_1. Their size distribution was all in the range of 50–80 nm, as shown in Fig. 4a_2–d_2, indicating the γ' particles exhibited a high thermal stability at the investigated temperatures.

Figure 5 shows the EBSD images and Kernel Average Misorientation (KAM) maps of the Ni-Co-based superalloy after different temperature tests. At 25 °C, the grains were elongated along the strain direction, and the extent gradually decreased when the test temperature was higher than 650 °C, as shown in Fig. 5a_1–a_4. The evolution of strain contours was related to the dislocation motion, multiplication, and their interactions with grain boundaries during plastic deformation. The KAM

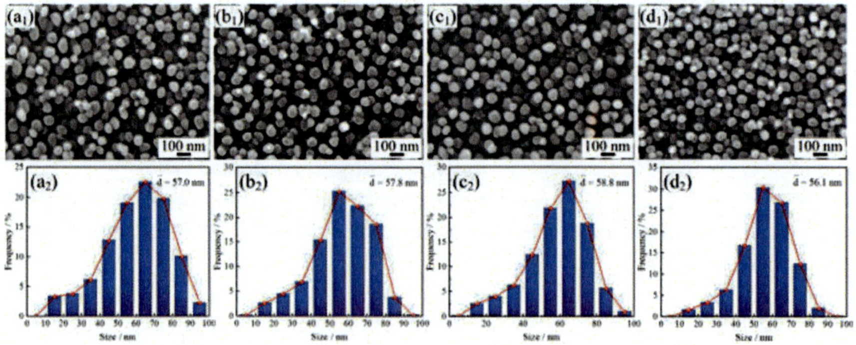

Fig. 4 Morphology and size distribution of γ' particles near the fracture region of the specimens tested at different temperatures: $\mathbf{a_1}$–$\mathbf{a_2}$ 25 °C; $\mathbf{b_1}$–$\mathbf{b_2}$ 650 °C; $\mathbf{c_1}$–$\mathbf{c_2}$ 700 °C; $\mathbf{d_1}$–$\mathbf{d_2}$ 760 °C

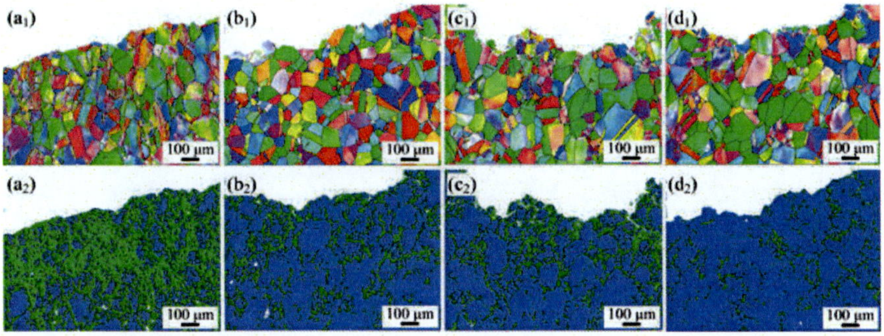

Fig. 5 The EBSD images and KAM maps near the fracture region of the specimens tested at different temperatures: $\mathbf{a_1}$–$\mathbf{a_2}$ 25 °C; $\mathbf{b_1}$–$\mathbf{b_2}$ 650 °C; $\mathbf{c_1}$–$\mathbf{c_2}$ 700 °C; $\mathbf{d_1}$–$\mathbf{d_2}$ 760 °C

is the average misorientation between each pixel and its surrounding pixels in EBSD measurements. At 25 °C, the high KAM at the grain boundaries and grain interior (Fig. 5a_2) indicated a high level of dislocation activity. For the specimens tested at 650 and 700 °C, the uniform distribution of KAM suggested there was no abundant dislocation multiplication (Fig. 5(b_2–c_2)). The distribution of KAM value was the most uniform in the specimen tested at 760 °C, the strain concentration at grain boundaries was reduced, compared to those samples tested at lower temperatures (Fig. 5d_2).

Figure 6 shows the fracture surfaces of the specimens tested at different temperatures. Many trans-granular cracks formed on the fracture surface of the room temperature specimen (Fig. 6a). At 700 °C, a larger number of cleavage facets and inter-granular cracks existed on fracture surfaces (Fig. 6c). At 760 °C, the fracture was composed of cleavage facets, shallow dimples and inter-granular cracks (Fig. 6d), indicating a mixture of inter-granular and trans-granular fracture.

Microstructure Changes After the Pre-straining Treatment

Figure 7 shows the EBSD images of the superalloy with a standard heat treatment (0#), PT3#, and PT6# (denote the samples with 3% and 6% pre-strain at room temperature, respectively). As shown by the inverse pole figure (IPF) of Fig. 7a_1–c_1, the original microstructure was composed of equiaxed grains with a random orientation and some annealing twins. After the pre-straining treatment, strain gradients were introduced into some grains of the PT3#. In PT6#, there was a clear strain gradient throughout the specimen. Based on the KAM diagram, there was no stress concentration in 0#. There was almost no obvious change in the grain size of PT3#, but stress concentrations were observed in some grain boundaries. PT6# showed stress concentration at the grain boundaries and grain interiors due to the larger deformation (Fig. 7a_2–c_2). Almost all the grains in 0# were recrystallized grains. About 9%

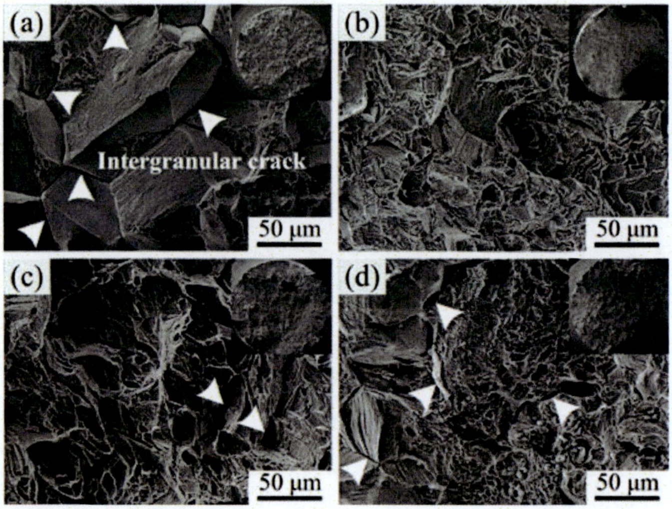

Fig. 6 SEM micrograph of the fracture surfaces of the specimens tested at different temperatures: **a** 25 °C; **b** 650 °C; **c** 700 °C; **d** 760 °C

of the equiaxed grains in PT3# had the formation of a substructure. About 67% of the grains of the PT6# had the formation of a substructure (Fig. 7a_3–c_3). For all the samples, the γ' phases were nearly spherical with the average diameter of 63.2 to 66.0 nm. It should be noted that some of γ' phase in PT3# and PT6 # were sheared.

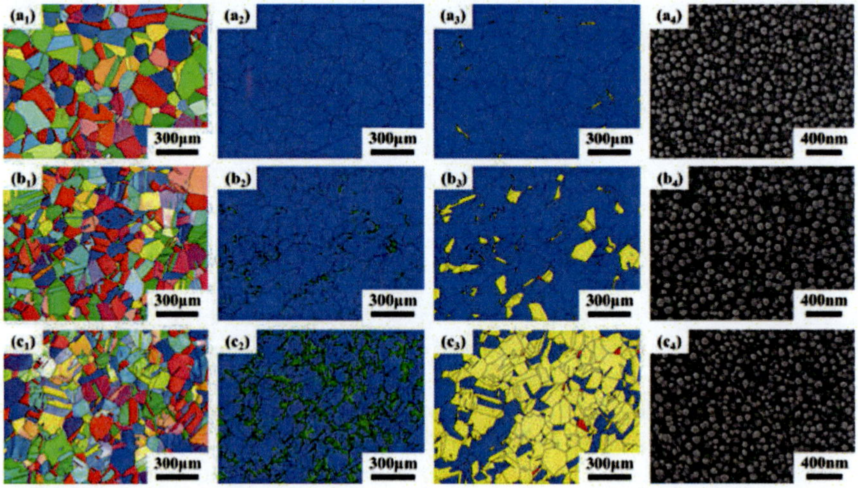

Fig. 7 EBSD analysis: IPF diagram, KAM diagram, recrystallized structure diagram, and SEM photographs: $\mathbf{a_1}$–$\mathbf{a_4}$ superalloy with a standard heat treatment without pre-straining 0#; $\mathbf{b_1}$–$\mathbf{b_4}$ PT3# and $\mathbf{c_1}$–$\mathbf{c_4}$ PT6#

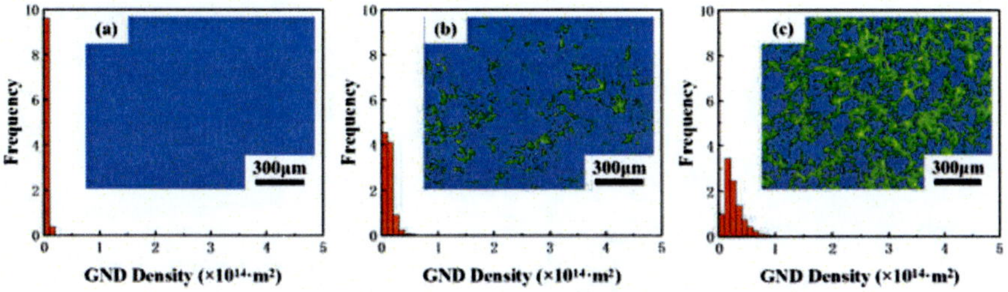

Fig. 8 GND density and distribution under different conditions: **a** original condition 0#; **b** PT3# and **c** PT6#

The Geometrically Necessary Dislocations (GNDs) values of the 0#, PT3#, and PT6# were statistically analyzed, and their distribution diagrams were plotted (Fig. 8). In order to study the evolution of dislocation density with the pre-strain degree, GND was used to characterize the dislocation density and its distribution in different grains. The closer the peak of the GND distribution is to the right side of the horizontal axis, the higher the GND density. The specimen with a standard heat treatment showed a very low dislocation density both inside the grain and near the grain boundary (Fig. 8a). For PT3#, a high density of dislocations was mainly observed at the grain boundary. Therefore, the peak position of the dislocation density distribution was almost the same as 0#, but the dislocation density distribution gradually moved to the left side of the horizontal axis (Fig. 8b). For PT6# with a larger deformation, the dislocation density in the grain interior or near the grain boundary increased rapidly, showing a typical single peak distribution (Fig. 8c).

In order to clarify the type of substructure induced by the pre-straining, Fig. 9 shows the TEM images of PT3# and PT6#. Substructures, such as dislocation lines, shear bands, and continuous stacking faults, can be clearly observed in PT3#, and dislocations in the matrix can cut the γ' phases. Isolated stacking faults were formed inside the γ' phase, many short isolated stacking faults can be observed (Fig. 9a–c). The amount of dislocations in PT6# increased significantly and entangled around the γ' phase, the larger deformation will assist the growth of continuous stacking faults. The obvious growth of isolated stacking faults was revealed by the high-resolution TEM image, and the stacking faults with different orientations intersected at 70.5° to form Lomer-Cottrell locks.

Tensile Behavior of Pre-Strained Specimens

Tensile Properties

The tensile properties of 0#, PT3#, and PT6# were shown in Fig. 10a, and the true stress-true strain curves were plotted (Fig. 10b). The yield strength and ultimate tensile strength of a novel Ni-based superalloy were improved after the pre-tensile

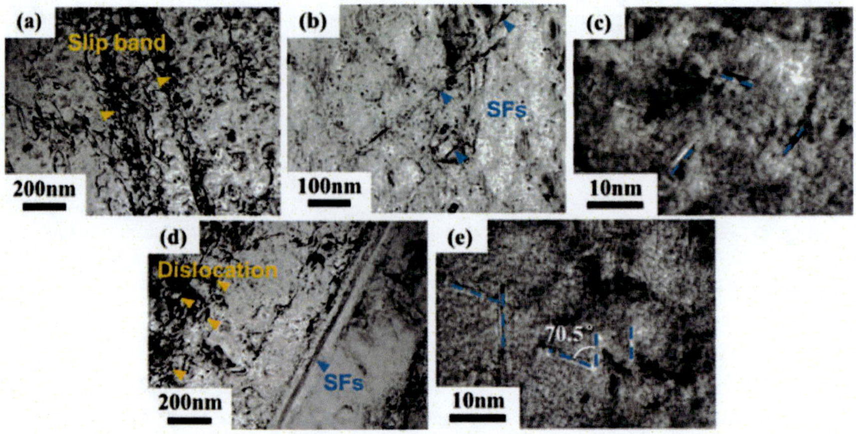

Fig. 9 TEM images of the pre-strained specimens: **a–c** PT3#; **d–e** PT6#

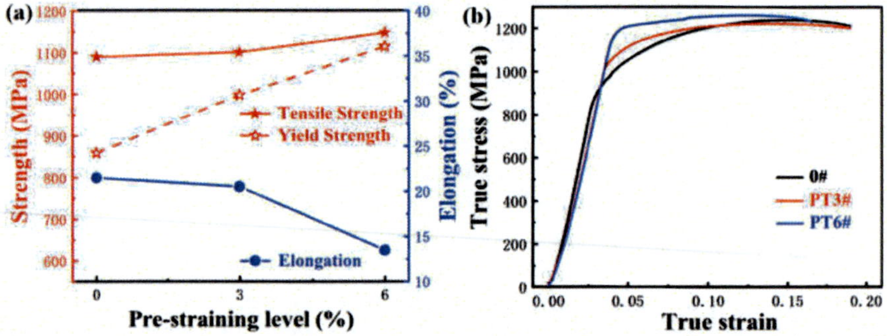

Fig. 10 **a** Strength and elongation tested at 760 °C versus the pre-straining level, **b** true stress-true strain profiles

treatments, but the elongation was gradually decreased. The yield strength of 0# at 760 °C was 859 MPa. The yield strength of PT3# and PT6# at 760 °C was 998 MPa and 1115 MPa, respectively, which was nearly 16.2% and 29.8% higher than that of 0#, respectively. The elongation of PT3# was equivalent to that of 0#, while the elongation of PT6# decreased slightly.

Deformed Microstructure After Tensile Tests

Figure 11 shows the tensile fracture morphology of 0#, PT3#, and PT6#. The fracture behavior was relatively the same: cracks originated from the edge, and the cracking area was mainly inter-granular fracture with a small number of dimples (Fig. 11a_1–c_1). The crack extended to the central area, where there were many cleavage planes

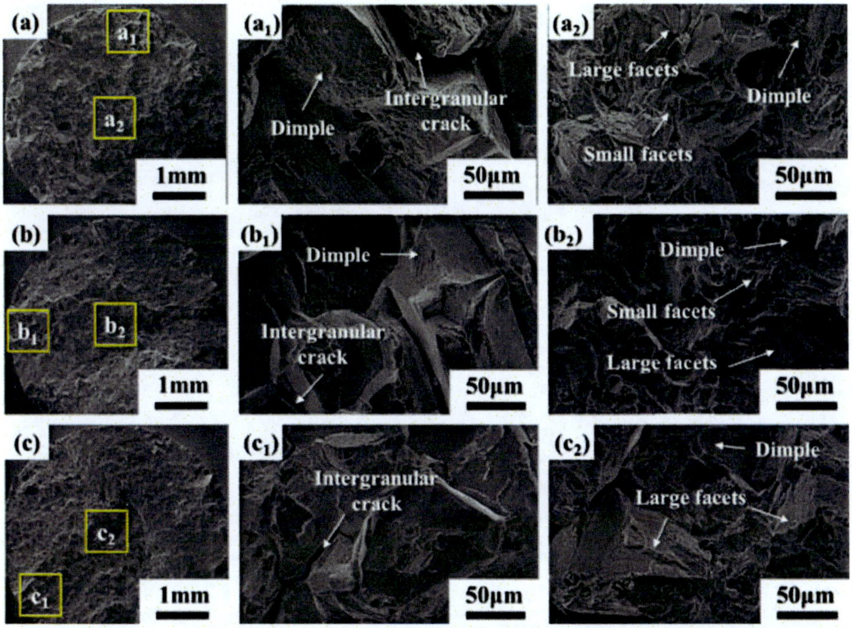

Fig. 11 Tensile fracture morphology of samples tested at 760 °C. **a–a$_2$** 0 #; **b–b$_2$** PT3#; **c–c$_2$** PT6#

and dimples (Fig. 11a$_2$–c$_2$). With the increase of pre-strain level, the amount of fine dimples in the cracking area gradually decreased, due to the stress concentration at the grain boundary. The amount of dimples in the central area of the fracture decreased, and the number of cleavage planes increased.

Figure 12 shows the TEM images of 0#, PT3#, and PT6# tensile tested at 760 °C. Continuous stacking faults and parallel deformation twins were observed in 0#, which can impede the motion of dislocations on both sides. The center dark field TEM image (Fig. 12b) revealed that the deformed twin could cut the γ' phase. The thickness of the deformation twin was about 6.1 nm, and there were isolated stacking faults around it (Fig. 12a–c). For PT3#, the amount of continuous stacking faults and parallel deformation twins increased. Deformation twins also cut into the γ' phase. The thickness of the deformation twin was about 5.9 nm, which was similar to that of 0#. There were parallel and continuous stacking faults near the deformation twins and the stacking faults of two different orientations can intersect at 70.5° to form Lomer-Cottrell locks (Fig. 12d–f). For PT6#, there was a significant increase in continuous stacking faults and parallel deformation twins. The γ' phase was cut by deformation twins. The thickness of the deformation twin was about 8.5 nm. There were continuous stacking faults in the same orientation as the deformation twin, and the continuous stacking faults can intersect with the deformation twin at 70.5° (Fig. 12d–f). With the increased amount of pre-strain, the number of deformation twins and Lomer-Cottrell locks increased,

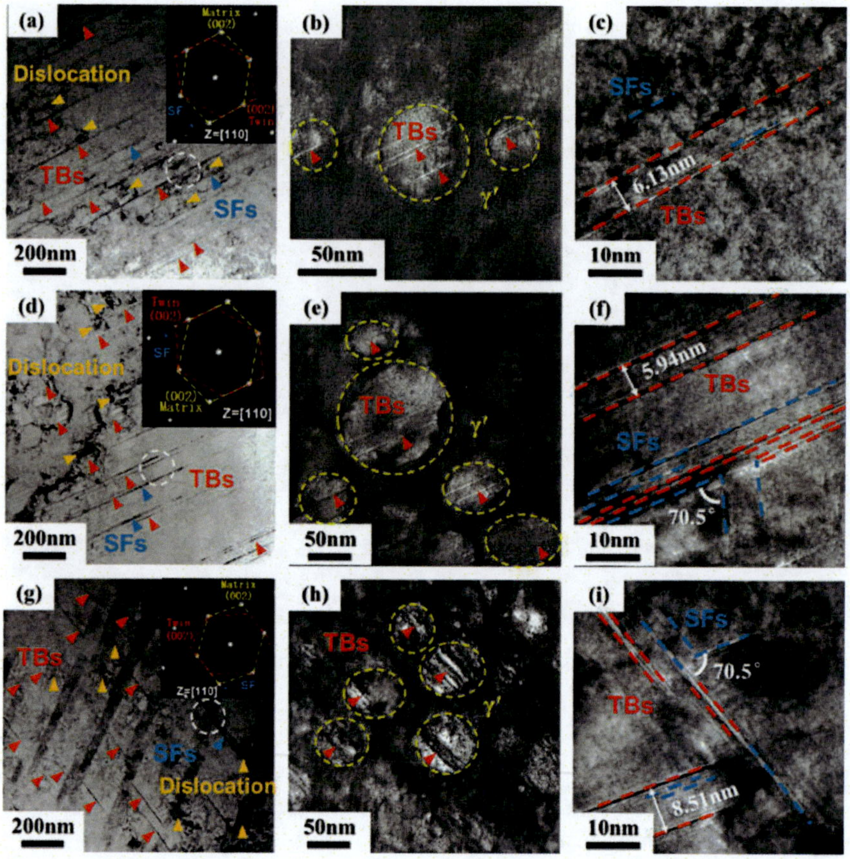

Fig. 12 TEM photos of tensile fractured specimens. **a–c** 0#; **d–f** PT3#; **g–i** PT6#

which improves the tensile strength. In addition, stacking faults formed near deformation twins can be used as the core of deformation twins. After atomic diffusion and rearrangement, stacking faults transformed into deformation twins, which increased the density of deformation twins.

Stress-Rupture Behavior of Pre-strained Specimens

Stress-Rupture Properties

The stress-rupture lifes of 0#, PT3#, and PT6# were measured at 760 °C/480 MPa (see Fig. 13). Pre-straining treatment reduced the stress rupture life of specimen with

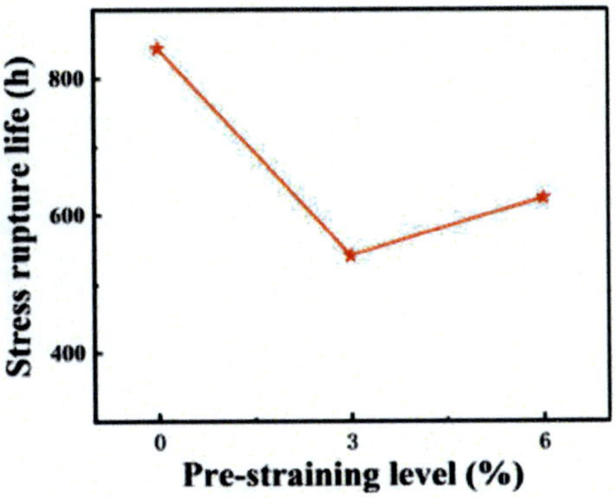

Fig. 13 Stress-rupture life of specimen versus pre-straining levels, measured at 760 °C/480 MPa

a standard heat treatment; however, the remaining stress-rupture life maintained a sufficiently high value.

Deformed Microstructure After Stress-Rupture Tests

Figure 14 shows the fracture surface of 0#, PT3# and PT6# tested at 760 °C/480 MPa (Fig. 14a–f).Their stress-rupture fracture mechanisms were relatively the same: cracks were generated from the central area of the sample with an inter-granular rupture manner. There were a few cleavage planes and dimples. There were many dimples in the cracking area of 0#, while there were no dimples in the cracking area of PT3# and PT6#. The ductility was reduced due to the stress concentration at the grain boundary. With the increase of pre-strain, the large cleavage facets in the central area of rupture decreased. However, both the small facets surface and the ductile rupture characteristics increased. The fracture morphologies were consistent with the stress rupture life revealed in Fig. 13.

Figure 15 shows the deformed microstructure of 0#, PT3#, and PT6# tested at 760 °C/480 MPa. The deformation mechanism was consisting of the formation of parallel deformation twins and stacking faults, which can obviously hinder the slip of dislocations on both sides. The $M_{23}C_6$ carbides formed at the grain boundary can prevent the slip of dislocations at the grain boundaries and stacking faults were formed in the $M_{23}C_6$ carbides after all dislocations cut into them (Fig. 15a–c). The amount of dislocation, deformation twins, and isolated stacking faults of the alloy with pre-straining treatments increased significantly (Fig. 15d, g and Fig. 15e, h). With the increase of pre-strain deformation, the number of deformation twins and isolated stacking faults in the pre-deformed structure increased. In addition, the $M_{23}C_6$ carbides at the grain boundaries played a more significant role in hindering dislocation movement, which was beneficial for improving the stress rupture life.

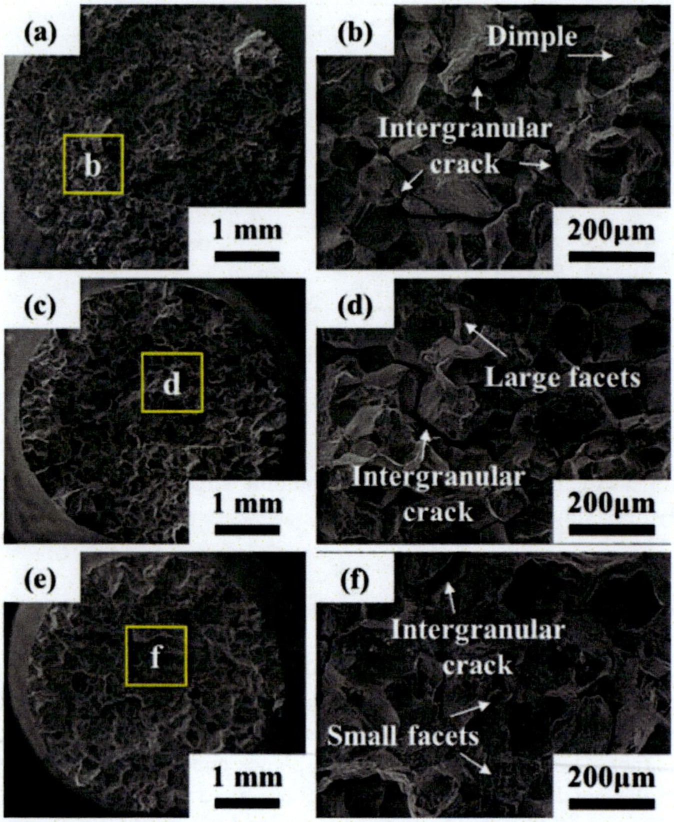

Fig. 14 Stress rupture morphology of specimens tested at 760 °C/480 MPa **a–b** 0#; **c–d** PT3#; **e–f** PT6#

Conclusions

The deformation behaviors from 25 to 760 °C were characterized and the impact of room temperature pre-straining on the tensile and stress-rupture properties at 760 °C was assessed. The major conclusions are summarized as follows.

(1) The strength maintained a sufficiently high level when the test temperature is lower than 650 °C. From 650 to 760 °C the tensile strength decreased with the test temperature, and the tensile strength decreased sharply at 760 °C
(2) The pre-straining treatment increased the yield strength at 760 °C. During the pre-straining process, the density of stacking faults and deformation twins increased, and the hindrance to dislocations increased, resulting in an increase in the yield strength at the elevated temperature.

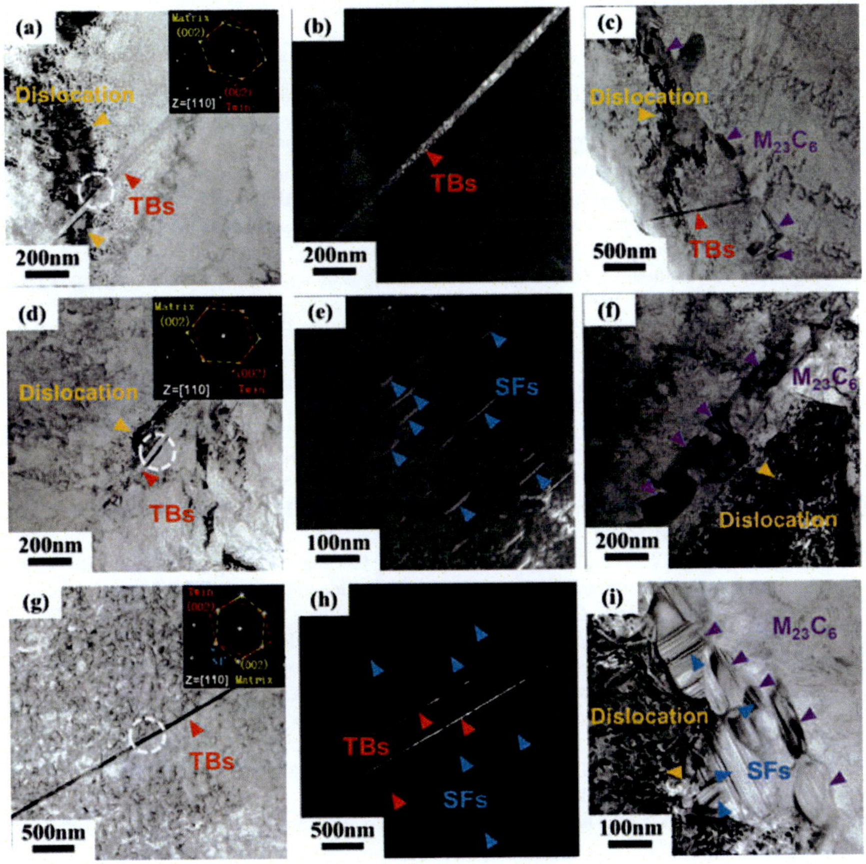

Fig. 15 TEM images of different samples tested at 760 °C/480 MPa **a–c** 0#; **d–f** PT3#; **g–i** PT6#

(3) After the pre-straining treatment, the stress rupture life of the alloy decreased. Pre-straining treatment reduced the stress rupture life of the un-stretched specimen. However, the remaining life maintained a sufficiently high value.
(4) With the increase of pre-strain deformation, the number of isolated stacking faults and deformation twins increased, and they can hinder the slip of dislocations on both sides. In addition, stacking faults and $M_{23}C_6$ hindered dislocation movement and grain boundary migration, which was beneficial for lengthening the stress-rupture life.

References

1. Pollock TM, Tin S, "Nickel-Based Superalloys for Advanced Turbine Engines: Chemistry, Microstructure, and Properties", *J. Propulsion & Power*, 2006, 22 (2), 361.
2. Donachie MJ, Donachie SJ, Superalloys: A Technical Guide, 2nd ed., ASM International, 2002.
3. Reed RC, the Superalloys Fundamentals and Applications, Cambridge University, 2006.
4. Kozar RW, Suzuki A, Milligan WW, Schirra JJ, Savage MF, Pollock TM, "Strengthening Mechanisms in Polycrystalline Multimodal Nickel-Based Superalloys", *Metall. Mater. Trans. A*, 2009, 40A, 1588–1603.
5. Argon AS, Strengthening Mechanisms in Crystal Plasticity, UK, Oxford University Press, 2007.
6. Hull D, Bacon DJ, Introduction to Dislocation, Butterworth-Heinemann, Oxford, UK, 2011.
7. Zhu T, Li J, "Ultra-strength Materials", *Prog. in Mater. Sci.*, 2010, 55, 710–757.
8. Kienl C, León-Cázares FD, Rae CMF, "Deformation Twinning During High Temperature Compression Tests of the Ni-based Superalloys ATI 718Plus", Acta Mater., 2022, 225, 115743.
9. Wei YJ, Li YQ, Zhu LC, Liu Y, Lei XQ, Wang G et al., "Evading the Strength-Ductility Trade-off Dilemma in Steel Through Gradient Hierarchical Nanotwins", *Nat. Comm.*, 2014, 5, 1–8.
10. Lu K, "Stabilizing Nanostructures in Metals Using Grain and Twin Boundary Architectures", *Nat. Rev. Mater.*, 2016, 1, 1–13.
11. Zhang ZJ, Sheng HW, Wang ZJ, Gludovatz B, Zhang Z, George EP, Yu Q, Mao SX, Ritchie RO, "Dislocation Mechanisms and 3D Twin Architectures Generate Exceptional Strength Ductility Toughness Combination in CrCoNi Medium-Entropy alloy", *Nat. Comm.*, 2017, 8, 14390.
12. Smith TM, Esser BD, Antolin N, Carlsson A, Willams REA, Wessman A et. al., "Phase Transformation Strengthening of High-Temperature Superalloys", *Nat. Comm.*, 2016, 7, 13434.
13. Barba D, Alabort E, Pedrazzini S, Collins DM, Wilkinson AJ, Bagot PAJ, Moody MP, et al., "On the Microtwinning Mechanism in a Single Crystal Superalloy", *Acta Mater.*, 2017, 135, 314–329.
14. Slaney JS, Nebiolo RA, "Development of Multiphase Alloy MP159 Using Experimental Statistics", *Metallography*, 1983, 16, 137–160。
15. Han GW, Feng D, Yin M, Yang NS, "Microstructure Variation of Cold Deformed MP159 Alloy During Aging", *J. Mater. Sci. Technol.*, 1996, 12, 109–113.
16. Han GW, Jones IP, Smallman RE, "Direct Evidence for Suzuki Segregation and Cottrel Pinning in MP159 Superalloys Obtained by FEG TEM/EDX", *Acta Mater.*, 2003, 51, 2731–2742.
17. Singh R, Doherty RD, "Strengthening in MULTIPHASE (MP35N) Alloy Part I. Ambient Temperature Defomation and Recrystallization", *Metall. Trans. A*, 1992,23, 307–319.
18. Gu J, Guo L, Gan B, et al., "Microstructure and Mechanical Properties of an MP159 Alloy Processed by Torsional Deformation and Subsequent Annealing", *Mater. Sci. Eng. A*, 2021, 802, 140676.
19. Pei B, Fan JP, Wang Z, Yang HJ, Qiao JW, Bi ZN, Gan B, "Excellent Combination of Strength and Ductility in CoNiCr-based MP159 Alloys at Cryogenic Temperature", *J. Alloys & Comp.*, 2022, 164144.
20. Gu J, Gan B, Bi ZN, Song M, "Microstructure and Mechanical Properties of a MP159 Superalloy after Pre-tensile Deformation and Subsequent Annealing", *Adv. Eng. Mater.*, 2021, 2100920.

21. Yuan Y, Gu YF, Cui CY, Osada T, Yokokawa T, Harada H, "A Novel Strategy for the Design of Advanced Engineering Alloys—Strengthening Turbine Disk Superalloys via Twinning Structures", *Adv. Eng. Mater.*, 2011,13, 296–300.
22. Yuan Y, Gu YF, Osada T, Zhong ZH, Yokokawa T, Harada H, "A new method to strengthen turbine disc superalloys at service temperatures", *Scripta Mater.*, 2012, 66, 884–889.
23. Tian CG, Cui CY, Xu L, Gu YF, Sun XF, "Dynamic Strain Aging in a Newly Developed Ni-Co Base Superalloy with Low Stacking Fault Energy", *J. Mater. Sci &Tech.*, 2013, 873–878.
24. Tian CG, Han GM, Cui CY, Sun XF, "Effects of Stacking Fault Energy on the Creep Behaviors of Ni-base Superalloy", *Mater. Des.*, 2014,316–323.
25. Tian CG, Han GM, Cui CY, Sun XF, "Effects of Co Content on Tensile Properties and Deformation Behaviors of Ni-based Disk Superalloys at Different Temperatures", *Mater. Des.*, 2015, 123–131.

Effect of Short-Term Isothermal Exposure on the Ductility Signature of Waspaloy in the Temperature Range of 750–950 °C: A Comparison with Haynes® 282®

Fabian Hanning, Abdul Khaliq Khan, Olanrewaju Ojo, and Joel Andersson

Abstract The evolution of microstructure and ductility has been investigated for Waspaloy after isothermal exposure between 5 and 1800s at 750–950 °C. Gamma prime (γ') with 1.7 nm diameter is found in the mill-annealed condition, while precipitate-growth following a $t^{1/3}$ relationship is observed for isothermal exposure. Grain boundary carbide networks are formed during isothermal exposure together with a rapid hardness increase. A drop in ductility is observed with the lowest values at 750 and 800 °C. Further ductility reduction during isothermal exposure correlates with the rapid hardness increase of Waspaloy. While grain boundary strengthening can compensate for the moderate age hardening observed for Haynes® 282®, the more rapid hardness increase due to γ' precipitation appears to be the dominating effect on ductility in Waspaloy. Carbide precipitation and growth kinetics are slower than those of Haynes® 282®, which further increases the relative effect of age hardening reactions on the ductility of Waspaloy.

Keywords Nickel-based superalloy · Waspaloy · Haynes 282 · Strain age cracking · Postweld heat treatment · Gleeble

Introduction

Hot structural components for aircraft engines are commonly manufactured using precipitation hardening nickel-based superalloys. The standard grade for such applications is Alloy 718, which has a favourable combination of high temperature strength and good fabricability [1]. The strength of Alloy 718 is based on the precipitation of

F. Hanning (✉) · J. Andersson
Department of Engineering Science, University West, Trollhättan, Sweden
e-mail: fabian.hanning@hv.se

A. K. Khan
Manitoba Institute for Materials, University of Manitoba, Winnipeg, MB, Canada

A. K. Khan · O. Ojo
Department of Mechanical Engineering, University of Manitoba, Winnipeg, MB, Canada

E. A. Ott et al. (eds.), *Proceedings of the 10th International Symposium on Superalloy 718 and Derivatives*, The Minerals, Metals & Materials Series,
https://doi.org/10.1007/978-3-031-27447-3_13

the gamma double prime γ" phase, which enables a maximum service temperature of 650 °C. The manufacturing of large structural components typically involves joining small parts together by welding [2]. This requires a good welding performance of the used material. Increased combustion temperatures in aircraft engines and the limited service temperature of Alloy 718 requires the use of higher temperature stable materials, which are typically gamma prime (γ') hardening alloys. Precipitation hardening superalloys require a post weld heat treatment (PWHT) to obtain uniform material properties. Fast aging kinetics of γ' hardening alloys can lead to strain age cracking (SAC), a cracking phenomenon characterized by low ductility intergranular fracture in the heat affected zone [3]. The general mechanism involves intragranular age hardening reactions, which localizes stress relaxation to grain boundaries. Different testing procedures have been developed to study SAC, where tests measuring ductility appear to be most suitable to compare different alloys [4]. The constant heating rate test (CHRT) is one of the few methods that have been used by several studies [5–8]. The test simulates slow furnace heating to PWHT, with samples being pulled to fracture in the temperature range of 700 to 1000 °C. A drop in ductility which is observed in this temperature range can be used to qualitatively rank different materials' susceptibility to SAC based on the severity of the drop in ductility. The CHRT is useful to create material rankings, it does however lack the ability to investigate the underlying mechanism of SAC. A controlled exposure time test has therefore developed in an earlier study [9]. It was found that isothermal exposure does not change the ductility of Haynes® 282®. This relatively new alloy has been developed as a potential complement of Alloy 718, above which it has a 150 °C maximum service temperature advantage [8]. Haynes® 282® was developed with a leaner composition as compared to Waspaloy or René 41 to improve weldability. Waspaloy has a very similar chemical composition but shows a faster age hardening response which is related to the higher content of γ' forming elements Al and Ti [10, 11]. This work involves a study of ductility signature and microstructural evolution of Waspaloy and provides a comparison with Haynes® 282® to investigate the effect of the former's faster age hardening kinetics on ductility after isothermal exposure.

Experimental

Material and Heat Treatments

Wrought Waspaloy is investigated in the form of 3.175 mm (1/8") rolled sheet with the chemical composition given in Table 1. The material is used in the as-received, mill-annealed condition, having a hardness of 301 ± 5HV. The microstructure with an average grain size of 24 ± 5 μm is shown in Fig. 1. The as-received microstructure shows no grain boundary carbide network, but primary MC-type carbides are visible in the microstructure.

Table 1 Chemical composition in wt.-% of the Waspaloy sheet used for the experiments

	Ni	*Cr*	*Co*	*Mo*	*Fe*	*Mn*	*Al*	*Ti*
wt.-%	Bal	19.08	13.56	4.52	1.57	0.08	1.41	3.00
	Zr	*B*	*C*	*Cu*	*P*	*S*	*Si*	
wt.-%	0.041	0.006	0.080	0.03	0.003	– [a]	0.05	

[a]Below detection limit

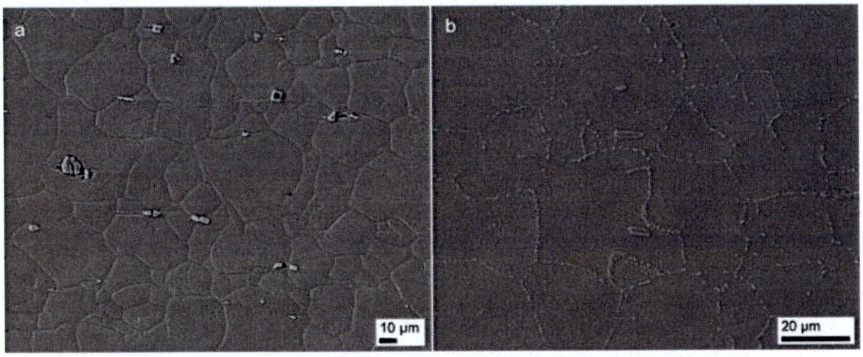

Fig. 1 **a** As received, mill-annealed microstructure with primary carbides present as stringers. **b** Age hardened microstructure with developed grain boundary carbide network

A conventional age hardening heat treatment has been carried out on some material. The heat treatment includes three aging steps, 995 °C 2 h, 845 °C 4 h, and 760° 16 h [10]. The heat treatment has been carried out on tensile test samples in a tube furnace under Argon gas with 100 l/h flow rate. Figure 1b shows the obtained microstructure which contains a developed grain boundary precipitate network. Grain size is not affected by the heat treatment, while the hardness increases to 411 ± 3 HV0.5.

Gleeble Thermomechanical Simulation

A Gleeble 3800D system is used for high temperature testing, with test temperatures ranging from 750 to 950 °C divided into steps of 50 °C. Short exposure times from 5 s are included to investigate the effect of early-stage γ' evolution. Fast heating with 1000 °C/s has been used to reduce the effect of heating on the test results. Similarly, the cooling rate after completed high temperature exposure was kept above 100 °C/s for temperatures above 500 °C to minimise microstructural changes during the cooling part of the test. Stroke rate has been found to affect the fracture mode in Haynes® 282® [9], with intergranular fracture being present only for low stroke rates. Two stroke rates are employed in this study, 0.055 and 55 m/s and results

Table 2 Test matrix and parameters

Temperature [°C]	750	800		850		900	950	
Exposure time isothermal tests [s]	5	10	30	60	180	600	1200	1800
Heating rate [°C/s]	1000							
Cooling rate (T > 500 °C) [°C/s]	100							
Stroke rate [mm/s]	0.055				55			
Chamber pressure [mbar]	0.1							
Thermocouple	Type K							

are compared to those observed for Haynes® 282®. All test specimens are prepared using abrasive waterjet cutting. Sample dimensions are the same as in [9], with 92 mm total sample length, a 26 mm gauge radius, and 7.5 mm minimum width in the gauge section. For microstructural analysis on undeformed material, reduced-size specimens with 55 × 7.5 × 3.15 mm are used. The test parameters are summarised in Table 2.

Three replicates are tested for each point, and the reduction in area (RA) has been measured using an Olympus SZX9 stereo optical microscope. The tests are randomized within repeats and only identifiable via their run number to minimize operational bias during evaluation.

Microstructural Characterization

Microstructural analysis is carried out on cross sections of the reduced-size specimens, cut out at the sample centre where the Gleeble control-thermocouple is located, and on fractured samples. Cross sectioned samples are mounted in hot mounting resin, followed by automated grinding and polishing. For general microstructural characterization, samples are etched using 10wt.-% oxalic acid at 3 V DC for 2–3 s. A Zeiss Evo 50 scanning electron microscope (SEM) and a Leo 1550 FEG SEM have been used for fracture and microstructure analysis. Selected samples have been further investigated using transmission electron microscopy (TEM). For this, cut-out cross sections have been thinned to 100 μm thickness, followed by dimple-grinding and electropolishing in 10:90 methanolic perchloric acid at −40 °C using a Struers Tenupol-3 twinjet electropolishing machine. A Jeol JEM-2100 field emission TEM with 200 kV acceleration voltage is used to acquire brightfield (BF) and darkfield (DF) images and selected area (SA) diffraction patterns end electron dispersive spectroscopy (EDS). Microhardness is measured using a Struers DuraScan-70 G5 with a force of 0.5kgf (HV0.5). 5 randomly located indents per sample were averaged and are presented with the respective standard deviation as error. Thermodynamic calculations are carried out using JmatPro v.11.2.

Results and Discussion

Microstructural Characterization

The as-received, mill annealed condition shows an average hardness of 301 ± 5HV0.5, which and can be attributed to the presence of γ' formed during cooling. SEM analysis did not reveal γ' precipitates, the presence of γ' superlattice reflections could however be observed in the TEM as shown in Fig. 2. The average particle diameter is measured to 1.7 ± 0.5 nm.

After age hardening heat treatment, the hardness increases to 411 ± 3 HV0.5. The first heat treatment at 996 °C leads to the precipitation of coarse γ' with a particle size of 133 ± 34 nm. This temperature is above the dissolution temperature of $M_{23}C_6$ carbides, which by the use of JmatPro is calculated to 965 °C. $M_{23}C_6$ carbides found on the grain boundaries are assumed to have formed during the subsequent steps of the aging treatment. The 3-step aging heat treatment lead to the formation of a complete network of $M_{23}C_6$ and γ' on the grain boundaries, as can be seen in Fig. 3a. Fine γ' that formed during the last heat treatment at 760 °C is evident in Fig. 3b. The particle size of this second species of γ' is measured to 34 ± 7 nm.

Particle size evolution during thermal exposure of mill annealed specimens in the Gleeble test are shown in Fig. 4 for 850 and 950 °C. Particle size data from TEM images is fitted to the Lifshitz-Slyozov-Wagner (LSW) theory [12], which predicts a linear relationship of particle radius and the cube root of time at a given temperature, assuming a constant γ' phase fraction. The fitting parameters R^2 are 0.9817 and

Fig. 2 TEM darkfield image using the < 2–1-1 > gamma prime diffraction spot on the gamma [102] zone axis, showing bright γ' precipitates with a particle diameter of 1.7 ± 0.5 nm

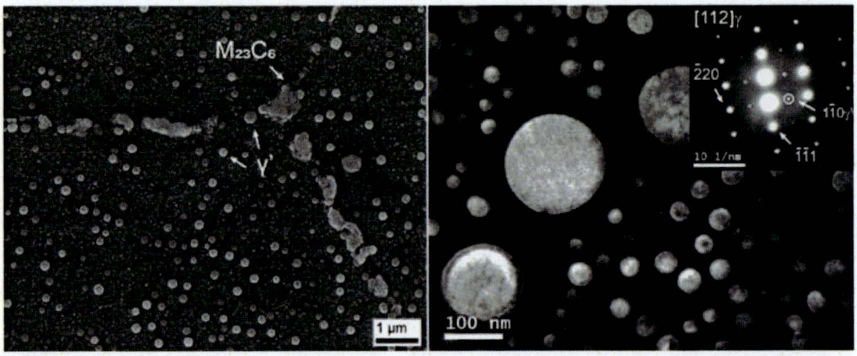

Fig. 3 **a** Waspaloy microstructure after full heat treatment showing $M_{23}C_6$ carbides on the grain boundaries and coarse γ' on grain boundaries and in the grain interior. **b** Gamma prime precipitates with 133 ± 34 nm and 34 ± 7 nm diameter in TEM dark field contrast using the <1–10> gamma prime diffraction spot on the gamma [112] zone axis

0.9891 for 850 and 950 °C, respectively. A similar relationship has been reported by Whelchel et al. for Waspaloy heat treated at temperatures above 800 °C [13].

Table 3 compiles the equilibrium γ' phase fractions and the exposure times necessary to reach equilibrium conditions. Note that these simulations assume a γ'-free microstructure as starting condition. The simulation data suggests that significant nucleation should occur during the initial stage of the heat treatments. The presence of γ' precipitates in the as-received material condition as shown in Fig. 2 however makes further nucleation unlikely especially for the higher test temperatures and diffusion-controlled growth can be anticipated to be the dominating mechanism for the investigated exposure times. This hypothesis is supported by the good fit of

Fig. 4 Particle radii measured at 850 and 950 °C for 180–1800s exposure time. Dashed lines show LSW fit, R2 > 0.98

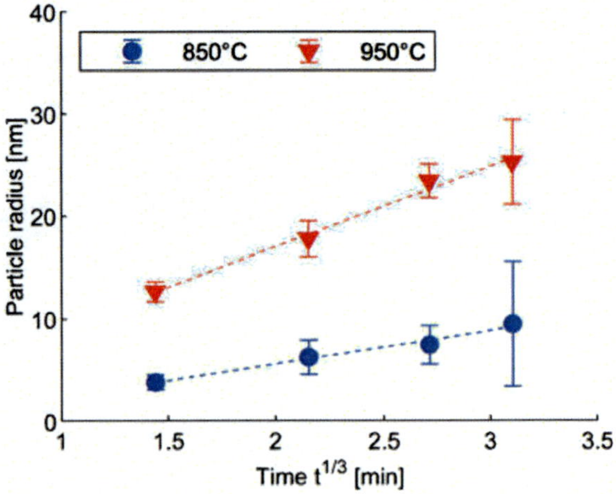

Table 3 Equilibrium γ' phase fraction and time to reach equilibrium at different temperatures for Waspaloy, calculated with JmatPro. Parentheses indicating amount of γ' formed after 1800s at 750 °C

Temperature [°C]	γ' phase fraction [%]	Exposure time [s]
750	22 (13.5)	4460 (1800)
800	20.24	1580
850	17.96	630
900	14.75	281
950	10.23	160

the measured particle size data to the LSW theory which otherwise should show a deviation from linear behaviour for shorter exposure times.

Direct isothermal exposure of mill annealed specimens during Gleeble testing produces the microstructure presented in Fig. 5. The precipitation of unimodal γ' can be observed for all investigated temperatures, while slower growth kinetics at the lower end of the investigated temperature window caused smaller precipitate sizes. Grain boundary carbide evolution is most pronounced at test temperatures up to 850 °C. Although some $M_{23}C_6$ carbides are present on grain boundaries after heat treatments at 950 °C, the close proximity to the $M_{23}C_6$ solvus temperature of 965 °C limited the carbide precipitation. Samples exposed to temperatures of 800 and 850 °C form a complete grain boundary carbide network with the carbides showing a connected, brick-like morphology. At 750 °C some grain boundary carbides are present after 1800s exposure, but not enough to form a connected network as shown in Fig. 6.

The hardness evolution is shown in Fig. 7, with the highest hardness of 409 ± 7 HV0.5 after 1200 s at 800 °C. This is close to the hardness obtained after a conventional age hardening heat treatment of 996 °C 2 h, 846 °C 4 h, and 760° 16 h (411 ± 3 HV0.5).

Fracture mode is intergranular for temperatures up to 850 °C when testing at stroke rates of 0.055 mm/s, with micro-voids on grain facets as shown exemplarily for a sample pulled after 1800s exposure at 800 °C. At 900 and 950 °C fracture mode transitioned towards ductile, which could be attributed to the increased temperature. The higher stroke rate of 55 mm/s did not lead to intergranular fracture at the lower test temperatures and instead grain rupture occurred. Figure 8 shows fracture surfaces after 1800s thermal exposure at 800 °C for 55 and 0.055 mm/s, respectively. The effect of stroke rate on the fracture morphology can be observed, which is discussed further below.

Comparison with Haynes 282

The hardness evolution and ductility response of Waspaloy is compared to that of Haynes® 282® [9]. The former is shown in the time–temperature-hardness (TTH) diagrams presented in Fig. 9, where the more rapid age hardening kinetics of Waspaloy is evident. The shape of hardening nose of Waspaloy is furthermore

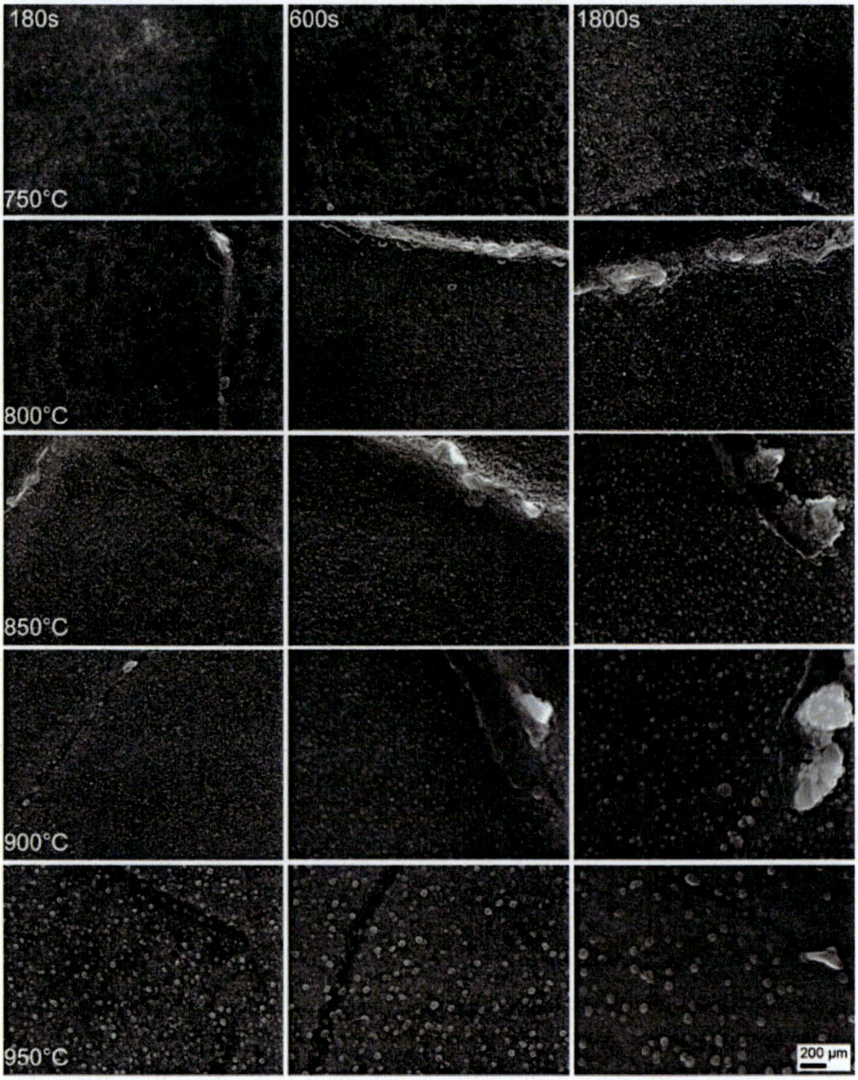

Fig. 5 Microstructure of Waspaloy after isothermal exposure in the temperature range of 750 to 950 °C, showing the evolution of γ' precipitation for 180, 600 and 1800s

different, with the nose location shifted towards higher temperatures, while the highest hardness is reached at 800–850 °C.

The rapid hardness increase of Waspaloy has a noticeable effect on the materials ductility when deformation is localized to the grain interior by a high deformation rate. The effect of the more rapid age hardening kinetics in Waspaloy leads to a clear decrease in ductility for long exposure times, as shown in Fig. 10. With ongoing

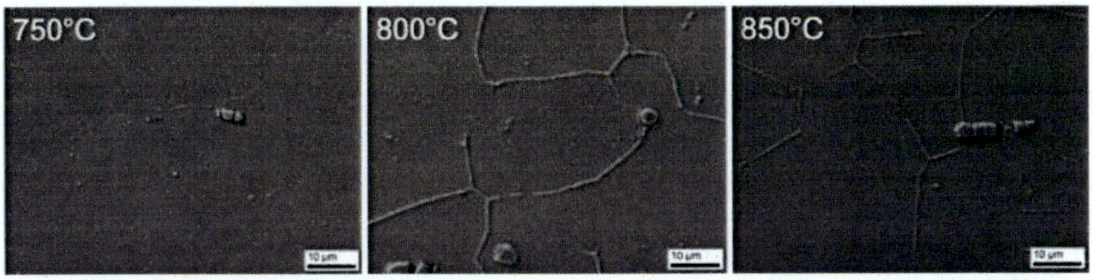

Fig. 6 Grain boundary carbide evolution during isothermal exposure. Minor formation at 750 °C, while a more continuous network of connected, brick-like carbides are formed at 800 and 850 °C

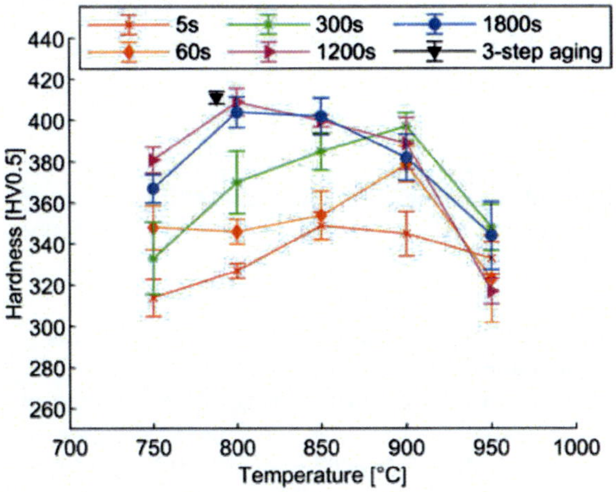

Fig. 7 Hardness evolution in Waspaloy during isothermal exposure. Hardness after conventional age hardening heat treatment is indicated for comparison

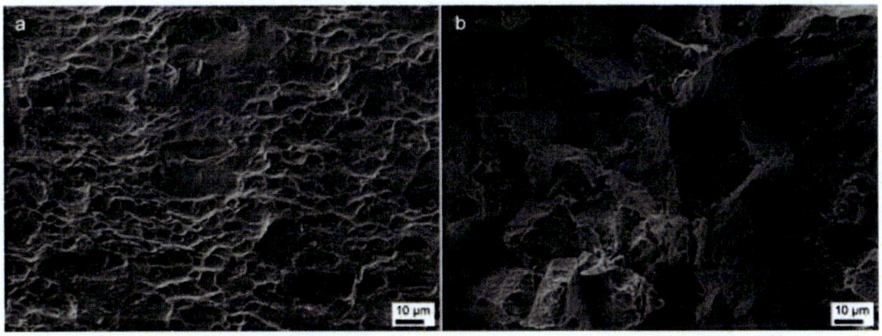

Fig. 8 Fracture surfaces of Waspaloy tested at 800 °C after 1800s. Grain rupture for 55 mm/s stroke rate (**a**), intergranular fracture, showing micro-voids on grain facets in (**b**) for a stroke rate of 0.055 mm/s

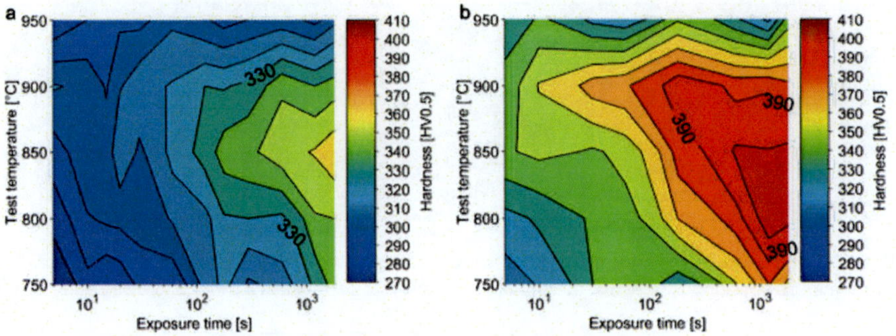

Fig. 9 Isothermal TTH-diagram of Haynes 282 (**a**) and Waspaloy (**b**) for exposure times up to 1800s. Contour spacing 10HV, average standard deviation ± 5 HV0.5

age hardening reactions, the ductility is also lower relative to Haynes® 282® which shows slower age hardening kinetics and lower overall hardness (cf. Fig. 9).

Using high stroke rates does not provide enough time for grain boundary slip to occur, which causes fracture in the grain interior. This in turn could explain the large effect of age hardening reactions in the grain interior on the ductility, where harder material is less ductile. Since strain age cracking is however a phenomenon that occurs on grain boundaries, testing with slower stroke rate is necessary to produce intergranular fracture as is characteristic for SAC. In the case of Haynes® 282® the effect of age hardening reactions was not found to significantly affect ductility when intergranular fracture is the dominating mode of failure [9]. Figure 11b shows ductility values measured for Waspaloy using a stroke rate of 0.055 mm/s, producing intergranular fracture. It can be observed that the three-step age hardening heat treatment leads to a higher ductility in the material as compared to the direct aging resulting from thermal exposure at a single temperature.

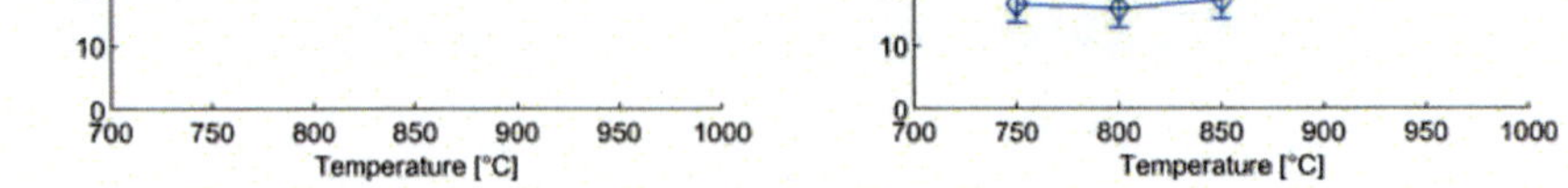

Fig. 10 Ductility as a function of temperature for samples tested with a stroke rate of 55 mm/s for Haynes® 282® (**a**) and Waspaloy (**b**). 3-step aging refers to furnace heat treatment followed by rapid heating to test temperature

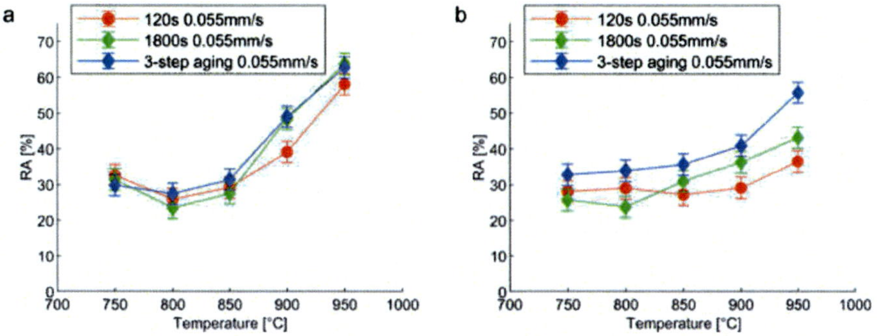

Fig. 11 Ductility as a function of temperature for samples tested with a stroke rate of 0.055 mm/s. Different ductility evolution for Haynes 282 (**a**) and Waspaloy (**b**), with isothermal exposure leading to reduced ductility for the latter. 3-step aging refers to furnace heat treatment followed by rapid heating to test temperature.

This seemingly counter-intuitive result can be explained by the difference in microstructure. The three-step heat treatment results in a bimodal γ' distribution, but also to the precipitation of grain boundary γ', as shown in Fig. 3a. Furthermore, the discrete particle distribution on the grain boundaries can be assumed to strengthen the grain boundaries more than direct aging at 750 and 800 °C. At those temperatures the grain boundaries develop a more continuous carbide network with large γ' precipitates being absent and the carbide morphology not being discrete. The combination of reduced grain boundary strength and the high hardness due to γ'-precipitation in the grain interior can explain the reduced ductility.

The effect of hardness increase is offset by the rapid formation of a grain boundary carbide network in Haynes® 282®, resulting in only minor variations in ductility. The significantly higher hardness increase in Waspaloy however appears to be the dominant mechanism affecting the ductility in the material, as 1800s exposure time led to a further reduction in ductility.

The secondary carbide species and their precipitation characteristics are different in the two alloys with Haynes® 282® containing both M_6C and $M_{23}C_6$ while Waspaloy only contains $M_{23}C_6$. Furthermore, Haynes® 282® forms a grain boundary carbide network at all investigated temperatures within the investigated range of exposure times [9], whereas no complete grain boundary coverage is observed for Waspaloy at 750 °C.

Two effects are observed to contribute to the ductility of Waspaloy. The formation of a grain boundary carbide network strengthens the grain boundaries, which is less pronounced at 750 °C, due to slower precipitation kinetics, while at 800 and 850 °C a relatively continuous network forms within 1800s. The second effect is the age hardening of the material due to γ' precipitation. Waspaloy shows a very rapid hardness increase, with the precipitation nose located at 800–850 °C. A hardness-increase of about 100 HV within the investigated time window of 1800s appears to

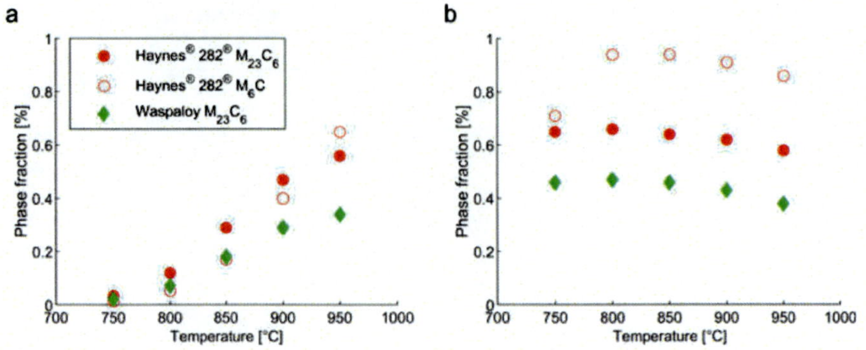

Fig. 12 Calculated secondary carbide phase fractions of Haynes® 282® and Waspaloy as a function of temperature after isothermal exposure of **a** 1800s and **b** 100000 s

have a larger effect than strengthening effects from the simultaneous formation of a grain boundary carbide network.

Haynes® 282® exhibits a more rapid formation of a grain boundary carbide network, which in combination with the presence of both $M_{23}C_6$ and M_6C carbides increases the total precipitate phase fraction as compared to Waspaloy. This visual observation is supported by the JmatPro simulations shown Fig. 12. The $M_{23}C_6$ phase fraction evolution after isothermal exposure of 1800 and 100000 s is systematically lower for Waspaloy. Note that the simulations assume a precipitate-free microstructure as a starting condition. The presence of low amounts of carbides in the as-received condition of both materials limits the accuracy of the simulation in terms of absolute values. The difference in precipitation kinetics and phase fraction can however clearly be seen.

At the same time, the alloy has a more moderate hardness increase with lower absolute hardness values and the location of the precipitation nose shifted upwards to 850 °C.

This difference in microstructural evolution can explain the higher SAC susceptibility of Waspaloy relative to Haynes® 282®. When comparing absolute ductility of the two materials, it becomes clear that the difference is only minor. Apart from the discussion above, considering the reported higher susceptibility towards SAC of Waspaloy [7, 11], Waspaloy was expected to show lower values than Haynes® 282®. A possible explanation is the difference in grain size of the two materials, which is 55 ± 5 μm for Haynes® 282® while that of the Waspaloy sheet investigated here is approximately half with 24 ± 5 μm. Large grain size is commonly seen as having a negative effect on SAC due to stress concentration on fewer grain boundaries [14, 15]. Having about half the grain size of Haynes® 282® could explain the high relative ductility measured for Waspaloy in this study, further investigations are however necessary to confirm this hypothesis.

Conclusions

The effect of age hardening reactions on the ductility of Waspaloy has been investigated in this study and compared to Haynes® 282®. Based on the results it can be concluded that.

- As-received Waspaloy contains fine γ' precipitates with a particle size of 1.7 ± 0.5 nm diameter
- Waspaloy shows rapid age hardening kinetics, resulting in a ~100 HV hardness-increase after 1800s isothermal exposure at 750–850 °C
- Ductility is affected by both age hardening via γ' precipitation and the formation of a network of grain boundary precipitates.
- Despite the high hardness, a conventional three-step age-hardening heat treatment produces the highest ductility, with a complete network of discrete grain boundary precipitates increasing grain boundary strength.
- The formation of a grain boundary carbide network during isothermal exposure cannot compensate for the hardness increase in the grain interior, leading to a ductility reduction at 750 and 800 °C.
- The slower age-hardening kinetics and lower absolute hardness of Haynes® 282® did not lead to a decrease in ductility for isothermal exposure up to 1800s.
- High absolute ductility of Waspaloy as compared to Haynes® 282® could be attributed to a difference in grain size.

Acknowledgements The support by the Consortium Materials Technology for Thermal Energy Processes (KME) through funding from Swedish Energy Agency and GKN Aerospace Sweden AB is highly appreciated. Haynes International is acknowledged for providing Waspaloy sheet material used in this study. Special thanks go to Asala Gbenga from the University of Manitoba for his help with TEM sample preparation and valuable discussions.

References

1. Sims CT, Stoloff NS, and Hagel WC (1987) Superalloys II. Wiley, New York
2. Sjöberg G (2010) Casting Superalloys for Structural Applications. 7th International Symposium on Superalloy 718 and Derivatives pp. 117–130
3. Lippold JC (2015) Welding metallurgy and weldability. John Wiley & Sons Inc Hoboken, New Jersey
4. Hanning F and Andersson J (2016) A Review of Strain Age Cracking in Nickel Based Superalloys. Conference Proceedings of the 7th International Swedish Production Symposium, Lund, Sweden
5. Fawley RW and Prager M. (1970) Evaluating the Resistance of René 41 to Strain-Age Cracking. WRC Bulletin, vol. 150, pp. 1–12
6. Rowe MD (2006) Ranking the resistance of wrought superalloys to strain-age cracking. Welding Journal 85(2):27s–34s
7. Metzler DA (2008) A Gleeble®-based method for ranking the strain-age cracking susceptibility of Ni-based superalloys. Welding Journal 87(10): 249s–256s

8. Pike LM (2006) HAYNES® 282™ Alloy - A New Wrought Superalloy Designed for Improved Creep Strength and Fabricability. Proceedings of ASME Turbo Expo 2006: Power for Land, Sea and Air, Barcelona, Spain vol. 05/2006, pp. 1031–1039
9. Hanning F, Khan AK, Steffenburg-Nordenström J, Ojo O, and Andersson J (2019) Investigation of the Effect of Short Exposure in the Temperature Range of 750–950 °C on the Ductility of Haynes® 282® by Advanced Microstructural Characterization. Metals 9(12): 1357
10. Haynes International Inc. (2008) Waspaloy alloy product brochure, Publication Number H-31228C
11. Pike LM (2008) Development of a Fabricable Gamma-Prime (γ') Strengthened Superalloy. Proceedings of the 11th International Symposium on Superalloys, pp. 191–200
12. Lifshitz IM and Slyozov VV (1961) The kinetics of precipitation from supersaturated solid solutions. Journal of Physics and Chemistry of Solids 19 (1): 35–50
13. Whelchel RL, Kelekanjeri VSKG, Gerhardt RA, and Ilavsky J (2011) Effect of Aging Treatment on the Microstructure and Resistivity of a Nickel-Base Superalloy. Metall. Mater. Trans. A 42(5):1362–1372
14. Carlton JB and Prager M (1970) Variables Influencing the Strain-Age Cracking and Mechanical Properties of René 41 and Related Alloys. WRC Bulletin, vol. 150 pp. 13–23
15. Berry TF and Hughes WP (1069) A Study of the Strain-Age Cracking Characteristics in Welded René 41 - Phase II. Welding Journal. 48(11): 505s–msms

Characterization of γ' Precipitation Behavior in Additively Manufactured IN738LC Superalloy via In-Situ Small-Angle Neutron Scattering

Hailong Qin, Hai Chi, Ying Tao, Mingzhao Xie, Songyi Shi, Hongyao Yu, Jinli Xie, Qing Tan, and Zhongnan Bi

Abstract In the present study, quantitative evaluation of the size and volume fraction of γ' particles have been studied by Small-Angle Neutron Scattering (SANS) and FE-SEM in additively manufactured IN738LC Superalloy employing isothermal heat treatment conditions at 850 °C. The results show that there is no observable γ' precipitation in the as-deposited alloy, while a large number of small cellular and striated sub-structures can be characterized with widths ranging from 0.5 to 1.5 μm. During the 850 °C isothermal heat treatment, the γ' phase rapidly precipitates in large quantities and shows a unimodal irregular sphere-like shape. The average size of the γ' particles increases with longer aging times. After 10 min of isothermal heat treatment the average γ' particle size is 72.5 nm. This average size increase to 137.3 nm after 120 min of ageing. However, the volume fraction (VF) of γ' precipitates does not change after 10 min of isothermal heat treatment time, at which point the VF % reaches about 40%. This result varies significantly from IN738LC that was prepared by traditional casting processes.

Keywords Small-Angle Neutron Scattering · IN738LC · Precipitation behavior · Additive manufacturing

---

Qin and Chi: These authors contributed equally to this work and should be considered co-first authors.

H. Qin (✉) · M. Xie · S. Shi · H. Yu · J. Xie · Z. Bi
Beijing Key Laboratory of Advanced High Temperature Materials, Central Iron & Steel Research Institute, Beijing 100081, China
e-mail: hailongqin@126.com

H. Qin · H. Chi · M. Xie · S. Shi · H. Yu · J. Xie · Z. Bi
Beijing GAONA Materials & Technology Co., LTD, Beijing 100081, China

Y. Tao
University of Science and Technology Beijing, Beijing 100083, China

Q. Tan
Department of Microstructure Physics and Alloy Design, Max-Planck-Institut Für Eisenforschung GmbH, Max-Planck-Straße 1, 40237 Düsseldorf, Germany

E. A. Ott et al. (eds.), *Proceedings of the 10th International Symposium on Superalloy 718 and Derivatives*, The Minerals, Metals & Materials Series,
https://doi.org/10.1007/978-3-031-27447-3_14

Introduction

Additive manufacturing (AM) technique is developing more and more quickly in the recent years, there are many types of manufacturing methods in this fabrication technology, among them, Selective laser sintering (SLS) and laser powder bed fusion (L-PBF) are the most prominent ones [1–3]. Compared with other traditional fabrication methods, this process has the following advantages: geometrical freedom which enable to build complicated morphology of products, shorter time of the process from design to product, mass customization, and material flexibility [4, 5]. Due to such advantages, laser powder bed fusion (L-PBF) technology is very suitable to fabricate complicated morphology and valuable products, especially shows its great potential in medical and aircraft industry [6, 7].

Nickel-based superalloys, which play a key role in aerospace and aviation industries as hot section components, are one of the first manufacturing metallic material. Nickel-based superalloys are generally composed of two phases: matrix γ phase and strengthening γ' phase. γ' phase is the main strengthening phase of nickel-based superalloy, which plays a decisive role in the mechanical properties of the alloy. With the increase of the mass fraction of γ 'phase, the dissolution temperature of γ' phase increases. When the mass fraction of γ 'phase is increased to 50%, and the mechanical properties at high temperature are significantly improved. Therefore, the volume fraction of γ 'phase in alloys applied in high temperature field is relatively high, and the volume fraction of γ' phase in high-performance superalloys often exceeds 50% [8, 9]. IN738LC is a typical high-end Ni-based superalloy developed to meet the needs for harsh working environment [9–11]. IN738LC has up to 51% mass fraction of γ 'phase at room temperature. The total amount of Al and Ti of its composition is over 6.8 wt%, which causes the susceptibility to strain age cracking (SAC) [12]. Particularly, γ' particles precipitate rapidly during the subsequent heating process, resulting in volume shrinkage and local stress concentration. An urgent need has to conduct a study of γ' precipitation behavior during the post-treatment process to solve the problem of component cracking caused by SAC.

In recent years, neutron small-angle scattering has been gradually used in the study of high-temperature alloys [13–17]. Small-Angle Neutron Scattering (SANS) technique is particularly suited to quantitative characterize the microstructure of two-phase systems [18–21]. In the present study, quantitative evaluation of the size and volume fraction of γ' particles have been studied by SANS and FE-SEM in additively manufactured IN738LC Superalloy employing isothermal heat treatment conditions of 850 °C.

Materials and Test Methods

In this experiment, samples are made from IN738LC powders that prepared by vacuum induction gas atomization. The diameter of the powders is 15~50 μm. The powders are cleaned by Ethanol and dried at 90°C for more than 6 h before used. The L-PBF building machine is SLM 125 with the laser source of single YAG (yttrium aluminum garnet) system. The building chamber is under argon atmosphere in order to protect from oxidation. The samples are built layer by layer through selective laser melting. In order to reduce the thermal gradient, the laser scanning direction in each layer is rotated 67°. The chemical composition of the alloy is shown in Table 1.

In order to study the precipitation behavior of γ 'after 850°C ageing treatment for different time periods, the ageing time was set as 1, 10, 30 min, 1H (one hour), 3H, 8H, 16H, 24H, and 36H. The heat treatment method is shown in Fig. 2. OM (Optical Microscope) and FE-SEM (Field Emission Scanning Electron Microscopy) were used to observe and characterize the microstructure of samples after ageing. The samples were subjected to electrolytic polishing followed by electrolytic corrosion. The electrolytic polishing agent was 20%H_2SO_4 + 80%CH_3OH, the voltage was about 18~22 V, and the polishing time was about 5~10 s. The electrolytic corrosion agent was 150 mlH_3PO_4 + 10mlH_2SO_4 + 15gCrO_3, the voltage was 3~5 V, and the corrosion time was about 1~5 s.

Small-Angle neutron scattering technology can give the microstructure information in the nanoscale range of materials, such as morphology, size, distribution, content, and interface information. It has the advantages of deep penetration, good statistics, and simple sample preparation. It has been widely used in polymer, biology, and materials research. Compared with other nanostructure characterization methods, small Angle scattering also has the advantages of no special sample preparation, a wide variety of samples, and abundant experimental results. The precipitation size statistic software (Precipitation-Pro software) was adopted to calculate the size of γ 'in this experiment. Then, Origin was used to calculate the frequency distribution of precipitation size and calculate the average value of precipitation size for comparison. The volume fraction and size change of the enhanced phase during heat treatment were measured by SANS. The in-situ 850 °C ageing heating process is shown in Fig. 1. After a rapid ramp-up to 500 °C, the scattering curve is collected by staying at a low temperature of 500 °C for 5 min, followed by a ramp-up to 850 °C at a rate of 20 °C/min for 2 h.

Table 1 Chemical composition of the IN738LC alloy (wt%)

Ni	C	Cr	Co	W	Ta	Mo	Al	Ti	Fe	Nb
Bal	0.1–0.2	15.7–16.3	8.9–9.0	2.4–2.8	1.5–2.0	1.5–2.0	3.2–3.7	3.0–3.5	≤ 0.5	0.6–1.1

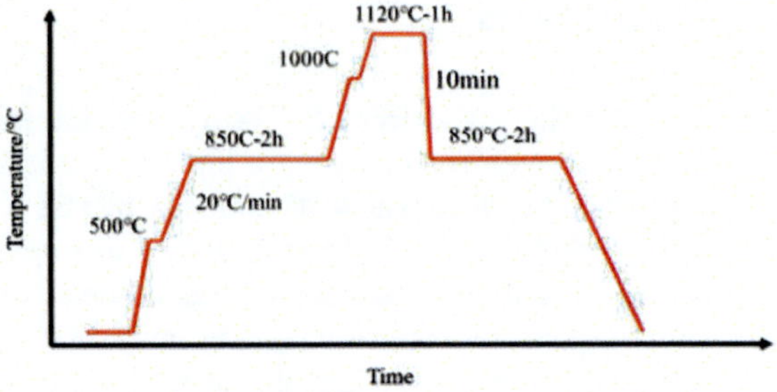

Fig. 1 Heat treatment regime of in-situ 850 °C ageing

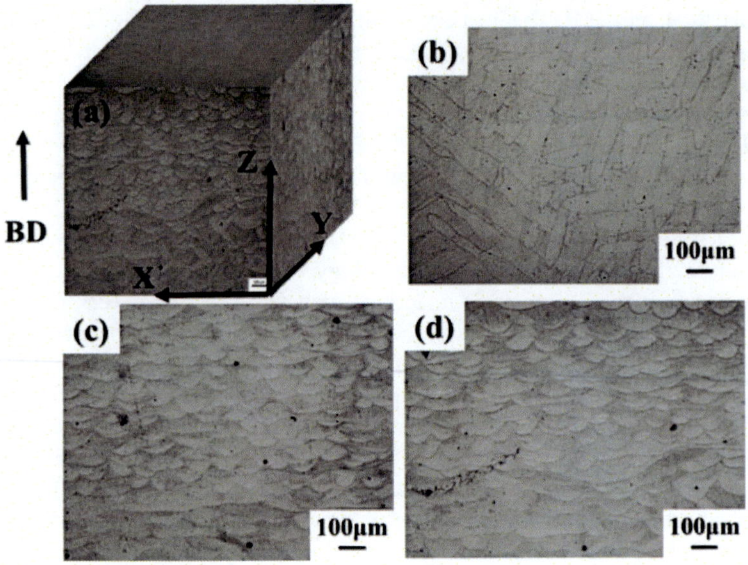

Fig. 2 OM image of the L-PBF as-fabricated products of IN738LC alloy **a** 3D image; **b** top view; **c** and **d** and two other side view

Results and Discussion

Characteristic Microstructure of the L-PBF As-Fabricated Sample

As mentioned above, the L-PBF process is a layer-wise building process. The heat is mainly dissipated through the already solidified part, which means downward to the base plate. Therefore, the grain growth direction will be mainly upward which is in the inverse thermal gradient direction. The internal structure of the sample in the initial state is shown in Fig. 2. It can be seen that channels (melt pool) overlapped with each other. The longitudinal section of the sample is shown in Fig. 2c and d, showing the horizontal arrangement of fish squamous molten pool structure. There were no obvious cracks that could be observed by an optical microscope.

In order to study the characteristic grain orientation in the L-PBF as-produced sample, EBSD for the sample are done, the results are shown in Fig. 3. Figure 3a and c is colored with respect to the Z direction separately from top view and side view. As can be seen, the Z direction (perpendicular to the building direction) shows a <001> cubic texture. It can be clearly observed that the grains in the transverse section are mainly equiaxed grains and the grains in the longitudinal section are mainly columnar grains, which further indicates that there are columnar grains growing along the forming direction in the longitudinal section. Figure 3b and d is colored with respect to KAM (Kernel Average Misorientation) figure separately from top view and side view, which can reflect the dislocation density and stress distribution in the alloy to a certain extent. As can be seen, there is a high-level local misorientation in the as-fabricated sample, especially at the grain boundary of each grain. The KAM value is high, indicating that the stress is concentrated here and the dislocation density is large. Besides, striated sub-structures can be characterized with widths ranging from 0.5 to 1.5 μm, as shown in Fig. 4.

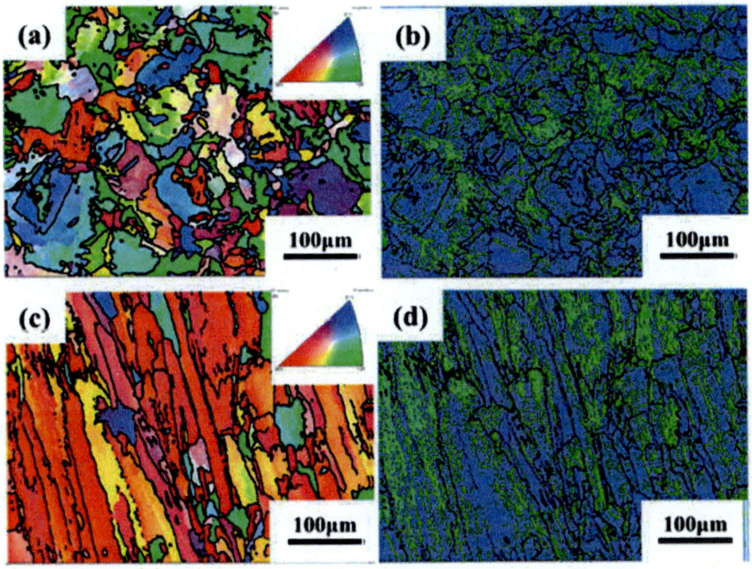

Fig. 3 EBSD measurement from side view of sample: **a** coloring with respect to the IPF in Z direction from top view; **b** KAM figure from top view; **c** coloring with respect to the IPF in Z direction from side view; **d** KAM figure from side view;

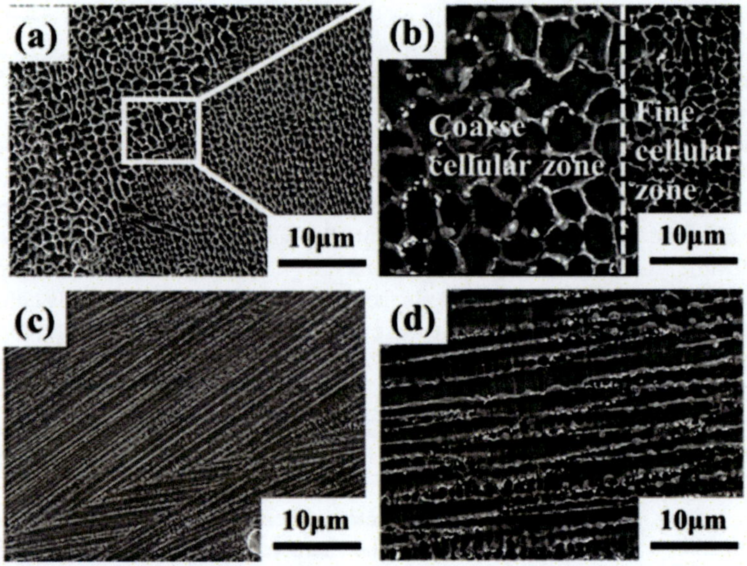

Fig. 4 Striated sub-structures of the L-PBF as-fabricated products of IN738LC alloy

Characteristic Microstructure After Different Time of Ageing Heat Treatment

Figure 5 shows the changes of the internal substructure of the alloy after different ageing times at 850 °C. When the ageing time is less than 1 h, the internal substructure of the alloy is uniformly distributed, while a small amount of carbide is attached to the boundary of the substructure. However, with further extension of the ageing time, the internal substructure of the alloy disappears, i.e., the high temperature and long ageing leads to dislocation recovery and reduction of small-angle grain boundaries, while a large number of submicron γ' phases can be observed.

The changes of γ' phase inside the alloy after different ageing times at 850 °C are showed in Fig. 6. A large number of nanoscale γ' precipitates with irregular curved short rods can be observed uniformly in the substructure at 1 min of ageing. During the 850 °C isothermal heat treatment, the γ′ phase rapidly precipitates in large quantities and shows a unimodal irregular sphere-like shape. The changes of γ' phase inside the alloy after different ageing times at 850 °C are showed in Fig. 6.

Figure 7 shows the size statistics of the γ' phase inside the alloy with different ageing times after analysis by the Precipitation-Pro software, and the average size of the γ' phase grows from about 40 nm at the initial ageing time of 1 min to 220 nm at 36 h. The average size of the γ′ particles increases as aging time increases: After isothermal heat treatment for 10 min, the average particle size is approx. 70 nm; and it reaches 150 nm when ageing time is 3 h. When the ageing time was extended to 8 h, the size of γ′ particles did not continue to grow, which was about 220 nm. The results

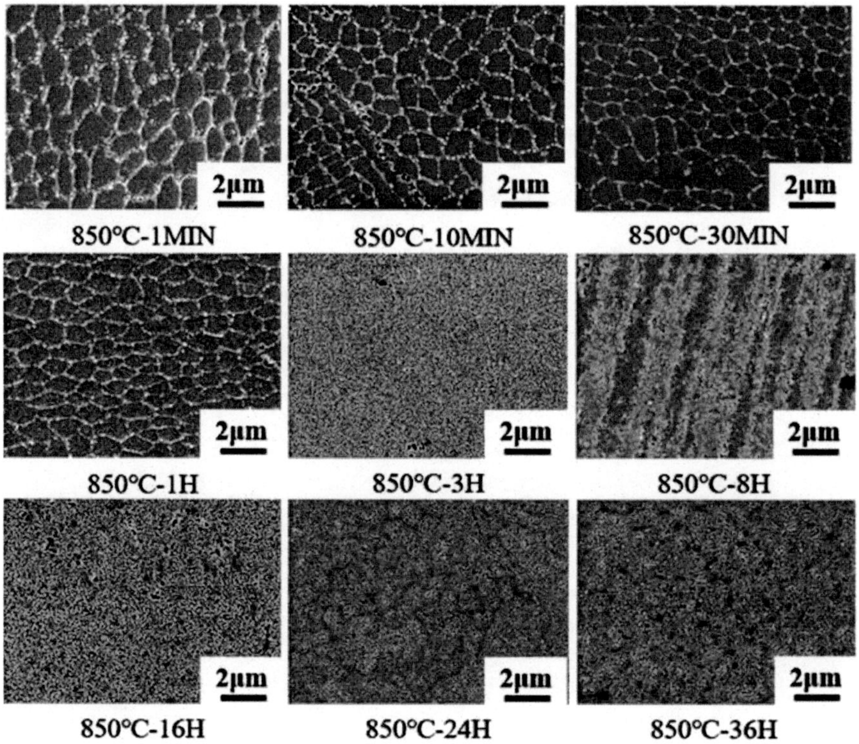

Fig. 5 Internal substructural changes of the alloy after different ageing times at 850 °C

of γ' phase precipitation show that the initial alloy is in a supersaturated state, and a large amount of γ' phase is precipitated rapidly during the heating process, which is completely precipitated at the early stage of ageing. The morphology of γ' phase shows an irregular rod shape with single-peak distribution, and the precipitation behavior is obviously different from the alloys formed by conventional nucleation growth.

In-Situ SANS Test

The quantitative information in the nanoscale range can be obtained by using the small-angle scattering by neutrons technique, and statistical information such as morphology, size, volume fraction of γ′ particle can be fitted by selecting appropriate model parameters. SANS is a method used to analyze the peak in scattered Neutron appearing at the scattering angle $2\theta = 0°$, although the data is usually expressed as a function of the scattering vector Q, defined as $|Q| = 4\pi\sin\theta/\lambda$. The SANS intensity, $I(Q)$ can be described as follows:

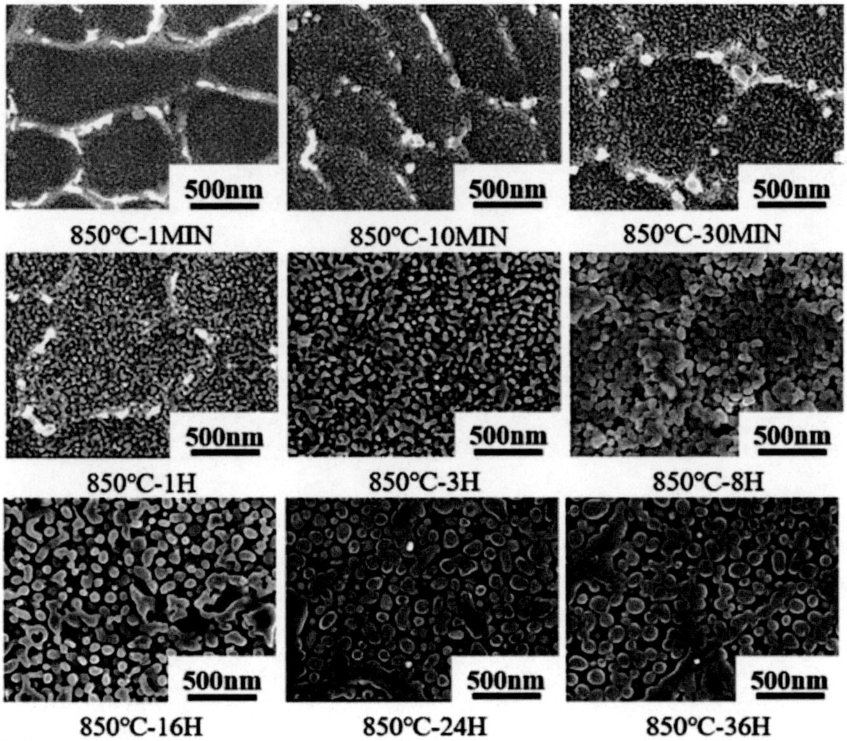

Fig. 6 Changes of γ' phase inside the alloy after different ageing times at 850 °C

$$I(Q) = \Delta\rho_p^2 N_p \int_0^\infty V_p(r) F_p{}^2(Q, r) f_p(r) dr \quad (1)$$

N_p is the number density factor, V_p is the volume of the particles, $\Delta\rho_p$ is the difference in scattering length density, F_p is the form factor, f_p is the size distribution function. Here, p indicates a group of the particles described by the size distribution. The typical scattering profile has two features, i.e., a relatively flat region in a log $I(Q)$-log Q plot, up to the point where Q attains one half of the intensity maximum.

As shown in Fig. 8, the scattering intensity versus scattering vector is plotted at different time nodes. During the heating process, when the temperature rises to 500 °C, the γ' phase has not yet precipitated, there is no peak shape of the scattering curve in the sample, which means no nanoscale inhomogeneous structure exists inside the alloy. When the temperature increased to 850 °C, the peak shape appears in the scattering curve, indicating the existence of nanoscale precipitation phase inside the alloy at this time. When the temperature was further increased to 1000 °C, the intensity of the scattering curve of the sample increased substantially and the size of the precipitated coarsened. When the temperature rises to 1120 °C-the precipitation phase back

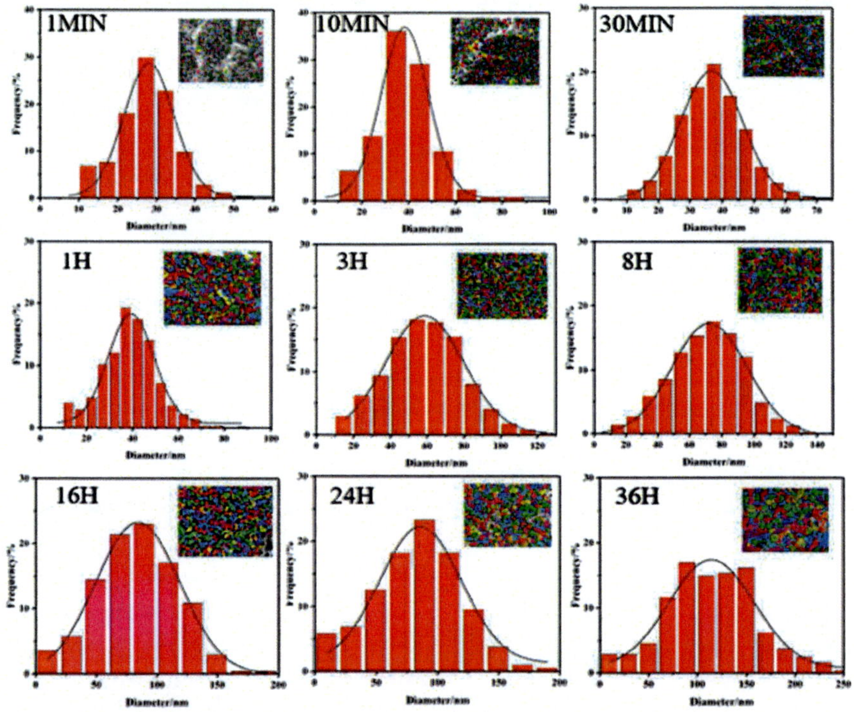

Fig. 7 Identification statistics of the internal γ' phase size of alloys with different ageing times at 850 °C

to the solubility point, the strength drops abruptly, at which time the internal precipitation phase of the alloy disappears. When the temperature drops again to 850 °C, precipitation phase is larger in size, resulting in the higher intensity of the scattering curve.

In order to specifically study the changes of precipitation phase size and volume fraction during direct ageing, the data at the first 850 °C ageing were fitted. As shown in Fig. 9, in the small scattering vector Q region, a clear peak shape appears after ageing at 850 °C, and the corresponding scattering intensity increases continuously with longer ageing time. The Hard Sphere model was used to analyze the relevant experimental data in the low Q region, and the multi-dispersion model LogNorm was selected. The distribution trends of size and volume fraction with time are shown in Fig. 10. With the ageing time increasing, the size of the scatter gradually increases and the volume fraction is nearly constant, fluctuating around 40%. The size distribution of the resolved phase shifts to the right and the number density decreases. The fit results are shown in Table 2. The fitted dimensions are slightly different from those observed under SEM and TEM, and are smaller than the morphological observations, which are considered to be due to the in-situ heating and interrupted observations as

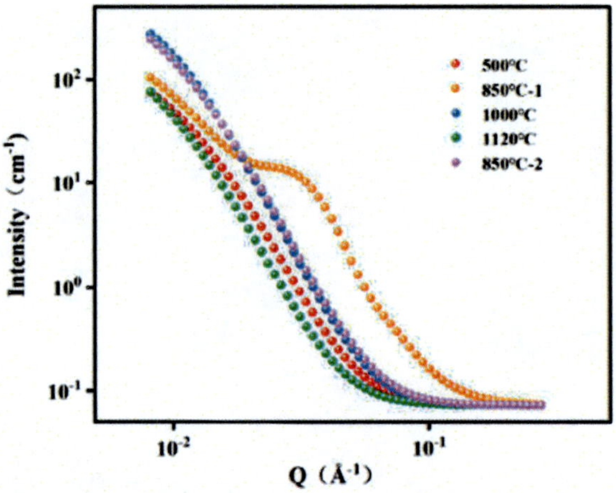

Fig. 8 Neutron small-angle scattering curves at different time points in in-situ heating; 850 °C-1: the first arrival at 850 °C during the heating process; 850 °C-2: solid solution at 1120 °C and returning to 850 °C for holding

well as the errors in the fitting function. The overall trend is consistent with the dimensional evolution observed under SEM at 850 °C. It is worth noting that precipitation of γ' phase is very fast and the volume fraction of γ' phase does not change with the extension of isothermal heat treatment time, reaching about 40% after 10 min, which is quite different from that prepared by traditional casting process [22].

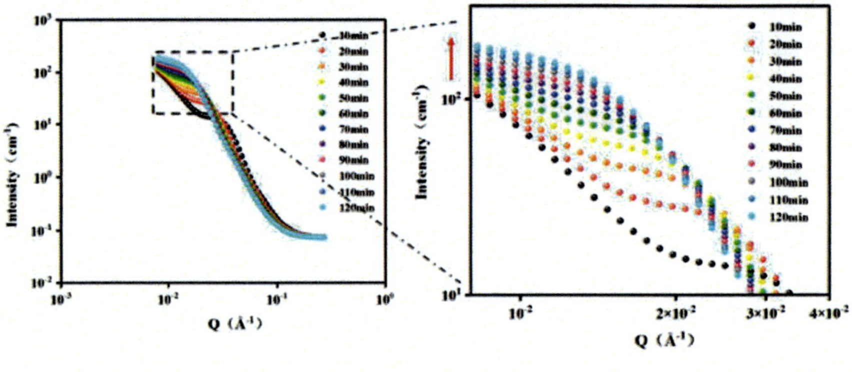

Fig. 9 Scattering curve of IN738LC under different ageing time at 850 °C

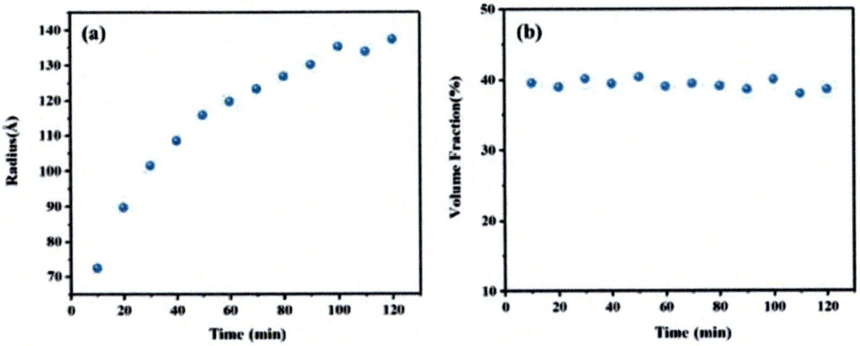

Fig. 10 Size and volume distribution of the precipitated phase of IN738LC after different ageing times at 850 °C **a** Precipitation phase size versus time; **b** Precipitation phase volume fraction versus time)

Table 2 Results of fitting the size and volume fraction of the precipitated phase for scattering curves after different ageing times at 850 °C

Time	Diameter (Å)	Volume fraction (%)
10 min	72.5	39.6
20 min	89.8	39
30 min	101.5	40.2
40 min	108.6	39.5
50 min	116	40.4
60 min	119.8	39.1
70 min	123.4	39.5
80 min	126.9	39.1
90 min	130.2	38.6
100 min	135.3	40.1
110 min	133.9	378
120 min	137.3	38.6

Conclusion

There is no observable γ′ precipitation in the as-deposited alloy, while a large number of small cellular and striated sub-structures can be characterized with widths ranging from 0.5 to 1.5 μm.

During the 850 °C isothermal heat treatment, the γ′ phase rapidly precipitates in large quantities and shows a unimodal irregular sphere-like shape. The average size of the γ′ particles increases on prolonging the time:

After isothermal heat treatment for 10 min, the average particle size is 72.5 nm; and it reaches 137.3 nm when ageing time is 120 min. However, the volume fraction of γ′ phase does not change with the extension of isothermal heat treatment time, reaching about 40% after 10 min.

Acknowledgements This work was supported by the National Natural Science Foundation of China (Grant No. 12205055) and the National Key Research and Development Program of China (Grant No. 2021YFB3702500).

References

1. Haines M P, Rielli V V, Primig S, et al. Powder bed fusion additive manufacturing of Ni-based superalloys: a review of the main microstructural constituents and characterization techniques[J]. Journal of Materials Science, 2022, 57(30): 14135–14187.
2. Fernandez Z P, Kirka M M, Rossy A M, et al. Nickel-based superalloy single crystals fabricated via electron beam melting[J]. Acta Materialia, 2021, 216: 117–133.
3. Trevor K, Greta L, Supriyo G, et al. Application of finite element, phase-field, and CALPHAD-based methods to additive manufacturing of Ni-based superalloys[J]. Acta Materialia, 2017, 139: 244–253.
4. Boswell J, Jones J, Barnard N, et al. The effects of energy density and heat treatment on the microstructure and mechanical properties of laser additive manufactured Haynes 282[J]. Materials & Design, 2021(6): 109725.
5. Yasa E, Craeghs T, Badrossamay M, et al. Rapid manufacturing research at the catholic university of Leuven[C]. Rapid Tech 2009: US-TURKEY workshop on rapid technologies, 2009: 1–10.
6. Zhang D, Feng Z, Wang C, et al. Comparison of microstructures and mechanical properties of Inconel 718 alloy processed by selective laser melting and casting[J]. Materials Science and Engineering: A, 2018, 724: 357–367.
7. Jiang X, Meng X, Song R, et al. Study on the Effects of Temperature Field of Material State Change in the SLM Process[J]. Applied Laser, 2015, 35(2): 155–159.
8. Olakanmi E O, Cochrane R F, Dalgarno K W. Densification mechanism and microstructural evolutionin selective laser sintering of Al-12Si powders[J], Journal of Materials Processing Technology, 2011, 221: 113–121.
9. Messé O M, Moreno R M, Illston T, et al. Metastable carbides and their impact on recrystallisation in IN738LC processed by selective laser melting[J]. Additive Manufacturing, 2018, 5: 394–404.
10. Ojo O A, Richards N L, Chaturvedi M C. Study of the fusion zone and heat-affected zone microstructures in tungsten inert gas-welded INCONEL 738LC superalloy[J]. Metallurgical and Materials Transactions A, 2006, 37(2): 421–433.
11. Grosdidier T, Hazotte A, Simon A. Precipitation and dissolution processes in γ/γ' single crystal nickel-based superalloys[J]. Materials Science and Engineering: A, 1998, 256(1–2): 183–196
12. Ojo O A, Richards N L, Chaturvedi M C. Contribution of constitutional liquation of gamma prime precipitate to weld HAZ cracking of cast Inconel 738 superalloy[J]. Scripta Materialia, 2003, 50(5): 641–646.
13. Mathon M H , Perrut M , Zhong S Y , et al. Small angle neutron scattering study of martensitic/ferritic ODS alloys[J]. Journal of Nuclear Materials, 2012, 428(1–3): 147–153.
14. Wang Y Q, Clark S J, Cai B, et al. Small-angle neutron scattering reveals the effect of Mo on interphase nano-precipitation in Ti-Mo micro-alloyed steels[J]. Scripta Materialia, 2020, 174: 24–28.
15. Karge L, Gilles R, Mukherji D, et al. The influence of C/Ta ratio on TaC precipitates in Co-Re base alloys investigated by small-angle neutron scattering[J]. Acta Materialia, 2017: 132.
16. Xin X, Joakim O, Magnus H C, et al. Effect of cooling rate after solution treatment on subsequent phase separation during ageing of Fe-Cr alloys: A small-angle neutron scattering study[J]. Acta Materialia, 2017, 134: 221–229.

17. Gilles R, Mukherji D, Eckerlebe H, et al. Investigations of early stage precipitation in a tungsten-rich nickel-base superalloy using SAXS and SANS[J]. Journal of Alloys & Compounds, 2014, 612: 90–97.
18. Pavel S, Debashis M, Oliver N, et al. Characterization of nanoporous superalloy by SANS[J]. Physica B: Physics of Condensed Matter, 2006, 385(part-P1): 626–629.
19. Zrník, J, Strunz P, Horňák P. et al. Microstructural changes in long-time thermally exposed Ni-base superalloy studied by SANS.[J]. Applied Physics A: Materials Science & Processing, 2002, 74: 1.
20. Brass A M, Chêne J. Sans analysis of γ' precipitation in the γ matrix of Ni base superalloy single crystals[J]. Scripta Materialia, 2000, 43(10): 913–918.
21. Fergin L A, Svergun D I. Structure Analysis by Small-Angle X-Ray and Neutron Scattering[M]. New York: Plenum Press, 1987: 68–90.
22. Rosenthal R, West D R. Continuous γ' precipitation in directionally solidified IN738 LC alloy [J]. Materials Science and Technology, 1999, 15: 1387–1394.

Chemical Mapping of Superalloys at the Nanoscale

Pritesh Parikh, Darshan Jaware, Jiangtao Zhu, Karol Putyera, and Rajiv S. Soman

Abstract Ni-based superalloys with applications to aerospace, nuclear, electrical, and automotive industries contain various alloying elements for enhanced thermal and corrosion resistance as well as improved mechanical strength. The underlying property and long-term stability of the superalloys is directly correlated to the processing conditions, materials structure, and alloying wt %. Hence, quantitative materials analysis of the alloying elements is essential for current and future superalloy development and failure investigations. Alloying elements introduced in superalloys can segregate to grain boundaries or form nanoscale clusters either as a result of processing conditions, secondary phase formation or failure from long-term use. Thus, suitable techniques at the nanoscale are necessary for a quantitative analysis and to understand the effect of clusters and segregation on material properties. APT (atom probe tomography) is the only analysis technique that can provide 3D elemental distribution with nm spatial resolution and up to 10 ppm chemical sensitivity to investigate both grain boundary segregation and clusters. Moreover, new processing techniques including additive manufacturing and nanoparticle-based alloys, further require suitable analytical techniques at the nanoscale, like APT, to identify chemical compositions. Herein, we use APT to better understand the influence of nanoscale effects in standard samples of a widely used Ni-based superalloy. Our analysis reveals differences in elemental distributions at low wt % that are responsible for material properties such as mechanical strength and oxidation resistance. We also compare bulk analysis using GDMS (glow discharge mass spectrometry), ICP-MS (inductively coupled plasma mass spectrometry) with APT to show localized fluctuations in the compositions.

The manuscript was written through contributions of all authors. All authors have given approval to the final version of the manuscript.

P. Parikh (✉) · D. Jaware · J. Zhu
Eurofins Nanolab Technologies, 1708 McCarthy Blvd, Milpitas, CA 95035, USA
e-mail: priteshparikh@eurofins-nanolab.com

K. Putyera · R. S. Soman (✉)
Eurofins EAG Materials Sciences LLC, 103 Commerce Boulevard, Liverpool, NY 13088, USA
e-mail: rajivsoman@eurofinseag.com

E. A. Ott et al. (eds.), *Proceedings of the 10th International Symposium on Superalloy 718 and Derivatives*, The Minerals, Metals & Materials Series,
https://doi.org/10.1007/978-3-031-27447-3_15

Keywords Atom probe tomography · Glow discharge mass spectrometry · Superalloys · Nanoscale composition

Introduction

Ni-based superalloys that were primarily developed by the aerospace industry have seen increased application due to their high strength and corrosion resistance in extreme temperature environments. Specially Inconel (Ni-based) alloys have replaced traditional stainless steel (Fe-based) alloys as the material of choice for gas turbine engine components, fan blades and parts for turbojet engines, exhaust systems in rocket, engine components in automotive parts, and as well in nuclear reactor cladding. This development has been made possible due to various alloying elements added to improve specific mechanical and thermal properties. Cr aids in corrosion resistance, Ta towards creep stress, oxidation resistance and casting processes, Al toward oxidation resistance, B for ductility and mechanical resistance and Co towards stacking fault energy mitigation [1].

However, stringent requirements and qualification on the material properties for Ni-based superalloys are necessary due to their industrial applications in extreme environments. Material analysis techniques such as GDMS, ICP-MS (inductively coupled plasma-mass spectrometry), and IGA (instrumental gas analysis) have traditionally been used to identify and quantify the compositions of the alloying elements (using GDMS and ICP-MS) and gas forming elements such as O, N, C, H, and S (using IGA). GDMS is a highly sensitive and robust analytical tool for monitoring elemental compositions and impurities in high-performance materials down to trace and ultra-trace mass fraction levels. Although these techniques investigate the bulk composition, enhanced performance of superalloys can be attributed to the influence of the microstructure on mechanical properties. In this respect, TEM (transmission electron microscopy) has been the technique of choice to understand dislocations, crystal structure, and associated changes at grain boundaries. However, TEM fails to accurately quantify nanoscale clusters coming from projection issues in 2D imaging and composition below 1 at%. APT (atom probe tomography) is thus a crucial technique towards analysis of superalloy microstructure to complement the structural information obtained from TEM. APT can provide analysis of segregating elements along grain boundaries and intergranular uniformity, analysis of clusters using data mining techniques and comparison of cluster versus matrix levels.

Based on extensive work in this field, a comprehensive understanding on the superior performance of Ni-based superalloy now exists as occurring from the formation of γ' (Ni_3Al/Ti) and γ'' (Ni_3Nb) precipitate phases after thermal ageing. The early work by Thomson et al. [2] demonstrated the application of APT towards high temperature materials including carbide precipitates in steels and nickel-based superalloys. The use of APFIM (atom probe field ion microscopy) and APT allowed analysis of precipitates and segregating elements along grain boundaries. To this effect, Miller et al. [1] and Blavette et al. [3] expanded on the application of APT to specifically

Ni-based superalloys. Kindrachuk et al. [4] investigated Inconel 706 using both HR-TEM and APT after direct ageing and modified stabilized heat treatment. Their work helped towards understanding γ' and γ'' precipitates and also observing Nb clusters around the precipitates. Bagot et al. [5] used APT for polycrystalline RR1000 Ni-based superalloy. Gardner et al. [6] Þ on Ni superalloys used for exhaust applications. Alam et al. [7] also demonstrated B, P, and C segregation in Inconel 718 and Rielli et al. [8] compared Inconel 718 with conventional and direct ageing for shape and size of nano precipitates of γ' and γ'' phase, which occur during ageing. Alternatively, Kulawik et al. [9] used EDX mapping and FIB-SEM-based tomography to get volume fraction of γ' and γ'' precipitates. First principles methods were used by Geng et al. [10] and Ebner et al. [11] and thermodynamic predictions by Miller et al. [12] to provide enhanced correlation between microstructure observed by APT and theoretical methods. Both Alam et al. [13] and Theska et al. [14] provided further understanding on formation and growth of the γ' and γ'' precipitates in early stages of the ageing process. Theska et al. [15] and Viskari et al. [16] explored the challenges on APT by exploring better analysis metrology with voltage and laser mode investigations as well as understanding the effect of the parameters on bulk composition and analysis of cluster sizes.

While the advantage of using APT in investigating nanoscale homogeneity allows observation and analysis of precipitates, one primary concern is the preferential evaporation of elements that affects accuracy of material compositions. Thus, to allow APT to become a tool for quantitative analysis for superalloys, comparison with standard samples as well as comparison with other techniques is necessary to establish suitable baselines. Herein, we use two commercially available Ni-superalloy reference materials to compare compositions and nanoscale homogeneity using bulk analysis techniques (GDMS, ICP-MS and IGA) with APT. The two reference materials used are from two different agencies, the first one from NIST (National Institute of Standards and Technology), which provides a standardized measurement infrastructure for US and the second one from ARMI (Analytical Reference Materials International) that is part of LGC Standards which is part of UK's designated National Measurement Institute. Through our analysis, we show that while bulk techniques are still necessary for determining accurate compositions, nanoscale clusters and segregations are observed with APT that can affect local composition. The clusters are rich in Al; however in each case, different sets of elements are also either rich or poor in the clusters. Possible reasons for deviations in composition obtained from APT compared to standard samples are also discussed. Our results demonstrate that standard samples would benefit from a nanoscale analysis to further understand onset of microstructure precipitates as well as bridge the gap in accurate quantification of composition using APT.

Experimental

Materials Used

Two commercially available Ni-based alloy reference materials obtained in disk form of size ~41 mm and ~19 mm thick were obtained for analysis; NIST SRM 1249 (referred to as Sample 1) and IARM 56H (referred to as Sample 2). NIST SRM standard sample was prepared by Inco Alloys, Inc. (Huntington, WV) and IARM 56H was prepared by ARMI/MBH. The reference standards have similar composition with different levels of low-concentration elements (below 1 wt %) which can provide a good comparison metric for APT and GDMS. The certified reference values for the composition are available for download [17, 18].

GDMS and ICP-MS

GDMS (glow discharge mass spectrometry) was carried out on the disks at the EAG Materials Science office in Syracuse. Multiple areas were chosen from the disks to get average values for the composition. GDMS is a mass spectrometric technique, which is based on glow discharge atomization/ionization source combined with a high-resolution sector-field mass analyzer. Argon is typically used as the discharge gas. Argon ions are accelerated towards the sample surface resulting in atomization of the analyzed area layer by layer. The sputtered species are diffused into the plasma where they are ionized. The ions are then extracted and subsequently accelerated into a mass spectrometer where they are separated according to their mass to charge ratio, permitting identification and measurement. The atomization and ionization processes in the analytical GD cells are largely separated in space and time, which is critical for the near matrix-independent quantification. Quantification is based on ion current ratios, and relative sensitivity factors (RSF).

ICP-MS (Inductively coupled plasma mass spectrometry) involves digestion of the sample to form a solution. The resulting sample solution is then nebulized, after which the generated aerosol is transported into the core of an inductively coupled argon plasma, where temperatures of approximately 9000 K are attained. At such high temperatures, the solution aerosol is vaporized, and the analyte species are atomized and ionized. In ICP-MS, ions generated in the high temperature argon plasma core are subsequently accelerated into a mass analyzer for both elemental and isotopic analysis. Intensity measurements are converted to elemental concentration by comparison with calibration standards. The analyses included in this study were performed on a Thermo Fisher Scientific Element GDMS, and a solution-based Perkin Elmer NexION 350D ICP-MS. Additional technical details are available [19].

IGA

A high temperature furnace is used to rapidly heat the sample and thereby convert gas elements into volatile forms in order to separate, detect and measure them. The analytical method is based on the complete oxidation of the solid sample by combustion in an oxygen plasma (above 2000 °C). The sample is placed in a ceramic crucible in a high-frequency induction furnace where it is heated. The combustion of the sample releases various gases, which are measured by appropriate detectors.

Sample Preparation for APT

Samples that were previously analyzed with GMDS were shipped from Syracuse, and used as is for APT sample preparation without any prior surface modification. The area chosen for the analysis was outside the GDMS analysis rings to ensure that an accurate representation of the standard sample itself was obtained. Standard FIB (focused ion beam) liftout procedures were used in order to prepare specimen tips for analysis, as described below.

Sample disks were loaded into the Helios 460 Dualbeam from Thermofisher Scientific. A thin ~200 nm Pt capping layer was placed on the area of interest using electron beam-assisted deposition followed by a thicker ~500–700 nm Ga ion beam-assisted Pt deposition. An 8 um long by 2 um wide section was carved into the sample using Ga ion beam milling at 30 kV accelerating voltage and ~2.5 nA beam current. The Omniprobe micromanipulator was used to lift out the wedge section and 2–3 tips for each sample were attached on Si micro posts using electron beam-assisted Pt in-fill. Subsequently, annular milling was carried out at 30 kV accelerating voltage and at reducing beam currents from 0.8 nA to 40 pA to create a conical shape on the specimen with an end radius less than 100 nm at the apex. A final low kV beam raster at 2 kV and 40 pA aids in removing the Pt and creating a specimen tip.

APT Analysis

APT technique uses a cryogenically cooled specimen tip with an apex radius <100 nm to field ionize (by application of a DC bias) and field evaporate (by AC pulsed voltage or laser) individual atoms that hit a 2D detector for spatial positioning. Concurrently, a time-of-flight measurement provides a chemical signature of the individual evaporating atoms. The LEAP 5000 XR at Eurofins Nanolab Technologies is used to acquire data on the specimen tips. A minimum of 2 tips for each sample were run at similar conditions to ensure direct comparison. The specimens were run at a temperature of 50 K with a laser pulse energy of 50 pJ and laser pulse rate of 200 kHz. The collected data was reconstructed and analyzed using AP suite 6.1.

While GDMS is a surface analysis technique, APT probes ~100 nm into the surface. Moreover, sample preparation forgoes the first ~5–10 nm of the surface. In this case, the APT analysis is expected to start ~10 nm below the surface. Additional technical details are available [20].

Results and Discussion

For a thorough analysis of Ni-based alloys that are known to have a microstructure, a nanoscale technique that can look within grain, at the grain boundaries and at nanoscale clusters is necessary. APT is a 3D technique that provides nanoscale spatial information along with ppm level chemical sensitivity that can meet this requirement. To understand the accuracy of APT towards quantitative analysis, the composition of the alloy is investigated first. Table 1 provides the compositions for Sample 1, comparing reference and certified values, GDMS results and composition analysis from 2 tips for APT. Since, APT composition is in at %, while certified, reference, and other analytical methods are reported in wt % the composition from APT is converted from at % to wt %, using total ranged ions for each data set and atomic masses for the most abundant isotope for that element from NIST database [21]. Table 1 shows that while certain elements have a close match between the certified values and APT data, not all elements are accurately quantified. Specifically, Al, S, C, and Zr from Table 1 are widely different from the certified and reference values. Moreover, differences between the 2 datasets for bulk composition with APT are also observed. GDMS provides composition of trace elements that matches with certified and reference values. From Table 2, a similar observation is obtained with Al, B, S, C, and P having widely different values compared to IARM certified values. GDMS, ICP-MS, and IGA in this case provide good reliability in composition. However, for the matrix elements in both cases, APT provides good agreement in the element composition (within 2 at %) ensuring that minor experimental/data analysis adjustments towards accurate values would improve reliability. Moreover, further observations on nanoscale composition changes are also valuable from APT.

Bulk analysis techniques like GDMS works with the assumption that the composition is uniform throughout the sample area analyzed. In order to investigate whether the compositions are homogenous, the 2D (two dimensional) ionic maps for each element as obtained from APT are shown in Fig. 1. 2D ionic maps are shown in the figures in lieu of 3D maps for ease of viewing which are also obtained using APT. Figure 1a shows all the ionic maps for Sample 1 and Fig. 1b shows the ionic maps for Sample 2. As seen from the maps, at the nanoscale, heterogeneity in the composition is seen in both samples. Both Sample 1 and Sample 2 show two types of non-uniformity. First is Al clusters that are nanoscale in size and contain O, while the second type is Si segregation either along boundaries or within the bulk. Additionally for Sample 2, the Al clusters are also rich in other elements, such as S, Ti, Cu, C, and Ga, which are uniform in Sample 1 (Fig. 1a). As seen in literature, the microstructure of alloys has a correlation with ageing, annealing processes, strength,

Table 1 Comparison of the overall bulk and low concentration composition analysis for Sample 1 using the reference values, two APT specimen tips, and GDMS. GMDS values are reported from measured values. Significant figures for APT values are reported in accordance with reference values from NIST. Certified values are provided for Al, Ti, Cr, Fe, Co, Ni, Cu, Nb and Mo. The remaining are reference values

Element	SRM 1249 (wt%)	APT Dataset 1 (wt %)	APT Dataset 2 (wt%)	GDMS (wt%)
Al	0.5682	0.1285	0.1263	Matrix
Cr	18.472	19.737	20.090	Matrix
Mg	0.0012	–	0.0025	0.0010
Nb	5.196	4.547	4.527	Matrix
Ti	0.959	0.944	0.930	Matrix
B	0.0023	0.0025	0.0025	0.0022
Cu	0.1402	0.1537	0.1494	-
Mn	0.108	0.125	0.120	0.1200
Ni	53.29	51.16	50.87	Matrix
S	0.00064	0.00554	0.02050	0.0008
V	0.0338	0.0671	0.0442	0.0360
C	0.0380	0.0008	0.0004	-
Fe	17.693	18.780	18.943	Matrix
Mo	3.112	3.427	3.445	Matrix
Si	0.120	0.156	0.151	0.1000
W	0.0846	0.0765	0.0670	–
Co	0.3371	0.5471	0.4382	Matrix
P	0.0134	0.0118	0.0301	0.0110
Ta	0.0027	–	–	0.0021 –
Zr	0.0029	0.0950	0.0187	
Ga	0.0019	0.0024	0.0096	0.0022
As	0.0013	0.0039	0.0026	0.0017
Sn	0.0024	0.0062	–	0.0021
Sb	0.00030	–	–	0.0003
N	–	–	–	–
O	–	0.0211	0.0103	–

and creep resistance [6–8, 13, 14]. The 2D maps reveal that even between the two samples compared, the composition at the nanoscale is not the same and thus can impact the microstructure.

Based on the 2D maps and compositional non-uniformity, the extent of the change in the elements is also important to understand. For the case of Al, the size of the clusters, the density, and the composition are parameters of interest to compare

Table 2 Comparison of the overall bulk and low concentration composition analysis for Sample 1 using the reference values, two APT specimen tips, and GDMS

Element	56H (wt%)	APT Dataset 1 (wt %)	APT Dataset 2 (wt%)	GDMS (wt%)
Al	0.51	0.20	0.11	0.4100
Cr[a]	17.86	19.03	18.52	17.7000
Mg	0.0010	0.0029	0.0033	0.0009
Nb[a]	5.20	4.66	5.20	5.1900
Ti[a]	0.96	0.97	0.98	0.9500
B	0.0045	0.0011	0.0011	0.0049
Cu	0.012	0.054	0.061	0.0082
Mn	0.040	0.048	0.048	0.0420
Ni[a]	53.8	51.5	52.0	53.9000
S[b]	0.0005	0.0583	0.0732	<0.0005
V	0.025	0.039	0.037	0.0300
C[b]	0.0248	0.0008	0.0062	0.0240
Fe[a]	18.6	19.5	19.0	18.2000
Mo[a]	2.90	3.26	3.36	2.8900
Si	0.054	0.087	0.083	0.0560
W	0.014	0.006	0.006	0.0059
Co	0.049	0.415	0.317	0.0520
P	0.0052	0.0156	0.0988	0.0059
Ta	0.006	0.003	0.003	0.0039
Zr	–	0.0562	0.0421	0.0001
Ga	–	0.0120	0.0167	0.0013
As	–	0.0052	0.0039	0.0011
Sn	–	0.0042	0.0083	0.0002
Sb	–	–	–	0.0001
N	0.0062	–	–	–
O	–	0.0575	0.0055	–

[a]These elements were determined by ICP-OES instead of GDMS and [b]were analyzed by IGA. GMDS, ICP-OES, and IGA values are reported from measured values. Significant figures for APT values are reported in accordance with reference values from IARM

between samples as well in aged samples. Figure 3 shows the analysis on the clusters for both samples. Sample 1 shows a smaller overall size of clusters within the analyzed volume. Moreover, the clusters are dispersed sporadically within the analysis volume indicating that a number density does not make sense to calculate for this case. Figure 2a–b shows the Al map and correspondingly all the clusters that were identified using a 1.5 at% iso-concentration surface for Al. Figure 2c shows the clusters that were selected for further analysis for the composition. Four clusters are chosen in this

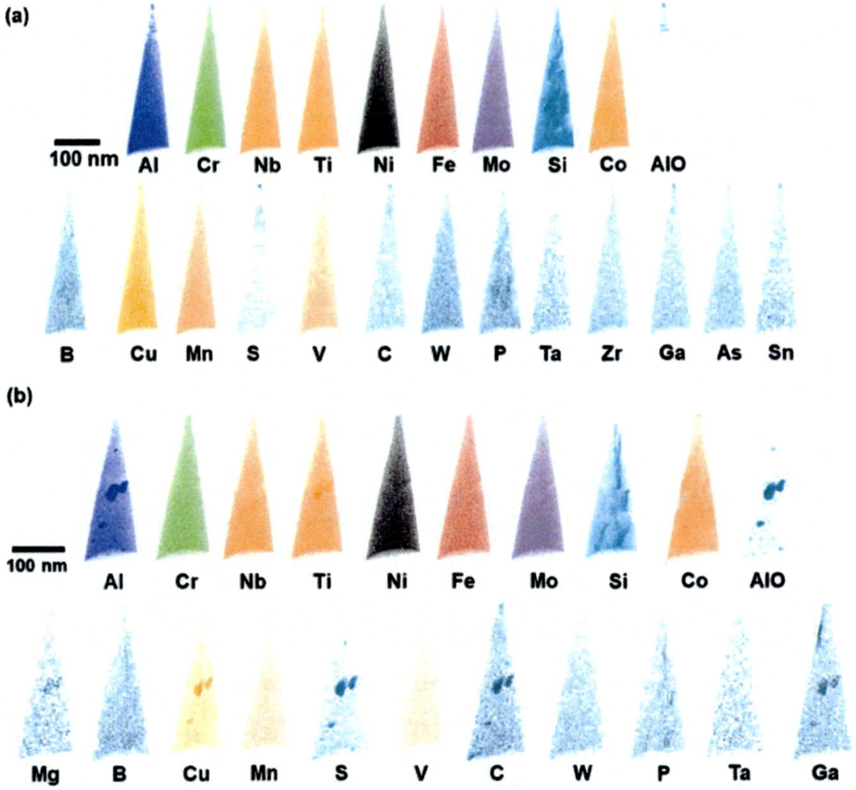

Fig. 1 2D ion map for each of the certified and reference elements in Ni-based superalloy for **a** Sample 1 and **b** and Sample 2. Non-uniform distribution is seen for Al, Si, and P for Sample 1 and for Al, Ti, Si, Cu, S, C, and Ga is seen for Sample 2

case to give an average value that would be more representative. The cluster extent in all three dimensions is shown in Fig. 2d. Figure 2e–f shows the elements that are either higher or lower within the cluster as compared to the matrix (right versus left of the 0 nm mark on the x-axis). Specifically, clustering of Al is accompanied by an increased concentration of O and P and a decrease in concentration of Ni and Fe. On the other hand, the clusters obtained in Sample 2 and the Al composition are shown in Fig. 2g–h as well as the three clusters used for the average composition are shown in Fig. 2i. The cluster extent as seen from Fig. 2j has sizes even over 10 nm. However, in this case the clustering of Al is not only accompanied by an increase in O but also P, C, Cu, S, and Ti. Concurrently a decrease in Cr, Fe, Ni, and Nb signals is seen in the cluster. The different elements that are clustering in Sample 1 versus 2 indicate that there are differences in the two samples from the onset that could likely carry forward in machined and prepared parts.

Compared to the changes in concentration seen in the clusters, Fig. 3 shows the changes in composition for Si. Figure 3a shows the cylindrical feature of size 25 nm

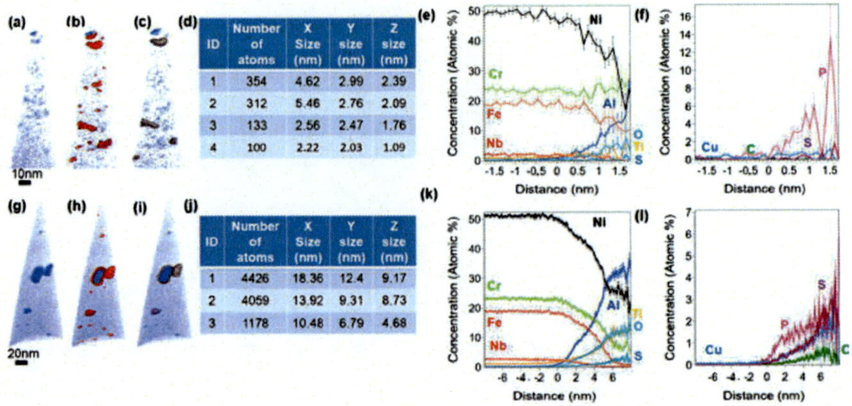

ID	Number of atoms	X Size (nm)	Y size (nm)	Z size (nm)
1	354	4.62	2.99	2.39
2	312	5.46	2.76	2.09
3	133	2.56	2.47	1.76
4	100	2.22	2.03	1.09

ID	Number of atoms	X Size (nm)	Y size (nm)	Z size (nm)
1	4426	18.36	12.4	9.17
2	4059	13.92	9.31	8.73
3	1178	10.48	6.79	4.68

Fig. 2 Analysis of clusters for Sample 1 and 2. **a** Shows the ionic map for Al with evidence of clusters for Sample 1 and (**g**) for Sample 2. **b** Shows all the clusters that were identified for an iso-concentration surface of 1.5 at% for Sample 1 and (**h**) Sample 2. **c, j** Shows the clusters that were chosen for analysis in **d–f** for Sample 1 and **k–l** for Sample 2. A difference in the elements that are clustering, and their concentration amounts are seen in **e, f** for Sample 1 versus **k–l** for Sample 2

by 25 nm and 45 nm long that is chosen for Sample 1 and (c) for Sample 2. Figure 3b and d show changes in the Si composition below 1 at% for both cases.

Based on the observations from Sample 1 and Sample 2, the difference in quantification for Al, B, S, C, P, and Zr could be explained as follows:

(a) for the case of Al and S in Sample 1 and Al, C, S, and P in Sample 2, the non-uniform distribution could indicate that while larger volumes have a constant value, the local value in nanoscale volumes can be different, which is evidenced in this case both by the total value for that data set and the clusters from Figs. 1 and 2. In contrast, for the case of Si, one of the reasons for systematic slightly higher composition in Tables 1 and 2 could be due to the much lower compositional changes as seen from Fig. 3.

(b) for the case of B, an underestimation which is also seen in Si-B films [22] could be due to the conditions needed to run the tips. One way to solve this would be to focus on B separately and run the samples at different conditions that are conducive to B and not the other elements. (c) With Zr and some of the other elements which are still different in composition, but are uniform in composition, the most likely explanation falls to the mass overlaps that APT faces in analysis for a multi-element system. Multiple overlaps are seen across the full list of elements and while the decomposed peak count for each sample is used in Tables 1 and 2, the full deconvolution for such a large set of elements is not possible to do directly without any additional post-processing and sample runs. However, comparison of data between samples provides a systematic analysis of the trends involved.

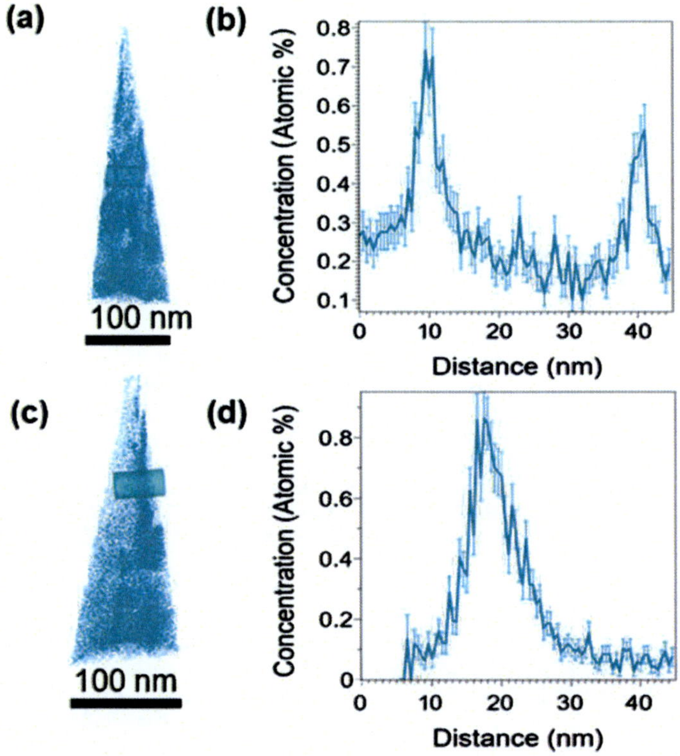

Fig. 3 Analysis of Si segregation for Sample 1 (**a**, **b**) and Sample 2 (**c**, **d**). A similar segregation level is seen in both cases

Conclusions

In conclusion, two different standard samples were evaluated with GDMS, ICP-MS, IGA, and APT to provide a comparison for the accuracy of the bulk composition. While GDMS provides better accuracy in overall composition, APT uniquely captures the potentially critical compositional heterogeneity at the nanoscale and can provide further analysis on the form and nature of it. The standard samples and comparative analysis will help to move APT from a qualitative to quantitative technique. Differences in standard samples are essential for qualification of feedstock and base materials used towards comparing new manufacturing approaches such as additive manufacturing. As such GDMS, ICP-OES, XRF, ICP-MS, and IGA will undoubtedly be essential in quantifying bulk compositions. However APT provides both a 3D visual analysis and understanding of the microstructure within grains and at nanoscale grain boundaries which can be linked to mechanical strength, fatigue, ageing, creep stress, and processing conditions.

Acknowledgements Robert Glaser performed recent GDMS measurements for NIST SRM 1249. The GDMS, ICP, and IGA teams carried out the respective analysis for IARM 56H.

References

1. Miller, M. K. Contributions of atom probe tomography to the understanding of nickel-based superalloys. 8 (2001).
2. Thomson, R. C. & Miller, M. K. Atom probe characterisation of high temperature materials. *Materials Science and Technology* 16, 1199–1206 (2000).
3. Blavette, D., Cadel, E. & Deconihout, B. The Role of the Atom Probe in the Study of Nickel-Based Superalloys. *Materials Characterization* 44, 133–157 (2000).
4. Kindrachuk, V. *et al.* Intragranular Precipitation in Inconel 706: 3D Atom-Probe and HRTEM Investigations. *steel research international* 75, 74–78 (2004).
5. Bagot, P. A. J. *et al.* An Atom Probe Tomography study of site preference and partitioning in a nickel-based superalloy. *Acta Materialia* 125, 156–165 (2017).
6. Gardner, H. *et al.* Atom Probe Tomography Investigations of Microstructural Evolution in an Aged Nickel Superalloy for Exhaust Applications. *Metall and Mat Trans A* 50, 1862–1872 (2019).
7. Alam, T. *et al.* Segregation of B, P, and C in the Ni-Based Superalloy, Inconel 718. *Metall and Mat Trans A* 43, 2183–2191 (2012).
8. Rielli, V. V. *et al.* Evolution of nanoscale precipitates during common Alloy 718 ageing treatments. *Materials & Design* 205, 109762 (2021).
9. Kulawik, K., Buffat, P. A., Kruk, A., Wusatowska-Sarnek, A. M. & Czyrska-Filemonowicz, A. Imaging and characterization of γ' and γ'' nanoparticles in Inconel 718 by EDX elemental mapping and FIB–SEM tomography. *Materials Characterization* 100, 74–80 (2015).
10. Geng, W. T., Ping, D. H., Gu, Y. F., Cui, C. Y. & Harada, H. Stability of nanoscale co-precipitates in a superalloy: A combined first-principles and atom probe tomography study. *Phys. Rev. B* 76, 224102 (2007).
11. Ebner, A. S. *et al.* Grain boundary segregation in Ni-base alloys: A combined atom probe tomography and first principles study. *Acta Materialia* 221, 117354 (2021).
12. Miller, M. K., Babu, S. S. & Burke, M. G. Comparison of the phase compositions in Alloy 718 measured by atom probe tomography and predicted by thermodynamic calculations. *Materials Science and Engineering: A* 327, 84–88 (2002).
13. Alam, T., Chaturvedi, M., Ringer, S. P. & Cairney, J. M. Precipitation and clustering in the early stages of ageing in Inconel 718. *Materials Science and Engineering: A* 527, 7770–7774 (2010).
14. Theska, F. *et al.* On the early stages of precipitation during direct ageing of Alloy 718. *Acta Materialia* 188, 492–503 (2020).
15. Theska, F., Ringer, S. P. & Primig, S. Atom Probe Microscopy of Strengthening Effects in Alloy 718. *Microsc Microanal* 25, 470–480 (2019).
16. Viskari, L. & Stiller, K. Atom probe tomography of Ni-base superalloys Allvac 718Plus and Alloy 718. *Ultramicroscopy* 111, 652–658 (2011).
17. https://www-s.nist.gov/srmors/certificates/1249.pdf
18. https://www.armi.com/hubfs/Certificates%20of%20Analysis/IARM-56H.pdf
19. https://www.eag.com/techniques/mass-spec/glow-discharge-mass-spectrometry-gdms/
20. https://www.eag.com/techniques/imaging/atom-probe-tomography-apt/
21. https://www.nist.gov/pml/atomic-weights-and-isotopic-compositions-relative-atomic-masses
22. Martin, A. J. & Yatzor, B. Examining the Effect of Evaporation Field on Boron Measurements in SiGe: Insights into Improving the Relationship Between APT and SIMS Measurements of Boron. *Microsc Microanal* 25, 617–624 (2019).

Part III
Environmental Behavior and Protection

Compatibility of Wrought Superalloys with Supercritical CO_2

B. A. Pint

Abstract Supercritical CO_2 (sCO_2) power cycles, particularly direct-fired cycles, have the possibility of revolutionizing clean fossil energy with peak temperatures above 700 °C and wrought precipitation strengthened alloys like Haynes 282™ for structural components. At temperatures <650 °C, it would be desirable to use less expensive alloys, however, steels are known to be susceptible to carburization. Laboratory 300 bar sCO_2 autoclave results were collected on a range of alloys including less expensive Ni-based alloys like 825 compared to advanced austenitic steels like alloy 709 at 600 °C. Both alloys 825 and 709 formed thin, protective Cr-rich oxides after 1,000 h. Alloy 825 also was exposed for 1,000 h in sCO_2 at 800 °C and compared to a range of Ni-based alloys. Comparing alloys 625, 825, and 282, the mass gain increased with increasing alloy Ti content under these conditions. High Al superalloys did not perform significantly better under these conditions at 800 °C.

Keywords Supercritical carbon dioxide · Environmental resistance · High temperature oxidation

Introduction

Supercritical CO_2 (sCO_2) is a potential working fluid for a number of power generation technologies including fossil, nuclear, geothermal, concentrating solar power (CSP), and waste heat recovery [1–4]. The supercritical state offers a unique set of properties and CO_2 has a relatively low critical point (31 °C/73.8 bar). Nickel-based alloys have been found to be very compatible with sCO_2 at 700–800 °C [5–13], where greater than 50% cycle efficiency is predicted [14]. Considerable development work was completed by the Advanced Ultra-supercritical (A-USC) consortium which had a goal of increasing steam temperatures to 760 °C/34.5 MPa (1400°F/5000 psi) in fossil-fired boilers [15]. The accomplishments included qualifying precipitate-strengthened (PS) γ/γ' alloys 740 (and 740H) and 282 [16, 17] into the ASME boiler

B. A. Pint (✉)
Oak Ridge National Laboratory, P.O. Box 2008, Oak Ridge, TN 37831-6156, USA
e-mail: pintba@ornl.gov

E. A. Ott et al. (eds.), *Proceedings of the 10th International Symposium on Superalloy 718 and Derivatives*, The Minerals, Metals & Materials Series,
https://doi.org/10.1007/978-3-031-27447-3_16

and pressure vessel code. Their high strength reduces the required wall thickness on pipes and tubes compared to conventional solid solution strengthened (SS) alloys (e.g., 625, 230, 617) [11].

However, for sCO_2-based power cycles to be commercially competitive, lower cost materials are needed in the lower temperature components in the cycle. There is concern about the use of steels in sCO_2 because of prior experience with Grade 9 (Fe-9wt.%Cr-1Mo) in the UK advanced gas cooled reactors (AGRs) operated with 42 bar CO_2 (sub-critical conditions) at $\leq$550 °C where severe internal carburization was observed [18]. A review concluded that creep-strength enhanced ferritic (CSEF) 9–12%Cr steels were limited to 450 °C in sCO_2 [19], significantly lower than the 580°-600 °C limit in supercritical steam [20]. A nuclear fuel cladding study found that a stainless steel such as type 316FR had much better oxidation resistance than 9–12%Cr CSEF steels in sCO_2 but 316FR began to show accelerated oxidation at 600 °C [21]. Carbon ingress in Fe-based alloys exposed to CO_2 environments has been observed for many years [22–24] but more recent studies by Young and co-workers [25, 26] have clearly explained the thermodynamic driving force for this behavior.

The current work is focused on identifying lower cost alloys for sCO_2 cycle components, particularly in the transition region where steels may not be sufficiently compatible with sCO_2. The current comparison included the high-Fe content, lower cost, alloy 825, and conventional and advanced austenitic steels at 600 °C. Alloy 825 also was exposed at 800 °C and compared to Ni-based PS and superalloys and one Co-based alloy that may be candidates for sCO_2 turbine components.

Experimental Procedure

The chemical compositions of the structural alloys studied are shown in Table 1. Alloy coupons (~12 × 20 × 1.5 mm) were polished to a 600 grit finish and ultrasonically cleaned in acetone and methanol prior to exposure. The specimens were exposed for 500-h cycles at 600° and 800 °C in 300 bar research grade (RG) sCO_2. Previously, the H_2O content of the RG CO_2 was measured by the vendor in 6 cylinders as 4.1 ± 0.7 ppm and the O_2 content is reported as <5 ppm [13]. The exposures were conducted in a vertically oriented autoclave (~266 mm × 83 mm inner diameter) made from alloy 282 and operated inside a three-zone furnace with an alloy 282 sample rack that sat on the bottom of the autoclave. The fluid flow rate was ~2 ml/min and additional details of the system have been provided elsewhere [6, 9, 12, 13]. The specimens were heated to temperature over several hours (~2 °C/min) in sCO_2, held at temperature ± 2 °C, and cooled in CO_2 to room temperature after exposure.

Because of variability observed in previous experiments, 4–6 coupons of each alloy were exposed in each condition with one specimen removed after 500 h. Before and after exposure, all specimens were weighed using a Mettler Toledo model XP205 balance with an accuracy of ± 0.04 mg or 0.01 mg/cm^2. For characterization, specimens were Cu-plated before being sectioned and mounted for light

Table 1 Alloy compositions determined by inductively coupled plasma and combustion analyses

Alloy	Fe	Ni	Cr	Al	Si	Other
316H	69.5	10.0	16.3	0.02	0.46	2.0 Mo, 0.84 Mn, 0.3 Co, 0.3 Cu, 0.041 C, 0.04 N
CF8C-Plus	61.1	12.8	19.3	0.01	0.57	3.9 Mn, 0.4 Mo, 0.5 Cu, 0.95 Nb, 0.09 C, 0.22 N
253MA	64.6	11.1	20.6	0.01	1.6	0.69 Mn, 0.3 Mo, 0.3 Cu, 0.17 Co, 0.03 Ce, 0.01 La, 0.31 C, 0.15 N
709	51.3	25.2	20.1	0.01	0.41	1.49 Mo, 0.89 Mn, 0.23 Nb, 0.02 Ti, 0.15 N, 0.06 C
825	30.8	39.5	22.7	0.18	0.34	3.0 Mo, 1.7 Cu, 0.97 Ti, 0.55 Mn, <5 ppmS, 0.02 C
617	1.2	54.6	22.3	1.0	0.05	11.9 Co, 8.2 Mo, 0.4 Ti, 0.12 W, 0.04 Mn, 0.06 C
282	0.16	57.1	19.6	1.6	0.04	10.6 Co, 8.6 Mo, 2.2 Ti, <3 ppmS
247	0.07	59.5	8.5	5.7	0.03	9.8 Co, 9.9 W, 0.7 Mo, 3.1 Ta, 1.0 Ti, 1.4 Hf
X4	0.05	60.8	6.4	5.8	0.02	9.5 Co, 6.5 Ta, 6.4 W, 2.9 Re, 0.96 Ti, 0.6 Mo, 0.08 Hf
NAFA	9.6	62.4	16.4	4.2	0.01	5.4 Mo, 0.5 Ti, 0.4 W, 0.6 Mn, 0.3 C, 0.06 Zr, 0.01 La
188	2.2	22.3	23.2	0.2	0.3	35.4 Co, 14.7 W, 1.0 Mn, 0.1 Ti, 0.04 La

microscopy and scanning electron microscopy (Tescan model MIRA3) equipped with energy-dispersive X-ray spectroscopy (EDS).

Results

Figure 1 shows the mass change data for specimens exposed to 300 bar RG sCO_2 for 500 and 1000 h at 600 °C. The conventional 316H austenitic steel specimens showed high mass gains under these conditions. Table 1 shows that it has the lowest combination of Cr and Ni, and it will not be further discussed here. Specimens of cast CF8C-Plus and wrought 253MA showed considerably lower mass gains under these conditions. This is an expected benefit of higher Cr and Ni contents in these alloys, Table 1. However, the mass gains were much higher than the 709 and 825 specimens, Fig. 1.

Figure 2 shows polished cross-sections of one of the specimens after 1000 h sCO_2 exposures at 600 °C. The 253MA and CF8C-Plus specimens showed more significant nodule formation than the 709 and 825 specimens, consistent with the mass change. Characterization using EDS revealed that the nodules were Fe-rich with the typical duplex structure of inward-growing Cr-rich spinel-type oxide and outward-growing Fe-rich oxide. The higher Ni and Cr contents in these alloys may prevent the nodules from growing and spreading to form a continuous duplex structure, like that formed on 316H in sCO_2 at 650 °C [27]. However, longer exposures are needed to determine to what extent the nodules will continue to grow. Perhaps of more importance is the C ingress that may have occurred during exposure. Those measurements, by bulk combustion analysis and surface measurements by sputtering glow discharge

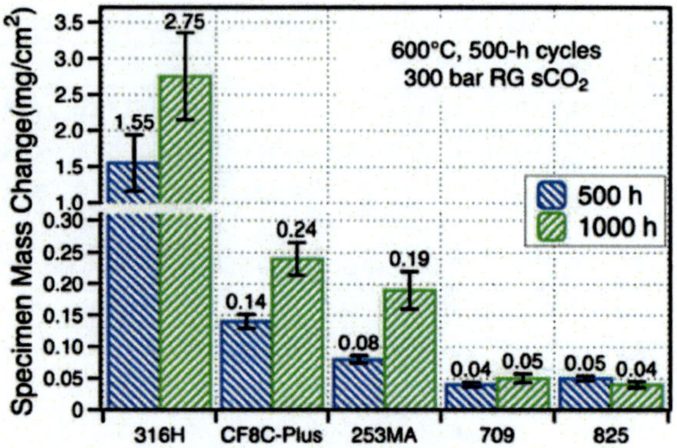

Fig. 1 Specimen mass change after 500 and 1000 h at 600 °C in 300 bar sCO_2. The whiskers show one standard deviation of 3–6 specimens exposed at each condition

optical emission spectroscopy (GDOES) or electron-probe microanalysis (EPMA) using wavelength dispersive spectroscopy, are in progress.

Figure 2c and d shows the thin reaction products formed on 709 and 825 specimens, respectively. Because the Cu plating separates at the scale interface, the scale on the 709 substrate looks thicker in Fig. 2c. Higher resolution SEM imaging was needed to characterize the scale. A few small oxide nodules also were observed on these specimens. Figure 3a shows a SEM image of the thin scale formed on alloy 709 at 600 °C. The associated EDS maps in Fig. 3 show that the oxide scale is enriched in Cr, Mn, and a small amount of Si. The Cr map in Fig. 3b shows Cr depletion near the metal-scale interface but the ~1 μm spatial resolution of EDS makes it difficult to quantify the depletion which appears to be on the order of 1 μm in depth. Figure 4 shows a similar SEM/EDS analysis of one alloy 825 specimen exposed for 1000 h at 600 °C. In this case, the Cr-rich oxide appears to contain Ti and Mn with some Mo-rich precipitates evident. Figure 4b also shows a small amount of Cr depletion

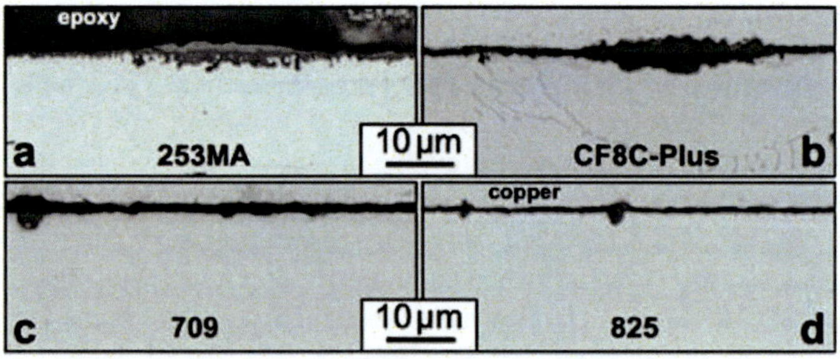

Fig. 2 Light microscopy of polished cross-sections of specimens after exposure to 300 bar sCO_2 for 1000 h at 600 °C **a** 253MA, **b** CF8C-Plus, **c** 709, and **d** 825

near the metal-scale interface that was measured as 15 wt.% (compared to 22.7%Cr measured in the bulk alloy, Table 1).

Figure 5 shows mass change measurements after 500 and 1000 h at 800 °C in 300 bar RG sCO_2. Despite the high Fe content (30.8%) in alloy 825, the average mass gains for this Cr_2O_3-forming alloy were similar to those for 282 and slightly higher than the specimens of alloy 617 and Co-based alloy 188. Figure 6a shows example cross-sections of the scales formed after 1000 h at 800 °C in sCO_2. The external scale

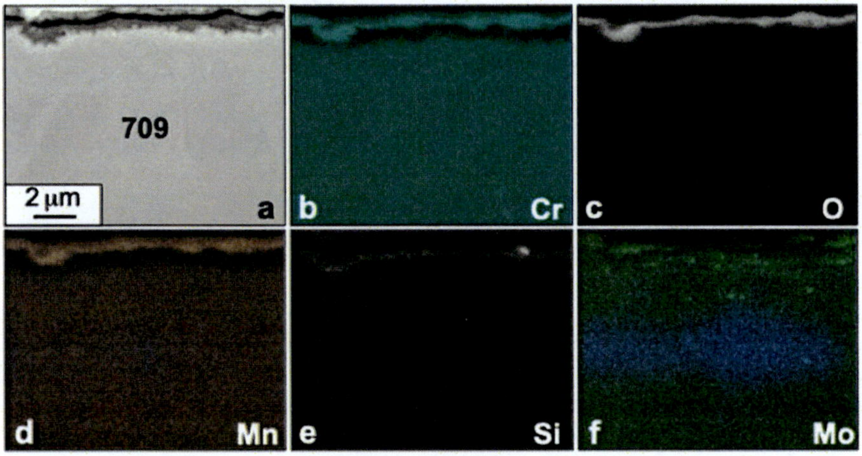

Fig. 3 **a** SEM secondary electron image of alloy 709 polished cross-section after 1000 h exposure to 300 bar sCO_2 at 600 °C and **b–f** associated EDS maps

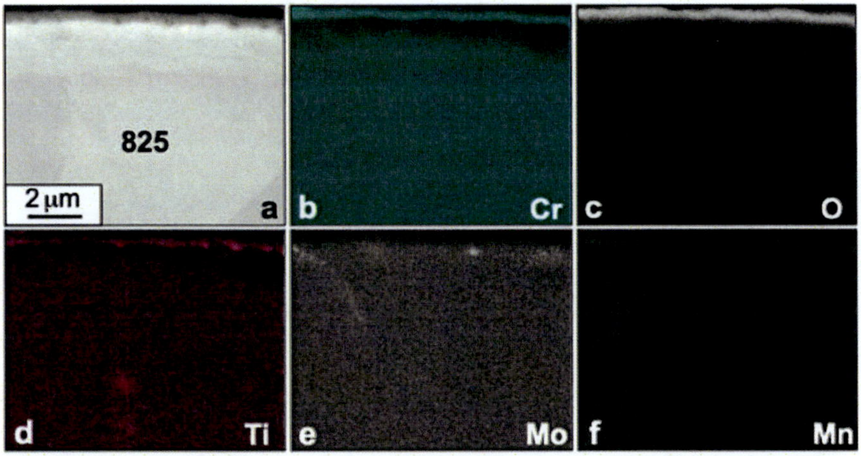

Fig. 4 **a** SEM secondary electron image of alloy 825 polished cross-section after 1000 h exposure to 300 bar sCO_2 at 600 °C and **b–f** associated EDS maps

formed on 825 (Fig. 6b) was slightly thicker than that formed on 282 (Fig. 6a) but with less internal oxidation. Additional characterization of the Cr_2O_3-forming alloys is in progress. A glow discharge optical emission spectroscopy (GDOES) sputter depth profile of the 825 specimen exposed for 1000 h at 800 °C in sCO_2 showed no evidence of C ingress, similar to other Ni-based alloys exposed for up to 10,000 h at 750 °C in sCO_2 [13].

Similar to observations at 750 °C [12] and 900 °C [28], the mass gains were higher for the specimens of 247, which does not form an Al-rich scale in CO_2 environments but shows significant internal oxidation of Al. An example cross-section from a 247 exposed at 800 °C is shown in Fig. 6c. For comparison, specimens were included of other Al_2O_3-forming alloys, superalloy X4 and a new wrought Ni-based alumina-forming austenitic alloy, designated NAFA, Table 1. Lower mass gains were observed for these alloys but not significantly different than the best Cr_2O_3-forming alloy 617; see Fig. 5. A cross-section of one X4 specimen is shown in Fig. 6d. It appeared to have less internal oxidation than the 247 specimen.

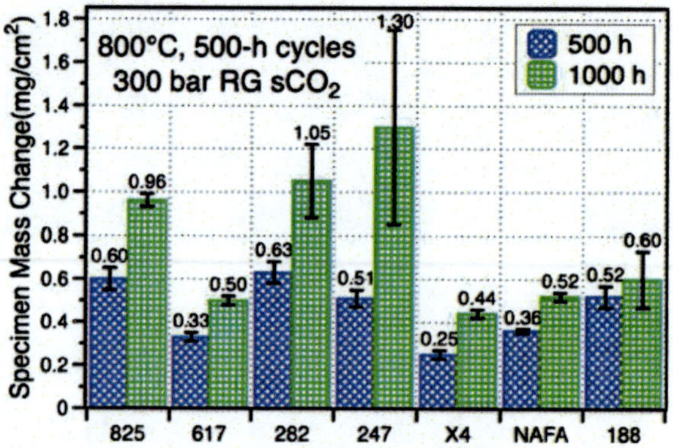

Fig. 5 Specimen mass change after 500 and 1000 h at 800 °C in 300 bar sCO_2. The whiskers show one standard deviation of 3–6 specimens exposed at each condition

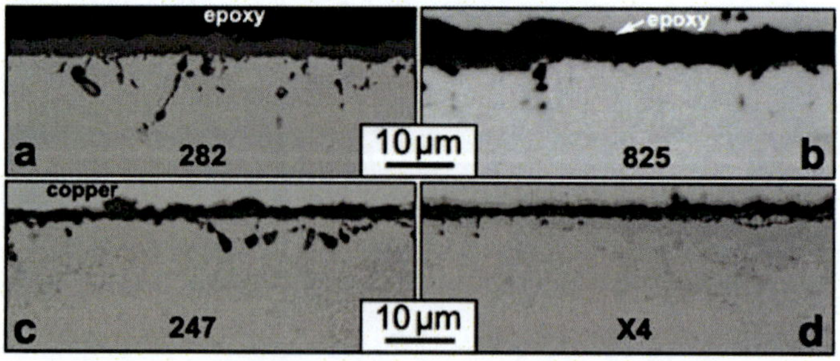

Fig. 6 Light microscopy of polished cross-sections of specimens after exposure to 300 bar sCO_2 for 1000 h at 800 °C **a** 282, **b** 825, **c** 247, and **d** X4

Discussion

The alloy 825 specimens performed well at both temperatures which suggests the alloy would be a good candidate for sCO_2 components in the 550–700 °C range. The high Fe content in this alloy did not appear to have a negative effect on sCO_2 compatibility but longer exposures may be needed to evaluate possible C ingress. The initial GDOES results suggest that the Cr-rich scale formed on alloy 825 acts as a C diffusion barrier [24].

Many of the observations appear to be well-explained by composition effects. At 600 °C, the best results for a Fe-based alloy were for 709, which contained the highest Cr and Ni contents. The results for CF8C-Plus and 253MA suggest that additions of high levels of Mn (CF8C-Plus) and Si or mischmetal (Ce,La) are not sufficient to prevent Fe-rich nodule formation in sCO_2. While Mn and Si can clearly become incorporated into the scale at 600 °C (Fig. 3), this may be too low a temperature for reactive elements like Ce and La to show a beneficial effect [29]. The relatively low Ni contents in these two alloys may be an issue in preventing nodule formation.

At 800 °C, for the Cr_2O_3-forming alloys, the importance of Al and Ti additions has been shown previously [13, 30]. The alloys 617, 825, and 282 with increasing Ti contents show increasing mass gains. Figure 6a and b comparing the Cr-rich scales formed on 282 and 825 suggests a thicker external scale on 825 with less internal oxidation than 282. This is likely due to the lower Al content in 825 [30]. For 247, the high Hf content (1.4%) may contribute to the internal oxidation, which was not as significant for X4, with less Hf (0.08%), Table 1. Both alloys have similar Al contents and would be expected to form a continuous Al_2O_3 scale at higher temperatures. Component fabrication might benefit from pre-oxidation to initiate a slow-growing α-Al_2O_3 scale at a higher temperature without CO_2.

Finally, an aspect that has not been addressed yet is the effect of high O_2 and H_2O impurities in the sCO_2 from combustion in the direct-fired Allam cycle [4]. At 750 °C, little difference in reaction rates was noted between RG sCO_2 and the low levels ($\leq$50 ppm) of impurities possible in industrial grade sCO_2 [12]. However, the higher impurity levels expected for Allam cycles, compared to indirect-fired or closed cycles, have been shown to affect reaction rates at 450–750 °C [12, 31]. Further work is needed to determine if alloy 825 is affected by high impurity levels. Only limited work has been conducted on the effects of impurities in sCO_2 at elevated temperatures [7, 12]. Despite decades of operation, it is not clear from the published literature how the additions (CH_4, H_2O, etc.) in the AGR CO_2 used to prevent oxidation of the graphite moderator in the reactor core may have affected alloy performance in that environment.

Summary

Alloy 825 was compared to a group of advanced austenitic steels at 600 °C and Ni-based alloys at 800 °C in 300 bar research grade supercritical CO_2. After 1,000 h at 600 °C, only the advanced austenitic steel alloy 709 performed as well as 825 under these conditions with both alloys forming thin, protective Cr-rich oxides. Alloy 825 also performed well after 1,000 h in sCO_2 at 800 °C compared to a range of Ni-based alloys. Among the Cr_2O_3-forming alloys, the performance of 825 may be explained by its Ti content compared to alloys 617 and 282. High Al content superalloys did not perform significantly better under these conditions at 800 °C.

Acknowledgements The experimental work was conducted by B. Johnston, J. Keiser, T. Lowe, M. Lance, V. Cox, and D. Newberry at ORNL. Y.-F. Su and S. Dryepondt provided useful comments on the manuscript. Research sponsored by the U.S. Department of Energy, Office of Fossil Energy and Carbon Management, Crosscutting Technology Program.

References

1. Dostal V, Hejzlar P, Driscoll MJ (2006) The supercritical carbon dioxide power cycle: Comparison to other advanced power cycles. Nucl. Technol. 154(3):283–301.
2. Chen H, Goswami DY, Stefanakos EK (2010) A review of thermodynamic cycles and working fluids for the conversion of low-grade heat. Renewable & Sustainable Energy Rev 14:3059–3067.
3. Iverson BD, Conboy TM, Pasch JJ, Kruizenga AM (2013) Supercritical CO_2 Brayton cycles for solar-thermal energy. Applied Energy 111:957–970.
4. Allam R, Martin S, Forrest B, Fetvedt J, Lu X, Freed D, Brown, Jr. GW, Sasaki T, Itoh M, Manning J (2017) Demonstration of the Allam Cycle: An Update on the Development Status of a High Efficiency Supercritical Carbon Dioxide Power Process Employing Full Carbon Capture. Energy Procedia 114:5948–5966.
5. Olivares RI, Young DJ, Marvig P, Stein W (2015) Alloys SS316 and Hastelloy-C276 in Supercritical CO_2 at High Temperature. Oxid. Met. 84:585–606.
6. Pint BA, Keiser JR, Initial Assessment of Ni-Base Alloy Performance in 0.1 MPa and Supercritical CO_2. JOM 67(11):2615–2620.
7. Mahaffey J, Adam D, Brittan A, Anderson M, Sridharan K (2016) Corrosion of Alloy Haynes 230 in High Temperature Supercritical Carbon Dioxide with Oxygen Impurity Additions. Oxid. Met. 86:567–580.
8. Dheeradhada V, Thatte A, Karadge M, Drobnjak M (2016) Corrosion of Supercritical CO_2 Turbomachinery Components. in Proceedings of the EPRI International Conference on Corrosion in Power Plants, EPRI, Charlotte, NC.

9. Pint BA, Brese RG, Keiser JR (2017) Effect of Pressure on Supercritical CO_2 Compatibility of Structural Alloys at 750°C. Mater. Corros. 68:151–158.
10. Oleksak RP, Tylczak JH, Carney CS, Holcomb GR, Dogan ON (2018) High-Temperature Oxidation of Commercial Alloys in Supercritical CO_2 and Related Power Cycle Environments JOM 70:1527–1534.
11. Pint BA (2018) Performance of Wrought Superalloys in Extreme Environments. in E. Ott et al. (Eds.), Proceedings of the 9th International Symposium on Superalloy 718 and Derivatives, TMS, Warrendale, PA, pp.165–178.
12. Pint BA, Lehmusto J, Lance MJ, Keiser JR (2019) The Effect of Pressure and Impurities on Oxidation in Supercritical CO_2. Mater. Corros. 70:1400–1409.
13. Pint BA, Pillai R, Lance MJ, Keiser JR (2020) Effect of Pressure and Thermal Cycling on Long-Term Oxidation in CO_2 and Supercritical CO_2. Oxid. Met. 94:505–526.
14. Feher EG (1968) The Supercritical Thermodynamic Power Cycle. Energy Conversion 8:85–90.
15. Viswanathan R, Shingledecker J, Purgert R (2010) Evaluating Materials Technology for Advanced Ultrasupercritical Coal-Fired Plants. Power 154(8):41–45.
16. Zhao SQ, Xie XS, Smith GD, Patel SJ (2003) Microstructural stability and mechanical properties of a new nickel based superalloy. Mater. Sci. Eng. A 355:96–105.
17. Pike LM (2008) Development of a Fabricable Gamma-Prime (γ') Strengthened Superalloy. In: Superalloys 2008, R. C. Reed et al. eds, TMS, Warrendale, PA, pp.191–200.
18. Gong Y, Young DJ, Kontis P, Chiu YL, Larsson H, Shin A, Pearson JM, Moody MP, Reed RC (2017) On the breakaway oxidation of Fe9Cr1Mo steel in high pressure CO_2. Acta Mater. 130:361–374.
19. Sarrade S, Férona D, Rouillard F, Perrin S, Robin R, Ruiz JC, Turc HA (2017) Overview on corrosion in supercritical fluids. J. Supercritical Fluids 120:335–344.
20. Shingledecker JP, Pint BA, Sabau AS, Fry AT, Wright IG (2013) Managing Steam-Side Oxidation and Exfoliation in USC Boiler Tubes. Adv. Mater. Processing, 171(1):23–25.
21. Furukawa T, Inagaki Y, Aritomi M (2011) Compatibility of FBR structural materials with supercritical carbon dioxide. Progress in Nuclear Energy 53:1050–1055.
22. McCoy HE (1965) Type 304 Stainless Steel vs Flowing CO_2 at Atmospheric Pressure and 1100–1800°F. Corrosion 21:84–94.
23. Fujii CT, Meussner RA (1967) Carburization of Fe-Cr Alloys During Oxidation in Dry Carbon Dioxide. J. Electrochem. Soc. 114:435–442.
24. Meier GH, Coons WC, R. A. Perkins RA (1982) Corrosion of Iron-, Nickel- and Cobalt-Base Alloys in Atmospheres Containing Carbon and Oxygen. Oxid. Met. 17:235–262.
25. Young DJ, Zhang J, Geers C, Schütze M (2011) Recent advances in understanding metal dusting: A review. Mater. Corros. 62:7–28.
26. Gheno T, Monceau D, Young DJ (2013) Kinetics of breakaway oxidation of Fe-Cr and Fe-Cr-Ni alloys in dry and wet carbon dioxide. Corrosion Science 77:246–256.
27. Pint BA, Pillai R, Keiser JR (2021) Effect of Supercritical CO_2 on Steel Ductility at 450°–650°C. ASME Paper #GT2021–59383, for Turbo Expo 2021, New York, NY.
28. Pint BA, Keiser JR (2022) Exploring Material Solutions for Supercritical CO_2 Applications above 800°C," Oxid. Met. 98:545–559.
29. Pint BA (1996) Experimental Observations in Support of the Dynamic Segregation Theory to Explain the Reactive Element Effect. Oxid. Met. 45:1–37.
30. Lehmusto J, Ievlev AV, Keiser JR, Pint BA (2021) A tracer study on sCO_2 corrosion with multiple oxygen-bearing impurities. Oxid. Met. 96 (2021) 571–587.
31. Pint BA, Lance MJ, Pillai R, Keiser JR (2022) Compatibility of Steels at 450°–650°C in Supercritical CO_2 with O_2 and H_2O Additions. AMPP (formerly NACE) Paper C 2022–18018, Houston, TX.

Effects of High-Temperature Oxidation on Fatigue Life of Additive-Manufactured Alloy 625

Grace de Leon Nope, Guofeng Wang, Juan Manuel Alvarado-Orozco, and Brian Gleeson

Abstract The effect of isothermal oxidation on the fatigue performance of differently processed Alloy 625 was studied (wrought 625, laser powder bed fusion, direct energy deposition). Uniaxial fatigue tests at room temperature were conducted after prior exposures at 800 °C for 24 h, 300 h, and 1000 h in either air or argon. Exposures in air resulted in chromia-scale formation, internal attack, and the formation of subsurface precipitates (i.e., δ-phase and σ-phase). Fatigue results indicated a consistent life reduction of up to 96% for the oxidized additive-manufactured test bars compared to their counterpart aged in back-filled argon. The fatigue life decreases as the oxidation exposure time increases. By contrast, any of the prior high-temperature exposures were not detrimental to the performance of the wrought Alloy 625. Microstructural analysis of the after-testing oxidized AM-processed bars indicated that the failure mode was attributed to the exacerbation of interfacial and subsurface defects from the oxidation exposure (i.e., internal attack, decohesion of the scale, and subsurface precipitates). These defects acted as preferential crack-initiation sites, leading to a reduction in fatigue life. On the other hand, the failure mode for the thermal-aged bars, without superficial degradation, involved fragmentation of ($\delta + \sigma$) precipitate clusters favored by the Nb and Mo segregation from the as-built microstructure of both AM processes (LPBF and DED). A large precipitate fraction depletes the matrix, facilitating crack formation.

Keywords Mechanical fatigue · Isothermal oxidation · Alloy 625 · Additive manufacturing

G. de Leon Nope (✉) · G. Wang · B. Gleeson
Department of Mechanical Engineering and Materials Science, University of Pittsburgh, Pittsburgh, PA 15261, USA
e-mail: gvd4@pitt.edu

J. M. Alvarado-Orozco
Centro de Investigación y de Estudios Avanzados del IPN, Unidad Querétaro, 76230 Querétaro, Mexico

E. A. Ott et al. (eds.), *Proceedings of the 10th International Symposium on Superalloy 718 and Derivatives*, The Minerals, Metals & Materials Series,
https://doi.org/10.1007/978-3-031-27447-3_17

Introduction

Alloy 625 is a nickel-based superalloy strengthened mainly by niobium and molybdenum in a solid solution. The alloy shows outstanding mechanical properties and good weldability [1] and can be used at temperatures ranging from cryogenic to about 900 °C. As a result, it is used for service in harsh environments such as those found in gas turbines, chemical reactors, and sea-water applications. Oxidation of Alloy 625 typically involves the formation of a continuous chromia (Cr_2O_3) scale along with internal oxidation. Even though a chromia scale can provide good oxidation and corrosion resistance, the internal attack has the potential to reduce fatigue life [2].

Several studies have assessed the effect of prior oxidation on fatigue performance, showing a potential for internal oxides to act as preferred crack-initiation sites in Ni-based alloys used in rotor disc applications [2–5]. A study of ME3 alloy, also known as Rene 104, showed that notch fatigue life reduction is proportional to the oxide scale thickness and the internal oxidation depth, where the testing was performed in air at 704 °C [4]. Specifically, life reduction was attributed to carbide dissolution in the depleted region and internal oxidation. Gabb et al. [5] also reported a reduction in the low-cycle fatigue life reduction of alloy U720 at 650 °C and ME3 alloy at 704 °C due to internal oxidation. Cracks in non-oxidized samples were shown to initiate at the subsurface (~50 μm below the surface), whereas oxidized samples showed initiation at the surface. To the best of the authors' knowledge, no reports have been published regarding the effects of oxidation on the fatigue behavior of Alloy 625.

Prior studies on the oxidation behavior of additively manufactured (AM) chromia-forming Ni-based alloys have shown differences in comparison to counterpart conventionally processed alloys that have a potential to affect fatigue performance. For example, oxidized LPBF(laser powder bed fusion)-Alloy 718 was found to exhibit deeper internal attack and a faster internal oxidation rate [6]; whereas studies on oxidized LPBF-Alloy 625 reported higher spallation [7], abundant subsurface voids [8], and an increase of subsurface δ-precipitates [9]. The current study compares the oxidation of as-printed LPBF, as-printed DED (directed energy deposition), and fully heat-treated wrought Alloy 625 and evaluates the impact of the oxidation-driven features on the subsequent fatigue performances.

Experimental

Materials

Alloy 625 fabricated by three different manufacturing methods was investigated. The two laser-assisted AM processes selected were LPBF and DED. Commercially available wrought Alloy 625 was used as a conventional manufacturing method for

Table 1 Measured chemical composition (wt.%) of the wrought and powder Alloy 625 used in this study

	Cr	Mo	Nb	Fe	Si	Al	Ti	Mn	Co	Ni
Wrought	21.1	8.52	3.58	4.67	0.21	0.17	0.13	0.05	0.03	Balance
Powder	21.6	9.18	3.72	4.79	0.16	0.27	0.26	0.04	0.05	Balance

comparison. Argon-atomized pre-alloyed powder provided by VDM Metals was used for both AM processes. The powder had a unimodal distribution, with d_{10} = 27.9 μm, d_{50} = 44.5 μm, and d_{90} = 63.5 μm. Table 1 presents the composition of the wrought and the pre-alloyed powder measured by inductively coupled plasma optical emission spectroscopy (ICP-OES). Regarding LPBF, 15 mm × 80 mm × 15 mm (height) samples were fabricated using an EOSINT M 280 system (EOS GmbH-Electro Optical Systems, Germany). The following LPBF laser parameters were used: 40 μm of layer thickness; 960 mm/s of scan speed; 110 μm of hatch distance; striped laser pattern with 67° rotation between layers. Regarding DED, 50 mm × 80 mm × 20 mm (height) samples were fabricated using a custom-built system, consisting of a three-port coaxial head BEO D70, with a laser source TruDisk6002 (Trumpf, Saalfeld, Germany), and a robotic arm IRB6620-120 (ABB Ltd., Zürich, Switzerland). The following DED laser parameters were used: 1 mm of layer thickness; 4 mm/s of deposition speed; 5 g/min of powder feed rate; 850 W of laser power; and 3 mm of hatch distance.

Static Oxidation of Flat Coupons

Oxidation of flat coupons was investigated as a reference to characterize the oxide-induced microstructure in the absence of loading. Coupons of approximate dimensions 10 mm × 10 mm × 2 mm were obtained using a low-speed saw from the LPBF and DED cubes, *both* in as-built conditions. The surface was prepared to a P600-SiC finish and then ultrasonically cleaned in ethanol. The flat coupons were oxidized in lab air at 800 °C for 24, 300, and 1000 h in an open furnace. After testing, cross-sections of the coupons were prepared metallographically, and the microstructure was characterized using scanning electron microscopy (Apreo system, ThermoFisher Scientific, Waltham, MA, USA) at 15 keV.

Fatigue Test Bars Preparation

For the fatigue tests, thermal aging (TA) and oxidation (OX) treatments were done for samples processed by each manufacturing method to independently evaluate the contributions from oxidation and the concurrent evolution of microstructure due to

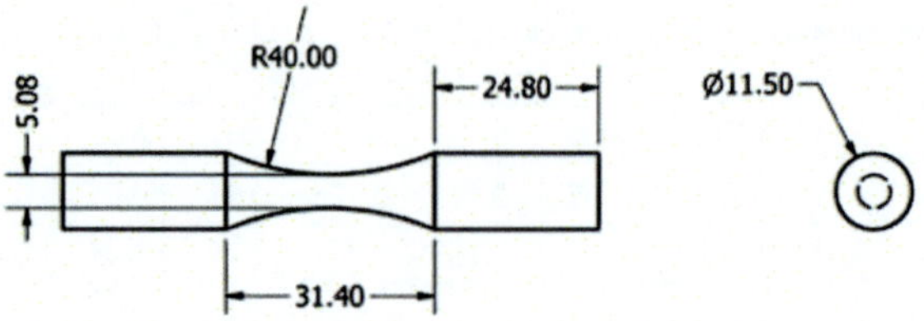

Fig. 1 Fatigue test bar dimensions following ASTM E466-07. Measurements shown are in mm

high-temperature exposure (i.e., phase transformation, grain growth, recrystallization). Both these exposures were conducted on as-printed test bars for 24 h, 300 h, and 1000 h at 800 °C in a horizontal furnace. Laboratory air was the atmosphere for the OX exposures, whereas argon was used for the TA exposures.

Cylinders were cut from the bulk AM samples by wire electrical discharge machining (EDM) and then machined by computer numerical control (CNC) milling to obtain the final geometry illustrated in Fig. 1. The standard hourglass-shaped test bar used for the fatigue testing is in accordance with ASTM E466-07. Each fatigue test bar had a gauge length of 31.4 mm and a minimum gauge diameter of 5.08 mm.

For the OX conditions, the test bar surface finishing was prepared using 320-grit SiC paper and the results of internal attack depth and oxide thickness were compared to previous work [9]. On the other hand, the TA test bars were polished to a mirror finish with a very fine non-woven alumina polishing pad to avoid detrimental interference of a high surface roughness on the fatigue life.

Fatigue Testing

Uniaxial fatigue testing was conducted at room temperature using an MTS servo-hydraulic universal test machine (MTS 880 landmark, MTS, Eden Prairie, Minnesota, USA) under load control mode. A stress ratio $R = -1$ was used ($R = \sigma_{min}/\sigma_{max}$), following a sinusoidal waveform at a frequency of 36–38 Hz. All experiments used a maximum stress level of 440 MPa and the fatigue run-out condition was set to 10^7 cycles. Every condition was performed in duplicate, except for the DED at 300 h in OX and TA conditions.

After testing, fractographic analysis was done using an optical microscope Smartzoom 5 (Carl Zeiss, Jena, Germany) to identify crack-initiation sites. A cross-section of the fractured sample, parallel to the load axis, was then cut from the main crack initiation and prepared metallographically to characterize secondary cracking and degradation. Eight micrographs at 5000× magnification per sample were examined via SEM at 15 keV. Microstructural features after fatigue testing, including oxide thickness and maximum depth of internal oxidation, were quantitatively measured using Gwydion software (Czech Metrology Institute, Brno, Czech Republic). Subsurface precipitate area and average precipitate aspect ratio were measured using Particleanalyzer_HD software [10] (University Hospital Heidelberg, Heidelberg, Germany), where precipitates consisted of δ-phase and σ-phase. Finally, the linear detachment fraction, the non-planar oxide/metal interface index, and the interfacial defects area were calculated based on metrics acquired using ImageJ software (National Institutes of Health, Bethesda, MD, USA). The term "interfacial defect" is used in this study to refer to the formation of free space at the oxide/alloy interface. The linear detachment fraction was calculated as the ratio between the interfacial length without oxide divided by the total interfacial length at the external oxide/metal interface. The non-planar interface index, a measurement related to roughness, was estimated as the ratio between the total interfacial length at the external oxide/metal interface divided by the length if that interface is perfectly flat.

Results and Discussion

Oxidation Behavior of Flat Coupons

Our previous study showed the as-built microstructure of the studied samples and their effect on oxidation at 800 °C for 24 h [9]. Summarizing, both AM samples possess a cellular-dendritic structure. Interdendritic regions exhibit high segregation of Nb and Mo, and dendritic regions are constituted by FCC γ(Ni–Cr–Fe) matrix depleted of Nb and Mo. The average grain size of the wrought (25.6 μm [95% CI, 23.9–27.3 μm]) and the LPBF (26.8 μm [95% CI, 24.5–29.0 μm]) are in the same order of magnitude, while DED grain size (150.1 μm [95% CI, 132.7–167.5 μm]) is six times greater. After oxidation at 800 °C in the air for 24 h, wrought, LPBF, and DED samples formed a protective external scale comprised mainly of chromia, and δ-phase precipitates beneath the chromia. The greater the Nb and Mo segregation in the as-built condition, the higher the δ-precipitate fraction [9]. Other studies on the oxidation of Alloy 625 report the same features, as well as some Mn–Cr spinel oxides at the scale surface [7, 11, 12].

In this study, exposures were extended to 300 h and 1000 h at 800 °C. Figure 2 shows cross-sectional images of the flat coupons after oxidation at 800 °C for 1000 h. At longer exposures, the chromia scales formed on LPBF and DED samples were slightly thicker than those formed on the wrought, in accordance with the kinetics trend reported at shorter times [9]. The main differences in the oxidation behavior between wrought and AM samples occur in the subsurface. Beneath the chromia scale, the composition of the bright precipitates corresponds to δ-phase [$(Ni, Cr)_3(Nb, Mo)$] and σ-phase [Mo–Ni–Si–Cr–Nb]. LPBF and DED showed a higher fraction of these subsurface precipitates than the wrought due to a higher Nb availability from the original segregation in the as-built state [7].

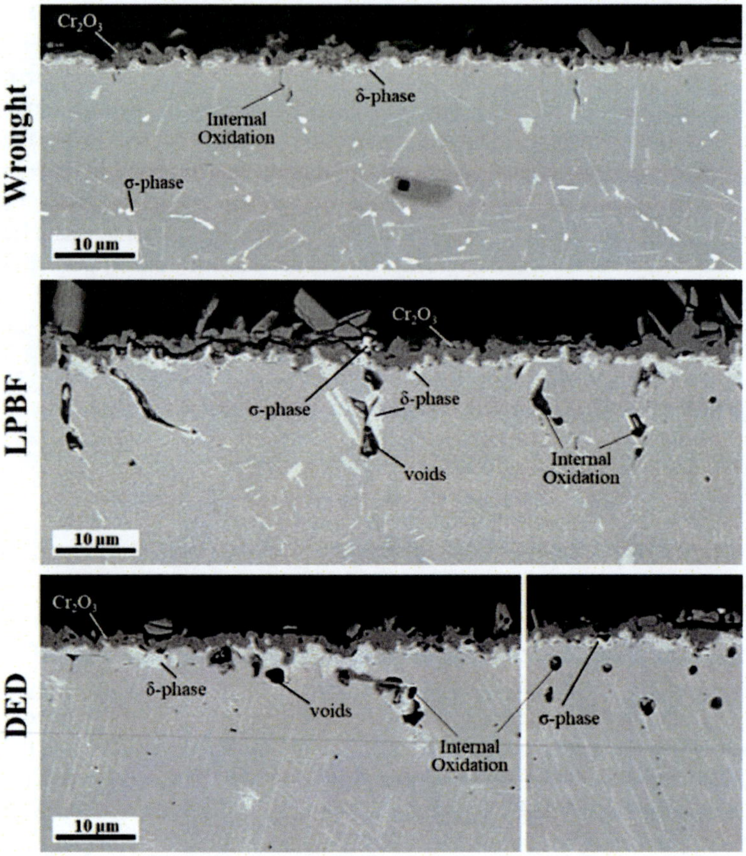

Fig. 2 SEM backscattered electron image of Alloy 625 oxidized at 800 °C for 1000 h in the air for the wrought, LPBF, and DED fabrication conditions

Internal oxidation and bulk transformation represented the most significant differences among the samples. The wrought alloy exhibited internal oxidation preferentially located at grain boundaries and consisting mainly of compact alumina. Internal oxidation in *both* AM alloys was deeper, coarser, and more abundant than that in the wrought alloy. As postulated by Sanviemvongsak et al. [6], this difference can be attributed to the improved oxidation resistance of twin boundaries present in the wrought. After oxidation, grain boundaries in the AM samples were found to be decorated by coarser δ-precipitates, internal oxidation, and a noticeable amount of free space among them. The exacerbation of voids in AM samples' grain boundaries has been reported previously and attributed to a faster local Cr consumption [8]. Additionally, internal oxides and voids form preferentially among the precipitate clusters located in the interdendritic regions of the DED-processed alloy, where the highest Nb–Mo segregation was in the as-built condition. Void formation in oxidized NiCr alloys has been attributed to the Kirkendall effect [6, 13].

On the other hand, the bulk transformation after the oxidation differs for the three conditions (e.g., distribution, morphology, and precipitates' size). Since the original microstructure varies among them (i.e., grain size, refractory segregation, and residual stress), different evolution after the heat treatment is expected. Therefore, to solely evaluate the influence of the oxidation behavior on fatigue performance, a counterpart thermal aging under a non-oxidizing atmosphere of argon was done for the same time as each oxidized sample.

Past studies regarding oxidized samples have shown that internal oxidation along grain boundaries [2, 3, 14] and the precipitates-free region [4] are critical factors in accelerating crack initiation during fatigue testing, showing that subsurface evolution during oxidation plays a vital role in fatigue performance. Therefore, since the main differences in the oxidation behavior of the additive samples occur in the subsurface, differences in the fatigue life may be expected. Other possible contributions to initiate and propagate cracks are localized interfacial defects (local detachment of the scale/alloy). Although pores in the scale may also be considered, crack propagation throughout the oxide layer is generally resisted by its compressive stresses. The molar volume difference between the oxide and the alloy and the mismatch between the thermal expansion coefficients of the oxide and the alloy contribute to the compressive stresses of the external oxide layer [15].

Finally, the long horizontal cracks in the scale are inferred to be artifacts caused during the cutting and polishing of the oxidized coupon. Thus, the integrity of the scale is expected to be intact before starting the fatigue testing.

Fatigue Results

Figure 3 shows the mean fatigue life as a function of thermal exposure conditions. The wrought test bars did not fail, reaching run-out (10^7 cycles), regardless of the exposure condition. In contrast, both the LPBF and DED test bars failed before 10^7 cycles with the exception of the as-built LPBF samples that were thermally aged for 24 h and 300 h. Apparently, the differences in the oxidation and thermal aging behaviors associated with the two additive processes are significant enough to affect fatigue performance.

The results in Fig. 3 indicate that both LPBF and DED consistently exhibit a higher fatigue life after thermal aging than oxidation for the same exposure times. The oxidation reduces the mean fatigue life in the LPBF test bars by 25%, 82%, and 91% for 24 h, 300 h, and 1000 h, respectively. For the DED test bars, the fatigue life reduction by oxidation was 46%, 96.8%, and 88% for 24 h, 300 h, and 1000 h, respectively. Therefore, since oxidized samples show the same bulk evolution as thermal aging, the fatigue life reduction between TA and OX is attributed to

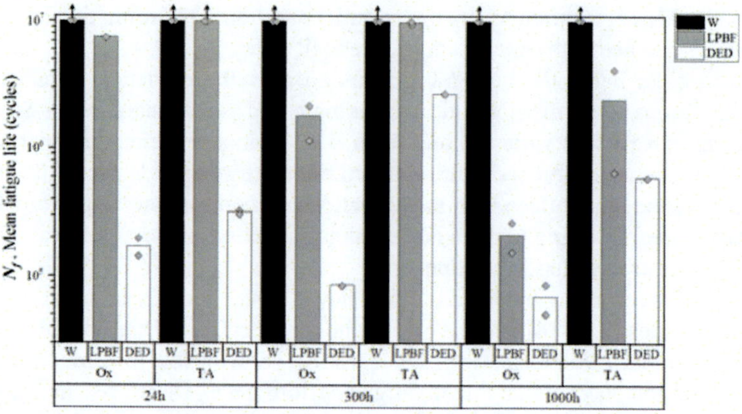

Fig. 3 Mean room temperature fatigue life of the wrought (W), LPBF, and DED Alloy 625 test bars that were oxidized (Ox) and thermal aged (TA) for 24 h, 300 h, and 1000 h at 800 °C prior to testing

the surface and subsurface evolution processes associated with oxidation. Several studies on Ni-based alloys have shown a reduction in fatigue life on exposure to high temperatures in air atmosphere, either during (e.g., N18 alloy [16]) or before the fatigue test (e.g., ME3 alloy [4, 5], IN100 [17]). In this study, the general trend shows that life reduction increases with increasing oxidation time. It is relevant to note that previous studies have shown that life reduction also increases with higher temperature exposures [4, 16]. Additionally, the DED-processed test bars exhibited a shorter fatigue life than those processed by LPBF when comparing oxidation at the same exposure times.

Microstructural Degradation Assessment of Oxidized Samples After Fatigue Test

Analyzing the microstructure evolution after testing can lead to a better understanding of the differences in fatigue performance. Figure 4 shows representative micrographs of the cross-sections after fatigue testing. The wrought microstructures do not show appreciable degradation after 10^7 cycles; the scales keep their integrity, and non-appreciable cracks are identified after the testing. By contrast, both the LPBF and DED test bars show increased subsurface degradation and breaches of the scale integrity. Quantitative measurements of the degradation state at the failure moment were therefore acquired to better correlate degradation with the fatigue performance. The degradation variables were roughly classified into the following three main types: (1) oxide products; (2) subsurface precipitates (δ- and σ-phase); and (3) oxide/metal interfacial detachment. Oxide products included maximum depth of internal oxidation and external oxide thickness. Subsurface precipitate measurements consisted of their total area and aspect ratio. Finally, the interfacial detachment included the detachment linear fraction, non-planar interface index, and interfacial defects area (empty space between the oxide scale and the alloy). Figure 5 summarizes the results for the wrought, LPBF, and DED alloys for various exposure conditions using spider diagrams, where all values are normalized relative to the maximum value of each variable; the complete data can be found in Appendix A.

The first general observation is that the bigger the area for a given variable, the smaller the mean fatigue life; hence, those variables that most affect the area play an important role in determining fatigue life. The second observation is the increase in degradation with the oxidation time for the LPBF and DED test bars; the lesser the degradation, the higher the fatigue life. A more detailed analysis is presented in the following.

Since wrought test bars withstand 10^7 cycles, neither the load nor the oxidation induced any critical-size crack. Therefore, the wrought values can be considered not detrimental, as a degradation state where failure is not likely. Accordingly, if the variables of an unknown sample are within the wrought range, failure by fatigue is not expected. For example, at 24 h the most significant difference between the wrought

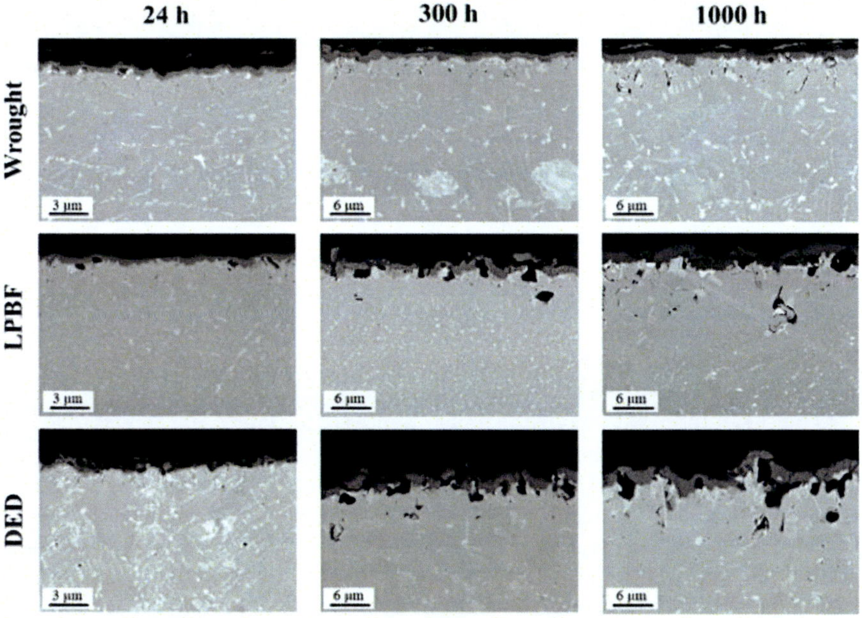

Fig. 4 SEM images of cross-sections of the oxidized test bars (OX) after fatigue testing. The oxide layer shown is parallel to the load axis

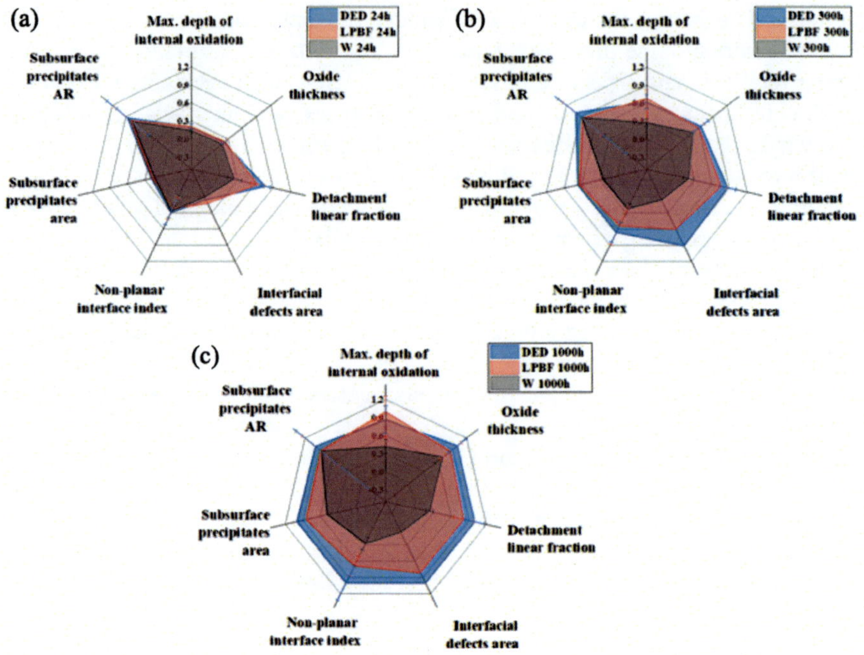

Fig. 5 Graphical representation of microstructural evolution and degradation measurements after fatigue testing of oxidized test bars. Comparison of wrought, LPBF, and DED alloys after different oxidation times: **a** 24 h, **b** 300 h, and **c** 1000 h

and the AM samples is the detachment linear fraction followed by the area of the interfacial defect. The detachment linear fraction of LPBF and DED is more than twice that of the wrought. This trend continues at 100 and 300 h, where the value of wrought keeps constant while the LPBF and DED values continue to increase with time. Thus, a linear fraction below 0.20 corresponds to non-failure and above 0.28 corresponds to failure. Additionally, the linear fraction increment is inversely proportional to the fatigue life for the LPBF and DED test bars, being the measured variable that best describes the fatigue life.

At the 300 h and 1000 h oxidation times, the area of the interfacial defects together with the detachment linear fraction shows the biggest difference between the non-failure region demarcated by the wrought and the AM test bars, showing that the exacerbation of interfacial defects is a key precursor to imminent failure. The maximum depth of internal oxidation, non-planar interface index, and subsurface precipitates area follow these in relevance.

In contrast, the oxide thickness and the precipitate aspect ratio do not directly correlate with the fatigue life. Regarding the aspect ratio (AR) of the subsurface precipitates, no substantial changes were observed, regardless of the exposure condition. Since the influence of the precipitates' AR will be in the same range in all studied samples, it is not directly contributing to differences in the reported fatigue

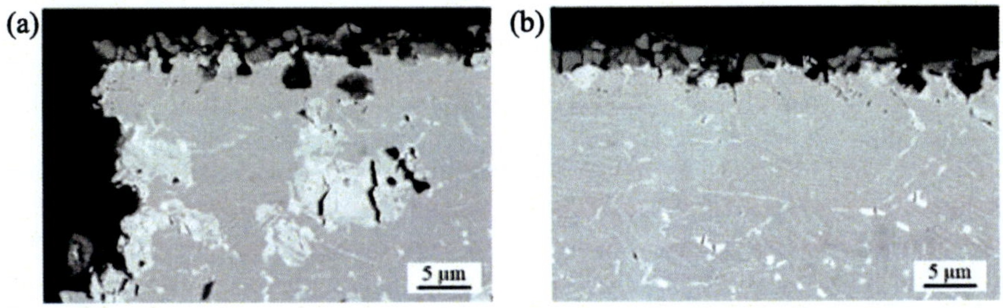

Fig. 6 SEM backscattered electron images of a cross-section of a failed wrought test bar oxidized after 1000 h. A fatigue test was stopped until failure (5 × 10^7 cycles) at different gauge lengths: **a** at the fracture (cross-sectional area ~20.2 mm^2) and **b** 5 mm from the fracture in the load axis (cross-sectional area ~25.6 mm^2)

life. The main difference in the subsurface precipitates among the samples was in the phase fraction, precipitate size, and spatial distribution. Regarding the oxide thickness, the standard deviation of the measurement can be interpreted as a scale homogeneity, showing that AM samples (high deviation) have a high variability in the oxide scale thickness compared with the wrought (low deviation) for the same exposure times. A highly variable oxide thickness can affect the compressive stress distribution throughout the scale, promoting local stress concentrators that lead to interfacial defects [18].

An additional fatigue test until failure was conducted for a wrought test bar oxidized after 1000 h. The sample failed at 5 × 10^7 cycles, and Fig. 6 shows the cross-section of the microstructure near the fracture region. The interfacial and subsurface degradations show similarities with the microstructures of the AM test bars (both LPBF and DED) oxidized at 1000 h after failure (Fig. 4). These similarities include the exacerbation of interfacial defects, an increase of linear fraction detachment, and external oxide scale cracking. As such, it indicates that the degradation prior to failure is similarly independent of the manufacturing process, but both LPBF and DED test bars underwent a higher degradation rate than the wrought, causing earlier failure in the former. The reasons behind the acceleration in the degradation during the fatigue testing in the AM test bars (both LPBF and DED) will be discussed in the next section.

Fatigue Crack-Initiation Mechanism of Oxidized Samples

The microstructural variables that best describe the degradation state at failure were discussed in the previous section. Since the characterization of the subsurface degradation in this study requires destructive techniques, it is difficult to assess the degradation evolution. This study proposes an alternative way to evaluate the degradation evolution by analyzing the cross-sections after testing at different distances from the center. Given that the cross-sectional area varies along the gauge length, the stress level also varies along the gauge. The bigger the cross-sectional area, the lower the stress, and as such it will result in a degradation decrease. Therefore, the degradation evolution with the applied stress can be analyzed. To that end, it is assumed that during fatigue at the same test time, a bigger cross-sectional area experiences an equivalent degradation to a smaller cross-sectional area at shorter times and higher stress.

Figure 7 shows representative cross-sections along the gauge of the oxidized LPBF sample after testing. Figure 7a shows the largest sample cross-section and, in turn, the area of lowest applied stress during fatigue testing. The microstructure is similar to the one shown for the oxidized flat coupons (Fig. 2). The external oxide scale is still intact, and there is a development of a more pronounced non-planar scale/alloy interface and an increase of interfacial local decohesion. Figure 7b corresponds to a higher stressed region than that shown in Fig. 7a and shows how interfacial defects grow and subsurface voids start to appear along the grain boundaries, located among internal oxides and δ-precipitates. As stress increases (Fig. 7c), interfacial defects continue to grow to the extent that the oxide scale cracks and eventually spalls, while a more pronounced development of subsurface voids also takes place. Finally, Fig. 7d shows deep and broad void formation in the vicinity of the fracture because of subsurface and surface defects coalescence. It is plausible that such void formation could serve to initiate crack development during fatigue loading. The subsurface voids associated with the internal oxidation process and the δ-precipitation could thus act as preferential crack-initiation sites. Despite that interfacial defects accelerate the degradation, observed microstructures after failure indicate that the sample can withstand abundant subsurface defect formation before failure. However, the same assumptions are no longer valid for subsurface voids since less of them are necessary to generate a critical-size defect that triggers crack propagation and failure.

Formation of Interfacial Defects

Interfacial defect formation starts as oxide buckling, but their growth continues toward the alloy as stress increases. The initial scale buckling is presumed to be in response to a compressive stress buildup in the scale during the oxidation in combination with relatively weak oxide/alloy adhesion. Specifically, if the oxide/alloy interface has a poor adhesion relative to the cohesive strength of the oxide, decohesion that leads to buckling will be favored before the oxide cracking (i.e., failure by shear) [19, 20]. Since the LPBF and DED test bars consistently exhibited buckling formation before oxide cracking, this indicates low adhesive interfacial forces in these systems. Secondly, the interfacial defect growth toward the alloy during the fatigue testing is promoted by the strength loss of the depletion region. The oxidation process promoted Cr and Nb depletion, making the subsurface alloy weaker than the bulk alloy. Moreover, the alloy subsurface is under tensile stresses as a compensation for compressive stresses in the oxide [21], which contributes to enhance the defect growth.

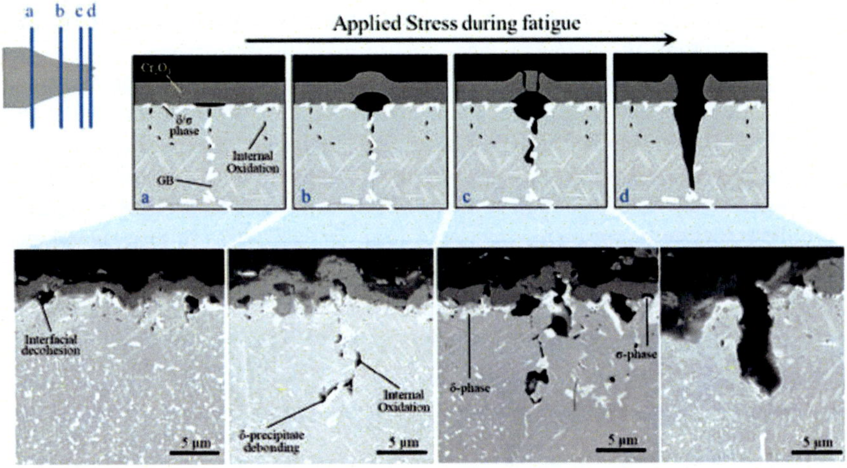

Fig. 7 Oxidized sample degradation evolution with applied stress. Bottom micrographs correspond to the cross-section of OX-LPBF for 300 h after fatigue testing

Several reported contributions that favor poor interfacial adhesion and accelerate scale spallation in the AM systems are as follows:

- *Convoluted scale/alloy interface*: Local heterogeneities that influence the diffusion through the oxide promote a non-planar interface (e.g., composition changes in oxide, subsurface precipitates, and grain boundaries) [22]. Composition changes in the oxide of the AM samples occur mainly during the first stages of oxidation before the establishment of the chromia layer. The as-built microstructure of the AM samples has regions with significant segregation of Nb, Mo, and Ti [9] that likely impact the local transient products due to the different oxygen affinity and different diffusion rates through the alloy and the oxide [22]. Another source of local heterogeneity comes from the grain boundaries. Enhanced diffusion promotes a faster chromia formation, resulting in oxides ridges above grain boundaries [8]. Additionally, the coarsening of subsurface precipitates and phase fraction increase contribute to convolute the interface.
- *Scale/alloy bond strengths:* Interfacial segregation of elements such as sulfur and phosphorus weakens the scale/metal bond strength [18]. Sulfur decreases the surface energy required to form a metal surface, favoring void nucleation [23]. Despite no measurements of sulfur segregation being acquired in this study, higher content of sulfur in alloys produced by LPBF in comparison with conventional methods has been reported. Son et al. [24] reported that the sulfur content of LPBF 625 powder (17 ppm) is twice the value of the wrought counterpart (34 ppm). Sanviemvongsak et al. [25] reported that the sulfur content in alloy 718 fabricated by LPBF (23 ppm) was more than one order magnitude higher than the wrought alloy (1 ppm), with greater scale spallation reported for the AM-718 samples in comparison with the wrought.

Furthermore, in the case of Alloy 625, the interface consists of oxide/δ-precipitate/alloy in several regions. Such a multiphase interface would necessarily have a greater amount of interphase boundaries in comparison to a stable scale/alloy interface. The degree of coherency at an interface depends on crystallographic differences between the phases (e.g., lattice misorientation and the ratio of molar volumes), with an incoherent interface providing a potent sink site for vacancies [26]. The formation of interfacial defects in the oxide/δ-precipitate due to the injection of vacancies stemming from the scale-growth process is likely. Although these defects may not be appreciable, they are prone to expand during the fatigue testing and compromise the adherence of oxide/δ-precipitate. Two experimental observations support this. First, there is a correlation between the precipitates fraction area and the detachment linear fraction. Second, most of the δ-precipitates remain within the alloy when there is a scale decohesion.

Formation of Subsurface Voids

Abundant voids form in the bulk and grain boundaries of the LPBF and DED alloys during oxidation in comparison to the wrought alloy (Fig. 2). The interaction of internal oxidation, voids, and large precipitates promoted localized deformation during fatigue (see Fig. 7). The inferred process for this is as follows. Nb and Mo segregation in the LPBF and DED alloys promotes a faster nucleation and favors the precipitate growth. The precipitate coarsening process resulted in relatively large precipitates surrounded by a denuded FCC matrix [27]. Subsurface voids were able to form and grow due, in part, to strength loss in these denuded zones.

This precipitate-associated subsurface void-formation process was most pronounced on grain boundaries, which in turn were associated with the deepest internal oxidation. The coalescence of subsurface voids along the grain boundaries likely triggered crack formation at least as deep as the internal oxidation. Furthermore, since AM grain boundaries were deeper, the crack propagation reached a critical size sooner in these alloys than in the wrought (see Appendix A).

Fatigue Crack-Initiation Mechanism of Thermally Aged Samples

Since the fatigue life generally increases as the grain size decreases [28, 29], the DED-processed alloy is expected to have a shorter fatigue life than the LPBF and wrought alloys. However, there is barely a difference in the LPBF and wrought grain size to explain the difference in their fatigue lives. Figure 8 shows representative cross-sections of the 1000 h TA-LPBF fractured test bar to evaluate the microstructural

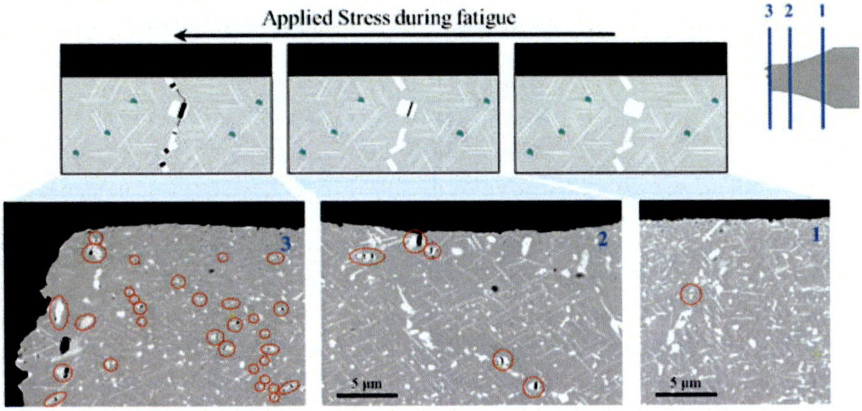

Fig. 8 Thermal-aged sample degradation evolution with applied stress. Bottom micrographs correspond to the cross-section of TA-LPBF for 1000 h after fatigue testing

degradation evolution with the stress applied, given that the cross-sectional area varies along the gauge length. Unlike the oxidized samples, no appreciable surface degradation occurred since thermal aging treatments were performed under back-filled argon. The degradation of the TA samples correlates mainly with the debonding and fracture of σ and δ- precipitates, where the fractures are mainly perpendicular to the loading axis. As the applied stress increases, the number of fractured precipitates increases. These fractured precipitates are more abundant closer to the surface.

During fatigue, in the absence of induced surface/subsurface degradation, fatigue life is mainly determined by the interactions of dislocations and precipitates. Phases predicted at equilibrium at 800 °C for the nominal composition of alloy 625 are mainly δ-[$(Ni,Cr)_3(Nb,Mo)$] and σ-[Mo–Ni–Si–Cr–Nb] [30]. The beneficial effect of δ-phase is reported in moderate amounts, where it provides grain-size control and grain refinement through a pinning effect below 990 °C [31, 32]. In coexistence with γ″, δ-phase increases mechanical properties below 426 °C for IN718 and Super Waspalloy [32]. This may explain the LPBF fatigue results after thermal aging for 24 h and 300 h. However, the coarsening of incoherent δ precipitates at longer times reduces the strengthening process effectiveness and lowers the ductility [33]. Since δ and σ precipitates act as stress and strain concentrators, they fracture once the stress concentration overcomes the precipitate strength. Then, these fractured precipitates can induce crack nucleation [34–36].

These precipitates/dislocation interactions depend on the precipitate's size [37], morphology, distribution [38], orientation to the tensile stress, location relative to the surface [39, 40], and the strength characteristics of the matrix [35, 36]. The heterogeneities from the initial microstructures (i.e., as-built condition) influence these factors.

Table 2 Predicted phase fractions at equilibrium at different segregated and depleted compositions

Sample		Nb (wt.%)	Mo (wt.%)	δ $Ni_3(Nb,Mo)$	σ (Mo–Ni–Si–Cr)	γ (Ni–Fe–Cr)
Wrought	–	3.45	8.40	0.06	0.10	0.83
LPBF	Depleted	3.14	8.19	0.03	0.08	0.89
	Segregated	6.44	10.05	0.17	0.12	0.71
DED	Depleted	2.70	8.91	0.04	0.04	0.92
	Segregated	16.20	12.72	0.48	0.32	0.20

The different Nb and Mo segregation levels in AM samples affect the phase fraction predicted at equilibrium. As an illustrative exercise, the phase fractions at equilibrium assuming an alloy with the measured composition of depleted (dendritic) and segregated (interdendritic) regions were calculated using ThermoCalc in combination with the TCNI8 database (Table 2). In agreement with the reported literature [30], the higher the Nb and Mo segregation, the bigger the difference between the (δ + σ)-phase fraction in the depleted and segregated regions.

Representative cross-sections of samples under thermal aging at 800 °C for 5000 h are shown in Fig. 9. Needle-like precipitates correspond to the δ-phase, while the bright small globular precipitates correspond to the σ-phase. The total precipitates fraction (δ + σ) in each sample was 12.1 ± 1.2% for the wrought, 16.1 ± 1.7% for LPBF, and 17.3 ± 2.3% for DED; demonstrating that the higher the precipitates fraction, the shorter the fatigue life. However, the difference in the precipitates fraction in the LPBF and DED alloys is not substantial enough to explain the difference in their fatigue life. The main microstructural differences between them are in the size and distribution of the precipitates. The histograms below the micrographs in Fig. 9 represent the area fraction of precipitates as a function of distance for a band of 5 μm width demarked in the image. The wrought sample shows the smallest variation between the peak and valleys, demonstrating a homogeneous (δ + σ) precipitate distribution. The LPBF sample exhibits an enrichment of precipitates area fraction along the grain boundaries. Intergranular precipitates are coarser than the intergranular ones, due very likely to the higher driving force for δ precipitation than in the bulk [27]. The DED samples show the highest heterogeneity of (δ + σ) precipitate distribution. The original highly segregated regions result in (δ + σ) precipitate clusters, while the original depleted region does not favor (δ + σ) precipitation, which is consistent with the ThermoCalc predictions.

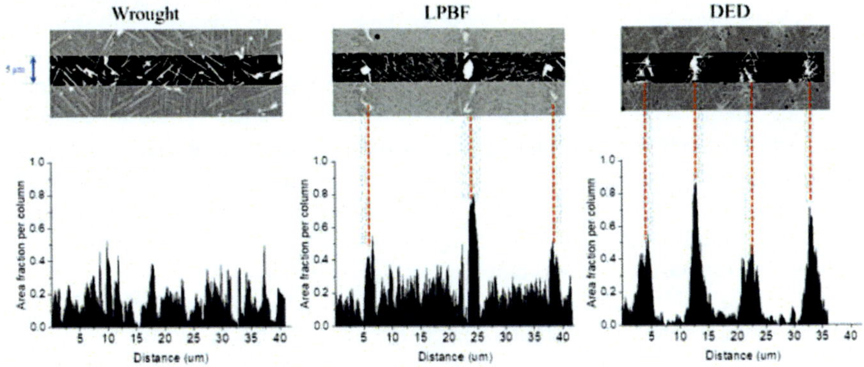

Fig. 9 Micrographs of wrought, LPBF, and DED after thermal ageing for 1000 h. Histograms represent the area fraction of precipitates as a function of distance. Each column in the histogram evaluates the precipitates fraction in 0.068 μm^2 (5 μm width by 0.014 μm length). The bright areas correspond to ($\delta + \sigma$) precipitates and the dark areas correspond to the matrix

The heterogeneities in chemical composition and ($\delta + \sigma$) precipitate distribution are detrimental to the fatigue performance. On one hand, highly segregated regions will promote a bigger area of precipitates that embrittle the alloy and favor crack nucleation. While on the other hand, highly depleted regions are not strengthened by solution or precipitates, so that cracks can propagate easily. The initial composition variability impacts the proportionality of the degree of embrittlement in the segregated regions and the degree of strength loss in the depleted regions. Consequently, the fatigue life trend among the TA samples exposed at the same time, i.e., wrought >> LPBF > DED, is explained by the heterogeneity of ($\delta + \sigma$) precipitates distribution.

Conclusions

This study assessed the influence of thermal- and oxidation-driven features on the uniaxial fatigue performance of Alloy 625 fabricated by LPBF, DED, and wrought processes. The following conclusions are drawn from the results:

- Oxidation is more detrimental than thermal aging on the fatigue life of the AM samples, where the life reduction increases with increasing oxidation exposure time. Oxidation induces interfacial defects (i.e., free space formation) and subsurface voids that accelerate crack initiation. Prior to failure, the sample withstands abundant interfacial defects formation before subsurface voids coalesce and initiate crack formation. Interfacial voids originate from local decohesion of

the oxide/alloy interface during cyclic loading. Subsurface voids result from the interaction of internal oxidation, δ precipitate coarsening, and δ/matrix weakening.
- Neither the oxidation-driven features nor the heat-treatment evolution was detrimental to the wrought 625 under the conditions studied. The AM oxidized samples showed an acceleration of degradation during fatigue testing compared to the oxidized wrought. Formation of interfacial defects in the AM samples was promoted by a convoluted scale/alloy interface and also the oxide/δ-precipitate/alloy interface, whereas grain boundary precipitates enhanced the extent of internal oxidation promoting subsurface voids.
- The fatigue failure mechanism in thermally aged AM samples (both LPBF and DED) involves the fragmentation and coalescence of δ and σ subsurface precipitates, leading to crack development throughout the sample. The higher the segregation from AM processing, the greater the heterogeneity of the (δ + σ) precipitate distribution and the shorter the fatigue life.

Acknowledgements This research was supported by the Division of Civil, Mechanical, and Manufacturing Innovation Division of the National Science Foundation (Alexis Lewis, Program Manager), award number 1662615. The authors thankfully acknowledge Rafael Rodriguez for the Thermo-Calc results and Mr. Brandon Blasko for assisting in the preparation of AM and fatigue test bars.

Appendix A

Quantitative measurements of the degradation state at the failure moment.

Sample ID	Oxide products				Subsurface precipitates (δ-phase and σ-phase)					Oxide/alloy interfacial detachment						
	Max. depth of internal oxidation [μm]		Oxide thickness [μm]		Precipitates size (ECD) [μm]			Precipitates AR		Detachment linear fraction [μm/ μm]		Non-planar interface index[2]		Defects area [μm^b]		
	Average	STD	Average	STD	Average	STD	Total sum[1]	Average	STD	Average	STD	Average	STD	Average	STD	Total sum[a]
Wrought 24h	1.10 ±	0.26	0.37 ±	0.09	0.29 ±	0.14	24.63	2.27 ±	0.98	0.13 ±	0.07	0.06 ±	0.02	0.17 ±	0.09	4.22
Wrought 300h	2.17 ±	0.48	1.09 ±	0.24	0.42 ±	0.18	43.19	2.60 ±	1.45	0.11 ±	0.04	0.05 ±	0.02	0.39 ±	0.43	16.12
Wrought 1000h	3.33 ±	0.90	1.56 ±	0.37	0.59 ±	0.37	84.94	2.53 ±	1.58	0.12 ±	0.03	0.06 ±	0.02	0.40 ±	0.50	16.12
LPBF 24h	1.67 ±	0.37	0.53 ±	0.13	0.29 ±	0.17	29.81	2.38 ±	1.04	0.34 ±	0.06	0.05 ±	0.04	0.32 ±	0.27	29.70
LPBF 300h	5.42 ±	1.19	1.37 ±	0.25	0.63 ±	0.42	112.78	2.55 ±	1.23	0.40 ±	0.05	0.11 ±	0.07	0.84 ±	1.18	143.97
LPBF 1000h	7.95 ±	2.04	1.95 ±	0.51	0.79 ±	0.55	146.65	2.55 ±	1.17	0.43 ±	0.09	0.14 ±	0.04	1.46 ±	1.84	193.21
DED 24h	1.61 ±	0.37	0.48 ±	0.11	0.30 ±	0.20	33.19	2.55 ±	1.33	0.41 ±	0.04	0.06 ±	0.05	0.21 ±	0.14	15.04
DED 300h	4.95 ±	0.99	1.46 ±	0.34	0.71 ±	0.51	116.88	2.96 ±	3.20	0.46 ±	0.07	0.14 ±	0.04	1.52 ±	1.67	218.59
DED 1000h	6.89 ±	1.82	2.27 ±	0.42	0.82 ±	0.67	174.83	2.88 ±	3.31	0.53 ±	0.09	0.21 ±	0.07	1.90 ±	2.29	239.61

[a]Total summations calculated along 331.2 μm (8 micrographs at 5000x)

[b] $Non-planar\ interface\ index = \frac{Real\ oxide/alloy\ interface}{flat\ interface} - 1$; a zero value indicates that the interface is flat

References

1. S. Floreen, G. E. Fuchs, and W. J. Yang, in *Superalloys 718, 625 and Various Derivatives* (Pittsburgh, Pennsylvania, USA, 1994), pp. 13–37.
2. S. Cruchley, M. P. Taylor, H. Y. Li, H. E. Evans, P. Bowen, D. J. Child, and M. C. Hardy, Materials at High Temperatures **32**, 68 (2015).
3. S. Cruchley, H. Y. Li, H. E. Evans, P. Bowen, D. J. Child, and M. C. Hardy, International Journal of Fatigue 265 (2015).
4. C. K. Sudbrack, S. L. Draper, T. T. Gorman, J. Telesman, T. P. Gabb, and D. R. Hull, in *Superalloys 2012* (Seven Springs, PA, USA, 2012), pp. 863–872.
5. T. P. Gabb, J. Telesman, P. Kantzos, J. Smith, and P. Browning, in *Superalloys 2004* (Seven Springs, PA, USA, 2004), pp. 269–274.
6. T. Sanviemvongsak, D. Monceau, C. Desgranges, and B. Macquaire, Corrosion Science **170**, 108684 (2020).
7. N. Ramenatte, A. Vernouillet, S. Mathieu, A. Vande Put, M. Vilasi, and D. Monceau, Corrosion Science **164**, 108347 (2020).
8. A. Chyrkin, K. O. Gunduz, I. Fedorova, M. Sattari, A. Visibile, M. Halvarsson, J. Froitzheim, and K. Stiller, Corrosion Science **205**, 110382 (2022).
9. G. de Leon Nope, G. Wang, J. M. Alvarado-Orozco, and B. Gleeson, JOM **74**, 1698 (2022).
10. S. Schroeder, S. Braun, U. Mueller, R. Sonntag, S. Jaeger, and J. P. Kretzer, Journal of Biomedical Materials Research - Part B Applied Biomaterials **108**, 225 (2020).
11. A. Chyrkin, P. Huczkowski, V. Shemet, L. Singheiser, and W. J. Quadakkers, Oxidation of Metals **75**, 143 (2011).
12. A. M. de Sousa Malafaia, R. B. de Oliveira, L. Latu-Romain, Y. Wouters, and R. Baldan, Materials Characterization **161**, 110160 (2020).
13. C. Desgranges, F. Lequien, E. Aublant, M. Nastar, and D. Monceau, Oxidation of Metals **79**, 93 (2013).
14. Y. Oshida and H. W. Liu, Grain Boundary Oxidation and an Analysis of the Effects of Pre-Oxidation on Subsequent Fatigue Life (Syracuse, New York, 1986).
15. D. J. Young, *High Temperature Oxidation and Corrosion of Metals: Second Edition*, Second (Elsevier, Amsterdam, 2016).
16. R. Jiang, S. Everitt, N. Gao, K. Soady, J. W. Brooks, and P. A. S. Reed, International Journal of Fatigue **75**, 89 (2015).
17. M. Reger and L. Remy, **19**, 2259 (1988).
18. F. H. Stott, Materials Science and Technology **4**, 431 (1988).
19. H. E. Evans, Materials at High Temperatures **3409**, 155 (2014).
20. W. Christl, A. Rahmel, and M. Schütze, Oxidation of Metals **31**, 1 (1989).
21. M. Schütze, in *Shreir's Corrosion*, edited by B. Cottis, M. Graham, R. Lindsay, S. Lyon, T. Richardson, D. Scantlebury, and H. Stott (Elsevier, Amsterdam, 2010), pp. 153–179.
22. S. Taniguchi, Transactions of the Iron and Steel Institute of Japan (1985).
23. H. J. Grabke, D. Wiemer, and H. Viefhaus, Applied Surface Science **47**, 243 (1991).
24. K. Son, M. E. Kassner, and K. A. Lee, Advanced Engineering Materials **1900543**, 1 (2019).
25. T. Sanviemvongsak, D. Monceau, M. Madelain, C. Desgranges, J. Smialek, and B. Macquaire, Corrosion Science **192**, 109804 (2021)
26. B. Pieraggi, R. A. Rapp, and J. P. Hirth, Oxidation of Metals **44**, 63 (1995).
27. S. Azadian, L. Wei, and R. Warren, **53**, 7 (2004).
28. D. J. Morrison and J. C. Moosbrugger, International Journal of Fatigue **19**, 51 (1997).
29. F. L. Liang and C. Laird, Materials Science and Engineering A **117**, 95 (1989).
30. G. Lindwall, C. E. Campbell, E. A. Lass, F. Zhang, M. R. Stoudt, A. J. Allen, and L. E. Levine, Metallurgical and Materials Transactions A: Physical Metallurgy and Materials Science **50**, 457 (2019).
31. C. Ruiz, A. Obabueki, and K. Gillespie, Superalloys 33 (1992).
32. J. F. Radavich and W. H. Couts, Jr., Superalloys 497 (1984).

33. M. Anderson, A. L. Thielin, F. Bridier, P. Bocher, and J. Savoie, Materials Science and Engineering A **679**, 48 (2017).
34. S. Suresh, in *Materials Science and Engineering: A*, edited by H. Muhrabi (VCH, Weinheim, Germany, 1993), pp. 509–563.
35. A. Luft, Progress in Materials Science **35**, 97 (1991).
36. D. Lambert, Department of Engineering Materials (1994).
37. Y. Murakami, in *Metal Fatigue: Effects of Small Defects and Nonmetallic Inclusions* (Elsevier Science Ltd, 2002), p. Elsevier Science Ltd,.
38. C. Qiu, Net-shape Hot Isostatic Pressing of a Nickel-based Powder Superalloy, University of Birmingham, 2010.
39. A. de Bussac, Fatigue & Fracture of Engineering Materials & Structures **17**, 1319 (1994).
40. J. A. Moore, D. Frankel, R. Prasannavenkatesan, A. G. Domel, G. B. Olson, and W. K. Liu, International Journal of Fatigue **91**, 183 (2016).

Long-Term Thermal Stability and Oxidation Resistance of HAYNES 233 Alloy

L. M. Pike and B. Li

Abstract HAYNES 233 alloy is an alumina-forming alloy developed for use in a variety of high temperature applications. It is unique in the marketplace for having a combination of high creep strength and excellent oxidation and high temperature corrosion resistance to temperatures of 2000°F (1093 °C) and above while still being readily fabricable (formable, weldable, etc.). The alloy may be used in the solution annealed conditions at these high temperatures, or age-hardened to provide high strength at intermediate temperatures while maintaining its oxidation/corrosion resistance. This study will explore the long-term performance of 233 alloy in terms of microstructural stability, tensile properties, and oxidation resistance. Thermal exposures of up to 8000 h were explored at temperatures ranging from 1200°F (649 °C) to 1800°F (982 °C). The effect of the initial material condition (solution annealed vs. age-hardened) was considered as well. It was found that prior age-hardening can provide the alloy with significantly improved ductility at intermediate temperatures before and after thermal exposure. A comparative year-long oxidation test at 2000°F (1093 °C) confirmed the excellent oxidation resistance of the alumina-forming 233 alloy compared to competing alloys (typically chromia formers).

Keywords Thermal stability · Microstructural stability · Oxidation resistance · HAYNES 233 alloy · Alumina-former

Introduction

HAYNES® 233™ alloy was recently developed to provide the market with a readily fabricable alloy combining the properties of high creep strength and excellent oxidation and high temperature corrosion resistance to temperatures of 2000°F (1093 °C) and above. This combination of properties is desirable for a number of high temperature applications, including hot gas components in aerospace and industrial gas turbines, industrial heating fixtures and sensors, and various structural components

L. M. Pike (✉) · B. Li
Haynes International, Inc., 1020 West Park Ave, Kokomo, IN 46904-9013, USA
e-mail: lpike@haynesintl.com

E. A. Ott et al. (eds.), *Proceedings of the 10th International Symposium on Superalloy 718 and Derivatives*, The Minerals, Metals & Materials Series,
https://doi.org/10.1007/978-3-031-27447-3_18

in the emerging power generation technology markets such as concentrated solar power, advanced-ultra supercritical (A-USC), and sCO_2 power cycle. The development and key features of 233 alloy have been detailed elsewhere [1]. While initially designed for use at very high temperatures in solution annealed condition, 233 alloy can also be age-hardened to provide excellent strength at intermediate temperatures, specifically those below the gamma-prime solvus of approximately 1767°F (964 °C).

An important feature of any high temperature alloy is its ability to withstand high temperatures for extended periods of time [2]. Long-term thermal exposures may have several effects on an alloy and its performance. For example, the microstructure may change significantly leading to changes in the mechanical properties which may or may not be acceptable for a given application. Similarly, the oxidation behavior of an alloy can also change during long-term thermal exposure as a result of oxide scale spallation and gradual depletion of oxide-forming elements. The purpose of this paper will be to examine the effects of long-term thermal exposures on the microstructure, mechanical properties, and oxidation resistance of 233 alloy.

Experimental

The material used in the thermal stability study was HAYNES 233 alloy in the form of cold-rolled and mill annealed sheet with a thickness of 0.063″ (1.6 mm) and an average ASTM grain size of 3.5. The nominal chemical composition of 233 alloy is given in Table 1. Two initial conditions were investigated: (1) solution (mill) annealed, and (2) age-hardened. The age-hardening heat treatment (labeled AHT2 in this paper) was 1650°F (899 °C)/4 h/AC + 1450°F (788 °C)/8 h/AC (where AC is air cooled). The solution annealed samples were subjected to thermal exposures of 8000 h in flowing air at four different temperatures: 1200°F (649 °C), 1400°F (760 °C), 1600°F (871 °C), and 1800°F (982 °C). While 233 alloy can be used at even greater temperatures, this intermediate range was selected as the most likely to be less thermally stable. For the samples initially in the age-hardened condition, thermal exposures of 1000, 4000, and 8000 h were applied at temperatures of 1200°F (649 °C), 1400°F (760 °C), and 1600°F (871 °C), which are all below the gamma-prime solvus.

The thermally exposed samples were tested for tensile properties and microstructural characteristics. For all thermal exposure conditions, duplicate tensile tests (using dog bone shaped samples) were conducted at both room temperature (RT) and the corresponding exposure temperature. The microstructure of each thermal exposure condition was examined using optical metallography and scanning electron microscopy/energy-dispersive X-ray spectroscopy (SEM/EDS). The microstructural samples were taken from the shoulder section of one of the RT tensile samples. Optical metallography samples were electrolytically etched using an oxalic/hydrochloric acid solution [3]. SEM samples were prepared either in as-polished or electropolished conditions for standard imaging or given the electrolytic "Radavich" etch [4] for high magnification imaging of the gamma-prime phase. Additionally, electrolytically extracted residue samples from selected thermal exposure

Table 1 Nominal chemical composition of the alloys in this study (wt.%)

Alloy	Ni	Co	Cr	Al	Mn	Si	Ti	Fe	Mo	W	C	B	Others
233	Bal.	19	19	3.3	0.4[a]	0.2[a]	0.5	1.5[a]	7.5	0.3[a]	0.1	0.004	0.03Zr, 0.5Ta, 0.025Y[a]
230	Bal.	5[a]	22	0.3	0.5	0.4	0.1[a]	3[a]	2	14	0.1	0.015[a]	0.02La
188	22	Bal.	22	–	1.25[a]	0.35	0.15[a]	3[a]	–	14	0.1	0.015[a]	0.03La
625	Bal.	1[a]	21	0.4[a]	0.5[a]	0.5[a]	0.4[a]	5[a]	9	–	0.1[a]	–	3.7(Nb + Ta)
617	Bal.	12.5	22	1.2	0.2[a]	0.2[a]	0.3	1	9	–	0.07	0.006[a]	

[a]Maximum

conditions were examined by X-ray diffraction (XRD) to assist in phase identification of the minor phases.

A one-year long-term oxidation test was performed to study and compare oxidation behaviors of 233 alloy and other commercial high temperature alloys, such as 230® alloy, 188 alloy, 617 alloy, and 625 alloy. Table 1 lists the nominal composition of the alloys tested, in which the 233 alloy is an alumina-forming alloy and the other alloys are chromia formers. To prepare for the oxidation test, alloy sheets with a thickness of ~0.125″ (3.2 mm) in the solution annealed condition were firstly cut into coupons with an approximate size of 1″ × 1″ (25 mm × 25 mm), and then polished with 120-grit silicon carbide paper. The test coupons were cleaned with acetone and weighed prior to the oxidation test. The oxidation test was performed for 360 days (8640 h) cycled each 30 days at 2000°F (1093 °C) in a tube furnace. During the test, the coupons were inserted directly into the furnace at the set temperature at the start of each cycle and removed from the furnace, and air cooled for weight measurement after each cycle. After the completion of the test, the test coupons were cross-sectioned for optical metallographic and SEM/EDS examination.

Microstructure of HAYNES 233 Alloy

Before looking at the effects of long-term thermal exposure it is beneficial to first examine the initial microstructure. In the solution annealed condition the microstructure of 233 alloy consists of primary carbides of three types, MC, $M_{23}C_6$, and M_6C, scattered within the gamma matrix as seen in Fig. 1a, b (also seen in the figures are cavities resulting from sample preparation). These carbides have been identified with a combination of SEM/EDS and XRD. The grain boundaries are seen to be essentially free of precipitation. In the age-hardened condition (Fig. 2a, b) the microstructure is similar except for the presence of fine intragranular precipitates and precipitation at the grain boundaries. The intragranular precipitates include scattered fine Cr-rich

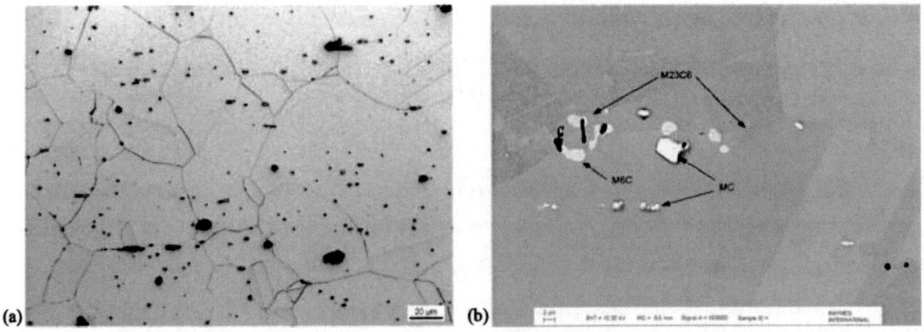

Fig. 1 The microstructure of solution annealed 233 alloy. **a** Optical micrograph, **b** SEM-BSE image

precipitates. Also, present both at grain boundaries and intragranularly is the gamma-prime phase. The shape of the intragranular gamma-prime precipitates is somewhat of a rounded cuboid with a single size distribution having an average cube length of about 65 nm (Fig. 2c). The grain boundaries are complex in nature (Fig. 2d) with fine Cr-rich and Mo-rich precipitates scattered amid heavy amounts of gamma-prime.

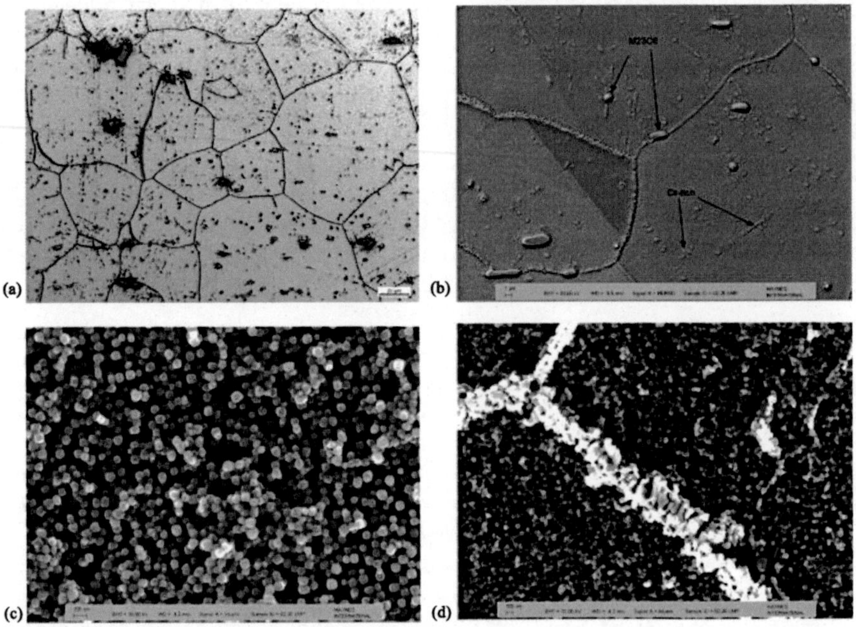

Fig. 2 The microstructure of age-hardened 233 alloy. **a** Optical micrograph, **b** SEM-BSE image, **c** SEM In-Lens SE image of typical gamma-prime, **d** SEM In-Lens SE image of gamma-prime and grain boundary precipitates

Thermal Stability

In service, HAYNES 233 alloy may be used either in the solution annealed or age-hardened condition depending on the requirements of the application. For applications where the entire component is operating at temperatures below the gamma-prime solvus temperature, it would often make sense to use the alloy in the age-hardened condition to provide high strength and to minimize the likelihood of dimensional changes in service. Similarly, when the component sees temperatures greater than the gamma-prime solvus it may seem best to start with the alloy in the solution annealed condition. However, often these high temperature components will be subjected to a range of temperatures and may age-harden in service in the cooler regions. Thus, an interesting question is whether or not to age-harden such components prior to service. Because of this, there is an added layer of complexity to this thermal stability study. We will present the results of thermal exposures given to 233 alloy in both initial conditions.

Solution Annealed + Thermal Exposure

Microstructure. SEM-BSE images of the microstructure of solution annealed 233 alloy subjected to 8000 h thermal exposures at 1200°F (649 °C), 1400°F (760 °C), 1600°F (871 °C), and 1800°F (982 °C) are shown in Fig. 3. After thermal exposure at 1200°F (649 °C) some fine grain boundary precipitation was observed as seen in Fig. 3a. Also present were 10 nm diameter spherical gamma-prime (note that the gamma-prime is present in all of the microstructures in Fig. 3, but not visible due to its small size and the surface preparation). The 1400°F (760 °C) thermal exposure resulted in larger grain boundary precipitates (Fig. 3b). Based on EDS and XRD data it is believed this is the Cr-rich sigma phase. Since further confirmation work is needed, we will hereby refer to this phase as the "intergranular Cr-rich phase". Gamma-prime is also seen at the grain boundary intermittently (darker phase). Also visible in Fig. 3b are intragranular acicular precipitates which EDS and XRD suggest is the Mo-rich mu phase. The gamma-prime precipitates formed during the 1400°F (760 °C) thermal exposure are "squarish" spheroids with a diameter of about 120 nm. It should be noted that the precipitation of gamma-prime, mu, and sigma phases at 1400°F (760 °C) was expected based on calculations from multiple commercial phase diagram software packages. Thermal exposures at 1600°F (871 °C) resulted in similar grain boundary phases (Fig. 3c) as the 1400°F (760 °C) thermal exposure. Both the intragranular Mo-rich and intergranular Cr-rich precipitates appeared to be less predominant at this temperature. The gamma-prime phase after the 1600°F (871 °C) thermal exposure was fully cuboidal and very coarse with an average cube length of 340 nm. Finally, the microstructure of the solution annealed 233 alloy after the 8000 h exposure at 1800°F (982 °C) is shown in Fig. 3d. M_6C and $M_{23}C_6$ carbides (verified by EDS and XRD) of various sizes (some quite large) can be seen in the figure. No strengthening gamma-prime was observed after exposures at 1800°F (982 °C) although there were some isolated pockets of very coarse gamma-prime. An interesting phenomenon observed after thermal exposures at all temperatures of 1400°F (760 °C) or greater was the partial decomposition of MC carbides into mostly M_6C and gamma-prime ($MC + \gamma \rightarrow M_6C + \gamma'$). Other phases sometimes observed in the region of decomposing MC carbides include MN and $M_{23}C_6$.

Tensile Properties. The tensile properties of solution annealed 233 alloy before and after 8000 h thermal exposure at the four temperatures are shown in Table 2. As would be expected, there were significantly different responses for each exposure temperature. The exposure at 1200°F (649 °C) resulted in the RT yield strength (YS) jumping from 52.5 ksi (362 MPa) to 142.3 ksi (981 MPa) due to the formation of fine gamma-prime (age-hardening) during the thermal exposure. A similarly large increase was observed in the 1200°F (649 °C) yield strength. The ultimate tensile strength (UTS) at RT and 1200°F (649 °C) also both increased significantly. A drop in elongation occurred at both test temperatures following the 1200°F (649 °C) thermal exposure. This was likely caused by the combination of high strength and the presence of fine grain boundary precipitation. The 1400°F (760 °C) thermal exposure was found to result in significant strengthening at RT, increasing up to 115.5 ksi (796 MPa) as a result of age-hardening. When the tensile test was conducted at

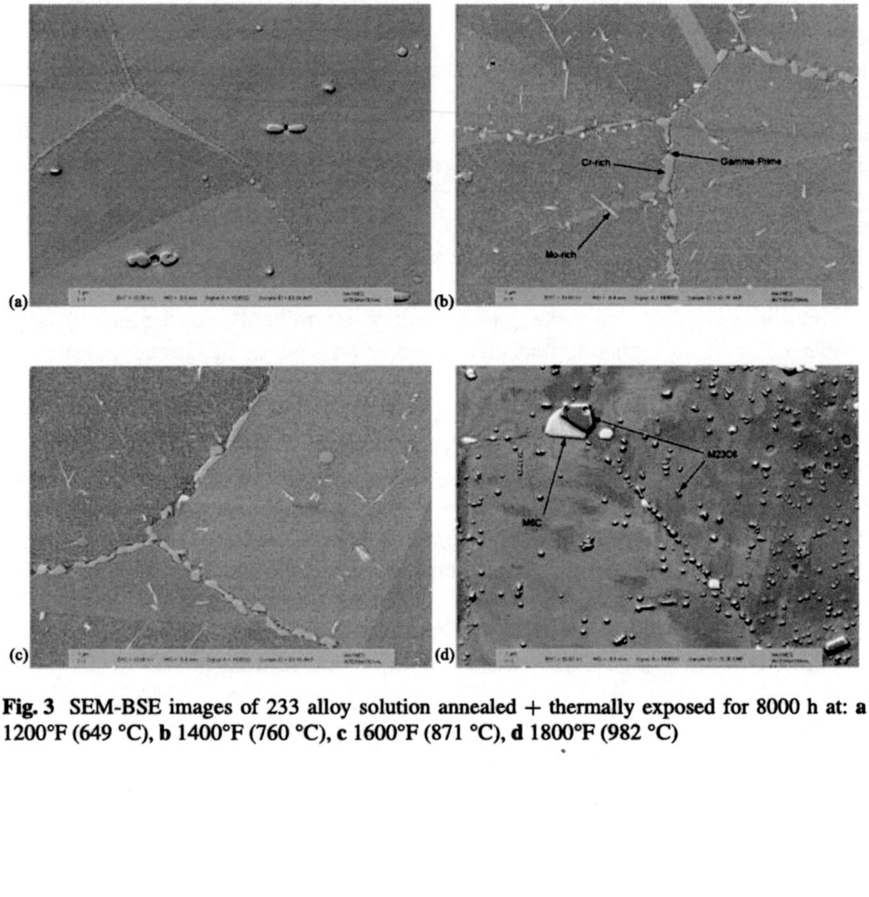

Fig. 3 SEM-BSE images of 233 alloy solution annealed + thermally exposed for 8000 h at: **a** 1200°F (649 °C), **b** 1400°F (760 °C), **c** 1600°F (871 °C), **d** 1800°F (982 °C)

the exposure temperature the yield strength was less dramatic, increasing from 78.1 ksi (539 MPa) to 87.7 ksi (605 MPa). It is believed that age-hardening of the as-solution annealed sample during the tensile test itself obscured the age-hardening effects of the thermal exposure. Similar thermal exposure effects were observed on the UTS at RT and 1400°F (760 °C). However, it is noted that the tensile elongation dropped from 57.3% to 7.7% at RT and from 14.2% to 9.5% at 1400°F (760 °C) as a result of the 1400°F (760 °C) thermal exposure. The intergranular Cr-rich and intragranular mu precipitates, particularly the former, are likely responsible for the lower elongation values in this condition. As will be discussed later in this manuscript, these drops in ductility can be mitigated somewhat by age-hardening prior to thermal exposure. At thermal exposures of 1600°F (871 °C) the effects were again mixed. The RT yield strength increased modestly up to 66.8 ksi (461 MPa), while the yield strength at 1600°F (871 °C) decreased from 61.9 ksi (427 MPa) down to 31.4 ksi (217 MPa) after the 1600°F (871 °C) thermal exposure. This apparent drop is again partly due to in-test age-hardening of the solution annealed sample, but is also a result of overaging of the gamma-prime phase during the thermal exposure. The elongation was found to decrease at RT following the 1600°F (871 °C) thermal exposure, while the elongation at the exposure temperature increased. Lastly, the thermal exposure at 1800°F (982 °C) was found to only modestly affect the tensile properties at either RT or the exposure temperature. Note that this temperature is above the gamma-prime solvus so no age-hardening during the thermal exposure would be expected.

Table 2 Tensile properties* of solution annealed HAYNES 233 alloy before and after thermal exposure

Initial Condition	Exposure Temperature		Exposure Duration	Test Temperature		Yield Strength		Ultimate Tensile Strength		Elong
	°F	°C	h	°F	°C	ksi	MPa	ksi	MPa	%
SA	–	–	–	RT	RT	52.5	362	118.6	818	57.3
SA	–	–	–	1200	649	52.0	359	102.0	703	59.5
SA	–	–	–	1400	760	78.1	539	100.9	696	14.2
SA	–	–	–	1600	871	61.9	427	73.3	505	24.5
SA	–	–	–	1800	982	12.1	84	16.3	112	99.3
SA	1200	649	8,000	RT	RT	142.3	981	192.8	1329	20.6
SA	1200	649	8,000	1200	649	122.4	844	166.0	1145	13.5
SA	1400	760	8,000	RT	RT	115.5	796	165.8	1143	7.7
SA	1400	760	8,000	1400	760	87.7	605	114.6	790	9.5
SA	1600	871	8,000	RT	RT	66.8	461	134.5	927	22.6
SA	1600	871	8,000	1600	871	31.4	217	52.5	362	47.8
SA	1800	982	8,000	RT	RT	49.1	339	116.8	806	46.2
SA	1800	982	8,000	1800	982	10.7	74	15.3	105	102.8

*Average of duplicate tests; SA = Solution Annealed

Age-Hardened + Thermal Exposure

Microstructure. The microstructure of age-hardened (using the AHT2 treatment described above) 233 alloy after thermal exposures of 1000, 4000, and 8000 h at 1200°F (649 °C), 1400°F (760 °C), and 1600°F (871 °C) was investigated as part of this study. For the sake of brevity, only the 8000 h microstructures will be described here. Optical micrographs of the 8000 h thermally exposed samples are shown in Fig. 4. SEM-BSE images of the 8000 h thermally exposed samples are shown in Fig. 5. Also shown in Fig. 5 are high magnification SEM images of the gamma-prime precipitates after thermal exposure. The microstructure of the 1200°F (649 °C) thermally exposed sample (Figs. 4a and 5a) was quite similar to the as-age-hardened condition discussed previously (Fig. 2a, b). Qualitatively it appears that the precipitation may be heavier in the thermally exposed sample, but not significantly. Moreover, the size (~70 nm) and shape of the gamma-prime (Fig. 5b) did not change much at all from the as-age-hardened condition (Fig. 2c). In contrast, thermal exposure at 1400°F (760 °C) resulted in several changes to the microstructure (Figs. 4b and 5c). Most noticeable was the formation of large intergranular Cr-rich precipitates (again EDS and XRD suggest this is the sigma phase). Also, intragranular acicular Mo-rich phase is formed. This has been confirmed by XRD to be mu phase. Additionally, the intergranular gamma-prime was found to be much heavier after the thermal exposure and took on a layer-like morphology (more about this in later section). Lastly, the intragranular gamma-prime (Fig. 5d) was found to grow during the 1400°F (760 °C) thermal exposure to an average cube length of about 130 nm, more than double that of the age-hardened condition. The microstructure of the sample thermally exposed at 1600°F (871 °C) had somewhat similar features (Figs. 4c and 5e) to the 1400°F (760 °C) thermal exposure with some differences. The Cr-rich intergranular phase is less predominant. The intragranular acicular Mo-rich phase precipitates are coarser and less common. And lastly, the intragranular gamma-prime precipitates (Fig. 5f) are now fully cuboidal and have grown to an average cube length of 420 nm (over 7 times larger than the as-age-hardened condition). Similar to the solution anneal + thermal exposure samples, the primary MC carbides in the age-hardened + thermal exposure samples were found to be fully or partially decomposed for exposure temperatures of 1400°F (760 °C) and above.

Tensile Properties. The tensile properties of age-hardened 233 alloy before and after thermal exposures of 1000, 4000, and 8000 h at 1200°F (649 °C), 1400°F (760 °C), and 1600°F (871 °C) are listed in Table 3. Tests of the thermally exposed samples were conducted at both RT and the corresponding exposure temperatures. The RT tensile properties are plotted in Fig. 6 as a function of exposure time, while the elevated exposure temperature tests are plotted in Fig. 7. In Fig. 6a it is seen that for all 3 exposure temperatures, the RT yield strength is essentially constant for exposure durations greater than 1000 h. For the exposure temperature of 1200°F (649 °C) the RT yield strength increases after the 1000 h thermal exposure but remains fairly constant after longer thermal exposures. The relatively modest changes in the tensile properties with 1200°F (649 °C) thermal exposure are consistent with

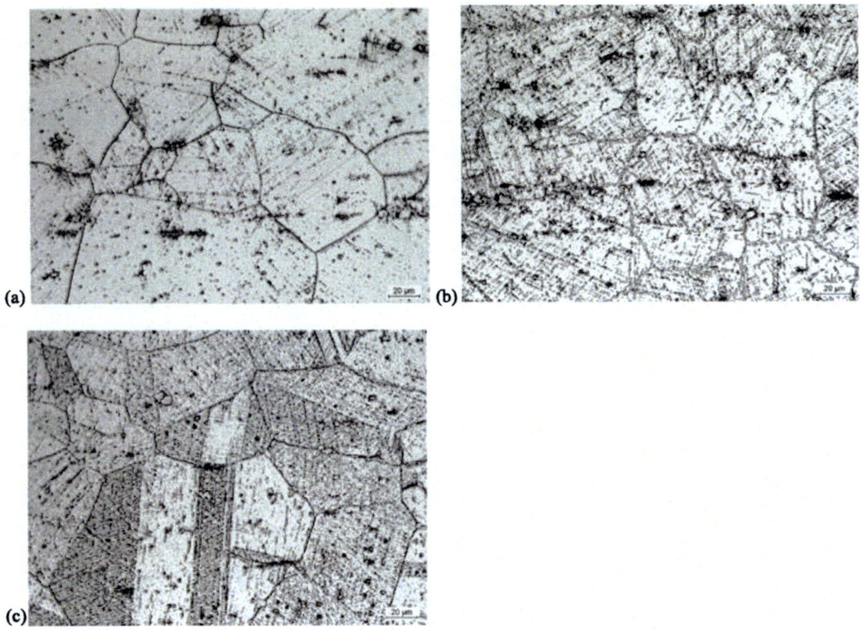

Fig. 4 Optical micrographs of 233 alloy age-hardened + thermally exposed for 8000 h at: a 1200°F (649 °C), b 1400°F (760 °C), c 1600°F (871 °C)

the observation that the microstructure is quite stable at that temperature. For the exposure temperature of 1400°F (760 °C) the RT yield strength is almost unchanged even after the full 8000 h thermal exposure. As mentioned in the previous section the size of the gamma-prime increased from 65 to 130 nm over that time, suggesting that even at 130 nm the gamma-prime phase is an effective strengthener. And for the thermal exposure temperature of 1600°F (871 °C) the RT yield strength initially decreases, before leveling off for longer thermal exposures. The drop in yield strength is presumably due to overaging of the gamma-prime which has grown much more significantly at this temperature as well as partial solutioning of the gamma-prime at this temperature (which is 150°F (83 °C) above the last step of the AHT2 age-hardening treatment). The RT UTS values were found to follow similar trends as the RT yield strength. (Fig. 6b). The RT elongation values (Fig. 6c) were found to decrease somewhat with exposure duration for thermal exposures at 1200°F (649 °C). However, they remained greater than 18% and appeared to have mostly leveled off after the first 1000 h. In contrast, the RT elongation of the samples thermally exposed at 1400°F (760 °C) was found to decrease with the exposure duration all the way to 8000 h (reaching as low as 8.6%), but the slope of the decrease did appear to be leveling off. The decrease in elongation is likely a result of the formation of the intergranular Cr-rich and intragranular mu phases, particularly the former. Loss of RT ductility with thermal exposure is not unusual in wrought superalloys, for example,

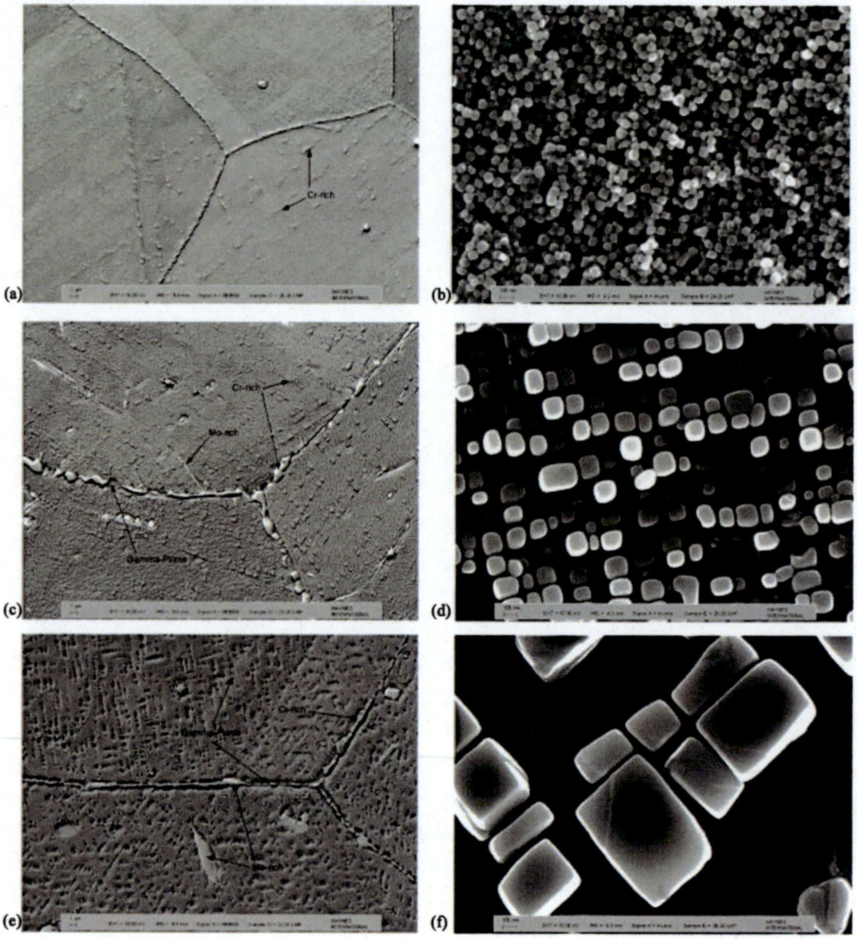

Fig. 5. 233 alloy age-hardened + thermally exposed for 8000 h. SEM BSE images and In-Lens SE images of typical gamma-prime, respectively: **a, b** 1200°F (649 °C), **c, d** 1400°F (760 °C), **e, f** 1600°F (871 °C)

the commonly-used 188 and 625 alloys [2, 5]. Similarly to 1200°F (649 °C), thermal exposures at 1600°F (871 °C) had only a modest effect on the RT tensile elongation of 233 alloy; it was observed to increase slightly after 1000 h and then drop off slightly with further exposure duration, always remaining greater than 23%. While the RT tensile elongation did decrease with 1400°F (760 °C) thermal exposure, for high temperature applications it is often more relevant to consider the effects of thermal exposures on the tensile properties *at the exposure temperature itself*. That is done in the following paragraph.

As with the RT yield strength, the exposure temperature yield strength (Fig. 7a) was found to remain essentially constant after the first 1000 h thermal exposure. This

Table 3 Tensile properties* of age-hardened HAYNES 233 alloy before and after thermal exposure

Initial Condition	Exposure Temperature		Exposure Duration	Test Temperature		Yield Strength		Ultimate Tensile Strength		Elong
	°F	°C	h	°F	°C	ksi	MPa	ksi	MPa	%
SA + AHT2	–	–	–	RT	RT	112.9	779	172.0	1186	27.7
SA + AHT2	–	–	–	1200	649	95.4	658	156.6	1079	24.7
SA + AHT2	–	–	–	1400	760	97.4	671	116.2	801	29.2
SA + AHT2	–	–	–	1600	871	56.8	392	66.1	456	24.1
SA + AHT2	1200	649	1,000	RT	RT	133.7	922	189.0	1303	21.8
SA + AHT2	1200	649	4,000	RT	RT	132.9	917	187.0	1289	18.3
SA + AHT2	1200	649	8,000	RT	RT	131.7	908	186.9	1289	18.5
SA + AHT2	1200	649	1,000	1200	649	113.2	781	170.3	1174	17.6
SA + AHT2	1200	649	4,000	1200	649	114.4	789	164.7	1135	14.4
SA + AHT2	1200	649	8,000	1200	649	119.3	822	172.5	1190	16.4
SA + AHT2	1400	760	1,000	RT	RT	116.4	803	177.0	1220	20.9
SA + AHT2	1400	760	4,000	RT	RT	114.6	790	174.4	1203	13.3
SA + AHT2	1400	760	8,000	RT	RT	110.1	759	165.2	1139	8.6
SA + AHT2	1400	760	1,000	1400	760	91.4	630	115.4	796	33.5
SA + AHT2	1400	760	4,000	1400	760	90.9	626	113.1	780	29.2
SA + AHT2	1400	760	8,000	1400	760	84.4	582	111.0	765	30.3
SA + AHT2	1600	871	1,000	RT	RT	76.8	529	147.6	1018	29.0
SA + AHT2	1600	871	4,000	RT	RT	68.5	472	137.2	946	27.7
SA + AHT2	1600	871	8,000	RT	RT	68.3	471	134.9	930	23.8

(continued)

Table 3 (continued)

Initial Condition	Exposure Temperature		Exposure Duration	Test Temperature		Yield Strength		Ultimate Tensile Strength		Elong
	°F	°C	h	°F	°C	ksi	MPa	ksi	MPa	%
SA + AHT2	1600	871	1,000	1600	871	33.4	230	56.8	391	43.1
SA + AHT2	1600	871	4,000	1600	871	33.2	229	53.4	368	49.1
SA + AHT2	1600	871	8,000	1600	871	29.7	205	52.7	363	46.9

*Average of duplicate tests; SA = Solution Annealed; AHT2 = 1650°F (871 °C) / 4 h + 1450°F (871 °C) /8 h

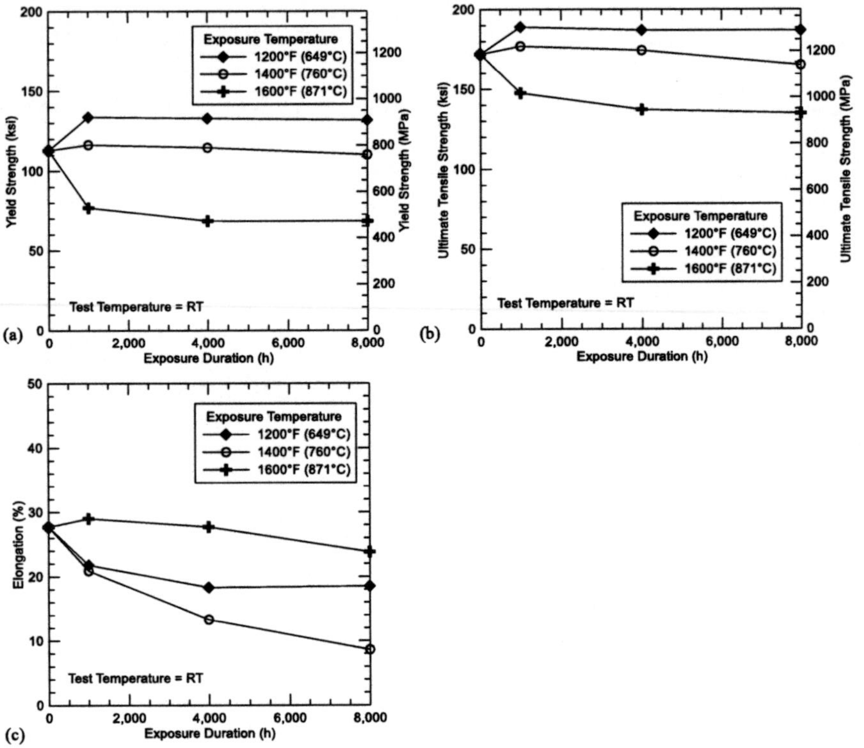

Fig. 6 Room temperature tensile properties of age-hardened + thermally exposed 233 alloy. **a** Yield Strength, **b** Ultimate Tensile Strength, and **c** Elongation

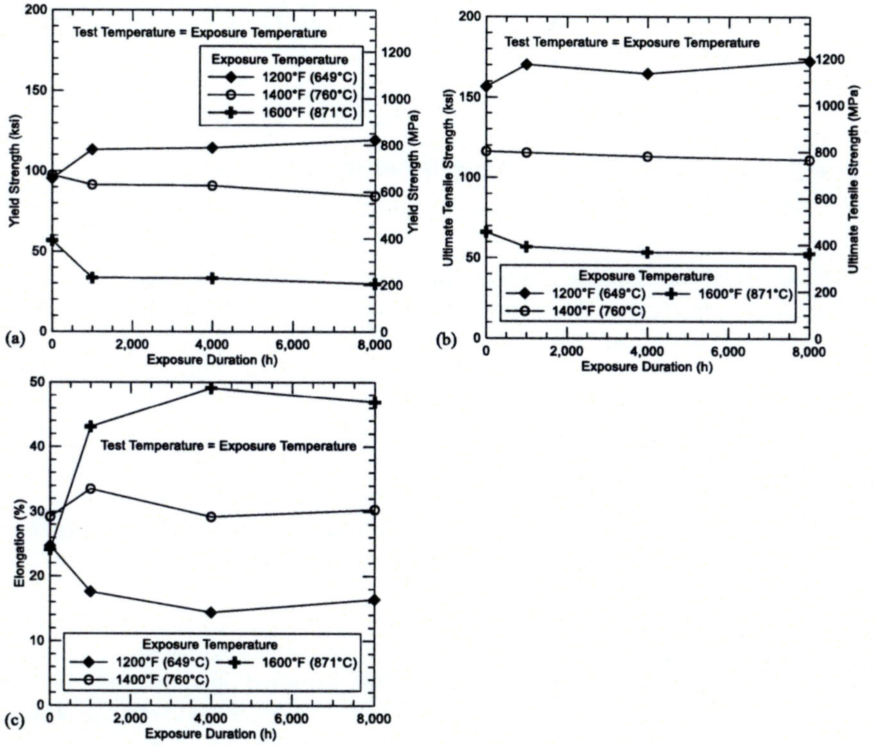

Fig. 7 Elevated temperature tensile properties of age-hardened + thermally exposed 233 Alloy. Tests performed at the corresponding exposure temperature. **a** Yield Strength, **b** Ultimate Tensile Strength, and **c** Elongation

was true for all three exposure temperatures. Again, the YS was found to initially increase with exposure duration for 1200°F (649 °C) thermal exposures, to remain essentially unchanged for 1400°F (760 °C) thermal exposures, and to initially slightly decrease for 1600°F (871 °C) thermal exposures. The effect of thermal exposures on the UTS (Fig. 7b) at the exposure temperature again followed similar trends to the RT data. However, the effect of thermal exposure on the tensile elongation at the exposure temperatures (Fig. 7c) did not always follow the same trends as the RT data. For 1200°F (649 °C) thermal exposures the exposure tensile elongation at the exposure temperature was found to drop within the first 1000 h, but remained fairly constant with further exposure duration, averaging a respectable 16% for thermal exposures between 1000 and 8000 h. In this case, the trends were similar to the RT data. A much different behavior was seen for 1400°F (760 °C) thermal exposures. Remarkably, the 1400°F (760 °C) tensile elongation remained quite high (~30%) regardless of the exposure duration. This was a pronounced difference in behavior from the RT data and will be discussed in more detail in the next section of this paper. Finally, for 1600°F (871 °C) thermal exposures it was found that the tensile elongation at the exposure temperature increased significantly out to exposure durations of 4000 h, but then leveled off. This trend also did not follow that of the RT data.

Beneficial Effects of Prior Age-Hardening on Thermal Stability

As discussed above, for applications which have components which see operating temperatures greater than the gamma-prime solvus of 1767°F (964 °C) but also have areas that would see lower temperatures, there is a question of whether to age-harden the component prior to service. The answer may depend in part on dimensional concerns. That is, how to best accommodate the density change (or shrinkage) which occurs as a result of age-hardening. However, another key factor is the effect of age-hardening on the ductility of the alloy at the critical intermediate temperatures—often around 1400°F (760 °C) where ductility minimums can occur in high-strength alloys. And furthermore, what is the effect of prior age-hardening on the intermediate temperature ductility after long-term thermal exposure? The results of this study can provide insight into these questions.

Consider the tensile elongation at the critical intermediate temperature of 1400°F (760 °C) before and after an 8,000 h thermal exposure at that same temperature (Fig. 8). For material originally in the solution annealed condition, the 1400°F (760 °C) elongation is seen to drop from 14.2% to 9.5% after the thermal exposure. However, when the starting material is age-hardened the 1400°F (760 °C) elongation is significantly higher, 29.2%, and actually increases following the thermal exposure to 30.3%. Prior age-hardening is clearly beneficial at this temperature. Looking at other exposure temperatures, there was a positive, but less pronounced, effect of prior age-hardening for 1200°F (649 °C) thermal exposures (1200°F tensile elongation after thermal exposure was 16.4% when given a prior age-hardening treatment vs. 13.5% without). For 1600°F (871 °C) thermal exposures there was little difference in 1600°F (871 °C) tensile elongation with or without prior age-hardening (47.8% and 46.9%, respectively). Overall, however, the data clearly suggest that age-hardening prior to thermal exposure is beneficial to intermediate temperature ductility both prior to and after thermal exposure.

To understand the benefits of prior age-hardening on the thermal stability of 233 alloy it is helpful to consider the grain boundary structure. As described earlier, the age-hardening treatment results in significant grain boundary precipitation of the gamma-prime phase (see Fig. 2d) along with some fine Mo-rich and Cr-rich precipitates. It is believed that this complex grain boundary structure is responsible for the improved ductility. More work is needed to better understand the exact mechanism. Now consider what happens after long-term thermal exposures. As described earlier, after 8000 h thermal exposures at 1400°F (760 °C) there is considerable precipitation of intergranular Cr-rich precipitates for both initial conditions (solution annealed and age-hardened). While this precipitation results in the drop of the 1400°F (760 °C)

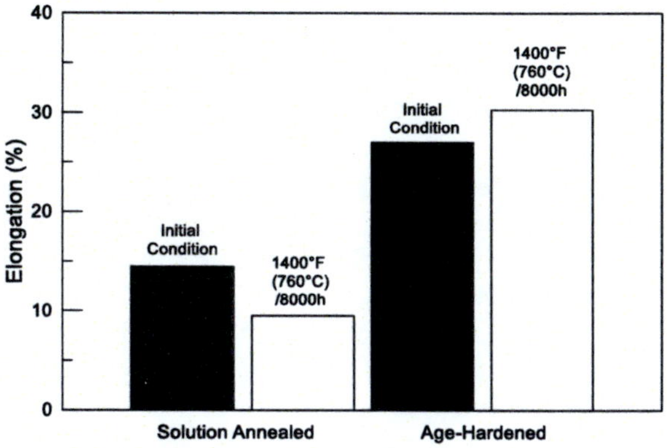

Fig. 8 Comparison of 1400°F (760 °C) tensile elongation of solution annealed and age-hardened 233 alloy before and after 8,000 h thermal exposure at that same temperature

tensile elongation following the thermal exposure of solution annealed samples (as might be expected), no such drop is observed for samples given the age-hardening treatment prior to thermal exposure. Again, the grain boundary structure may provide insight. Higher magnification (20Kx) SEM images of 1400°F (760 °C)/8000 h thermally exposed samples of both initial conditions are shown in Fig. 9a, b. Note that both the Cr-rich phase and gamma-prime can be seen at the grain boundaries in both images. However, in Fig. 9a (solution anneal + thermal exposure) the gamma-prime is discontinuous and sits adjacent to the blocky Cr-rich precipitates. However, in Fig. 9b (age-harden + thermal exposure) the gamma-prime phase is seen to be wider, continuous, and fully envelops the Cr-rich phase. It is likely that this difference in grain boundary structure is related to the significant improvement in 1400°F (760 °C) ductility, but work is needed to better understand this phenomenon.

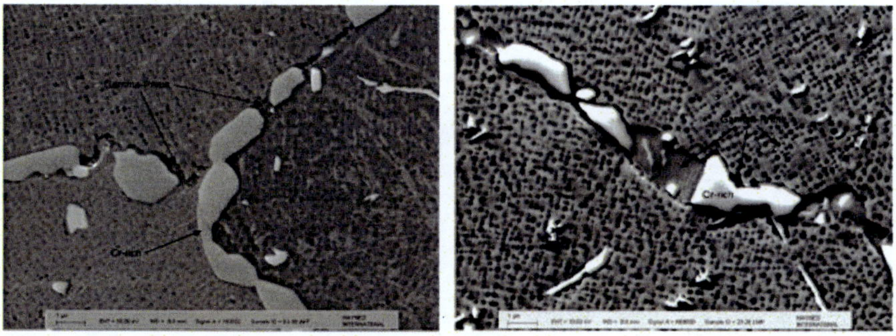

(a) (b)

Fig. 9 Higher magnification SEM-BES images of the grain boundaries of 233 alloy thermally exposed for 8000 h at 1400°F (760 °C). Initial condition: **a** Solution Annealed, **b** Age-hardened

Long-Term Oxidation Resistance

Another key property of HAYNES 233 alloy is its excellent oxidation resistance. Since the focus of this paper is on long-term performance, a long-term oxidation test (1-year) will be presented. Shorter term oxidation studies of 233 alloy have been presented elsewhere [1, 6]. The alloys selected for comparison were 230, 617, 188, and 625 alloys—all are readily fabricable alloys, like 233 alloy, used commonly in gas turbine engine applications. Figure 10 shows the weight changes of the alloys tested over the one-year test duration, in which 625 alloy and 188 alloy coupons were removed from the test after eight cycles (240 days/5760 h) due to severe metal loss, exhibiting breakaway oxidation behaviors after only two 30-day cycles at 2000°F (1093 °C). The other chromia forming alloys, 230 and 617, showed better high temperature oxidation resistance than 625 and 188 alloys without experiencing breakaway oxidation. Among all tested alloys, the alumina forming 233 alloy shows superior oxidation resistance without weight loss over the test period. The results clearly exhibit the significance to resist high temperature oxidation attack by the formation of a protective alumina scale on 233 alloy, and this protection is maintained over the full year-long test.

Besides metal loss due to high temperature oxidation attack, internal oxidation penetration is another important material degradation metric. Internal oxidation attack on the alloys tested was further evaluated from optical cross-sectional examination, and the results are tabulated in Table 4. Metal Loss was calculated from the initial coupon weight prior to the oxidation test and its final weight after the oxidation test and descale treatment. Average Metal Affected (Avg. Met. Aff.) and Maximum Metal Affected (Max. Met. Aff.) were calculated by summing Metal

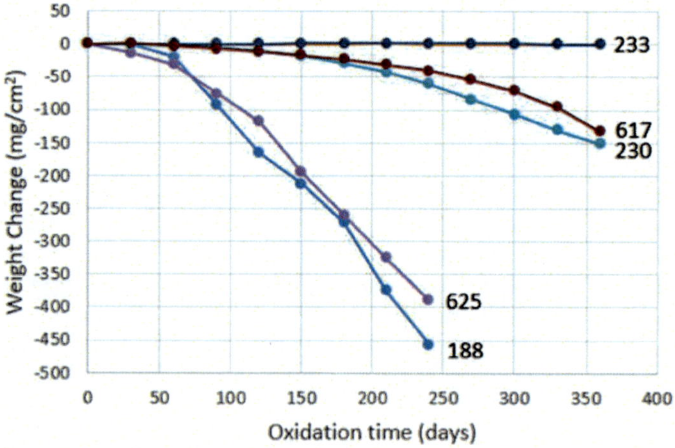

Fig. 10 Long-term oxidation behaviors of several high temperature alloys exposed for one-year cycled each 30 days at 2000°F (1093 °C)

Table 4 Oxidation attack measurement on the alloys oxidized for one year at 2000°F (1093 °C)

Alloy	Metal Loss (mils/μm)	Avg. Met. Aff. (mils/μm)	Max. Met. Aff. (mils/μm)
233	0.2/5	1.3/33	1.6/41
617	7.1/180	12.4/315	12.7/323
230	7.7/196	16.0/406	18.3/465
625 [a]	19.0/483	22.1/561	22.9/582
188 [a]	21.8/554	25.1/638	26.2/665

[a]The results of 625 and 188 alloys were obtained after only eight cycles (240 days/5760 h)

Loss and average internal penetration and maximum internal penetration, respectively, measured by cross-sectional examination. The average internal penetration was calculated from eight measurements taken across the full coupon. The oxidation attack results are consistent with the weight change results, but provide more accurate oxidation attack information. Among the alloys tested, 233 alloy showed minimal metal loss and internal penetration, *i.e.*, superior oxidation resistance, when compared to the chromia forming alloys over the one-year oxidation period.

The excellent long-term oxidation resistance of 233 alloy is attributed to the formation of an exclusive alumina scale, which is shown in Fig. 11. The alumina scale formed was about 15 μm in thickness with good integrity and adherence to the substrate. There is a local area in Fig. 11 exhibiting thinner alumina scale, ~5 μm, which is likely due to spallation of the top of the alumina scale. It is noted that the partial scale spallation did not result in oxidation attack on the substrate, which implies that the remaining alumina scale still provided good oxidation protection. Through the cross-section, no internal oxidation was observed, confirming the excellent oxidation resistant behavior of 233 alloy. The top of the oxide scale showed a small amount of local spinel with Al, Ni, and Cr. Some particles (bright phases in the alumina scale on the image) were observed in the alumina scale, and they were detected to be (Ta, Zr, Ti)-rich oxides. These oxide particles in the alumina scale could be related to original carbides existing in the alloy matrix and imply inward growth of the alumina scale. Although not visible in the image, some voids co-existed around these oxide particles. The existence of the oxide particles and/or voids in the alumina scale likely enhanced the scale growth rate by increasing diffusion paths in the alumina scale [7]. As similar oxide particles and voids were observed on 233 alloy oxidized for 1008 h at 1149 °C [6],these results indicate that they are not detrimental to the oxidation resistance of 233 alloy.

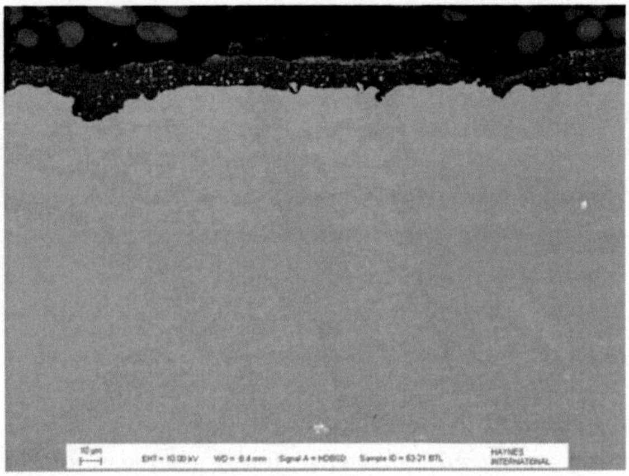

Fig. 11 Cross-sectional SEM-BSE image of 233 alloy showing the formation of an exclusive alumina scale after the one-year oxidation test at 2000°F (1093 °C)

Summary and Conclusions

1. The microstructure of solution annealed 233 alloy consists of primary carbides of three types, MC, $M_{23}C_6$, and M_6C, scattered within the gamma matrix. The grain boundaries are free of precipitation.
2. After the recommended two-step age-hardening treatment (AHT2), gamma-prime (~65 nm) is formed intragranularly and greatly strengthens the alloy. Additionally, a grain boundary complex is formed including gamma-prime and isolated Cr-rich and Mo-rich precipitates. This greatly improves the ductility of the alloy at the critical 1400°F (760 °C) temperature.
3. Thermal exposure of age-hardened 233 alloy at 1200°F (649 °C) for up to 8,000h had little effect on the microstructure. The strength (YS and UTS) and elongation at both RT and the exposure temperature were comparatively stable as well.
4. Thermal exposures of age-hardened 233 alloy at 1400°F (760 °C) for up to 8,000h resulted in several microstructural changes including the precipitation of both intragranular mu and an intergranular Cr-rich phase, coarsening of the gamma-prime phase to ~130 nm, and further growth of the gamma-prime layer at the grain boundary.
5. Despite these microstructural changes, the RT and 1400°F (760 °C) strength (YS and UTS) remained very stable over the duration of the exposure. While the RT ductility was found to drop as a result of the thermal exposure, the ductility at the exposure temperature remained quite high for the duration of the exposure.
6. The high 1400°F (760 °C) ductility of the age-hardened 233 alloy even after long-term thermal exposure is likely attributable to the presence of a gamma-prime grain boundary layer which completely envelops the Cr-rich phase.
7. Thermal exposures of age-hardened 233 alloy at 1600°F (871 °C) for up to 8000h produced microstructures similar to those of the 1400°F (760 °C) exposure, but

the amounts of the intergranular Cr-rich and intragranular mu phases were much less. The intragranular gamma-prime coarsened to an average size of ~420 nm.

8. The coarsening of the gamma-prime at 1600°F (871 °C) led to a drop in strength at RT and at the exposure temperature which levels off after 1000h. The 1600°F (871 °C) thermal exposure did not negatively affect the ductility at either temperature.
9. The effect of thermal exposures of 1200°F (649 °C), 1400°F (760 °C), 1600°F (871 °C), and 1800°F (982 °C) on the microstructure and tensile properties of solution annealed 233 alloy were also presented.
10. Overall, it was concluded that prior age-hardening of 233 alloy provides improved thermal stability, particularly with respect to ductility at the critical 1400°F (760 °C) temperature.
11. The long-term oxidation resistance of 233 alloy was found to be excellent in comparison to other readily fabricable alloys commonly used in gas turbine engines. This can be attributed to the formation of an exclusive alumina scale which remained protective through the full 1-year oxidation test at 2000°F (1093 °C).

Acknowledgements The efforts of laboratory personnel Jennifer Schoolcraft, Ryan Markley, Rupinder Sharma, Mark Richeson, John Ryan, and Bridget North to produce the data in this paper are gratefully acknowledged. Thanks also to Michael Fahrmann for careful review of the paper. HAYNES, HASTELLOY, and 230 are registered trademarks and 233 is a trademark of Haynes International, Inc.

References

1. Pike LM, Srivastava SK, Forsythe, AF (2019) A New Alumina-Forming Ni-Co-Cr Base Alloy for Service in Gas Turbine Engine Combustors and Other High-Temperature Applications. Journal of Materials Engineering and Performance 28(4):1929–1935.
2. Wlodek ST (1999) The Stability of Superalloys. In: Fuchs GE et al. (ed) Long Term Stability of High Temperature Materials. The Minerals, Metals & Materials Society, Warrendale, PA p 3–40.
3. Haynes International, Metallographic Preparation of HASTELLOY® and HAYNES® alloys. https://www.haynesintl.com/tech-briefs/general-haynes-international-information/metallographic-preparation-of-hastelloy-and-haynes-alloys-(H-1107). Accessed 4 January 2023.
4. Radavich JF (1997) Electron Metallography of Alloy 718. In: Loria EA (ed) Superalloys 718, 625, 706, and Various Derivatives. The Minerals, Metals & Materials Society, Warrendale, PA p 17–26.
5. Rowe MD, Ishwar VR, Klarstrom DL, (2006) Properties, Weldability, and Applications of Modern Wrought Heat-Resistant Alloys for Aerospace and Power Generation Industries. Transactions of the ASME 128(April):354–361.
6. Li B, Pike LM, Deodeshmukh VP, and Srivastava SK (2021) High Temperature Oxidation Resistance of a Newly Developed Ni-Based Alumina-Forming NiCoCrMoAl Alloy. Paper (#16537) presented at Corrosion 2021, NACE & AMPP, Houston, TX, April 2021.
7. Wessel E, Kochubey V, Naumenko D, Niewolak L, Singheiser L, Quaddakkers WJ (2004) Effect of Zr addition on the microstructure of the alumina scales on FeCrAlY-alloys. Scripta Materialia 52:987–992.

Subcritical Crack Growth of Alloy 718 in Marine Exposure Conditions and Microstructural Modeling

A. Arcari, D. J. Horton, M. Zikry, and M. Chen

Abstract UNS N07718 is widely used in marine service applications under a variety of conditions: alternate immersion, different levels of cathodic protection, and freely corroding galvanic couples. Environmentally assisted cracking can significantly affect the performance of this alloy and constrains design as it needs to account for subcritical crack growth in service. We measured subcritical crack growth rates and thresholds in different environmental conditions for two different heat treatments of UNS N07718. The first heat treatment, following AMS 5664 is widely used in the aircraft industry and for marine fasteners, and the second, following API 6A, is principally used in the marine and oil and gas industries. The material environmentally assisted cracking was studied under alternate immersion to natural seawater and under cathodic protection in natural seawater. Microstructural modeling is presented to understand and predict how precipitates, their volume fraction, morphology, and properties, affect the evolution and accumulation of dislocation densities within the microstructure, influencing the fracture process at different physical scales.

Keywords Natural seawater · Hydrogen embrittlement · Alternate immersion

Introduction

Nickel based alloy Inconel 718 is a high strength alloy extensively used in marine environments due to its combination of high general corrosion resistance, good strength, and high resistance to chloride and sulfide stress corrosion cracking [1, 2]. The most commonly used condition in marine applications, particularly in the Oil and Gas sector, is produced to meet the American Petroleum Institute (API)

A. Arcari (✉) · D. J. Horton
Center for Corrosion Science and Engineering, Chemistry Division, U.S. Naval Research Laboratory, 4555 Overlook Avenue SW, Washington, DC 20375, USA
e-mail: attilio.arcari@nrl.navy.mil

M. Zikry · M. Chen
Department of Mechanical and Aerospace Engineering, North Carolina State University, Raleigh, NC 27695, USA

E. A. Ott et al. (eds.), *Proceedings of the 10th International Symposium on Superalloy 718 and Derivatives*, The Minerals, Metals & Materials Series,
https://doi.org/10.1007/978-3-031-27447-3_19

standard 6A 718 specification, which has a minimum yield strength of 120 ksi [3, 4]. Inconel 718 API 6A has a unique chemistry, heat treatment, and microstructure to enhance its resistance to hydrogen embrittlement (HE) and stress corrosion cracking (SCC). The API specification requires compliance with NACE MR0175 for all cases involving exposure to H_2S, which limits the maximum hardness to Rockwell Hardness (HRC) 40. Variations in material's microstructure are a function of the manufacturing process [5] and there is considerable interest in understanding the role of the microstructure in the HE performance of the material.

The annealed condition (AMS 5662) is commonly sold for general use in the aerospace and oil and gas industries, which can then be heat treated to meet various other conditions: aged per AMS 5663 to meet aerospace requirements or aged to meet API 6A and/or NACE MR0175 requirements for use in the oil and gas industry for marine applications. The AMS 5663 condition requires a minimum of 150 ksi yield strength and 180 ksi strength. Typical 718 bar for Navy use are procured per AMS 5664, or if that is not available, per AMS 5662 or 5663 by applying an additional heat treatment to effectively yield a 5664 material. The requirement is similar to AMS 5663 in terms of strength, but with a higher solution annealing temperature and longer first and second heat treatment steps.

The use of Inconel 718 (API 6A) in naval applications is increasing, particularly as a fastener material, due to its resistance to SCC and HE under cathodic protection systems that maintain an applied potential equal to or more electronegative than −850 mV vs Ag/AgCl [6, 7]. This work compares the behavior of these two commonly used forms of 718, API 6A and AMS 5664, in two typical marine environments: natural seawater under alternating wet and dry conditions and under cathodic protection at −850 mV vs Ag/AgCl. Modeling of the material microstructure to study the effects of precipitated phases within the metal grains is also presented.

Materials and Experimental Methods

The AMS 5664 material was in the form of a round bar of 4.25 in diameter and the minimum yield strength was 150 ksi. The API 6A material was in the form of a round bar of 3.75 in diameter and with minimum yield strength of 120 ksi. The samples were both CT samples, 1.25 in thickness, and 2.5 in width, and loaded using a bolt, as shown in Fig. 1. The samples were tested in natural unfiltered seawater environment at open circuit potential under alternate immersion conditions and under cathodic protection. The latter was achieved by connecting the samples to a low voltage aluminum anode (LVAA) also immersed in seawater to maintain the nominal potential of −850 mV vs Ag/AgCl.

We subjected fatigue pre-cracked samples to a constant crack mouth opening displacement (CMOD) while exposed to the environment, and monitored the progression of the crack over time by measuring the sample compliance with the use of a strain gauge attached to the back face of the sample [8]. This technique allows to study the environmental assisted cracking (EAC) behavior of the material when the

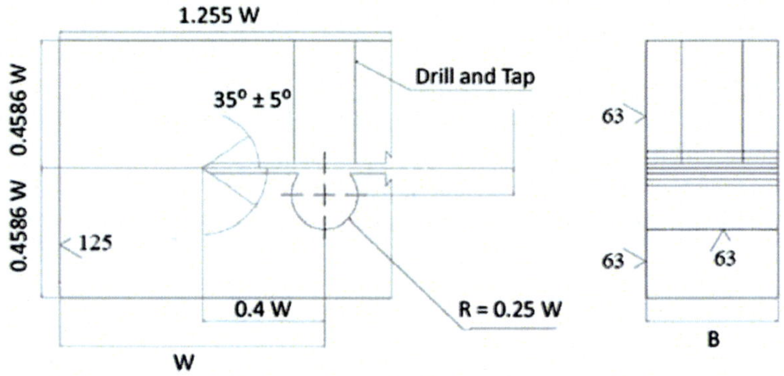

Fig. 1 Bolt loaded Compact Tension (CT) geometry

driving force progressively reduces over time, with the advancement of the crack. All samples were fatigue pre-cracked according to ASTM E647 guidelines and using the load shedding method [9]. The fatigue pre-crack length was approximately 0.5 W for all samples. The final maximum of the Stress Intensity Factor (SIF), K_{max}, during fatigue cycling was lower than the applied stress intensity factor K at the beginning of the test per ASTM standard guidelines.

A clip gauge on the front face of the sample was used to monitor the applied bolt pre-load before exposure to the testing environment. The SIF was calculated by knowing the initial crack length and the applied CMOD. Immediately after loading, the clip gauge was removed, the sample was deployed in the test environment, and then monitored continuously using a strain measurement on the back face of the sample [8]. The strain was monitored using a quarter Wheatstone bridge configuration with completion resistors within the data recording system. The temperature of one of the samples was monitored to allow temperature correction of the strain recording. The resistance of the cables from the location of testing to the location of data recording (~30 feet) was included in the gauge factor as an additional correction. Previous work showed that the resolution of this technique is 25–50 μm/day. The strain gauge was protected using a polyurethane rubber compound for OCP conditions.

Results

For alloy 718 AMS 5664 a total of 3 samples were tested under OCP conditions and 5 under CP conditions. Table 1 shows for each sample: the environment of testing, open circuit potential under alternate immersion, abbreviated OCP-A_{Im}, or cathodic protection under full immersion conditions, abbreviated CP_{-850}, the initial K, the average crack growth rate measured for the sample, the final recorded value of K, and the total test time.

Table 1 Test conditions for nickel alloy 718 AMS 5664 and average measured crack growth rates. The two exposure environments are OCP-A_{Im}, Open Circuit Potential—Alternate Immersion, and CP, Cathodic Protection with reference electrode potential of −850 mV vs Ag/AgCl

	Environment	Initial K (ksi√in)	Average da/dt	Final K (ksi√in)	Total Test Time (hrs)
Sample 1	OCP—A_{Im}	106.0	1.015 10^{-5}	94.75	>10,000
Sample 2	OCP—A_{Im}	104.9	No cracking	*	>10,000
Sample 3	OCP—A_{Im}	106.1	No cracking	*	>10,000
Sample 4	CP_{-850}	103.7	1.792 10^{-4}	82	1,800
Sample 5	CP_{-850}	103.0	3.560 10^{-5}	94.5	>5,000
Sample 6	CP_{-850}	82.4	4.672 10^{-5}	78	>5,000
Sample 7	CP_{-850}	82.0	1.891 10^{-5}	75	>5,000
Sample 8	CP_{-850}	81.3	No cracking	*	>5,000

The results for Samples 1, 4, 5, 6, and 7 of AMS 5664 are shown in Fig. 2. All these samples showed measurable crack growth by the end of the test. The general trend for crack progression was of a very low incubation time upon exposure to the environment, and of high crack growth rates in the early phases of the test. The growth rates progressively reduced and crack arrest occurred between 800 and 1,800 h of testing. The samples were left exposed for over 10,000 h in the case of OCP testing and over 5,000 h for the tests under CP and did not show any further crack progression. Sample 4 showed a shorter incubation time and generally faster growth rates than the corresponding Sample 5, tested at nominally the same conditions. Figure 2b shows a direct comparison of these two tests for the initial phases of cracking. We do not currently know the reason for this behavior and can only assume it represents the scatter in EAC behavior of this material. Analysis of incubation times and stress intensity among all tests did not show a clear trend or correlation, suggesting that there is significant variability in the EAC behavior of this material and that a much larger test matrix is needed.

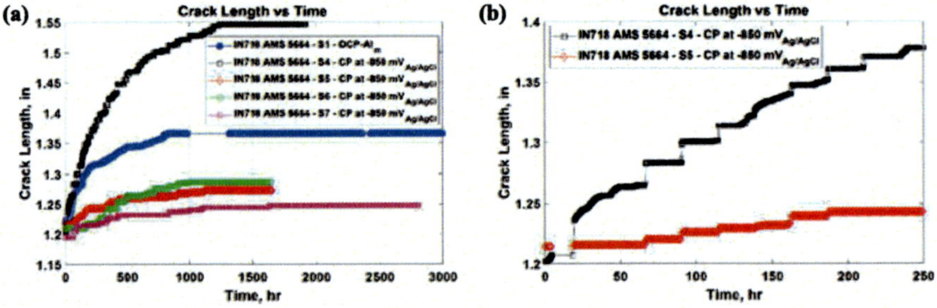

Fig. 2 Resulting EAC progression for samples that showed susceptibility during long term exposure testing

Table 2 Test conditions for nickel alloy 718 API6A and average measured crack growth rates

	Environment	Initial K (ksi$\sqrt{}$in)	Measurable da/dt ?	Final K (ksi$\sqrt{}$in)	Total Test Time (hrs)
Sample 1	OCP—A_{Im}	92.9	4.472 10^{-7}	89.0	>10,000
Sample 2	OCP—A_{Im}	91.4	No	*	>10,000
Sample 3	OCP—A_{Im}	92.1	No	*	>10,000

Alloy 718 API(6A) was not susceptible to EAC under OCP alternate immersion testing conditions. The initial stress intensity for the three samples tested was near 92 ksi$\sqrt{}$in and only Sample 1 showed some initial indication of cracking. This was recorded during the initial few hours of exposure, followed by a crack arrest for the remaining test duration. An average crack growth rate is calculated for comparative purposes only. Samples 2 and 3 did not show any significant crack progression. The data are summarized in Table 2. The data under OCP alternate immersion conditions are compared to literature results for CP.

Figure 3 shows a summary of all the results regressed as SIF vs crack growth rate expressed in inches per hour (in/hr). The downward arrows indicate the lowest SIF that showed cracking for the material and experimental condition tested. The AMS 5664 variant showed significant cracking under OCP alternate immersion conditions. The crack growth rates progressively reduced until crack arrest was observed at a SIF of 94.75 ksi$\sqrt{}$in, here identified as the observed threshold. The same material under CP conditions showed measurable crack growth at significantly lower K. As shown, Samples 4, 6, and 7 all showed crack growth rates above 10^{-5} in/hr for K higher than 80 ksi$\sqrt{}$in. The behavior of Sample 4 and Sample 5 however were not in good agreement, with the first showing significant crack progression and a decrease in K of almost 20 ksi$\sqrt{}$in, while the second showed crack growth rates decreasing below 10^{-5} in/hr at a K = 94.5 ksi$\sqrt{}$in, followed by crack arrest. Two out of the three materials tested at an initial K of 80 ksi$\sqrt{}$in showed measurable crack progression. The initial fatigue crack growth rates were in agreement with the measurements from Sample 4 at the beginning of the test and then progressively reduce below 10^{-5} in/hr when K reduced below 75 kis$\sqrt{}$in.

Under OCP alternate immersion conditions 718 API6A did not show significant crack progression and an average crack growth rate is calculated and shown in Fig. 3 for reference only. Data from [10] show that the crack growth rate for 718 API6A in constant immersion conditions is near 10^{-6} in/hr for an applied stress intensity of 100 ksi$\sqrt{}$in, and near 10^{-7} in/hr for an applied stress intensity of 80 ksi$\sqrt{}$in. The data from the literature are for a constant-K type test rather than a K-shedding test and represent a conservative upper bound.

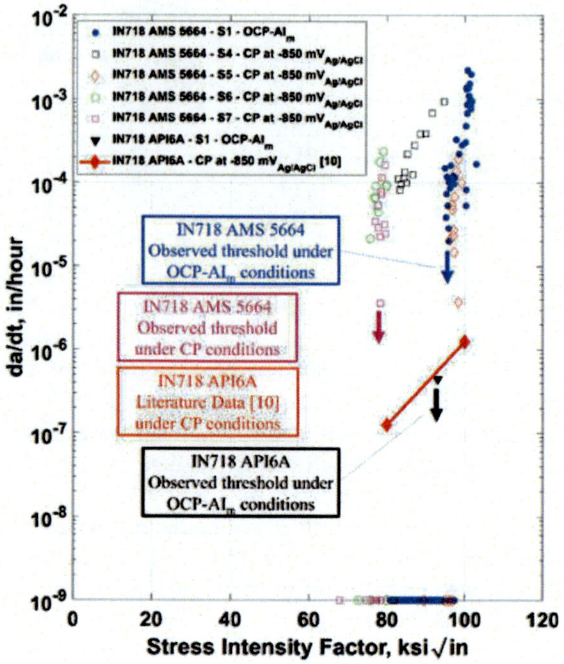

Fig. 3 Summary of results from long term exposure testing of alloy 718 (IN718) produced according to AMS 5664 and API6A

Modeling

While the microstructure of alloy 718 has been extensively studied and specific compositional limits and thermal processing steps are standardized to reduce HE susceptibility, the role of precipitates in deformation and hydrogen mechanism behavior is not well understood. A microstructural meso-scale model was developed to study the mechanical response of the elements of the microstructure for a constant applied displacement. Scanning Electron Microscopy (SEM) was performed to identify the parts of the microstructure to be modeled. A portion of the cross section of the 718 AMS 5664 round bar where the samples were extracted from was encapsulated in conductive epoxy, ground, and polished to progressively finer grit. A final step of vibratory polishing in colloidal silica allowed the visualization of the grain boundaries and of the δ phase. The microstructure showed metal grains with average dimensions close to 10 microns (ASTM E114 number near 8) and the presence of δ phase at the grain boundary. The δ phase is generally considered to have an orthorhombic structure, with stoichiometry Ni_3Nb. Other precipitates, specifically γ' and γ'', are smaller and mostly coherent with the FCC matrix and are not visible under SEM. Direct characterization using transmission electron microscopy (TEM) is planned in future work.

The grain structure was reproduced by using Voronoi tessellation: a set of 16 grains in a 16 μm × 16 μm microstructure in a 2-dimensional plane-strain finite element analysis. The number of grains was defined based on a compromise between good representation of the variations in the material microstructure and a manageable computational time. As our initial goal is to understand the roles of the different parts of the microstructure, computational time is minimized to allow an expansion of modeling parameters. The grain size, overall similar to the observed grain size from polished samples of AMS 5664, was also a compromise between reflecting the expected size in the materials studied and the computational burden given by the modeling of the phases within each grain. The larger the grain, the higher the number of precipitated phases inside the grains and therefore the computational size of the model. The δ phase was modeled at the grain boundaries as a circular phase with diameter ranging from 0.1 μm to 0.2 μm. The model allows for different options for δ phase shape, size, and distribution on the grain boundary. The option for the results presented in this work was of a fixed number of two δ phases on each grain boundary, for grain boundaries that were long enough to accommodate them. Their number on different grain boundaries also varied as to avoid intersections with other δ phases or parts of the microstructure. The δ phase was modeled as an ordered tetragonal crystal with elastic constants C_{11} = 264 GPa, C_{12} = 175 GPa, C_{13} = 155 GPa, C_{33} = 261 GPa, C_{44} = 142 GPa, C_{66} = 150 GPa [11]. The material orientation of the δ phase was simplified as the average of the Euler angles of the two adjacent grains.

The γ'' phase particles were modeled at the interior of the grain boundary as ellipses with major to minor axis ratio of 2. The major axis dimension was varied between 0.1 μm to 0.2 μm. The former value is representative of an upper bound limit for the expected size of this phase for materials with comparable thermal history to the material tested in this work (AMS 5664 and API 6A) [5, 12, 13]. The larger size is used for comparative purposes only and it is not meant to be a representative microstructure. It is a useful reference microstructure to study the trends in the model. The γ'' phase particles were randomly distributed at the interior of each grain with constraints dictated by distance from the grain boundary, to avoid intersection with δ phase or grain boundaries, and by distance to the other γ'' phase particles. Although a characteristic set of orientations was observed for this phase within the γ microstructure [14], this model randomizes the orientation of the major axis with respect to the orientation of the metal grain. Future work will include the expected misorientations of the γ'' phase. Constant volume fractions of γ'' phase were compared by varying their size and number within the microstructure. The γ'' phase was also modeled as an elastic crystalline material with the same constants as the δ phase. The material orientation of the γ'' phase was the same as the grains that includes it.

A crystal plasticity formulation was implemented into a User Material subroutine (UMAT) in Abaqus based on the work by Huang [15] to model the mechanical response of the metal grains. Before plastic deformations occur, the FCC γ-matrix of the material is modeled as an elastic crystal with constants: C_{11} = 233 GPa, C_{12} = 147 GPa, C_{44} = 112 GPa [16]. The formulation accounts for shear strain deformation on all 12 slip planes when plastic deformation occurs. The constitutive equations were modified to account for the distribution of mobile and immobile dislocations within the microstructure. Following the work by Zikry [17, 18], the strength of each slip system was modeled as a function of the mobile and immobile dislocation densities developed upon plastic deformations. The main step of the implementation of the theory is in the following evolution equations and their coupling to the shear strain rate and strength equations:

$$\dot{\gamma}^{(\alpha)} = \dot{\gamma}_{ref}^{(\alpha)} \left(\frac{\tau^{(\alpha)}}{\tau_{ref}^{(\alpha)}} \right) \left(\frac{|\tau^{(\alpha)}|}{\tau_{ref}^{(\alpha)}} \right)^{(1/m-1)}$$

$$\tau_{\mathrm{ref}}^{(\alpha)} = \left(\tau_{\mathrm{y}}^{(\alpha)} + G \sum_{\beta=1}^{nss} b^{(\beta)} \sqrt{\alpha_{\alpha} \beta \rho_{im}^{(\beta)}} \right)$$

where $\dot{\gamma}^{(\alpha)}$ is the shear strain rate and $\dot{\gamma}_{ref}^{(\alpha)}$ is a reference shear strain rate, $\tau^{(\alpha)}$ is the resolved shear stress, $\tau_{\mathrm{y}}^{(\alpha)}$ is the initial strength, G is the shear modulus, $b^{(\beta)}$ is the Burger's vector, and all refer to the slip system α. The variable *nss* is the number of slip systems, $a_{\alpha\beta}$ represents the influence of dislocation accumulation along a β slip system, 0.2 for cross-coupling and 1.0 for self-coupling, $\rho_{im}^{(\beta)}$ is the immobile dislocation density for the β slip system. The evolutions of the mobile and immobile dislocation densities are coupled through a separate set of differential equations [17, 18]. The material properties for the crystal plasticity model were developed by simulating the tensile response of a grain aggregate and comparing with the stress–strain response of the material.

The model also included a sharp crack-like defect at one side, and it was loaded in tension by applying a constant displacement on the top boundary with the bottom boundary being fixed. The area near the crack is modeled as an elastic crystal. The mesh was refined around all the δ and γ″ phases, for a minimum element size of 0.0125 μm. The largest model, named model C1, included 16 grains, 41 δ phase particles, and 6100 γ″ phase particles; the number of nodes was 1,132,375. The model was developed as parametric, allowing all the inputs to be changed systematically and to observe the variations in mechanical response. Results for three variations of the largest model developed are presented here. A summary of the input variations of model C1 is shown in Table 3, each model is given a name for ease of comparison.

The first set of results compares the effects of size of γ″ phase particles at the same volume fraction. The volume fractions translate into an area ratio in the 2D model developed. The total area of the γ″ phase particles is 23.9 μm^2 for the first model, named C1a, and 23.5 μm^2 for the second model, named C1b. The difference between the two models is in the size of the γ″ phase: the major diameter is 0.1 μm for model C1a and 0.2 μm for model C1b, all γ″ phase particles have the same major to minor diameter ratio of 2.0. Their distribution is different, and it is determined by randomly positioning particles within the interior of the grain, leaving a distance equal to the diameter of the δ phase from the grain boundary. The models also have the same number of grains, grain orientation, δ phase position, size, and orientation.

The resulting total accumulated shear strains for models C1a and C1b are shown in Fig. 5. The maximum in the contour plot refers to 0.5% shear strain, while the minimum refers to no accumulated shear. The accumulated shear strain is slightly

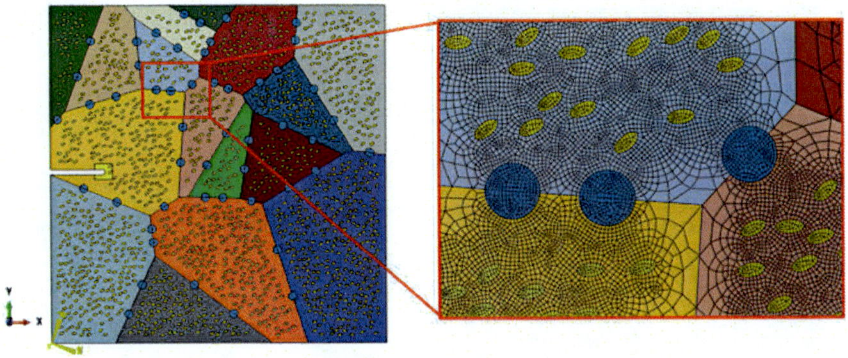

Fig. 4 Representative microstructure 2D plane-strain model for alloy 718; detail view of the phases in the finite element model: grain boundary δ phase and intergranular γ″ phase

Table 3 Summary of the differences in δ and γ″ phase modeling between the three models presented in this work

Model Name	γ″ number of particles	γ″ major diameter, μm	γ″ total area, μm²	δ number of particles	δ diameter, μm	δ total area, μm²
C1a	6100	0.1	23.9	41	0.4	5.15
C1b	1500	0.2	23.5	41	0.4	5.15
C1c	1500	0.2	23.5	41	0.2	1.29

higher for model C1a with the smaller γ″ major diameter. Some differences are visible near a grain boundary towards the bottom of the microstructure. A higher accumulated shear strain occurs there because the two adjacent grains have high misorientation. The contours for model C1a are less fragmented and more homogenous throughout the model. Near the grain boundary this results in a higher accumulated shear strain for model C1a.

Some differences are also observed in the distribution of immobile dislocation densities, with dislocation accumulation on a (111) plane in the [−110] direction, shown in Fig. 6. A small increase in dislocation density for the model with the larger size of γ″ particles can be appreciated in several parts of the microstructure. Higher level of dislocation activity is shown at grain boundaries, particularly at locations within δ phase particles. A simple quantitative metric can be obtained from the figures to have a better sense of their visual qualitative difference. The images are translated into an 8-bit RGB representation and the pixels with red component higher than 200/255, and with blue and green component equal or lower than 1/255 are counted. Results show that the number of red pixels for model C1a is lower than for model C1b.

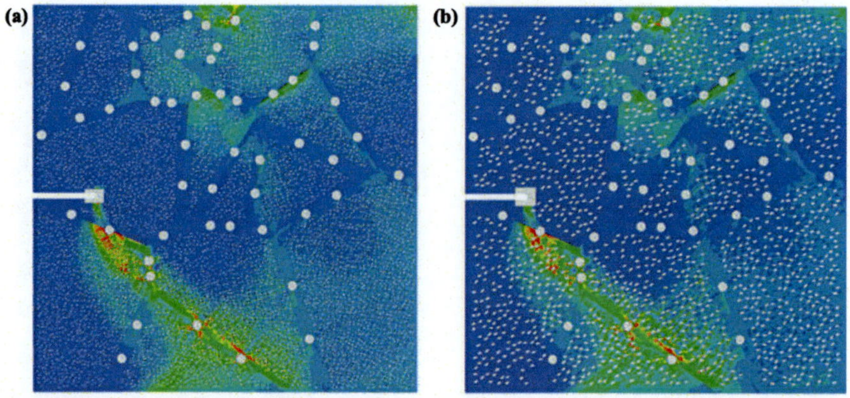

Fig. 5 Total accumulated shear strain for models C1a and C1b. A higher accumulation of shear strain for model C1b occurs at a grain boundary towards the bottom of the microstructure, while higher accumulated shear occurs within the grains of model C1a

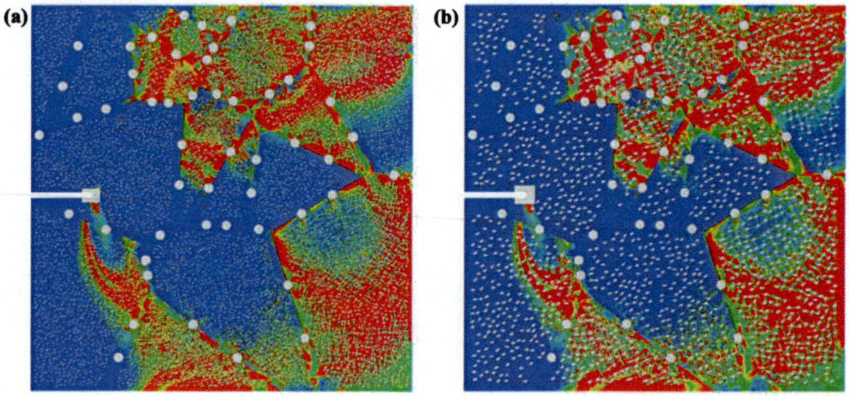

Fig. 6 Dislocation density along a (111) plane in the [−110] direction, for model C1a and model C1b

The third model, named C1c, uses the same size, distribution, and orientation of the γ'' phase particles in model C1b and compares the effects of the δ phase volume fraction by reducing the size of the δ phase in the model to 0.1 μm in diameter. The results of accumulated slip, in Fig. 7, identify the same critical grain boundary visualized previously as the location of highest accumulated slip. The maximum in the contour plot refers to 0.5% shear strain, while the minimum refers to no accumulated shear. For both models, the accumulated shear strain contour plots are similarly fragmented by the network of γ'' phase particles. The comparison also shows a trend of increasing shear strain with decreasing δ phase diameter. Figure 8 shows the difference in immobile dislocation densities along the (111) plane in the [−110] direction as shown previously. There is a very significant impact of the δ phase size along the grain boundaries, with significantly decreased activity throughout the microstructure. Particularly significant is the reduction in dislocation density at the critical grain boundary earlier identified. The same metric calculated previously confirms these qualitative observations and the red pixel number in Fig. 8 is reduced with respect to both models C1a and C1b.

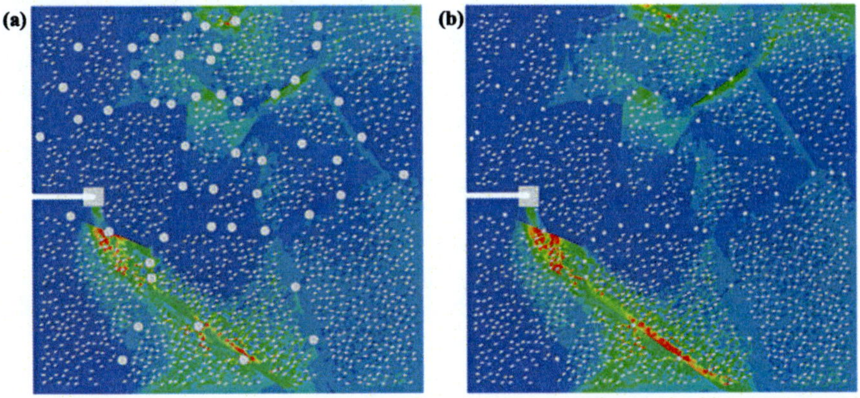

Fig. 7 Total accumulated shear strain for models C1a and C1c. A higher accumulation of shear strain for model C1c occurs at a grain boundary towards the bottom of the microstructure

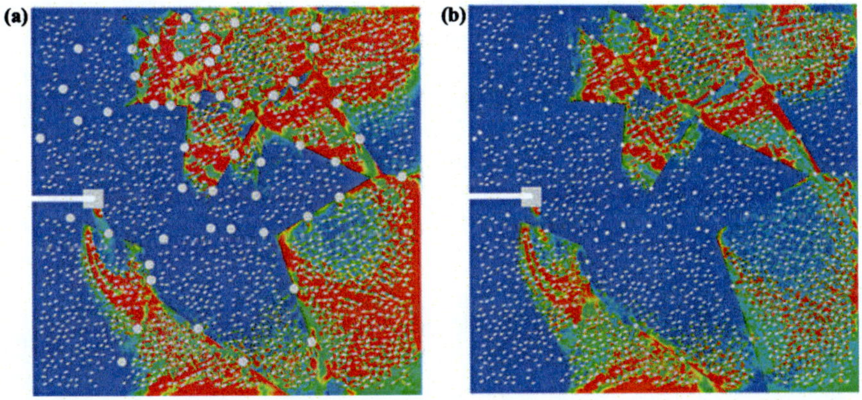

Fig. 8 Dislocation density along a (111) plane in the [−110] direction, for model C1b and model C1c

Discussion

Experimental

The experimental results reflect a trend of higher growth rates and lower threshold for higher hydrogen-producing environments. Both materials show a degree of susceptibility to SCC and HE, with the 5664 variant being affected the most. All the results presented, with the exception of the results from [7], were obtained for K-shedding conditions. These results are noteworthy, as shedding-K testing conditions are considered less aggressive than constant- or rising-K. Fasteners in stiff joints are more likely to experience a decreasing load when the stiffness of the fastener reduces while the joint displacement remains constant or in redundant structures if the load redistributes to other parts of the bolted connection. Additionally, because of the constant applied bolt displacement in the tests, the K decreases faster over time for the samples showing higher crack growth rates. The higher the initial rate, the higher the reduction in *dK/da* experienced by the sample per unit time, which is known to affect the response of the materials to SCC and HE [7, 19–22].

It is interesting to note that despite the higher rate of reduction of *dK/da* for Sample 4 when compared to Sample 5, the latter shows crack arrest at higher K. We can infer that the sensitivity to a reduction in K is not as strong in natural seawater under the application of cathodic protection at this potential. Trends for nickel alloy 718 and other precipitation hardened nickel alloys in comparable environmental conditions [7] show lower crack growth rates for K-shedding conditions. Constant- or rising-K tests are therefore expected to show higher growth rates and lower thresholds. It is possible that the long exposure time in unfiltered natural seawater causes local crack tip conditions to be significantly more aggressive and reduce the sensitivity of the material to a decreasing K condition. This practically means that crack arrest may not readily occur and cracks may continue to progress for AMS 5664 material even after a quick reduction in K. The observed differences may also be related to the intrinsic material variabilities. Additional experiments are needed to clarify which factor affects the results more strongly.

Modeling

The models developed represent a microstructure with coarse γ'' precipitates and grain boundaries decorated with δ precipitates of regular size. The coarse size of the γ'' precipitates is an upper bound limit of observed distributions for the materials modeled [13]. The δ phase distribution is representative of the occasional precipitation of regularly shaped δ phase within the microstructure, which meets the acceptance criteria of the API 6A specification [4]. The C1a model more closely represent the microstructure for a material that underwent a two-phase heat treatment, like AMS 5664, with coarser and less coherent precipitates present within the matrix, discrete δ phase, and a precipitate-free zone near the grain boundary [13, 23]. The microstructure for 718 API6A is tailored to minimize the precipitation of δ and the coarsening of the γ'' phase. Recent work showed that the API6A could still result in precipitation of δ phase at the grain boundary, albeit in less regular, more elongated shape, and γ'' precipitates with a length near and above 50 nm within the microstructure [7]. As small differences in microstructures of precipitation hardened alloy 718 were shown to determine significant differences in EAC behavior, two models that better represent the materials tested in this work will be developed based on SEM and TEM examinations in future work.

The model developed in this work also simplifies the deformation behavior of the microstructure of the material, in particular, it simplifies the interaction of the matrix and the precipitates without considering the shearing of precipitates [24]. Because of the choice in constitutive equations, the precipitates are considered fully coherent with the matrix during elastic deformations, while they act like obstacles to deformations when the matrix plastically deforms. The intent is to idealize the behavior of the interaction to reduce the computational burden for a numerical simulation of a larger portion of the microstructure ahead of the crack tip. The results from the study of mechanical response are being currently used to develop a model for hydrogen transport through the microstructure, coupled with the deformation and dislocation density results presented.

Improvements to the model are underway to represent a more realistic distribution and orientation of the metal grains, and of the γ'' precipitates, as well as to include γ' precipitates, along with more realistic shapes of δ phase at the grain boundary. A tailored two-step heat treatment process able to yield a 150 ksi API6A-compliant material was shown to decrease the size of the γ'' precipitates, increase the co-precipitation of γ', and obtain precipitate-free grain boundaries [7, 10, 25]. The modeling presented will focus next on these relatively small changes in microstructure to study their impact on mechanical behavior.

Despite its shortcomings, the model is able to capture the main trends that have been identified by experimental work on nickel alloy 718 microstructures. The modeling results in this work indicate that coarsening of the γ'' phase results in increased dislocation density accumulation at the critical areas of the microstructure in the simulated deformation. Larger γ'' phase particles in the model likely act by enhancing defined paths of shear deformation in the model, while obstructing shear deformations on other paths. Increased dislocation density for other slip planes modeled at the same location (not shown) supports the theory that intersection of dislocation slip bands and interaction of dislocation motion may result in increased strain localization and increased susceptibility to EAC. An increased dislocation activity is also observed in between the two δ precipitates modeled on one of the critical grain boundaries in Figs. 5 and 7.

The reduction in δ phase size also translates into a decreased activity comparing model C1b and C1c in Fig. 8. This seems consistent with observations of increased occurrence of intergranular cracking in the presence of δ phase at the boundaries, particularly for grain boundaries with higher coverage of δ precipitates. The change in dislocation activity is significant in all of the areas of the microstructure, where the δ phase acts as a magnet for increased dislocation activity. The results also indicate increased strain localization in the precipitate-free zone.

Conclusions

This study compared the EAC behavior of two commonly used variants of a alloy 718. Two different exposure conditions, OCP under alternate immersion conditions and CP at −850 mV vs Ag/AgCl highlight the susceptibility to hydrogen-producing environments of AMS 5664 variant. A significant degree of variability in the results at high applied SIF was also observed. The API 6A variant was significantly less susceptible to EAC in OCP alternate immersion conditions. A meso-scale microstructural model was developed to study the role of precipitates within the material's microstructure. An initial set of models was developed to represent the susceptible microstructure of the AMS 5664 material studied. The model uses a dislocation density based crystal plasticity formulation and simplifies the matrix-precipitate interaction to represent partial coherence with the matrix. Results of the model show trends in line with literature observations on HE sensitivity of material microstructures. Increased dislocation activity for larger γ'' and δ precipitates at the grain boundary are interpreted as increased hydrogen transport paths to weak areas of the microstructure.

References

1. T. Chen, J. Nutter, J. Bai, J. Hawk, and X. Liu, "Corrosion fatigue crack growth behavior of oil-grade nickel-base alloy 718. Part 2: Effect of aging treatment," *Corrosion Science,* vol. 98, pp. 280–290, 2015/09/01/ 2015.
2. T. Chen, J. Nutter, J. Hawk, and X. Liu, "Corrosion fatigue crack growth behavior of oil-grade nickel-base alloy 718. Part 1: Effect of corrosive environment," *Corrosion Science,* vol. 89, pp. 146–153, 2014/12/01/ 2014.
3. American Petroleum Institute, NACE MR0175, *Petroleum and natural gas industries - Materials for use in H_2S-containing environments in oil and gas production.*
4. *Age-Hardened Nickel-Based Alloys for Oil and Gas Drilling and Production Equipment* API 6A718, 2013.
5. B. Saleem, H. B. Dong, and V. Patel, "Alloy 718 Subsea Bolt in Relation to Surface Cracking: A Microstructural Perspective," *Materials Sciences and Applications,* vol. 11, pp. 787-816, 2020.
6. K. A. Esaklul and T. M. Ahmed, "Prevention of failures of high strength fasteners in use in offshore and subsea applications," *Engineering Failure Analysis,* vol. 16, no. 4, pp. 1195–1202, 2009/06/01/ 2009.
7. R. Thodla, "Environmentally Assisted Cracking of High Strength Nickel Based Alloys Under Cathodic Protection," presented at the CORROSION 2018, Phoenix, Arizona, USA, 2018/7/20/, 2018.
8. A. Arcari, J. P. Moran, D. J. Horton, E. J. Lemieux, R. McCoy, and T. M. Newbauer, "The Effect of Alternate Immersion on Stress Corrosion Cracking Behavior of Steel and Nickel Alloys in Natural Seawater," in *CORROSION 2021,* Day 8 Mon, April 26, 2021, D081S029R012.

9. *Standard Test Method for Measurement of Fatigue Crack Growth Rates*, ASTM E647–15E01, 2015.
10. X. Li, G. B. Viswanathan, and R. Thodla, "Hydrogen Embrittlement Study of Three Heats of UNS N07718 in Subsea Applications," in *CORROSION 2019*, 2019, vol. All Days, NACE-2019–13057.
11. Z. Y. Xiong *et al.*, "Accelerating optimization of IN718 by mapping alloying effects on phase stabilities and mechanical properties using high-throughput calculations," *Materials & Design*, vol. 217, p. 110603, 2022/05/01/ 2022.
12. C. Slama and M. Abdellaoui, "Structural Characterization of the Aged Inconel 718," *Journal of Alloys and Compounds* vol. 306, pp. 277-284, 2000.
13. G. C. Obasi, Z. Zhang, D. Sampath, R. Morana, R. Akid, and M. Preuss, "Effect of Microstructure and Alloy Chemistry on Hydrogen Embrittlement of Precipitation-Hardened Ni-Based Alloys," *Metallurgical and Materials Transactions A*, vol. 49, no. 4, pp. 1167–1181, 2018/04/01 2018.
14. R. Y. Zhang *et al.*, "Evolution of Lattice Spacing of Gamma Double Prime Precipitates During Aging of Polycrystalline Ni-Base Superalloy: An In-Situ Investigations," *Metallurgical and Materials Transactions A*, vol. 51A, pp. 574-585, 2019.
15. Y. Huang, *A User-Material Subroutine Incorporating Single Crystal Plasticity in the ABAQUS Finite Element Program*. Unknown Publisher, 1991.
16. S. Gupta and C. A. Bronkhorst, "Crystal plasticity model for single crystal Ni-based superalloys: Capturing orientation and temperature dependence of flow stress," *International Journal of Plasticity*, vol. 137, p. 102896, 2021/02/01/ 2021.
17. Q. Wu and M. A. Zikry, "Prediction of diffusion assisted hydrogen embrittlement failure in high strength martensitic steels," *Journal of the Mechanics and Physics of Solids*, vol. 85, pp. 143–159, 2015/12/01/ 2015.
18. M. A. Zikry and M. Kao, "Inelastic microstructural failure mechanisms in crystalline materials with high angle grain boundaries," *Journal of the Mechanics and Physics of Solids*, vol. 44, no. 11, pp. 1765–1798, 1996/11/01/ 1996.
19. P. L. Andresen and M. M. Morra, "Effect of Rising and Falling K Profiles on SCC Growth Rates in High-Temperature Water," *Journal of Pressure Vessel Technology*, vol. 129, no. 3, pp. 488-506, 2006.
20. K. Chen, J. Wang, D. Du, P. L. Andresen, and L. Zhang, "dK/da effects on the SCC growth rates of nickel base alloys in high-temperature water," *Journal of Nuclear Materials*, vol. 503, pp. 13–21, 2018/05/01/ 2018.
21. Z. D. Harris, S. K. Lawrence, D. L. Medlin, G. Guetard, J. T. Burns, and B. P. Somerday, "Elucidating the contribution of mobile hydrogen-deformation interactions to hydrogen-induced intergranular cracking in polycrystalline nickel," *Acta Materialia*, vol. 158, pp. 180–192, 2018/10/01/ 2018.
22. K. A. Nibur and B. P. Somerday, "Identification of Conservative Hydrogen Assisted Cracking Fracture Threshold Measurements," in *International Hydrogen Conference (IHC 2012): Hydrogen-Materials Interactions*, B. P. Somerday and P. Sofronis, Eds.: ASME Press, 2014, p. 0.
23. X. Lu, J. Du, Q. Deng, and J. Zhuang, "Stress rupture properties of GH4169 superalloy," *Journal of Materials Research and Technology*, vol. 3, no. 2, pp. 107–113, 2014/04/01/ 2014.
24. C. H. Zenk, L. Feng, D. McAllister, Y. Wang, and M. J. Mills, "Shearing mechanisms of co-precipitates in IN718," *Acta Materialia*, vol. 220, p. 117305, 2021/11/01/ 2021.
25. R. Thodla, A. Chandra, X. Li, C. D. Taylor, H. Ke, and N. Sridhar, "Hydrogen embrittlement of 718 under cathodic polarization," *Corrosion Science*, vol. 165, p. 108361, 2020/04/01/ 2020.

Hot Corrosion Behavior of a GH4720Li Disk Superalloy at 700 °C

Teng An, Fangzhen Duan, Yu Gu, Yuting Shi, Di Wang, Jinglong Qu, Zhongnan Bi, and Jinhui Du

Abstract The nickel-based superalloy disk components in the turbine sections were subjected to Type II hot corrosion damage in the sulfur-containing salt contaminants at 650 °C–750 °C. The alloy GH4720Li with different grain sizes was corroded in a mixture of sulfates (25% NaCl + 75% Na_2SO_4) at 700 °C for 200 h, and the microstructure evolution was investigated. The experimental results showed that when the grain size increased from 15.9 to 127 μm, the mass loss decreased by 96%, and the corrosion layer thickness decreased by 44%. The hot corrosion resistance increased with the increasing grain size, and the corrosion failure mechanisms changed from pitting corrosion to uniform corrosion. The corrosion layer comprised $NiCr_2O_4$, Al_2O_3, CoO, TiO, Ni_3S_2, and CoS_2. The oxide layer, Ni/Co-rich layer and S-rich layer were stratified and sequentially located on the alloy GH4720Li surface. The corrosion behavior was accelerated by the triangular grain boundaries (GBs) and γ' phase, the segregation behavior of Cr elements in the GBs, as well as the γ' phase formation promoted the tendencies for pit nucleation in the fine-grained structure. In contrast, the Ni/Co-rich layer provided better resistance to hot corrosion and was easier to form on the surface of the coarse-grained structure. The sulfide-oxidation cycle mechanism could well describe the hot corrosion behavior of the disk superalloy.

Keywords Disk superalloy · Grain size · Hot corrosion · Microstructure evolution

T. An (✉) · F. Duan · Y. Gu · Y. Shi · D. Wang · J. Qu · Z. Bi · J. Du
Beijing GAONA Materials and Technology Co., LTD, Beijing 100081, China
e-mail: anteng.1009@163.com

Beijing Key Laboratory of Advanced High Temperature Materials, Central Iron and Steel Research Institute, Beijing 100081, China

E. A. Ott et al. (eds.), *Proceedings of the 10th International Symposium on Superalloy 718 and Derivatives*, The Minerals, Metals & Materials Series,
https://doi.org/10.1007/978-3-031-27447-3_20

Introduction

The nickel-based superalloys are widely used to fabricate the turbine components of jet engines due to their excellent mechanical properties, such as high-temperature strength, toughness, and corrosion resistance at high temperatures. The turbine disks and blades suffer from hot corrosion in the marine environment [1], resulting in the degradation of mechanical properties and catastrophic failure of turbine components [2, 3].

The hot corrosion is often categorized into high-temperature hot corrosion (Type I, 850 °C–950 °C) and low-temperature hot corrosion (Type II, 650 °C–800 °C) depending on the temperature regime and the salt composition. Type I hot corrosion in nickel-based superalloy has been extensively studied due to the increasing jet engine turbine blade operating temperature, in which the degradation of material mechanical properties was induced by deposited molten salts above approximately 800 °C. However, the turbine disk operating temperature increased above approximately 700 °C, which met the Type II hot corrosion temperature regime. Therefore, an indepth understanding of the disk material attack due to the presence of eutectic salt mixtures is required, which occurs in a stable molten form. The Type II hot corrosion databases need to be generated over a wide range of time, temperature, and environmental conditions to support the disk material applications. The present study focused on the effect of Type II hot corrosion on the GH4720Li alloys at 700 °C.

Type II hot corrosion mechanisms have been described by researchers [4]. The eutectic liquid Na_2SO_4–MSO_4 formed on the surface of materials is essential for Type II hot corrosion occurrence. The melting point of Na_2SO_4–$NiSO_4$ and Na_2SO_4–$CoSO_4$ is 671 °C and 576 °C, respectively [5]. The formation of Na_2SO_4–MSO_4 eutectic solutions leads to a high corrosion rate with a maximum at about 700 °C for the Ni–Cr–Co-based GH4720Li alloy. The attack form of hot corrosion usually generates pitting damage with internal sulfidation in the front of the corrosion pit, which affects fatigue behavior [6]. The corrosion pits in Rene104 serve as stress concentration sites where fatigue cracks initiate and grow under cyclic loads [7]. The corrosion damage evolution of Rene 104 was represented as selective carbide oxidation leading to a pitted surface, followed by sulfur-related accelerated hot corrosion damage [8]. Li et al. [9] proposed a hot corrosion fatigue life estimation of a PM superalloy, in which the corrosion pit due to hot corrosion was considered the micro notch that decreased the fatigue life. The pit damage due to the hot corrosion has been extensively studied in the liberate, and this study focused on the grain size effects on the hot corrosion.

The alloying element [10, 11], grain boundary (GB) [12], grain size [13], and γ' phase affect the corrosion attack evolution. Taylor et al. [14] investigated Type II hot corrosion of the CMSX-4 alloy at 700 °C, and the results showed that the outer layer was rich in Co and Ni, which transformed into mixed oxides of Co and Ni, and the inner layer was rich in Cr, Al, and S. Cr promoted the continuous Cr_2O_3 scale formation which could be self-healed under hot corrosion circumstance [10]. Co and Ti also can provide better hot corrosion resistance of the FGH4096 alloy

[15]. The corrosion mechanism of nickel-based superalloy changed from pitting corrosion to uniform corrosion with the grain size increment. The driving force for the Cr diffusion from grain to boundary in the coarse grain was lower than the fine grain [16]. Penetration of S and depletion/enrichment of alloying elements can be confined by the sample boundaries which significantly reduces hot corrosion of 617 alloys [12]. The grain size and GM of GH4720Li alloy significantly affect the hot corrosion behavior. It is of great importance to study the hot corrosion behavior at disk operating temperatures for a longer duration.

This study aims to study the Type II hot corrosion behavior of a GH4720Li disk superalloy with different grain sizes at 700 °C for 200 h. The microstructural development of the corrosion scale and subscale microstructure were studied by scanning electron microscopy (SEM), energy-dispersive spectroscopy (EDS), Electron probe X-ray microanalyzer (EPMA), and X-ray diffraction(XRD).

Materials and Experimental Procedures

GH4720Li is a γ' precipitation-hardened Ni-based superalloy used for aero-engine turbine disk applications. The ingot of as-received material was produced by vacuum induction melting, electro-slag refining, and vacuum arc refining. The chemical composition of the investigated alloy is shown in Table 1. The forged bar was obtained after the ingot homogenization treatment, fast forge, and radial forge. Then, the forged bar was forged to form a GH4720Li disk at 1100 °C, and the fine-grained microstructure was observed. The disk diameter was 220 mm, and the rim thickness was 45 mm. The samples for hot corrosion tests were cut from the edge of the disk using electro-discharge machining along the tangential direction, as shown in Fig. 1. The circular and square samples were $\varphi 10 \times 10$ mm and $10 \times 20 \times 10$ mm, respectively.

The grain size effects on Type II hot corrosion performance of forged GH4720Li superalloy were investigated. The different grain sizes of samples were gained by controlling the heat treatment in a chamber furnace. The samples were produced by heat treatment at 1130 °C–1160 °C with three different holding times (1, 2, or 4 h), followed by air cooling, and then treated by two-step aging heat treatment: 650 °C/24 h/air quenching (AQ) + 760 °C/16 h/AQ. Three circular samples and three square samples were used in hot corrosion tests to ensure experimental data accuracy. Circular samples were used for surface observation after corrosion tests, and the square samples were used for cross-section observation. The circular samples and square samples with different heat treatments were named H1–H4 samples and

Table 1 Chemical composition of the as-received GH4720Li alloy (in wt.%)

Cr	Al	Ti	Co	Mo	W	B	C	Ni
16.17	2.56	4.94	14.73	2.92	1.27	0.014	0.015	Bal

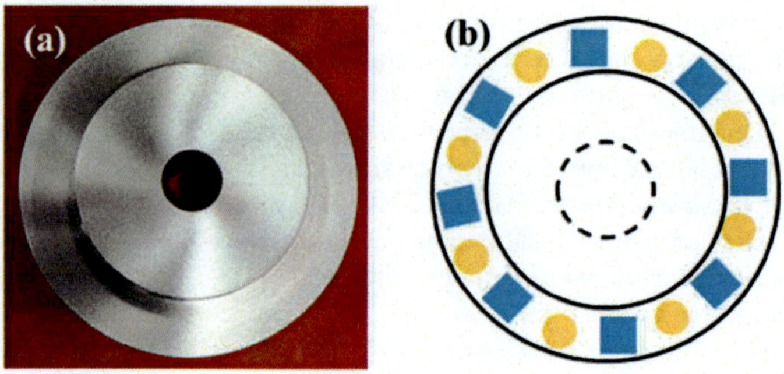

Fig. 1 Schematic diagram of the GH4720Li disk (**a**) and the samples used for hot corrosion tests (**b**)

Table 2 Heat treatment system of the GH4720Li superalloy

Samples		Temperature/°C	Hold time/h	Cooling method	Aging
Circles	Squares				
Z1	H1	1130	1	Air	650 °C/24 h/ AQ + 760 °C/16 h/AQ
Z2	H2	1140	4		
Z3	H3	1150	2		
Z4	H4	1160	1		

Z1–Z4, respectively. The detailed heat treatment process and sample number are shown in Table 2.

All surfaces of samples were ground on SiC paper (1200 grit) with a mean particle diameter of 15 μm. The samples were ultrasonically cleaned in ethanol and dried in hot air to remove atmospheric contaminants before the hot corrosion tests. Type II hot corrosion tests were carried out to investigate the hot corrosion behavior of the GH4720Li superalloy. The samples with different grain sizes were buried in alumina crucibles under the 25% NaCl + 75% Na_2SO_4 salt mixtures. Then, the crucibles were heated in a chamber furnace at 700 °C for 200 h. After the hot corrosion tests, all samples were desalted in boiling water at 100 °C. The mass of each sample and the collected corrosion products were recorded before and after hot corrosion exposure. The specific mass per surface area was also calculated.

The corrosion products on the surface of the samples were identified via XRD using a Bruker D8 Focus diffractometer. For the metallographic examinations, the corroded samples were embedded in epoxy resin and ground in cross-section on SiC papers from 400 to 2000 grit and chemically polished with a 0.02 μm silica suspension to obtain a smooth surface (Ra < 0.1 μm). The section samples were cleaned in acetone anhydrous before metallographic observations. The corroded samples were observed by optical light microscopy (OM, OLYMPUS GX71 microscope). The samples for SEM observation were electrolytically polished in a solution of 20% H_2SO_4 + 80% CH_3OH under the voltage of 25–30 V, followed by electrolytic etching in the solution of 170 ml H_3PO_4 + 10 ml H_2SO_4 + 15 g CrO_3 under a voltage of 5 V. The microstructure observations were conducted by SEM (JSM-7200F microscope) coupled with an EDAX detector for EDS analyses. Element mappings with a spatial resolution of 1 μm were obtained by electron beam microprobe (EPMA, JXA-8350F).

Results and Discussion

The initial microstructures of the as-received materials are shown in Fig. 2. The GH4720Li alloy consisted of the γ matrix with equiaxed grain and a high fraction of γ' precipitates. The γ' precipitates were formed in the forging process and existed on the GBs and within the grains. The area fraction of γ' precipitates was about 45%. The average grain size of this alloy was about 10 μm. The primary γ' precipitates (1–3 μm) pinned the GBs or triple-junction (Fig. 2a). The grains could maintain uniform and fine structure during the hot deformation at sub-solvus temperatures due to the pinning effect by primary γ' precipitates in GBs. The secondary γ' precipitate (90–200 nm) and tertiary γ' precipitate (15–50 nm) were distributed in the interior of grains (Fig. 2c, d).

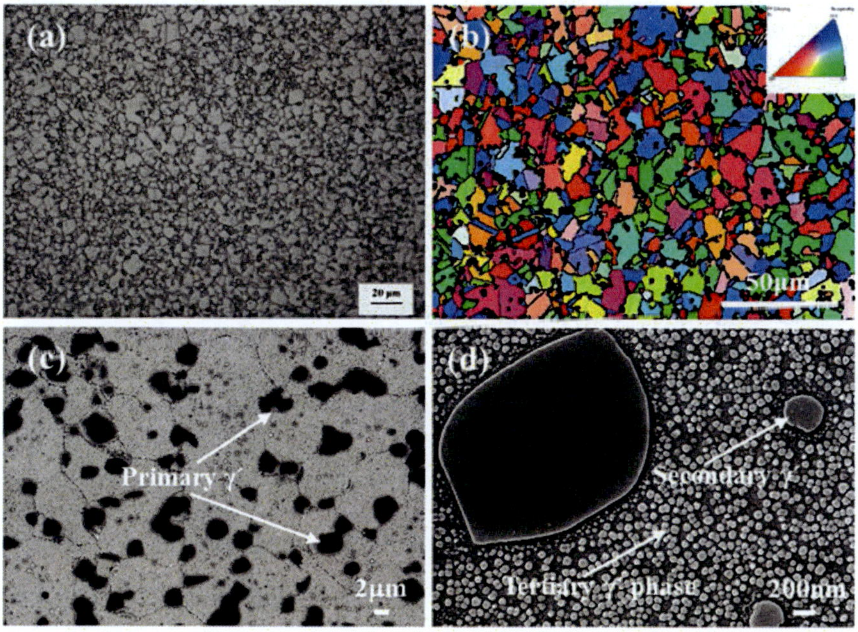

Fig. 2 Initial microstructure of the GH4720Li alloy: **a** the OM image, **b** the IPF map, **c, d** SEM images

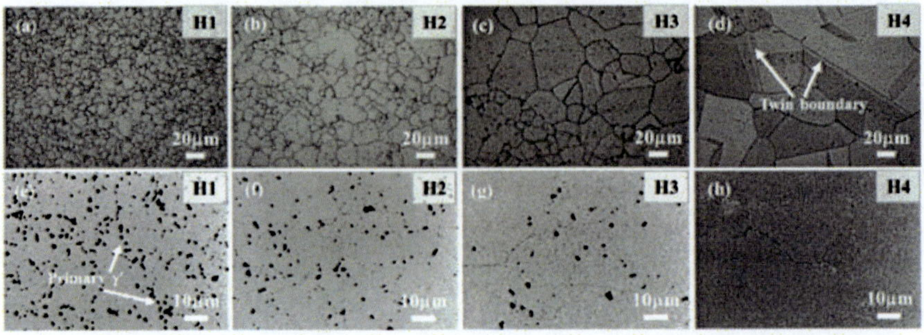

Fig. 3 Micrographs of the grain structure and primary γ′ precipitate distribution of the GH4720Li alloy: **a, e** H1 samples; **b, f** H2 samples; **c, g** H3 samples; and **d, f** H4 samples

The microstructures of H1–H4 samples are given in Fig. 3. The results showed that the grain grew gradually with the solution temperature increase. The volume fraction of primary γ′ precipitate, which was mainly distributed on the GBs, decreased from 35 to 0% with increasing solution temperature.

It can be seen from Fig. 4a that the particle size of primary γ′ precipitates was 0.5–8.5 μm in the H1 sample. The equilibrium volume fraction of γ′ precipitates decreased with increasing temperature [17]. Hence, the area fraction of primary γ′ precipitates with the sizes of 3–5 μm in the H2 sample decreased compared with the H1 sample. This suggests that the γ′ precipitates with a size larger than 3 μm were gradually dissolved when the solution treatment temperature increased to 1140 °C. In addition, as shown in Fig. 4b, the volume fraction of the primary γ′ precipitate of the H1 sample was approximately 5%, and the average grain size was 15.9. The average grain size slightly increased from 15.9 to 127.0 μm with the solution temperature increase from 1100 °C to 1160 °C. The dissolving temperature of the γ′ precipitates of GH4720Li was 1160 °C [18]. The grain growth behavior during sub-solvus temperature heat treatment suggested that the restriction of the grain growth process could mainly be attributed to the pinning effect of γ′ precipitates [17, 19]. The volume fraction of γ′ precipitates decreased significantly to 0.03% for the H4 sample, which would eliminate the particle-pinning forces and quickly result in grain growth [20, 21].

Figure 5 shows the macroscopic morphologies of the GH4720Li alloy samples with different microstructures after hot corrosion at 700 °C for 200 h. The edges of the samples were corroded, resulting in the formation of corrosion pits and corrosion products. Samples were easily subjected to corrosion due to the large stress concentration at their edge locations. The uncorroded area of the corroded samples gradually increased, corrosion layer and corrosion pits gradually reduced with the grain size increase. The corrosion pits were not observed on the surfaces of the H4 samples. This indicates that the larger grain size improved the corrosion resistance of materials.

Figure 6 shows the mass change comparison of the H1–H4 samples after hot corrosion at 700 °C for 200 h. The weight gain per unit area (Δw in mg/cm^2) as

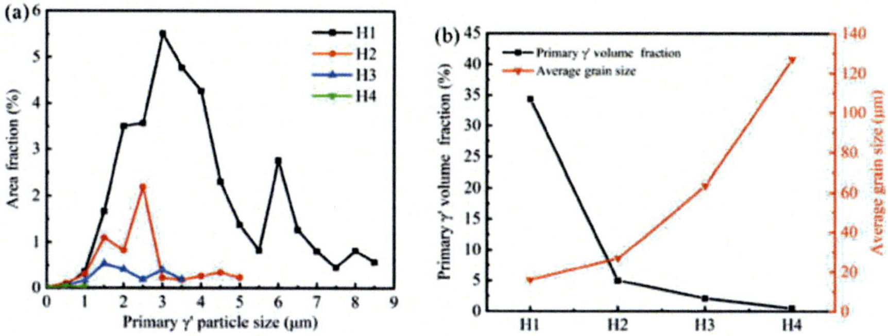

Fig. 4 Primary γ′ precipitate size distribution (**a**) and average grain size (**b**) for H1–H4 samples

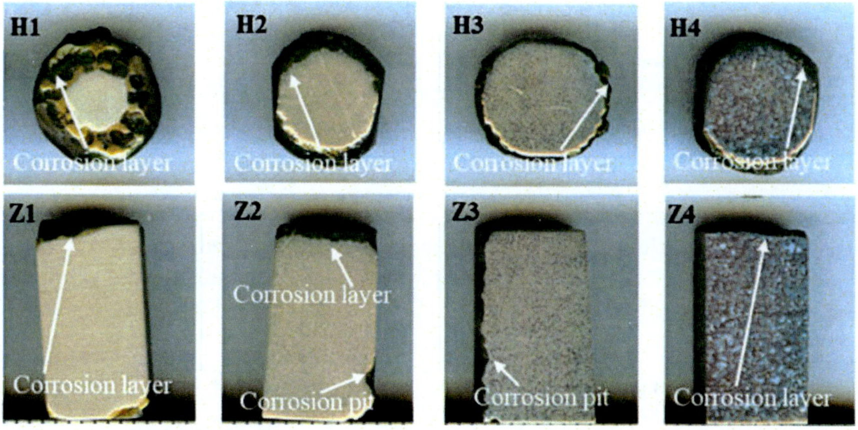

Fig. 5 Macroscopic morphologies of the samples with different microstructures after hot corrosion at 700 °C for 200 h. The surface morphology of H1–H4 samples after hot corrosion and the cross-sectional morphology of Z1–Z1 samples after hot corrosion are shown. The corrosion layer of the samples was removed

a function of grain size showed that no spallation or evaporation occurred during exposure tests for all samples, and no sudden drop was observed. It was clear that weight gain increased gradually as the grain size decreased. The weight gain per unit area corrosion of H4 samples (0.454 mg/cm^2) was about 27 fold the H1 samples (0.017 mg/cm^2). This indicates that the coarse grains of the GH4720Li superalloy had better hot corrosion resistance than the fine grains.

The XRD tests were carried out to detect the phase constitution of the corrosion layer. The XRD patterns of the corrosion products on the surface of the H1–H4 samples after hot corrosion at 700 °C for 200 h are illustrated in Fig. 7. The XRD patterns of the GH4720Li alloy after hot corrosion showed similar characteristics. The corrosion products were composed of Ni, $NiCr_2O_4$, TiO, CoO, Ni_3S_2, and CoS_2, indicating that the change of microstructures had no effects on the types of corrosion

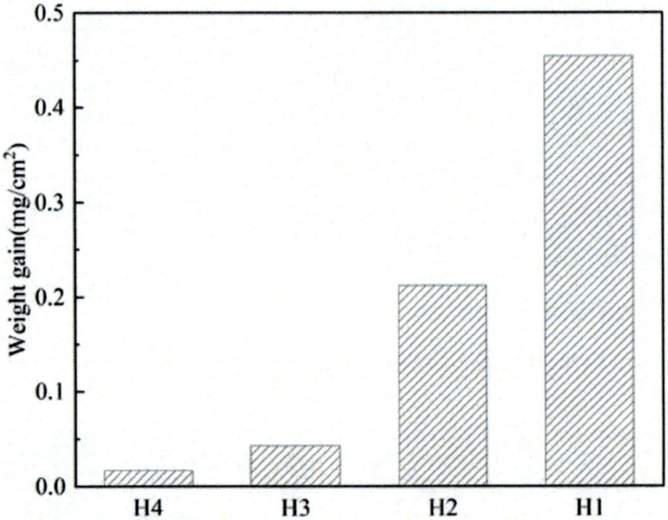

Fig. 6 Mass change of the Z1–Z4 samples after hot corrosion at 700 °C for 200 h

products. Furthermore, the products demonstrated that the Type II hot corrosion of GH4720Li alloy included two processes: oxidation and sulfuration.

The corrosion layers and corrosion pits were formed on the surface of the samples after hot corrosion of GH4720Li at 700 °C for 200 h. The results were consistent

Fig. 7 XRD patterns of the corrosion products of the H1–H4 samples after hot corrosion at 700 °C for 200 h

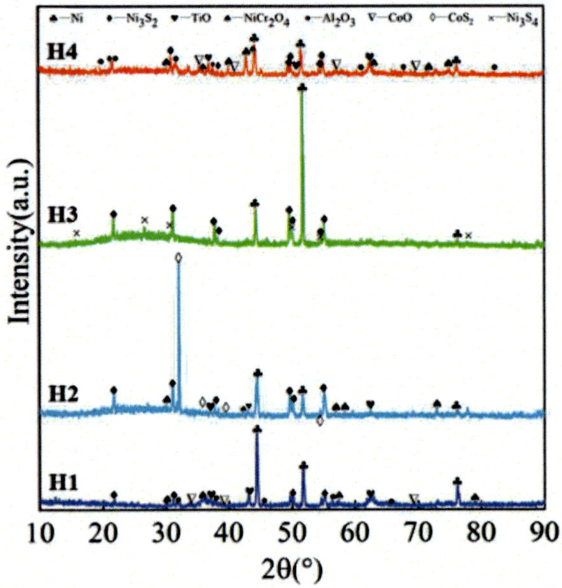

with those reported by Jiang H et al. [15] and Zheng L et al. [22]. Chloride in the form of volatiles leads to the formation of voids and pits, which provides an easy path for the flow of corrosive agents in the sodium chloride-coated samples [23, 24]. The metal matrix would be corroded rapidly when the protective layer is broken. However, the corrosion rates of GH4720Li were influenced by grain size, γ' phase, and GB. Figures 5 and 6 indicate that the samples with larger grain sizes had better hot corrosion resistance. The corrosion products demonstrated that hot corrosion included two processes: oxidation and sulfuration. $NiCr_2O_4$ was formed from NiO and Cr_2O_3 spontaneously [22, 25]. The outward diffusion of elements such as Cr, Ni, and Ti interacted with oxygen to form oxides [26]. Cr_2O_3 can play an important role in protecting materials' substrates from further corrosion [27]. The formation of sulfides resulted from the interaction between the metal matrix and the molten salt of sodium sulfate, and several reports proved that sulfur could only be obtained from the dissociation of sulfates [5, 28].

Figure 8 shows morphologies of the corrosion pit on the surface of GH4720Li samples subjected to hot corrosion at 700 °C for 200 h. It can be confirmed that corrosion pitting was the main form of corrosion attack. It can be observed that intergranular corrosion occurred around the corrosion pit on the surface of H1–H3 samples, as shown in Fig. 8a–c. However, the GBs on the surface of the H4 sample were corroded slightly, as shown in Fig. 8d. The GBs around the corrosion pit were also corroded, accelerating the corrosion process. The attack form of hot corrosion usually generates pitting damage with internal sulfidation in the front of the corrosion pit. Figure 9 displays the surface micrographs around the corrosion pit and EDS elemental mapping analysis on the surface of the H1 sample after hot corrosion at 700 °C for 200 h. The elemental maps demonstrated that the major element in area A was O, S, Cr, Co, and Ni. Distributions of Cr, Co, and Ni overlapped that of O, which proved that Cr, Co, and Ni were preferentially oxidized during hot corrosion, and the corrosion layer was mainly composed of Cr_2O_3 and NiO. Due to the high stability, Al_2O_3 can form a protective scale to avoid sulfur and oxygen attack at elevated temperatures by inhibiting the diffusions or penetrations of oxygen and sulfur. Cr_2O_3 was also protective to protect the substrate from further corrosion. At the same time, sulfides were detected in area B. The S element penetrated the region near the interface along the GB.

The corrosion attack severity investigation was carried out by cross-section microstructure analysis. The cross-sections micrographs of GH4720Li samples after hot corrosion at 700 °C for 200 h with different grain sizes are shown in Fig. 10. The cross-section morphologies showed that the lamellar distribution characteristic was mainly composed of a white layer and a gray layer. The average thickness of corrosion layer decreased from 955.65 to 540.86 μm with the grain size increase. Meanwhile, the corrosion layer tended to become dense, and the coarse-grained samples had better hot corrosion resistance, consistent with the results observed in Figs. 5 and 8.

The interface micrographs between the substrate and corrosion layer on the cross-section of the Z1 and Z2 samples after hot corrosion at 700 °C for 200 h are shown in Fig. 11. Figure 11b, c shows the higher magnification images of areas A and B.

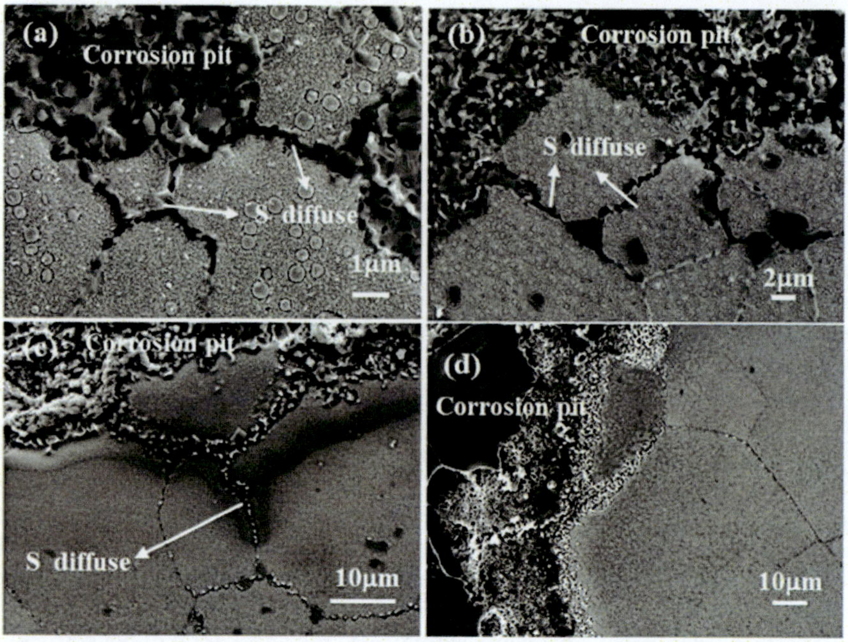

Fig. 8 Surface morphologies of the corrosion pit for the GH4720Li samples after hot corrosion at 700 °C for 200 h: **a** H1 sample; **b** H2 sample; **c** H3 sample; and **d** H4 sample. The corrosion layer of the samples was removed

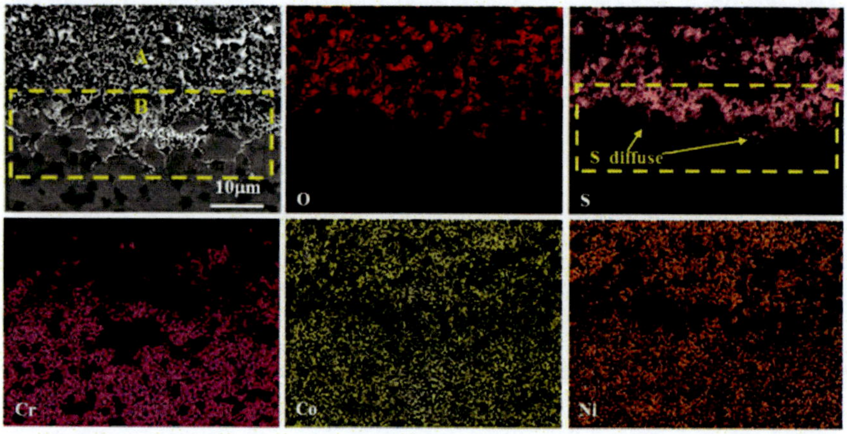

Fig. 9 Surface micrographs around the corrosion pit and EDS elemental mapping analysis of the H1 sample after hot corrosion at 700 °C for 200 h. The corrosion layer of the samples was removed

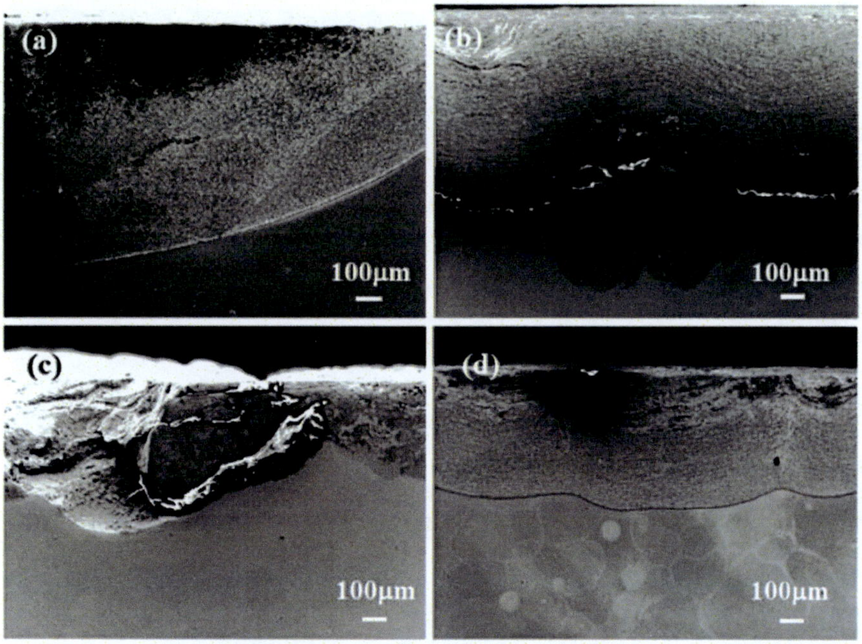

Fig. 10 Cross-section micrographs of the Z1–Z4 samples after hot corrosion at 700 °C for 200 h: **a** Z1 sample; **b** Z2 sample; **c** Z3 sample; and **d** Z4 sample

Figure 11e, f show the higher magnification images of areas C and D. Areas A and C showed a sparse corrosion layer on the cross-section of the corroded sample. The elemental distribution of O, S, Na, Al, Ti, Cr, Co, and Ni was given by EPMA in Area A (Fig. 12). The analysis results confirmed that the corrosion layer was mainly enriched with Na, O, S, Ni, and Co elements. This layer was vulcanized but not fully oxidized. S diffused into the substrate through the porous oxide scales. The GBs in the matrix were corroded, and S penetrated into the substrate along the primary γ' precipitates and crystal boundaries.

The GBs at areas B and D were uncorroded. The EPMA elemental mapping analysis of area B is shown in Fig. 13. The corrosion layer was divided into two layers, with the outer layer being sparse and the inner layer comparatively dense. Elemental analysis results revealed that the outer layer was mainly enriched with Cr, Al, and Ti oxides. The inner layer was enriched in Co and Ni elements and partially sulfided to form CoS_2 and Ni_3S_2. A small amount of Na_2SO_4 molten salt solution expanded downward along the loose layer.

The interface micrographs between the substrate and corrosion layer on the cross-section of the Z3 and Z4 samples after hot corrosion at 700 °C for 200 h are shown in Fig. 14. The interface corrosion morphology of the Z3 sample was similar to that of Z1 and Z2 samples, but the GB was corroded slightly. The interface corrosion morphology of the Z4 sample showed GBs were not corroded. Figure 14f shows that

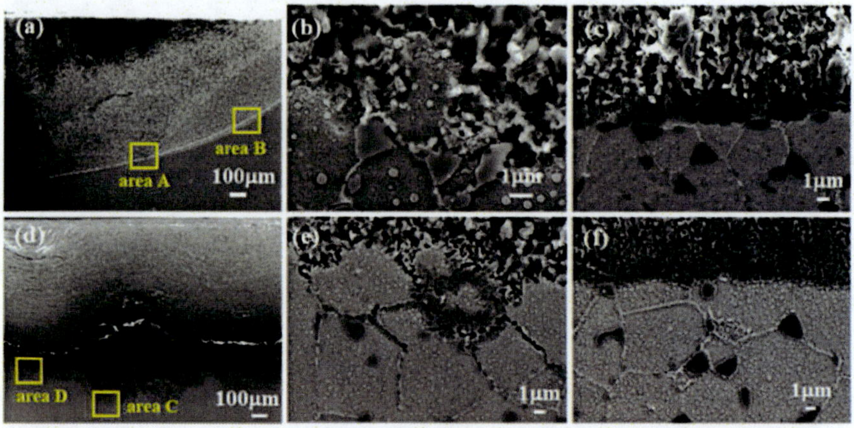

Fig. 11 Interface micrographs between the substrate and corrosion layer on the cross-section of the Z1 sample (**a**) and Z2 sample (**b**) after hot corrosion at 700 °C for 200 h. Figure 12b, c shows the higher magnifications of areas A and B, and Fig. 12e, f shows the higher magnifications of areas C and D

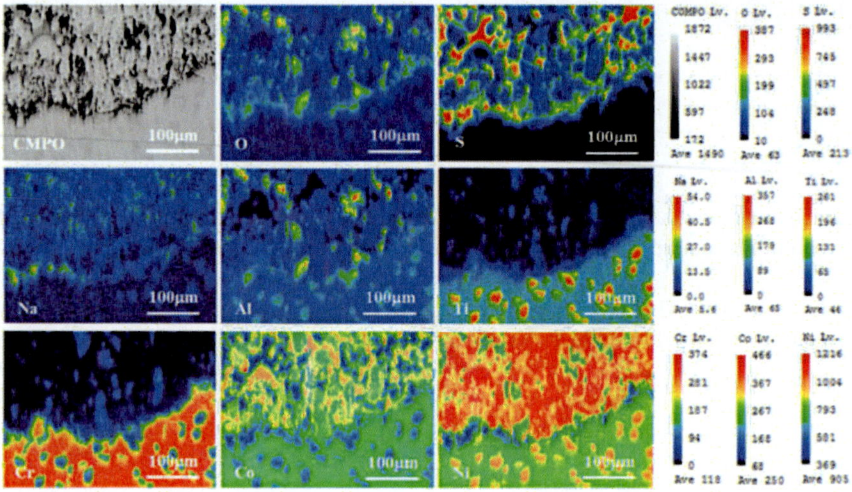

Fig. 12 EPMA elemental mapping analysis of area A on the cross-section of the Z1 sample

the corrosion layer located in area 3 was denser and mainly enriched with the Co and Ni elements (Table 3). The corrosion layer gradually became loose along the direction from matrix to surface. Area 1 was the oxide layer, where Cr, Co, Ni, Al, Ti, and other elements were oxidized, as shown in Table 3. Area 2 was the sulfide layer, where Ni, Co, and other elements were gradually sulfided.

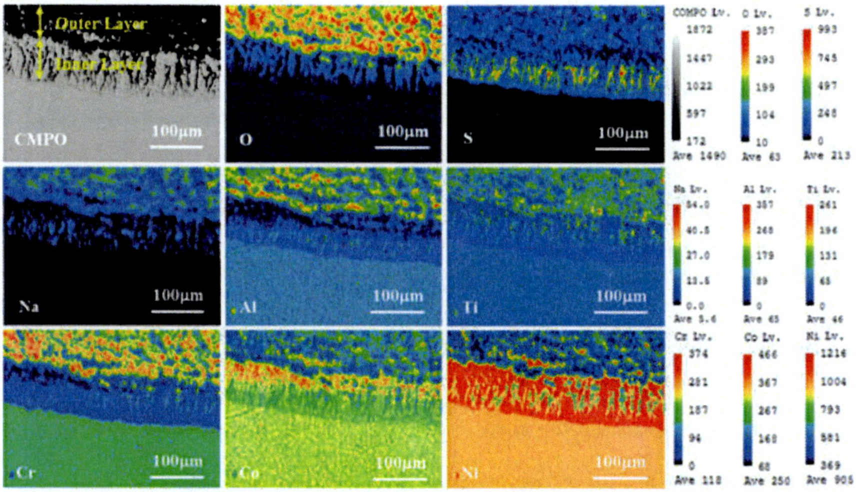

Fig. 13 EPMA elemental mapping analysis of area B on the cross-section of the Z1 sample

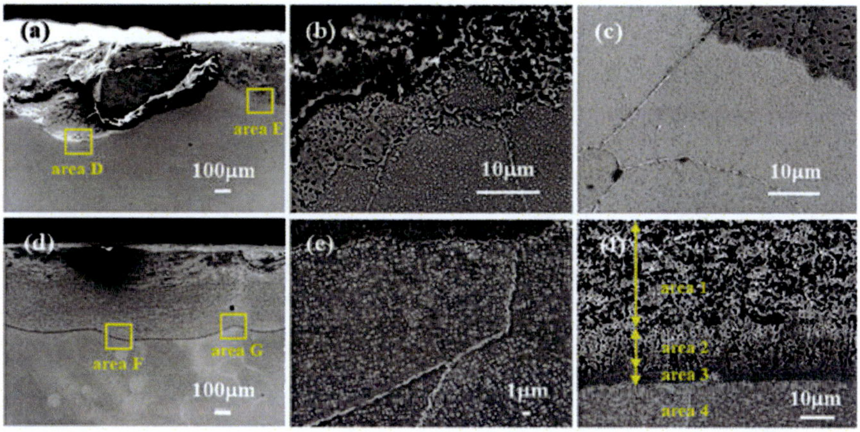

Fig. 14 The interface micrographs between the substrate and corrosion layer on the cross-section of the Z3 sample (**a**) and Z4 sample (**d**) after hot corrosion at 700 °C for 200 h. Figure 15b, c shows the higher magnifications of areas D and E, and Fig. 15e, f shows the higher magnifications of areas F and G

Table 3 Typical elemental composition of the corrosion layer (in wt.%). The location is shown in Fig. 14f

Element	O	S	Na	Al	Ti	Cr	Co	Ni
Area 1	38.63	0.86	5.19	2.87	17.15	19.30	6.83	8.94
Area 2	18.77	19.46	8.61	0.84	2.13	14.62	5.91	25.61
Area 3	–	1.34	–	1.82	0.82	5.95	15.07	74.99
Area 4	–	–	–	2.89	5.93	13.37	12.73	58.16

The enrichment of certain alloying elements (viz. Cr, Ni, Co, Al, and Ti) along with sulfur and oxygen shows the formation of both internal sulfides and oxides in the GH4720Li alloy (Figs. 9, 12, and 13). These results were consistent with the literature concerning hot corrosion studies of numerous grades of superalloys [15, 22, 29]. The released S coming from the Na_2SO_4 molten salt solution corroded the matrix when the protective scale was broken down [30]. With further time extension, sulfides were oxidized to form a loose oxide layer (Fig. 9). Oxidation of the sulfide released S atoms, diffusing into the matrix and accelerating the corrosion attack. The sulfides penetrated deep into the matrix, and the corrosion of GH4720Li alloy repeated sulfidation and oxidation processes [31].

The corrosion resistance of GH4720Li samples with different microstructures was investigated after hot corrosion at 700 °C for 200 h. Coarse-grained samples had better hot corrosion resistance than fine-grained samples (Fig. 11). The major characteristic was the difference between the corrosion layer and GB. Large grain size samples (H4) showed uniform corrosion, while relatively small grain size samples (H1–H3) showed pitting corrosion. The major reason was that Ni/Co-rich layers had better hot corrosion resistance and were more easily developed on the surface of coarse-grained structures (Figs. 11 and 14). On the other hand, high-angle GBs (HAGBs) are important in promoting intergranular corrosion [12]. The H1–H3 samples had a smaller grain size and higher GBs than the H4 sample. The triangular GBs and primary γ' phases accelerated the corrosion behavior and promoted the formation of corrosion pits. S atoms were more likely to corrode the matrix through the GB, and the H4 sample showed uniform corrosion. In addition, the GBs of the coarse grain (H4) had longer element diffusion paths and lower interfacial energy with a significant impact on impeding the diffusion of the percolating elements (Cr, S, and O) [12]. The longer GBs in the coarse grain seem to be resistant to S diffusion, which prevented S atoms from percolating deeper into the matrix and eventually delayed the pit core formation from a hot corrosion attack.

Conclusions

In this study, the microstructure effect with different grain sizes on the high-temperature hot corrosion behavior of the studied alloy was critically evaluated by immersing some of the processed samples in the salt mixture of 75% Na_2SO_4 + 25% NaCl at 700 °C for 200 h. Some of the important conclusions that can be drawn from this study are as follows:

(1) The hot corrosion process of GH4720Li consisted of a cooperating process of sulfidation and oxidation. The corrosion products were similar in the samples with different grain sizes and mainly consisted of $NiCr_2O_4$, Al_2O_3, TiO, CoS_2, and Ni_3S_2.

(2) Increasing the grain size improved the hot corrosion resistance of the GH4720Li alloy, resulting in the corrosion failure characteristic change from pitting corrosion to uniform corrosion. The triangular GBs and primary γ' phase in fine grains accelerated the corrosion behavior. The Ni/Co-rich layer and long GBs in the coarse grains prevented S penetration along the GBs, which reduced the corrosion rate.

(3) The H4 sample exhibited a significantly lower corrosion depth (~540 μm). This clearly highlights that the coarse-grained microstructure of GH4720Li has practical potential in the hot corrosion environment, although it does not completely terminate it.

Acknowledgements This work was financially supported by the National Key R&D Program of China (Nos. 2017YFA0700703) and the National Natural Science Foundation of China (No. 52074092).

Declaration of competing interest The authors declare that they have no known competing financial interests or personal relationships that could have appeared to influence the work reported in this paper.

References

1. HENDERY M L, WHITTAKER M T, COCKINGS B J, et al. The effect of salt composition on the stress-free and corrosion-fatigue performance of a fine-grained nickel-based superalloy [J]. Corrosion Science, 2022, 198:110–3.
2. CHAN K S, ENRIGHT M P, MOODY J, et al. HOTPITS: The DARWIN approach to assessing risk of hot corrosion-induced fracture in gas turbine components [J]. Engineering Fracture Mechanics, 2020, 228:106889.
3. GABB T P, TELESMAN J, HAZEL B, et al. The Effects of Hot Corrosion Pits on the Fatigue Resistance of a Disk Superalloy [J]. Journal of Materials Engineering and Performance, 2009, 19(1): 77-89.
4. LUTHRA K L, SHORES D A. Mechanism of Na_2SO_4 Induced corrosion at 600°–900°C [J]. Journal of The Electrochemical Society, 2019, 127(10): 2202–10.
5. GRéGOIRE B, MONTERO X, GALETZ M C, et al. Mechanisms of hot corrosion of pure nickel at 700°C: Influence of testing conditions [J]. Corrosion Science, 2018, 141:211–20.
6. BACHE M, BALL C, HARDY M, et al. Corrosion fatigue and damage tolerance in the nickel-based superalloy RR1000 subjected to SO_2 environments [J]. Fatigue & Fracture of Engineering Materials & Structures, 2022, 45(5): 1537–49.
7. CHAN K S , ENRIGHT M P , MOODY J P. Development of a Probabilistic Methodology for Predicting Hot Corrosion Fatigue Crack Growth Life of Gas Turbine Engine Disks[C]// ASME Turbo Expo 2013: Turbine Technical Conference and Exposition. 2013.
8. BIRBILIS N, BUCHHEIT R G. Measurement and Discussion of Low-Temperature Hot Corrosion Damage Accumulation upon Nickel-Based Superalloy Rene 104 [J]. Metallurgical and Materials Transactions A, 2008, 39(13): 3224–32.
9. LI Z, LI S, XU G, et al. The framework of hot corrosion fatigue life estimation of a PM superalloy using notch fatigue methodology combined with pit evolution [J]. International Journal of Fatigue, 2021, 153:106483.
10. CHANG J X, WANG D, ZHANG G, et al. Interaction of Ta and Cr on Type-I hot corrosion resistance of single crystal Ni-base superalloys [J]. Corrosion Science, 2017, 117:35–42.

11. HAN F F, CHANG J X, LI H, et al. Influence of Ta content on hot corrosion behaviour of a directionally solidified nickel base superalloy [J]. Journal of Alloys and Compounds, 2015, 619:102-8.
12. K D, MANDAL S, C.N A, et al. Implication of grain boundary engineering on high temperature hot corrosion of alloy 617 [J]. Corrosion Science, 2016, 106:293–7.
13. LIU L, LI Y, WANG F. Influence of grain size on the corrosion behavior of a Ni-based superalloy nanocrystalline coating in NaCl acidic solution [J]. Electrochimica Acta, 2008, 53(5): 2453–62.
14. TAYLOR M, DING R, MIGNANELLI P, et al. Oxidation behaviour of a developmental nickel-based alloy and the role of minor elements [J]. Corrosion Science, 2022, 196:110002.
15. JIANG H, DONG J-X, ZHANG M-C, et al. Hot corrosion behavior and mechanism of FGH96 P/M superalloy in molten NaCl–Na_2SO_4 salts [J]. Rare Metals, 2016, 38(2): 173–80.
16. KAITHWAS C K, BHUYAN P, PRADHAN S K, et al. 'Hall-Petch' type of relationship between the extent of intergranular corrosion and grain size in a Ni-based superalloy [J]. Corrosion Science, 2020, 175:108868.
17. MONAJATI H, JAHAZI M, BAHRAMI R, et al. The influence of heat treatment conditions on γ' characteristics in Udimet® 720 [J]. Materials Science and Engineering: A, 2004, 373(1–2): 286–93.
18. VAUNOIS, JEAN-ROCH, CORMIER J , VILLECHAISE P , et al. Influence of Both γ' Distribution and Grain Size on the Tensile Properties of UDIMET 720Li At Room Temperature[J]. John Wiley & Sons, Inc. 2010.
19. BHUYAN P, PRADHAN S K, MITRA R, et al. Evaluating the efficiency of grain boundary serrations in attenuating high-temperature hot corrosion degradation in Alloy 617 [J]. Corrosion Science, 2019, 149:164–77.
20. WAN Z, HU L, SUN Y, et al. Effect of solution treatment on microstructure and tensile properties of a U720LI Ni-based superalloy [J]. Vacuum, 2018, 156:248–55.
21. CHARPAGNE, MARIE-AGATHE, FRANCHET, et al. Overgrown grains appearing during sub-solvus heat treatment in a polycrystalline gamma-gamma ' Nickel-based superalloy [J]. Materials & Design, 2018, 144: 353–60.
22. ZHENG L, MAICANG Z, JIANXIN D. Hot corrosion behavior of powder metallurgy Rene95 nickel-based superalloy in molten NaCl–Na_2SO_4 salts [J]. Materials & Design, 2011, 32(4): 1981-9.
23. MAHOBIA G S, PAULOSE N, SINGH V. Hot Corrosion Behavior of Superalloy IN718 at 550 and 650°C [J]. Journal of Materials Engineering & Performance, 2013, 22(8): 2418-35.
24. GURRAPPA I. Hot Corrosion Behavior of CM 247 LC Alloy in Na_2SO_4 and NaCl Environments [J]. Oxidation of Metals, 1999, 51(5–6): 353–82.
25. ROBINO C V. Representation of mixed reactive gases on free energy (Ellingham-Richardson) diagrams [J]. 1996, 27(1): 65–9.
26. EL-AWADI G A, ABDEL-SAMAD S, ELSHAZLY E S. Hot corrosion behavior of Ni based Inconel 617 and Inconel 738 superalloys [J]. Applied Surface Science, 2016, 378: 224–30.
27. DHANANJAY P, SHANKAR M G, KAUSIK C, et al. Severe Hot Corrosion of the Superalloy IN718 in Mixed Salts of Na_2SO_4 and V_2O_5 at 700°C [J]. Journal of Materials Engineering and Performance, 2018, 27:4235–43.
28. GHENO T , GLEESON B . On the Hot Corrosion of Nickel at 700 °C[J]. Oxidation of Metals, 2015, 84(5–6):1–18.
29. YUAN L, WANG H M. Hot corrosion behaviors of a $Cr_{13}Ni_5Si_2$-based metal silicide alloy in Na_2SO_4 + 25 wt.% K_2SO_4 and Na_2SO_4 + 25 wt.% NaCl molten salts [J]. Intermetallics, 2010, 18(3): 324–9.
30. ZHAO S, XIE X, SMITH G D. The oxidation behavior of the new nickel-based superalloy Inconel 740 with and without Na_2SO_4 deposit [J]. Surface & Coatings Technology, 2004, 185(2–3): 178–83.
31. LI M, SUN X, HU W, et al. Hot Corrosion of a Single Crystal Ni-Base Superalloy by Na-Salts at 900°C [J]. Oxidation of Metals, 2006, 65(1): 137–50.

Part IV
Modelling and Data Analytics

Application of Computational Materials and Process Modeling to Current and Future Aero-Engine Component Development and Validation

David Furrer

Abstract Computational materials and process modeling capabilities have evolved over the past several decades. More recently, the Materials Genome Initiative (MGI) has provided focus on this technology and its application for rapid and lower cost materials and process development and implementation. Integrated Computational Materials Engineering (ICME) is now part of many organizations' engineering and design approaches and associated infrastructures. Nearly all current new and future materials and process technology developments do or will involve application of modeling and simulation. The evolution of materials modeling and application to mainstream and emerging supply-chain processes will be reviewed with some perspectives on what the future might hold. Special emphasis will be made to review application of ICME Alloy 718 and derivative materials and components. The use of computational modeling and simulation to material and process development is being extended to component qualification and certification. There are significant opportunities and prospects for materials and process modeling to enable further advancements in alloy design and definition, materials processing methods development, and enablement of enhanced utilization of material capabilities to new product application spaces.

Keywords Computational modeling · Simulation · Integrated computational materials engineering · ICME · Alloy 718 · Superalloys · Industry 4.0

Introduction

There has been and continues to be great promise for the use of computational modeling tools and methods to support the materials science and engineering discipline and the Materials Genome Initiative (MGI) [25]. This goal is being realized through many specific applications, from materials design, process design, component design, and component and process qualification and certification. These

D. Furrer (✉)
Pratt & Whitney, 400 Main Street, East Hartford, CT 06118, USA
e-mail: David.Furrer@PrattWhitney.com

E. A. Ott et al. (eds.), *Proceedings of the 10th International Symposium on Superalloy 718 and Derivatives*, The Minerals, Metals & Materials Series,
https://doi.org/10.1007/978-3-031-27447-3_21

advances are happening as a result of the confluence of the maturity of several critical technologies, including materials characterization, process sensors and controls, materials and process definitions, physics-based computational models, efficient computer technologies and infrastructure, data management plans and curation systems, and data analytics. Each of these technologies are critical enabling elements that are increasing the rate of application and adoption of computational material and process modeling. [10, 22].

The aerospace industry and aero-engines products specifically, have been driving the need for advanced materials capabilities for their demanding application environments. The evolution of turbine engine architectures along with the continued drive for increased fuel efficiency and durability has continued to push the demand for greater performance materials and components. These enhancements in capabilities have come from both new alloy systems and a greater understanding and control of processes. Material and ultimately component properties are controlled by path-dependent processing or exposure from initial material manufacture, component production to system operation, repair, and retirement. Traditional methods to design materials and processes by empirical Edisonian trial-and-error methods have served the industry well but are too expensive and too time consuming for the commercial demands and constraints of today. Computational modeling methods are enabling new discoveries in material science and engineering and are providing for rapid insertion of new materials and processes in the production of new products.

There are several main modeling types and applications that have been and continue to be applied to Alloy 718 and derivative alloys. Alloy design activities have been supported by thermodynamic and kinetics modeling. Process modeling tools have been matured to the point where they are invaluable for casting, forging, heat treating, and emerging manufacturing processes, such as additive manufacturing. These tools have been utilized by engineers for specific applications, but are now being applied at a much broader, system level to support product design optimization. Each of these main applications will be the focus of the following sections.

Thermodynamic and Kinetics Modeling

Phase diagrams are critical tools that materials engineers utilize to guide their research in the pursuit of new, more capable materials, and to understand how microstructures and associated phases can be manipulated by thermal or thermo-mechanical processing. Historically, such phase diagrams were developed by careful experimentation and ex situ or in situ physical characterization of samples of varying chemistry that were exposed to various thermal processes. While such careful experimentation is still needed and is even becoming more exacting with the development and maturation of new materials characterization and sensor technologies, thermodynamic modeling has become a significant and powerful tool that is being used to predict both equilibrium and non-equilibrium phased diagrams. Figure 1 shows an example of a predicted phase diagram for Alloy 7189. These predictive capabilities provide newfound knowledge regarding stable and meta-stable phase equilibria. The use of computational modeling and simulation tools can rapidly provide for prediction and assessment of the sensitivity of alloying elements to the phase equilibria and phase stability of Alloy 718 and derivative alloys.

Thermodynamic modeling is also being applied to the calculation of thermophysical properties. These predicted properties have been validated against careful physical measurements and provide a more complete picture of a material's properties as a function of temperature as compared to limited experimentally available data. Figure 2 shows examples of predictions of Alloy 718 Heat Capacity and Thermal Conductivity as a function of temperature. These types of predicted thermophysical properties are being incorporated into other types of modeling applications to enable prediction of such things as microstructure, mechanical property, and residual stress evolution.

The understanding of phase equilibria in alloys is important in the development and optimization of alloys, but knowledge of the kinetics of phase reactions is equally important. The competitive nucleation and growth of stable and meta-stable phases, and their associated volume fractions, sizes, and distributions ultimately control the mechanical behavior of alloys such as Alloy 718. Considerable work has been previously conducted to physically measure the kinetics of various reactions in Alloy 718 and derivatives [32] Oradei-Basile [30], Carlson [5]. Figure 3 shows predictions of precipitate formation and growth during cooling from two different solution temperatures and starting microstructures.

Long-time stability of alloys is also of great importance when designing new alloys and associated manufacturing processes. Alloy 718 for example will precipitate alpha-Cr during long exposures. The instability and application-based transformation have been successfully modeled and utilized for both alloy chemistry optimization and defining component application space.

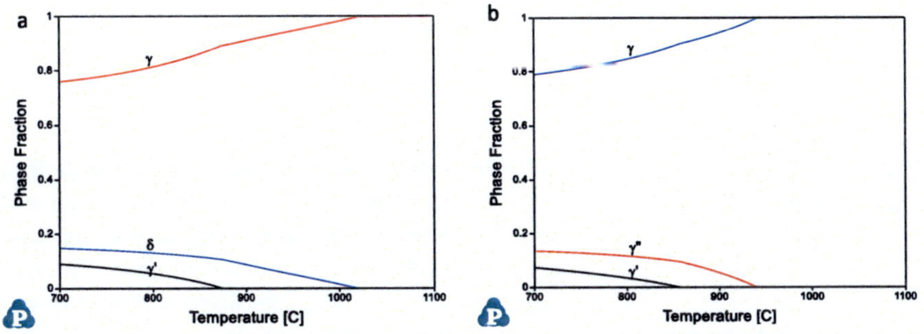

Fig. 1 Thermodynamic predictions of constituent phase fraction versus temperature for an Alloy 718 material with a nominal chemistry of: 18.24% Cr; 53.02% Ni; 3.07% Mo; 4.94% Nb; 0.97% Ti; 0.52% Al; 19.24% Fe in weight percent; where (A) shows predictions for the equilibrium gamma, δ, and γ′ phases and (B) for suspension of δ phase and inclusion of the meta-stable γ″ phase. Courtesy of F. Zhang and CompuTherm LLC

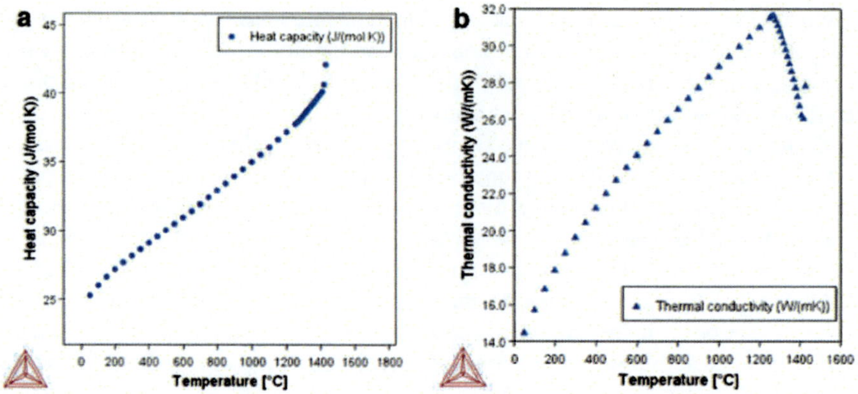

Fig. 2 Computed **a** heat capacity and **b** thermal conductivity for Alloy 718 as a function of temperature. Courtesy of P. Mason and Thermo-Calc Software

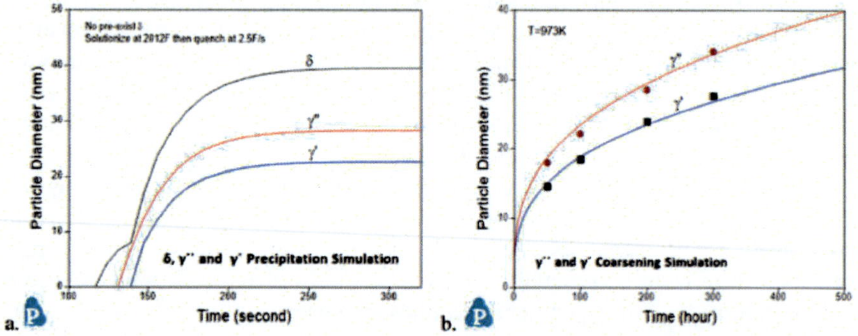

Fig. 3 Predicted particle diameter for δ, γ' and γ" during precipitation from a super-delta solvus solution heat treatment (**a**), and coarsening of γ' and γ" during isothermal aging at 700 °C. Courtesy of F. Zhang and CompuTherm LLC

The desired mechanical properties of Alloy 718 are largely controlled by the precipitation of gamma-prime and gamma-double-prime which are in competition with delta phase formation. A balance of mechanical properties can be realized with a controlled volume fraction, spatial distribution, and morphology of each of these phases for a given gamma grain size and alloy chemistry. Models have been developed to enable the prediction of mechanical properties of nickel-base superalloys as a function of microstructure [17, 23, 31] Such computational tools are being further developed to include further understanding of the physics of the mechanical property behavior. Precipitate shearing is a significant mechanism through which alloys are strengthened. The mechanism of precipitate shearing includes the formation and transmission of dislocation partials. Drag on these dislocations increases the strength of the material. Anti-phase boundary (APB) energy is one of the driving forces that hinder the ability of dislocations to penetrate and move through ordered precipitates which are present in Alloy 718 and other nickel-based superalloys. The ability to predict APB energies as a function of alloy and precipitate phase chemistry is aiding the design and optimization of new alloys [15].

Model-Based Alloy Design and Development

Computational materials models are being used more frequently to design and optimize new alloys for specific sets of mechanical properties and component performance [28, 29, 38]. Model-enabled alloy design has shown the ability to design a nickel-based alloy that has greater capability over that of the highest strength variants of Alloy 718 [18]. The approach used included ab initio calculations of APB energies as a function of chemistry, thermodynamic and kinetics modeling of phase formation, chemical portioning and microstructure evolution, mechanical property predictions, and associated optimization of chemistry and processing parameters. Figure 4 shows the overall computational workflow for an advanced Alloy 718 derivative alloy. Figure 5 shows measured yield strength properties for the computationally designed superalloy compared with the properties of a high-grain, high-strength version of Alloy 718.

Deformation and Microstructural Evolution Modeling

Forging modeling has been utilized to study, design and optimize deformation processes to achieve required microstructures throughout the volume of aerospace components, including Alloy 718 rotors [8, 9, 24, 34, 36] These modeling and simulation efforts have demonstrated the capability of prediction of recrystallization and grain growth. Figure 6 shows predicted average grain size contours for example, Alloy 718 disk [7, 35]. This example is unique in that it shows the as-heated grain size within the central low strain, die-lock areas. The as-heated grain size for the billet material was predicted to be ASTM average 8. The prediction also shows a significant reduction in grain size with increasing strain away from the top and bottom die-lock regions. The refinement in grain size is a result of dynamic recrystallization. The finest grain size predicted is ASTM 13. There is another region within the dynamically recrystallized volume, which has a "teardrop" shape in cross-section. This region has an increased grain size compared to the dynamically recrystallized region. The teardrop region underwent significant strain at a high strain rate. This resulted in significant adiabatic heating and subsequent grain growth to the predicted sizes shown. This simulation effort required a series of models to predict the initial grain growth, dynamic recrystallization, and local grain growth due to locally high temperatures during the deformation process. These models required a detailed understanding and control of the forging process parameters and process

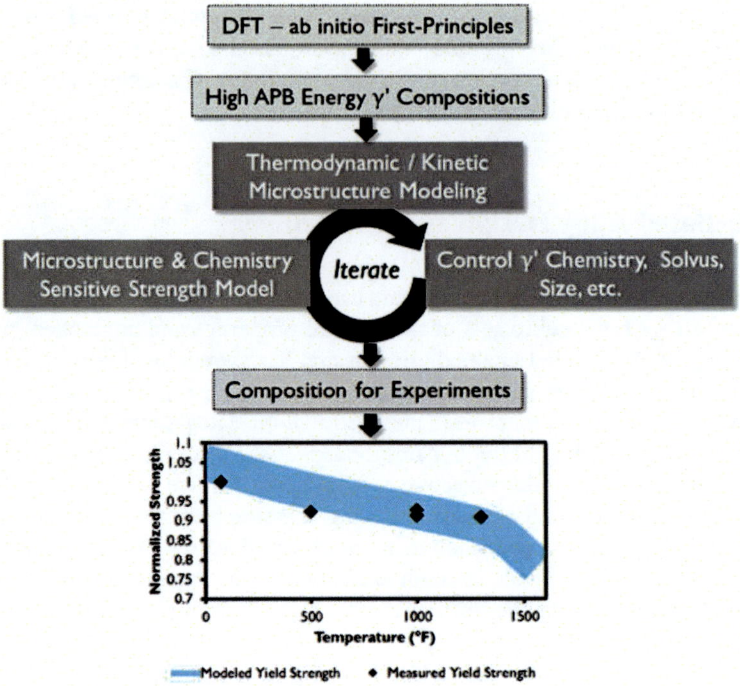

Fig. 4 The high-level framework for the computational modeling workflow for the development of a new nickel-base superalloy based on mechanistic strengthening elements also present in Alloy 718

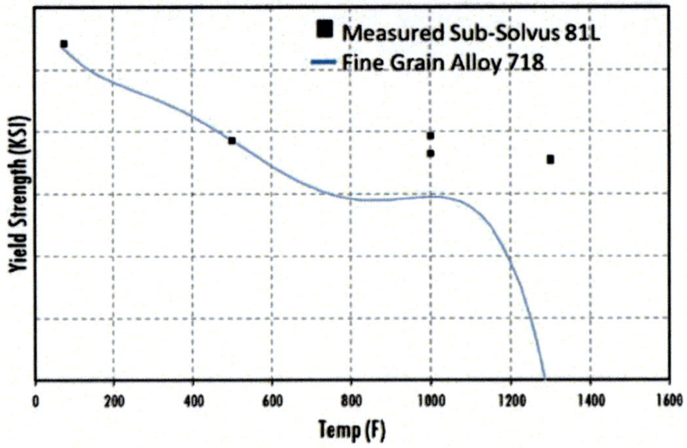

Fig. 5 Yield strength versus temperature for a high-strength, fine-grain Alloy 718 material compared to the measured tensile strength capability of the new nickel-base superalloy developed through the model guided framework shown in Fig. 4

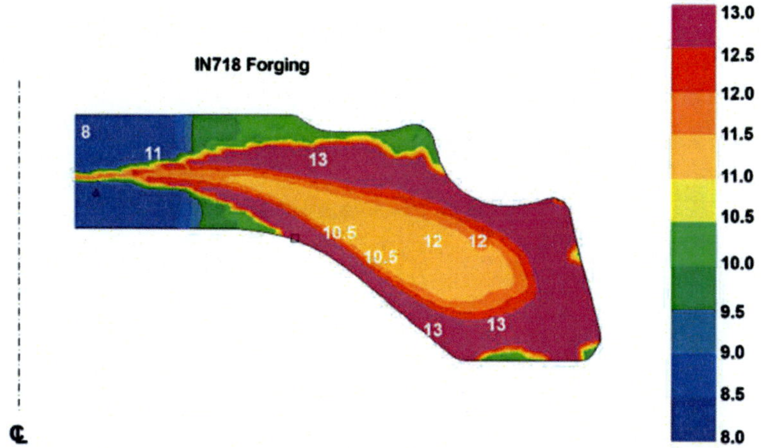

Fig. 6 Predicted grain structure for an Alloy 718 forging (contours) along with location-specific ASTM grain size measurements (inserted values). Copyright 2000 The Minerals, Metals & Materials Society. Used with permission.

boundary conditions. The predicted final grain structure is shown to compare very well with actual metallographic measures.

Thermal Process and Residual Stress Modeling

Thermal treatments of Alloy 718 and derivative components are critical to achieve desired microstructures, which in turn provides for optimal mechanical properties of final components and products. Models for the heat treatment processes for Alloy 718, such as solution heat treatment, stabilization, and aging are very mature. These computational modeling tools provide accurate time–temperature histories throughout the volume of processed components.

The thermal history during heat treatment can drive the development of dynamic thermal stresses, thermal strains, and subsequent residual stress in final processed castings, forgings, or fabrications. Efforts to predict residual stresses in IN718 turbine disks have been conducted and reported [21]. Figure 7 shows an example prediction of the bulk residual stress profile for a forged and heat treated superalloy pancake forging [[16].

Considerable effort has been undertaken to study, model, and validate predictions from linked material, thermal processing, and residual stress evolution models under a prior USAF sponsored program called the Foundational Engineering Problems (FEP) program [6, 39]. This prior effort showed that the accurate prediction of residual stress within Alloy 718 components is possible, but the dynamic material properties needed for such predictions, including flow stress and creep rates as

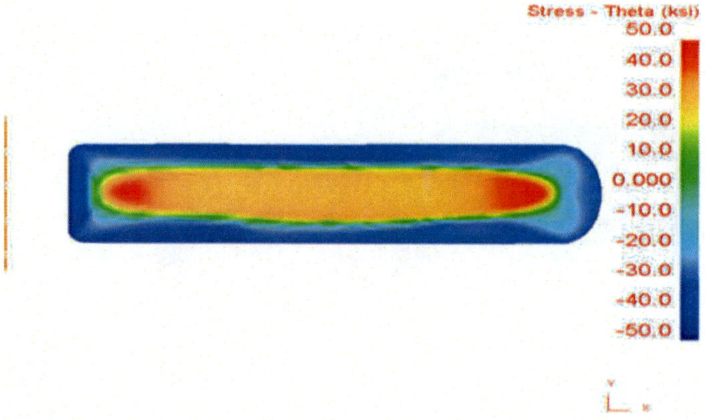

Fig. 7 Predicted residual hoop stress as a result of heat treat quench, and aging processes

a function of temperature and microstructure are the major sources of uncertainty in such predictions for Alloy 718. This effort also showed that understanding the spatial distribution of residual stresses within a component a priori can readily enable "smart testing" where specific locations can be defined that are most useful in validating the overall magnitude of residual stresses throughout the entire part volume. This model-enabled "smart testing" approach can provide for greater process and component knowledge, a quantitative reduction in residual stress uncertainty, and a reduction in the overall cost of quality testing. Figure 8 shows an example residual stress prediction contour with specific locations defined that will provide the most beneficial results from physical testing.

Learning from this prior program and other related efforts are being applied to other processes that produce thermally-induced residual stresses. Related efforts to model and predict local plastic strain from peening, burnishing, and cold expansion/compression methods are also demonstrating capabilities to accurately predict near-surface and local bulk residual stress development in final components. Machining processes can also provide for local surface plastic strains which can produce both tensile or compressive residual stresses, which drives the need for tight control of finish machining operations for Alloy 718 and derivative components [2].

Casting and Solidification Modeling

There has been significant progress and utilization of solidification modeling and simulation for casting processes. Virtual casting initiative activities have been "pathfinder' examples on how components and casting processes can be designed and optimized to produce highly capable, high-yield, high performance manufacturing processes and cast products [1]. Casting modeling has been and continues to be applied to Alloy 718 materials and components. Prediction and elimination of

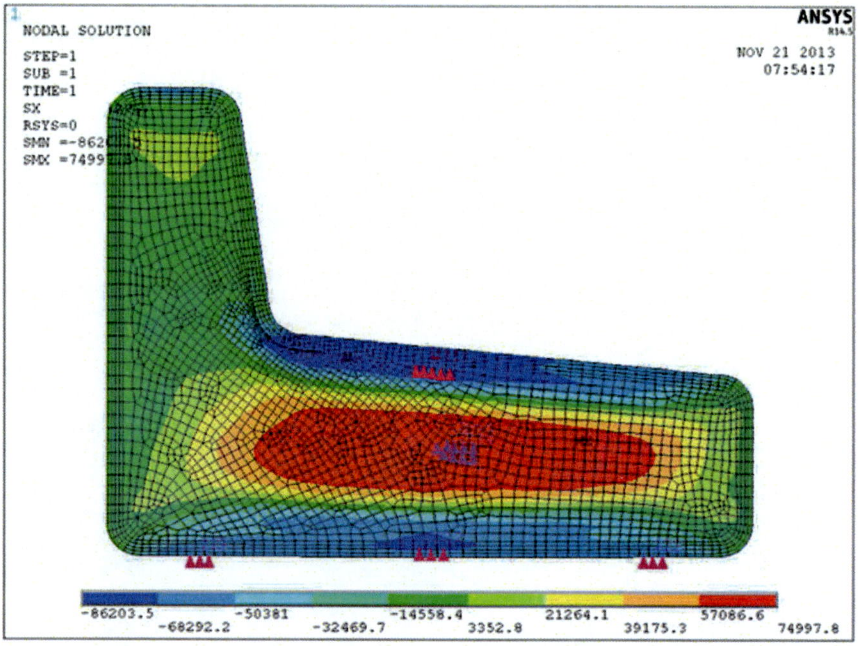

Fig. 8 Predicted residual stress contour plot for an Alloy 718 forging processed through heat treatment. The marked locations have been identified through simulation and uncertainty quantification analysis to be the most critical locations for testing

shrinkage porosity is one of the major goals of such modeling. Figure 9 shows an example of a casting solidification simulation where the sensitivity of casting parameters is shown to provide information regarding the occurrence of shrinkage porosity and its associated spatial distribution.

Casting and solidification models of this type are commonly being applied to casting geometry and process design. Predictive capabilities are such that this is an invaluable tool in the design of casting processes by enabling "design for manufacture" capabilities. Casting models, like other types of models, can be utilized in a deterministic or probabilistic framework. The propensity or probability for shrinkage porosity to occur at a given location within a part volume for a given casting process and defined range of process controls can be analyzed through Monte Carlo type analysis methods where knowledge of the range of variations in critical processing parameters can be utilized to determine the effect of individual or combined parameter ranges for a defined manufacturing process. Linking measured distributions of porosity characteristics with casting simulations that include process parameter variability can help to more clearly define the critical parameters that are driving the occurrence and occurrence rates observed in production castings.

Integrating modeling and data management methodologies enable fusion of simulation information with physically measured data. Superimposing captured porosity

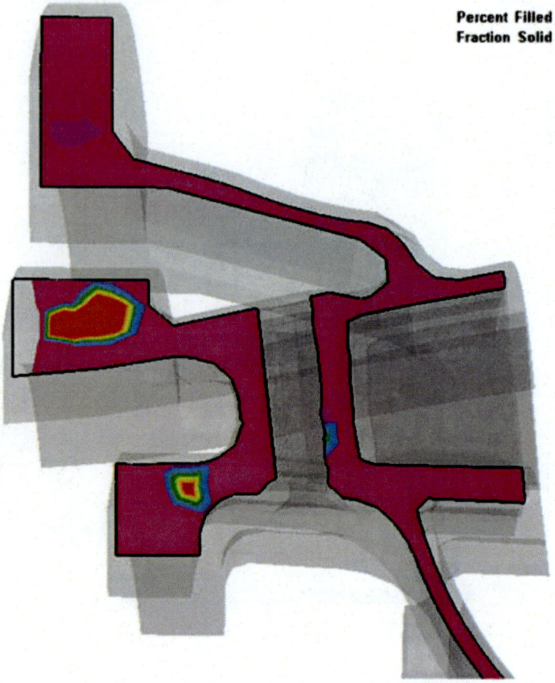

Fig. 9 Casting solidification prediction for an Alloy 718 component using a commercial casting solidification software showing hot-spot regions in locations where there is a propensity for solidification shrinkage porosity

occurrence rate, size, and spatial location onto 3D solid models are providing significant information about process control and capabilities. Capturing and superimposing of final part information from computed tomography, radiography, or periodic cross-sectional cut-up tests in the form of location-specific rates of porosity occurrence onto the cross-section of predicted part solid models provides for the ability to analyze probabilistic predictions with operational process statistics to guide identification of critical sources of combined variation and enhancement of subsequent manufacturing process control plans [20].

Data capture and integration with process modeling provides for a means to further understand the true capability of specific casting processes for specific component configurations, and whether the defined stacking-tolerances from the casting process could provide occurrences of deleterious feature or increased rates of occurrence that might not be acceptable from an economic or component performance perspective. Upfront modeling provides an opportunity for optimal design-for-manufacturing, while the capture and utilization of production data enables further validation and improvement of the uncertainty of models and process boundary conditions, resulting in an effective means of capturing knowledge in a re-usable modeling framework.

Additive Manufacturing Process Modeling

Additive manufacturing (AM) is becoming an attractive approach to produce complex components in a completely disruptive manufacturing process. Recent instantiations of AM processes have focused largely on geometric capabilities and ease of printing, so machine designers and builders have focused on these attributes. The issue of understanding and controlling the microstructure evolution and macroscopic process defect formation are now largely driving the development, optimization, and final implementation of this process for actual components. AM is an adaption of welding processes, where powders, wires, or other feedstocks are melted and added to a substrate in a layer-by-layer approach, similar to processes like submerged arc cladding or overlaying method, but with a number of significant differences. In most AM processes, such as laser powder bed fusion processes or powder feed processes, a very small, controlled melt pool is created, which is on the order of a couple of hundred microns in diameter, and which traverses a component surface at one or two orders of magnitude faster rates as compared to conventional welding processes. These macroscopic boundary conditions result in unique opportunities and challenges. The small fusion zone and powder sizes can enable very fine features and geometry control. The layer-by-layer construct of the process requires significant control of the "weld bead" at every point of every layer, as any anomaly of a deposit bead has the ability to propagate into subsequent layers or trap-in a defect feature that can greatly reduce the mechanical property capability of a final component.

Additive manufacturing is not an inexpensive process but can provide for an overall most efficient process as compared to other processes for specific component applications with complex geometries, difficult to process materials, and critical mechanical property requirements. These economic constraints often drive AM toward unique components that require tight control of often "organic" geometries and with high-levels of mechanical property capabilities, such as low-cycle and high-cycle fatigue. Figure 10 shows the impact of AM pedigree, and associated microstructure and geometric feature control on the fatigue capabilities of AM Alloy 718. This shows the very clear need to understand and control the features that can form during additive manufacturing, and which degrade the mechanical property capabilities.

Modeling and simulation of AM processes have been extensively worked on for a range of purposes, including defect prediction, microstructure evolution, residual stress development, and final component mechanical properties [4, 11, 19]. The ability to rapidly analyze and predict the formation of build defects in AM components is an enabling technology that can provide needed guidance in component design, process design, process control, and component and process qualification and certification. An effort has been undertaken to develop an advanced AM modeling and simulation framework to predict the formation of the main AM processing defects, including key-hole formation, un-melt-balling, and surface roughness, using fast-acting, physics-based analytical models [3].

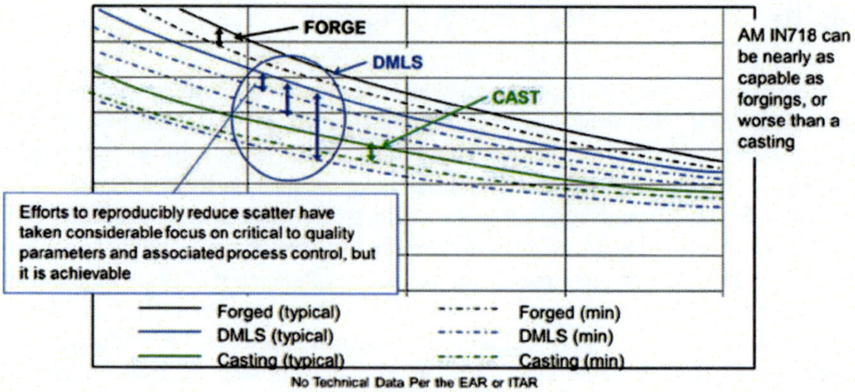

Fig. 10 Comparison of fatigue curves from Alloy 718 forgings, castings, and various pedigrees of additive manufactured material. It can be seen that the fatigue capabilities of Alloy 718 significantly depend on the microstructure, and the size, frequency, and type of build defects within each specific pedigree of material

The established model is a system of sub-models that consist of physics-based analytical expressions for heating from the laser, thermal conductivity, and dynamic thermal profile within the powder bed, melt pool formation, thermal profile within the melt pool, melt pool shape, key-hole formation, and fluid flow within the melt pool via Marangoni flow. The model contains many critical parameters ranging from the thermophysical properties of the powder material, size and shape of the powder, powder bed density, part geometry and build orientation, and defined or parametric ranges of laser power and scan speed as a function of volumetric location, to list a few.

This simulation process enables predicting the potential of each defect to form as a function of the local processing boundary conditions and processing path and history. It is well known that two parameters that impact defect formation potential are laser energy and laser traverse speed. Many have developed processing maps based on physical trials to describe the parameter space where each defect feature will form. Though this has been demonstrated by physical trials, these maps do not take all of the required processing information into account to assess defect formation potential. Empirically derived processing maps are often thought of as a single map description for a given material, which is a gross over simplification. AM defects, like those that can form in welding or casting, are holistically path-dependent and depend on a range of parameters and boundary conditions. Conductivity of powder beds and temperature of the substrate and powder bed layer at the time of laser processing are also major boundary conditions that are often overlooked or considered as secondary issues.

The developed AM modeling tools can provide for a dynamic prediction of "processing maps" in multi-dimensional parameter space as a function of build sequence. As the build process starts and progresses to produce the entire part, the processing

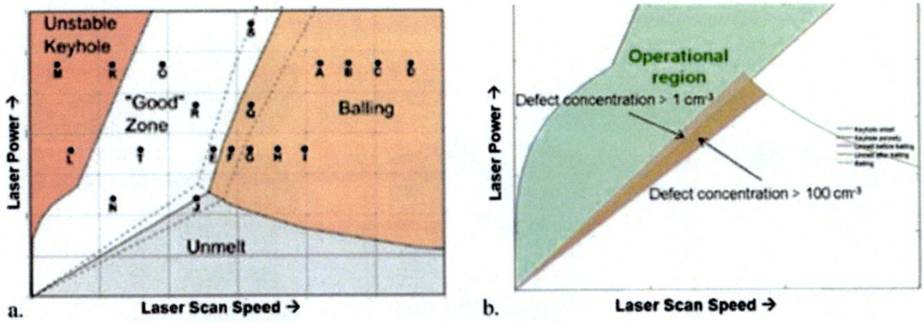

Fig. 11 Process maps for Alloy 718 for the processing parameters of laser power and scanning speed with all other processing parameters held constant. Measured validation points are shown in (**a**), while the uncertainty of the mechanism boundaries is shown in both (**a**) and (**b**) as a function of known secondary parameter control uncertainty

boundary conditions change, which shift the boundaries within "processing maps" that separate stable and unstable regions. The distribution of processing boundary conditions can also be included in such analysis efforts to enable the treatment of the process parameter boundaries to all of the probability of defect formation. Figure 11 shows an example predicted Alloy 718 processing map for a set of processing parameters, along with measured values. Incorporation of uncertainty in process boundary conditions also provides for a means to include uncertainty in the boundaries within the processing map.

Computational efficiency is of critical importance when developing modeling and simulation tools. The current model provides for extremely fast calculations that can provide input to designs for the optimization of processing paths for any arbitrary geometry. Simulation of components can be completed in seconds as compared to weeks for other modeling approaches. This modeling approach enables true design for manufacture and highlights the need for laser control and monitoring capabilities. Figure 12 shows an example of a series of Alloy 718 coupons that was modeled through the developed AM modeling and simulation framework. The predictions Fig. 12a show that the propensity to form defects as a function of build volume location can change for a constant set of build path parameters, such as those that might be provided as default machine settings for a specific build location type. As builds evolve, the thermal conditions change, which shift the boundaries in processing space where stable, defect-free material can be produced.

Figure 12b shows example of μ-CT images and physical cross-section metallography data for three different controlled build path parameters. These comparisons show that the established modeling framework can accurately predict various types of defects and spatial distributions for a wide range of processing conditions.

Complex real-world component can be analyzed, and defects can be predicted within the 3D volume for any set of build parameters. This computational tool allows engineers to manually or with automated algorithms optimize the build parameters on a build schedule and location-specific basis. Figure 13 shows the capability to

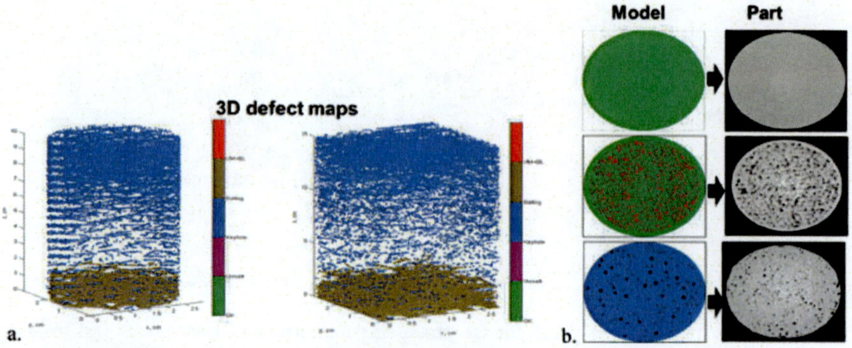

Fig. 12 Prediction of additive manufacturing defect formation within a 3D cross-section of an Alloy 718 coupon (**a**), and a comparison of defect structure within predicted and physically characterized test bars (**b**)

predict volumetric defect formation, surface defect formation and associated surface roughness as defined by depth of surface connected porosity for a complex component configuration.

This type of efficient modeling and simulation tool does provide a means to develop optimal manufacturing processes, assessment of post-build inadvertent variations, and could potentially be used for adaptive closed-loop control systems that incorporate process sensor data and provide for any required process parameter adjustments to ensure the build zone is kept within the most stable, defect-free conditions. Additionally, the ability to predict the probability of forming defects along with local microstructure can enable the prediction of location-specific mechanical properties, similar to the capabilities for parts produced by other manufacturing processes. This combined modeling capability can directly support the mechanical design and structural analysis of components during design and qualification processes.

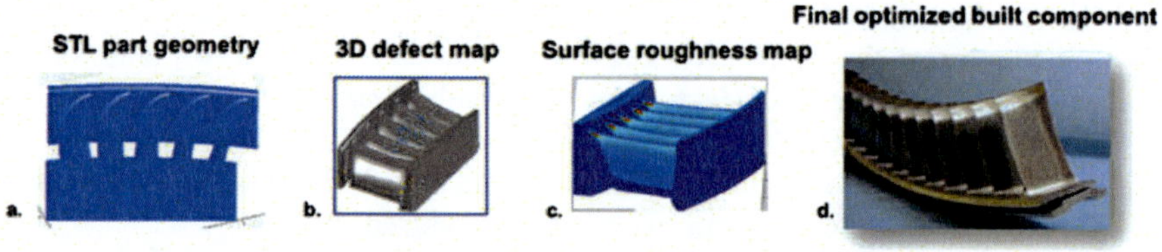

Fig. 13 Example of an Alloy 718 component that was analyzed for AM process defect formation and subsequent optimization, where **a** shows an input component STL file geometry, **b** prediction of 3D defect formation by location, **c** predicted surface roughness as a function of location, and **d** the final realized optimized AM component

Model-Based Material Definitions and Model-Enabled Qualification and Certification

Local microstructure (e.g., grain size and precipitate type, quantity, and size distributions) can be utilized to predict location-specific mechanical properties throughout a component, whether manufactured by forging and heat treating, casting, fabrication, or additive manufacturing. Figure 14 shows an example of an Alloy 718 forging showing contours of local mechanical properties. It can be seen that the properties throughout the volume of this forging are not constant and vary due to mechanistic reasons, such as thermo-mechanical processing history and grain size evolution, and specific cooling paths that impact the precipitation kinetics and final precipitate structure locally.

With this understanding of location-specific microstructure and property evolution, it is understandable that resultant test data will vary throughout the cross-section of this component. This example illustrates the issue with current static material specifications that provide single minimum property values. Within the current construct of material specifications, e.g., AMS5662, a minimum property value is given that is aimed at representing all material that is defined by the specification. The minimum property value for traditional static specifications is defined as the -2, or -3 sigma value from a population of tested samples. If these samples are taken from a part with a wide variation in microstructure due to large local strain and/or cooling rate histories, then a wide distribution of properties will be observed. The treatment of the resultant physically tested samples assumes the properties should be the same and are analyzed as a single continuous distribution of properties with scatter due to the material's inherent capabilities. The −2, or −3 sigma deviation values are utilized to capture the "natural" variation of this alloy and to provide a simple means to define the minimum property that can be achieved when this specification is applied to a design and structural analysis. As can be seen from the model example in Fig. 14, scatter in measured properties is not an alloy inherent feature, but rather a combined alloy and processing-path variation. Model-based material definitions (method of using a series of microstructure and property models) along with processing path simulation and prediction can provide for enhanced understanding of the true capability of a material and manufactured component on a location-specific basis [10].

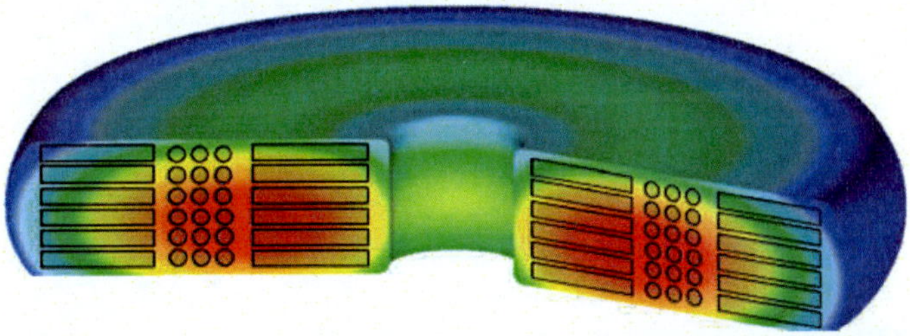

Fig. 14 Predicted mechanical property contours and physical test specimen layout plan

This approach is being applied to many new materials and processes, as workflow-based applications of model-based definitions are being understood and accepted across engineering, manufacturing, and quality disciplines.

Materials data analytics are rapidly blending with previous physics-based or phenomenological modeling approaches. Microstructure is an unquestionable critical feature of all engineered materials and components. The ability to characterize microstructures is rapidly changing with the maturation and increased availability of new methods, such as μ-CT and EBSD (electron backscatter diffraction) pattern imaging. Figure 15 shows an example of a complex AM Alloy 718 microstructure using EBSD imaging and analysis methods. This enhanced imaging capability provides for large quantity of digital data that can be analyzed to show recrystallized and un-recrystallized grain structure and the variations of crystallographic misorientation within and between grains to enable quantification of retained strain and texture respectively.

Machine learning (ML) and other artificial intelligence methods are being used to analyze and establish models from large datasets, such as microstructures [26]. ML tools have been utilized effectively to classify microstructure through principal components analysis (PCA) as shown in Fig. 17. Such analysis efforts are like established human facial recognition tools and can be used to separate and classify very complex microstructures such as that seen in Fig. 16. PCA methods have been applied to a wide range of materials, such as titanium and nickel-base superalloys to help guide investigation and understanding of complex microstructures.

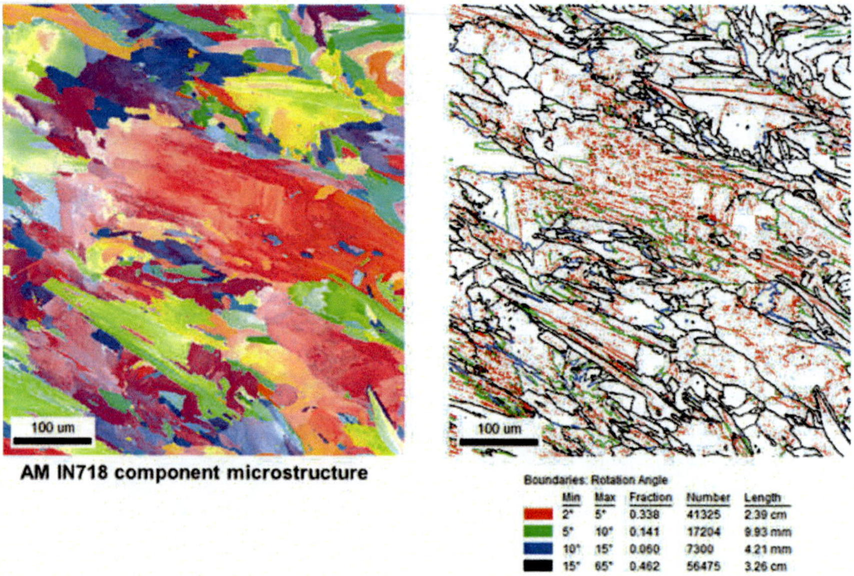

Fig. 15 EBSD image of an additively manufactured Alloy 718 sample (**a**), and subsequent analysis of grain and sub-grain misorientation angles (**b**)

Complex microstructures are very difficult or impossible for humans to assess and subjectively determine differences between samples or to photographic standards. ML models can provide more objective means of characterizing the microstructure for both alloy development and component quality control.

Digital characterization of microstructure can also enable objective, statistical representations that can be further used to model and simulate probabilistic mechanical property capabilities of a material. Statistically equivalent representative volume elements (SERVEs) are a new computational tool that defines the statistics of the microstructure and associated minimum length scale that can be used to define specific mechanical properties for a specific material [14, 37] Fig. 17 shows an example assemblage of microstructures that have been used to establish a SERVE for a nickel-base superalloy.

The approach of using SERVE definition of microstructure for a given material and component allows for assessing the potential range of microstructures possible that meet the statistics of the SERVE. ML models and computational tools such as

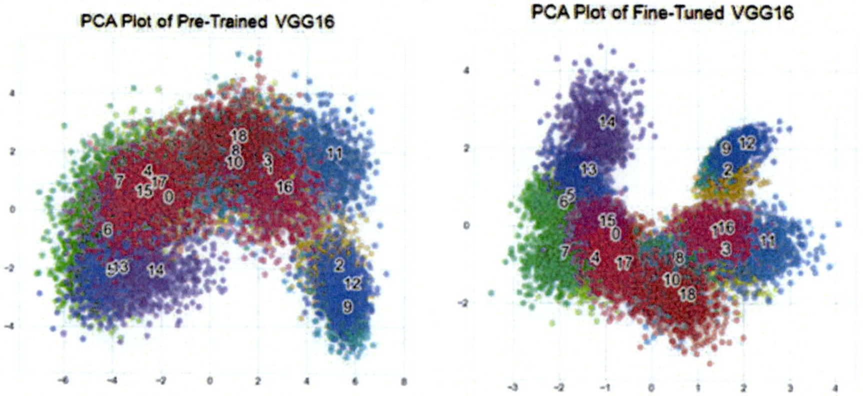

Fig. 16 PCA analysis examples for several trained models for a range of alloy microstructures[27].

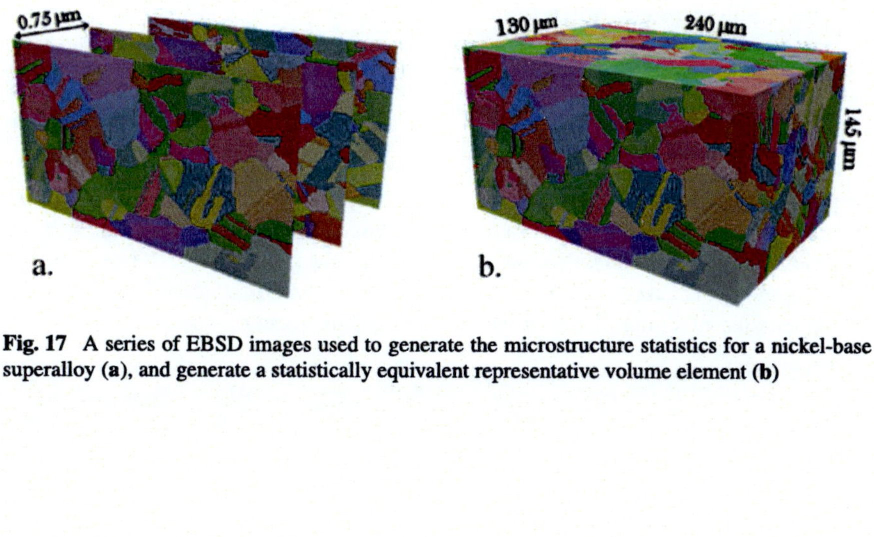

Fig. 17 A series of EBSD images used to generate the microstructure statistics for a nickel-base superalloy (**a**), and generate a statistically equivalent representative volume element (**b**)

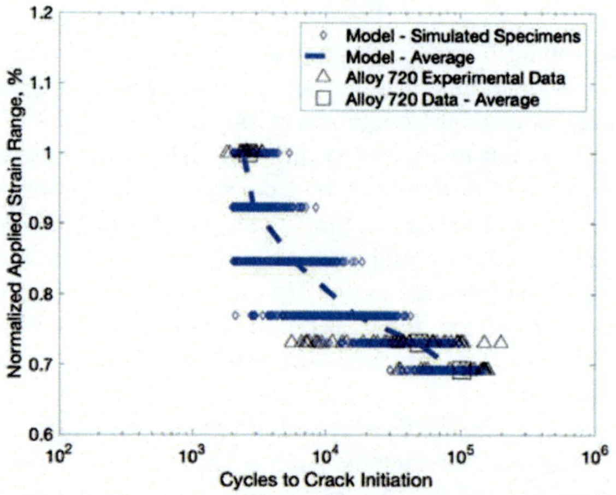

Fig. 18 Crystal-plasticity based model prediction of fatigue properties for Alloy 720 compared with measured data. The model well predicts the observed scatter of the measured values. It is also of interest to note the range of virtual test specimens that were predicted as compared to the limited number of actual physical specimens, owing to the speed and cost effectiveness of virtual material property testing and evaluation. Courtesy of Prof. M. Sangid, Purdue University

Dream3D allow for creation of artificial microstructures over the entire spectrum of the statistically defined microstructure. This provides for the ability to create the microstructures that make up the tales or rare event instantiations. Measuring and capturing tales or rare events are by definition statistically challenging, but ML models and data analytics make it possible to assess such structures, which are often the source of minimum property capabilities. Linking statistically defined microstructures with crystal-plasticity modeling provides for an approach to computationally predict the distribution of mechanical properties for a material. This approach has been applied to full-scale component predictions [12, 13], as well as for the assessment, augmentation [33], or complete development of traditional minimum capability design curves. Figure 18 is an example of crystal-plasticity based fatigue design curve compared with measured properties for Alloy 720 [33].

Conclusions

The development and application of computational models for Alloy 718 and derivatives have been successfully applied to define the thermodynamic and kinetic behavior of these materials. Process models of various types have also been created and utilized to accurately predict the evolution of microstructure and mechanical properties of components on a path-dependent, location-specific basis. New alloys and manufacturing processes are being developed and optimized through model-guided engineering methods.

The challenge and opportunity for the continued utilization of computational materials and process models is the development of model-based definitions that can provide the needed path-dependent predictive capability for material and process design, and qualification and certification. Model-based definitions are continuing to mature. They are comprised of formalized workflows that provide for the prediction of location-specific processing paths, associated microstructure evolution, and corresponding local mechanical properties. Minimum property design curves, though still readily used, are starting to give way to model-based material definitions that provide for more exacting description of properties throughout the volume of a component.

Statistical material science and engineering is a major element of the evolution of model-based material definitions, where predicted variations in chemistry and microstructure inform property prediction models which in turn provide predictions of the range of variation of mechanical properties on a spatial or location-specific basis.

Complex processes, such as additive manufacturing, must be analyzed and controlled relative to location-specific, path-dependent processing parameters. This capability has been shown and utilized in other, more mature processes, such as forging and casting, but is vital for rapid process development, qualification, and certification of additive manufactured components.

The future is bright for computational material and process modeling, as tools are developed, and workflows merged with those of other engineering, manufacturing, and quality disciplines.

References

1. Allison J, Li M, Wolverton C, and Su XM (2006) "Virtual Aluminum Castings: An Industrial Application of ICME", JOM, Vol. 58, No.11, 2006, pp. 28–35.
2. Arrazola PJ, Kortabarria A, Madariaga A, Esnaola JA, Fernandez E, Cappellini C, Ulutan D, Özel T (2014), "On the machining induced residual stresses in IN718 nickel-based alloy: Experiments and predictions with finite element simulation", Simulation Modelling Practice and Theory 41 (2014) 87–103.
3. Burlatsky SF, Novikov D, Brindley WJ, Furrer D (2017) "System and process for evaluating and validating additive manufacturing operations", U.S. Patent 10254730, Feb 24, 2017.
4. Burlatsky S, Sharon J, El-Wardany T, and Furrer D (2019) "Application and Validation of Laser Powder Bed Fusion (LPBF) Defect Prediction Models to Guide Optimal Process Windows", RAPID Conference Presentation, Jan. 7, 2019.
5. Carlson RG, and Radavich J (1989) "Microstructure Characterization of Cast 718", Proceedings of the Int. Symp. on Metallurgy and Applications of Superalloys 718, 625 and Various Derivatives, Ed. E.A Loria, June 12–14, 1989, TMS, pp. 70–85.
6. Cernatescu I, Venkatesh V, Glanovsky JL, Landry LH, R. N. Green RN, Gynther D, Furrer DU (2015) "Residual Stress Measurements for Model Validation As Applied in the United States Air Force Foundational Engineering Problem Program on ICME of Bulk Residual Stress in Ni Rotors", 56th AIAA/ASCE/AHS/ASC Structures, Structural Dynamics, and Materials Conference, AIAA SciTech Forum, (AIAA 2015–0387), 2015.

7. Furrer D, Groppi G, and Bunge G (2004) "Current and Future Trends in the Manufacture of Turbine Engine Disks", ASM Materials Solutions Conference, Oct. 18–21, 2004, Columbus, OH.
8. Furrer D, Goetz R, and Shen G (2010) "Modeling and Simulation of Alloy 718 Microstructure and Mechanical Properties", proceedings of the 7th International Symposium on Superalloy 718 & Derivatives, October 10–13, 2010, Pittsburgh, Pennsylvania, TMS, 2010.
9. Furrer D, Venkatesh V, Zhang F, Gynther D, Asare T, Novikov D, and Burlatsky S (2014) "Computational Modeling and Simulation of Alloy 718", 8th International Symposium on Superalloy 718 and Derivatives, Eds: Ott E, et.al., TMS, 2014. pp. 81–94.
10. Furrer DU, Dimiduk DM, Cotton JD and Ward CH (2017) Making the Case for a Model-Based Definition of Engineering Materials. Integ Mater Manuf Innov, 6:249–263 https://doi.org/10.1007/s40192-017-0102-7.
11. Furrer D and Burlatsky S (2022) "Model-Assisted Validation and Certification of AM Components", AM Bench Conference Presentation, TMS, Aug. 19, 2022, Bethesda, MD.
12. Gopalakrishnan S, Bandyopadhyay R, Sangid MD (2022) "A Framework to Enable Microstructure-Sensitive Location-Specific Fatigue Life Analysis of Components and Connectivity to the Product Lifecycle", Int. J. of Fatigue, 165, 2022, 107221, https://doi.org/10.1016/j.ijfatigue.2022.107211
13. Ghosh S, Dimiduk D, Furrer D (to be published) "Statistically Equivalent Representative Volume Elements (SERVE) for Material Behavior Analysis and Multiscale Modeling".
14. Ghosh S, and Venkatesh V (Unpublished Research).
15. Gorbatov OI, Lomaev IL, Gornostyrev YN, Ruban AV, Furrer D, Venkatesh V, Novikov DL, and Burlatsky SF (2016) "Effect of composition on antiphase boundary energy in Ni3Al based alloys: Ab initio calculations", Physical Review B 93, 224106 (2016).
16. Gostic W (2012) "Application of Materials and Process Modeling to the Design, Development and Sustainment of Advanced Turbine Engines", Superalloys 2012, Eds. Huron, E, Reed RC, Hardy MC, Mills MJ, Montero RE, Portella PD, Telesman J, TMS, 2012, pp. 3–12.
17. Jou H-J, Vorhees P, Olson GB (2004) "Computer Simulation for the Prediction of Microstructure/Property Variation in Aeroturbine Disks", Superalloys 2004, Eds. Green KA, Pollock TM, Harada H, Howson TE, Reed RC, Schirra JJ, Walston S (TMS), pp 877–886.
18. Kaplan MA, Liu X, and Furrer DU (2020) "Composition and method for enhanced precipitation hardened superalloys", U.S. patent: US10793934B2.
19. Khairallah SA, Martin AA, Lee JRI, Guss G, Calta NP, Hammons JA, Nielsen MH, Chaput K, Schwalbach E, Shah MN, Chapman MG, Willey TM, Rubenchik AM, Anderson AT, Wang YM, Matthews MJ, King WE (2020) "Controlling Independent Meso-Nanosecond Dynamics and Defect Generation in Metal 3D Printing", Science, Vol 368, Issue 6491, May 2020.
20. Kopper, A., Karkare, R., Paffenroth, R.C. and Apelian D (2020) "Model Selection and Evaluation for Machine Learning: Deep Learning in Materials Processing", Integrating Materials and Manufacturing Innovation, volume 9, 287–300 (2020). https://doi.org/10.1007/s40192-020-00185-1.
21. Krempaszky C, Werner EA, Stockinger M (2005), "Residual stresses in IN718 Turbine Disks", Superalloys 718, 625, 706 and Derivatives 2005, Ed. Loria EA, TMS 2005, pp. 527–538.
22. Liu X, Furrer DU, Kosters J, and Holmes J (2018) Vision 2040: A Roadmap for Integrated, Multiscale Modeling and Simulation of Materials and Systems. NASA/CR-2018-219771, E-19477, GRC-E-DAA-TN52454.
23. McDowell D, and Backman D (2010) "Simulation-Assisted Design and Accelerated Insertion of Materials", Computational Methods for Microstructure-Property Relationships, Springer, 2010, pp 617–647.
24. Mosser PE, Leconte G, Leray J, Lasalmonie A, and Honnorat Y (1989) "Metallurgical Aspects of Forge Modeling in Alloy 718", Proceedings of the Int. Symp. on Metallurgy and Applications of Superalloys 718, 625 and Various Derivatives, Ed. E.A Loria, June 12–14, 1989, TMS, pp. 179–188.
25. National Research Council (2008) "Integrated Computational Materials Engineering: A Transformational Discipline for Improved Competitiveness and National Security", Washington, DC: The National Academies Press. https://doi.org/10.17226/12199.

26. Noraas R, Somanath N, Giering M, and Oshin O (2020a) "Structural Material Property Tailoring of Dual Phase Titanium Alloy Microstructures Using Deep Neural Networks", AIAA Scitech 2020 Forum, 6-10 January 2020, Orlando, FL. https://doi.org/10.2514/6.2020-1151.
27. Noraas R, Venkatesh V, Rettberg L, Somanath N, Oshin O and Giering M (2020b) "Recognition and Quantification of Dual Phase Titanium Alloy Microstructures Using Convolutional Neural Networks", The 14th World Conference on Titanium, MATEC Web of Conferences 321, 11084. https://doi.org/10.1051/matecconf/202032111084.
28. Olson GB (1997) Computational Design of Hierarchically Structured Materials. Science 277:1237–1242.
29. Olson (2000) "Materials by Design", Science, 12 May 2000, Vol 288, Issue 5468, p. 995. https://doi.org/10.1126/science.288.5468.995.
30. Oradei-Basile A and Radavich J (1991) "A Current T-T-T Diagram for Wrought Alloy 718", Proceedings of the Int. Symp. on Metallurgy and Applications of Superalloys 718, 625 and Various Derivatives, Ed. E.A Loria, June 23–26, 1991, TMS, pp. 325–335.
31. Parthasarathy TA, Rao SI, and Dimiduk D (2004) "A Fast Spreadsheet Model for the Yield Strength of Superalloys", Superalloys 2004, TMS (The Minerals, Metals & Materials Society), 2004, pp. 887–896.
32. Radavich, J (1989) "The Physical Metallurgy of Cast and Wrought Alloy 718", Proceedings of the Int. Symp. on Metallurgy and Applications of Superalloys 718, 625 and Various Derivatives, Ed. E.A Loria, June 12–14, 1989, TMS, pp. 229–240.
33. Sangid MD, Maier HJ and Sehitoglu H (2011) "An energy-based microstructure model to account for fatigue scatter in polycrystals", Journal of the Mechanics and Physics of Solids,https://doi.org/10.1016/j.jmps.2010.12.014
34. Shen G, Rollins J, and Furrer D (1996) " Microstructure Modeling of Forged Waspaloy Discs", Superalloys 1996, Eds. Kissinger RD, Deye DJ, Anton DL, Cetel AD, Nathal MV, Pollock TM, and Woodford DA, TMS, 1996, pp. 613–626.
35. Shen G (2000) "Microstructure Modeling of Forged Components of Ingot Metallurgy Nickel Based Superalloys", Advanced Technologies for Superalloy Affordability, Eds. Chang K-M, Srivastava SK, Furrer DU and Bain KR, TMS, 2000, pp 223–231.
36. Shen G, Kahlke D, Denkenberger R and Furrer D (2001) "Advances in the State-of-the-Art of Hammer Forged Alloy 718 Aerospace Components", Superalloys 718, 625, 706 and Various Derivatives, ed. E.A. Loria, TMS, 2001, 237–247.
37. Swaminathan S, Ghosh S (2006) Statistically equivalent representative volume elements for composite microstructures, Part I: With interfacial debonding. J Comp Mater 40(7):605–621.
38. Tang YT, Panwisawas C, Ghoussoub JN, Gang Y, Clark JWG, Nemeth AAN, McCartney DG, Reed RC (2021) "Alloys-by-design: Application to new superalloys for additive manufacturing", Acta Materialia, Volume 202, 1 January 2021, Pages 417–436.
39. Venkatesh V, Green R, O'Connell J, Cernatescu I, Goetz R, Wong, T, Streich B, Saraf V, Glavicic M, Slavik D, Sampath R, Sharp A. Song B, Bocchini P (2018) "An ICME Framework for Incorporating Bulk Residual Stresses in Rotor Component Design" Integrating Materials and Manufacturing Innovation, Springer, https://doi.org/10.1007/s40192-018-0119-6, 2018.

Applied Calphad to Cast and Wrought Successors to IN718: A Physics-Based Approach with Implications for Phase Stabilities, Precipitation, and Microstructural Modeling

Erwin Povoden-Karadeniz and Nicolas Garcia Arango

Abstract This comparative computational study of Applied Calphad aims at the simulative analysis of mechanical properties and microstructures of recent C&W IN718 successors with a similar application at elevated temperatures.

Predictions are presented for candidates AD-730, and M647. Moreover, Rene88DT and derivatives GH4096, as well as Co-free EXP-G27, which, due to a new C&W technology involving electro-slag remelting continuous directional solidification [1], enter the list of promising C&W candidates, are researched. The different alloying concepts are presented, and their influence on microstructures during thermal processing is shown. Using the assessed MatCalc [2] multi-component Calphad thermodynamics and diffusion mobilities databases "mc_ni" (thermodynamic databases for fcc γ and γ', as well as the diffusion mobilities database used here are added at the end of this paper) for Ni-base superalloys within the system Ni–Fe–Al–Ti–Co–Cr–Mo–Nb–W, phase stabilities of Ni_3Al-based strengthening precipitates, affected by different alloying contributions, are presented. The evaluated solubility behaviour of elements, and the thermodynamic dissolution behaviour of second phases in the Ni-fcc matrix are discussed. By Scheil-Gulliver simulations of as-cast microstructures, segregation trends during casting are described. The kinetic phase evolutions during supersolvus heat treatment conditions are evaluated with nucleation and growth modeling by using MatCalc (version 6.04 rel0.127), predicting particle distributions, sizes, and densities.

We show that fully predictive, computational strengthening trends, employing simple Applied Calphad and computational thermokinetics of precipitation, combined with strengthening models for solid solution strengthening and order strengthening for γ', are a valuable tool for decisions on appropriate thermal operation conditions of IN718 successors.

Keywords C&W · Thermokinetics · Gamma prime · Bimodal

E. Povoden-Karadeniz (✉)
Institute of Materials Science and Technology, TU Wien, Vienna, Austria
e-mail: erwin.povoden-karadeniz@tuwien.ac.at

E. Povoden-Karadeniz · N. G. Arango
Christian Doppler Laboratory Interfaces and Precipitation Engineering, Institute of Materials Science and Technology, CDL-IPE, TU Wien, Vienna, Austria

E. A. Ott et al. (eds.), *Proceedings of the 10th International Symposium on Superalloy 718 and Derivatives*, The Minerals, Metals & Materials Series,
https://doi.org/10.1007/978-3-031-27447-3_22

Introduction—Applied Calphad

The core of Calphad is constituted by

(1) physically appropriate phase descriptions, including crystallographic structure, stoichiometry, and stability range of a phase, coupled with the thermodynamic state variables that define the temperature- and composition-dependent molar Gibbs energy of a phase,
(2) the optimization of model parameters taking into account selective, weighted/assessed input data from all available experimental data that are directly or indirectly connected to the model, i.e. thermodynamic properties and phase diagram data,
(3) the extension from low-order systems to high-order systems, with (optional) non-ideal mixing added to the purely mechanical combinatorial base of all low-order descriptions. In the case of quaternary and higher-order phases that occur, their individual thermodynamic description is indeed further required.

One key towards Applied Calphad for multi-component and multi-phase alloys is an iterative validation process of (1) and (2), which results in the refinement of (3) and the associated improvement of the Calphad-based modeling application, such as the prediction of thermokinetic phase evolution, precipitation strengthening, and recrystallisation.

While conventional thermodynamics tend to remain in a thermodynamic equilibrium consideration, allowing predictions on phase stabilities typically after sufficiently long annealing times, there is a scientific and technological demand on the understanding of transient states in terms of the evolving phases and microstructures in a system, often consisting of multi-component, multi-phase assemblies. For this applied computation case, nucleation and growth and inevitably their rates—time dependence—need consideration. In this context, the simplest way to start, in a mean-field modeling of kinetics, is to implement diffusion mobilities, nucleation theory, and growth kinetics with interfacial energies between an alloy matrix and an evolving second phase as a function of temperature *and* time described by the broken bond approach. In fact, such a modeling setup does not take into account local microstructural effects, such as dislocation dynamics, stacking faults, and deformation twinning, which represent just a few examples of the microstructural entities during real processing of Ni-base superalloys. Nevertheless, in this study, we show how the simplified approach of basic Applied Calphad [3] can be used to predict precipitation trends, and even trends of mechanical properties in different promising candidates acting as successor C&W developments to conventional IN718.

Studied Material

We split our study into two sections—Computations with materials where experimental data on precipitation are available—these are the alloys Rene88DT and

Table 1 Compositions of different IN718 successors

	Cr	Co	W	Mo	Ti	Al	Fe	Nb	*C*	*B*	*Zr*	Ni
Rene88DT	16	13	4	4	3.7	2.1	–	0.7	*0.03*	*0.015*	–	bal
GH4096	16	13	4	4	3.8	2.2	–	0.7	*0.05*	*0.015*	*0.05*	bal
AD-730	16	8.5	2.7	3	3.5	2.3	4	1.1	*0.015*	*0.01*	*0.03*	bal
M647	16	20	4	2.5	1.8	3.3	1	2.7	–	*0.015*	*0.03*	bal
EXP-G27	15	–	–	4	1.8	2	15.3	3.7	*0.03*	*0.005*	*0.03*	bal

GH4096—and fully predictive computations for those alloys where no experimental precipitation data are available. We selected AD-730, M647, and EXP-G27 for this purpose. The nominal compositions of studied alloys are presented in Table 1 (wt.%).

In part, quite different metallurgical strategies are evident between these superalloys. While Rene88DT and GH4096 represent the case of relatively balanced alloying of Cr and Co, as well as W, Mo, and Ti, with a bit less Al, in AD-730 considerably less Co is used, and instead Nb alloying is a bit increased. Its Fe amount is significant. AD-730 thus seems to represent a somewhat cheaper C&W opportunity—in the context of reduced alloying complexity and cost savings—with lack of both Co and W in the nominal composition, but considerable Fe concentration, and significantly increased Nb (which can mentionably represent an economic and political issue of its availability). The Co-free EXP-G27 variant is of particular interest due to its applicability in nuclear radiation settings, as well as the strong cost saving and humanity in the raw materials supply.

Here, we do not consider the minor elements wit italic letters in Table 1, C, B, and Zr.

Modeling

Computational Thermodynamics

Development of mc_ni Multi-component Database for Applications in Cast and Wrought Ni-Base Superalloy

Our multi-component Ni-base system of interest is Ni–Cr–Co–W–Mo–Ti–Al–Nb–Fe, a nonary system, which thus seems to be awfully complex, when setting up the appropriate Calphad database "mc_ni" (mc stands for MatCalc) for equilibrium phase stability computations. However, Calphad multi-component databases are *simply* based on the merging of ternary subsystems (an exception representing systems with a genuine quarternary phase), which strongly reduces the required model parameters of intermetallic compounds and alloying elements' interactions within the alloy matrix fcc (face-centered cubic) phase. Moreover, in particular for ternary combinations with the minor alloying elements W, Mo, Ti, Al, Fe, and particularly Nb, just Ni-containing binaries and ternaries are relevant. Unary descriptions are adopted from compiled SGTE (Scientific Group Thermodata Europe) data by Dinsdale 1991. These considerations/reliable simplifications lead to a reduction of the overall huge number of subsystems by recombinations to a manageable number of required Calphad sub-assessments of 35. These are listed in Table 2.

The most important strengthening phases in Ni-base superalloys, Ni_3Al-based γ', and Ni_3Nb-based γ'' (in the IN718 reference alloy) show chemical ordering, which in a physically correct modeling needs special attention, introducing ordering parameters in a "split model" (energetics of an fcc-phase with ordered site occupancies of substitutional elements among different sublattices is in each calculation step compared with the energetics of a disordered fcc-phase with random occupancies, and the same model applies for both) for chemically ordered phases. However, this modeling is only relevant when particular substitutions within the crystal lattices of the respective phases are studied and their effects within the framework of Ni-base processing and Ni-base properties are clear, which is not the focus of this study, and is open for discussion. This led us to the decision about another simplification in the database—we simplify the modeling of γ' and γ''; instead of ordering parameters and phase descriptions which allow for the disordered/ordered comparison described above, we view γ' and γ'' as simple intermetallic solid solution phases with their individual models that are different than the model of the fcc-alloy phase. Our modeling leads to almost the same molar Gibbs energies and chemical potentials as for the computations with split modeling. Moreover, when the computational costs are lower, the numerical stability is higher for the solid solution model used in Applied Calphad than for the split model.

Table 2 Assessed subsystems for mc_ni

NiCr	NiCo	NiW	NiMo	NiTi	NiAl	NiFe
NiCoCr	NiCoW	NiCoMo	NiCoTi	NiCoAl	NiCoFe	NiCoNb
NiCrW	NiCrMo	NiCrTi	NiCrAl	NiCrFe	NiCrNb	
NiWMo	NiWTi	NiWAl	NiWFe	NiWNb		
NiMoTi	NiMoAl	NiMoFe	NiMoNb			
NiTiAl	NiTiFe	NiTiNb				
NiAlFe	NiAlNb					
NiFeNb						

What Do We Learn from Computational Equilibrium Thermodynamics Using mc_ni—Implications on Solvi, Subsolvus, and Supersolvus Solution Treatment, and Phase Fractions Trends as a Function of Temperature

Understanding of equilibrium phase stabilities and metastabilities (γ''!) is an important prerequisite at the beginning of any alloy design. In the simple case of a second intermetallic phase occurring in an alloy matrix, phase fraction plots as a function of temperature tell the researcher the temperature region of steady volume fractions of the intermetallic phase (i.e. the maximum plateau of the respective phase fraction), in the first approximation representing the growth region of a precipitate. In the temperature region of descending phase fraction curve, a phase may start to coarsen as a function of time, and above its solvus temperature the phase dissolves within the matrix. While in the case of strengthening γ' and γ'' the temperature region above the solvus temperature leads to rapid material degradation in operation due to a drop of yield strength, the information of solvus temperature is during processing highly relevant for the knowledge of single-phase fcc-Ni, which represents the starting (solution treatment) of any design of a superalloy with typically homogeneously distributed nano-particles in the matrix by quenching and aging. The highest temperature limit of solution treatment is typically formed by the solidus temperature, where incipient melting of the material is observed. When a heterogeneous phase such as Ni_3Nb-Delta at grain boundaries is observed, the question arises whether it may be beneficial to maintain a "solution" temperature or intermediate heat treatment step before the main aging right below the delta-solvus. Even a subsolvus treatment concerning γ' is possible. This can lead to larger particles pinning grain boundaries and thus influencing recrystallisation and controlling grain growth, at the cost of less supersaturation for subsequently forming smaller precipitates in the scope of further heat treatment.

Microsegregation at Casting—Relevance for Homogenisation Issues—The Role of Proper Matrix Phase Thermodynamics in the Liquid and Solid States

Thermodynamically, microsegregation occurring during casting in accordance to Gulliver (1913) and Scheil 1942 [4] assumes that no diffusion takes place in the solid and that solute redistribution in the liquid is infinitely fast. The corresponding differential Eq. (1) and its analytical solution (2) read

$$(C_L - S_L)df_s = (1 - f_s)dC_L, \tag{1}$$

and

$$C_s = kC_0(1 - f_s)^{k-1}. \tag{2}$$

Concentrations in (1) and (2) are related solely to thermodynamics, in particular the chemical potentials in the liquid and the solid. When these are correct, the description of microsegregation by the Scheil-Gulliver approach will be internally consistent.

Thermokinetic Modeling and Simulation

Computational Method

Thermokinetic modeling of precipitation involves the transient nucleation rate, J, [5]

$$J = N_0 Z\beta exp\left(\frac{-G'}{k_B T}\right)exp\left(\frac{-\tau}{t}\right) \tag{3}$$

where N_0 is the number of potential nucleation sites, Z is the Zeldovich factor, β is the atomic attachment rate, G' is the critical nucleation energy, k_B is the Boltzmann constant, τ is the incubation time, t is time, and T is temperature. The critical nucleation energy, G', strongly determines J. It is commonly derived from the extremum value of the nucleation energy

$$\Delta G_{nucl}^{CNT} = \frac{4}{3}\pi\rho^3\Delta G_{vol} + 4\pi\rho^2\Delta G_{intf}, \tag{4}$$

where ΔG_{nucl}^{CNT} is the CNT free energy of nucleus formation.

In developing the Generalized Broken Bond (GBB) model, Sonderegger and Kozeschnik [6] utilize the direct relation between the specific interface energy and the solution enthalpy, ΔH_{sol}. In their treatment of the planar, sharp interface energy, γ_0, this quantity is written as

$$\gamma_0 = \frac{n_S \Delta z_{S,eff}}{N \Delta z_{L,eff}} \Delta\Delta H_{sol}, \tag{5}$$

where n_S is the number of atoms on unit interfacial area and N is the Avogadro number. The quantities $z_{S,\text{eff}}$ and $z_{L,\text{eff}}$ are the effective number of broken bonds across the interface and the effective coordination number, respectively. The structural factor, $z_{S,eff}/z_{L,eff}$, is obtained, after averaging over several interfacial directions, with approximately 0.329 for fcc crystal structure and 0.328 for bcc.

In follow-up work, the same authors investigate the impact of interfacial curvature [7] and temperature [8] on the effective interface energy, γ, of a small precipitate (nucleus), with the result that

$$\gamma = \alpha(\rho)\Delta\beta(T)\Delta\gamma_0(T, X_i), \tag{6}$$

where the functions $\alpha(\rho)$ and $\beta(T)$ are given as

$$\alpha(\rho) = 1 - \frac{20}{11}\cdot\frac{r_c}{\rho} + \left(\frac{45}{44} - \frac{1}{22}ln2\right)\cdot\left(\frac{r_c}{\rho}\right)^2 + \frac{1}{2}\cdot ln\left(\frac{\rho}{r_c}\right)\cdot\left(\frac{r_c}{\rho}\right)^2 \tag{7}$$

with $r_c = 0.3 \cdot r_1$, r_1 being the nearest neighbor atomic distance, and

$$\beta(T) \approx 8.4729\cdot\left(\frac{T}{T_c}\right)^6 - 26.691\cdot\left(\frac{T}{T_c}\right)^5 + 32.717\cdot\left(\frac{T}{T_c}\right)^4$$
$$- 17.674\cdot\left(\frac{T}{T_c}\right)^3 + 2.2673\cdot\left(\frac{T}{T_c}\right)^2 - 0.09\cdot\left(\frac{T}{T_c}\right) + 1.000476320. \tag{8}$$

Combining particle growth kinetics by the SFFK (Svoboda-Fischer-Fratzl-Kozeschnik) [9, 10] with CNT including modeling of interfacial energy, in a Kampmann-Wagner framework of numerical simulation (thermodynamic equilibrium is calculated with MatCalc for each time step and in a "phase loop" the precipitation behaviour of all phases is examined), the precipitation kinetics without further calibration in the simulation setup is *in essence predictive,* and the predictive power strongly depends on the quality of the thermodynamic and diffusion mobilities databases used. Here, we test the feasibility of predictive trends from alloy cases with validated precipitation.

Previous Findings—Udimet 720 LI as Example for the Appearance of Tertiary γ'—Their Explanation by Simple Thermodynamic and Diffusion Considerations

Radis et al. [11] have described the underlying mechanism of well-known bimodal to multimodal γ' distribution, which occurs during heat treatment of many Ni-base superalloys and for various heat treatments, based on their findings in Udimet 720 LI. When cooling the material below the γ' solvus temperature, the difference of volume free energy between matrix and precipitate increases, until the critical nucleation barrier is overcome, and relatively large nuclei start to form at a relatively low nucleation rate close to the thermodynamic γ' solvus temperature. Within the diffusion field of the growing particle, diffusional transport of particle-forming elements from the matrix, due to the element diffusion mobilities as a function of cooling rate, takes place. Eventually, in the regions of the depleted matrix, nucleation stops. At further cooling, the diffusion fields continuously shrink, and in the remaining zones that are no longer affected by the diffusional element transport to the first growing particles, supersaturation increases rapidly, leading to a second wave of precipitation. This scenario means that bi- to multimodal precipitation is governed by thermodynamics and their descriptions of the matrix phase and the precipitate, as well as diffusion mobilities, and database quality is responsible for the soundness of simulated particle distributions in respective Ni-base superalloys. Over decades, the role of tertiary γ' for strengthening stood for debate. The precipitate distribution has in fact important consequences on material properties, including yield strength, and it has been very recently proposed by Zhang et al. [12] that yield strength and ultimate strength clearly increase by several hundreds of MPa when the average particle size decreases by an order of magnitude. This eventually indicates that the appropriate thermodynamic Calphad Ni-base and diffusion mobilities have some direct consequence on the strengthening behavior.

Mechanical Modeling

For the modeling of strengthening in Ni-base superalloys, the comprehensive work by Ahmadi et al. [13] is implemented in MatCalc, which allows for the evaluation of different contributions to the yield strength, such as order strengthening (which takes into account the antiphase boundary energy created by the shearing of a γ' particle by a dislocation pair) and solid solution strengthening, in its classical formulation [14].

Results and Discussion

Computational Thermodynamics

Temperature-Dependent Phase Fractions

It should be noted that one of the authors of this paper is the developer of all MatCalc databases, its beginning dating back into the year 2009. In the following, we postulate the acceptable quality of mc_ni, since the assessment of it goes beyond the frame of this contribution. An extensive comparison of thermodynamic computations and comparisons among different databases can be found elsewhere [15]. Figure 1 shows that computed equilibrium solvus temperatures for γ' match experimental values and vary well for more than 50 tested superalloys.

As noted above, a comparison between the equilibrium phase fractions—temperature computations for the studied materials—is giving first indications on the expected differences of suitable process heat treatment conditions and operation temperatures (different solvus temperatures and T-dependent phase fraction evolution of γ' in the different studied materials). These are plotted in the figure series 2 (a) to (f). Carbon, boron, and Zr are neglected in the computations.

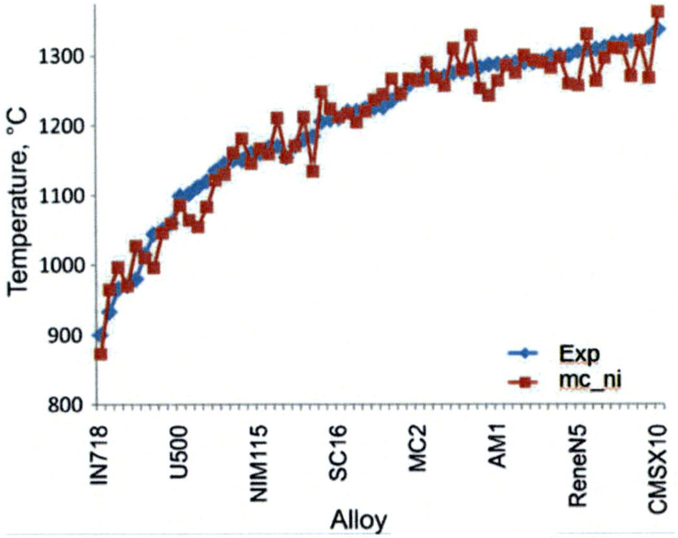

Fig. 1 γ' solvi, experimentally versus simulated. Each data point represents a specific Ni-base superalloy. More than 60 of them are arranged as function of increasing solvus temperature. The solid connection lines between data points are just guides to the eye

The computed thermodynamic equilibrium phases in Rene88DT (Fig. 2a) comprise, in addition to liquid, γ phase and γ' also topologically close-packed (TCP) μ-phase (MU_PHASE) and Ni_7Ti-based eta-phase. Eta-phase is however not prominent. It was also reported in Rene88DT in an early experimental study on those alloys [16]. In the same study, also the appearance of mu-phase after long-term aging was reported. The thermodynamic equilibrium phase fraction of the mu-phase is thus reasonable, but in the alloy processing the phase will not appear. The situation changes at elevated temperatures during operation, where topologically close-packed structures, mu-phase, and (minor amounts of) P-phase may evolve. This may indeed indicate a critical aspect to be considered for the application of the alloy. Long-time creep testing is needed for further discussion. Minor amounts of Cr-rich bcc-phase at low temperatures do likely not play an important role in both the strengthening or embrittlement of the material.

While AD-730 (Fig. 2b) reveals a very similar solvus temperature of γ' as Rene88DT, lying at around 1100 °C, this material seems to be a little less prone to μ-phase formation than Rene88DT (13 mol.% max. phase fraction of μ-phase, compared to around 16 mol.% in Rene88DT), and also to the eta-phase fraction (even less). With the comparable γ' solvus of AD-730 with Rene88DT, but less alloying leading to less tendency for other possibly detrimental phases, less Co, and only a little higher Nb than Rene88DT, AD-730 is indeed an interesting alternative. In fact, the questionable role of lower Co for microstructure issues—homogeneity of the microstructure—is not discussed here.

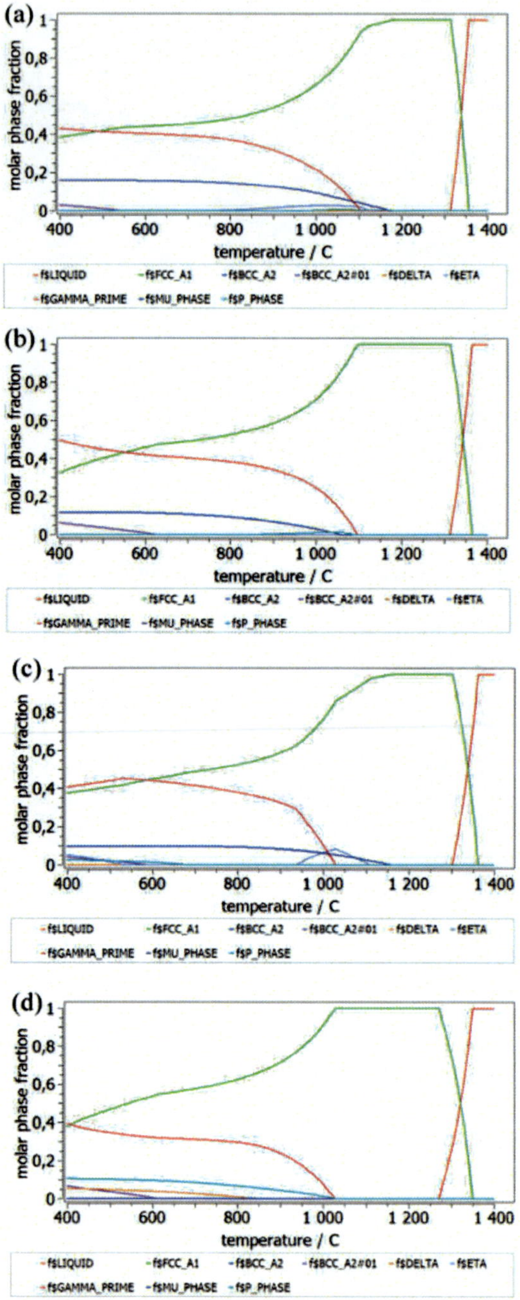

Fig. 2 Computed equilibrium phase fractions versus temperature, Rene88DT (**a**), AD-730 (**b**), and M647 (**c**). Note that equilibrium phase fractions of GH4096 are not shown due to its very small alloying difference from Rene88DT. GAMMA_PRIME denotes the γ' phase. BCC_A2#01 is Cr-rich bcc-phase

The third alloy of the present comparison, μ-phase, is even less; the γ' solvus temperature is around 15 °C higher. When one keeps the small eta-phase stability range around 1000 °C in the process out of consideration, suggested purely by Calphad computation, this alloy seems to be the most promising one in terms of its expected strengthening potential, creep performance, and high-temperature limitation of its use. Indeed, the microsegregation behavior due to elevated Nb-content needs to be discussed, which is the topic of the next section.

Finally, in our phase stabilities comparisons by Calphad equilibrium computations, we shift to the Ni–Fe alloyed material. Little is known about phase stabilities of EXP-G27, however, computation indicates that no potential eta-phase "problem" arises, due to missing Co (which is an eta-phase stabiliser). In terms of the long-term behavior of the material, one needs to critically observe the appearance and evolution of Ni–Cr–Mo P-phase (around 10 mol.% max.), as well as μ-phase (complex Cr–Mo-Ni–Co–W based). Both phases are TCP and are expected to behave similarly, and there is no indication whether one of them is better in terms of slower growth or lower embrittlement. The high Nb-alloying is an interesting alloying concept, since it stabilizes a *small* amount (max. 4.5 mol%.) of Ni*3*Nb delta-phase in the microstructure, which can be indeed used as grain boundary pinner. At the same time, embrittlement due to uncontrolled growth of delta-phase during operation is avoided. The γ' solvus temperature of EXP-G27 lies in the range of M647.

Next, we shortly compare the potential of metastable γ'' phase formations in test simulations removing TCP phases and delta-phase. It appears that γ'' does not form in any of the studied alloys.

Comparing the phase compositions of fcc-Ni and γ' phase from the calculation at a typical application temperature, 800 °C, we obtain information about the partitioning of different elements in different superalloys, as shown in Table 3.

Partitioning to γ phase is strongest for Cr (or Fe), followed by Co and Mo, whereas partitioning to γ' is strongest for Ti, followed by Nb and Al. W is only slightly enriched in γ. Due to the strong alloying similarity between Rene88DT and GH4096, the partitioning of these two alloys is also almost the same.

Interestingly, Co, even though it prefers the matrix phase, is also dissolved quite significantly in γ', where it also, as in the matrix, will play a role in the stacking fault energy. Since Nb strongly tends to dissolve in γ', remaining segregation dynamics after γ' formation will be likely affected. Actually, nucleating γ' may act as a "competing trap" to interfaces for the Nb segregation. This consideration may be taken into account when a subsolvus heat treatment step is planned prior to the main aging event. Microsegregation during casting is the topic of the next section.

Microsegregation

Evaluating microsegregation gives some indications on different relevance of alloyed elements for chemically heterogeneous starting conditions for the processing subsequent to casting and the necessity/feasibility of a homogenisation step. Tailoring of homogenisation is thereby supported by finite element diffusion simulations (not

Table 3 Equilibrium partitioning of elements between fcc-Ni γ phase and γ', wt.% at 800 °C

	$Cr_\gamma/Cr_{\gamma'}$	$Co_\gamma/Co_{\gamma'}$	$W_\gamma/W_{\gamma'}$	$Mo_\gamma/Mo_{\gamma'}$	$Ti_\gamma/Ti_{\gamma'}$	$Al_\gamma/Al_{\gamma'}$	$Fe_\gamma/Fe_{\gamma'}$	$Nb_\gamma/Nb_{\gamma'}$	$Ni_\gamma/Ni_{\gamma'}$
Rene88DT	27.63/1.32	17.26/5.3	1.6/0.69	3.51/0.58	0.39/11.53	1.36/9.96	–	0.04/1.11	48.2/69.5
GH4096	28.17/1.33	17.5/5.31	1.62/0.69	3.55/0.62	0.37/11.46	1.38/10.16	–	0.04/1.08	47.4/69.3
AD-730	28.49/1.52	11.56/3.37	1.03/0.56	2.7/0.44	0.31/10.85	1.16/10.4	6.57/0.44	0.06/1.59	48.1/70.8
M647	26.26/2.42	25.95/7.94	1.07/0.26	1.82/0.92	0.11/5.72	3.4/13.29	1.55/0.01	0.12/4.36	39.7/65.0
EXP-G27	23.1/2.4	–	–	3.1/0.91	0.14/6.48	1,56/10.0	22.0/2.0	0.49/6.13	49.5/72.1

treated here). In the following, we compare the computation results of microsegregations of different elements after casting, i.e. the chemical difference between the first solid fcc-structured matrix phase and the last crystallising liquid before complete solidification. Moreover, phase fractions of primary particles can be assessed. Evaluation results of element partitioning between liquid and γ alloy-matrix are given in Table 4.

Eta-phase is predicted to be a quite prominent primary phase, which—in addition to the Ni_3Ti-base—dissolves Cr and particularly Nb in its lattice, at temperatures above possible stabilisation of delta-Ni_3Nb. Al and Ti remain to accumulate in the residual liquid, which are forming the centers of the interdendritic zone, even though they are at the same time dissolved in primary eta-phase at the high temperatures of completed crystallisation after the casting. Also, Nb tends to accumulate in the residual liquid zones after casting. Fe enriches in the crystallising γ phase. Computation suggests the requirement of a homogenisation step prior to further processing for all studied candidates.

The decrease of liquid fraction and increase of the γ′ phase fraction is increased as a function of cooling, which is shown in Fig. 3a–d for four of the studied alloys. The strongest stabilisation of primary eta-precipitates occurs during the casting of EXP-G27 (see Fig. 3d; with a higher amount than 6 mol.%, its influence of further processing and materials response is still to be researched in detail).

Thermokinetic Precipitation Simulation—from Validated Cases to Prediction

For the time being, in the thermokinetic simulations we assume that the material has been fully homogenized after casting, which is the ideal case of erasing effects that were described in the previous section. The first thermokinetic simulations focus on the nucleation and growth of γ′ at the cooling stage from the homogenization temperature. To validate the simulations, experimental results on the diameter of secondary γ′ at various cooling rates, obtained by Wlodek et al. [16], are compared to thermokinetic predictions.

The mean sizes of secondary and tertiary γ′ as a function of ageing temperature in alloy GH4096 are simulated and also compared to the experimental results published in 2020 by Li et al. [17].

Case Studies with Experimental Input Data

The inverse relationship between cooling rate and size of secondary γ′ for 4 different alloys (RENE 88DT, AD-730, M647, and EXP-G27) is predicted and, in the case of RENE 88DT, compared to experimental values (Wlodek et al. [16]) in Fig. 4. The applied heat treatment consists of an initial holding stage at 1120 °C followed by a

Table 4 Element partitioning between liquid and γ alloy-matrix during casting

	Cr_{γ}/Cr_{liq}	Co_{γ}/Co_{iq}	W_{γ}/W_{iq}	Mo_{γ}/Mo_{iq}	Ti_{γ}/Ti_{iq}	Al_{γ}/Al_{iq}	Fe_{γ}/Fe_{iq}	Nb_{γ}/Nb_{iq}	Ni_{γ}/Ni_{iq}
Rene88DT	16.43/22.9	13.94/0.72	1.3/0.56	1.52/2.23	3.42/9.16	4.62/8.22	–	0.23/2.28	58.54/47.5
AD-730	16.84/16.4	10.14/0	0.91/0.23	1.31/1.35	3.08/9.82	4.52/9.6	4.98/4.34	0.29/4.3	57.93/53.3
M647	16.1/20.05	21.72/7.5	1.12/0.07	0.78/1.11	1.66/2.79	6.55/14.14	1.42/0.6	0.88/8.24	49.77/45.5
EXP-G27	17.08/18.1	–	–	2.33/1.71	1.0/6.54	3.65/5.07	19.64/8.14	0.8/7.86	55.5/52.62

Fig. 3 Computed increasing solid Scheil phase fractions ("PHASE_S" in MatCalc) and decreasing liquid fraction as function of continuous casting simulation for alloy **a** RENE88DT, **b** AD-730, **c** M647, and **d** EXP-G27. The largest primary precipitate fraction of eta-phase is expected for EXP-G27. Eta-phase is forming in all alloys during casting, whereas a small amount of primary γ' phase is only predicted for AD-730

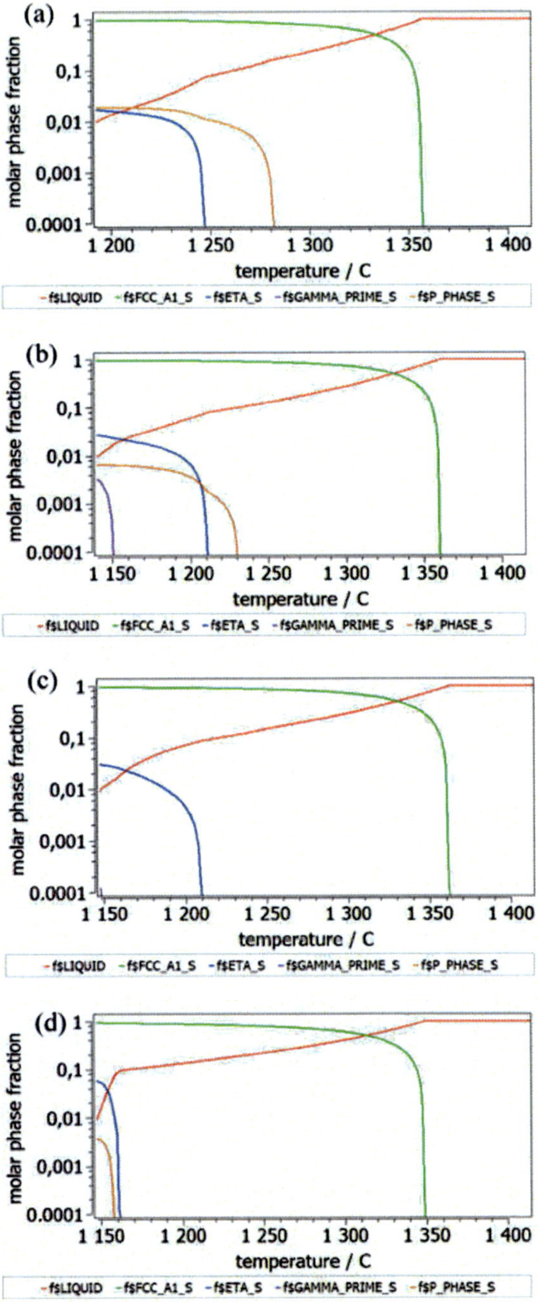

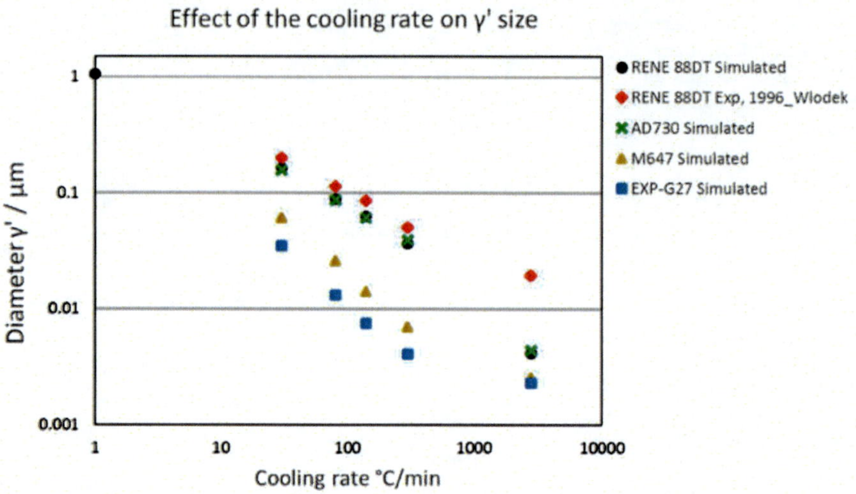

Fig. 4 Effect of the cooling rate on secondary γ′ size. Experimental results from RENE 88DT and simulated values from RENE 88DT, AD-730, M647, and EXP-G27 are included

cooling step at different rates (1 °C /min, 30 °C /min … 2800 °C /min). No primary γ′ is assumed in the microstructure as the cooling starts from 1 h at a supersolvus temperature, where primary γ′ is considered to be fully dissolved.

The simulated results predict a very similar behaviour between RENE 88DT and AD-730, undermining the impact Co has on the nucleation and growth of secondary γ′. Furthermore, for higher Nb contents, as in M647 and EXP-G27, a large effect in the precipitation of secondary γ′ is predicted, as the mean size of the precipitated γ′ at the same cooling rates is strongly reduced.

Following the results of 2020 by Li et al. [17] for the alloy GH4096, an age-hardening treatment where sizes of both secondary and tertiary γ′ were characterized. As explained for the case of Udimet 720 Li (Sect. 3.2.2, from [11]), the second wave of γ′ (tertiary γ′) nucleates during the latest part of the cooling; this event can be observed in the simulation when analysing the precipitate size distribution at two different stages of the process, as seen in Fig. 5.

The simulation on the nucleation and growth of secondary and tertiary γ′ is validated by comparing the mean diameter of the precipitates at different aging temperatures (700 °C, 730 °C, 760 °C, and 790 °C) for 8 h with results published in Li et al. [17], shown in Fig. 6. The main feature experimentally observed of secondary γ′ was its coarsening resistance to the applied heat treatments, an effect also seen in the simulations. On the contrary, the population of smaller γ′ displayed both, through experiments and simulations, a size increase after the applied age-hardening treatment, and the growth was more prominent at higher temperatures.

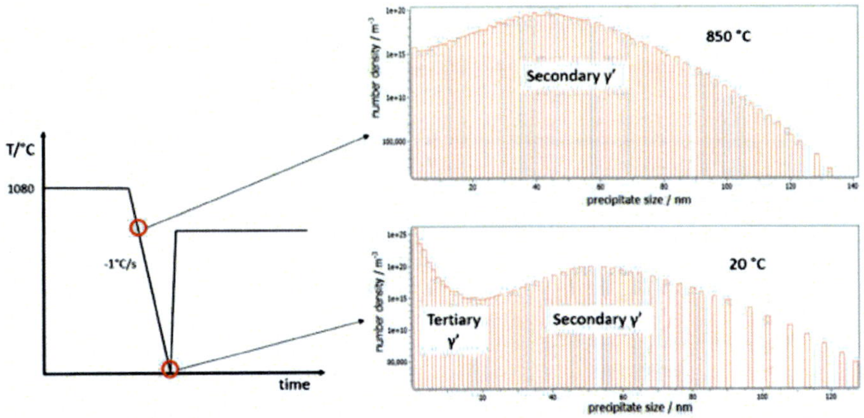

Fig. 5 Precipitate size distribution at two different stages of the cooling step (850 °C and 20 °C), initial precipitate population, termed as secondary γ′ nuclei at higher temperatures, and the second precipitate wave (tertiary γ′) is distinguished at the end of the cooling

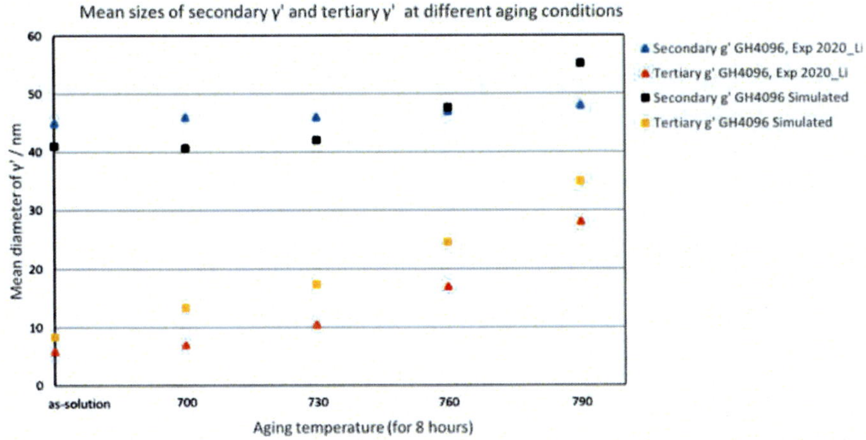

Fig. 6 Simulated and experimental results on the mean size of secondary and tertiary γ′ after aging for 8 h at different temperatures; the initial condition is also shown and labeled as "as-solution"

Fully Predictive Applied Calphad for New Alloy Design

The present simulation approach can be used to predict the response of newly designed alloys to heat treatments. In Fig. 7, the precipitate size distribution at the end of the cooling step for alloys GH4096, EXP-G27, AD-730, and M647 is displayed; in all 4 cases, both secondary and tertiary γ′ are present, although their sizes and number densities differ.

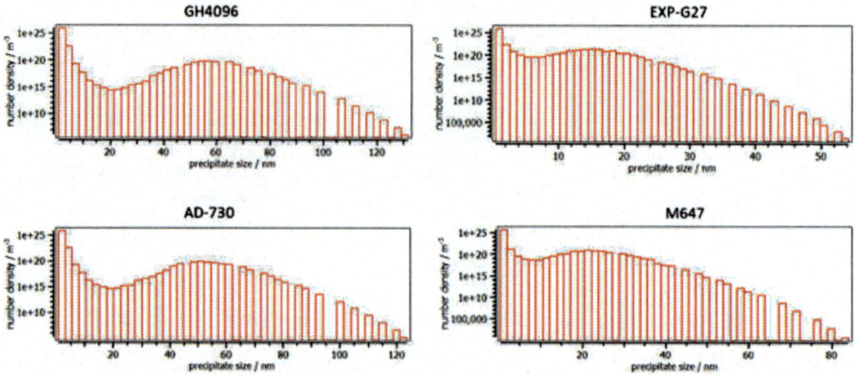

Fig. 7 Precipitate size distribution at the end of the cooling step for 4 alloys, GH4096, EXP-G27, AD-730, and M647. Both secondary and tertiary γ' are predicted in the 4 cases

Mechanical Properties

Validated Yield Strength Evaluation by Experiments

A preliminary analysis on the mechanical properties is done after comparing the yield strength of the GH4096 alloy obtained by Li et al. [17], as a function of the aging temperature to the total yield strength contribution from the precipitates (Fig. 8). To model the influence of γ' to the strength of the material, two different volumetric misfits were tested (1 and 2%—both lying in fact within the realistic range). Assuming that the total yield strength contribution from precipitates is the most relevant term to the strength of the material (around 1 order of magnitude larger than the solid solution contribution to the yield strength), the effect of increasing aging temperature is not huge on the mechanical properties of the material, given the high coarsening resistance of the secondary γ' precipitates. Simulations considering a volumetric misfit of 1% follow this trend more accurately.

Prediction of Mechanical Properties

A similar approach can be used to predict the response of newly designed alloys to various heat treatments and the effect they have on the strength of the alloy, introducing a volumetric misfit of 1%; yield strength contribution from precipitates in alloys AD-730, GH4096, and M647 as a function of aging temperature is shown in Fig. 9.

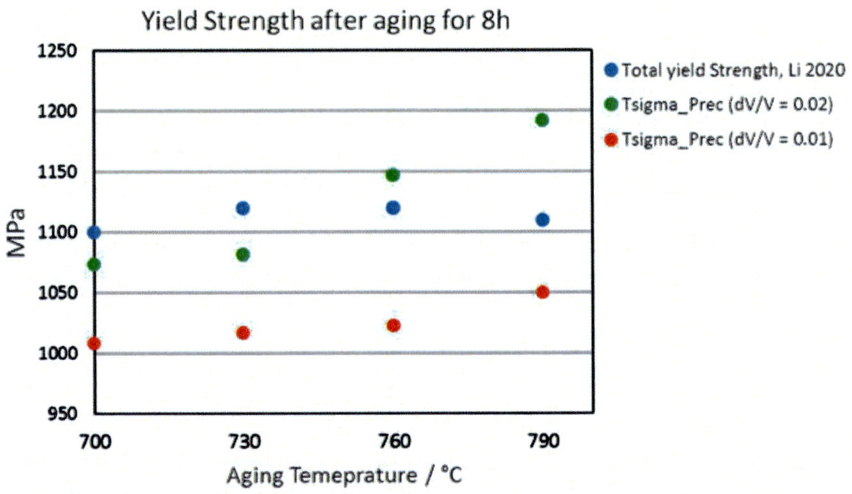

Fig. 8 Total yield strength contribution from γ' precipitates for two different volumetric misfits 1% and 2%, compared to the experimental values of total yield strength as a function of aging temperature

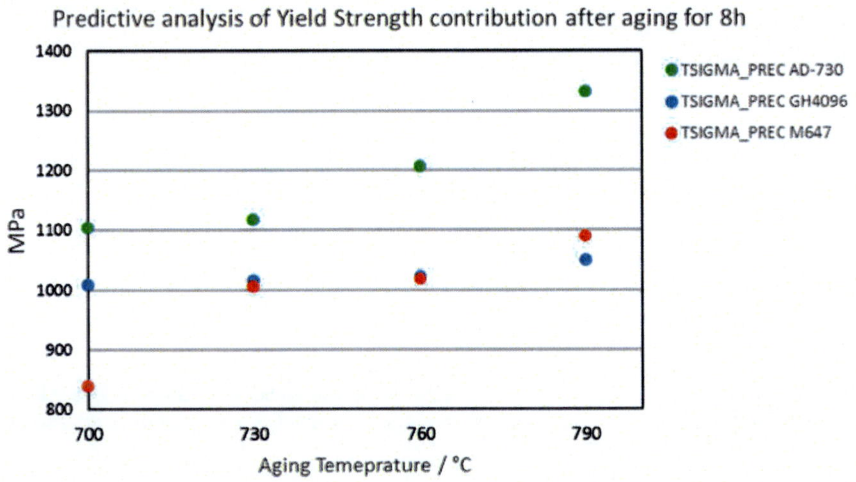

Fig. 9 Yield strength contribution from γ' precipitates in alloys AD-730, GH4096, and M647 as a function of aging temperature

Conclusion and Outlook

The kinetic stabilisation of topologically close-packed structures in the studied alloys requires experimental investigation and confirmation by long-term aging simulations.

This is also required for the primary precipitates being suggested by the Scheil calculation. Moreover, in a more comprehensive kinetic precipitation simulation, through process modeling, combining the casting step, the homogenisation and subsequent aging will be performed. In fact, a more complete comparative experimental validation of mechanical properties, which is not available to date, will consolidate the computational prediction. Moreover, the kinetic simulation setup can be designed to be more complex, in particular by including a potentially diffuse interface to model for it being available [18]. This will shift the nucleation start temperatures, however, currently there is a lack of experimental data, and thus calibration in the modeling will not be very reliable. Our next modeling and simulation step will involve particularly heterogeneous recrystallisation.

By deforming the material, new dislocations appear in the microstructure; these linear defects interact with each other creating substructures that evolve with time into new recrystallized grains. To simulate the process of discontinuous dynamic recrystallization in Ni-based superalloys, the multi-class grain distribution model included in MatCalc will be used. The onset and evolution of recrystallization consider dislocation kinetics, strain, strain rate, the initial grain configuration, temperature, and subgrain growth, which combined with precipitation kinetics allow to model and predict the microstructure evolution of the selected IN718 successor alloys.

References

1. Fu R, Chen XC, Feng D. The microstructure and hot deformation behavior of ESR-CDS Rene 88 DT alloy. J. Aeronaut. Mater. 31 (3), 2011, 8.
2. E. Kozeschnik, Mean-Field Microstructure Kinetics Modeling, Encycl. Mater. Met. Alloys. 4, 2022, 521–526.
3. P. Retzl, Y.V. Shan, E. Sobotka, M. Vogric, W. Wei, E. Povoden-Karadeniz, E. Kozeschnik, Progress of physics-based mean-field modeling and simulation of steel.
4. (2022) Berg Huettenmaenn Monatsh, 167 (1), 2022. 15–22.
5. E.Scheil, Z.Metallkd. 34, 1942, 70–72.
6. E. Kozeschnik, Modelling Solid-State Precipitation, Momentum Press, LLC, New York, New York, 2013. https://doi.org/10.5643/9781606500644.
7. B. Sonderegger, E. Kozeschnik, Generalized Nearest-Neighbor Broken-Bond Analysis of Randomly Oriented Coherent Interfaces in Multicomponent Fcc and Bcc Structures, Metall. Mater. Trans. A Phys. Metall. Mater. Sci. 40, 2009, 499–510.
8. B. Sonderegger, E. Kozeschnik, Size dependence of the interfacial energy in the generalized nearest-neighbor broken-bond approach, Scr. Mater. 60, 2009, 635–638.
9. J. Svoboda, F.D. Fischer, P. Fratzl, E. Kozeschnik, Modelling of kinetics in multi-component multi-phase systems with spherical precipitates I: Theory, Mater. Sci. Eng. A. 385, 2004, 166–174.
10. E. Kozeschnik, J. Svoboda, P. Fratzl, F.D. Fischer, Modelling of kinetics in multi-component multi-phase systems with spherical precipitates II: Numerical solution and application, Mater. Sci. Eng. A. 385, 2004, 157–165.
11. R. Radis, M. Schaffer, M. Albu, G. Kothleitner, P. Pölt, E. Kozeschnik, Multimodal size distributions of γ‘ precipitates during continuous cooling of UDIMET 720 Li, Acta Mater. 57 (19), 2009, 5739–5747.

12. H. Zhang, Y. Li, T. Ma, T. Chang, P. Zhang, X. Fang, K. Huang, Tailoring of nanoscale γ' precipitates and unveiling their strengthening mechanisms in multimodal nickel-based superalloy GH4720Li, Mater. Charact. 188, 2022, 11918.
13. M.R. Ahmadi, E. Povoden-Karadeniz, K.I. Oksuz, A. Falahati, E. Kozeschnik, A model for precipitation strengthening in multi-particle systems, Comput. Mater. Sci. 91, 2014, 173–186.
14. M.R. Ahmadi, E. Povoden-Karadeniz, L. Whitmore, M. Stockinger, A. Falahati, E. Kozeschnik, Yield strength prediction in Ni-base alloy I718plus based on thermo-kinetic precipitation simulation, Mater. Sci. & Eng. A 608, 2014, 114–122.
15. Ritter, NC; Sowa, R; Schauer, JC; Gruber, D; Goehler, T; Rettig, R; Povoden-Karadeniz, E; Koerner, C; Singer, RF, Effects of Solid Solution Strengthening Elements Mo, Re, Ru, and W on Transition Temperatures in Nickel-Based Superalloys with High γ'-Volume Fraction: Comparison of Experiment and CALPHAD Calculations, Metal. Mater. Trans. A – Physical Metal. Mater. Sci. 49A (8), 2018, 3206–3216.
16. S.T. Wlodek, M. Kelly, D.A. Alden, The structure of RENE 88 DT, Superalloys 1996, Eds. R.D Kissinger et al., the Minerals, Metals & Materials Society, 1996, 129–136.
17. F. Li, Y. Bai, L. Meng, R. Fu, Y. Zhong, J. Du, F. Yin, D. Feng, Impact of aging heat treatment on microstructure and mechanical properties of a newly developed GH4096 disk superalloy, Mater Charact. 161, 2020, 110175.
18. B. Sonderegger, E. Kozeschnik, Interfacial energy of diffuse phase boundaries in the generalized broken-bond approach, Metall. Mater. Trans. A Phys. Metall. Mater. Sci. 41 (2010) 3262–3269. https://doi.org/10.1007/s11661-010-0370-8.

Multi-variate Process Models for Predicting Site-Specific Microstructure and Properties of Inconel 706 Forgings

Nishan M. Senayake, Tiffany A. Dux, and Jennifer L. W. Carter

Abstract High throughput measurements of structure from simulations and microstructural analysis were developed and combined into an adaptable data-analytics processing–structure–property modeling framework to gain insight into the design envelope for forgings of Inconel 706. We highlight how thermal profiles from finite element simulations (DEFORM) can establish time–temperature boundary conditions for CALPHAD predictions of the combined $\gamma' \gamma''$ precipitate distribution in Inconel 706 (structure predictors). Experimental observations of these precipitate distributions allow for the tailoring of the CALPHAD interfacial energy. In this manner, a $25\times$ reduction in the number of physical observations of γ' and γ'' distribution (78 to 4) results in site-specific processing–structure, and processing–property models of forged parts with 80% predictive power. Additionally, the gradient boost modeling provides an opportunity to interpret feature importance in structure–property models to provide engineers with design insights for future development efforts. The adaptable framework would enable future DEFORM and CALPHAD simulations to be added to the dataset so that engineers can interpolate within the existing experimental dataset.

Keywords Inconel 706 · CALPHAD · Precipitate distributions · Data-enabled process modeling

Introduction

The performance of superalloy forgings hinges on the careful design of the thermomechanical history to promote distributions of often dependent microstructural features. Establishing predictive process–structure–property (PSP) models to tailor

N. M. Senayake · J. L. W. Carter (✉)
Department of Materials Science and Engineering, Case Western Reserve University, Cleveland, OH 44116, USA
e-mail: jwc137@case.edu

T. A. Dux
Howmet Aerospace Forgings (Formerly Arconic/Alcoa), Newburgh Heights, OH 44105, USA

E. A. Ott et al. (eds.), *Proceedings of the 10th International Symposium on Superalloy 718 and Derivatives*, The Minerals, Metals & Materials Series,
https://doi.org/10.1007/978-3-031-27447-3_23

manufacturing routes requires immense cost due to the time-consuming task of quantifying statistically significant observations of different predictors and performance metrics across multiple lengthscales [1]. Power analysis indicates that a simple multivariate linear model of one performance metric, P, dependent on six input process predictors ($k = 6$) (i.e., $P = f(k_1, k_2, \ldots k_6)$) with 80% predictive power, would require 120 observations; to predict n performance metrics ($P_1, P_2, \ldots P_n$), a statistical study protocol would require 120 observations of each n performance metric (ANOVA would require 175n).

Measuring a statistical number of observations of predictors and metrics is challenging at both ends of the lengthscale spectra. Nanoscale precipitates, like the distribution of the γ' and γ'' precipitates in Inconel 706 have required transmission electron microscopy which does not readily lend itself to transition from qualitative analysis to statistical measures [2]. Conventional mechanical testing, on the other hand, requires relatively large volumes of material for a single observation. For example, the PSP model work in aerospace disc alloys conducted by Semiatin et al. which separates the different processing steps to related thermomechanical processing to grain size [3] and supersolvus heat treatment to precipitate distribution [4], took years of development. This is because in many cases, predictors and metrics are observed/measured destructively resulting in development efforts that take many years to establish the database necessary for PSP models [5, 6]. Since PSP models still require an iterative design paradigm, this motivates the development of high-throughput measurements for both experimental approaches [7, 8] and physics-based predictions [9, 10] if the dramatic design paradigm shift predicted by the Materials Genome Initiative is to be fulfilled [11].

In this paper, high-throughput measures of the precipitation distributions from automated image processing of micrographs are used to calibrate CALPHAD simulations of precipitate distributions. DEFORM simulations provide the processing and structural predictors necessary to establish the boundary conditions of CALPHAD simulations of previously unobserved precipitation distributions. In this regard, we leverage high-throughput experiments and physics-based simulations to reduce the necessary observation space for predictive PSP models for forgings of Inconel 706. We demonstrate how a data-driven modeling framework, shown schematically in Fig. 1, can combine different data types to understand PSP linkages in real forgings.

Approach

The heritage data for this work was obtained over a period of 10 years by the Howmet Aerospace Forgings (formerly Arconic/Alcoa) during the requalification of the 50 k-ton press after it was rebuilt in 2012. The dataset includes both categorical data with discrete classes (i.e., heat treatment process location), and continuous numerical data (e.g., grain size and yield strength). Data were acquired from ten 2–2.25 m diameter (8620–12,920 kg) IN706 following OEM certification standards.

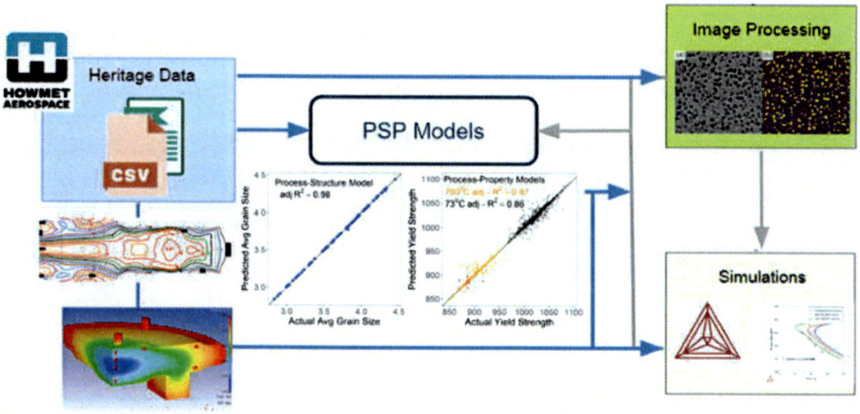

Fig. 1 The data-driven framework accommodates ingesting heritage data retained by Howmet Aerospace Forgings to create PSP models for Inconel 706 forgings. High-throughput imaging processing and physics-based simulations enable us to augment the heritage data with new observations and metrics

Howmet sampled variables from five different forging configurations, of two different diameter types, with cylindrically symmetric disc/wheel geometry from four concentric ring cutouts, from the bore (center) to the rim (the outer surface where blades would attach), are shown in Fig. 2. Test samples were also obtained at four angular locations (0°, 90°, 180°, and 270°), and tensile samples were loaded parallel to two orientations (tangential and axial). For each disc, variables were acquired from a maximum of 16 distinct locations, each with a different strain/thermal history. This amounts to 667 unique observations of tensile performance and 78 unique observations of microstructure. All of the variables were measured experimentally following ASTM standards, E930, E112, and E8-04, or simulated using commercial software (DEFORM, with optimized inputs) and are categorized into one of three different metrics, processing, structure, or properties, as shown in Table 1. The non-descriptive variables are maintained in the generalized framework to allow Howmet to continue to build the dataset as new forgings are processed. The blue-bold processing data in Table 1 are attributed to the tensile testing metadata.

The heritage data from Howmet did not include measured variables for the γ' and γ'' distributions. These were added to the dataset following procedures similar to those used to calibrate models of precipitation kinetics in Inconel 718 [12]. The CALPHAD interfacial energy was calibrated from observations of the precipitate distribution from electron micrographs from samples with known thermal profiles. After calibrating the interfacial energy, DEFORM simulations were utilized to input boundary conditions for predicted distribution variables in unmeasured locations. In this manner, the number of physical observations of γ' the γ'' distributions are reduced by 20x (78–4); and relying on more readily available experimental and computational approaches allowing for industrial use with limited available technical resources.

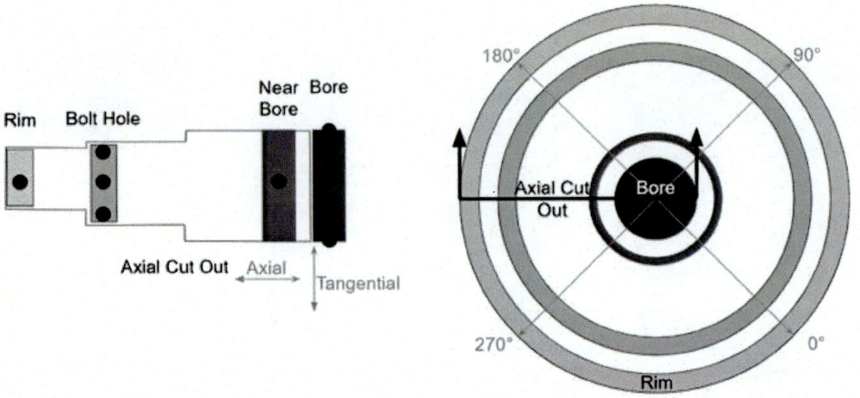

Fig. 2 Forgings have radial symmetry; the schematic on the right shows the sampling ring locations over a whole disc. A cut-out on the left highlights the individual sampling locations (black dots) within each ring

Table 1 Data frame of a subset of the descriptive categorized variables. The number in parentheticals indicates the total number of variables in the metric category

Processing (21)	Structure (7)	Properties (4)
Solution Temp	Avg. Grain Size	Tensile Strength
First Aging Time	Largest Grain Size	Yield Strength
Second Aging Time	Strain	Elongation
Configuration	Mean Radius of γ'	Reduction Area
Radial Location	Mean Radius of γ''	
Angular Location	Area Fraction of γ'	
Orientation	Area Fraction of γ''	
Temperature		

Precipitate distributions were measured in grains nominally oriented with <111> crystallographic direction normal (±8°) to the sample surface as determined by electron backscatter diffraction (EBSD) and located using the platinum marker as a reference. The total area fraction of γ' and γ'' was quantified from over 300 observations of individual particles from each of the four heat-treated samples in a Thermo Fisher Apreo scanning electron microscope (SEM) at 5 kV, in secondary electron mode, similar to the method discussed in Senanayake et al. [7].

The image analysis algorithm was developed in Python 3.9, using NumPy, pandas, and scikit-learn packages. First, the micrographs were denoised by removing objects smaller than 25 pixels (2 nm) in diameter, the scale bar was cropped, and the input pixel intensities were normalized to ensure robust image-to-image processing with the standard image processing algorithms. The next step of the image segmentation process is calculating the minimum threshold of the cross-entropy between the foreground(precipitates) mean intensity and, and the background (matrix) mean intensity [13]. The pixels in the area of interest were labeled as two classes, 1 for the background and 0 for the precipitates, as shown in Fig. 3. For the area fraction calculation, all the pixels of each type of precipitate are counted and divided by the area of the

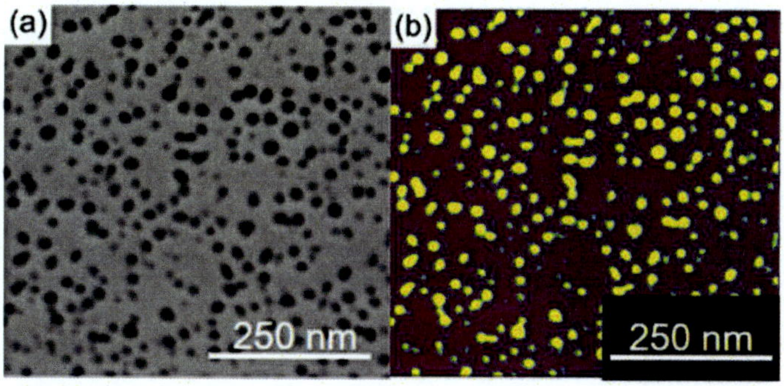

Fig. 3 An etched (with by volume of 50% glycerin, 30% Lactic acid, 17.5% Nitric acid, and 3% hydrofluoric acid in DI water) surface of superalloy IN706 using a secondary detector at 5 kV and 50 pA current **b** segmented image with two classes (background: purple, precipitates: yellow)

image. The uncertainty of the area fraction was calculated using the Monte Carlo method randomly sampling 300 particles. Precipitate sizes were calculated by the circular-equivalent radius method.

CALPHAD simulations were conducted in Thermo-Calc/PRISMA (TCNI9:Ni-Alloys v9.1 and MOBNI5:Ni-Alloys Mobility v9.1 databases) of the resulting γ' and γ'' precipitate distributions. The boundary conditions were the local time/temperature profiles from four different samples provided by Howmet. Starting with the γ' and γ'' precipitate interfacial energies for Inconel 718 [13] and understanding that chemistry is a small contributor [14], six combinations of γ' and γ'' interfacial energies, maintaining constant ratio (between 0.021 and 0.026 and 0.02175 and 0.02675 J/m^2, respectively) were compared to the experimental observations from high-throughput microscopy. As seen in Fig. 4, the trends of the simulation data align with the experimental data. Interfacial energy S4 (0.0240 J/m^2 for γ' and γ'' 0.02475 J/m^2 for γ'') combination gives the best fit of the simulation results to the experimental data for both assumptions. The values S4 range between the uncertainty values calculated for each experimental data point. Therefore, the interfacial values of 0.0240 J/m^2 for γ' and 0.02475 J/m^2 for γ'' were used for the further CALPHAD simulations to obtain the missing microstructure data for γ' and γ''.

Once interfacial energy was selected, the DEFORM simulations of thermal profiles from thermocouple positions of real forging geometries were utilized to subsequently approximate other 74 local time–temperature conditions at discrete locations within thermomechanically processed Inconel 706 forgings that were not experimentally observed. This was done by interpolating between locations following the one-dimensional, radially symmetric Fourier–Kirchhoff equation. The processing variables added to the database by interpreting DEFORM simulations are indicated in Table 1 by green-italic text, while the CALPHAD simulated microstructure variables added to the dataframe are indicated by red-italic text.

Finally, these high-throughput experimental observations of variables were added to the heritage dataset to enable machine learning modeling. This work used a gradient

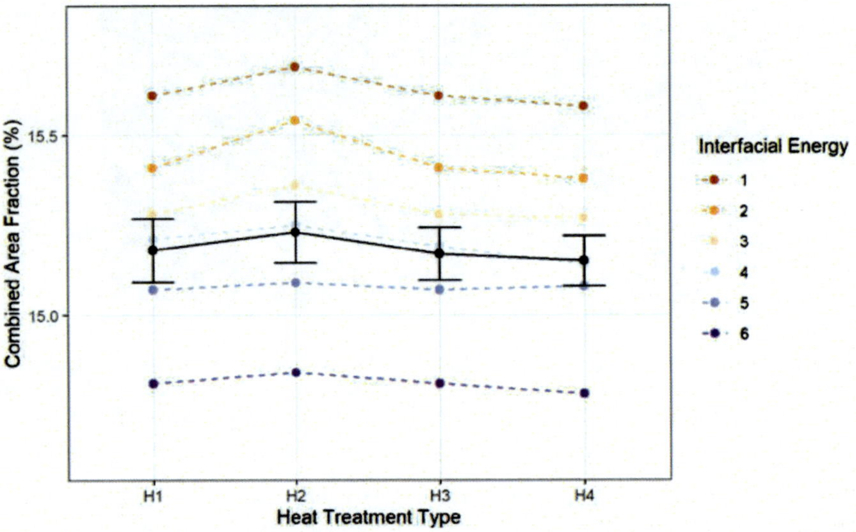

Fig. 4 The interfacial energy can be tailored to match the simulated total volume fraction of γ' and γ'' of samples within the uncertainty of the experimentally measured (black points and error bard) combined area fraction from four unique heat treatments

boost machine learning method of combining multiple models into a single composite model based on a constructive strategy of ensemble formation. In boosting, models in the ensemble that perform slightly better than random guessing (also known as weak learners) are integrated one at a time. At each iteration, a weak learner is trained with respect to the error of the entire ensemble model. The basic idea is to construct new learners to be maximally correlated with the negative gradient of the loss function [15, 16]. The efficacy of models was estimated by calculating the adj-R^2 value on the testing data following the conventional 80:20 train: test subsetting. The importance of different variables as features in a given model was assessed using a score derived from the Gini impurity [17]. The analytical framework was developed using Python version 3.9.2. The NumPy and Pandas Python packages were used for data ingestion and cleaning. XGBoost and scikit-learn Python packages were used for data imputation and modeling. The approach follows an iterative set of steps: data preprocessing, hypothesis generation, algorithm selection, and feature importance.

Results and Discussion

Exploratory Data Analysis

Exploratory data analysis (EDA) was conducted on the data to visualize trends and develop model hypotheses. Simple, pair-wise visualizations indicate that all tensile property metrics except percent elongation show a clustering with respect to test temperature, Fig. 5. The behavior for both yield strength and tensile strength are consistent so only one is visualized in Fig. 5. The data also indicates statistically insignificant trends in tensile properties with respect to the radial position (indicated in Fig. 2). Tensile anisotropy was only measured at room temperature, highlighted triangles in Fig. 5, a repeat of the yield strength figure with tangentially oriented samples of the yield strength grayed out. This data suggests a statistically insignificant trend in properties as a function of tensile loading parallel to the radial or tangential directions.

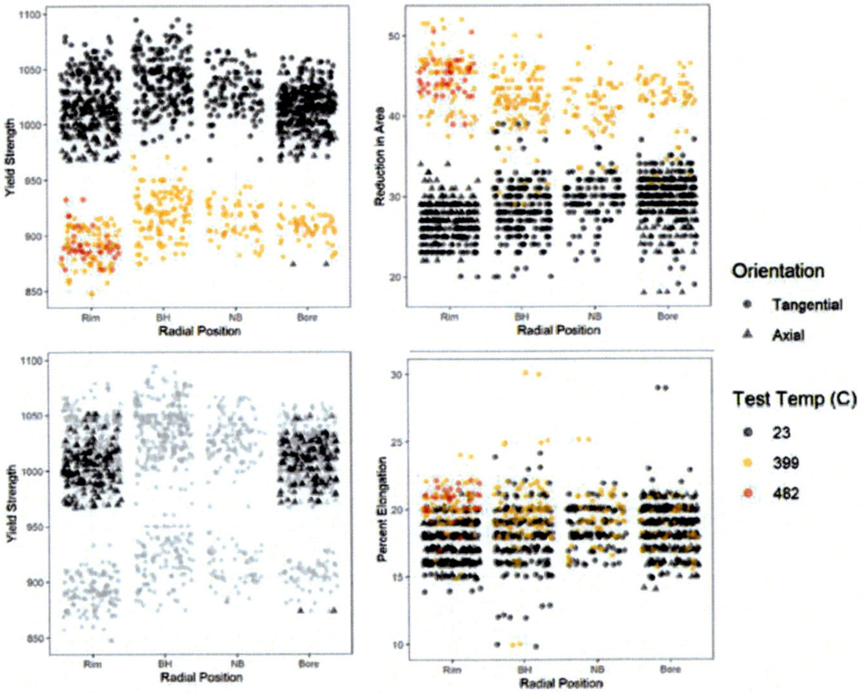

Fig. 5 Exploratory data analysis of the performance data indicates that there is a clustering of tensile performance with test temperature, which is consistent with the literature. The yield strength from the radially oriented samples collected at room temperature is highlighted with triangles, while all gray points are tangentially oriented tensile observations

The pair-wise plots indicate that there might be two methods of accommodating tensile test temperatures within a model paradigm: (1) include temperature as a numerical, continuous variable, or (2) include it as a factor in selecting which model to utilize to predict the performance. Paradigm one is appropriate if there is only one mechanism responsible for tensile properties or if the multiple mechanisms that are responsible for tensile properties are all correlatively dependent on test temperature. Paradigm two is more appropriate if we hypothesize that the mechanism responsible for tensile properties each varies independently with temperature.

Pair-wise plots are a biased method of exploring the data because the domain expert chooses which pairs of data to review. An unbiased method of EDA is to conduct cluster analysis, which are unsupervised machine learning algorithms that utilize different metrics to divide normalized, multidimensional datasets into homogeneous groups within a two-dimensional space. Collections of similar numerical data exist as a single cluster of points, while dissimilar data cluster in separate groups. The spacing between clusters depends on which numerical variables are included in the similarity analysis. For example, when test temperature and angular location are not considered a variable for similarity, the data in Fig. 6a separate into two clusters and the impact of sample orientation and radial position are conflated into a single cluster. When test temperature and angular location are considered a variable of similarity, the independent impacts of sample orientation and radial location become more evident in Fig. 6b. This is because now all tensile data were collected at all temperatures but samples at particular angular locations were only tested in one orientation, so the impact of temperature is reduced when included as a similarity variable. A similar separation by orientation and location occurs if the t-SNE analysis is only conducted on the tensile data collected at room temperature. No other macro-level processing metric provides an obvious cluster in the tensile performance. This indicates (1) that Howmet did a good job at designing the bulk processing conditions such that the five different forging configurations are statistically equivalent, and (2) that further insight into the tensile performance is going to require incorporation of the localized processing and or microstructural data for each of the five different forging configurations.

CALPHAD Prediction of Precipitate Distributions

The calibration of the CALPHAD interfacial energy was consistent with literature results for the slight impacts of chemistry. The slightly larger interfacial energy of γ/γ'' of Inconel 706 compared to Inconel 718 is consistent with the literature. Devaux et al. found that alloys that contain more nickel in the matrix favor lower interfacial energy with γ'' particles [18], while Liu et al. noted that Mo additions had the largest effect on decreasing the interfacial energy of γ/γ' [19]. The larger interfacial energies in IN706 due to the absence of Mo, and lower Ni content compared to Inconel 718 are aligned with previous studies.

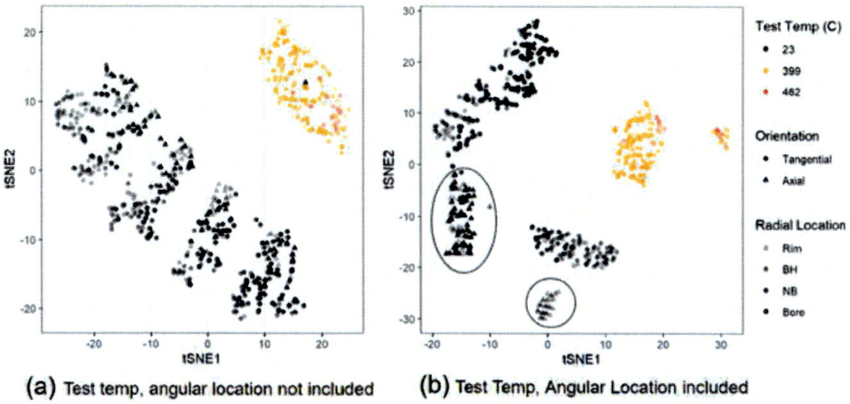

Fig. 6 The t-SNE visualization of the clustering of multidimensional datasets changes with variables considered. The impact of bulk processing variable radial location (saturation) and tensile test orientation (shape) is **a** less obvious when test temperature was not considered a variable of similarity and **b** more obvious when test temperature was included as a similar variable

Processing–Structure–Property Models

Processing–structure (P-S) gradient boost models were accurate (adj-$R^2 > 0.94$) for grain size and strain variables. The feature importance maps, shown in Fig. 7, indicate that radial location, configuration, and solutionizing temperature are near equivalent top-three contributors. The contribution of angular location/orientation and type on the P_S models is an not actual insight but arises from bias in the sampling; there are no angular locations for bore samples, so configuration and type are correlated. Additionally, though structure variables were measured at each angular location, all microstructure variables were only measured from one orientation.

The feature importance map provides insight into how model interpretation is dependent on how we select variables. For example, there is continual debate among the co-authors on how to categorize the strain variable: processing or structure.

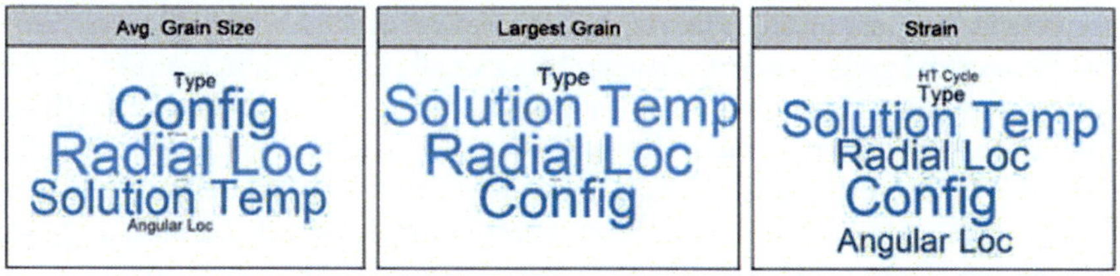

Fig. 7 The feature importance word map, where larger/lighter feature texts are more important, shows that the same top three processing variables are responsible for predicting the local grain size and strain

Table 2 Comparison of the predictive power (adj-R^2) of the P_P and S_P models at the two different temperature regimes

	P_P		S_P	
Property variable	**23°C**	**399°C**	**23°C**	**399°C**
Yield strength	0.86	0.87	0.62	0.80
Tensile strength	0.92	0.90	0.74	0.58
Reduction of area	0.78	0.83	0.60	0.53
Elongation	0.76	0.62	0.59	0.56

When creating P_S models or processing–properties (P_P) models using the categories defined in Table 1, the impact of strain on grain size or properties is inferred through the other processing variables. It is established knowledge that high dislocation density (strain) enhances the recover-recrystallization kinetics and exacerbates grain growth. The processing process does not occur all at once but in a series of steps: forging, solutionizing, aging, etc. Therefore, adding insights or the ability to extrapolate the P_S or P_P models to new processing conditions could be gleaned by stringing models along in a logical order of operations.

Gradient boost P_S models of the precipitate variables are inappropriate since only the local heat treatment parameters are considered in the CALPHAD predictions, and these parameters are perfectly correlated. The value in adding the precipitate values occurs as we transition to gaining insight from the structure–property models.

The processing–property (P_P) models for the property variables were created by segmenting the data by test temperature into room temperature (23 °C) and elevated temperature (399 °C). The data for the elevated temperature models included data collected at 399 °C and 482 °C. The temperature-specific P_P models for strength were more accurate (adj-$R^2 > 0.86$) than the P_P models for ductility variables, as summarized in Table 2.

The predictive power of the structure–property (S_P) models is lower than that of the P_P models, as shown in Table P_P&S_P, but the value of the S_P models is in the interpretability of the models to design new routes in an inverse fashion. The feature importance word maps, Fig. 8, provide insights into how the gradient boost model identifies how structural features that dominate the mechanical performance change with temperature. The fact that grain size is fairly unimportant is consistent with the range of grain sizes in the dataset being small (2.8–4.45), so even though the data are normalized from zero to one prior to modeling the importance is small. At low temperature, the strain processing parameter, which is a proxy for geometrically necessary dislocation density (GND) is the most important feature for metrics of ductility than strength. At elevated temperature, diffusion-mediated cross slip can alleviate the stored GND, and the precipitate features become more important to resisting further dislocation motion.

The predictability and insights from the models need to be considered within the context of the data provided. The data are from forgings that are all designed, using DEFORM, to manufacture material that all the various local conditions still

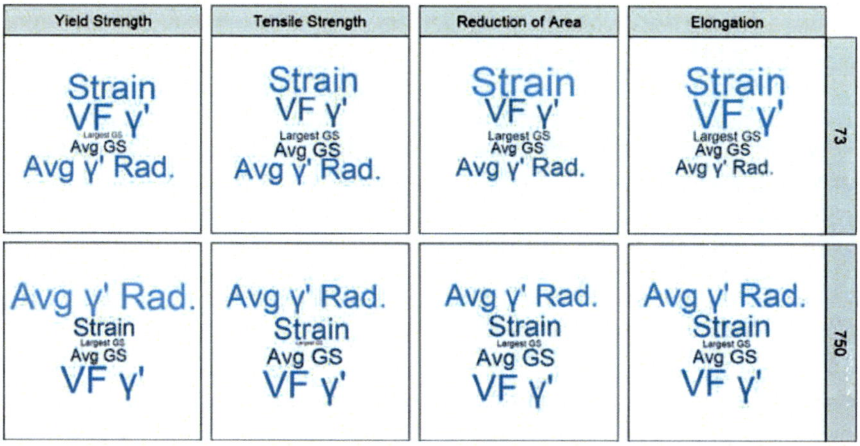

Fig. 8 The feature importance word map, where larger/lighter feature texts are more important, indicates that the importance of particular features is temperature-dependent

result in material that meets the OEM specification. Therefore, there is little variation in the dataset (consistent with the interpretation of the t-SNE cluster analysis in Fig. 6). Therefore, it is not surprising that the insights are limited but it is reassuring that these insights are consistent with common interpretation of physical mechanisms. The advantage of this approach is the data framework provides a means of combining heritage and new data sources with minimal redesign. Therefore, as the manufacturing engineers begin the interactive design process for new tooling, geometries, or process enhancements, new data points derived from DEFORM simulations of processing variables and the geometries to directly predict strain and thermal profiles can be added. These can then be used to predict precipitate distributions in CALPHAD and then the trained gradient boost models can ultimately predict grain sizes and properties for different geometries of the same material and expectedly reduce the process development time.

Conclusions

In this work, a data analytics framework was demonstrated that links physics-based simulations and experimental observations to develop predictive Process–Structure, Process–Property, and Structure–Properties models of forged Inconel 706. The framework enables multiple phases of analysis to efficiently explore data trends and build interpretable models. As part of this framework, we developed a novel two-part high-throughput approach to approximating distributions of nano-scale precipitates. A machine learning image algorithm of SEM images that replaces conventional, tedious analysis of transmission electron micrographs was established. This imaging approach and algorithm are implemented to calibrate interfacial energies of IN706 for CALPHAD simulations. Directly linking experimental and physics-based simulations enables a 25 × reduction in the micrograph collection and analysis time.

The implemented set of gradient boost PSP models for IN706 predicts the microstructure and the mechanical properties based on the processing and microstructure with an accuracy of up to 90%. The models and importance of features are capable of not only predicting the target variables with significant accuracy but also guiding to understanding and explaining the physical insights.

Acknowledgements This work was funded by the National Science Foundation, on the CAREER: A Systematic Data-Analytics Approach to the Design of Interface-Rich Materials (1552716). Scanning electron microscopy was conducted at the Center for Electron Microscopy and Analysis (CEMAS) at The Ohio State University. The CALPHAD calculations were conducted using the CWRU educational site license of Thermo-Calc awarded through the ASM Materials Genome Toolkit program.

References

1. S. Li, U. R. Kattner, and C. E. Campbell, "A Computational Framework for Material Design," *Integrating Mater. Manuf. Innov.*, vol. 6, no. 3, pp. 229–248, Sep. 2017. https://doi.org/10.1007/s40192-017-0101-8.
2. S. Zhang *et al.*, "Comparison study of microstructure and mechanical properties of standard and direct-aging heat treated superalloy Inconel 706," *Mater. Sci. Eng. A*, vol. 839, p. 142836, Apr. 2022. https://doi.org/10.1016/j.msea.2022.142836.
3. S. L. Semiatin *et al.*, "Plastic Flow and Microstructure Evolution during Thermomechanical Processing of a PM Nickel-Base Superalloy," *Metall. Mater. Trans. A*, vol. 44, no. 6, pp. 2778–2798, Jun. 2013. https://doi.org/10.1007/s11661-013-1675-1.
4. S. L. Semiatin, F. Zhang, J. S. Tiley, and D. U. Furrer, "A comparison of the precipitation behavior in PM γ-γ' nickel-base superalloys," *Mater. High Temp.*, vol. 33, no. 4–5, pp. 301–309, Jun. 2016. https://doi.org/10.1080/09603409.2016.1165449.
5. A. K. Verma, J. A. Hawk, L. S. Bruckman, R. H. French, V. N. Romanov, and J. L. W. Carter, "Mapping Multivariate Influence of Alloying Elements on Creep Behavior for Design of New Martensitic Steels," *Metall. Mater. Trans. A*, vol. 50, no. 7, pp. 3106–3120, 2019. https://doi.org/10.1007/s11661-019-05234-9.
6. J. L. W. Carter and A. K. Verma, "Informatics-Enabled Design of Structural Materials," *JOM*, vol. 73, pp. 3323–3325, Oct. 2021. https://doi.org/10.1007/s11837-021-04896-4.
7. N. M. Senanayake and J. L. W. Carter, "Computer Vision Approaches for Segmentation of Nanoscale Precipitates in Nickel-Based Superalloy IN718," *Integrating Mater. Manuf. Innov.*, vol. 9, no. 4, pp. 446–458, Dec. 2020. https://doi.org/10.1007/s40192-020-00195-z.
8. Senanayake N, Mukhopadhyay S, Carter JLW (2020) High-throughput approaches to establish quantitative process-structure-property correlations in Ni-base superalloy. In: Superalloys 2020, Seven Springs, PA, 2020, p 10. https://doi.org/10.1007/978-3-030-51834-9_66
9. C. J. Kuehmann and G. B. Olson, "Computational materials design and engineering," *Mater. Sci. Technol.*, vol. 25, no. 4, pp. 472–478, Apr. 2009. https://doi.org/10.1179/174328408X371967.
10. Semiatin SL et al (2021) A fast-acting method for simulating precipitation during heat treatment of superalloy 718. Metall Mater Trans A 1–17. https://doi.org/10.1007/s11661-020-06092-6

11. D. L. McDowell and S. R. Kalidindi, "The materials innovation ecosystem: A key enabler for the Materials Genome Initiative," *MRS Bull.*, vol. 41, no. 4, pp. 326–337, Apr. 2016. https://doi.org/10.1557/mrs.2016.61.
12. Zhang F, Cao W, Zhang C, Chen S, Zhu J, Lv D (2018) Simulation of co-precipitation kinetics of γ' and γ'' in superalloy 718. In: Proceedings of the 9th international symposium superalloy 718 derivatives: energy, aerospace, and industrial applications, pp 147–161. https://doi.org/10.1007/978-3-319-89480-5_8
13. C. H. Li and C. K. Lee, "Minimum cross entropy thresholding," *Pattern Recognit.*, vol. 26, no. 4, pp. 617–625, Apr. 1993. https://doi.org/10.1016/0031-3203(93)90115-D.
14. Q. Chen, K. Wu, G. Sterner, and P. Mason, "Modeling Precipitation Kinetics During Heat Treatment with Calphad-Based Tools," *J. Mater. Eng. Perform.*, vol. 23, no. 12, pp. 4193–4196, Dec. 2014. https://doi.org/10.1007/s11665-014-1255-6.
15. Natekin A, Knoll A (2013) Gradient boosting machines, a tutorial. Front Neurorobot 7. https://doi.org/10.3389/fnbot.2013.00021
16. Schapire RE (2003) The boosting approach to machine learning: an overview. In: Denison DD, Hansen MH, Holmes CC, Mallick B, Yu B (eds) Nonlinear estimation and classification. Springer, New York. pp 149–171. https://doi.org/10.1007/978-0-387-21579-2_9
17. S. Nembrini, I. R. König, and M. N. Wright, "The revival of the Gini importance?," *Bioinforma. Oxf. Engl.*, vol. 34, no. 21, pp. 3711–3718, Nov. 2018. https://doi.org/10.1093/bioinformatics/bty373.
18. A. Devaux *et al.*, "Gamma double prime precipitation kinetic in Alloy 718," *Mater. Sci. Eng. A*, vol. 486, no. 1, pp. 117–122, Jul. 2008. https://doi.org/10.1016/j.msea.2007.08.046.
19. X. L. Liu, S.-L. Shang, Y.-J. Hu, Y. Wang, Y. Du, and Z.-K. Liu, "Insight into γ-Ni/γ'-Ni3Al interfacial energy affected by alloying elements," *Mater. Des.*, vol. 133, pp. 39–46, Nov. 2017. https://doi.org/10.1016/j.matdes.2017.07.028.

Linking Stress-Rupture Properties to Processing Parameters of HAYNES® 718 Nickel Superalloy Using Machine Learning

David E. Farache, George M. Nishibuchi, Sebastian Elizondo, John G. Gulley, Alex Post, Kyle Stubbs, Keith Kruger, Arun Mannodi-Kanakkithodi, and Michael S. Titus

Abstract Requirements of stress-rupture life and elongation of nickel alloy 718 are often prescribed by specification AMS5596™ or AMS5662™, which broadly state that the stress-rupture life and elongation must exceed 23 h and 4% at 649 °C (1200 °F), respectively. Variability in product stress-rupture life can range from less than 2 h to more than 1000 h depending on test load, which can cause significant delays for testing, shipping, and delivery of a product. In this work, we predict the stress-rupture life and elongation of HAYNES® 718 sheet product utilizing machine learning models. The models were trained on data from 448 lots of material and inputs including composition, room temperature mechanical property data, processing data such as finish gauge, total reduction, final reduction, rule of mixture averaged properties, and environmental factors. Different sets of input features were chosen from the highest absolute Pearson correlation values, one-way ANOVA analysis, random forest (RF) model analysis methods, and generated compound features, and two separate RF models were trained using an 80–20% split between training and testing data. The resulting mean squared errors of best performing models of stress-rupture life and elongations were 102 h and 7.2%, respectively. Input features of the highest importance were observed to be room temperature tensile properties, finish gauge, and tramp elements such as Co, P, and Si. These models can be utilized to accelerate acceptance testing of 718 products by selecting the highest testing load that will still guarantee passing test life and elongation results.

Keywords Creep · Elongation · Machine learning

D. E. Farache, G. M. Nishibuchi—They are co-first authors.

D. E. Farache · G. M. Nishibuchi · S. Elizondo · J. G. Gulley · A. Mannodi-Kanakkithodi · M. S. Titus (✉)
School of Materials Engineering, Purdue University, 701 West Stadium Ave, West Lafayette, IN 47907, USA
e-mail: titus9@purdue.edu

A. Post · K. Stubbs · K. Kruger
Quality & Manufacturing Engineering, Haynes International, 2000 W Deffenbaugh St, Kokomo, IN 46904-9013, USA

E. A. Ott et al. (eds.), *Proceedings of the 10th International Symposium on Superalloy 718 and Derivatives*, The Minerals, Metals & Materials Series,
https://doi.org/10.1007/978-3-031-27447-3_24

Introduction

Nickel alloy 718 is commonly considered as the most significant nickel superalloy produced in the world. Often this important technological alloy is specified using either SAE AMS5596™ (flat products) or AMS5662™ (round products) specifications [1, 2], which outline requirements by product form for composition, microstructure, and mechanical properties. One of the longest tests required by specification is the stress rupture test, which can take in excess of 1000 h to complete and poses a risk for suppliers in delivering products on time.

At 649 °C (1200 °F) and a minimum of 655–689 MPa (95–100 ksi), AMS 5596 and AMS 5662 indicate minimum stress-rupture life and elongation of 23 h and 4%, respectively. Upon reaching minimum specification, the test may be accelerated using incremental loading of 34.5 MPa (5 ksi) every 8 h after the initial 23 h specification has been met. Incremental loading offers suppliers a route to accelerate rupture tests, but cannot always be used when other concurrent specifications prohibit uploading. However, if the initial stress can be increased significantly above 689 MPa (100 ksi) while meeting minimum life and elongation requirements, testing can be drastically accelerated, ensuring on-time deliveries to customers.

Significant work has recently been reported on predicting stress-rupture and creep properties in steels using machine learning techniques [3–7]. Due to the ubiquity of steel, very large datasets exist to aid in training machine learning models and include the NIMS databases [8], and highly accurate predictions of stress-rupture lives are able to be made when using more complex methods such as gradient boost algorithms [6, 7]. However, more limited data exists for Ni-based superalloys [8]. Some recent work has focused on predicting mechanical properties in additively manufactured components and single crystal Ni-based superalloys [9–13].

In this work, based on an extension of a Purdue University Senior Design project, we utilize machine learning to predict the stress-rupture life and elongation of HAYNES® 718 at 649 °C and 724 MPa (105 ksi). These models can be used to predict mechanical behavior prior to testing, and thus aid in initial stress selection to minimize testing time while ensuring minimum specification requirements are met.

Methodology

Data Selection and Initial Input

Data from 448 lots of HAYNES® 718 alloy were used to train machine learning models. The HAYNES® 718 data were drawn from an electroslag remelted (ESR) sheet product ranging from 0.38 to 5.2 mm (0.015 in. to 0.205 in.) in thickness and produced from 2018 to 2020. The resulting stress-rupture life distribution at 649 °C (1200 °F) and 724 MPa (105 ksi) ranged from 3.6 h to 1202 h, as shown in Fig. 1. The initial data included primary melt composition data, cold and hot rolling

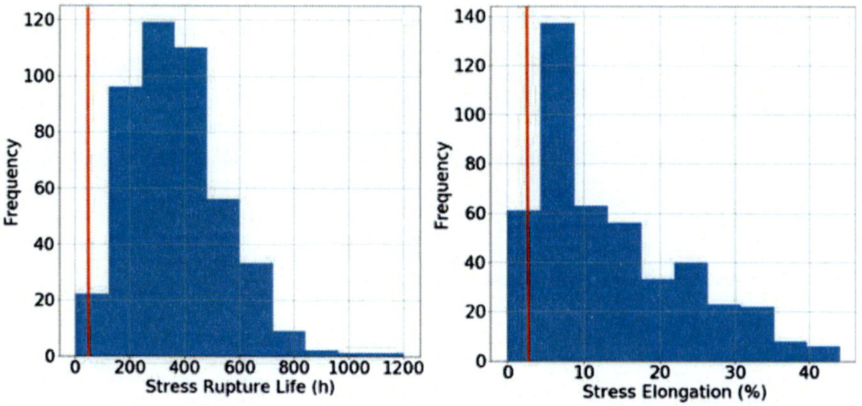

Fig. 1 Histogram of the stress-rupture life (left) and stress rupture elongation (right). The red line indicates the required specifications

data, ESR furnace and crucible number, and room temperature tensile testing data. Additional physical, mechanical, and environmental data were utilized to expand the dataset using a combination of rule of mixture averages for select composition data, tabulated physical properties of chemical elements collected from *Pymatgen* [14], as well as historical weather data.

The features selected were then categorized into 5 groups: composition, processing, physical, mechanical, and environmental. Composition refers to an encoding of the number and type of elements in a sample, physical refers to chemical or thermodynamic related elemental properties, mechanical refers to those material properties determined by the application of forces, environmental refers to weather data in the location of production on the day an ingot was processed, and processing refers to methods and specifications utilized to manufacture the ingot. In terms of the number of data points, most data could be compartmentalized into the compositional dataset as shown in Table 1. This table summarizes the groupings of data, but does not include all features included in the analyses. In total, 57 input parameters were considered for this work.

After preliminary data cleaning, an initial analysis of the Pearson correlation coefficients between input (Table 1) and output (stress-rupture life and elongation, Fig. 1) data was performed to determine any associations in the property groupings.

Exploration of non-linear relations between features in the dataset to creep rupture life and elongation was accomplished by performing one-way ANOVA analysis with (1) features combined additively to determine new compound features (e.g. nickel + boron), and (2) generation of new, higher-order compound features via a sure independence screening and sparsifying operator (SISSO)-inspired algorithm. One-way ANOVA determines how statistically significant independent variables are to one another [15]. *Statsmodels* library returns a p-value, which, if below 0.05, considers the variables to have statistical significance. Statistically significant compound features

Table 1 Input parameters used as features and their designated groupings. Features marked with asterisk are those added to the original Haynes (R) 718 alloy dataset

Input parameters	Compositional	Al	B	Co	Cr	Mo	Ni	P	Si
		Ti	W	Nb	Fe	Cu	Mn		
	Processing	Finish gauge		Total reduction		Final reduction		ESR furnace	
	Physical*	Electronegativity		Electrical resistivity		Melting point			
	Mechanical*	Bulk modulus		Young's modulus		Density			
	Environmental*	Daily max. temp		Daily min. temp		Humidity		Barometric pressure	

*Additional Features added to dataset

with $p < 0.05$ were identified, tested with newly trained models, and assessed based on model error.

For higher-order compound features, an implementation inspired by the SISSO compound feature generation algorithm [16] was created to expand the available features in the training set. The process for this feature generation can be seen in Fig. 2. These compound features were intended to capture higher-order physical relations by mathematically combining the features provided by Haynes, generating orders of magnitude more features than originally existed, and choosing the highest correlated features to be added to the machine learning, random forest (RF) model. This further identified compound features of high importance, typically via highly non-linear functions of multiple primary features.

After expanding the number of features, a RF machine learning model was utilized for its unique ability to select the most important features. The RF machine learning models were trained with the *scikit-learn* Python package utilizing 800 initial estimators (trees), a max depth (of tree) of 80, minimum samples per leaf of 8, bootstrapped samples to build trees, and mean squared error as the loss function [17]. This optimization was reached utilizing the *grid_search* function which performed tenfold cross validation upon the model until the lowest mean absolute error (MAE) was reached. The most relevant features were captured from the model via its Gini

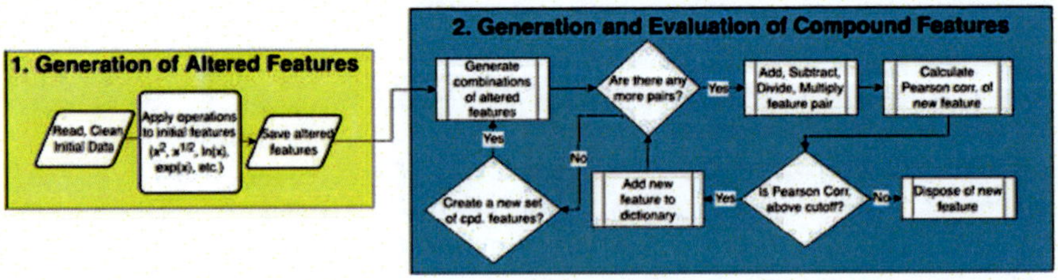

Fig. 2 Overview of compound feature generation

Importance, Mean Accuracy Decrease, and Shapley additive explanations (SHAP) analysis.

The Gini Importance (mean decrease impurity) reduces either the impurity or the variance: how consistent results are from between nodes in a decision tree, or the amount of change the model will experience from changes in the dataset, respectively. High variance is undesirable as that indicates overfitting and lack of ability for the model to learn trends. Mean Accuracy Decrease, or permutation importance, functions by determining the decrease of accuracy score based on the removal of a feature [17]. SHAP utilizes game theory to determine the marginal contributions of all permutations of the features [18]. The features then determined to be most important were fed into a different RF model with the same parameters from the *lolopy* Python package, to obtain uncertainty as well as the predicted value.

A classification model was also trained to identify whether a given sample would pass the requirements given by the AMS5596 and AMS5662 specifications as the regressor models have difficulty making predictions at the extreme ranges of the dataset. Several popular classifiers, including Gaussian, Bernoulli, Random Forest, Logistic-Regression, and K-nearest neighbors, were called from the *scikit-learn* library, and models were trained for stress-rupture life (4 to 400 h) and stress rupture elongation for (1 to 40% strain). Since different classifiers utilize substantially different operational logic, the effect of classifier choice on performance was evaluateable.

The Gaussian classifier attempts to classify the labels via a Gaussian distribution. The Bernoulli classifier uses the simplest version of Bayes' theorem to determine class from the input only of a discrete dataset such as the one used here. The random forest classifier works similarly to a random forest regressor, except the trees return a label value and perform majority voting for selection. Logistic-Regression instead computes the probability of the class via a matrix of maximum likelihood loss. K-nearest neighbors has each point vote for its class, and those nearest to one another are grouped and classed based on the majority vote. This was performed while utilizing the same inputs for features found by the random forest regressor analysis, since all classifiers (except random forest) are highly dependent on feature selection and utilization of default hyperparameters from *scikit-learn* [16].

Results

Summary of 718 Dataset and Statistical Analysis of Inputs to Outputs

The stress-rupture life data can be described as a normal distribution with outlier values past 800 h and the majority of the data points being past the required threshold of 24 h, as shown in Fig. 1. The stress rupture elongation has a right skewed distribution with the majority of the results satisfying or exceeding the requirement of

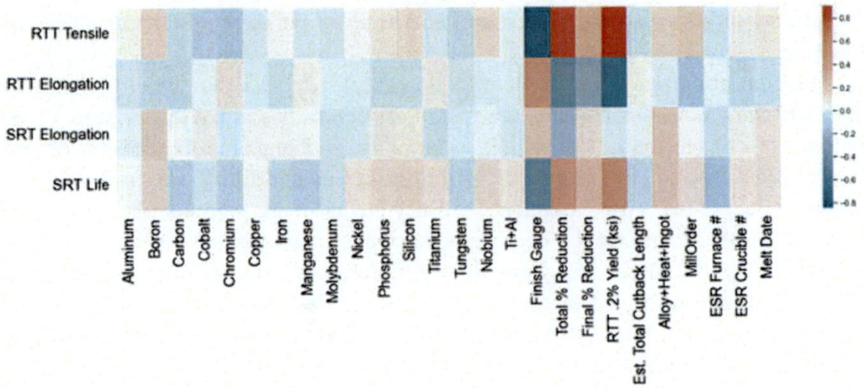

Fig. 3 Map of Pearson correlations between initial features and properties being predicted, with red being positively correlated and blue being negatively correlated

4% strain. The feature set was expanded with elemental parameters via *pymatgen* and the addition of weather data to identify contributing factors to the shape of the distributions.

Initial analysis of the Pearson correlation coefficients between input (Fig. 3) and output (stress-rupture life and elongation) data revealed that stress-rupture life shows moderate positive correlation with room temperature mechanical properties and total percent reduction, and moderate negative correlation with the finish and intermediate gauge and room temperature tensile elongation. In general, poor correlations below absolute values of 0.3 were observed for all input data with stress-rupture life and elongation.

Analysis of Compound Features and ANOVA (Expansion of Dataset)

We performed one-way ANOVA analysis, represented in Fig. 4, where the rows and columns are features combined to determine the influence of any two features together on the output variables, stress-rupture life and elongation. It becomes clear that a significant portion of the data now appears statistically significant, and Table 2 indicates that a combination of any feature and Finish_MO (Mill Order) has the most influence on the output. Because Finish_MO appeared to have the most influence on the output, the pairwise combinations of Finish_MO and the other features were tested, but were found to increase the mean absolute error without improving fit. Other feature combinations were shown to reduce the mean absolute error, such as Molybdenum + Iron for stress-rupture life, and Nickel + Boron and TempMin + Titanium for stress rupture elongation.

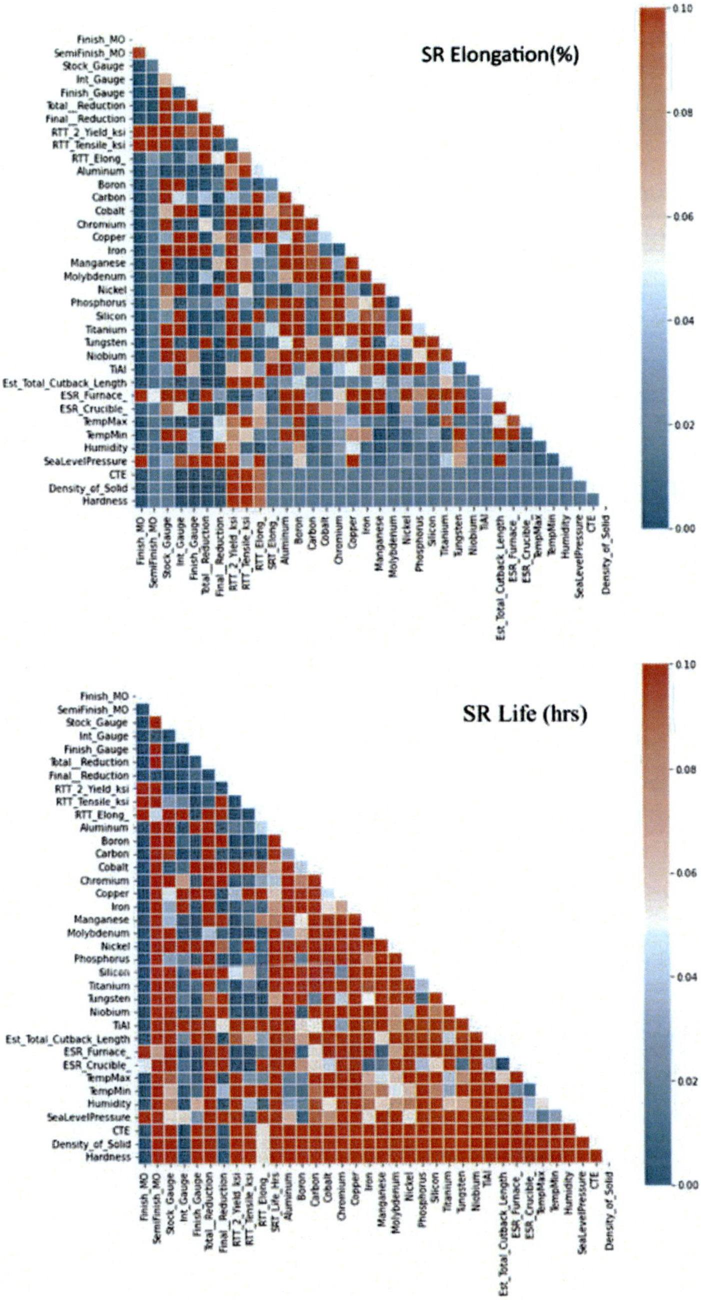

Fig. 4 The plot represents the p-value given between one-way ANOVA of the column and row features with SR life and SR elongation. P-values lower than 0.05 are considered statistically significant and are represented in blue (more blue means better) while those in red are less statistically significant

Table 2 Top 5 one-way ANOVA features by lowest p-value

SR life		SR elongation	
Feature	PR(>F)	Features	PR(>F)
Finish_MO + Humidity	2.73e-39	Finish_MO + Humidity	9.47e-40
Stock_Gauge + Finish_MO	3.06e-39	Finish_MO + TempMax)	1.66e-39
Molybdenum + Finish_MO	3.06e-39	Molybdenum + Finish_MO	1.85e-39
Finish_MO + CTE	3.06e-39	Stock_Gauge + Finish_MO	1.85e-39
Niobium + Finish_MO	3.06e-39	Finish_MO + CTE	1.85e-39

To further capture and identify feature importance as well as to expand the features available to train the model, the initial features were mathematically combined to create compound features intended to capture physical relations. Pearson Correlations for the most positive compound features for SRT Life and RTT Tensile strength are shown in Tables 3 and 4, respectively.

Compound features generated for stress-rupture life (Table 3) resulted in an overall increase in the Pearson correlation by ~0.06 compared to the highest correlated non-compound feature (~0.52) from the initial dataset. It is clear that phosphorus content, total reduction, yield strength, and titanium composition were the most important

Table 3 Top 5 compound features generated for SR Life

Compound feature	Pearson correlation
$\left([P]^4 * TotalReduction^4\right) + (Yield^{\frac{1}{4}} * [Ti]^{\frac{1}{4}})$	0.5938
$\left([P]^4 * TotalReduction^4\right) + (Yield^{\frac{1}{4}} + [Ti]^4)$	0.5934
$\left([P]^3 * TotalReduction^3\right) + (Yield^{\frac{1}{4}} + [Ti]^4)$	0.5916
$\left([P]^3 * TotalReduction^3\right) + (Yield^{\frac{1}{4}} * [Ti]^{\frac{1}{4}})$	0.5915
$([P] * TotalReduction) + (Yield^{\frac{1}{4}} * [Ti]^4)$	0.5840

Table 4 Top 5 compound features generated for RT tensile strength

Compound feature	Pearson correlation
$\left([B]^{\frac{1}{4}} * TotalReduction\right) * (FinishGauge^{\frac{1}{4}} * TotalReduction)$	0.8817
$\left(FinishGauge^{\frac{1}{2}} * TotalReduction^4\right) * ([Ni]^{\frac{1}{3}} - TotalReduction^{\frac{1}{4}})$	0.8801
$\left([B]^{\frac{1}{2}} * TotalReduction^{\frac{1}{2}}\right) * (FinishGauge^{\frac{1}{4}} * TotalReduction)$	0.8801
$\left([Cr]^{\frac{1}{4}} - TotalReduction^{\frac{1}{4}}\right) * (FinishGauge^{\frac{1}{4}} * TotalReduction)$	-0.8792
$\left([Cr]^{\frac{1}{3}} - TotalReduction^{\frac{1}{2}}\right) * (FinishGauge^{\frac{1}{4}} * TotalReduction)$	-0.8791

features for stress-rupture life given that all of the top five generated features for stress-rupture life were variations of the same set of initial features.

The top five compound features for RT tensile strength (Table 4) showed a marked increase in Pearson correlation over the next highest correlated non-compound feature (~0.52), yielding nearly 70% increases in the magnitude of Pearson correlation for the best compound features. These marked increases in Pearson correlation provided confidence in the chosen methods used for generating compound features.

Analysis of Features with Scikit-learn and SHAP

The *scikit-learn* RF model for stress-rupture life appears to have difficulties with outliers in the dataset. Both parity plots, shown in Fig. 5, exhibited similar trends of fitting similar values along a line that did not match the line of best fit, indicating that the models were unable to demonstrate learning, especially at the extremes. The model's attempt to minimize the variation did result in maintaining a low error of 103 ± 9.4 h (less than 10% of the total range of values) for the rupture life, while the stress rupture elongation had an error of 7.43 ± 0.51%. For both, most predictions fit within two standard deviations.

Given the model was able to maintain the predictions mostly within two standard deviations, we deem the model reliable enough to extract important features. The mean accuracy decrease results indicated that for stress rupture time, the features RT Tensile (ksi) and RT 0.2% Yield (ksi) are most important as seen from the Pearson correlation map. Interestingly, the compositional, environmental, and processing features begin to appear: Titanium for compositional, minimum day temperature

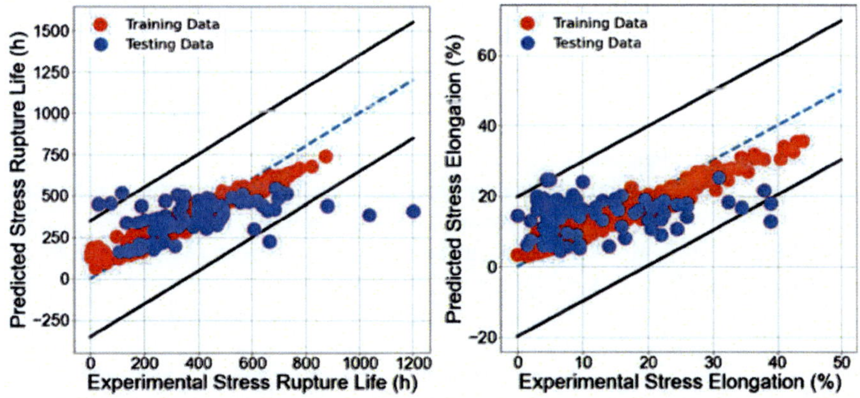

Fig. 5 The image is a parity plot of the *scikit-learn* RF model trained on all features with output stress-rupture life (left) and stress rupture elongation (right). Red dots indicate training, blue dots indicate testing data, black lines are two standard deviations from the line of best fit, and the dashed line is the line of best fit

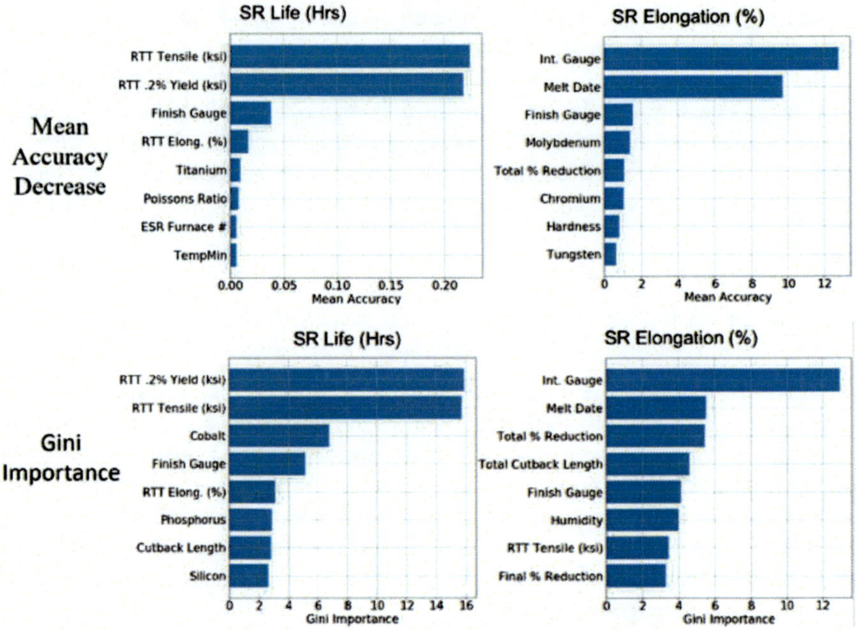

Fig. 6 Bar charts indicate features of greatest correlation in decreasing order based on Mean Accuracy Decrease analysis (top row) compared to Gini Importance (bottom row) for stress-rupture life (left) and stress rupture elongation (right)

for weather, and ESR Furnace # for processing features (Fig. 6). The Gini Importance analysis also selected the same top features but instead placed priority on the compositional features of Co, P, and Si as shown in Fig. 6.

The stress rupture elongation results from mean accuracy minimization placed high importance on Intermediate Gauge and the environmental feature of melt date. The high importance of melt date could be attributed to seasonal variation in environmental conditions. The Gini Importance assigned greater relevance to processing features seen in the Pearson correlation map but chose another environmental feature of humidity as seen in Fig. 6. The regressor further selected compositional features of Mo, Cr, and W and the first mechanical feature, hardness.

SHAP analysis leads to similar results to both Gini Importance and Mean Accuracy based analysis, with the addition of ESR Crucible # for stress-rupture life and the addition of Mo and P for stress elongation (Fig. 7). All features were compiled along with compound features and ANOVA features that had high correlation, or those features that reduced the mean absolute error of the model when added to the dataset.

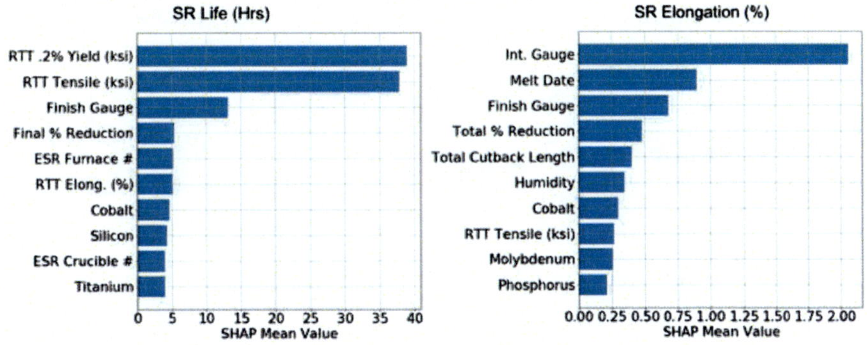

Fig. 7 Bar charts of greatest correlation of important/influential features using SHAP Mean value for stress-rupture life (left) and stress rupture elongation (right)

Machine Learning of Elongation and Stress-Rupture Life Lolopy and Scikit-learn Classifier

Based on the above analyses, a feature set comprising 23 features for stress-rupture life and 16 features for stress rupture elongation was used to train a lolopy random forest to not only predict features but also extract uncertainty. These results are summarized in Fig. 8.

The stress-rupture life results indicate that learning was achieved without overfitting, as observed by the strong correlation in the training data and similar correlation of the testing data. The testing data indicated that very low (<150 h) and very high (>500 h) stress-rupture life predictions were anomalously high and low, respectively, as shown in Fig. 8. However, the predicted values between 150 and 500 h show tight clustering around the parity line, well within two standard deviations of the dataset and indicating good predictability.

In contrast, the stress rupture elongation model indicated good training, but the resulting test set showed poor learning as shown by the relatively horizontal values in the stress rupture elongation plot in Fig. 8.

The classification models for the stress-rupture life perform well at low value cutoffs, with the Random Forest, Gaussian, K-Nearest, and Logistic classifiers having similar results, and only the Bernoulli failing at larger thresholds as seen in Fig. 9. The classification models for stress rupture elongation had greatest accuracy at the edge thresholds with the random forest classifier outperforming or matching all other models. It is also apparent that all classifier models tend to have worse performance as the threshold is set more towards the center of the dataset, making equivalent sections of the two classes. This is conspicuously apparent for the Bernoulli classifier stress-rupture life output (Fig. 9 at left), even though it is nearly equivalent in accuracy to all other classifiers for the prediction of rupture elongation. The stress-rupture life threshold was cut off at Haynes International's request, as the intention of this work was to apply findings to the manufacturing process. Batches above

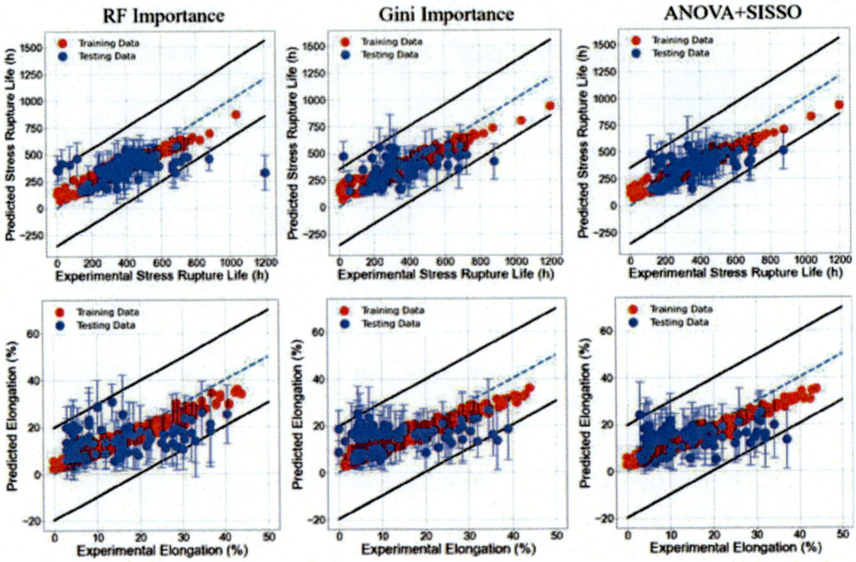

Fig. 8 Parity plots are shown for Predicted Life (top row) and Predicted Elongation (bottom row) based on training with Gini Importance features only (left plots), training with Gini Importance, Mean Accuracy Decrease, and SHAP features (middle plots), and training with all previous features including SISSO (compound) and ANOVA features (right plots). Red points represent training data, and blue points testing data with uncertainty of the prediction. The blue dashed line is the line of best fit, while the black lines are two standard deviations from that line of best fit

the 400 h threshold were judged to have negligible impacts on processing time. The observed correlation between the model accuracy and the sample frequency displayed in Fig. 1 is attributed to the difficulty in distinguishing differentiating features as sample frequency increases for a given threshold. As more samples are available for a given threshold, distinctive features are less uniform across a larger set of samples than a smaller set. Thus, we see predictive accuracy fall when a larger set of less uniform samples are passed to the binary classifier.

Discussion

Analysis of Feature Importance

The random forest features of highest importance determined from the Gini Importance method (mean decrease impurity) were the room temperature yield and tensile strength stress-rupture life, and intermediate gauge for stress rupture elongation. Other features of importance to stress-rupture life included phosphorous and silicon, which have been shown to dramatically affect stress-rupture life in nickel alloy 718

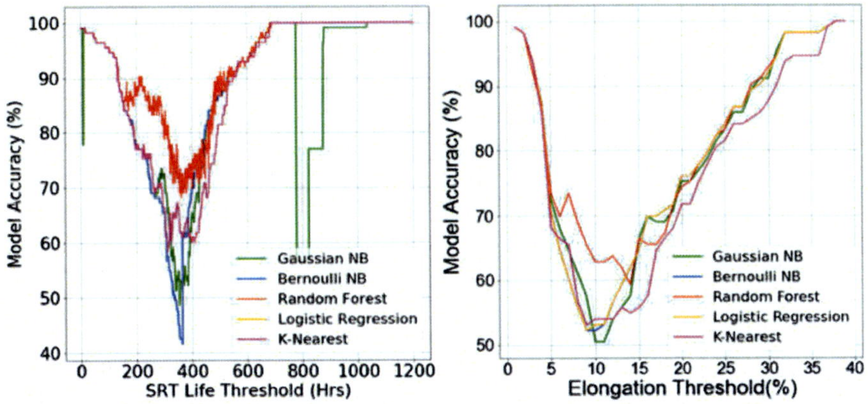

Fig. 9 Plots comparing model accuracy for various classifier tests and their error for each using an interval of 1 for stress-rupture life (SRT) at values from 4 to 400 h (left) and stress rupture elongation at values from 1 to 40% (right)

[19, 20]. As can be seen in Fig. 10, the inclusion of features, including complex ones, improves the model as they are added with the MAE of stress rupture time decreasing from 109 ± 10.3 (h) to 102 ± 8.5 (h) and stress rupture elongation from 7.23 ± 0.49 (%) to 7.19 ± 0.41 (%).

In comparison to the *scikit-learn* model with all features used for training and the *lolopy* models with selected features, the *lolopy* models outperformed the *scikit-learn* regressors as expected. The error for stress-rupture life dropped from 103 ± 9.4 to 102 ± 8.5 (h) and the stress elongation error dropped from 7.43 ± 0.51 to 7.19 ± 0.41 (%). Furthermore, the *scikit-learn* parity plots did not exhibit good

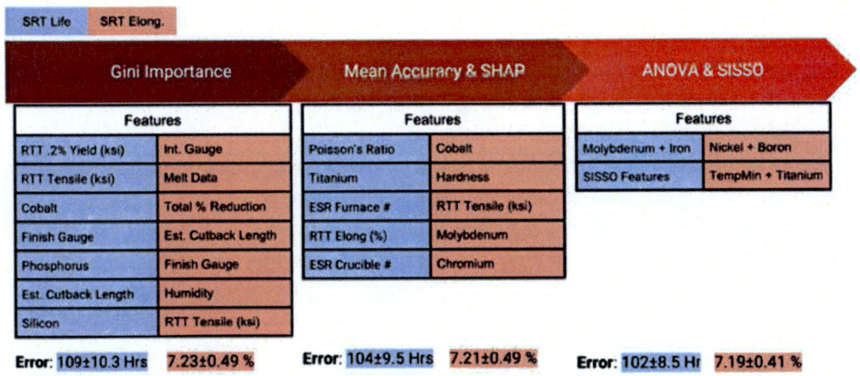

Fig. 10 Schematic showing the reduction of Mean Absolute Error (MAE) with increasing incorporation of compound features, starting with Gini Importance, then adding those from Mean Accuracy Decrease shortened to Mean Accuracy and SHAP, and finally adding ANOVA and SISSO features (compound features)

fitting behavior, and had difficulty with outliers, while the feature selected *lolopy* models were capable of learning, exhibiting better conformance to the line of best fit and keeping outliers within two standard deviations from it. In addition to this, *lolopy* provides an uncertainty value for predictions, granting further confidence in the prediction value and future values that can be utilized for diagnostics.

Classification of Elongation and Stress-Rupture Life

The classifier proved most effective at the extremes of the dataset that were the true focus of this investigation and suggests the strong potential to inform whether products would meet the requirements of the AMS5596 or AMS5662 standard. The classifier would work in combination with the regressor as a prerequisite to making a prediction of stress-rupture life using the regressor.

Classification model accuracy is highest at the two extremes of the dataset, that is, when < 4 h or $>> 400$ h is used as the threshold, because the data almost entirely belongs to one category, and it is easier for the classifier to place all data points in that category. On the other hand, it can be seen from the histogram plot that 400 h is closer to the center of the data distribution. In other words, there are a large number of data points above and below the 400 h threshold, making it harder for the classifier to predict with high accuracy. We thus see a model accuracy trend which inversely correlates with the frequency of occurrence of SRT values.

Further Work

The models have displayed interesting correlations, especially with features related to composition and environmental effects. Due to their high potential impact, further experimentation is recommended to better understand the effects of trace elements Co, P, and Si, and the environmental effect of humidity. Of further interest would be to perform the same analysis with the addition of microstructural features such as grain size, delta phase fraction and morphology, and twin boundary fraction. Additionally, more processing data including mean strains across rolling steps, rolling temperatures before and after each step, and potentially remelting melt rates could provide additional insight into creep performance based on processing history.

Conclusions

The regressor models were shown to be able to reasonably predict stress-rupture life within two standard deviations of a line of best fit for HAYNES® 718 sheet, with the most accurate predictions for the middle of the distribution of the dataset. The models did perform worse at predicting values at the extreme ends of the dataset, which was addressed by the use of a classifier to inform whether or not ingots at the extremes of the training set would pass specification. From the feature analyses, physical processing features including finish gauge, total % reduction, and RT tensile and yield were found most important. For stress rupture elongation, environmental features of humidity and melt date were also flagged as highly impactful. Trace elements of Co, P, and Si further appeared, which correlated well with the literature that suggested small compositional changes of such elements could discernably affect these types of output parameters.

To further enhance the accuracy of the model, it is believed that microstructural data could improve the feature analysis methods used. In order to obtain higher Pearson correlations in our compound feature creation, the addition of microstructural features to the dataset would allow for the discovery of compound features that would connect better with physical intuition rather than simply the importance of a given compound or mechanical property. Similar results would be expected for ANOVA analysis.

Acknowledgements This work was supported by a generous gift from Haynes International, Inc., and through computational resources provided by the Rosen Center for Advanced Computing within Information Technology at Purdue, West Lafayette, Indiana.

References

1. SAE International, "AMS5596™: Alloy Sheet, Strip, And Plate, Corrosion And Heat Resistant Nickel Base–19Cr–3.1Mo–5.2(Cb+Ta)–0.90Ti–0.60Al Consumable Electrode or Vacuum Induction Melted, Annealed," https://www.sae.org/standards/content/ams5596/.
2. SAE International, "AMS5662™: Alloy Bars, Forgings, And Rings, Corrosion And Heat Resistant Nickel Base–19Cr–3.1Mo–5.1(Cb+Ta)–0.90Ti–0.50Al Consumable Electrode or Vacuum Induction Melted, Solution Treated," https://www.sae.org/standards/content/ams5662/, (2022).
3. Wang J, Fa Y, Tian Y, Yu X. A machine-learning approach to predict creep properties of Cr–Mo steel with time-temperature parameters. Journal of Materials Research and Technology. 2021 Jul 1;13:635–50.
4. He JJ, Sandström R. Creep rupture prediction using constrained neural networks with error estimates. Materials at High Temperatures. 2022 Jun 4:1–3.
5. Zhang XC, Gong JG, Xuan FZ. A deep learning based life prediction method for components under creep, fatigue and creep-fatigue conditions. International Journal of Fatigue. 2021 Jul 1;148:106236.
6. Mamun O, Wenzlick M, Sathanur A, Hawk J, Devanathan R. Machine learning augmented predictive and generative model for rupture life in ferritic and austenitic steels. npj Materials Degradation. 2021 Apr 16;5(1):1–0.
7. Mamun O, Wenzlick M, Hawk J, Devanathan R. A machine learning aided interpretable model for rupture strength prediction in Fe-based martensitic and austenitic alloys. Scientific Reports. 2021 Mar 9;11(1):1–9.
8. National Institute for Materials Science (NIMS), "Creep Data Sheet." https://smds.nims.go.jp/creep/

9. Han H, Li W, Antonov S, Li L. Mapping the creep life of nickel-based SX superalloys in a large compositional space by a two-model linkage machine learning method. Computational Materials Science. 2022 Apr 1;205:111229.
10. Liu Y, Wu J, Wang Z, Lu XG, Avdeev M, Shi S, Wang C, Yu T. Predicting creep rupture life of Ni-based single crystal superalloys using divide-and-conquer approach based machine learning. Acta Materialia. 2020 Aug 15;195:454–67.
11. Polimetla V, Gangolu S. Predictive Approach to Creep Life of Ni-based Single Crystal Superalloy Using Optimized Machine Learning Regression Algorithms. InSoft Computing in Materials Development and its Sustainability in the Manufacturing Sector (pp. 21–35). CRC Press.
12. Zhu Y, Duan F, Yong W, Fu H, Zhang H, Xie J. Creep rupture life prediction of nickel-based superalloys based on data fusion. Computational Materials Science. 2022 Aug 1;211:111560.
13. Sanchez S, Rengasamy D, Hyde CJ, Figueredo GP, Rothwell B. Machine learning to determine the main factors affecting creep rates in laser powder bed fusion. Journal of Intelligent Manufacturing. 2021 Dec;32(8):2353–73.
14. Jain A, Ong SP, Hautier G, Chen W, Richards WD, Dacek S, Cholia S, Gunter D, Skinner D, Ceder G, Persson K. The Materials Project: A Materials Genome Approach to Accelerating Materials Innovation. APL Materials, 1 (1): 011002, 2013.
15. Seabold S, Perktold J. Econometric and statistical modeling with Python skipper seabold 1 1. InProc 9th Python Sci Conf 2010 Jun 28 (Vol. 57, p. 61).
16. Ouyang R, Curtarolo S, Ahmetcik E, Scheffler M, Ghiringhelli LM. SISSO: A compressed-sensing method for identifying the best low-dimensional descriptor in an immensity of offered candidates. Physical Review Materials. 2018 Aug 7;2(8):083802.
17. Pedregosa F, Varoquaux G, Gramfort A, Michel V, Thirion B, Grisel O, Blondel M, Prettenhofer P, Weiss R, Dubourg V, Vanderplas J. Scikit-learn: Machine learning in Python. the Journal of machine Learning research. 2011 Nov 1;12:2825–30.
18. Lundberg SM, Lee SI. A unified approach to interpreting model predictions. Advances in neural information processing systems. 2017;30.
19. Sun WR, Guo SR, Lee JH, Park NK, Yoo YS, Choe SJ, Hu ZQ. Effects of phosphorus on the δ-Ni3Nb phase precipitation and the stress rupture properties in alloy 718. Materials Science and Engineering: A. 1998 Jun 1;247(1–2):173–9.
20. Sun WR, Guo SR, Guo JT, Tong BY, Yang YS, Sun XF, Guan HR, Hu ZQ. The common strengthening effect of phosphorus, sulfur, and silicon in lower contents and a problem of a net superalloy. Superalloys 2000;2000:467–76.

Competitor Ti-Comprising Refractory High Entropy Alloys to Superalloy 718 for Aeroengine Applications

Tanjore V. Jayaraman and Ramachandra Canumalla

Abstract Superalloy 718 and its derivatives are ubiquitous to aeroengine applications owing to their excellent formability, ultra-high strength, good thermal stability, adequate weldability, and so forth. However, currently, the relatively lighter Ti-comprising refractory high entropy alloys (Ti-comprising RHEAs), having a unique combination of ambient and elevated temperature mechanical properties and corrosion resistance, are projected as potential competitors to superalloy 718 and their derivatives. We analyzed the current literature data of, relatively lighter, Ti-comprising RHEAs by a novel combination of multiple attribute decision making (MADM), hierarchical clustering (HC), and principal component analysis (PCA)—to identify the probable competitors to superalloy 718 for aeroengine applications. The ranks assigned by six MADMs, chosen for the investigation, including ARAS (additive ratio assessment), MEW (multiplicative exponent weighing), OCRA (operational competitiveness ratio), ROVM (range of value method), SAW (simple additive method), and WEDBA (weighted Euclidean distance-based approach), were concordant. PCA consolidated the MADM ranks of the alloys, while HC identified similar top-ranked alloys. The analyses identify three Ti-comprising RHEAs, viz., ONS-BCC–Ti17.8 ($Al_{20.4}$–$Mo_{10.5}$–$Nb_{22.4}$–$Ta_{10.1}$–$Ti_{17.8}$–$Zr_{18.8}$), EF-BCC-Cr20–Ti20 (Ti_{20}–Zr_{20}–Hf_{20}–Nb_{20}–Cr_{20}), and ONS-BCC–Ti27.9 ($Al_{11.3}$–$Nb_{22.3}$–$Ta_{13.1}$–$Ti_{27.9}$–$V_{4.5}$–$Zr_{20.9}$), having properties comparable or superior to superalloy 718 and reveal their potential to substitute critical parts in aeroengines.

Keywords Superalloy 718 · Ti-comprising refractory high entropy alloys · Multiple attribute decision making · Material selection

T. V. Jayaraman and R. Canumalla—Equal contribution.

T. V. Jayaraman (✉)
Department of Mechanical Engineering, University of Michigan-Dearborn, Dearborn, MI 48128, USA
e-mail: tvjraman@umich.edu

R. Canumalla (✉)
Weldaloy Specialty Forgings, Warren, MI 48089, USA
e-mail: r.canumalla@weldaloy.com

E. A. Ott et al. (eds.), *Proceedings of the 10th International Symposium on Superalloy 718 and Derivatives*, The Minerals, Metals & Materials Series,
https://doi.org/10.1007/978-3-031-27447-3_25

Introduction and Background

Superalloy 718 and its derivatives, having an excellent combination of properties, have been extensively used in aeroengine applications. For the past few decades, there have been prominent efforts to reduce the weight of the aeroengines by using lighter, stronger, and corrosion-resistant materials [1]. The use of conventional titanium alloys, pronounced for their significantly low density (about half that of steel and superalloys), good mechanical properties (both at room and elevated temperatures up to about 600 °C), corrosion resistance, and forgeability have gone up from about 0% in 1950 to beyond 30% at present in various aeroengine fan and compressor (shafts, discs, blades, casings, etc.) parts [2–5]. In low-pressure turbine blades, the intermetallic TiAl alloys (Ti–48Al–2Cr–2Nb and other variants) with even lower density and superior elevated temperature properties compared to the conventional Ti alloys have replaced heavier superalloys [6–8].

With growing interest in replacing the heavier superalloys, the sustained research over more than a decade in the new class of alloys, the high entropy alloys (HEAs), has presented opportunities for refractory high entropy alloys (RHEA). The RHEAs have an excellent combination of basic properties (reported in the current literature) including density, yield strength at room and elevated temperatures, good structure that meets the thermal stability requirements, and great potential for elevated temperature applications [1, 9–20]. To narrow down the RHEAs (comprising refractory elements Ti, Zr, Hf, V, Nb, Ta, Cr, Mo, and W) for the aeroengine applications, where density is a crucial parameter for weight reductions, it is apt to focus on the RHEAs that comprise the lighter refractory element Ti (i.e., Ti-comprising RHEAs) in the current literature, and, subsequently, compare them with the current industry benchmark of superalloy 718 and its derivatives, Inconel 718 [21–23].

The number of such Ti-comprising RHEAs is limited in the current literature. It is imperative to devise a methodology to rank the available Ti-comprising RHEAs using the basic properties mentioned above (density, yield strength at room temperature, and yield strength at elevated temperature). Once the alloys are ranked, the focus could be shifted to a few top-ranked Ti-comprising RHEAs with equivalent or superior properties compared to the benchmark and, subsequently, generate extensive data of various other properties, viz., stress rupture, creep, fracture toughness, thermal stability, etc., based on the application requirements. Such an approach will save time, effort, and cost compared to generating extensive data for numerous alloys. Material selection is a holistic approach to selecting an optimal material from a list of materials, which typically involves trade-offs between various properties, cost, availability, environmental effects, etc. [24]. Multiple attribute decision making (MADM) finds wide applications in many industries, including manufacturing, logistics, construction, transportation, and material selection, which involves making preference decisions over the available alternatives characterized by multiple and usually conflicting attributes or properties [25–27].

In the present investigation, we applied multiple attribute decision making (MADM) coupled with advanced statistics to rank the Ti-comprising RHEAs in the literature and identify the competitors to superalloy 718 for elevated temperature aeroengine applications. The paper ranks the Ti-comprising RHEAs in the literature by MADM, consolidates the ranks evaluated by diverse MADMs by applying mean-based and principal component analysis (PCA)-based techniques, assesses the relative similarities (or differences) among the alloys by hierarchical clustering (HC), and identifies/recommends the top three Ti-comprising RHEAs for further evaluation for the potential replacement of superalloy 718 and its derivatives in aeroengine applications.

Materials and Methodology

We applied a novel methodology of data-driven sorting and selection of Ti-comprising RHEAs from the literature, which consisted of the following routines: (i) compile literature data, (ii) apply six multiple attribute decision making (MADM) methods, viz., ARAS (additive ratio assessment), MEW (multiplicative exponent weighing), etc., (iii) consolidate the ranks by advanced statistical techniques, and (iv) identify Ti-comprising RHEAs that are competitors to superalloy 718 for aeroengine applications. Figure 1 presents the flowchart of the novel methodology for data-driven sorting and selection of Ti-comprising RHEAs from the literature.

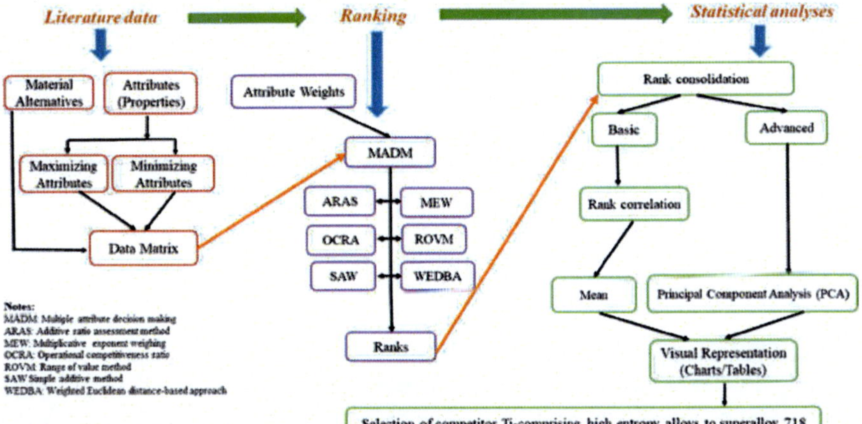

Fig. 1 The flowchart of data-driven sorting and selection of Ti-comprising refractory high entropy alloys

Literature Data

We compiled a list of Ti-comprising RHEAs and their properties from the literature, including conference proceedings and peer-reviewed journals [1, 9–19]. Table 1 presents the alloy chemistry (in at. %), processing conditions, imminent microstructures of the alloys, and unique identifier assigned for the current study—alloy designation, while Table 2 presents their properties. The properties density (ρ), yield strength at room temperature (*0.2% YS-RT*), and yield strength at 800 °C (*0.2% YS-800 °C*) reported in the current literature were analyzed. For the targeted aeroengine turbine applications, a combination of low density and high yield strengths at ambient and elevated temperatures is desirable. Hence, in the parlance of MADM, ρ is a minimizing attribute (lower the better), while *0.2% YS-RT* and *0.2% YS-800 °C* are maximizing attributes (higher the better). Thus, the alternatives (*Alloy designation*) and the attributes (ρ, *0.2% YS-RT*, and *0.2% YS-800 °C*) form the data matrix for the study.

Ranking and Analyses

We evaluated the ranks of the decision matrix (columns *Alloy designation*, ρ, *0.2% YS-RT*, and *0.2%YS-800* °C in Table 2) by several MADM methods. Making preference decisions over the available alternatives that are often characterized by multiple and usually conflicting attributes is MADM [25, 26]. Distinct components of MADM are (i) the decision matrix, which comprises alternatives and attributes, and (ii) attribute weights that quantify the relative importance of the attributes [25, 26, 28]. The attribute weights are of two types: (a) objective—applies a mathematical model to quantify the relative weights of the attributes; and (b) subjective—takes experts' opinions (sound judgment based on the intended application) and designers' opinions (design constraints) to quantify the relative weights of the attributes. We adopted subjective attribute weights wherein we assigned equal weights (1/3) to each attribute (ρ, *0.2% YS-RT*, and *0.2% YS-800* °C) in this investigation. The six MADM methods identified for the investigation are as follows: Additive ratio assessment method (ARAS) [29–31], Multiplicative exponent weighting (MEW) [25, 28, 32], Operational competitiveness ratio (OCRA) [33–35], Range of value method (ROVM) [36, 37], Simple additive weighting (SAW) [25, 26, 38, 39], and Weighted Euclidean distance-based approach (WEDBA) [26, 40, 41]. The modus operandi of the MADMs was soft-coded in Microsoft Excel.

Each MADM method applies a unique mathematical aggregation procedure to rank the alternatives; consequently, the ranks evaluated by the methods are likely to deviate. We evaluated Spearman's correlation coefficients to quantify the similarities (or differences) among the ranks from the six MADMs [42, 43]. The ranks obtained by various MADMs were consolidated by basic and advanced statistical techniques. In the former, ranks were consolidated by taking the mean (average), while, in the

Table 1 Ti-containing refractory high entropy alloys from the literature; includes alloy chemistry/composition, processing conditions, imminent microstructures, and unique identifier assigned for the current study

S#	Alloy chemistry in at. %	Processing step 1	Processing step 2	Microstructure	Alloy designation	References
1	$Ni_{47.9}$–$Al_{10.2}$–$Co_{16.9}$–$Cr_{7.4}$–$Fe_{8.9}$–$Mo_{0.9}$–$Nb_{1.2}$–$W_{0.4}$–$C_{0.4}$–$Ti_{5.8}$	Vac. arc melting followed by DS to produce columnar microstructure	Solution Treated (ST) at 1210 °C/10 h to Homogenize; Aging at 800 °C/20 h	γ' + γ'—L_{12}-Ni_3(Ti,Al) (69% 290 nm avg. size)	TKT-HESA1-FCC–Ti 5.8	[1, 9]
2	$Al_{6.25}$–C_1–Co_{15}–Cr_{13}–$Fe_{4.5}$–$Mo_{1.75}$–$Nb_{0.6}$–Ni_{48}–V_5–Ti_5	Vac. arc melting (5 times) and suction casting	ST 1175 °C/2 h-850°C/8 h-650°C/8 h-Water Quenched (WQ)	60% vol. fraction of γ'—L_{12}; 450 nm sized γ'	KWG-HESA2-FCC–Ti 5	[9]
3	Al_{10}–Co_{25}–Cr_8–Fe_{15}–Ni_{36}–Ti_6	Vac. induction melted and solidified directionally	Homogenized at 1220 °C/20 h/Furnace Cooled (FC)-900 °C/5 h/Air Cooled (AC)	L_{12} γ'-Ni_3(Ti,Al) (45% Vf/450 nm) in γ FCC solid sol. and B2/NiAl (needle like—up to 50 µm long) (< 5% V_f) precipitates	HMD-HESA3-FCC–Ti 6-5H	[10]
4	Al_{10}–Co_{25}–Cr_8–Fe_{15}–Ni_{36}–Ti_6	Vac. induction melted and solidified directionally	Homogenized at 1220 °C/20 h/FC-900 °C/50 h/AC	L_{12} γ'-Ni_3(Ti,Al) (46% Vf/460 nm) in γ FCC solid sol. & B2/NiAl (needle like, up to 50 µm long) (<5% V_f) ppts	HMD-HESA3-FCC–Ti 6-50H	[10]
5	$Al_{20.4}$–$Mo_{10.5}$–$Nb_{22.4}$–$Ta_{10.1}$–$Ti_{17.8}$–$Zr_{18.8}$	Vac. arc melting-remelted 5 times	AC, Hot Isostatic Pressing (HIP) at 1400 °C/207 MPa/2 h, 1400 °C/24 in Ar	BCC1 + BCC2; 75 µm avg. grain size; nanolamellar structures of the two phases	ONS- BCC–Ti17.8	[11, 12]
6	$Al_{21.9}$–Nb_{32}–Ta_9–$Ti_{26.7}$–$Zr_{10.3}$	Vac. arc melting-remelted 5 times	AC, HIP at 1400 °C/207 MPa/2 h, 1400 °C/24 h in Ar	BCC; 2000 µm avg. grain size; nanophases	ONS-BCC–Ti 26.7	[11, 12]

(continued)

Table 1 (continued)

S#	Alloy chemistry in at. %	Processing step 1	Processing step 2	Microstructure	Alloy designation	References
7	$Al_{7.9}$–$Hf_{12.8}$–Nb_{23}–$Ta_{16.8}$–$Ti_{18.9}$–$Zr_{20.6}$	Vac. arc melting-remelted 5 times	AC, HIP at 1200 °C/207 MPa/2 h, 1200 °C/24 h in Ar	BCC; 140 μm avg. grain size; nanophases	ONS- BCC–Ti18.9	[11, 12]
8	$Al_{5.7}$–$Nb_{23.5}$–$Ta_{17.6}$–$Ti_{27.2}$–Zr_{26}	Vac. arc melting-remelted 5 times	AC, HIP at 1200 °C/207 MPa/2 h, 1200 °C/24 h in Ar	BCC1 + BCC2; 200 μm avg. grain size; nanolamellar structures of the two phases	ONS-BCC–Ti 27.2	[11, 12]
9	$Al_{5.2}$–$Nb_{23.4}$–$Ta_{13.2}$–$Ti_{27.7}$–$V_{4.3}$–$Zr_{26.2}$	Vac. arc melting-remelted 5 times	AC, HIP at 1200 °C/207 MPa/2 h, 1200 °C/24 h in Ar	BCC; 180 μm avg. grain size; nanophases	ONS-BCC–Ti 27.7	[11, 12]
10	$Al_{11.3}$–$Nb_{22.3}$–$Ta_{13.1}$–$Ti_{27.9}$–$V_{4.5}$–$Zr_{20.9}$	Vac. arc melting-remelted 5 times	AC, HIP at 1200 °C/207 MPa/2 h, 1200 °C/24 h in Ar	BCC1 + BCC2; 100 μm avg. grain size; nanolamellar structures of the two phases	ONS-BCC–Ti 27.9	[11, 12]
11	$Al_{26.6}$–$Nb_{23.8}$–$Ti_{25.1}$–$V_{24.5}$	Vac. arc melting-remelting 5 times—and casting	Homogenized at 1200 °C/24 h	BCC SS (300 to 400 μm grain size)	NDS-BCC–Ti 25.1	[13]
12	Nb_{20}–Cr_{20}–Mo_{10}–Ta_{10}–Ti_{20}–Zr_{20}	Vac. arc melting and re-melted 5 times and casting	AC, HIP at 1450 °C/207 MPa/3 h	BCC1 (67% V_f) + BCC2 (16% V_f) + Laves (FCC) (17% V_f)	ONS-BCC–Ti 20	[14]
13	$Nb_{28.3}$–$Ti_{24.5}$–V_{23}–$Zr_{24.2}$	Vac. arc melting and re-melted 5 times and casting	AC, HIP at 1200 °C/207 MPa/2 h, 1200 °C/24 h	BCC + submicron ppts	DBM-BCC–Ti 24.5	[15, 16]
14	$Nb_{22.6}$–$Ti_{19.4}$–$V_{37.2}$–$Zr_{20.8}$	Vac. arc melting and re-melted 5 times and casting	AC, HIP at 1200 °C/207 MPa/2 h, 1200 °C/24 h	BCC1 + BCC2 + BCC3	DBM–BCC–Ti19.4	[15, 16]

(continued)

Table 1 (continued)

S#	Alloy chemistry in at. %	Processing step 1	Processing step 2	Microstructure	Alloy designation	References
15	$Cr_{24.6}$–$Nb_{26.7}$–$Ti_{23.9}$–$Zr_{24.8}$	Vac. arc melting and re-melted 5 times and casting	AC, HIP at 1200 °C/207 MPa/2 h, 1200 °C/24 h	BCC + Laves (ordered FCC)	DBM-BCC–Ti 23.9	[15, 16]
16	$Cr_{20.2}$–Nb_{20}–$Ti_{19.9}$–$V_{19.6}$–$Zr_{20.3}$	Vac. arc melting and re-melted 5 times and casting	AC, HIP at 1200 °C/207 MPa/2 h, 1200 °C/24 h	BCC + Laves (ordered FCC)	DBM- BCC–Ti19.9	[15, 16]
17	Hf_{20}–Mo_{20}–Nb_{20}–Ti_{20}–Zr_{20}	Vac. arc melting and re-melted 5 times and casting	AC, Homogenized at 1100 °C/10 h/SC	BCC (dendritic)-no ordered phases	NNG-BCC–Ti 20	[17]
18	$Ta_{19.68}$–$Nb_{18.93}$–$Hf_{20.46}$–$Zr_{21.23}$–$Ti_{19.7}$	Vac. arc melting and casting	AC, HIP at 1200 °C/207 MPa/2 h, 1200 °C/24 h (only homogenized)	BCC (dendritic and nonuniform); equiaxed grains about 100 μm bottom to 200 μm at the top	ONS- BCC–Ti19.7	[18]
19	Ti_{20}–Zr_{20}–Hf_{20}–Nb_{20}–V_{20}	Vac. induction melting and cast	As-Cast	BCC + U (unknown intermetallic phase)	EF-BCC-V20–Ti 20	[19]
20	Ti_{20}–Zr_{20}–Hf_{20}–Nb_{20}–Cr_{20}	Vac. induction melting and cast	As-Cast	BCC + Laves (L1 + L2)	EF-BCC–Cr20–Ti 20	[19]

Table 2 The properties—density (ρ), yield strength (0.2% strain offset) at room temperature (*0.2% YS-RT*), and yield strength (0.2% strain offset) at 800 °C (*0.2% YS-800 °C*)—of the Ti-containing refractory high entropy alloys (from the literature) identified for the data-driven analyses

Alloy designation	Density (ρ) (g/cm^3)	0.2% YS-RT (MPa)	0.2% YS-800 °C (MPa)	References
TKT-HESA1-FCC–Ti 5.8	7.44	875	800	[1, 9]
KWG-HESA2-FCC–Ti 5	7.60	1000	700	[9]
HMD-HESA3-FCC–Ti 6-5H	7.38	568	535	[10]
HMD-HESA3-FCC–Ti 6-50H	7.38	596	581	[10]
ONS- BCC–Ti17.8	7.21	2000	1597	[11, 12]
ONS-BCC–Ti 26.7	6.74	1280	728	[11, 12]
ONS- BCC–Ti18.9	8.86	1841	796	[11, 12]
ONS-BCC–Ti 27.2	7.91	1965	362	[11, 12]
ONS-BCC–Ti 27.7	7.49	1965	678	[11, 12]
ONS-BCC–Ti 27.9	7.27	2035	796	[11, 12]
NDS-BCC–Ti 25.1	5.46	1020	685	[13]
ONS-BCC–Ti 20	8.19	1595	983	[14]
DBM-BCC–Ti 24.5	6.50	1105	187	[15, 16]
DBM- BCC–Ti19.4	6.44	918	240	[15, 16]
DBM-BCC–Ti 23.9	6.66	1260	300	[15, 16]
DBM- BCC–Ti19.9	6.53	1298	615	[15, 16]
NNG-BCC–Ti 20	8.62	1575	825	[17]
ONS- BCC–Ti19.7	9.79	929	535	[18]
EF-BCC-V20–Ti 20	8.00	1170	1148	[19]
EF-BCC-Cr20–Ti 20	8.17	1375	1325	[19]

latter, the ranks were consolidated by principal component analysis (PCA) [44–46]. PCA, a multivariate technique, reduces the dimensionality of the data set consisting of several variables to a new set of variables by orthogonal transformation. The new set of variables, commonly termed principal components (PC), are ordered such that the first few PCs (usually one or two) retain the most variations in the original data. Additionally, we evaluated the relative similarities (or differences) among the alloys by hierarchical clustering (HC) [47]. The statistical analyses were carried out on the commercial software Minitab® 20.3.

Results and Discussion

Table 3 presents the descriptive statistics of the Ti-comprising RHEAs in the literature. Inconel 718, the current benchmark for aeroengine turbine applications, is a conventional alloy (not a high entropy alloy) whose ΔS_{config}, ρ, *0.2% YS-RT,* and

Table 3 Descriptive statistics of the properties of the Ti-comprising refractory high entropy alloys from the literature

Attribute	Minimum	Maximum	Range	Mean ± stdev
$\Delta S_{config}/R$ (mol^{-1})	1.35	1.76	0.41	1.58 ± 0.12
ρ *(g/cm^3)*	5.46	9.79	4.33	7.47 ± 0.98
0.2% YS-RT (MPa)	568	2035	1467	1319 ± 464
0.2% YS-800 °C (MPa)	187	1597	1410	721 ± 351

0.2% YS-760 °C are ~10.81 J/Kmol, 8.28 g/cm^3, 1034 MPa, and 758 MPa (at 760 °C), respectively [21–23]. Comparing the properties of the Inconel 718 with the descriptive statistics of the literature data of Ti-comprising RHEAs reveals the following: (i) the mean of ρ (~7.47 g/cm^3) is less than the benchmark, (ii) the mean *0.2% YS-RT* (~1319 MPa) is greater than the *0.2% YS-RT* of the benchmark, (iii) the yield strength for Inconel 718 (~758 MPa) at a relatively lower temperature of 760 °C is greater than the mean of *0.2% YS-800* °C (~721 MPa) of the literature data, which is at a higher temperature, i.e., 800 °C; it is reasonable to assume that *0.2% YS* for Inconel 718 at 800 °C is likely to be similar to the mean of the *0.2% YS-800* °C of the Ti-comprising high entropy alloys. Hence, the combination of properties of certain Ti-comprising RHEAs is likely to be better than the benchmark based on the three attributes chosen (and available in the literature) for the investigation.

Figure 2 shows the ranks of the alloys evaluated by the six MADMs. The alloys ranked 1, 2, and so on are considered top or best alloys. Since each MADM method applies a unique mathematical aggregation procedure to sort the alternatives, the ranks evaluated by various methods are likely to deviate, as evident from the figure. For example, all the MADMs identify ONS- BCC–Ti17.8 (shaded in green) as the top-ranked alloy (rank#1). On the contrary, the rank evaluated by the various MADMs to NDS- BCC–Ti25.1 (colored in red) differs significantly. Table 4 shows Spearman's correlation coefficients (S_ρ) that quantify the similarities (or differences) among the ranks evaluated by the six MADMs. However, of the 15 combinations, 11 combinations (~73%) have Spearman's correlation coefficients (S_ρ) > 0.80, while the rest have S_ρ > 0.60. Such a strong correlation of ranks elicits that it is reasonable to consolidate the ranks from various MADMs.

Figure 3a presents the mean-based consolidation of the ranks of Ti-comprising RHEAs from the six MADMs. The consolidated rank of the alloys is superimposed (solid yellow points and dashed green lines) over the individual MADM ranks as in Fig. 2. The three top-ranked alloys are ONS-BCC–Ti17.8 (Rank#1), EF-BCC-Cr20–Ti 20 (Rank#2), and ONS-BCC–Ti 27.9 (Rank#3). Figure 3b shows the PCA-based consolidation (score plot) of the ranks of Ti-comprising high entropy alloys from the six MADMs. The score plot presents the first two components (*PC1* and *PC2*), post-reduction of the data dimension (6, i.e., ranks from six MADMs) into a two-dimensional space. Table 5 presents the eigenvalues (and their proportion) that capture the variation of the distribution of each principal component. The new axes capture ~98% of the variation in the original data. Thus, this way of presentation qualifies to be called a rank chart. The first principal component (*PC1*) captures

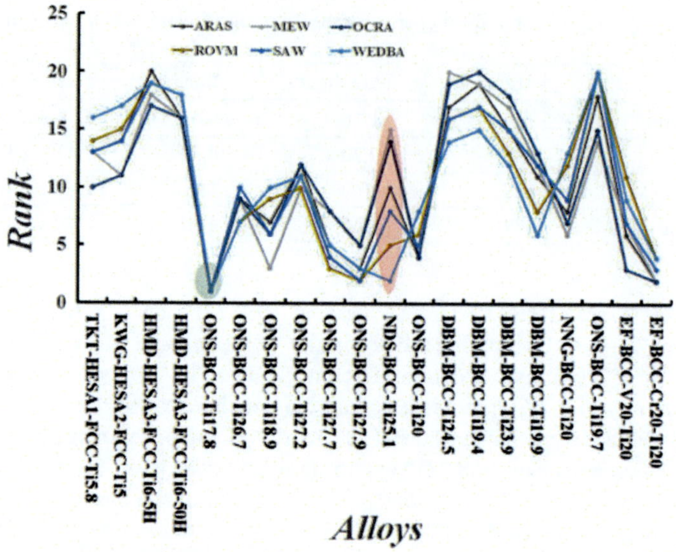

Fig. 2 The ranks of Ti-comprising refractory high entropy alloys evaluated by the six multiple attribute decision making (MADM) methods. For example, all the MADMs assign similar rank #1 (green ellipse) to the alloy ONS- BCC–Ti17.8. On the other hand, MADMs give a diverse rank (pink ellipse) to the alloy NDS-BCC–Ti 25.1

Table 4 The Spearman rank (S_ρ) correlation among the six multiple attribute decision making (MADM) methods

	ARAS	MEW	OCRA	ROVM	SAW
MEW	0.920				
OCRA	0.929	0.965			
ROVM	0.910	0.741	0.728		
SAW	0.979	0.872	0.877	0.941	
WEDBA	0.836	0.618	0.617	0.965	0.877

~88% of the variation or scatter in the original data, while the second principal (*PC2*) describes ~10% of the variation. Since *PC1* captures nearly 88% of the variation in the initial six dimensions (i.e., sets of ranks), it approximates the consolidated ranks of Ti-comprising RHEAs. An imaginary reference line (perpendicular to *PC1*) traversing from left to right (–4 to 4) indicates the overall ranks of the alloys. The alloys ONS- BCC–Ti17.8 ($Al_{20.4}$–$Mo_{10.5}$–$Nb_{22.4}$–$Ta_{10.1}$–$Ti_{17.8}$–$Zr_{18.8}$), EF-BCC-Cr20–Ti 20 (Ti_{20}–Zr_{20}–Hf_{20}–Nb_{20}–Cr_{20}), and ONS-BCC–Ti 27.9 ($Al_{11.3}$–$Nb_{22.3}$–$Ta_{13.1}$–$Ti_{27.9}$–$V_{4.5}$–$Zr_{20.9}$) are the top three alloys in that order. The three top-ranked alloys by PCA-based consolidation are strikingly similar to the top-ranked alloys by mean-based rank consolidation. The dendrogram, by hierarchical clustering (HC), in Fig. 4 revealed that the top three alloys are ~85% similar to one another and are only somewhat similar to the rest of the alloys. Apparently, the similarities among the top three Ti-comprising RHEAs are that the configurational entropy is greater than ~13.3 J/K mol and the alloys have a body-center cubic crystal structure. Specifically, the top-ranked alloy ONS- BCC–Ti17.8 ($Al_{20.4}$–$Mo_{10.5}$–$Nb_{22.4}$–$Ta_{10.1}$–$Ti_{17.8}$–$Zr_{18.8}$) is lighter than superalloy 718 by ~13% and has significantly superior room temperature and elevated temperature yield strengths. The potency of the data-driven methodology could further be tapped by effectively and appropriately choosing the weights of the properties for specific aeroengine turbine applications.

From the materials engineering perspective, refractory elements Ti, Zr, Hf, V, Nb, Ta, Cr, Mo, and W in RHEAs extend the application of these alloys to elevated temperatures [11–20] due to their intrinsic nature—high melting point promoting higher thermal stability and service temperature, and the ease to form single-phase microstructures. Ti, the lightest among the refractory elements, would contribute to a significant reduction of density. Further, to reduce the density, Al and Si are added that improves other properties as well. With a proper combination of the refractory alloying elements along with the Al or Si or both, Ti-comprising RHEAs having lower densities and superior properties to superalloys could be achieved. Two different approaches appear to emerge. In the first approach, the solid solutions of the body-centered-cubic microstructures dominate since the majority of the refractory metals used to make these RHEA alloys are essentially BCC elements (viz., ONS- BCC–Ti17.8, ONS-BCC–Ti 26.7, ONS- BCC–Ti18.9, ONS-BCC–Ti 27.2, ONS-BCC–Ti 27.7, ONS-BCC–Ti 27.9, NDS-BCC–Ti 25.1, ONS-BCC–Ti 20, DBM-BCC–Ti 24.5, DBM- BCC–Ti19.4, DBM-BCC–Ti 23.9, DBM- BCC–Ti19.9, NNG-BCC–Ti 20, ONS- BCC–Ti19.7, EF-BCC-V20–Ti 20, and EF-BCC-Cr20–Ti 20 shown in Table 1). Each composition of these RHEAs is unique and needs to be developed and studied for their microstructures, like a single BCC phase or multiple BCC phases of solid solutions to provide the required properties based on the solid solution strengthening mechanism; in general, being single or a two-phase with no precipitates or very minimal are expected to have better thermal stability and oxidation resistance for use for the intended high-temperature applications. The second approach is similar to the physical metallurgy of superalloys (viz., TKT-HSEA1-FCC–Ti 5.8, KWG-HESA2-FCC–Ti 5, HMD-HESA3-FCC–Ti 6-5H, and HMD-HESA3-FCC–Ti 6-50H shown in Table 1) where the RHEAs have significant volume fractions of stable, coherent γ'' to impart high-temperature strength [1, 9, 10]. This approach is evolving, and more alloys and data and understanding of the mechanisms are underway by various groups.

The literature data is predominantly cast alloys subjected to thermal treatments (to reduce chemical segregation), and only a limited few were thermomechanically processed for microstructure evolution. Wrought microstructures with grain refinement and other strengthening mechanisms should be investigated to further improve properties in a targeted way. Further, most of the data in the literature are for compression testing, and tensile data is required, especially for a clearer picture of the ductility. Furthermore, creep, fatigue, fracture toughness, and oxidation resistance studies are also desirable. While there is a lot of data that is desired, the present effort would assist in selecting the top-ranked alloys using the ranking methodology, and thus, one

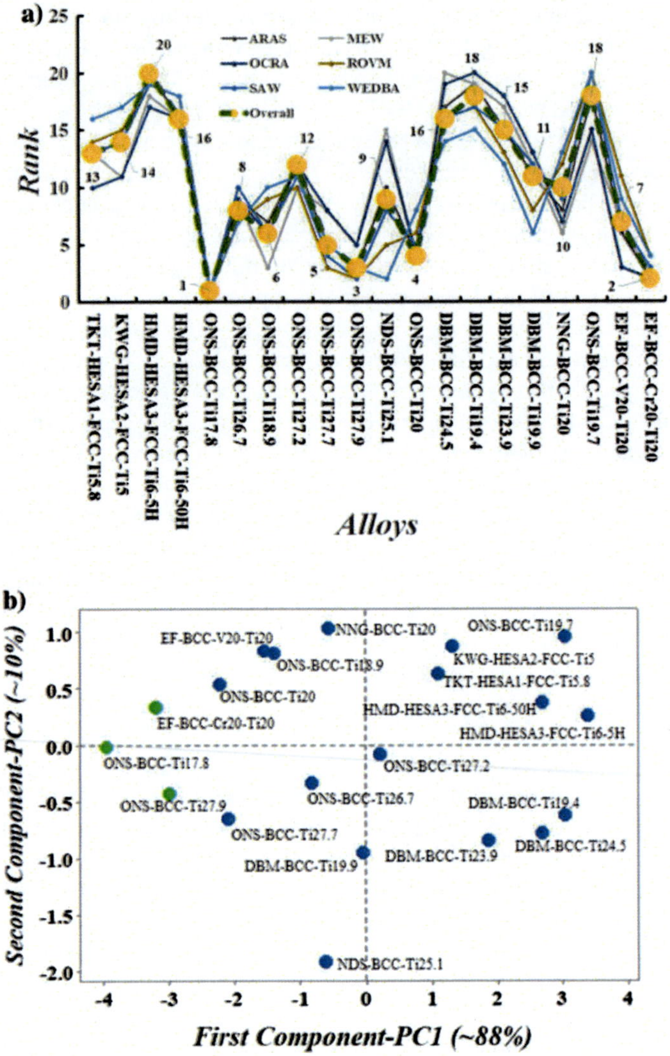

Fig. 3 The rank consolidation by **a** mean and **b** principal component analysis (PCA) of Ti-comprising refractory high entropy alloys evaluated by the six multiple attribute decision making (MADM) methods. The ranking of the top three alloys by both methods match

could concentrate on some selected alloys to generate extensive data in the desired direction. The investigation identifies a few Ti-comprising RHEAs that match the current benchmark, and provides guidelines and directives to focus on the further evaluation and development of the identified Ti-comprising RHEAs. Future work will include additional diverse MADM methods for robust analyses of a broader data set, additional properties/attributes, and Ti-comprising RHEAs in the literature.

Table 5 The eigenvalues (and their proportion) by principal component analysis (PCA) of the ranks of the Ti-comprising refractory high entropy alloys evaluated by the six multiple attribute decision making (MADM) methods

	PC1	PC2	PC3	PC4	PC5	PC6
Eigenvalue	5.2696	0.6314	0.0415	0.0316	0.0153	0.0106
Proportion	0.878	0.105	0.007	0.005	0.003	0.002
Cumulative	0.878	0.983	0.990	0.996	0.998	1.000

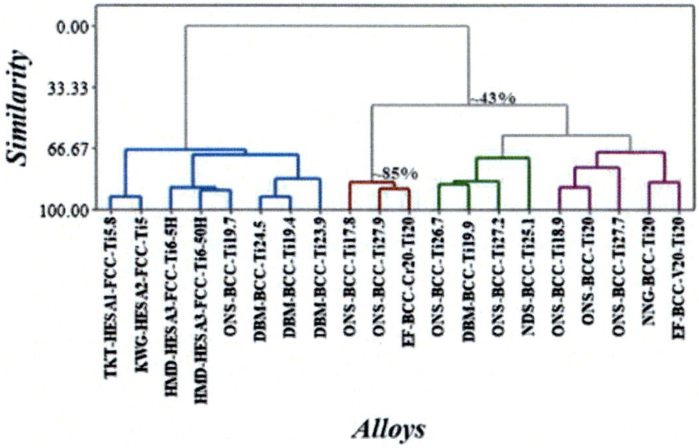

Fig. 4 The dendrogram, by hierarchical clustering (HC), of the ranks Ti-comprising refractory high entropy alloys evaluated by the six multiple attribute decision making (MADM) methods

The focus and objective of the present work have been to rank the available research alloys based on the limited available properties/attributes using the MADM methodology, and recommend extensive study and generation of desired data on the top-ranked alloys for the intended use effectively and efficiently saving time, effort, and cost. The methodology has been demonstrated as a useful tool in this paper and iterations of this procedure could be done as more and more research data becomes available on the limited basic attributes on more research Ti-comprising RHEAs as discussed.

Summary, Conclusions, and Future Work

We analyzed the basic properties of Ti-comprising RHEAs available in the current literature by a novel combination of multiple attribute decision making (MADM)

methods and advanced statistics—hierarchical clustering (HC) and principal component analysis (PCA), to identify the probable competitors to superalloy 718 for aeroengine applications. The ranks assigned by six MADMs (ARAS, MEW, OCRA, ROVM, SAW, and WEDBA) were concordant. Simple and advanced (PCA) statistical techniques consolidated the ranks of the alloys and identified the following top three Ti-comprising RHEAs: ONS-BCC–Ti17.8 ($Al_{20.4}$–$Mo_{10.5}$–$Nb_{22.4}$–$Ta_{10.1}$–$Ti_{17.8}$–$Zr_{18.8}$), EF-BCC-Cr20–Ti 20 (Ti_{20}–Zr_{20}–Hf_{20}–Nb_{20}–Cr_{20}), and ONS-BCC–Ti 27.9 ($Al_{11.3}$–$Nb_{22.3}$–$Ta_{13.1}$–$Ti_{27.9}$–$V_{4.5}$–$Zr_{20.9}$), in that order, as competitors to superalloy 718 and their derivatives. Hierarchical clustering (HC) revealed that the top three alloys are ~85% similar—alloys have not only configurational entropy greater than ~13.3 J/K mol but also a body-center cubic crystal structure. The top-ranked alloy ONS- BCC–Ti17.8 ($Al_{20.4}$–$Mo_{10.5}$–$Nb_{22.4}$–$Ta_{10.1}$–$Ti_{17.8}$–$Zr_{18.8}$) is lighter than superalloy 718 by ~13% and has significantly superior room temperature and elevated temperature yield strengths. The potency of the methodology could further be tapped by effectively and appropriately choosing the weights of the properties for specific aeroengine turbine applications. Future work will include additional diverse MADM methods for robust analyses, other properties, and more Ti-comprising RHEAs as they appear in the literature. From the materials engineering point of view, the ranking could apparently suggest the direction to be pursued to develop body-centered cubic structures if the extensive and rapid data generation on other desired properties on the top-ranked alloys also falls in line with the intended application.

Acknowledgements T. V. Jayaraman thanks the College of Engineering and Computer Science (grant# 049150) and the Institute of Advanced Vehicle Systems (grant# 052349) at the University of Michigan-Dearborn for their support. Ramachandra Canumalla thanks the Weldaloy Specialty Forgings management for all the support.

References

1. Tsao TK, Yeh AC, Kuo CM, Kakehi K, Murakami H, Yeh JW, Jian SR (2017) The high temperature tensile and creep behaviors of high entropy superalloy, Scientific Reports 7:
2. Eylon D, Fujishiro S, Postans PJ, Froes FH (1984) High-temperature titanium alloys—a review, Journal of Metals 1:55–62
3. Gogia AK (2005) High-temperature titanium alloys, Defense Science Journal 55:149–173
4. Rao MN (2011) Materials for gas turbines–An overview, in Advances in Gas Turbine Technology, Intechopen, 2011:293–314.
5. Canumalla R (2020) On the low tensile ductility at room temperature in high temperature titanium alloys, SCIREA Journal of Metallurgical Engineering 4(2):16–51
6. Bewlay B, Nag S, Suzuki S, Weimer M (2016) TiAl alloys in commercial aircraft engines, Materials at High Temperatures 33(4–5):549–559
7. Janschek P (2015) Wrought TiAl Blades, in Materials Today: Proceedings
8. Sallot P, Martin G, Knittel S (2017) Implementation of γ-TiAl Alloys for Low Pressure Turbine Blades: Opportunities and New Challenges, in *TMS 2017*
9. Gaber KW (2020) The design and characterization of HEA for high-temperature applications, New Mexico, New Mexico: New Mexico Institute of Mining and Technology

10. Daoud HM, Manzoni AM, Wanderka N, Glatzel U (2015) High-temperature tensile strength of Al10Co25Cr8Fe15Ni36Ti6 compositionally complex alloy (high entropy alloy), Journal of Materials (JOM).
11. Senkov ON, Senkova SV, Woodward C (2014) Effect of aluminum on the microstructure and properties of two refractory high entropy alloys, Acta Materialia 68:214–228
12. O. N. Senkov ON, C. Woodward (2014) Microstructure and properties of aluminum-contaning refractory high entropy alloys, Journal of Materials (JOM) 66(10):2030–2042
13. Stepanov ND, Shaysultanov DG, Salishchev GA, Tikhonovsky MA (2015) Structure and mechanical properties of a lightweight AlNbTiV high entropy alloy, Materials Letters 142:153–155
14. Senkov ON, Woodward CF (2011) Microstructure and properties of a refractory NbCrMo0.5Ta0.5TiZr, Materials Science and Engineering A 529:311–320
15. Senkov ON, Senkova SV, Woodward C, Miracle DB (2013) Low-density refractory multi-principal element alloys of CrNbTiVZr system: microstructure and phase analysis, Acta Materialia 61:1545–1557
16. Senkov ON, Senkova SV, Miracle DB, Woodward C (2013) Mechanical properties of low-density refractory multi-principal element alloys of CrNbTiVZr system, Materials Science and Engineering A 565:51–62
17. Guo NN, Wang L, Luo LS, Li XZ, Su YQ, Guo JJ, Fu HZ (2015) Microstructure and mechanical properties of refractory MoNbHfZrTi high entropy alloy, Materials & Design 81:87–94
18. Senkov ON, Scott JM, Senkova SV, Meisenkothen F, Miracle DB, Woodward CF (2012) Microstructure and elevated temperature properties of a refractory TaNbHfZrTi alloy, Journal of Materials Science 47:4062–4074
19. Fazakas E, Zadorozhnyy V, Varga LK, Inoue A, Louzguine-Luzgin DV, Tian F, Vitos L (2014) Experimental and theoretical study of Ti20Zr20Hf20Nb20X20 (X 1/4 V or Cr) refractory high entropy alloys, International Journal of Refractory Metals and Hard Materials 47:131–138
20. Miracle DB, Senkov ON (2017) A critical review of high entropy alloys and related concepts., Acta Mater. 122:448–511
21. *INCONEL alloy 718 Data Dheet,* Special Metals Corporation, 2007.
22. "https://www.hightempmetals.com/techdata/hitempInconel718data.php," 2015. [Online].
23. *ALLOY BARS, FORGINGS, AND RINGS, CORROSION AND HEAT RESISTANT Nickel Base-19Cr-3.1Mo-5.1(Cb + Ta)-0.90Ti -0.50Al Solution and Precipitation Heat Treated Consumable Electrode or Vacuum Induction Melted AMS5663,* SAE International, 2022.
24. Jahan A, Ismail MY, Sapuan SM, Mustapha F (2010) Materials screening and choosing methods–A review, Materials and Design 31:696–705
25. G. H. Tzeng GH, J. J. Huang JJ (2011) Multiple attribute decision making methods and applications, CRC Press
26. Rao RV (2007) Decision making in the manufacturing environment, Using graph theory and fuzzy multiple attribute decision making methods, Springer
27. Zavadskas EK, Turskis Z, Kildiene S (2014) State of art surveys of overviews on MCDM/MADM methods, Technol. Econo. Dev. Economy. 20)(1):165–179
28. Jahan A, Edwards KL, Bahraminasab M (2016) Multi-criteria Decision Analysis—For supporting the selection of engineering materials in product design, 2 ed., Butterworth-Heinemann, Elsevier
29. Zavadskas EK, Turskis Z (2020) A new additive ratio assessment (ARAS) method in multicriterial decision making, Technol. Economic Dev. Of Economy (Baltic Journal of Sustainability) 16(2):159–172
30. Zavadskas EK, Turskis Z, Vilutiene T (2010), Multiple criteria analysis of foundation installment alternatives by applying additive ratio assessment (ARAS) method, Archives of Civil and Mechanical Engineering 10(11):123–141
31. Stanujic S, Jovanovic R (2012), Measuring a quality of faculty website using ARAS method, Contemporary Issues in Business Management and Education, 2012:545–554
32. Windarto AP, Muhamad A (2017) Comparison of weighted sum model and multi-attribute decision making weighted product methods in selecting the best elementary school in Indonesia, Int. J. Software Eng. App. 11(4):69–90

33. Chatterjee P, Chakraborty S (2012), Material selection using preferential ranking methods, Mater. & Design 35:384–393
34. Erdogan S, Aydin S, Balki MK, Sayin C (2020) Operational evaluation of thermal barrier coated diesel engine fueled with biodiesel/diesel blend by using MCDM method base on engine performance, emission and combustion characteristics Renewable Energy 151:698–706
35. Darji VP, Rao RV (2014) Intelligent multi-criteria decision making methods for material selection in sugar industry, Procedia Materials Science 5:2585–2594
36. Madic M, Radovanovic M, Manic M (2016) Application of the ROV method of selection of cutting fluids, Decision Science Letters 5:245–254
37. Jha GK, Chatterjee P, Chatterjee R, Chakraborty S (2013) Suppliers selection in a manufacturing environment using a range of value method, J. of Mech. Eng. 3(3):16–22
38. Afshari A, Mojaahed M, Yusuff RM (2010), Simple additive weighting approach to personal selection problem, Int. J. Innov. Mgt. Technol. 1(5):511–515
39. Memariani A, Amini A, Alinezhad A (2009) Sensitivity analysis of simple additive weighting method (SAW): the results of change in the weight of one attribute on the final ranking alternatives, J. Indust. Eng. 4:13–18
40. Rao RV, Singh D (2012) Evaluating flexible manufacturing systems using Euclidean distance-based integrated approach, Int. J. Risk. Manag., 3:32–53
41. Rao RV, Singh D (2012) Weighted Euclidean distance-based approach as a multiple attribute decision making method for plant or facility layout design selection, Int. J. Indust. Eng. Comp. 3:365–382
42. Levine DM, Ramsey PP, Smidt RK (2001) Applied Statistics for Engineers and Scientists, Upper Saddle, NJ: Prentice-Hall
43. Navidi W (2010) Statistics for Engineers and Scientists, 3 ed., McGraw-Hill Science/Engineering
44. Rajan K (2005) Materials Informatics, Materials Today 8(10):38–45
45. Cadima J, Jolliffle IT (2016) Principal Component Analysis: A review and recent developments, Phil. Trans. R. Soc. A 374:20150202
46. George L, Hrubiak R, Rajan R, Saxena SK (2009), Principal component analysis on properties of binary and ternary hydrides and a comparison of metal versus metal hydride properties, J. Alloys and Compds. 478(1–2):731–735
47. Arabie P, Hubert LJ, De Soete G (1999) Clustering and Classification, River Edge, NJ: World Scientific

An ICME Framework to Predict the Microstructure and Yield Strength of INCONEL 718 for Different Heat Treatments

Taiwu Yu, Thomas Barkar, Carl-Magnus Lancelot, and Paul Mason

Abstract The superalloy INCONEL 718 stands out for its excellent manufacturability and strength at ambient temperature. In most recent studies people tried to improve the mechanical properties of the alloy through adjusting different processing conditions such as solution annealing temperature, aging temperature and holding time, and the amount of intermediate cold work. Such studies could be expensive and time consuming. This study aims to build a CALPHAD-based ICME framework to investigate the microstructural stability and mechanical properties using the Thermo-Calc software. The evolution of precipitates is characterized using the TC-PRISMA precipitation module paired with the TCNI12 and MOBNI6 databases. The microstructure is simulated in terms of the nucleation and growth of the precipitates. The precipitation of the secondary phases$\gamma\prime$, $\gamma\prime\prime$ and δ is simulated under different aging temperatures and their contributions to yield strength are quantified. A simplified yield strength model is applied to predict the precipitate, grain boundary and solid solution strengthening. The quantified results show good agreement with the experiment.

Keywords INCONEL 718 · Thermo-Calc software · Yield strength · Microstructure · Precipitation

Introduction

With the dramatic achievements and significant progress in the last decade, Integrated Computational Materials Engineering (ICME) provides an unparalleled design and manufacturing paradigm in accelerating the R&D of advanced materials. In recent

T. Yu (✉) · P. Mason
Thermo-Calc Software Inc, 4160 Washington Road, McMurray, PA, USA
e-mail: taiwu@thermocalc.com

T. Barkar · C.-M. Lancelot
Thermo-Calc Software AB, Råsundavägen 18, 169 67 Solna, Sweden

E. A. Ott et al. (eds.), *Proceedings of the 10th International Symposium on Superalloy 718 and Derivatives*, The Minerals, Metals & Materials Series,
https://doi.org/10.1007/978-3-031-27447-3_26

years, the ICME tools have been widely used in the development of Ni-based superalloys [1–4]. These tools provide the advantage of predicting mechanical properties such as creep life [1, 2], fatigue life [4] as well as oxidation resistance [1] while saving time and effort. As an example, Conduit et al. [5] applied a neural network to obtain desired properties. However, there are a limited number of alloys for whose measured properties agree with the theoretical predictions within acceptable uncertainties. These alloys include some commercial alloys such as Udimet720, LSHR, Rene104 and RR1000 alloys. For other alloys, the low accuracy can often be attributed to the lack of sufficient information related to microstructure (e.g., particle density, distribution, grain size, etc.) which is used to predict the mechanical properties. There are limited studies on the correlation between microstructure and yield strength during heat treatment through ICME. However, with a CALPHAD-based ICME framework exemplified in this work, it is possible to simulate the precipitation over the allowable composition range for these alloys. This works presents the simulation of the microstructure and a corresponding yield strength (YS) of INCONEL 718 (IN718 for short throughout the entire paper).

Modeling and Results

The following sections begin by discussing the CALPHAD-based thermodynamic description that provides the foundation of this framework, followed by a simulation of the developing microstructure in IN718 during heat-treatment, and the prediction of mechanical properties based on this microstructure.

Thermodynamic Calculation with the CALPHAD Method

Calculation of Phase Diagrams (CALPHAD) is a proven methodology for predicting thermodynamic, kinetic, and other properties of multicomponent material systems [6]. The theoretical basis of phase diagram calculation rests on the fact that the location of a phase boundary is the result of stability competition between two or more phases. The essence of this approach is to develop thermodynamic databases of multi-component systems using the available experimental data of lower order systems, i.e., binaries and ternaries, using physics-based models to extrapolate and interpolate the descriptions into the unknown parameter space. Thus, the predictive capacity of a thermodynamic database depends on the quality of the thermodynamic modeling and assessments of low-order systems.

Combined with the Thermo-Calc software version 2022b [7], TCNI12 was used in this work to predict phase stabilities in IN718. Compared to previous version, TCNI12 includes 371 binary and 431 ternary systems. TCNI12 is a self-consistent

Table 1 The composition range of IN718 from ASM Specialty Handbook [8]

	Ni	Cr	Nb	Mo	Ti	Al	Fe
Range (wt.%)	50–55	17–21	4.75–5.5	2.8–3.3	0.65–1.15	0.2–0.8	Bal
This work (wt.%)	54.6	19.0	5.0	3.0	0.9	0.5	17.0

thermodynamic database developed using the CALPHAD approach for the application of Ni-based superalloys. With the CALPHAD technique an internally consistent thermodynamic description can be obtained to describe phase equilibria through the underlying thermodynamics of a system. Additionally, various thermophysical properties such as molar volume and thermal expansivities can also be critically evaluated for most phases of importance to Ni-based superalloys. Some of these phases crucial to IN718 are the primary matrix phase γ (labeled as FCC_L12#1 in the database), the strengthening phase γ' (labelled as FCC_L12#2), the main strengthening phase γ'' (labelled as BCT_D022) and the often unwanted phase δ (NI3TA_D0a).

The chemistry of IN718 can be found within the limits as specified in ASM specialty handbook [8] (unit: wt.%): 50.0 to 55.0 min Ni + Co; 17.0 to 21.0 Cr; 4.75 to 5.50 Nb; 2.80 to 3.30 Mo; 0.65 to 1.15 Ti; 0.20 to 0.80 Al; 1.00 max Co; 0.08 max C; 0.35 max Mn; 0.35 max Si; 0.015 P; 0.015 S; 0.006 max B; 0.30 max Cu; balance Fe. The composition ranges of the major elements and the composition used for CALPHAD calculation performed during this work is shown in Table 1. Note that the impurity elements were excluded in order to avoid minor phases and to reduce the computational costs for the simulations.

Prior to performing any kinetic simulations, it is necessary to determine which phases may be present. To achieve this, an equilibrium calculation was first performed with the thermodynamic database TCNI12, varying the composition shown in Table 1 and temperature ranging from 600 to 2000 °C. The result is shown in Fig. 1a. This shows the formation of secondary phases δ, σ and γ'. However, as can be seen, under equilibrium conditions, the γ'' phase, which is often observed in the literature [9–12], is not present. The reason for this, is that γ'' is a metastable phase. By excluding the δ, P, σ, and η-Ni_3Ti phases which are kinetically less favored in the calculations shown in Fig. 1b, it can be seen that γ'' will then appear. The calculated solvus temperatures of γ' and δ in a full equilibrium are 820 °C and 1002 °C, respectively. The calculated solvus temperature of δ phase shows good agreement with the experiment (around 1000 °C) [13, 14]. However, it is reported that at temperature 900 ~ 920 °C, γ' precipitates still exist with γ'' precipitates [15]. Figure 1b indicates the solvus temperature of γ' and γ'' are 886 °C and 927 °C, respectively, which agrees well with the experiments [15, 16].

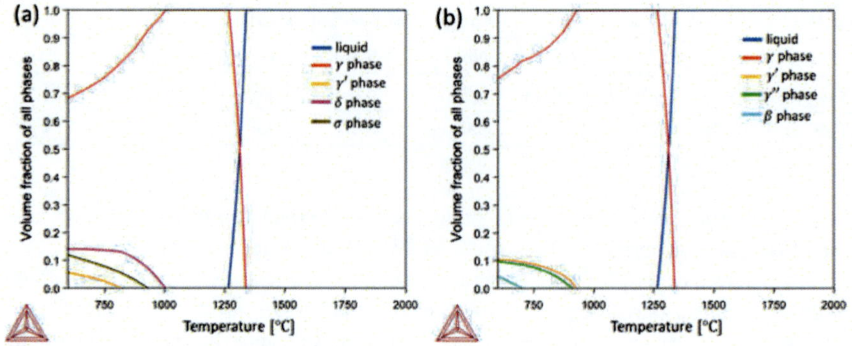

Fig. 1 Volume fraction of different phases as a function of temperature **a** all phases available; **b** with suspension of δ, P, σ, and η phases

Precipitation Kinetics Simulations

In addition to the thermodynamic database referenced above, the kinetic simulations also requires a mobility database for Ni-based superalloys to be selected. In this case, MOBNI6 is used as this is compatible with TCNI12. Here compatibility means that the descriptions of elements and phases are identical in both databases and the mobility parameters have been optimized using TCNI12, i.e., the thermodynamic factors required during the optimization of mobility parameters against experimental interdiffusion or intrinsic diffusion data were calculated based on the thermodynamics in TCNI12. MOBNI6 contains descriptions for the atomic mobility of the γ (FCC), $\gamma\prime$ (FCC ordered) and liquid phases.

The precipitation add-on module to Thermo-Calc, TC-PRISMA, has been developed based on Langer-Schwartz theory [17], and adopts the Kampmann-Wagner numerical (KWN) method [18] for concurrent nucleation, growth, and coarsening of dispersed precipitate phases. Employing particle size distribution (PSD) $f(R,t)$ in terms of particle size (as radius R) and time (t), the approach proceeds by simultaneously solving the continuity equation

$$\frac{\partial f(R,t)}{\partial t} = -\frac{\partial}{\partial R}[v(R)f(R,t)] + j(R,t) \quad (1)$$

where $v(R)$ is the growth rate and $j(R,t)$ is the nucleation rate. The mass balance equation between matrix and precipitates is described as

$$C_{0i}^{m} = C_i^m(t) + \left[C_i^p(t) - C_i^m(t)\right]\int_0^\infty \frac{4\pi}{3} f(R,t)R^3 dR \quad (2)$$

where C_{0i}^{m} is the initial alloy composition (mole per unit volume) of solute element i, and $C_i^{p}(t)$ and $C_i^{m}(t)$ are the instantaneous precipitate and matrix composition of solute element i, respectively.

The nucleation model is based on classical nucleation theory [19, 20] and in this work, a spherical morphology of the $\gamma\prime$ phase and plate morphology of $\gamma\prime\prime$ is assumed for the simulation.

The oblate spheroid (plate) morphology is described as

$$\frac{x_1^2}{l^2}+\frac{x_2^2}{l^2}+\frac{x_3^2}{r^2}\leq 1 l > r \tag{3}$$

with an aspect ratio $\alpha=\frac{l}{r}>1$, where x_1, x_2, x_3 are the coordinates of point on the surface and l is the horizontal radius of the oblate and r is the thickness of the spheroid.

Several studies have shown that aspect ratio varies as a function of particle size [21, 22]. The ratio becomes larger with increasing particle size. According to Moore et al. [22], for the $\gamma\prime\prime$ particles with an average particle length larger than 50 nm, a good approximation of the average aspect ratio is 5. Therefore, the aspect ratio was here assumed to be 5 and held constant during the precipitation. For the plate morphology of $\gamma\prime\prime$, Eshelby's theory [23, 24] was adopted to calculate the elastic strain energy efficiently. The calculation of elastic strain energy is also derived by Wu et al. [25]. According to Devaux et al. [21], the coherency strain of $\gamma\prime\prime$ is calculated as $\varepsilon_{11}^{T}=\varepsilon_{22}^{T}=6.67\times 10^{-3}$ and $\varepsilon_{33}^{T}=2.86\times 10^{-2}$.

By combining the thermodynamic (TCNI12) and kinetic (MOBNI6) databases available for Thermo-Calc with additional property data (e.g., interfacial energies and nucleation site densities), we have the necessary data input for TC-PRISMA, through these simulations, the temporal evolution of mean radius, number density, volume fraction, and size distribution of secondary phase precipitates is acquired. The nucleation rate and precipitate composition can also be obtained during the simulation. Coupled with relevant microstructure-property models, the Thermo-Calc Software can be used to estimate the change of mechanical properties of alloys upon aging or tempering.

There are many studies in which TC-PRISMA has been applied to study the precipitation kinetics of steels [26–29] and Ni-based superalloys [25, 30–34]. In a study by Fahrmann and Metzler [30], only $\gamma\prime$ is simulated and discussed, whereas in IN718, there are $\gamma\prime$, $\gamma\prime\prime$, and δ precipitates formed during the aging heat treatment. This work presents simulations of the precipitation of $\gamma\prime$, $\gamma\prime\prime$ and δ phases in the alloy, including co-precipitation of $\gamma\prime$ and $\gamma\prime\prime$, using TC-PRISMA. The simulations adopts a mean-field approach that assumes uniform elemental distributions in the pre-existing matrix phase, and carbides and other minor phases have not been considered.

The precipitates $\gamma\prime$ and $\gamma\prime\prime$ form during aging of the supersaturated γ matrix, where increased solute mobility allows for localized elemental segregation. The $\gamma\prime\prime$ phase, however, is metastable at room temperature and can transform from bct into an orthorhombic δ phase (Ni3(Nb,Ti)) during periods of prolonged heating or exposure to sufficiently high temperatures. Therefore, to optimize the interfacial energy at

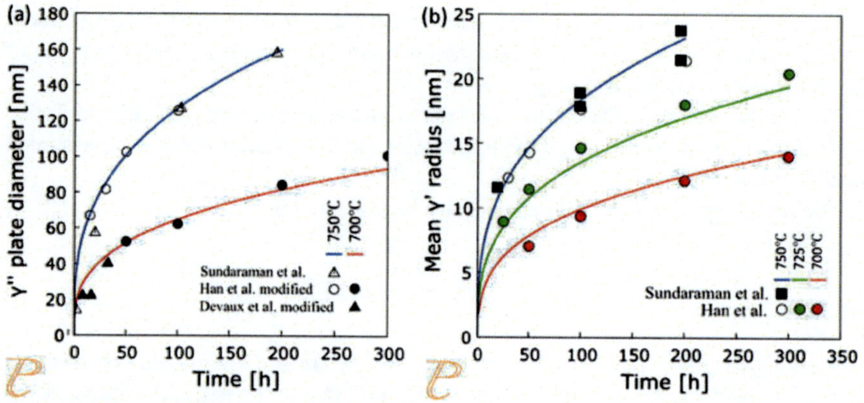

Fig. 2 Relationship between aging time and **a** the diameter of γ'' circular plate at 700 °C and 750 °C; and the **b** mean radius of γ' during aging at 700 °C, 725 °C and 750 °C, compared with literature data [21, 39, 40]

aging temperature between 700 ~ 750 °C, only the γ' and γ'' precipitates have been considered.

Figure 2 compares the experimental measurement of mean particle sizes and simulations by TC-PRISMA. The figure presents the variation of mean plate diameter of γ'' as a function of aging time at temperature 700 °C and 750 °C (Fig. 2a), and mean radius of γ' during aging at 700, 725 and 750 °C (Fig. 2b). The fitted interfacial energy between γ and γ' is $-9.43 * 10^{(-5)} * T + 0.11131\ J/m^2$ and the interfacial energy between γ and γ'' is a constant 0.014 J/m^2 which is within the range of values in literature [21, 35].

The time-temperature-precipitation (TTP) diagram of the system was then calculated. To conduct the simulation, the interfacial energy between γ and δ is -0.0003*T+0.438567 J/m^2 where T is the temperature. The TTP diagrams of two different systems are simulated: 1. the system with γ' and γ'' precipitates is shown in Fig. 3a; 2. The system with γ', γ'' precipitates in bulk and δ precipitates in the bulk and at grain boundaries is shown in Fig. 3b. In both systems, the nucleation site density is set as $10^{20} m^{-3}$. In the second system, the wetting angle of grain boundary is set as 70 °. As a comparison, Brooks and Bridges quantified experimental TTP diagrams [36] as shown in dashed line in Fig. 3; there are a few more empirical TTP diagrams available in literature, but this one was selected because it is the only one to separate the curves for onset of γ' and γ'' formation. The simulation shows good qualitative agreement with the empirical diagram. The upper limits to TTP curves of γ'', γ' and δ correspond to 825 °C, 850 °C and 875 °C (for bulk), or 825 °C (for grain boundary). The quantified limit of γ'' shows good agreement with the experiment, while the limit of γ' is overestimated (800 °C in the experiment [36]) and the limit of δ is underestimated (925 °C for bulk and grain boundary in the experiment [36]) for the bulk and grain boundary.

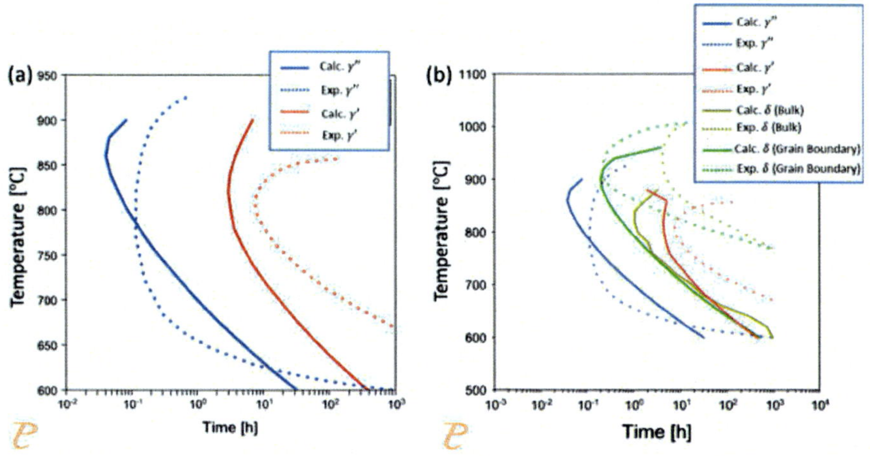

Fig. 3 Calculated TTP diagram which **a** γ' and γ'' precipitation start; **b** γ', γ'' and δ precipitation start. Both with experimental curves reproduced from Brooks and Bridges [36]

Figures 4 and 5 show the variation of volume fraction of precipitates as a function of aging time. In Fig. 4, at relatively low temperature (720 °C), the volume fraction of γ' and γ'' after aging is predicted to be 5.7% and 11.3%, respectively, after 200 h, which is consistent with the results measured by Oblak et al. [37], e.g. 4.75% for γ' and 14.25% for γ''. The difference could be attributed to the heterogeneity of the microstructure and composition in the alloy or error in experimental measurements (e.g. 2D projection through TEM foil thickness).

Moreover, for relatively high temperatures (800 ~ 875 °C), predicted volume fractions of γ', γ'' and δ are shown in Fig. 5. As can be seen from Fig. 5a at 800 °C, γ'' phase is predicted to precipitate first, and its volume fraction reaches over 9% within 1 h. During the aging process, γ'' precipitates decompose in favor of the γ' and δ phase. The required time for decomposition decreases with increasing aging temperature to 520 h and 150 h at 825 and 850 °C respectively (Fig. 5b and c). At 875 °C, only δ phase is predicted to precipitate (Fig. 5d). This result shows good agreement with experiments [14, 38], which shows that γ' and γ'' may dissolve at 900 °C.

Calculation of Yield Strength

The yield strength (YS) model implemented within Thermo-Calc estimates the yield strength and hardness of a material for a given temperature T that is used to evaluate the equilibrium state of the system, and the resulting compositions and phase fractions which are then used in the evaluation of the mechanical properties. It returns the calculated yield strength at room temperature for the specified material using the phase equilibria evaluated at temperature T (i.e., the state at T is frozen-in).

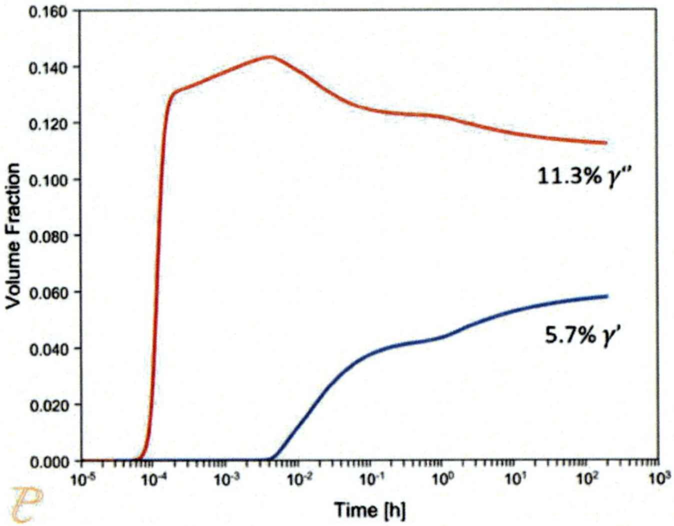

Fig. 4 Relationship between aging time and the volume fraction of $\gamma\prime$ (red line) and $\gamma\prime\prime$ (blue line) at 720 °C

The yield strength model, as implemented considers the following contributions:

- Intrinsic strength (σ_i) for the pure elements
- Intrinsic strength, σ_i, is the base strength as evaluated from linear combination of the pure elements' strength. The value is derived from the handbook values of the pure elements or related reference [41] which is implemented in Thermo-Calc software.
- Grain boundary strengthening (σ_{gb})
- The grain boundary strengthening, σ_{gb}, is the contribution to the total strength stemming from the grain boundary's ability to hinder dislocation movement described by the Hall–Petch relation [42]. The parameters are the Hall–Petch constant k_{h-p}, in MPa $\cdot$ $\mu\text{m}^{\frac{1}{2}}$, and the average grain diameter is assumed as 100 μm. In this study, the matrix grain growth was not considered.
- Solid solution strengthening (σ_{ss})
- The solution strengthening, σ_{ss}, is the contribution to total strength due to the elastic strains in the crystal lattice caused by alloying elements of a lattice parameter differing from the main constituent. For the solid solution strengthening the option of evaluating the strength at any temperature is available. The result is normalized with respect to the sum of all mole fractions for FCC, BCC and HCP. This option uses the phase fractions and compositions calculated at the overall equilibrium temperature. The model by Walbrühl et al. [43] is implemented.
- Precipitation strengthening (σ_p)
- The precipitation strengthening, σ_p, is the contribution to total strength due to precipitation of a secondary phase, e.g., gamma prime in Ni-base alloys. This

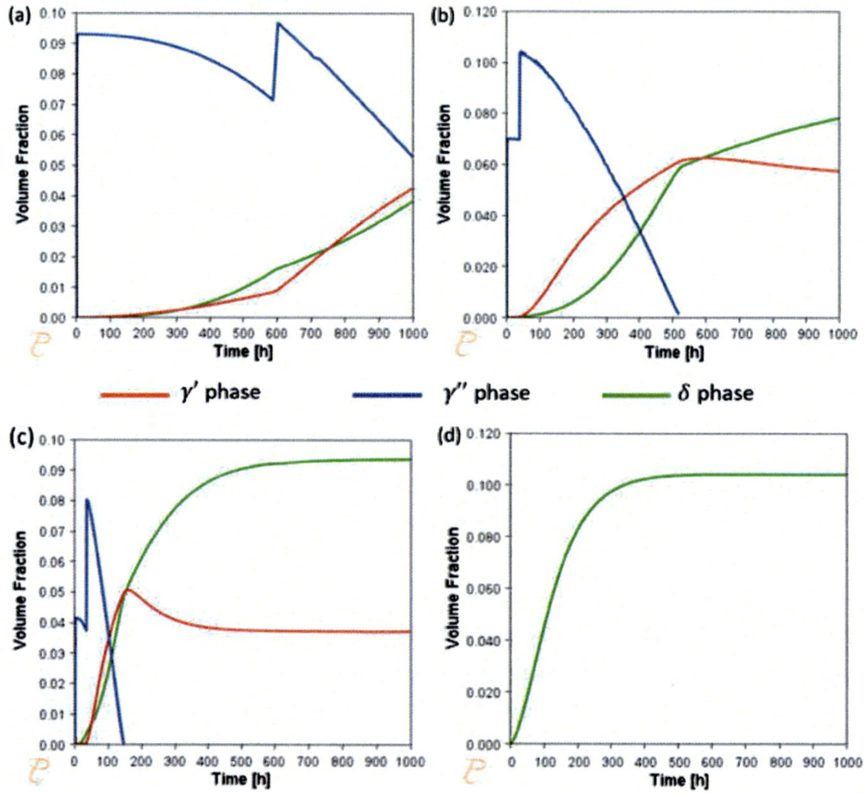

Fig. 5 Relationship between aging time and the volume fraction of $\gamma\prime$, $\gamma\prime\prime$ and δ at **a** 800 °C **b** 825 °C; **c** 850 °C; **d** 875 °C, respectively

model is available for any number of precipitating phases. The total particle strength is calculated by the square mean of the individual particle strength contributions: $\sigma_p = \sqrt{\sigma_{p1}^2 + \sigma_{p2}^2 + \cdots \sigma_{pn}^2}$

where $\sigma_{pi}(i = 1, 2, \ldots n)$ is the precipitation strengthening of precipitate i.

For precipitation strengthening, the simplified model was adopted in this work, which is based on the model developed by Zander et al. [44]. The model considers the general mechanisms of cutting and looping without regard to any detailed dislocation mechanisms. The relationship between precipitation hardening σ_p and r is given by

$$\sigma_{pi} = \frac{M}{b} \cdot \sqrt{\frac{3f}{2\pi}} \cdot \left(2\beta G b^2\right) \cdot \left(\frac{1}{r_c}\right)^{\frac{3}{2}} \cdot \sqrt{\bar{r}}, r < r_c \quad (4)$$

$$\sigma_{pi} = \frac{M}{b} \cdot \sqrt{\frac{3f}{2\pi}} \cdot \left(2\beta G b^2\right) \cdot \frac{1}{\bar{r}}, r \geq r_c \quad (5)$$

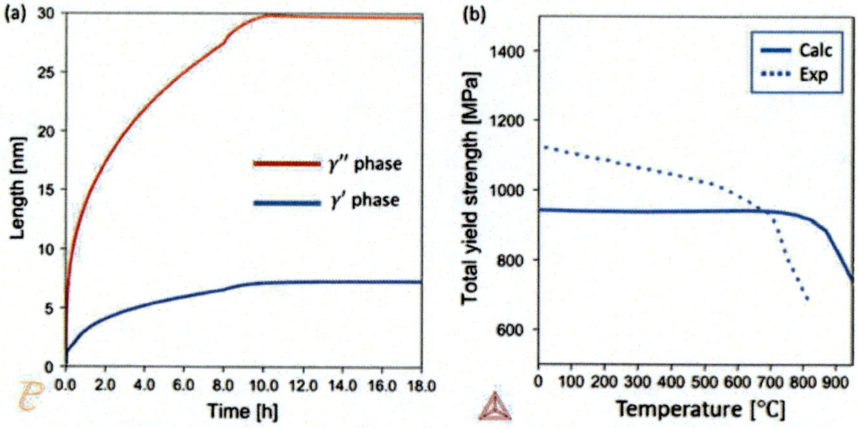

Fig. 6 Results of **a** Simulated average length for $\gamma\prime$ phase (radius) and $\gamma\prime\prime$ phase (circular-equivalent diameter) as a function of aging time and **b** YS with optimized parameters strengthening constant: k_p and critical radius: r_c. The experimental data is from Special Metals IN718 datasheet [46]

where M is the Taylor factor, $\overline{r}$ is the mean particle radius, G is the shear modulus, b is the Burgers vector and f is the volume fraction of the precipitates. β is a constant. Here $\frac{M}{b} \cdot \sqrt{\frac{3f}{2\pi}} \cdot \left(2\beta G b^2\right)$ is treated as strengthening constant k_p and critical radius r_c. The value of k_p and r_c can be optimized using the experimentally measured YS values.

Here we take one example as follows: aging at 718 °C or 8 h, furnace cool to 620 °C then hold for total aging time of 18 h. The mean particle sizes of $\gamma\prime$ and $\gamma\prime\prime$ as function of aging time is shown as Fig. 6a. After 18 h' aging, the average sizes of $\gamma\prime\prime$ and $\gamma\prime$ are around 30 nm and 7.5 nm, respectively. It is reported that in Ni base superalloy the strengthening benefit of the $\gamma\prime\prime$ outweighs that of $\gamma\prime$ on a normalized volume fraction basis [45]. Therefore, only the strengthening benefit of $\gamma\prime\prime$ is considered for precipitation strengthening. The fitting parameters k_p and r_c of the simplified model are 9.55×10^{-6}MPa $\cdot$ m and 2.88×10^{-9} m for $\gamma\prime\prime$. The grain size is assumed as 100 μm. The results with optimized fitting parameters are shown in Fig. 6b.

The predicted yield strength shows good with the experiment [46]. At low temperatures (0 ~ 600 °C), the YS is underestimated, which could be caused by (1) the strengthening effects of $\gamma\prime$ precipitates; (2) possible defects induced by hot roll according to the reference; (3) experimental measured error.

Summary

In this study, the precipitation of $\gamma\prime$, $\gamma\prime\prime$ and δ in the alloy IN718 has been simulated for different aging temperatures to predict a corresponding yield strength using a CALPHAD-based ICME framework accessible in the Thermo-Calc software package.

With this framework, the equilibrium volume fraction and solvus temperature of the precipitates $\gamma\prime$, $\gamma\prime\prime$ and δ was predicted. Interfacial energies were optimized to reproduce coarsening data for $\gamma\prime$ and $\gamma\prime\prime$ from literature. This was used together with adjusted nucleation site densities to calculate a TTP diagram for the formation of $\gamma\prime$, $\gamma\prime\prime$ and δ. The provisional results show a promising agreement with the experiments. There is ongoing work to further improve the simulated precipitation of $\gamma\prime$, $\gamma\prime\prime$ and δ phases, in order to predict the optimal precipitation strengthening conditions.

To calculate the yield strength, a simplified model was used which includes contributions from the intrinsic strength, grain boundary strengthening, solid solution strengthening, and precipitation strengthening. The predicted yield strength shows good agreement with handbook data. In the future the effects of processing on IN718 will be investigated.

References

1. C.D. Taylor, B. Tossey, T. Yu, Y. Wang, S. Niezgoda, ICME for Advanced Manufacturing of Ni Superalloy Heat Exchangers with High Temperature Creep + Oxidation Resistance for Supercritical CO2, (2022). https://doi.org/10.2172/1844170.
2. C. Fu, Y. Chen, S. He, S. Antonov, L. Li, W. Zheng, Q. Feng, ICME Framework for Damage Assessment and Remaining Creep Life Prediction of In-Service Turbine Blades Manufactured with Ni-Based Superalloys, Integr Mater Manuf Innov. 8 (2019) 509–520. https://doi.org/10.1007/S40192-019-00161-4/FIGURES/9.
3. M. Zou, W. Li, L. Li, J.C. Zhao, Q. Feng, Machine Learning Assisted Design Approach for Developing γ′-Strengthened Co-Ni-Base Superalloys, Minerals, Metals and Materials Series. (2020) 937–947. https://doi.org/10.1007/978-3-030-51834-9_92/COVER.
4. M. Detrois, J. Rotella, M. Hardy, S. Tin, M.D. Sangid, Tailoring the Properties of a Ni-Based Superalloy via Modification of the Forging Process: an ICME Approach to Fatigue Performance, Integr Mater Manuf Innov. 6 (2017) 265–278. https://doi.org/10.1007/S40192-017-0103-6/FIGURES/13.
5. B.D. Conduit, N.G. Jones, H.J. Stone, G.J. Conduit, Design of a nickel-base superalloy using a neural network, Mater Des. 131 (2017) 358–365. https://doi.org/10.1016/J.MATDES.2017.06.007.
6. N. Saunders, A.P. Miodownik, CALPHAD (Calculation of Phase Diagrams) : a Comprehensive Guide., (1998) 497.
7. J.O. Andersson, T. Helander, L. Höglund, P. Shi, B. Sundman, Thermo-Calc & DICTRA, computational tools for materials science, Calphad. 26 (2002) 273–312. https://doi.org/10.1016/S0364-5916(02)00037-8.
8. J.R. Davis, Nickel, cobalt, and their alloys, ASM international, 2000.
9. S. Mahadevan, S. Nalawade, J.B. Singh, A. Verma, B. Paul, K. Ramaswamy, Evolution of δ phase microstructure in alloy 718, in: Proceedings of the 7th International Symposium on Superalloy, 2010: pp. 737–750.
10. M.K. Miller, S.S. Babu, Phase compositions in alloy 718: A comparison between APT/APFIM measurements and thermodynamic predictions, in: Advanced Technologies for Superalloy Affordability, Proceedings of the 129th TMS Annual Meeting, TMS2000, 2000: p. 63.
11. W.E.I. Xian-ping, Z. Wen-jie, S. Zhi-gang, L.E.I. Ting, Y. Qi-long, X.I.E. Qing-cheng, Elemental Partitioning Characteristics of Equilibrium Phases in Inconel 718 Alloy at 600-1100, O ▲ (Ĭ İ). 20 (2013) 88–94.

12. S. Azadian, Aspects of precipitation in alloy Inconel 718, (2004). http://urn.kb.se/resolve?urn=urn:nbn:se:ltu:diva-25931 (accessed September 14, 2022).
13. N.C. Ferreri, S.C. Vogel, M. Knezevic, Determining volume fractions of γ, γ', γ'', δ, and MC-carbide phases in Inconel 718 as a function of its processing history using an advanced neutron diffraction procedure, Materials Science and Engineering: A. 781 (2020) 139228. https://doi.org/10.1016/J.MSEA.2020.139228.
14. J. Lacaze, M. Dehmas, A. Niang, B. Viguier, TEM study of high-temperature precipitation of delta phase in inconel 718 alloy, Advances in Materials Science and Engineering. 2011 (2011). https://doi.org/10.1155/2011/940634.
15. A. Niang, B. Viguier, J. Lacaze, Some features of anisothermal solid-state transformations in alloy 718, Mater Charact. 61 (2010) 525–534. https://doi.org/10.1016/J.MATCHAR.2010.02.011.
16. S.L. Semiatin, N.C. Levkulich, R. Larsen, J.S. Tiley, K.N. Wertz, F. Zhang, T.M. Smith, R.Y. Zhang, H.B. Dong, P. Gadaud, J. Cormier, The Application of Differential Scanning Calorimetry to Investigate Precipitation Behavior in Nickel-Base Superalloys Under Continuous Cooling and Heating Conditions, Metall Mater Trans A Phys Metall Mater Sci. 52 (2021) 3706–3726. https://doi.org/10.1007/S11661-021-06362-X/FIGURES/18.
17. J.S. Langer, A.J. Schwartz, Kinetics of nucleation in near-critical fluids, Phys Rev A (Coll Park). 21 (1980) 948. https://doi.org/10.1103/PhysRevA.21.948.
18. R. Kampmann, R. Wagner, KINETICS OF PRECIPITATION IN METASTABLE BINARY ALLOYS -THEORY AND APPLICATION TO Cu-1.9 at % Ti AND Ni-14 at % Al, (1984) 91–103. https://doi.org/10.1016/B978-0-08-031651-2.50018-5.
19. D. Kashchiev, Nucleation, Elsevier, 2000.
20. D. Kashchiev, G.M. van Rosmalen, Nucleation in solutions revisited, Crystal Research and Technology: Journal of Experimental and Industrial Crystallography. 38 (2003) 555–574.
21. A. Devaux, L. Nazé, R. Molins, A. Pineau, A. Organista, J.Y. Guédou, J.F. Uginet, P. Héritier, Gamma double prime precipitation kinetic in Alloy 718, Materials Science and Engineering: A. 486 (2008) 117–122. https://doi.org/10.1016/J.MSEA.2007.08.046.
22. I.J. Moore, M.G. Burke, N.T. Nuhfer, E.J. Palmiere, Evaluation of classical precipitation descriptions for γ''(Ni3Nb-D022) in Ni-base superalloys, J Mater Sci. 52 (2017) 8665–8680. https://doi.org/10.1007/S10853-017-1091-9.
23. J.D. Eshelby, The determination of the elastic field of an ellipsoidal inclusion, and related problems, Proc R Soc Lond A Math Phys Sci. 241 (1957) 376–396.
24. J.D. Eshelby, The elastic field outside an ellipsoidal inclusion, Proc R Soc Lond A Math Phys Sci. 252 (1959) 561–569.
25. K. Wu, Q. Chen, P. Mason, Simulation of Precipitation Kinetics with Non-Spherical Particles, J Phase Equilibria Diffus. 39 (2018) 571–583. https://doi.org/10.1007/S11669-018-0644-1/FIGURES/7.
26. K. Krishna Kumar, J. Anburaj, T. Satish Kumar, A. Chandrasekar, R. Subramanian, Thermodynamic Simulation of Niobium-Containing Austenitic Stainless Steels: Microstructure and Corrosion Properties, Transactions of the Indian Institute of Metals. 72 (2019) 1627–1630. https://doi.org/10.1007/S12666-019-01642-Y/FIGURES/4.
27. S. Jin, Q. Chen, J. Bratberg, Thermodynamic calculations and precipitation simulations of HSLA steels, 7th International Conference on High Strength Low Alloy Steels, HSLA Steels 2015, International Conference on Microalloying 2015, Microalloying 2015 and International Conference on Offshore Engineering Steels 2015, OES 2015. 1 (2015) 173–177. https://doi.org/10.1007/978-3-319-48767-0_16/COVER.
28. J.P. Sanhueza, D. Rojas, O. Prat, J. Garcia, R. Espinoza, C. Montalba, M.F. Melendrez, Precipitation kinetics in a 10.5%Cr heat resistant steel: Experimental results and simulation by TC-PRISMA/DICTRA, Mater Chem Phys. 200 (2017) 342–353. https://doi.org/10.1016/J.MATCHEMPHYS.2017.07.083.
29. O. Prat, J. García, D. Rojas, J.P. Sanhueza, C. Camurri, Study of nucleation, growth and coarsening of precipitates in a novel 9%Cr heat resistant steel: Experimental and modeling, Mater Chem Phys. 143 (2014) 754–764. https://doi.org/10.1016/J.MATCHEMPHYS.2013.10.010.

30. M.G. Fahrmann, D.A. Metzler, Simulation of γ' Precipitation Kinetics in a Commercial Ni-Base Superalloy, JOM. 68 (2016) 2786–2792. https://doi.org/10.1007/S11837-016-2097-5/FIGURES/9.
31. K. Vattappara, V. A Hosseini, C. Joseph, F. Hanning, J. Andersson, Physical and thermodynamic simulations of gamma-prime precipitation in Haynes® 282® using arc heat treatment, J Alloys Compd. 870 (2021) 159484. https://doi.org/10.1016/J.JALLCOM.2021.159484.
32. K. Vattappara, Understanding the effect of temperature and time on Gamma prime coarsening for Nickel-base superalloy Haynes 282, (2019). http://urn.kb.se/resolve?urn=urn:nbn:se:kth:diva-259774 (accessed September 18, 2022).
33. F. Zhang, J. Ilavsky, G. Lindwall, M.R. Stoudt, L.E. Levine, A.J. Allen, Solid-state transformation of an additive manufactured inconel 625 alloy at 700◦C, Applied Sciences (Switzerland). 11 (2021). https://doi.org/10.3390/APP11188643.
34. G. Lindwall, C.E. Campbell, E.A. Lass, F. Zhang, M.R. Stoudt, A.J. Allen, L.E. Levine, Simulation of TTT Curves for Additively Manufactured Inconel 625, Metall Mater Trans A Phys Metall Mater Sci. 50 (2019) 457–467. https://doi.org/10.1007/S11661-018-4959-7/FIGURES/10.
35. C.K. Sudbrack, R.D. Noebe, D.N. Seidman, Compositional pathways and capillary effects during isothermal precipitation in a nondilute Ni–Al–Cr alloy, Acta Mater. 55 (2007) 119–130. https://doi.org/10.1016/J.ACTAMAT.2006.08.009.
36. J.W. Brooks, P.J. Bridges, Metallurgical stability of Inconel alloy 718, Superalloys. 88 (1988) 33–42.
37. J.M. Oblak, D.F. Paulonis, D.S. Duvall, Coherency strengthening in Ni base alloys hardened by DO22 γ' precipitates, Metallurgical and Materials Transactions B 1974 5:1. 5 (1974) 143–153. https://doi.org/10.1007/BF02642938.
38. S.A. Nalawade, M. Sundararaman, J.B. Singh, A. Verma, R. Kishore, Precipitation of γ' phase in δ-precipitated Alloy 718 during deformation at elevated temperatures, Materials Science and Engineering: A. 527 (2010) 2906–2909. https://doi.org/10.1016/J.MSEA.2010.01.006.
39. M. Sundararaman, P. Mukhopadhyay, S. Banerjee, Some aspects of the precipitation of metastable intermetallic phases in INCONEL 718, Metallurgical Transactions A 1992 23:7. 23 (1992) 2015–2028. https://doi.org/10.1007/BF02647549.
40. V.F. Han, P. Deb, M.C. Chaturvedi, Coarsening behaviour of γ''- and γ'-particles in Inconel alloy 718. 16 (2013) 555–561. https://doi.org/10.1179/030634582790427118.
41. S. Metals, Product Handbook of High Performance Nickel Alloys, Product Handbook. (2008).
42. N. Hansen, Hall–Petch relation and boundary strengthening, Scr Mater. 51 (2004) 801–806. https://doi.org/10.1016/J.SCRIPTAMAT.2004.06.002.
43. M. Walbrühl, D. Linder, J. Ågren, A. Borgenstam, Modelling of solid solution strengthening in multicomponent alloys, Materials Science and Engineering: A. 700 (2017) 301–311. https://doi.org/10.1016/J.MSEA.2017.06.001.
44. J. Zander, R. Sandström, One parameter model for strength properties of hardenable aluminium alloys, Mater Des. 29 (2008) 1540–1548. https://doi.org/10.1016/J.MATDES.2008.02.001.
45. P.M. Mignanelli, N.G. Jones, E.J. Pickering, O.M.D.M. Messé, C.M.F. Rae, M.C. Hardy, H.J. Stone, Gamma-gamma prime-gamma double prime dual-superlattice superalloys, Scr Mater. 136 (2017) 136–140. https://doi.org/10.1016/J.SCRIPTAMAT.2017.04.029.
46. INCONEL ® alloy 718, (n.d.).

Part V
High Temperature Fe-, Ni, and, Co-based Alloys

Factors Influencing Propensity for Stress Relaxation Cracking in Inconel® Alloy 740H® and Practical Guidance for Applications

John Shingledecker, John Siefert, Tapasvi Lolla, John Dupont, Jack deBarbadillo, and Ronnie Gollihue

Abstract Inconel® alloy 740H® (UNS N07740) was the first age-hardenable nickel-based alloy approved by the ASME Boiler and Pressure Vessel Code for use in pressure-boundary applications. In recent years, advanced energy systems such as supercritical CO_2 power cycles have utilized alloy 740H in large demonstration projects driven by the requirement for higher fluid temperatures and pressures. Stress relaxation cracking (SRxC) following post weld heat treatment (PWHT), also known as strain age cracking (SAC), has been identified in a limited number of weldments during these industrial builds resulting in focused research to further clarify factors influencing this cracking tendency. This paper will summarize some of the findings from shop and field fabrication leading to successful welds and characteristics of observed SRxC. Laboratory experiments supported by microstructural characterization will be presented to highlight the importance of variables such as strain, material starting condition, and PWHT temperatures. Finally, the results will be summarized within the context of practical guidance for industry to successfully weld the material in boiler, heat exchanger, and piping applications.

Keywords Relaxation cracking · Strain age cracking · Supercritical CO_2 power cycles · Advanced ultrasupercritical · Concentrated solar power

---

J. Shingledecker (✉) · J. Siefert · T. Lolla
Electric Power Research Institute, Charlotte, NC 28262, USA
e-mail: jshingledecker@epri.com

J. Dupont
Lehigh University, Bethlehem, PA 18015, USA

J. deBarbadillo · R. Gollihue
Special Metals Corporation, Huntington, WV, USA

E. A. Ott et al. (eds.), *Proceedings of the 10th International Symposium on Superalloy 718 and Derivatives*, The Minerals, Metals & Materials Series,
https://doi.org/10.1007/978-3-031-27447-3_27

Introduction

Inconel® alloy 740H® (UNS N07740), herein referred to as alloy 740H (nominal composition in Table 1), was the first age-hardenable nickel-based alloy approved by the ASME Boiler and Pressure Vessel Code for use in pressure-boundary applications. It was originally developed as a coal-fired boiler tubing alloy for advanced ultrasupercritical (A-USC) steam applications. Over the past ~20 years, the alloy has been optimized for weldability and high-temperature stability, approved for use in different applications, and the number of product forms and manufacturing processes demonstrated now includes the required product forms needed to manufacture complex high-temperature components or systems. These product forms include thin-wall tubing, heavy wall piping, cold-formed and induction bends, seam welded piping, fittings, forging, sheet, etc. [1]. Figure 1 shows a timeline of the alloy development including key research topics, technical development milestones, and successful code qualification activities. The development of a supply chain combined with the advantageous properties of the alloy (high-temperature creep strength, oxidation and corrosion resistance, etc.) have resulted in the alloy being specified for use in new high-temperature power cycle demonstration projects as shown in Table 2.

Nickel-based alloys may be susceptible to a number of weldability challenges [2, 3]. A 2018 review of the weldability and performance of high-strength nickel-based alloys to enable advanced power systems proposed that alloy 740H was generally not

Table 1 Alloy 740H nominal composition (wt%)

C	Ni	Fe	Cr	Mn	Si	Mo	Co	Al	Ti	B	Nb + Ta
0.06	Bal.	0.7	24.5	0.3	0.15	0.1	20	1.35	1.35	0.001	1.5

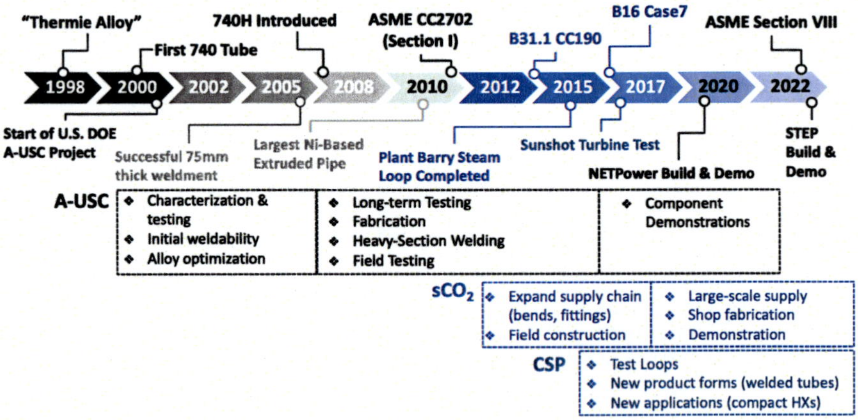

Fig. 1 Development timeline for alloy 740H including major milestones and applications (CC = code case, A-USC = advanced ultrasupercritical, sCO_2 = supercritical carbon dioxide, CSP = Concentrating Solar Power)

Table 2 Demonstrations utilizing 740H

Years	Name	Description	Product forms	Approx. Scale
2011–15	Plant Barry Steam Loop	760 °C + Steam + 2 yr operation	Tubing (~50 mm diameter)	<20 m of tubing
2014–16	Sunshot Turbine Test	700 °C + microtube HX and connecting piping	6.35 mm dia. tubing, piping	25 m of piping, 6,000 m of tubing
2017-current	NET Power Demonstration	700 °C direct-fired sCO_2 test facility	Piping, bends, forgings	~10,000 kg overall
2018-current	STEP Heater	50 MW_{th} natural gas fired sCO_2 heater	Tubing, piping, forgings	~25,000 kg overall ~6,100 m of tubing
	STEP Piping	715 °C, 250 bar sCO_2 piping system	Heavy wall piping (200–250 mm dia.), bends and elbows	~17,000 kg overall ~150 m piping 26 bends/elbows
2019–2022	A-USC Fabrication Demonstrations	Shop and field erection study for full-scale A-USC header assembly	Heavy-wall piping, header, induction bends	~39,000 kg ingot

susceptible to fusion zone solidification cracking, and had minimal risk of liquation cracking if proper alloy chemistry control and welding practices were used, but the risk for stress relaxation cracking (SRxC) during post weld heat treatment (PWHT) or high-temperature service also known as strain age cracking (SAC) was a potential concern which required further evaluation [4]. The complexities of the SRxC damage mechanism require a specific combination of a susceptible microstructure, the presence of stress (typically a residual stress from welding), and high-temperature exposure in a specific temperature range and time to cause relaxation of the imposed stress by cracking. The resulting SRxC is thus sensitive to the component design, fabrication practice, and application [5]. In 2018, when this review was written, SRxC had not been observed in 740H, but at that time, most of the practical welding experience on the alloy was restricted to laboratory studies, industrial qualification activities, and relatively limited demonstrations resulting in 10's of welds. Today, as shown in Table 2, tons of 740H products including km of tubing and piping have now been produced and welded as part of several completed and ongoing large-scale demonstrations with thousands of welds now successfully executed. This paper provides insights into the occurrence of the SRxC in some of these projects, highlights recent laboratory work to explore some of the variables leading to SRxC, and provides recommendations to reduce the risk of SRxC in the future for alloy 740H.

Industry Welding Experience

Overall, the experience with alloy 740H welding has been very good. A large variety of section thicknesses from 1.65 mm (0.065-in.) autogenous laser welds to a 94 mm (3.70-in.) thick multi-pass narrow-groove hot-wire gas tungsten arc weld (GTAW) and all sizes in-between have been successfully fabricated in industrial fabrication shops [6]. Field welding of nickel-based alloys requires proper welder training and skill, and with training and oversight, multiple organizations have demonstrated field welding of tubing and piping using GTAW processes. To capture these experiences and identify challenges, EPRI hosted an industry workshop in 2021 for alloy 740H users to share experiences [7]. While the majority of the experience was positive, a few instances of cracking after PWHT had been identified by users which are described in more detail below.

The U.S. DOE Supercritical Transformational Electric Power Project (STEP Demo) is constructing a multi-user facility as a 10 MW electric advanced pilot project to advance the development and deployment of supercritical carbon dioxide (sCO_2) power cycles [8]. A key component of the overall facility is the gas-fired heater which is designed to enable sCO_2 cycle conditions of 276 bar (4,000 psig) at >700 °C (1,292 °F). The high-temperature heater coil is the largest known application of alloy 740H designed, fabricated, and stamped to ASME Boiler and Pressure Vessel Code Section I. The heater required ~25,000 kg (55,000 lbs.) of alloy 740H including ~6,100 m (~20,000 feet) of tubing [9].

After post weld heat treatment of the coil assembly, cracking was identified in a small percentage of tube butt welds. A detailed metallurgical investigation was undertaken on a population of samples removed from the heater to determine the failure mechanism with more details in [10]. A comprehensive characterization of the cracking involving non-destructive evaluation and advanced microscopy methods indicated SRxC during post weld heat treatment was the operative failure mechanism. Table 3 provides a list of the key alloy 740H components in the STEP Demo fired heater, the number of welds conducted, and the number of SRxC cracks observed through field failures and non-destructive evaluation. All failures occurred after PWHT in the thin-wall tubing and Fig. 2 provides a summary of the observed damage in this application. The cracks characterized in the laboratory showed initiation at the weld toe region, intergranular propagation, minimal branching or extensive damage removed from the main crack, and propagation through the base metal, heat affected zone (HAZ), and/or weld metal depending on the orientation. Detailed characterization confirmed high strain was present at the crack initiation site, Fig. 3, leading to the rapid formation of precipitate free zones (PFZs) during the ~4-h PWHT, Fig. 4. Prior observations of PFZs had only been previously observed in long-term creep tested weldments after exposure in the 100 s to 1,000 s of hours [11].

Other incidences of SRxC have been identified including multiple transverse cracks in a pipe-to-pipe weld near a pipe wall thickness transition, Fig. 5, and extensive cracking in the socket weld, forged block, and tubing of an instrumentation port in a small diameter piping system, Fig. 6. In both instances, the cracking occurred

Table 3 Alloy 740H welding experience for the STEP Demo fired sCO_2 heater [10]

Type	Description	Cracking/nde indications	Total number fabricated	Failure rate (%)
Thick-wall	End Plates	0	4	0
	Flange-to-header	0	2	0
Thin-to-thick	Tube-to-header	0	292	0
Dissimilar	740H–347H	0	4	0
Thin-to-thin	Tube-to-tube butt welds	39	1,296	3.0
	Repairs[a]	0	87	0
Total		39	1,685	2.3

[a] Repairs included pup piece installations for 39 cracks/indications (2X repair welds), 3 segments removed for additional NDE verification (2X repair welds) plus 3 sacrificed welds during repairs (1X repair welds)

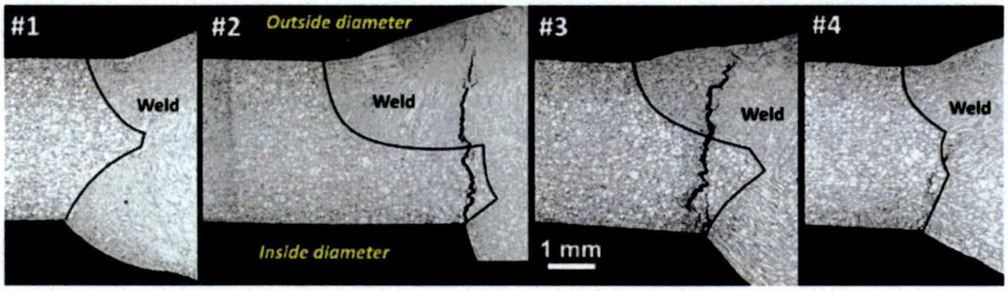

Fig. 2 Cross-sectional etched micrographs for one side of a SRxC in an alloy 740H tube-to-tube weldment around the circumference showing crack initiation at the weld toe (#4), crack growth from weld toe through the HAZ and weld (#3 and #2), and no cracking (#1)

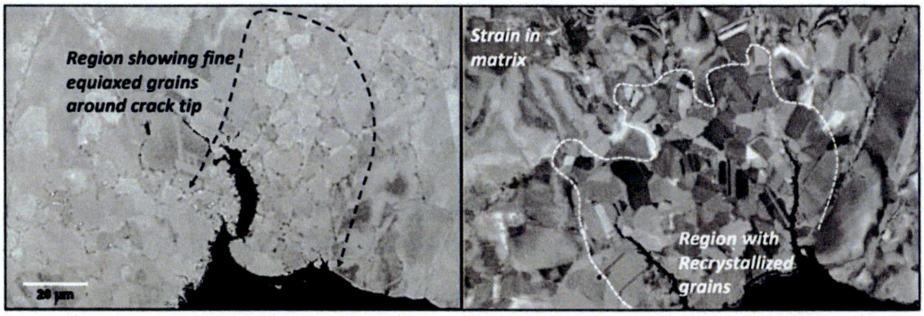

Fig. 3 SEM image near weld toe showing a region of fine equiaxed (recrystallized) grains coincident with crack tip surrounded by strained grains in the matrix (channeling contrast)

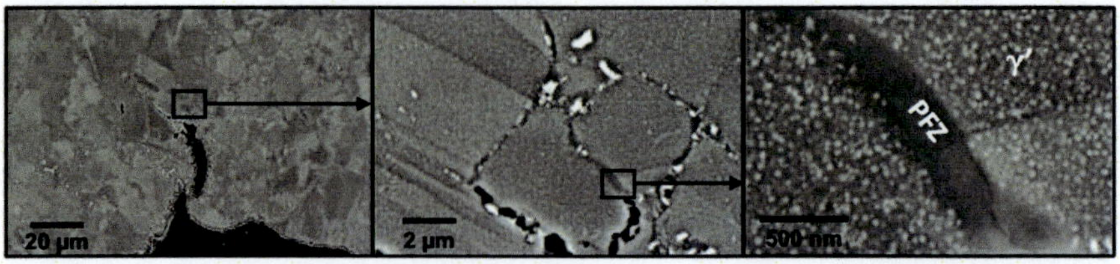

Fig. 4 SEM images at 1,500X (left), 15,000X (middle), and 100,000X (right) showing microstructure ahead of SRxC initiation site including precipitate free zones (PFZ) along the grain boundaries

Fig. 5 ~325 mm dia., 7.8 mm wall thickness pipe weld near a pipe wall thickness transition showing axial cracking in the weldment consistent with SRxC

after PWHT, was intergranular with minimal branching, and microstructural investigations found the occurrence of discontinuous coarsening along grain boundaries leading to PFZ regions where relaxation damage and cracking accumulates.

Assessment of Contributing Factors for SRxC

There are many contributing factors that need to be considered for SRxC. These can largely be classified as material, thermal history, and stress state. Based on the alloy 740H experience and failure analysis reports, available laboratory data, and knowledge of similar alloys, a group of experts were asked to rank their opinion on the most important variables to consider leading to SRxC of alloy 740H. Table 4 shows these results with general agreement that the stress state (from residual stresses, constraint, and deformation) was playing a significant role in the observed failures with some disagreement on the importance of the welding process, PWHT conditions, bulk alloy chemistry (within the 740H specification range), and overall weld quality. Based

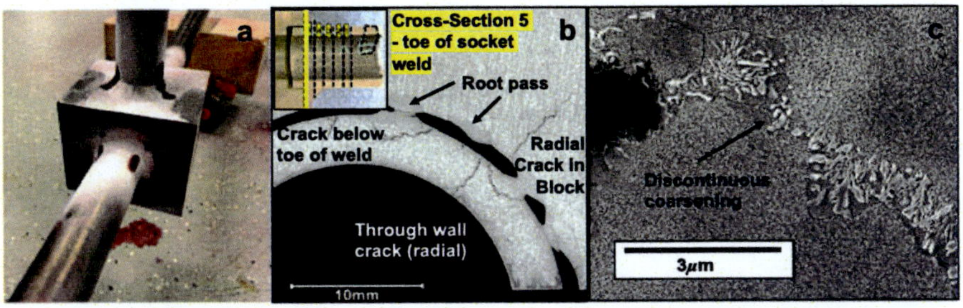

Fig. 6 Die penetrate inspection **a** shows multiple cracks in the socket weld, block forging, and tubing of an instrumentation port for all three welds. The metallographic cross-section **b** shows cracking in the weld toe, tubing through-wall cracks, and cracks extending into the block forging. The scanning electron microscopy images in **c** show the discontinuous coarsening reaction ahead of a crack tip

on this feedback, a limited laboratory study using a Gleeble™ thermo-mechanical simulator was conducted on 3 different heats of 740H evaluating variables including strain (stress) level, PWHT temperature, and starting material condition [12]. While a detailed discussion on the specific test methods and materials is beyond the scope of this paper, the methodology utilized followed previous work by Kant and DuPont's study establishing general alloy susceptibility to SRxC including alloy 740H [13].

Figure 7 (left) shows a typical result confirming the expert review where a strong dependency on the stress state was observed. In this experiment, for a single heat of the material, 0–4% strain did not cause sample failure (macro cracking) but strain levels above 4% induced sample fracture during relaxation. Heat-to-heat variability was identified in the laboratory tests. As an example, Fig. 7 (right) shows five different heats of alloy 740H subjected to the same relaxation testing method. Only two of the heats exhibited macro cracking and failure. In this test method, microcracking can occur in the sample which may not lead to fracture or failure of the test so there may be relaxation cracks in other samples which can only be assessed via post-test investigation. In comparing the composition of the five heats, it appears sulfur (S) is of particular importance to relaxation cracking. It is well established that high-temperature ductility, in the range of 700 to 900 °C (1,292 to 1,652°F), of nickel-based alloys can be greatly reduced due to segregation of S to grain boundaries even at levels as low as 9 ppm [14–18]. Minimizing the segregation of S to grain boundaries has been shown to improve weldability and reduce cracking occurrences in nickel-based alloys [19]. Magnesium (Mg) is one element commonly added to control S in nickel-based alloys [20, 21]. For the study conducted, one of the two heats which cracked had higher S resulting in a very low Mg/S ratio (less than 1), whereas other heats, more representative of today's commercial practices for alloy 740H, had ratios >6. It should also be noted that an optimally designed melting/remelting practice is necessary to avoid the potentially embrittling effects of excessive retained and uncombined Mg and S and that the processing of different product forms may have a role in the local grain boundary structures and cracking susceptibility. Overall, these results

Table 4 Results of an expert solicitation on SRxC variables in Alloy 740H

			Expert Solicitation (1–4 ranking, 1 = highest to 4 = lowest)				
Factor	Variable		A	B	C	D	Sum
Material	Composition	Bulk Composition	4	2	2	2	10
		Local Composition	3	1	2	3	9
	Grain size		4	3	3	4	14
	Processing and heat treatment history		3	2.5	3	2	10.5
Thermal history	Welding process	Filler Metal	3	4	4	4	15
		Heat-Input	2	2.5	4	2	10.5
		Multi-pass welding	3	2.5	1	2	8.5
	Pwht	Temperature	2	2	2	2	8
		Heating Rate and/or Control	2	2	4	3	11
	Service conditions	Temperature	4	2	4	4	14
		Time	4	2	4	4	14
Stress	Residual stress		1	1	4	1	7
	Deformation/working process		1	1	1	1	4
	Constraint		1	1	1	1	4
	Service induced		4	1	4	4	13
Welding quality	Techniques		1	4	2	3	10
	Training		1	4	1	3	9

suggest Mg and S may account for some of the heat-to-heat differences in SRxC and deserve further attention in future work. Currently, the chemistry requirements are generally controlled by the fact that the alloy is still patented and manufactured by a single producer. Thus, the control of stress state appears to be the most critical controllable factor for SRxC risk mitigation. However, as more suppliers with different melting practices produce the alloy, chemistry and processing may need to be more thoroughly assessed beyond the comments noted regarding the importance of the Mg and S content.

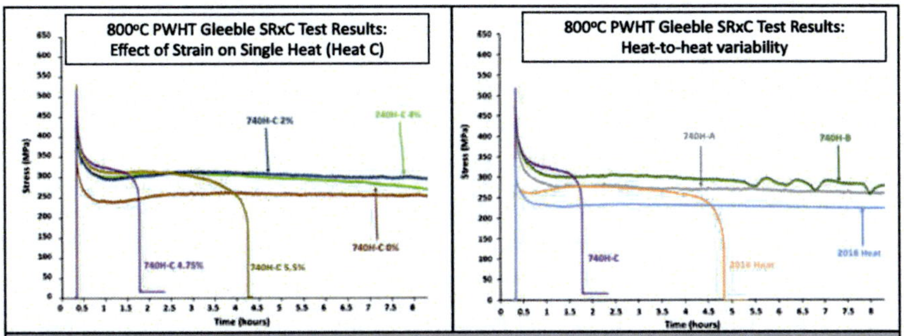

Heat ID	2016	2018	A	B	C
Product Form	Plate	Plate	Tubing	Plate	Extruded Pipe
Mg (wt%)	0.0022	0.0053	0.0062	0.0032	0.0013
S (wt%)	0.0025	0.0004	0.001	0.0005	0.0003
Mg/S	0.88	13.3	6.2	6.4	4.3

Fig. 7 Gleeble™-based SRxC experiment for a single heat of Alloy 740H (left) showing relaxation without failure for 0, 2, and 4% strain but sample failure at 4.75 and 5.5% strain and (right) the heat-to-heat variability

Guidance

EPRI is creating technical requirements and guidelines to provide purchasers and end users the opportunity to enhance purchase specifications for alloy 740H and to reduce the variability in the fabrication or operation of alloy 740H components or systems [22]. A key part of these recommendations includes comprehensive welding guidance. The qualification of a welding process and welder to ASME Section IX will not guarantee success. Welding of nickel-based alloys and alloy 740H require guidance that goes beyond the minimum requirements given in the code of construction. The basic guidance for alloy 740H can be found in [23] which addresses welding technique and training including the following:

- Cleanliness
- Fluidity & penetration
- Bead shape
- Shielding gas selection
- Heat input, pre-heating, and interpass
- Weld reinforcement

The field findings and research suggest additional practical considerations are required to reduce the risk of SRxC. These are as follows:

1. *Avoid excessive reinforcement and other local geometric changes which lead to high localized stresses.*

Figure 8 shows an example of excessive reinforcement resulting in a high stress location on the tube ID. When combined with the residual stresses from welding, this location initiated the crack which propagated through the wall thickness. For manual welding, providing favorable access and appropriate training and/or experience with nickel-based welding is critically important. Alternatively, automated welding may reduce the propensity for such features.

2. *Joint location and fabrication sequence should be considered in the design. This is especially true for manual welding to provide ease of access, reduce local constraints, minimize alignment requirements, and/or the introduction of excessive welding residual stress.*

In the STEP Demo heater example (Table 3), the failed tube butt welds were biased to the return end of the heater. All of the non-return end welds were conducted on a table with favorable welder access. Conversely, as shown in Fig. 8, welder access was limited, especially as welding progressed. Furthermore, as shown in the figure insert, the closure welds often required 'jacking tubes into place' to align the two tubes. Such practices should be avoided and/or prohibited.

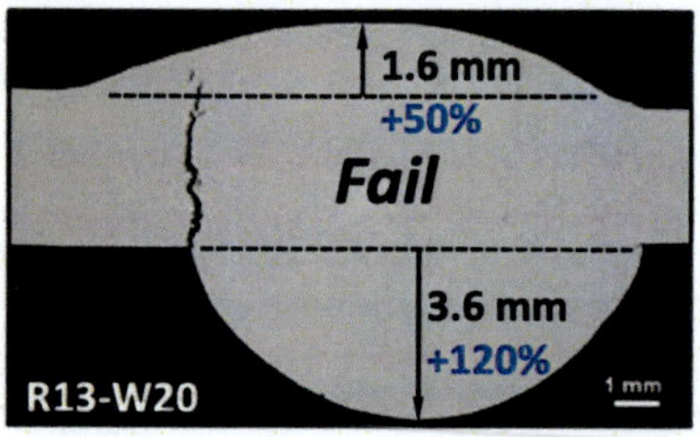

Fig. 8 Example of excessive reinforcement leading to a stress concentration at the toe of the weld

Fig. 9 Example of challenging welder access and highly restrained closure welds which superimpose additional stresses to the welding residual stresses during the welding process [10]

3. *Do not place welds at thickness transitions which can lead to high localized stresses in the weldment*

Transition pieces should be utilized to avoid placing welds at thickness transitions where the difference in stiffness can lead to increased localized stresses. This is true for tubing or piping applications. Special design or fabrication considerations must be given to transitions that occur between dissimilar materials to alloy 740H.

4. *Block fittings, intersections or valves must be avoided.*

It is relatively commonplace to procure block forgings that are subsequently machined into fittings, intersections or valves with minimal OD machining or contouring. Figure 10(a) shows a different view of the failure in Fig. 6. While a detailed model was not constructed, it is likely the large block tee piece acted as a heat sink during welding, thus requiring the welder to compensate with excessive heat input. Additionally, the non-symmetric shape of a thin tube into a solid block likely led to additional stresses upon cooling. Figure 10(b) shows an improved design where OD machining provided a more uniform circumferential thickness and eliminated the socket weld in favor of a full penetration butt weld. Figure 10(c) shows the completed new instrument attachment which did not crack after PWHT and is now in operation.

5. *The sequence for volumetric inspection*

In many cases of SRxC, the fabricator or end user performed inspections after welding was complete only to discover failures that are believed to have originated after PWHT during a hydro acceptance test. While some inspection after welding should be performed to identify if gross defects are present and more easily remediated prior to PWHT, it is recommended to perform a comprehensive inspection after PWHT as the origination of the majority of recent cracking has been attributed to SRxC.

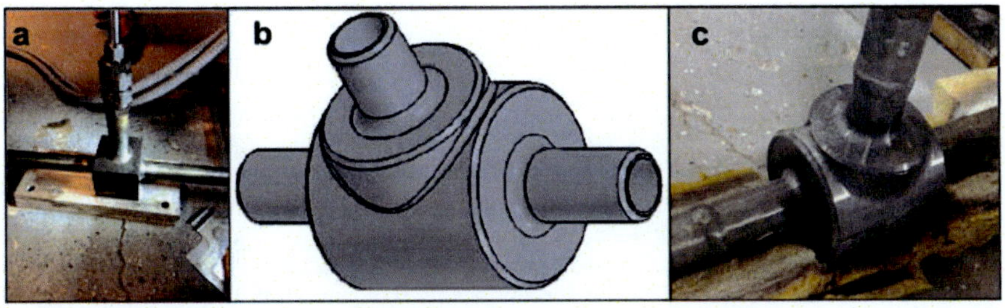

Fig. 10 Original instrument 'block tee' design (**a**), a re-designed contoured tee (**b**), and the final fabricated tee which did not show any cracking (**c**)

Summary

740H is a commercial nickel-base alloy attracting strong interest to enable the design, fabrication, and operation of advanced, high-temperature power cycles and technologies. Its recent application in a number of demonstration facilities has confirmed that the alloy can be welded and fabricated into complex components of varying sizes. Stress relaxation cracking (SRxC) during PWHT has been identified through these demonstrations, and, similar to other austenitic materials, is a legitimate concern. The cracking case studies reviewed in this manuscript provided supporting metallurgical investigations and targeted laboratory studies. This information, in addition to industry workshops and collation of recent experience, is leading to the first-of-a-kind document to disseminate the necessary practical guidance to reduce the uncertainty and variability in future installations requiring this material.

Acknowledgements This manuscript and research were supported by the U.S. Department of Energy's Office of Energy Efficiency and Renewable Energy (EERE) under the Solar Energy Technologies Office Award Number DE-EE0009378. The guidance of Kamala Raghaven, DOE, is appreciated.

References

1. J.J. deBarbadillo, "INCONEL alloy 740H", Materials for Ultra-Supercritical and Advanced Ultra-Supercritical Power Plants, ed. A. Di Gianfrancisco, Elsevier, London, 2017, 469–506. ISBN: 978-08-100552-1
2. J. Dupont, J. C. Lippold and S. D. Keiser: 'Welding metallurgy and weldability of nickel base superalloys'; 2009, New York, John Wiley and Sons Inc.
3. S.A. David, J.A. Siefert, J.N. Dupont, J.P. Shingledecker. "Weldability and weld performance of candidate nickel base superalloys for advanced ultrasupercritical fossil power plants part I: fundamentals." *Science and Technology of Welding and Joining*. Vol. 20, Issue 7 (October 2015), pp. 532–552.
4. John P. Shingledecker, John A. Siefert. "Age Hardenable Nickel-based Alloy Developments and Research for New High Temperature Power Cycles." In Ott E. et al. (eds) *Proceedings of the 9th International Symposium on Superalloy 718 & Derivatives: Energy, Aerospace, and Industrial Applications*, © 2018 The Minerals, Metals & Materials Series. Springer, Cham. https://doi.org/10.1007/978-3-319-89480-5_1
5. J.A. Siefert, J.P. Shingledecker, J.N. Dupont, S.A. David. "Weldability and weld performance of candidate nickel base superalloys for advanced ultrasupercritical fossil power plants part

II: weldability and cross-weld creep performance." *Science and Technology of Welding and Joining*. Available online October 2015. https://doi.org/10.1179/1362171815Y.0000000094
6. John deBarbadillo, Ronald Gollihue, Brian Baker. "Recent Development in Welding 740H" DOE-NETL. available at: https://netl.doe.gov/sites/default/files/netl-file/21WELD_Debarbadillo.pdf
7. "740H Fabrication and Welding Experience", *EPRI Industry Workshop (virtual)*, December 13, 2021.
8. J. Marion et. al. "The STEP 10 MWe sCO2 Pilot Demonstration Status Update." *ASME Turbo Expo 2020*: September 21–25, 2020 (Virtual). GT2020–14334. https://doi.org/10.1115/GT2020-14334
9. M. Hauth, D. Stanley, D. Washburn. "ASME BPVC Code-Compliant Heater Design & Fabrication Using Inconel 740H." *Proceedings – Evaluation of Welding Issues in High Nickel and Stainless Steel Alloys for Advanced Energy Systems*. DOE-NETL: March 10, 2021. https://netl.doe.gov/21WELD-proceedings
10. John Shingledecker, John Siefert, Tapasvi Lolla, Matthew Hauth, Mark Stevens, Trenton Cook. "Investigation of Weldment Cracking During Fabrication of a 700°C Fired sCO2 Heater." *The 7th International Supercritical CO2 Power Cycles Symposium*. February 21 – 24, 2022, San Antonio, Texas. Paper #170. https://sco2symposium.com/proceedings2022/170-paper.pdf
11. Daniel H. Bechetti, John N DuPont, John A Siefert, John P Shingledecker. "Microstructural Evolution and Creep-Rupture Behavior of A-USC Alloy Fusion Welds." Metallurgical and Materials Transactions A. pp. 1–17. Available online June 2016: http://dx.doi.org/ https://doi.org/10.1007/s11661-016-3603-7
12. *Innovative Method for Welding in Generation 3 CSP to Enable Reliable Manufacturing of Solar Receivers to withstand Daily Cycling at Temperatures Above 700°C.* DE-EE0009378
13. R. Kant, J. DuPont. "Stress Relief Cracking Susceptibility in High-Temperature Alloys." *Welding Journal*, AWS, February 2019, 29-s. https://doi.org/10.29391/2019.98.003
14. R.A. Mulford: *Treatise Mater. Sci. Technol.*, 1983, vol. 25, pp. 1–19.
15. T. Hu, S. Yang, N. Zhou, Y. Zhang, and J. Luo: *Nat. Commun.*, https://doi.org/10.1038/s41467-018-05070-2.
16. D. McLean: Grain Boundaries in Metals, Clarendon Press, Oxford, 1957.
17. R.A. Mulford: Metall. Trans. A, Phys. Metall. Mater. Sci., 1983, vol. 14 A, pp. 865–70.
18. R.T. Holt and W. Wallace: Int. Met. Rev., 1976, vol. 21, pp. 1–24.
19. J.N. Dupont, J.C. Lippold, and S.D. Kiser: Welding Metallurgy and Weldability of Nickel-Base Alloys
20. J.J. deBarbadillo: Superalloys, 1976, pp. 95–107.
21. K. Banerjee: Mater. Sci. Appl., 2011, vol. 02, pp. 1243–55.
22. Technical Requirements for Procurement of Components Fabricated from INCONEL ® Alloy 740H®. EPRI, Palo Alto, CA: 2022. *3002025494*.
23. A Practical Guide on Welding INCONEL® Alloy 740H®. *Special Metals Brochure*

Mechanical and Microstructural Properties of Brazed Honeycomb Liner Material Haynes 214

Jonas Vogler, Jieun Song, Jakob Huber, Rainer Völkl, and Uwe Glatzel

Abstract Abradable honeycomb sealing systems are widely used in turbines to improve efficiency and thus reduce carbon dioxide emissions. The honeycomb sealing systems are produced by brazing together nickel-based superalloy sheets with a nickel-based brazing metal. This investigation aimed to generate data on the mechanical performance of Haynes 214 metal sheets brazed with the nickel–chromium-silicon filler metal BNi-5 (71 wt.% Ni, 19 wt.% Cr, 10 wt.% Si). Tensile properties of brazed metal sheet composites are tested. Interdiffusion zones and hard particles with high chromium contents are observed along the brazed joint. Even a very thin brazing layer reduces the ductility considerably.

Keyword Brazing · Haynes 214 · Honeycomb Nickel-based superalloys

Introduction

Honeycomb sealing systems are used in aircraft turbines to minimize air leakage in the gaps between rotating parts and the turbine casing to improve efficiency and thus reduce carbon dioxide emissions [1]. In addition, the honeycomb structure protects the fins of the rotating turbine blades from critical damage in the case of a contact caused by thermal or mechanical expansion [2].

The honeycombs themselves are point welded thin metal sheets of a nickel-based superalloy brazed onto a substrate usually made too of a nickel-based superalloy (Fig. 1). During the brazing process, capillary forces draw braze filler alloy into gaps

J. Vogler · R. Völkl · U. Glatzel (✉)
Metals and Alloys, University of Bayreuth, Bayreuth, Germany
e-mail: uwe.glatzel@uni-bayreuth.de

J. Song
Institute of Thermal Turbomachinery, Karlsruhe Institute of Technology, Karlsruhe, Germany

J. Huber
Institute of Material Science and Mechanics of Materials, Technical University of Munich, Munich, Germany

E. A. Ott et al. (eds.), *Proceedings of the 10th International Symposium on Superalloy 718 and Derivatives*, The Minerals, Metals & Materials Series,
https://doi.org/10.1007/978-3-031-27447-3_28

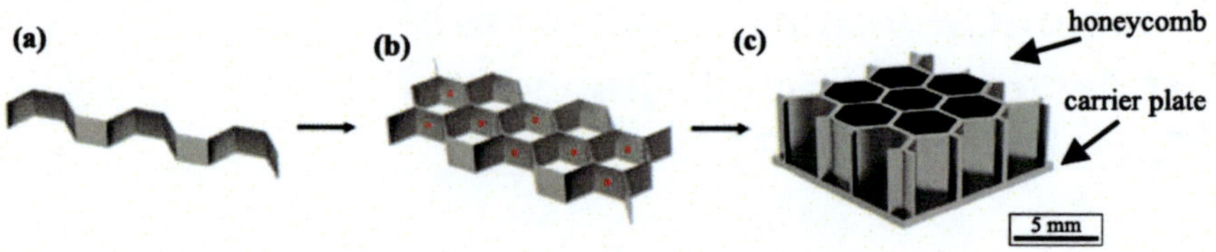

Fig. 1 Honeycomb seal manufacturing schematic: **a** corrugated half-hexagon structure, **b** spot-welding and **c** honeycomb structure brazed onto the carrier plate

between the metal sheets. Additionally, the braze filler alloy can travel up to the rubbing surface at the outer faces of the honeycomb cells [3].

The formation of hard and brittle phases due to the chemical composition of the braze filler alloy can deteriorate the mechanical properties of the abradable sealing [3]. Additionally, the original chemistry of the base alloy can be altered, thus influencing the oxidation behavior [4]. A detrimental influence of the increased contact surface area with the rotor fin can also be expected.

For the low-pressure turbine with temperatures of up to 1100 °C, Haynes 214 is a commonly used honeycomb liner material [2]. At these temperatures, alumina-forming alloys are more suitable than chromia-forming alloys. For Haynes 214 metal sheets, thermo-physical properties can be found in the literature [4–7]. However, for the honeycomb sealing systems, the mechanical properties of brazed Haynes 214 metal sheets are highly relevant. Therefore, this study compares the microstructure and the mechanical properties of Haynes 214 metal sheets before and after brazing with the braze filler alloy BNi-5. Compared to other nickel-based braze filler alloys BNi-5 has a higher solidus temperature, hence it is very well suited for brazed parts operating at high temperatures.

Experimental Methods

In this work, the nickel-based superalloy Haynes 214 is selected as the base material since it is widespread for turbine seal honeycomb applications [2]. Haynes 214 is a nickel–chromium-aluminum-iron alloy with excellent oxidation resistance developed for high-temperature applications. It can be age hardened at temperatures below 925 °C [5].

Honeycomb seals are typically brazed with nickel- or cobalt-based braze filler alloys since the sealing has to withstand high temperatures of up to 1100 °C [2]. A commonly used braze filler alloy is BNi-5 (AMS4782). The liquidus temperature of BNi-5 is 1135 °C [8]. The nominal compositions of Haynes 214 and braze filler alloy BNi-5 are given in Table 1.

The rolled and annealed tape of Haynes 214 alloy was supplied by Elgiloy Specialty Metals. The alloy was cold-rolled and solution-annealed at 1095 °C for 2 h

Table 1 Nominal chemical composition (wt.%) of the Haynes 214 metal sheets and the brazing filler alloy BNi-5 used [5, 8]

Element	Ni	Cr	Si	Al	Fe	Mn, Mo, Ti, W
Haynes 214	75 Bal	16	<0.2	4.5	3	<0,5
BNi-5	71 Bal	19	10			

and rapid air-cooled. For the tensile tests metal sheets with a thickness of 0.6 mm were used.

For the brazed samples two Haynes 214 metal sheets with the dimension of 145 × 28 mm and an average thickness of 0.29 mm were brazed together by the Listemann AG. A BNi-5 brazing foil with a thickness of 0.04 mm is used to join two Haynes 214 metal sheets in a lap joint geometry. The samples were placed on level ground. A graphite plate was placed horizontally on top of the samples to ensure a uniform brazing clearance. The brazing is performed under vacuum ($<10^{-2}$ Pa) at a temperature of 1170 °C with a holding time of 15 min. Subsequently the samples were furnace cooled to 900 °C (~5 °C/min) and then rapidly cooled to room temperature with a cooling rate of 20 °C/min.

Mechanical testing of the brazed sandwich structure is performed on a Zwick Z2.5 universal testing machine at a crosshead speed of 0.05 mm/s. Strain is measured with a video extensometer with an accuracy of ~ ± 0.0025% [9]. Specimens were taken with electron discharge machining. The gauge length of the specimen is 5 mm with a width of 0.9 mm. Microhardness (Fischerscope HM 2000) on the polished cross-sections is measured by applying 50 mN force. At least 30 hardness indents were recorded for each specimen tested.

Microstructure studies were carried out using scanning electron microscopy (SEM), element distribution analysis was performed by energy-dispersive x-ray spectroscopy (EDS), and grain sizes were determined using electron backscattering diffraction (EBSD).

Results

The as-delivered Haynes 214 metal sheets with a thickness of 0.29 mm have an equiaxed grain structure with an average grain size of 28 μm. After the brazing, the average grain size increases to ~ 330 μm. The grain size distribution is shown in Fig. 2. The number of grains throughout the thickness of the metal sheet is strongly reduced to less than six grains.

Figure 3 shows the as-brazed microstructure. The dashed red lines indicate the initial thickness of the brazing foil (40 μm). In Fig. 3a, an annular BSE detector was used to identify phases with different chemical content. The brazing layer almost completely dissolved during the brazing process since the elements from the brazing filler alloy diffuse into the base material. Thus, a dark appearing phase can be seen

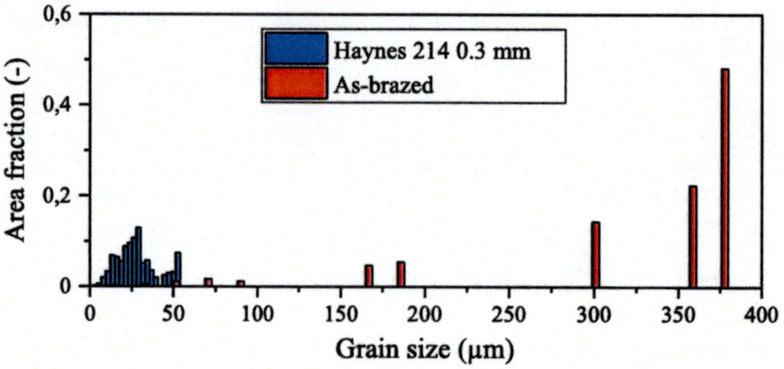

Fig. 2 Grain size distribution before (blue) and after brazing (red)

in the brazing region and at the grain boundaries. No cracks and pores are detected. Hence the BNi-5 brazing foil is very well suited to join samples with the geometry used in this study. The EBSD orientation map in Fig. 3b shows grains and twinning. It also becomes clear that grain growth is interrupted at the centerline.

The EDS results of the chemical composition in the brazing region are shown in Fig. 4. A silicon-rich diffusion affected zone with a width of ~ 50 µm can be identified in the EDS mapping. Furthermore, precipitates in the middle of the joint region and along the grain boundaries are clearly visible. The chemical composition shown in Table 2 was obtained by averaging of seven EDS point measurements in the precipitates. The chromium content is high with about 88 wt.%, whereas aluminum and iron content are low compared to the Haynes 214 base metal.

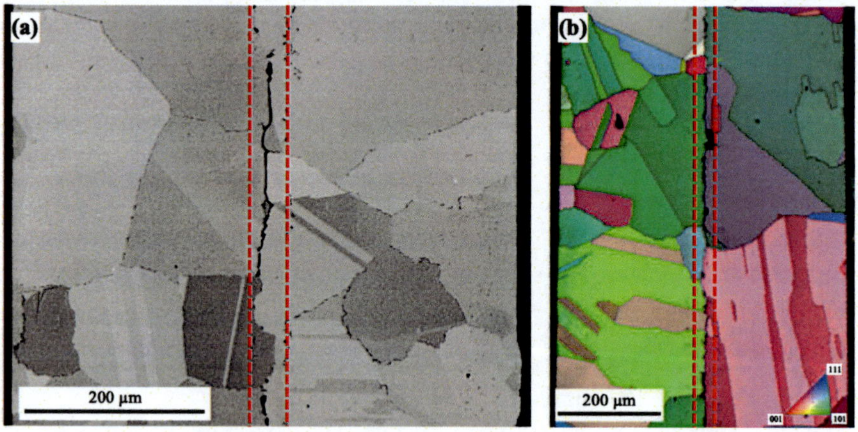

Fig. 3 Microstructure of the Haynes 214 alloy after brazing with BNi-5 **a** Backscattered electron image and **b** EBSD orientation map

Table 2 Mean value from the EDS point measurements (wt.%) of the phase or precipitate respectively marked in Fig. 4

Element	Ni	Cr	Si	Al	Fe	B, C
Wt.%	7	88	<0.2	<0.2	<0.2	<2

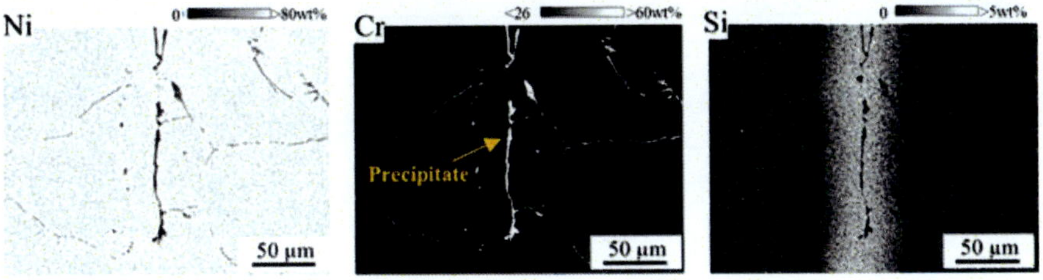

Fig. 4 EDS mapping in the brazing joint region

The microhardness of the brazed Haynes 214 metal sheets slightly increases from 3062 N/mm^2 to 3337 N/mm^2. The chromium-rich phase in the joint region has a high hardness of over 6500 N/mm^2.

The stress–strain curves for the Haynes 214 metal sheets with a thickness of 0.3 mm and for the Haynes 214 metal sheets brazed with BNi-5 are given in Fig. 5.

The data obtained for the Haynes 214 metal sheets are slightly lower than the values provided by Haynes International, Inc. [5]. The values for the yield strength, the ultimate tensile strength and the elongation at fracture are summarized in Table 3.

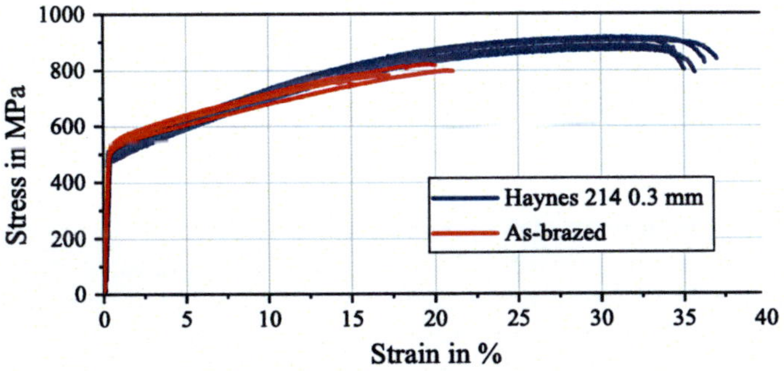

Fig. 5 Stress–strain curves of Haynes 214 metal sheets and lap brazed metal sheets

Table 3 Tensile properties of Haynes 214 metal sheets and lap brazed metal sheets at room temperature

	Yield strength at 0.2% offset (MPa)	Ultimate tensile strength (MPa)	Elongation at fracture (%)
As-Brazed	521	794	18
Haynes 214 (this study)	497	903	36
Haynes 214 [5]* *Cold-rolled and solution-annealed sheet with a thickness of 2.0 to 3.0 mm	577	975	37

Discussion

A silicon rich diffusion affected zone, typical when BNi-5 is used as braze filler alloy, is identified by EDS mappings. The width of the diffusion affected zone is influenced by the brazing temperature and time [10]. In literature [10–12] the eutectic reaction $L \leftrightarrow (Ni) + Ni_5Si_2 + Cr_3Ni_5Si_2$ is confirmed in the Cr-Ni-Si ternary system. According to Han et al. [10] the typical microstructure of BNi-5 consists of γ-Ni solid solution, Ni_5Si_2, G-phase, Ni_3Si and $Cr_3Ni_5Si_2$. In this study highly chromium-rich precipitates at the grain boundaries and in the brazing region are detected. The chemical composition of the chromium-rich precipitates is not consistent with the previously published literature. Reasons for this may be the use of a very thin braze metal foil and a different brazing strategy (temperature and time) in this study. Furthermore, the base metal too influences the microstructure in the brazing region. For Haynes 214 brazed with BNi-5 no published literature can be found.

Interestingly, the yield strength for the brazed Haynes 214 metal sheets is increased, meaning the resistance against plastic deformation is higher for the brazed metal sheets. Additionally, the hardness of the brazed Haynes 214 metal sheets is slightly increased. This is due to the brazing heat treatment causing residual stresses and a change in the chemical composition due to elemental diffusion from the braze filler alloy into the base material. In contrast to the yield strength, the ultimate tensile strength is, on average, reduced by roughly 100 MPa. The reduced ultimate tensile strength is mainly attributed to the precipitates forming in the brazing region and along the grain boundaries. These chromium-rich precipitates cause stress concentrations in the lap brazed metal sheets, which strongly decrease the elongation at failure from 36 to 18%.

To obtain optimal mechanical properties, it is crucial to prevent the formation of a hard chromium-rich phase in the brazing region. The precipitate volume fraction can be reduced by reducing the amount of brazing filler alloy or subsequent aging heat treatment. However, in honeycomb sealing systems, a sufficient amount of brazing filler alloy is necessary to guarantee good structural stability between the honeycomb and the carrier plate (see Fig. 1). Nevertheless, it is possible to reduce the amount of brazing material at the contact surface with the rotor by modifying the component geometry or by using a pack aluminide coated base material [13].

Summary and Conclusion

In this work, the influence of the filler alloy BNi-5 on the microstructural and mechanical properties of lap-brazed Haynes 214 superalloy metal sheets was studied. Near the joint region, a silicon diffusion affected zone with a thickness of about 50 μm is formed. Furthermore, particles of hard chromium-rich phases are observed in the joint region and along grain boundaries. After the brazing process, the grain size is drastically increased.

The yield strength of the brazed metal sheet is higher than the yield strength of the Haynes 214 base metal. However, the elongation to failure of the brazed metal sheet is considerably reduced from 36 to 18%.

Despite using an extremely thin brazing foil with 0.04 mm thickness, the mechanical properties, particularly the elongation to failure, are negatively affected. This is especially unfavorable when the brazed part, like a honeycomb abradable sealing system, is exposed to mechanical loads.

References

1. Stocker HL, Cox DM, Holle GF (1977) Aerodynamic performance of conventional and advanced design labyrinth seals with solid-smooth abradable, and honeycomb lands. NASA, Indianapolis
2. Sporer D, Fortuna D (2014) Selecting Materials for Brazing a Honeycomb in Turbine Engines. Weld J. 93(2):44–48
3. Ulan kyzy S, Völkl R, Munz O, Fischer T, Glatzel U (2019) The Effect of Brazing on Microstructure of Honeycomb Liner Material Hastelloy X. J Mater Eng Perform 28:1909-1913. https://doi.org/10.1007/s11665-019-03910-w
4. Sporer DR, Shiembob LT (2004) Alloy Selection for Honeycomb Gas Path Seal Systems. ASMEDC 4:763–774. https://doi.org/10.1115/GT2004-53115
5. Haynes International (2021) Datasheet Haynes 214. https://haynesintl.com/docs/default-source/pdfs/new-alloy-brochures/high-temperature-alloys/brochures/214-brochure.pdf?sfvrsn=bf7229d4_30. Accessed 8 August 2022
6. Ulan kyzy S, Völkl R, Munz O, Fischer T, Welzenbach S, Glatzel U (2020) Thermo-physical properties of Hastelloy X and Haynes 214 close to the melting range. Mater Sci Tech 36(10):1012–1019. https://doi.org/10.1080/02670836.2020.1753154
7. Deevi SC, Sikka VK (1996) Nickel and iron aluminides: an overview on properties, processing, and applications. Intermetallics 4:357–375. https://doi.org/10.1016/0966-9795(95)00056-9
8. The Prince & Izant Company (2022) Datasheet AMS 4782. https://princeizant.com/uploads/Tech_Data_Sheets/AMS_4782_-_BNi-5.pdf. Accessed 8 August 2022
9. Völkl R, Fischer B, Beschliesser M, Glatzel U (2008) Evaluating strength at ultra-high temperatures - Methods and results. Mater Sci Eng A 587–589. https://doi.org/10.1016/j.msea.2006.09.171

10. Han WP, Wan M, Tan JF, Zhao R, Kang H, Qu P (2022) Study on mechanical properties and microstructure development of Inconel 718 ultrathin-walled capillary-and-plate brazed structure using BNi-5 filler metal. Weld World. https://doi.org/10.1007/s40194-021-01219-8
11. Schuster JC, Du Y (2000) Experimental investigation and thermodynamic modeling of the Cr-Ni-Si system. Metall and Mater Trans A 31:1795–1803. https://doi.org/10.1007/s11661-006-0248-y
12. Han W, Wan M, Zhao R, Kang H, Rao Y (2021) Microstructural evolution and mechanical properties of brazed IN718 ultrathin-walled capillary structure using different particulate reinforced filler alloy. Chinese J Aeronautics 35:550-564. https://doi.org/10.1016/j.cja.2021.03.001
13. Shiembob LT (1975) Continued development of abradable gas path seals. Pratt & Whitney Aircraft, Cleveland

Effect of Heat Treatment on the Mechanical Property and Deformation Mechanism of a Novel Cast Nickel-Based Superalloy

Pengfei Zhao, Min Wang, Meiqiong Ou, Yingche Ma, and Kui Liu

Abstract The microstructure, mechanical properties, fracture behavior, and deformation mechanisms of a novel cast nickel-based superalloy subjected to various aging treatments were investigated. The microstructure of this new alloy K4800 consists of γ, MC, $M_{23}C_6$, and the γ' in two sizes after conducting a heat treatment of solution annealing and double-stage aging. It is found that an initial higher temperature aging process is beneficial and somewhat necessary to optimize the tensile and creep properties of the alloy. The strength of alloy K4800 rises with the volume fraction of small-sized γ' phase; meanwhile, the precipitation of large-sized γ' phase has a desirable impact on the elevated temperature ductility of the alloy. However, the existence of large-sized γ' phase may accelerate the ripening process of γ' phase and decrease of creep life of alloy K4800. Therefore, accurate control of the content and proportion of the γ' phases in two sizes is the key to obtaining an optimal mechanical property of K4800. TEM microstructure investigations show that the main strengthening mechanisms of the alloy are APB cutting at room temperature and Orowan bypassing at high temperature. Additionally, based on the results of a long-term aging experiment at 800 °C up to 2000 h, the material is not expected to precipitate any undesirable phase like σ or η, exhibiting an outstanding microstructural stability.

Keywords Nickel-based superalloy · Precipitation · Mechanical property · Deformation mechanism · Heat treatment

Introduction

Nickel-based high-temperature alloys have become the material of choice for vital hot-end components in aero-engines and gas engines because of their excellent high-temperature creep resistance, decent high-temperature oxidation and corrosion resistance, and good microstructural stability [1–3]. With the increasing thrust-to-weight ratio of engines and the development of integral casting technology, the structures

P. Zhao · M. Wang (✉) · M. Ou · Y. Ma · K. Liu
Shi-Changxu Innovation Center for Advanced Materials, Institute of Metal Research, Chinese Academy of Sciences, Shenyang 110016, China
e-mail: minwang@imr.ac.cn

E. A. Ott et al. (eds.), *Proceedings of the 10th International Symposium on Superalloy 718 and Derivatives*, The Minerals, Metals & Materials Series,
https://doi.org/10.1007/978-3-031-27447-3_29

of aero-engine hot-end components are becoming increasingly complex. Therefore, nickel-based high-temperature alloys with better temperature-bearing capacities, especially the ones also possessing good welding and casting properties, are needed in the development of advanced aero-engines [4]. In the present work, a newly-developed cast nickel-based superalloy K4800 is studied, which has excellent mechanical properties and microstructural stability at 800 °C.

The mechanical properties of cast nickel-based high-temperature alloys depend highly on their microstructural features, such as grain size, and the size, volume fraction, morphology, and distribution of precipitates [5–8]. γ' phase is the main strengthening secondary phase of nickel-based superalloys. The matrix γ phase and γ' phase have lattice constants slightly different from each other, and an elastic stress field can be easily generated at the interface of the two phases [1, 3], which hinders dislocation movement. Unlike many other nickel-based high-temperature alloys which are strengthened by the γ' phase with uniform particle size, the strength of alloy K4800 is provided by the γ' particles in a bimodal size distribution. Heat treatment is an important process to control the precipitation characterization of γ' phase. Through carefully designing the parameters of aging heat treatment, a microstructure with the γ' phase in various sizes can be obtained. It is well-known that the size of γ' phase increases with aging temperature and time, and coarse γ' particles can form in an over-aging regime or at high aging temperatures. Xu et al. found that the large size of the γ' phase precipitated at high temperature was beneficial to dislocation movement and improved alloy plasticity. Sharma et al. also found that multiple ageing treatments could obtain multi-scale γ' phases, improving the tensile strength, tensile plasticity, and high-temperature stress ruptured life of alloy [9–10].

As a new alloy, the effects of heat treatment regime on the microstructure and mechanical properties of alloy K4800 are not fully understood. Besides, the strengthening mechanism of multiple-scale γ' phase in this alloy is yet to be investigated in detail. Therefore, in this work, alloy K4800 was aged at different temperatures, and tensile and creep tests were conducted. Microstructures of heat-treated and fractured specimens were investigated and the strengthening mechanism of the alloy was analyzed.

Experimental

Experiments were conducted using cast alloy K4800 test bars, which were fabricated by the investment casting method in a vacuum induction melting furnace. The chemical composition of alloy K4800 was determined by an ICP6300 spectrometer, as listed in Table 1. The Thermo-Calc software was used to depict the precipitation

Table 1 The chemical composition of alloy K4800 (wt.%)

wt.%	C	Cr	Co	W	Mo	Al	Ti	Nb	B	Ni
K4800	0.11	17	9.5	2.0	1.5	1.5	3.2	2.7	0.01	bal

Table 2 The equilibrium volume fraction of γ' phase at different temperatures

Temperature, °C	1100	1090	1060	1030	800
Volume fraction, vol.%	0	0.1	7.8	13.8	35.4

behaviors of secondary phases in alloy K4800 between 500–1500 °C. Four heat treatment conditions (T1–T4) were designed based on the calculated equilibrium phase diagram (Table 2):

T1:1180°C × 4 h.AC + 1160°C × 4 h.AC + 1030°C × 2 h.AC + 800°C × 16 h.AC (Air Cool).

T2:1180°C × 4 h.AC + 1160°C × 4 h.AC + 1060°C × 2 h.AC + 800°C × 16 h.AC.

T3:1180°C × 4 h.AC + 1160°C × 4 h.AC + 1090°C × 2 h.AC + 800°C × 16 h.AC.

T4:1180°C × 4 h.AC + 1160°C × 4 h.AC + 1100°C × 2 h.AC + 800°C × 16 h.AC.

All of these four heat treatment conditions consisted of a homogenization process, a solution process, and the first and second step aging processes. Heat treatments T1 to T4 were only different in the temperature of the first-step aging, ranging from 1030 to 1100°C. Names of the specimens heat-treated following those four methods were designated as T1–T4 directly.

Heat-treated bars were machined into specimens with a gauge length of 30 mm and a gauge diameter of 5 mm. At least two samples were prepared for each mechanical test. Tensile tests were carried out at both room temperature (RT) and 800 °C, which is the highest normal service temperature of this alloy. During the tests, the temperature variation was maintained within ± 2 °C, and the initial strain rate before yielding was controlled to $3 \times 10^{-4}\ s^{-1}$. Microstructures were characterized by a ZEISS MERLIN Compact scanning electron microscope (SEM) and an FEI Talos 200X scanning transmission electron microscope (STEM) equipped with energy dispersive spectroscopy. TEM discs no thinner than 450 μm in thickness were cut from the gauge section of specimens perpendicular to the loading direction. After being mechanically ground to about 45 μm, thin foils with a diameter of 3 mm were punched out and prepared by a twin-jet electropolishing machine with a solution of 10 vol.% $HClO_4$ and 90 vol.% C_2H_5OH at 19 V and − 22 °C. Most of the dislocation configuration observations were conducted in a STEM bright field (BF) mode. The size and volume fraction of γ' phase in the alloy were calculated using the ImageJ-Pro software.

Results and Discussion

Microstructure

Thermodynamic calculation software was employed to predict the equilibrium phase diagram of alloy K4800, as shown in Fig. 1. In this diagram, the weight fractions of phases are given as a function of temperature. The γ phase and liquid phase coexist within the temperature range of 1270–1340 °C, and their weight fractions change with temperature. Figure 1a indicates that no γ' phase can form at a temperature higher than 1100 °C, and when below 1100 °C, the weight fraction of γ' phase increases with the decrease in temperature.

Furthermore, according to the calculated diagram Fig. 1b, the volume fraction of γ' phase increases dramatically at the beginning of its precipitation at the temperature near 1100 °C. The equilibrium precipitation volume fraction of γ' phase is about 0.1 vol.% at 1090 °C. However, as the temperature descends to 1060 or 1030 °C, the volume fraction of γ' phase increases significantly to 7.8 and 13.8 vol.%, respectively. During an aging process above 1000 °C, γ' phase is expected to precipitate in large-sized particles due to the low nucleation rate in this condition [8]. At 800 °C, the γ' phase volume fraction is calculated to be 35.4%. As it can be seen from Fig. 1a, at temperatures below 800 °C, the equilibrium precipitation content of γ' phase becomes roughly steady, and at these relatively low temperatures, the precipitation size of γ' phase is generally small.

The heat-treated microstructures of specimens T1–T4 are quite similar, and the microstructure of specimen T3 is given in Fig. 2 as an example. It can be seen that the grains are generally equiaxed and a large number of MC carbides precipitate within the grains in the forms of block, rod, or skeleton. In addition, blocky MC and granular $M_{23}C_6$ carbides are found on the grain boundaries, as presented in Fig. 2d. The grain size of the alloy was calculated to be 640 ± 175 μm, using the intersection method on the Disc IAS image analysis software.

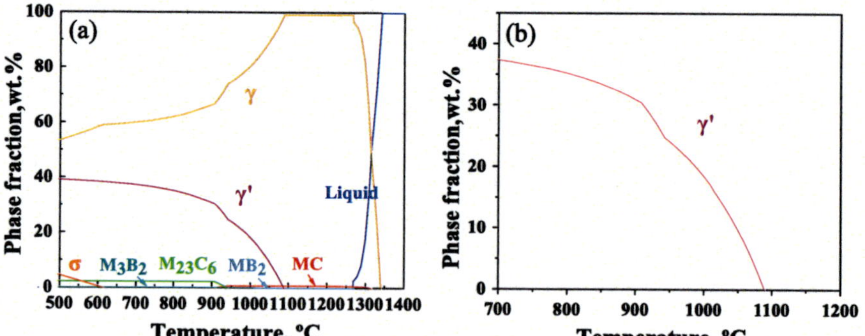

Fig. 1 The equilibrium phase diagram of alloy K4800 for **a** all phases and **b** γ' phase

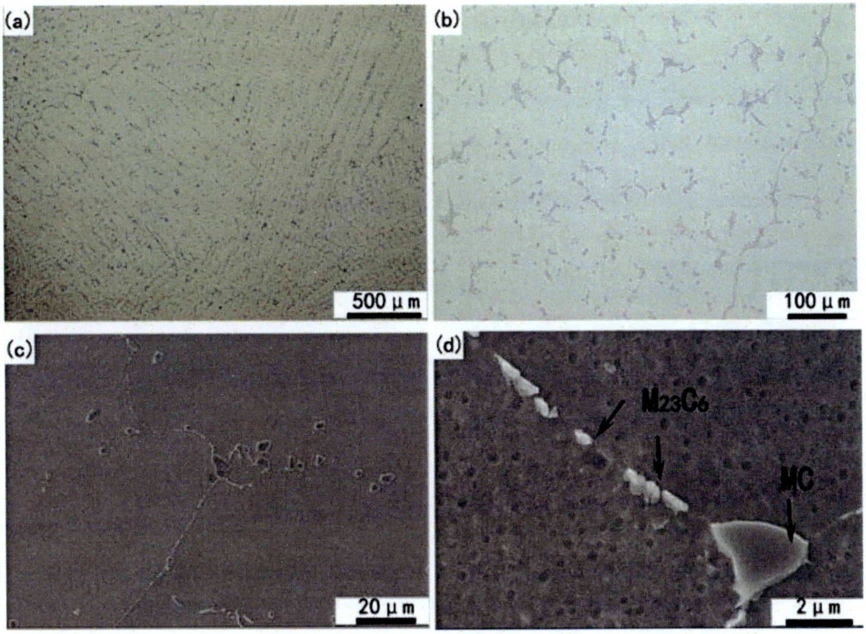

Fig. 2 Grain and carbide morphology of the heat-treated specimen T3 observed by **a-b** OM and **c-d** SEM

Figure 3 shows the precipitations of γ′ phase in the specimens T1–T4. It can be seen from the micrographs that the precipitation characterizations in these specimens are quite different. The γ′ particles in specimens T1–T3 precipitate in a bimodal size distribution, whereas in specimen T4, they are in a uniform size. The heat treatment conditions of specimens were only different in the temperature of the first-step aging process in which the large γ′ phase generate. A prior solution was employed to produce a supersaturated matrix preparatory to the subsequent aging by dissolving the minor phases precipitating during casting. The γ′ phase with a large size precipitated during the first-step aging process at a temperature lower than the prior solution but higher than 1000 °C. In contrast, the γ′ phase precipitated in the following lower-temperature aging step was generally small. For the convenience of explanation, the γ′ phase with a larger size formed at high temperatures is named as the γ_1' phase, and the small γ′ particles precipitated at 800 °C are named as the γ_2' phase.

In order to accurately acquire the size and content of γ′ phase, TEM micrographs of the specimens in four different heat treatment conditions were employed in measurements by using the ImageJ-Pro software, and the results are listed in Table 3. Higher the aging temperature, the larger the size and less the precipitation content of γ_1' phase. As the first-step aging temperature increases from 1030 to 1090 °C, the volume fraction of γ_1' phase decreases from 32.5 to 4.3 vol.%, while the volume fraction of γ_2' phase increases from 9.2 to 30.4 vol.%. In addition, the average size of γ_1' phase increases

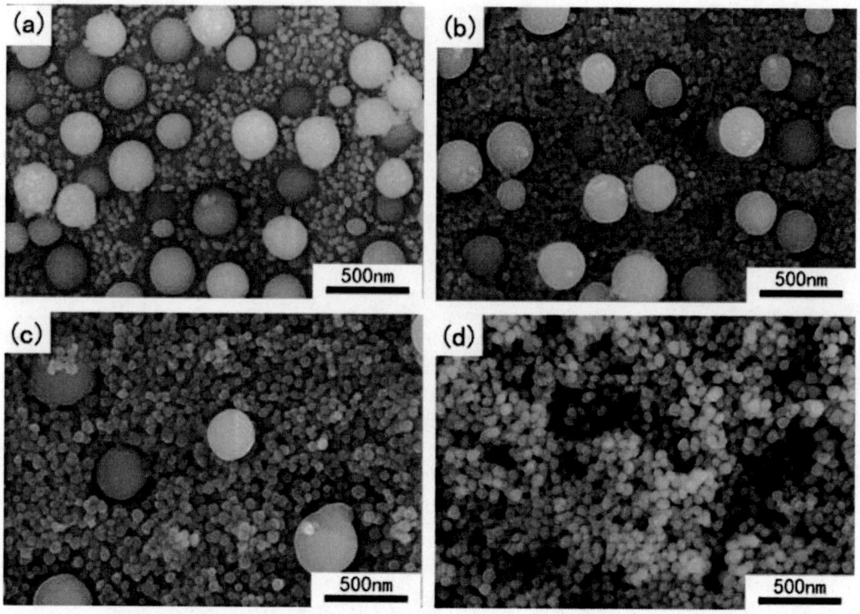

Fig. 3 SEM observations on γ' phase in the specimens **a** T1, **b** T2, **c** T3, and **d** T4

from about 273 to 356 nm and the average size of γ_2' phase rises from 45 to 52 nm. As to specimen T4, no γ_1' phase forms, and the volume fraction and average size of γ_2' phase is 37.1 vol.% and 63 nm, respectively.

At a temperature slightly lower than the solution temperature of γ' phase, the precipitation undercooling is small and consequently, the nucleation rate of γ' phase is low. Besides, at a high temperature, the alloying elements are rich in the matrix and their diffusion rates are high. As a result, the average size of γ_1' phase in specimen T1 is the largest among the four kinds of specimens, meanwhile, the content of γ_1' phase in T1 is the lowest. The precipitation of γ_1' phase will consume a certain number of alloying elements and then decrease the precipitation content of γ_2' phase. If the precipitation of γ_1' phase is numerous, accordingly, the amount of the small-sized γ_2' phase that precipitates in the subsequent aging process will be insufficient due

Table 3 The volume fraction and average particle size of γ' phase in specimens

Specimen	Volume fraction, vol.%		Average particle size, nm	
	γ_1' phase	γ_2' phase	γ_1' phase	γ_2' phase
T1	32.5	9.2	273	45
T2	17.7	17.1	28	49
T3	4.3	30.4	356	52
T4	/	37.1	/	63

to the lack of γ' forming elements. For specimen T4, because there is no γ_1' phase precipitates, the size of γ_2' phase is the largest among specimens under the most sufficient solute supply.

Tensile Property and Strengthening Mechanism

Figure 4 shows the tensile strength and ductility of specimens T1–T4 tested at room temperature and 800 °C. It can be seen that the room temperature (RT) yield strength (YS) and ultimate tensile strength (UTS) of specimen T1 reach 835 and 1090 MPa, respectively, while the RT YS and UTS of specimens T2, T3, and T4 increase gradually. The RT YS of specimen T3 increases to 942 MPa, while specimen T4 exhibits the highest YS of 958 MPa. It seems that the RT YS of the alloy increases with the first-step aging temperature of the specimen. However, the aging temperature has little effect on the RT tensile plasticity of the material. The difference in elongation and reduction in the area among specimens is not significant.

When the tensile temperature is increased to 800 °C, the effect of heat treatment on the strength of the alloy is similar to that at room temperature. The best-elevated

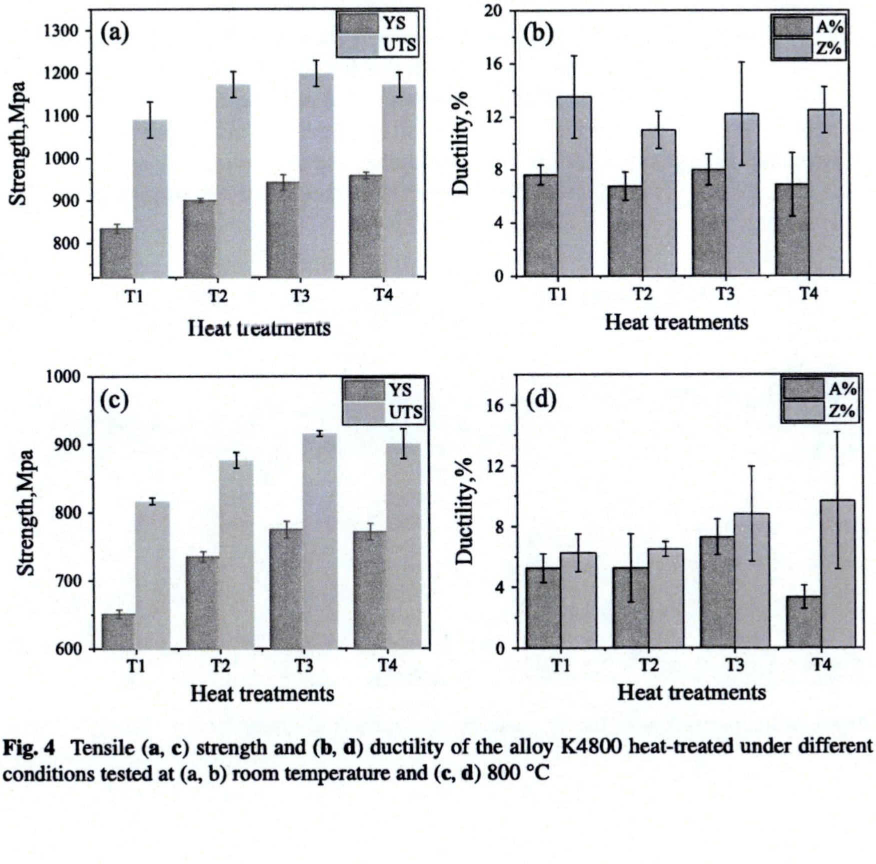

Fig. 4 Tensile (**a**, **c**) strength and (**b**, **d**) ductility of the alloy K4800 heat-treated under different conditions tested at (a, b) room temperature and (**c**, **d**) 800 °C

temperature mechanical performance was exhibited by specimen T3, with the YS and UTS of 775 and 915 MPa, respectively. Besides, unlike that at RT, the first-step aging temperature also has an obvious effect on the high-temperature tensile ductility of the specimen. The tensile elongations of specimens T1–T3 obtained at 800 °C are no lower than 6.0%. In contrast, the elongation of specimen T4 is only about 3.3%, which is lower than that of the other three kinds of specimens. The difference in the high-temperature reduction in the area of material is insignificant.

Despite different aging regimes were employed, fracture surface morphologies of the tensile specimens tested at room temperature are similar, as shown in Fig. 5. Shallow dimples and a certain number of tearing edges can be observed on all the fracture surfaces. In addition, no obvious secondary crack can be found on those fracture surfaces, indicating that the primary RT tensile fracture mode of alloy K4800 can be classified as transgranular.

The fracture surface morphologies of specimens tested at 800 °C are presented in Fig. 6. Cleavage facets and dimples can be found on these fracture surfaces, and basically, the morphologies do not show much difference from that of the specimens tested at room temperature. Many broken MC carbides are also witnessed on the facture surfaces but no obvious crack is found. This result suggests that the specimens ruptured in a transgranular mode at 800 °C as they did at room temperature, indicating that alloy K4800 still holds a good grain-boundary strength at elevated temperature.

However, Fig. 4d demonstrates that the tensile elongation of specimen T4 is obviously lower than that of T3 at 800 °C. Then after being given a more detailed examination, it is discovered that the number of flat cleavage facets on the fracture surface of specimen T4 is higher than that on T3, and the number and depth of dimples on the fracture surface are larger for the latter as well. This implies that to some extent intergranular failure occurred during the tensile test of specimen T4 at 800 °C, resulting in the apparent high-temperature ductility drop compared with that of specimen T3.

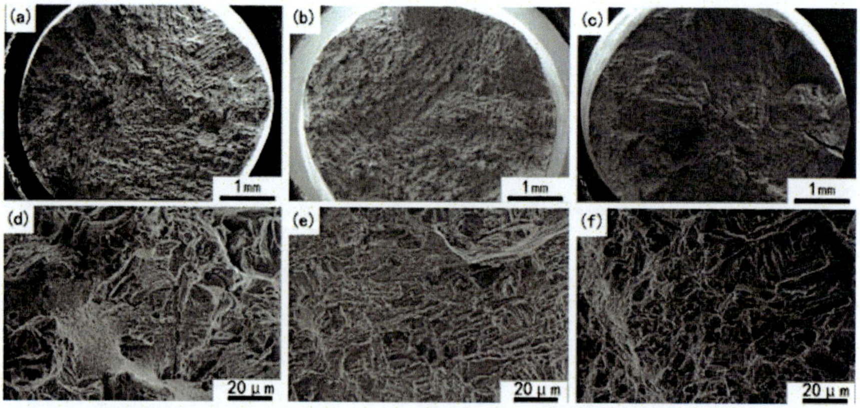

Fig. 5 SEM observations on the RT tensile fracture surfaces of specimens **a**, **d** T1, **b**, **e** T3, and **c**, **f** T4

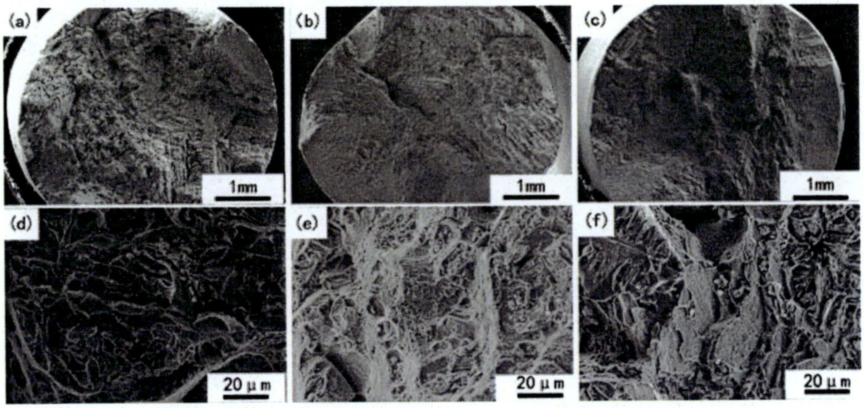

Fig. 6 SEM observations on the 800 °C tensile fracture surfaces of specimens **a, d** T1, **b, e** T3, and **c, f** T4

The longitudinal section view of tensile fracture was revealed and displayed in Fig. 7 Since the facture sections of specimens are quite similar in morphology, only the micrographs of specimen T3 are given. As it can be seen from Fig. 7a, d, the fracture surfaces of specimens are fairly flat at both room temperature and 800 °C, and no obvious intergranular secondary cracks can be found, indicating that the alloy ruptured in a transgranular mode at both room temperature and 800 °C. Inside the grains, cracks can be found in the interior and vicinity of MC carbide. It is because the MC/γ interface is a favorable location for dislocation accumulation. When dislocations are substantially accumulated, they can induce stress concentration and subsequently lead to the initiation of micropores and microcracks. Figure 7c shows that there are multiple micropores at the MC/γ interface. When dislocations shear into MC carbide, they will promote the formation of internal microcracks, in which the surrounding micropores play the role of cracking source. With the enlarging of uniaxial tensile stress, the internal microcracks of MC carbides continue to expand resulting in the fragmentation and spalling of MC carbides, which eventually leads to the formation of intragranular cracks and the final rupture.

Figure 8 demonstrates the dislocation configurations in the specimens fractured at room temperature, and the slip band is clearly the main deformation feature in all specimens. The existence of slip bands instead of widespread dislocations indicates that the deformation in the material is not uniform and concentrates within these band-like areas. Due to the ordered structure of γ′ phase, dislocations pass the γ/γ′ interface in pairs. During the tensile tests of the four kinds of specimens, the deformation mechanism of the alloy is identical (dislocation-pair cutting). Figure 3 shows the relative contents of γ_1' and γ_2' phases in specimens are different, but it seems like that this variation did not affect the deformation mechanism of the alloy. Despite that, some subtle differences still can be noticed. For instance, the slip bands in specimens T1 and T2 are all parallel to each other, and the degree of dislocation entanglement is relatively low. In contrast, the number of slip bands in specimens T3 and T4 is larger, and obviously,

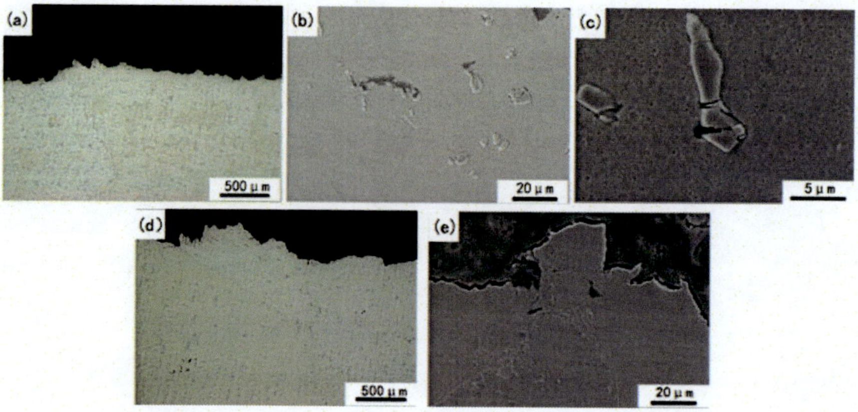

Fig. 7 OM and SEM observations on the longitudinal tensile fracture morphology of specimen T3 tested at **a-c** RT and **d**, **e** 800 °C

more than one slip system is activated. The dislocation density in slip bands becomes higher as well. It is well-known that a greater extent of dislocation accumulation and entanglement is conducive to hindering dislocation movement and improving the strength of the alloy, which is consistent with the fact that the room temperature tensile strengths of specimens T3 and T4 are higher than that of T1 and T2.

Figure 9 shows the morphology and distribution of dislocation in the specimens tested to rupture at 800 °C. It can be found that dislocations are also confined within slip bands as they do in the microstructure fractured at room temperature. However, at 800 °C, dislocations tend to move forward in a way of bypassing γ' phase particles and leaving dislocation loops around them. This phenomenon indicates that the deformation mechanism is Orowan bypassing during the testing of the alloy at 800 °C.

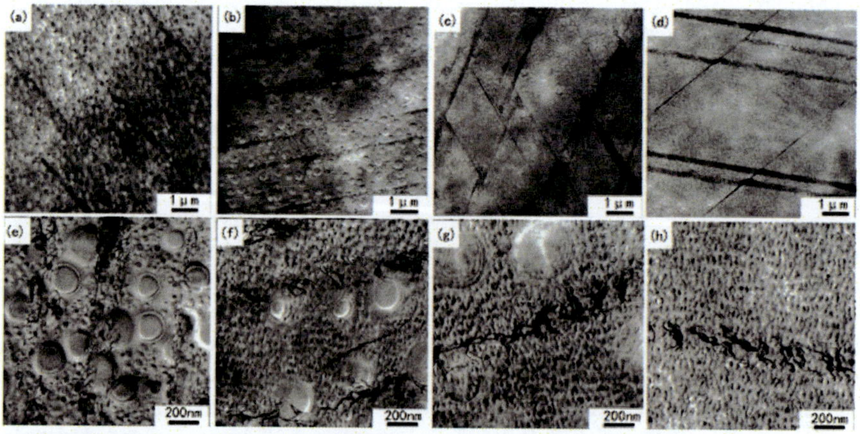

Fig. 8 TEM observations on dislocation configuration in the tensile specimens of **a**, **e** T1, **b**, **f** T2, **c**, **g** T3, and **d**, **h** T4 tested at room temperature

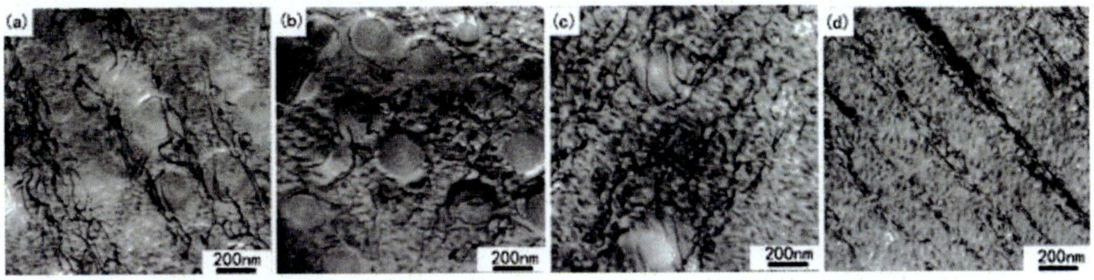

Fig. 9 TEM observations on dislocation configuration in the tensile specimens of **a** T1, **b** T2, **c** T3, and **d** T4 after 0.2% plastic strain at 800 °C

According to the analyses above, the primary deformation mechanism of alloy K4800 during the tensile process is dislocation-pair cutting at room temperature, and Orowan bypassing at 800 °C. With the increase of γ_2' phase content, the extent of dislocation accumulation and entanglement within slip bands become more and more severe at both room and elevated temperatures. Besides, it can be seen in Fig. 9 that most dislocations distribute around γ_2' particles and seldom interact with the γ_1' phase whose size is much larger. The same phenomenon can be found in Fig. 8e–h as well. Therefore, the room-temperature and high-temperature tensile strengths of alloy K4800 are both closely related to the small-sized γ_2' phase in the microstructure. It is obvious that the strength of the alloy rises with the temperature of first-step aging, which decreases the content of γ_1' phase and enhances the precipitation of γ_2' phase.

On the other hand, although the excessively coarsened γ_1' precipitates are detrimental to the strength of the alloy, specimens T1–T3 with γ' phase in various sizes exhibit better tensile ductility compared with specimen T4 at elevated temperature. It is believed that the large spacings between γ_1' particles are conducive to dislocation movement, thus improving the high-temperature ductility of the alloy. This is consistent with the work of Wang et al. [10]. Besides, it is well acknowledged that the intergranular strength of superalloy becomes inferior to the intragranular strength at elevated temperatures. The existence of γ_1' phase mildly lowers the intragranular strength of the alloy and narrows the strength rap between the grain and grain boundary. Under this circumstance, micropores and cracks not only initiate on the grain boundaries, but also within grains, thus extensive grain-boundary cracking is prevented and the high-temperature ductility of material is improved.

Creep Properties and Microstructural Stability

The creep properties of alloy K4800 were tested at 870 °C to investigate the over-temperature performance of this alloy. Since in an over-temperature condition, generally γ' particles in the matrix tend to coarsen in a very fast rate, which can lead to an obvious decreae in alloy strength. The coarsening rate of γ' particles is closely related to the precipitation characteristic of them, which depends on the heat treatment of

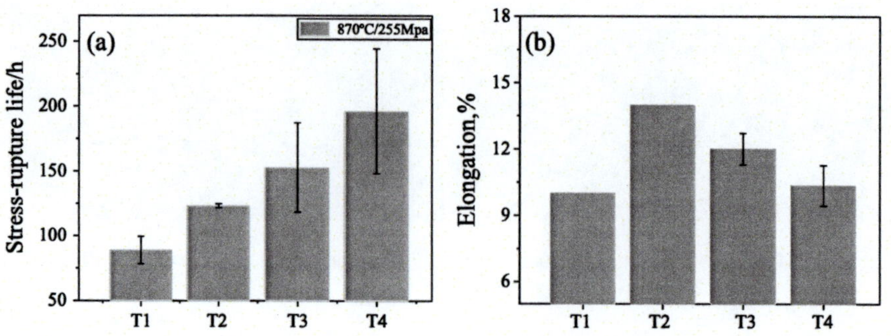

Fig. 10 Stress-rupture **a** life and **b** elongation of alloy K4800 heat-treated under different conditions

the alloy. The specimens heat-treated under four different conditions were all tested at 870 °C with a constant loading stress of 255 MPa, as demonstrated in Fig10. The creep elongations of specimens are generally good, but the creep lives of them are quite different. The life to rupture of specimen T1 is merely 88.9 h, while the creep lives of T2 and T3 reach 123.3 and 152.8 h, respectively. The best creep performance is presented by specimen T4 with a creep life as high as 196.2 h. Therefore, it is clear from the results above that the relative precipitation content of the large-sized γ_1' phase and the small-sized γ_2' phase significantly affects the creep properties of alloy K4800.

Figure 11 shows the morphology of γ' near the fracture of the stress ruptured specimens T2 and T3. It can be found that the γ' particles observed have significantly deformed and coarsened. The γ_2' phase in specimen T2 has completely disappeared, while specimen T3 still retains a large amount of γ_2' phase. All the γ' particles regardless of size in T3 present the trace of being cut by dislocations. During the creep test, γ' particles with a larger size constantly grow at the expense of smaller ones. Under this circumstance, the existence of γ_1' phase would dramatically accelerate the ripening process of γ_2' phase. As a result, the distribution of γ_2' phase becomes more and more sparse, so the strengthening effect of γ_2' phase is greatly weakened, leading to a reduction in alloy creep life. Therefore, it can be concluded that the relatively short stress rupture lives of specimens T1 and T2 generally stem from the evident coarsening and ripening of γ'phase. The γ' phase in specimen T3 also coarsened to some extent, but due to the limited precipitation of γ1' phase, a lot of γ_2' particles still remain with a dense distribution till the end of the creep test. According to that, the stress rupture life of specimen T3 is higher than that of T1 and T2. Since no γ_1' phase precipitates in specimen T4, so this specimen has the slowest γ' coarsening rate and the longest stress rupture life (190 h).

Stress ruptured specimen T3 was longitudinally sectioned near its fracture surface and investigated by using OM and SEM, as shown in Fig. 12. The creep fracture morphology of specimens T1–T4 are similar, so the micrographs of other specimens are omitted. As it can be seen from Fig. 12, specimen T3 presents a mixed fracture feature. Specifically, many intragranular cracks and pores can be observed near the

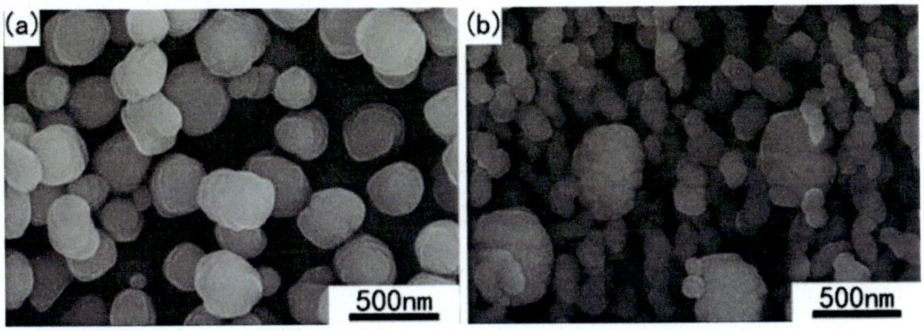

Fig. 11 SEM observations on γ' phase in the stress ruptured specimen **a** T2 and **b** T3

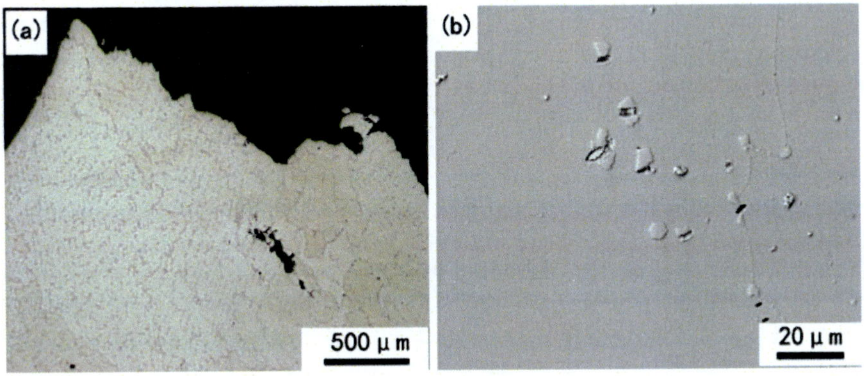

Fig. 12 Longitudinal microstructure of stress ruptured specimen T3

fracture surface, whereas no serious intergranular cracking is detected, which indicates that alloy K4800 has a good inter-/intragranular strength balance during the test.

The strength of alloy K4800 increases inversely with the volume fraction of γ_1' phase, and specimen T4 shows the highest tensile and creep strengths. However, for specimen T3, although its tensile and creep strengths are lower than T4, it has a better elevated temperature ductility. Therefore, from a comprehensive aspect, the co-existence of γ_1' and γ_2' phases can provide the alloy with a good combination of strength and ductility.

Alloying elements such as Co, Cr, Mo, and W are added to alloy K4800. The addition of many alloying elements enables alloy K4800 to obtain higher strength. However, during the long-term aging process of superalloys, those alloying elements can cause or accelerate the precipitation of harmful TCP phases like σ and η, which will lead to the deterioration of alloy properties. Considering this, the chemical composition of alloy K4800 was carefully designed and verified by calculations and experiments to prevent the precipitation of TCP phase. The microstructural evolution of alloy K4800 was investigated by conducting a long-term age at 800 °C. Figure 13

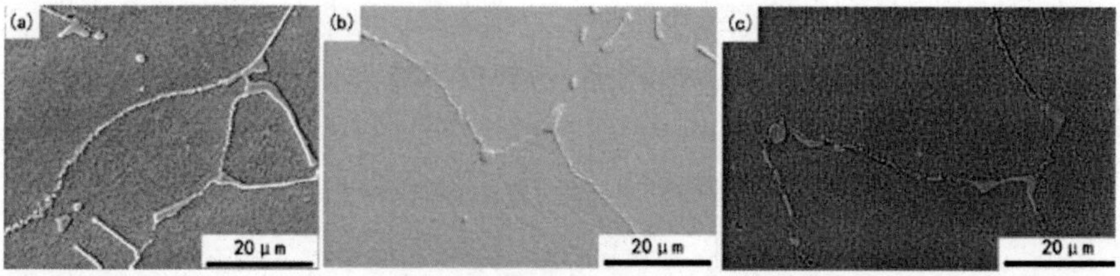

Fig. 13 SEM observations on the specimen T2 aged at 800 °C for **a** 500 h, **b** 1000 h, and **c** 2000 h

shows SEM observations on specimen T2 aged at 800 °C for 500, 1000, and 2000 h. It can be seen that no harmful phase precipitates during aging, which confirms the excellent microstructural stability of alloy K4800. Besides, it is also found in the aged specimen that the granular $M_{23}C_6$ carbides on the grain boundary exhibit a tendency of growing and linking together with time. Investigations on the evolution of microstructure and mechanical property of this alloy during long-term aging are in progress.

For the two-stage aging strategy used in this work, the γ' phase with a bimodal size distribution can form in the microstructure of alloy K4800. As aforementioned, the strength of alloy K4800 increases with the volume fraction of the small-sized γ_2' phase. The precipitation of a large-sized γ_1' phase can enhance the high-temperature ductility of the alloy, but too much of them will accelerate the ripening of γ_2' phase and deteriorate the creep property of the alloy. In other words, the large-sized γ_1' phase precipitated during high-temperature aging is necessary for alloy K4800 to obtain a good combination of strength and ductility, but the content of them needs to be controlled. Accordingly, the heat treatment regime T3, with a prior stage aging conducted at a temperature merely 10 °C lower than the solution temperature of γ' phase, is the most suitable one for alloy K4800.

Conclusions

The impacts of various aging treatments on the microstructural evolution, mechanical properties, and strengthening and fracture mechanisms of a newly-developed alloy K4800 were investigated in the present work. Based on the present study, a two-stage aging strategy is recommended for the microstructural and property optimization of alloy K4800. The main results are listed as follows:

1. A prior aging at 1090 °C can induce a large-sized γ_1' phase in the microstructure of alloy K4800 optimizing the general precipitation characterization of γ' phase and enabling this alloy to possess balanced strength and ductility.
2. The existence of the γ_1' phase can improve the high-temperature ductility of the alloy. After the tensile and creep tests conducted at 800 °C and above,

no obvious intergranular cracking can be found on the fracture surface and specimens ruptured in a transgranular mode.

3. The content of the γ_1' phase needs to be accurately controlled because that too much of them would restrict the precipitation of the small-sized γ_2' phase and increase the ripening rate of γ_2' phase, impairing the tensile strength and stress rupture life of alloy.
4. The deformation mechanism of alloy K4800 is mainly the dislocation-pair cutting at room temperature and the Orowan bypassing at 800 °C insensitive to the presence of the γ_1' phase.

References

1. Reed R.C. (2008) The Superalloys: Fundamentals and applications [M]. Cambridge University Press, Cambridge, UK.
2. Williams J.C., Starke E.A.. (2003) Progress in structural materials for aerospace systems[J]. Acta Materialia, 51: 5775-5799.
3. Schulz U., Leyens C., Fritscher K.. et al. (2003). Some recent trends in research and technology of advanced thermal barrier coatings[J]. Aerospace Science and Technology, 7: 73-80.
4. Xu J, Zhang M J,Tang J, et al. (2018) Influence of solution treatment on microstructure and stress rupture properties of K439 nickel-base superalloy [J]. Atlas. J. Mater. Sci.,3(6):78
5. Cheng K Y, Jo C Y, Jin T, Hu ZQ. (2010) Influence of applied stress on the γ' directional coarsening in a single crystal superalloy. Mater Des;31:968–71.
6 Moshtaghin RS, Asgari S. (2003) Growth kinetics of γ' precipitates in superalloy In 738LC during long term aging. Mater Des;24:325–30
7. Sun F., Gu Y.F., Yan J.B., et al. (2016) Tensile deformation-induced dislocation configurations at intermediate temperatures in a Ni-Fe-based superalloy for advanced ultra-supercritical coal-fired power plants[J]. Journal of Alloys and Compounds, 657: 565-569.
8. Xu Y.L., Yang C.X., Ran Q.X., et al. (2013). Microstructure evolution and stress-rupture properties of Nimonic 80A after various heat treatments[J]. Materials and Design, 47: 218-226.
9. Sharma K.K., Banerjee D., Tewari S.N. (1988) Effect of reverse-aging treatment on the microstructure and mechanical properties of Nimonic alloys[J]. Materials Science and Engineering A,104: 131-140.
10. Wang J, Zhou L, Sheng L, Guo J. (2012) The microstructure evolution and its effect on the mechanical properties of a hot-corrosion resistant Ni-based superalloy during long-term thermal exposure. Mater Des;39:55–62.

Microstructural Stability and Strengthening Mechanism of a Ferritic Fe–Cr–Ni–Al Superalloy Containing Cuboidal B2 Nanoparticles

Zhenhua Wang, Beibei Jiang, Haiyang Liu, Ben Niu, Hongyao Yu, and Qing Wang

Abstract Development of Fe-based superalloys is highly demanded for both scientific interests and broad applications. Generally, the mechanical properties of traditional alloys strengthened by carbides and Laves phase, etc., are degraded after long-term operation at HTs (>923 K) due to microstructural instability. Considering that the superiority of Ni-based superalloys benefits from their unique coherent microstructure, the coherent precipitation of cuboidal B2 nanoparticles will certainly improve the HT microstructure and mechanical properties of Fe-based superalloys. In the present work, we developed a new Fe-based superalloy with cuboidal B2-NiAl nanoparticles coherently-precipitated into BCC matrix, which has not been reported. This alloy exhibits an excellent microstructural stability at 973 K with a slow particle coarsening rate, which is ascribed to the moderate lattice misfit (ε = 0.24 ~ 0.67%) between BCC/B2. Also, it is the cuboidal B2 nanoprecipitation that renders the current alloy with yield strength of σ_{YS} = 238 ~ 258 MPa at 973 K.

Keywords Ferritic stainless steels · Coherent precipitation · Lattice misfit · Microstructural stability · Strengthening mechanism

---

Z. Wang · H. Liu · B. Niu · Q. Wang (✉)
Key Laboratory of Materials Modification By Laser, Ion and Electron Beams (Ministry of Education), Engineering Research Center of High Entropy Alloy Materials (Liaoning Province), School of Materials Science and Engineering, Dalian University of Technology, Dalian 116024, China
e-mail: wangq@dlut.edu.cn

B. Jiang
Analysis and Test Center, Guangdong University of Technology, Guangzhou 510006, China

H. Yu
Beijing Key Laboratory of Advanced High Temperature Materials, Central Iron and Steel Research Institute, Beijing 100081, China

E. A. Ott et al. (eds.), *Proceedings of the 10th International Symposium on Superalloy 718 and Derivatives*, The Minerals, Metals & Materials Series,
https://doi.org/10.1007/978-3-031-27447-3_30

Introduction

Ferritic stainless steels (FSSs) have been widely used into high-temperature (HT) applications due to their high thermal conductivity, low thermal expansion, prominent mechanical properties, as well as good economy [1–4]. Theses conventional FSSs are generally strengthened by carbides ($Cr_{23}C_6$ and MC, M = Nb, Ti, V, etc.), Z-CrNbN, and Laves phase (Fe_2M) [1–5]. However, the creep resistance of those alloys could degrade in long-term operation at HTs (above 923 K), which is ascribed to the microstructural instability, i.e., the rapid coarsening of carbides induced by the non-coherency with body-centered-cubic (BCC) ferritic matrix [1, 2]. Luckily, the outstanding superiority of Ni-based superalloys at HTs is primarily attributed to the unique microstructure of ordered γ'-Ni_3Al ($L1_2$-Cu_3Au type) nanoparticles coherently-precipitated in the FCC-γ matrix, resulting in unparalleled HT mechanical strength and creep resistance even at the temperature exceeds 1073 K [6, 7]. In particular, the cuboidal $L1_2$-γ' nanoprecipitation could enhance the service temperature of single crystalline superalloys up to 1373 K (~0.9 of melting point) [8]. Similarly, the coherent precipitation of ordered B2 nanoparticles in the BCC matrix would undoubtedly improve the HT microstructural stability and mechanical properties of FSSs.

Recently, it has reported that spherical B2-NiAl nanoparticles could be coherently-precipitated in the BCC ferritic matrix in Fe-10Cr-10Ni-6.5Al-3.4Mo-0.25Zr-0.005B (weight percent, wt. %, designated as FBB8) alloy by intentionally tailoring the composition, exhibiting a higher microstructural stability than non-coherent precipitation strengthened FSSs at 973 K [9–11]. Intriguingly, the further addition of 2 wt. % Ti into FBB8 could produce cuboidal B2-NiAl/$L2_1$-Ni_2AlTi hierarchical nano-precipitates, in which the $L2_1$-Ni_2AlTi is another highly-ordered phase of BCC solid solution [12–14]. This significantly enhances the tensile yield strength from original 120 up to 280 MPa at 973 K due to the cuboidal nanoprecipitation, which is much higher than that (160 MPa at 973 K) of conventional F82H FSS (Fe-8Cr-2W-V) strengthened by incoherent carbides [14, 15]. It is emphasized that the formation of spherical or cuboidal nanoprecipitates is closely related to the lattice misfit ε between these two coherent phases, in which a small $\varepsilon(\varepsilon < 0.2\%)$ corresponds to a spherical nanoprecipitation, a lager ε ($\varepsilon > 1.0\%$) leads to a basket-like microstructure, and only an appropriate ε promotes the cuboidal B2 nanoprecipitation [16–18]. Thus, the cuboidal B2/$L2_1$ and spherical B2 is mainly ascribed to the larger $\varepsilon = 0.7\%$ and the smaller $\varepsilon = 0.06\%$ in FBB8 alloys with and without Ti, respectively [10, 12, 19]. However, the addition of Ti would deteriorate the resistance to oxidation since Ti accelerates the internal oxidation due to high oxygen permeability at HTs [20]. Therefore, to develop novel FSS strengthened by cuboidal B2-NiAl precipitates would be fascinating for many structural applications at HTs.

Actually, it is difficult to tailor the BCC/B2 lattice misfit to a moderate value for cuboidal nanoprecipitation due to a much larger composition difference between them, where BCC/B2 always exhibits a weave-like microstructure [21, 22]. In previous work, we designed a series of multi-component alloys in light of the cluster composition formula of Al_2M_{14} (M presents different combinations of Fe, Co, Ni, and Cr), showing a coherent microstructure with cuboidal B2 nanoprecipitates in the BCC matrix [16, 23]. Here, by setting the M as $M_{14} = Fe_{10}Ni_2Cr_2$, we can get the composition of $Al_2(Fe_{10}Ni_2Cr_2)$ (= Fe-12.5Cr-14.1Ni-6.5Al, wt. %), which is an FSS alloy according to the Cr- and Ni-equivalence in stainless steels [24]. Moreover, Mo and W as slowly diffused element could inhibit the coarsening of B2 particles due to segregate to the BCC/B2 interfaces [25]. The addition of Mo could increase the lattice constant of the BCC matrix, leading to a favorable reduction of the BCC/B2 lattice misfit [25, 26]. Moreover, as slow diffusing elements, Mo and W could inhibit the coarsening of B2 particles due to their segregation on the BCC/B2 interfaces [10]. Hence, a minor amount of Mo and W is substituted for Cr in above composition with a specific ratio of Cr to (Mo + W) = 8/1 (in molar fraction). A trace amount of Zr and B should also be added to improve the ductility of high-strength steels through enhancing grain boundary cohesion for prevention from the intergranular fracture [27, 28]. Based on all above considerations, the optimal composition is final determined as Fe-10.9Cr-13.9Ni-6.4Al-2.2Mo-0.5W-0.04Zr-0.005B (wt. %) to achieve BCC/B2 coherent microstructure with cuboidal B2 nanoprecipitates. Then, the BCC/B2 microstructural evolution with the aging time at 973 and 1073 K will be investigated, in which the coarsening behavior of cuboidal B2 nanoparticles will be discussed. The mechanical properties of this alloy at both room temperature and 973 K will be studied, where the high strength will be discussed via the strengthening mechanism.

Experimental

The current alloy ingots with a weight of 100 g were prepared by arc melting and suction cast into a 6 mm-diameter cylindrical copper mold under an argon atmosphere. The purities of raw metals are 99.999 wt. % for Al, 99.99 wt. % for Fe, Ni, Zr, and B, 99.9 wt. % for Cr, Mo, and W. The chemical homogeneity of the ingot is improved by electromagnetic stirring. The cast ingot was solid-solutioned at 1473 K for 2 h in a muffle furnace, during which they were sealed in a vacuum tube to prevent oxidation. Finally, these solid-solutioned specimens were aged at 973 K for different periods of time (up to 200 h). Each heat-treatment was followed by water quenching. The crystalline structures of alloy specimens at different treatments were identified using a Bruker D8 X-ray diffractometer (XRD) with the Cu K_α radiation (λ = 0.15406 nm). The microstructure was examined using Olympus optical microscopy (OM), Zeiss Supra 55 scanning electron microscopy (SEM), JEM2100F FEG scanning transmission electron microscopy (STEM). The etching solution for OM and SEM observation was composed of 5 g $FeCl_3 \cdot 6H_2O$ + 25 ml HCl + 25 ml C_2H_5OH, and the TEM specimens were prepared by twin-jet electro-polishing in a solution of 10% $HClO_4$ + 90% C_2H_5OH (volume fraction) at a cryogenic temperature of 243 K. The chemical composition was analyzed using SHIMADZU electronic probe micro-analyzer (EPMA) and Super-X energy dispersive. The melting temperature of the current alloy was determined on NETZSCH STA 449F3 Differential Scanning Calorimetry (DSC) at 10 K/min. The statistical analysis of the volume fraction and size of precipitated particles and the grain sizes at different heat-treated states were measured from the SEM/OM morphology images (at least 6 images) employing the ImageJ software [29]. The mean radius (r) of precipitates is calculated from the traced areas using a circular-equivalent, i.e., $r = \sqrt{(\text{area}/\pi)}$.

Uniaxial tensile tests were conducted on a UTM5504 Material Test System (MTS) with a heating furnace. The tensile tests were performed at a nominal strain rate of $1 \times 10^{-3}\ s^{-1}$ at 973 K, in which the gauge dimension of tensile samples is $5 \times 1.5 \times$ 26 mm (width $\times$ thickness $\times$ length), and three samples for each heat treatment were tested. Microhardness of different heat-treated samples was tested with a HVS-1000 Vickers hardness tester under a load of 500 g for 15 s, in which at least 10 indents were taken to obtain an average value.

Results

Microstructural Characterizations

The chemical composition by EPMA is Fe-(11.08 ± 0.15)Cr-(11.85 ± 0.08)Ni-(5.90 ± 0.08)Al-(2.19 ± 0.02)Mo-(0.51 ± 0.03)W-(0.04 ± 0.02)Zr-(0.004 ± 0.003)B (wt. %), which is close to the nominal composition. As shown in Fig. 1, the melting temperature of the current alloy is determined to be ~ 1679 K by DSC, which is comparable to that (1698 K) of the austenitic stainless steel 304 (Fe-18Cr-9Ni-0.07C wt. %) [30]. Figure 2 shows the XRD patterns of the current alloy at different heat-treated states, in which all the alloy specimens consist of primary BCC solid solution plus B2 phase (characterized by a weak (100) diffraction peak). The lattice constants of two phase were then calculated, being $a_{BCC} = (0.2877 \pm 0.0042)$ nm and $a_{B2} = (0.2887 \pm 0.0047)$ nm, respectively. Moreover, the lattice misfit calculated with the equation of $\varepsilon = 2 \times (a_{B2} - a_{BCC})/(a_{B2} + a_{BCC})$ [31] could increase from $\varepsilon = 0.24\%$ in 0.5 h-aged sample to $\varepsilon = 0.67\%$ in 1000 h-aged sample with prolonging aging time, in which the average value is about $\varepsilon \sim 0.35\%$.

The OM image (Fig. 3 in Supplementary materials) of the alloy in solid-solution state indicates that the alloy matrix is constituted of coarse ferritic grains with a size of 300 ~ 500 μm, where no second-phase particles appear on grain boundaries. The precipitation of B2 nanoparticles in different heat-treated states are observed by SEM and presented in Fig. 4a-f, from which it is found that cuboidal nanoparticles with a radius of ~ 30 nm are uniformly distributed in the ferritic matrix in the solid-solution state. Besides, there still exist ultra-fine spherical nanoparticles with a size of 3 ~ 5 nm (Fig. 4a), which might be caused by secondary precipitation during water-quenching [32]. After aging at 973 K, the ultra-fine nanoparticles disappear. Moreover, these cuboidal nanoparticles are gradually regular with prolonging the aging time. In particular, the radius of these cuboidal B2 precipitates increases

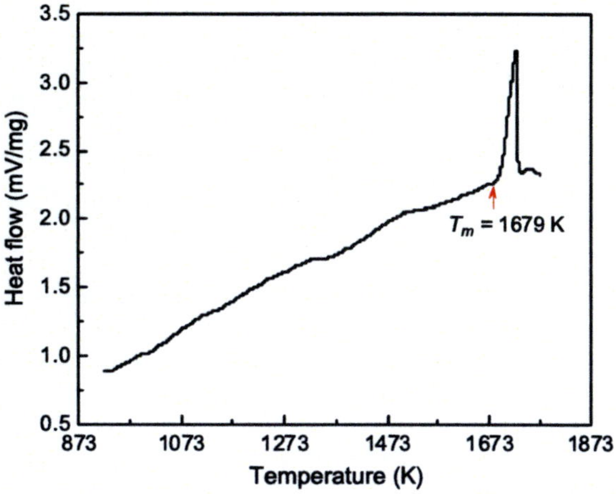

Fig. 1 DSC curve for the current alloy, in which the solidus temperature is T_m ~ 1679 K

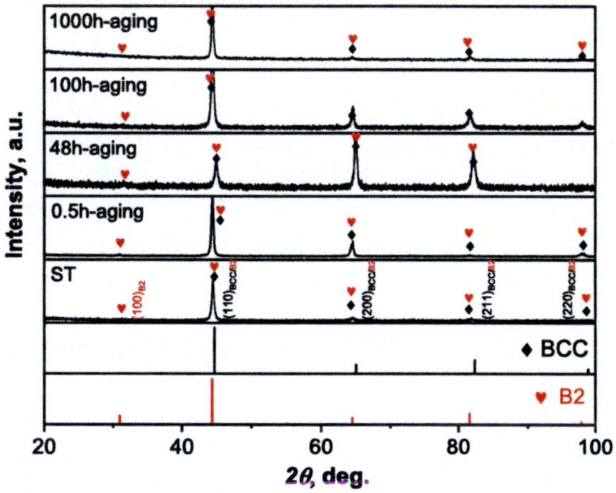

Fig. 2 XRD patterns of the current alloy at different heat-treated states

gradually form ~ 37 nm for 0.5 h-aging to ~ 78 nm for 100 h-aging (Fig. 4b–c). With further aging time extension, the B2 particle size increases more slowly from ~ 88 nm for 200 h-aging to ~ 114 nm 500 h-aging, then to ~ 136 nm 1000 h-aging, as presented in Fig. 4d–f. Moreover, the volume fraction of cuboidal B2 precipitates increases slightly from ~ 20.4% in 0.5 h-aged state to ~ 24.8% in 48 h-aged state, and then does not vary with the increase of the aging time. Furthermore, the coarsening of coherent precipitates in ferritic alloys would occurs by agglomeration during aging, and whereas the extra nucleation and growth hardly happen, thus resulting in a very small change of the volume fraction [33]. Similar to the 973 K-aged alloy, the B2 nanoparticles in 1073 K-aged alloy maintained a cuboidal shape. The radius of B2 particles coarsened from r ~ 141 nm for 24 h-aging to r ~ 158 nm for 48 h-aging, then

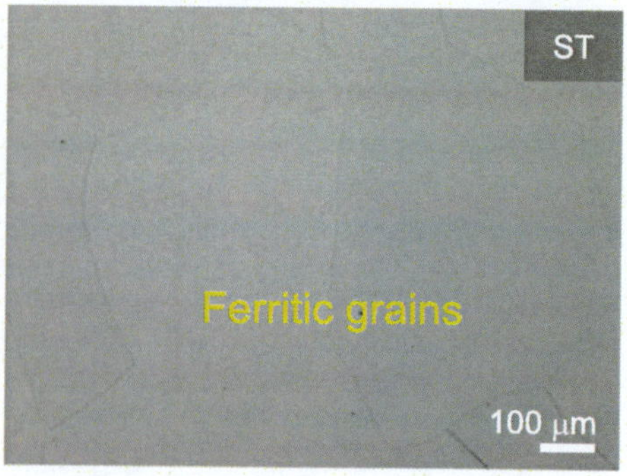

Fig. 3 OM image of solid-solutionized alloy at 1473 K for 2 h

to $r \sim 175$ nm for 100 h-aging (Fig. 5), which is obviously faster than the coarsening of B2 particles for 973 K-aging.

The 48 h-aged alloy samples were further analyzed by the TEM to verify the crystalline structure of precipitates, as shown in Fig. 6. This shows the cuboidal B2 nanoprecipitates embedded into the BCC matrix with a volume fraction of ~ 36.7%, as demonstrated by the selected area electron diffraction (SAED) pattern along the $[110]_{BCC}$ direction in both bright-field (BF) and dark-field (DF) images (Fig. 6a, b). The high-resolution TEM (HRTEM) image and fast Fourier transform (FFT) patterns presented in Fig. 6c, c-1, c-2 indicate that the B2 precipitate are fully coherent with the BCC matrix, exhibiting a coherent interface, where the schematic crystalline structures of BCC and B2 phases are also inserted in Fig. 6c. The elemental distribution in the 48 h-aged alloy was also analyzed with the EDS equipped in DB-FIB, as shown in Fig. 7. It is found that Ni and Al elements are enriched in B2 nanoparticles, and Fe, Cr, Mo, and W are mainly segregated to the BCC matrix.

Mechanical Properties

Microhardness (*HV*) measurements were initially conducted to evaluate the age-hardening response of the current alloy. The variation of *HV* with the aging time at 973 K is presented in Fig. 8a, in which the tendency of the radius of B2 precipitates is also shown. It is found that the *HV* decreases rapidly from ~ 467 *HV* in the solid-solutionized state to ~ 419 *HV* after 0.5 h-aging, which might result from the dissolution of ultra-fine spherical B2 nanoparticles into the BCC matrix since the particle size (60 ~ 70 nm) of cuboidal B2 nanoparticles in both states are comparable (Fig. 4a, b). Then, it keeps almost constant till 24 h-aging, followed by a slight decrease to ~ 344 *HV* (48 h-aged), to ~ 304 *HV* (300 h-aged), and lastly to ~ 284 *HV* after 1000 h-aging. The slow decrease of *HV* in the late-period aging is ascribed to the relatively-high thermal stability of B2 nanoparticles.

High-temperature tensile tests at 973 K of the aged alloy specimens for 24, 48, and 100 h were subsequently performed, and the engineering stress–strain curves are

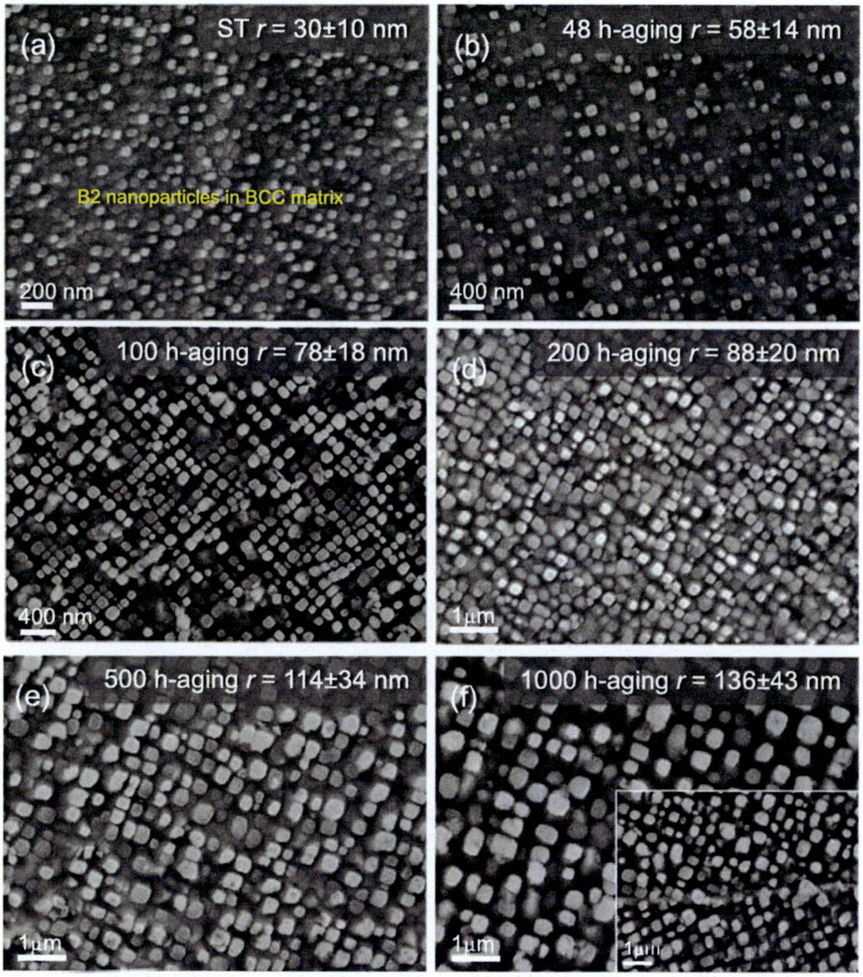

Fig. 4 SEM observations of the current alloy after different heat-treatments. **a**: Solid-solutionized at 1473 K for 2 h, **b** – **f**: aged at 973 K for 48 h, 100, 200, 500, and 1000 h, respectively

shown in Fig. 7b, from which the yield strength (σ_{YS}) and elongation to fracture (δ) are measured. It is found that the mechanical properties of these aged samples are comparable, with a yield strength of $\sigma_{YS} = 238 \sim 258$ MPa and a ductility of $\delta = 50 \sim 56\%$ at 973 K. Obviously, the σ_{YS} of this alloy is much higher than that ($\sigma_{YS} =$ 120 MPa at 973 K) of FBB8 alloy containing spherical B2 precipitates with a similar particle size of $r \sim 53$ nm [10, 14], and is comparable to that ($\sigma_{YS} = 280$ MPa at 973 K) of Ti-modified FBB8 alloy containing cuboidal B2/L2$_1$ precipitates with $r \sim$ 57 nm [12, 14].

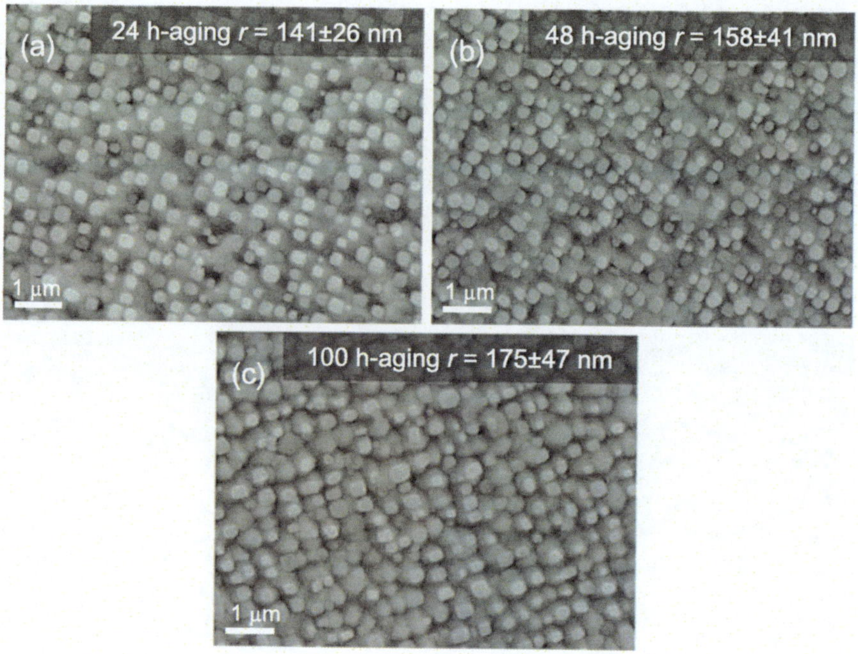

Fig. 5 SEM images of 1073 K-aged alloy for 24 h (**a**), 48 h (**b**), and 100 h (**c**)

Discussion

Microstructural Stability of Coherent BCC/B2

Since the volume fraction of B2 nanoprecipitates remains almost unchanged with the aging time prolonging in the current alloy, the classic Ostwald ripening theory proposed by Philippe and Voorhees (i.e., the PV theory) could be applied to study the coarsening behavior of B2 nanoparticles in the current alloy, which has been extensively used in the multicomponent alloys [34, 35]. Such a time-dependent coarsening process can be described with the following Eq. (1):

$$r^3(t) - r^3(t_0) = k(t - t_0) \tag{1}$$

$$k = \frac{8V_m \sigma D c_i^M}{9RT(c_i^P - c_i^M)^2} \tag{2}$$

where $r(t)$ is the mean radius of precipitates at the aging time of t, in which r is calculated from the traced areas using a circular-equivalent, i.e.,$r = \sqrt{}(\text{area}/\pi)$; t_0 = 0.5 h, and k is the coarsening rate constant, which is largely dependent on the

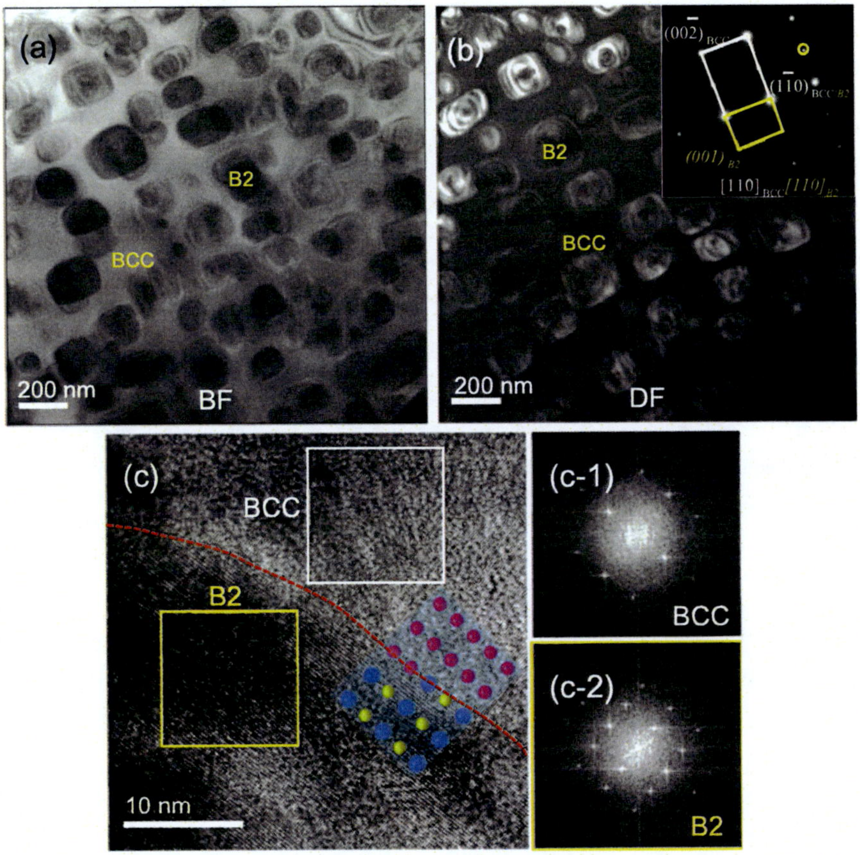

Fig. 6 TEM characterization of the current alloy aged at 973 K for 48 h. The bright-field (BF) image (**a**) and the corresponding dark-field (DF) image (**b**) show that the cuboidal B2 nanoprecipitates are dispersed into the BCC matrix; HRTEM image and FFT patterns (**c**, **c-1**, **c-2**) exhibit that the B2 particle is coherent with the BCC matrix

diffusion of the solute elements and the interfacial energy between the precipitates and matrix [36]. Especially, the coarsening rate constant of the coherent precipitates can be expressed with Eq. (2), in which V_m is the molar volume of the precipitates in the current alloy; $\sigma = 0.03\ \text{J}\cdot\text{m}^{-2}$ is the interfacial energy of coherent BCC/B2 phases [37]; c_i^M and c_i^P are the equilibrium composition (in mole fraction) of the element i in the matrix and precipitates, respectively; R is the gas constant (=8.314 J·mol^{-1} ·K^{-1}); T is the absolute temperature; and D is the diffusion coefficient of the slow diffusing element in the matrix [36]. Figure 9 gives the variation of particle size r^3 of B2 nanoprecipitates with the aging time, which could be well fitted by the PV theory. Thus, the coarsening rate constant is obtained as $k = 6.9 \times 10^{-28}\ \text{m}^3\cdot\text{s}^{-1}$. For comparison, we also plotted the variation of r^3 of particles with the aging time at 973 K in existing FBB8 and Ti-modified FBB8 alloys, from which the coarsening rate

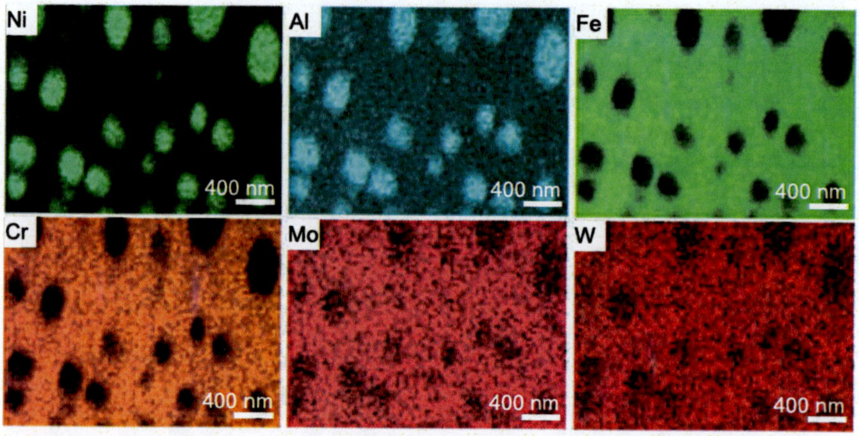

Fig. 7 Elemental distribution of the 48 h-aged alloy mapped with EDS equipped in DB-FIB

constants are 4.9×10^{-28} $m^3 \cdot s^{-1}$ and 5.9×10^{-28} $m^3 \cdot s^{-1}$, respectively, comparable to that of the current alloy. Obviously, the coarsening rate constant of coherent particles in BCC steels is much lower than that (~ 1.7×10^{-26} $m^3 \cdot s^{-1}$ at 873 K) of $Cr_{23}C_6$ carbides in Fe-9Cr-1Mo-0.2 V-0.1C-0.08Nb-0.05N (wt. %, T91) steel [38], showing a much slower coarsening behavior and a higher microstructural stability. More importantly, the current alloy contains a minor amount of slow diffusion elements Mo and W, which are mainly partitioned into the BCC matrix (Fig. 7). As W has a very slow diffusion coefficient in BCC α-Fe [24], W could stabilize the BCC matrix to an extent. Thus, the coarsening of the B2 nanoparticles for the current alloy is relatively slow.

It is known that the equilibrium morphologies of precipitates are strongly affected by both strain energy and interfacial energy at the precipitate/matrix interface [39, 40]. Thompson et al. [41] proposed a characteristic parameter $L = \varepsilon^2 C_{44} r/\sigma$ to describe the particle morphology, where $C_{44} = 116$ GPa is an elastic constant of the BCC matrix [42]; $\varepsilon = 0.35\%$ is the lattice misfit between BCC and B2 phases in the current alloy; the interfacial energy is $\sigma = 0.03$ $J \cdot m^{-2}$ [37]. Since the parameter L is proportional to both the lattice misfit ε and the mean radius r of particles, the ε will be crucial in determining the particle shape when the particle sizes are comparable ($r = 30$ ~ 90 nm). Thus, the calculated L value for this current alloy is $L = 1.4$ ~ 4.3, which is less than $L^* = 5.6$, the critical value for the morphology of precipitates transforming from cuboid to plate-like or needle-like shape [41]. This moderate lattice misfit ($\varepsilon = 0.24$ ~ 0.67%) contributes for the formation of cuboidal B2 nanoprecipitates in the current alloy.

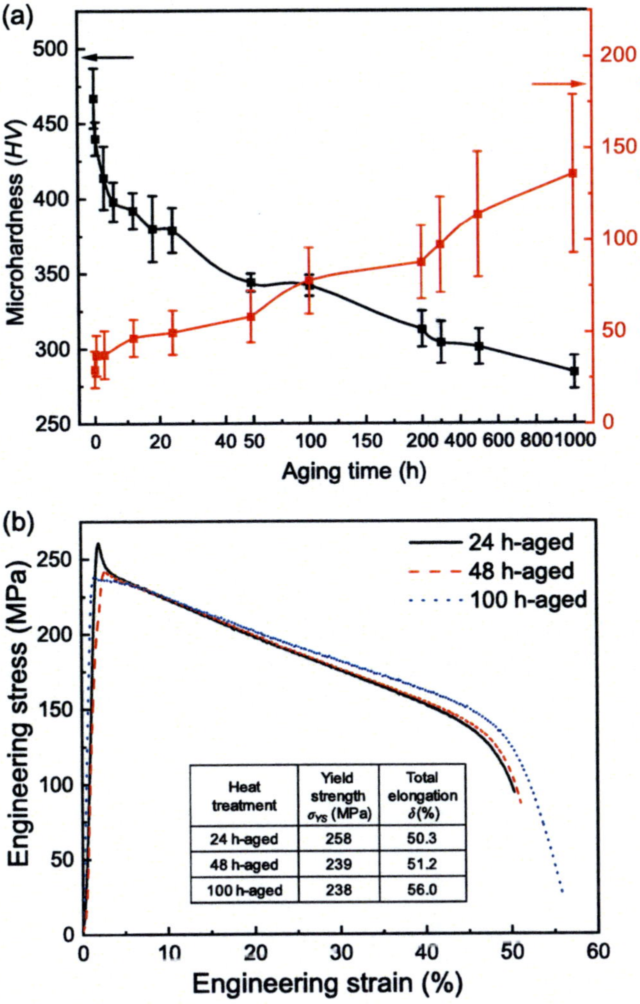

Fig. 8 **a**: Variation tendency of the microhardness *HV* and B2 particle radius *r* in the current alloy with aging time at 973 K; **b**: Engineering tensile stress–strain curves at 973 K of aged alloy for 24 h, 48 h, and 100 h; **c**: Computations of $(\Delta\sigma_{CS} + \Delta\sigma_{MS})$ and $\Delta\sigma_{orowan}$ as a function of particle size, *r*, for the current alloy

Strengthening Mechanism

The contributions in yield strength of the current alloy primarily originate from the solid solution strengthening of solutes and the precipitation strengthening of B2 nanoprecipitates. Thus, the strength increment could be calculated with the formula of $\sigma_{YS} = \sigma_0 + \sigma_{ss} + \sigma_p$ [31, 37], where $\sigma_0 = 50$ MPa is the lattice fictional

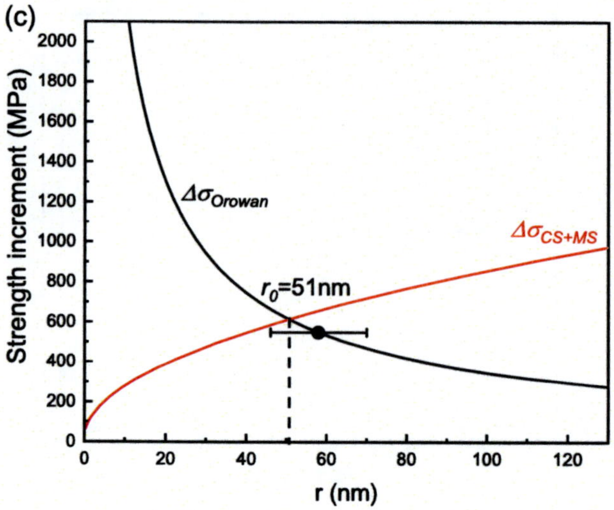

Fig. 8 (continued)

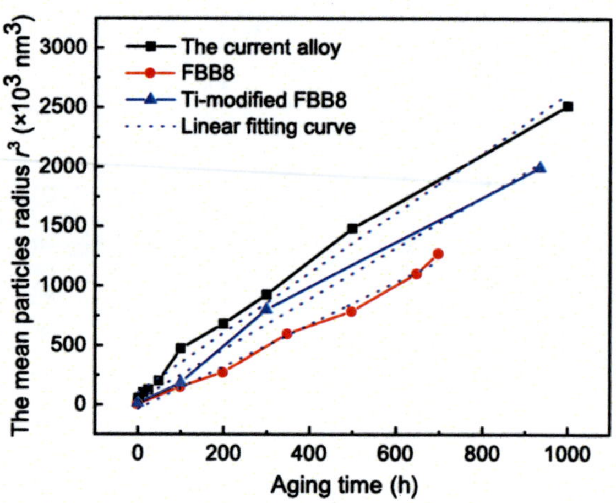

Fig. 9 Variation of the mean particle radius r^3 with the aging time at 973 K for the current alloy, FBB8, and Ti-modified FBB8, where the corresponding linear fitting curves are also shown

stress of BCC α-Fe [43]; and σ_{ss} is the contribution from the solid-solution strengthening, which is taken as $\sigma_{ss} = 509$ MPa in FBB8 alloy since both have similar compositions [44]. For the precipitation strengthening (σ_p) of B2 nanoparticles, the strengthening can be divided into two categories, the shearing mechanism and the Orowan bypassing mechanism [37]. The strength increment from the former results from both the coherency strengthening ($\Delta\sigma_{CS}$) and modulus mismatch strengthening ($\Delta\sigma_{MS}$) since the B2 nanoparticles are too much larger to be cut by dislocations [31, 37, 45–48]. They could be expressed with the Eqs. (3), (4), and (5), respectively:

$$\Delta\sigma_{CS} = M\alpha_{\varepsilon}(G\varepsilon_C)^{\frac{3}{2}}(\frac{rf}{0.5Gb})^{\frac{1}{2}} \quad (3)$$

$$\Delta\sigma_{MS} = 0.0055M(\Delta G)^{\frac{3}{2}}\left(\frac{2f}{G}\right)^{\frac{1}{2}}(\frac{r}{b})^{\frac{3m}{2}-1} \quad (4)$$

$$\Delta\sigma_{orowan} = M\frac{0.4Gb}{\pi\sqrt{1-\nu}}\frac{\ln(2\sqrt{\frac{2}{3}}r/b)}{\lambda_P}$$

$$\lambda_p = 2\sqrt{\frac{2}{3}}r(\sqrt{\frac{\pi}{4f}} - 1) \quad (5)$$

where $M = 2.73$ is the Taylor factor for the BCC structure [31]; $\alpha_{\varepsilon} = 2.6$; $m = 0.85$; $\varepsilon_c = 2\varepsilon/3$ is the constrained lattice misfit. $G = 83$ GPa and $\Delta G = 3$ GPa are the shear modulus of the matrix and the shear modulus mismatch between precipitates and matrix, respectively [31]; b is the Burgers vector; f is the volume fraction of the precipitates; $\nu = 0.3$ is the Poisson ratio, and λ_p is the inter-precipitates spacing. Figure 8c plots the strength increment of ($\Delta\sigma_{CS} + \Delta\sigma_{MS}$) or $\Delta\sigma_{orowan}$ as a function of the particle radius r in the current alloy, in which the lattice misfit and volume fraction are taken as $\varepsilon = 0.35\%$ and $f = 24.8\%$, respectively, for the 48-aged alloy. It is found that the maximum strength increment of 601 MPa reaches at the critical $r_0 = 51$ nm, which is within the range of experimental value of $r = 58 \pm 14$ nm. Thus, the total yield strength increment is $\sigma_{YS} = \sigma_0 + \sigma_{ss} + \sigma_p = 50 + 509 + 588 = 1147$ MPa, which is in a good agreement with the measured value of 1032 MPa that was converted from the $HV =$ ~ 344 in 48 h-aged state according to σ_{YS} ~ 3 * HV in steels [49]. Therefore, the precipitation of cuboidal B2 particles with the present particle size and volume fraction could achieve the maximum strengthening effect.

Moreover, the yield strength $\sigma_{YS} = 239$ ~ 258 MPa of the present alloy at 973 K is much higher than that ($\sigma_{YS} = 120$ MPa at 973 K) of FBB8 alloy containing spherical B2 precipitates with a particle size of ~ 53 nm, which is mainly ascribed to the fact that the present lattice misfit $\varepsilon = 0.35\%$ of BCC/B2 is significantly higher than that ($\varepsilon = 0.06\%$) of FBB8 alloy [10, 14]. Moreover, its yield strength is comparable to that ($\sigma_{YS} = 280$ MPa at 973 K) of Ti-modified FBB8 alloy containing cuboidal B2/L2$_1$ precipitates with a particle radius of ~ 57 nm, even though the lattice misfit $\varepsilon = 0.24$ ~ 0.67% is slightly lower than that ($\varepsilon = 0.7\%$) of Ti-modified FBB8 [12, 14]. Previous research indicated that the dominant creep mechanism of FBB8 alloy is dislocation climb bypass of B2 nanoprecipitates with repulsive elastic interaction strain field due to the lattice misfit (ε) [26, 50]. The threshold strength (σ_{th}) needed for dislocations to bypass precipitate particles is greatly affected by the lattice misfit ε between precipitates and matrix. Generally, the magnitude of threshold strength (σ_{th}) is about 0.4 ~ 0.5 times of the strength increment from the Orowan mechanism ($\Delta\sigma_{orowan}$) [26, 50]. And the threshold strength (σ_{th}) of the current alloy is roughly estimated to be ~ 175 MPa with the Eq. (2-c), where $G = 57$ GPa at 973 K [51]. Evidently, its threshold strength is greater than that ($\sigma_{th} = 69$ MPa) of FBB8 alloy, and comparable to that ($\sigma_{th} = 172$ MPa) of Ti-modified FBB8 alloy [14, 26]. A relatively larger lattice misfit ($\varepsilon = 0.24 \sim 0.67\%$) of the current alloy as well as Ti-modified alloy could also result in a strain field that more effectively impedes dislocations on the arrival or departure side of the precipitates, compared to the FBB8 alloy [52, 53]. It is emphasized that an exaggerated lattice misfit ε ($\varepsilon > \sim 1.3\%$) could induce a weaker elastic interaction with dislocations because of the disappearance of coherency [44, 52, 54]. Therefore, a moderate lattice misfit ($\varepsilon = 0.24 \sim 0.67\%$) of the current alloy induced to the formation of cuboidal B2 nanoprecipitates would also certainly improve the creep property at HTs.

Conclusions

In this study, a new ferritic Fe–Cr–Ni–Al series ferritic stainless steel (Fe-10.9Cr-13.9Ni-6.4Al-2.2Mo-0.5W-0.04Zr-0.005B wt.%) strengthened by cuboidal B2-NiAl precipitates was successfully designed via the cluster formula approach. In the solution state, coarse B2 particles with a size of ~ 60 nm and high-density fine B2 particles with a size of 3 ~ 5 nm are precipitated in BCC ferritic matrix. After aging, only one kind of cuboidal B2 particles coherently precipitated due to a moderate $\varepsilon = 0.69\%$. The coarsening behavior discussed by the LSW theory is a time dependent process controlled by the diffusion of the solute and interfacial energy between the precipitates and matrix. The rate coarsening constants is $k = 4.08 \times 10^{-29}$ $m^3 \cdot s^{-1}$. The higher strength ($\sigma_s = 258$ MPa) of the 24 h-aged alloy at 973 K is primarily attributed to the coherent precipitation of cuboidal nanoprecipitates in BCC matrix.

Acknowledgements This work was supported by the National Natural Science Foundation of China [52171152 and U1867201], and the Science and Technology Program of Guangzhou (No. 202102021123).

References

1. Xu YT, Li W, Wang MJ, Zhang XY, Wu Y, Min N, Liu WQ, Jin XJ (2019) Nano-sized MX carbonitrides contribute to the stability of mechanical properties of martensite ferritic steel in the later stages of long-term aging. Acta Mater. 175:148–159.
2. Aghajani A, Somsen C, Eggeler G (2009) On the effect of long-term creep on the microstructure of a 12% chromium tempered martensite ferritic steel. Acta Mater, 57:5093-5106.
3. Kostka A, Tak KG, Hellmig RJ, Estrin Y, Eggeler G (2007) On the contribution of carbides and micrograin boundaries to the creep strength of tempered martensite ferritic steels. Acta Mater. 55:539–550.
4. Pesicka J, Kuzel R, Dronhofer A, Eggeler G (2003)The evolution of dislocation density during heat treatment and creep of tempered martensite ferritic steels. Acta Mater. 51:4847–4862.
5. Niu B, Wang ZH, Wang Q, Pan QF, Dong C, Zhang RQ, Liu HQ, Liaw PK, Xu W (2020) Dual-phase synergetic precipitation in Nb/Ta/Zr co-modified Fe-Cr-Al-Mo alloy. Intermetallics 124:106848.

6. Reed RC (2008), The Superalloys: Fundamentals and Applications. Cambridge University Press, Cambridge.
7. Pollock TM, Argon AS (1992) Creep resistance of CMSX-3 nickel base superalloy single crystals. Acta Metall. Mater. 40:1–30.
8. Zhang JX, Harada H, Koizumi Y, Kobayashi T (2010) Dislocation Motion in the Early Stages of High-temperature Low-stress Creep in a Single-crystal Superalloy with a Small Lattice Misfit. J. Mater. Sci. 45:523–532.
9. Teng ZK, Liu CT, Ghosh G, Liaw PK, Fine ME (2010) Effects of Al on the microstructure and ductility of NiAl-strengthened ferritic steels at room temperature. Intermetallics 18:1437–1443.
10. Teng ZK, Miller MK, Ghosh G, Liu CT, Huang S, Russell KF, Fine ME, Liaw PK (2010) Characterization of nanoscale NiAl-type precipitates in a ferritic steel by electron microscopy and atom probe tomography. Scripta Mater. 63:61–64.
11. Sun Z, Liebscher CH, Huang S, Teng Z, Song G, Wang G, Asta M, Rawlings M, Fined ME, Liaw PK (2013) New design aspects of creep-resistant NiAl-strengthened ferritic alloys. Scripta Mater. 68:384–388.
12. Song G, Sun ZQ, Poplawsky JD, Gao YF, Liaw PK (2017) Microstructural evolution of single Ni2TiAl or hierarchical NiAl/Ni2TiAl precipitates in Fe-Ni-Al-Cr-Ti ferritic alloys during thermal treatment for elevated-temperature applications. Acta Mater. 127:1-16.
13. Song G, Hong SJ, Lee JK, Song SH, Hong SH, Kim KB, Liaw PK (2019) Optimization of B2/L21 hierarchical precipitate structure to improve creep resistance of a ferritic Fe-Ni-Al-Cr-Ti superalloy via thermal treatments. Scripta Mater. 161:18–22.
14. Song G, Sun Z, Li L, Xu X, Rawlings M, Liebscher CH, Clausen B, Poplawsky J, Leonard DN, Huang S (2015) Ferritic alloys with extreme creep resistance via coherent hierarchical precipitates. Sci. Rep. 5:16327.
15. Kohyama A, Hishinuma A, Gelles DS, Klueh RL, Ehrlich K (1996) Low-activation ferritic and martensitic steels for fusion application. J. Nucl. Mter. 233-277:138–147.
16. Ma Y, Wang Q, Jiang BB, Li CL, Hao JM, Li XN, Dong C, Nieh TG (2018) Controlled formation of coherent cuboidal nanoprecipitates in body-centered cubic high-entropy alloys based on $Al_2(Ni,Co,Fe,Cr)_{14}$ compositions. Acta Mater. 147:213–225.
17. Wang Q, Han JC, Liu YF, Zhang ZW, Dong C, Liaw PK (2021) Coherent precipitation and stability of cuboidal nanoparticles in body-centered-cubic $Al_{0.4}Nb_{0.5}$ $Ta_{0.5}TiZr$ $_{0.8}$ refractory high entropy alloy. Scripta Mater. 190:40–45.
18. Li CL, Ma Y, Hao JM, Yan Y, Wang Q, Dong C, Liaw PK (2018) Microstructures and mechanical properties of body-centered-cubic $(Al,Ti)_{0.7}(Ni,Co,Fe,Cr)_5$ high entropy alloys with coherent $B2/L2_1$ nanoprecipitation. Mater. Sci. Eng. A 737:286-296.
19. Teng ZK, Ghosh G, Miller MK, Huang S, Clausen B, Brown DW, Liaw PK (2012) Neutron-diffraction study and modeling of the lattice parameters of a NiAl-precipitate-strengthened Fe-based alloy. Acta Mater., 60:5362–5369.
20. Yamamoto Y, Brady MP, Lu ZP, Maziasz PJ, Liu CT, Pint BA, More KL, Meyer HM, Payzant EA (2007) Creep-resistant, Al_2O_3-forming austenitic stainless steels. Science, 316:433–436.
21. Wang WR, Wang WL, Yeh JW (2014) Phases, microstructure and mechanical properties of $Al_xCoCrFeNi$ high-entropy alloys at elevated temperatures. J. Alloys Compd. 589:143–152.
22. Chen RR, Qin G, Zheng HT, Wang L, Su YQ, Chiu YL, Ding HS, Guo JJ, Fu HZ (2018) Composition design of high entropy alloys using the valence electron concentration to balance strength and ductility. Acta Mater. 144:129–137.
23. Wang Q, Ma Y, Jiang BB, Li XN, Shi Y, Dong C, Liaw PK (2016) A cuboidal B2 nanoprecipitation-enhanced body-centered-cubic alloy $Al_{0.7}CoCrFe_2Ni$ with prominent tensile properties. Scripta Mater., 120:85–89.
24. Bhadeshia H, Honeycombe R (2017) Steels-Microstructure and Properties. 4th ed., Butterworth Heinemann, Oxford.
25. Calderon HA, Fine ME, Weertman JR (1987) Coarsening and morphology of β' particles in Fe-Ni-Al-Mo ferritic alloys. Metall. Trans. A 19:1135–1146.
26. Vo NQ, Liebscher CH, Rawlings MJS, Asta M, Dunand DC (2014) Creep properties and microstructure of a precipitation-strengthened ferritic Fe-Al-Ni-Cr alloy. Acta Mater. 71:89–99.

27. Briant CL, Banerji SK (1978) Intergranular failure in steel: the role of grain boundary composition. Int. Metal. Rev. 23:164–199.
28. Baker I, Munroe PR (1988) Improving intermetallic ductility and toughness, JOM 40:2.
29. Schneider CA, Rasband WS, Eliceiri KW (2012) NIH Image to ImageJ: 25 years of image analysis. Nat. Methods 9:671–675.
30. Shackelford JF, Han YH, Kim S (2016) CRC Materials science and engineering handbook. Boca Raton (FL): CRC Press
31. Hosford WF (2005) Mechanical Behavior of Materials. first ed., Cambridge University Press, New York.
32. Sun ZQ, Song G, Ilavsky J, Liaw PK (2015) Duplex precipitates and their effects on the room-temperature fracture behaviour of a NiAl-strengthened ferritic alloy. Mater. Res. Lett. 3:128–134.
33. Baik SI, Wang SY, Liaw PK, Dunand DC (2018)Increasing the creep resistance of Fe-Ni-Al-Cr superalloys via Ti additions by optimizing the $B2/L2_1$ ratio in composite nanoprecipitates. Acta Mater. 1571142–154
34. Philippe T, Voorhees PW (2013) Ostwald ripening in multicomponent alloys. Acta Mater. 61:4237–4244.
35. Orthacker A, Haberfehlner A, Taendl J (2018) Diffusion-defining atomic-scale spinodal decomposition within nanoprecipitates. Nat Mater. 17:1101–1107.
36. Philippe T, Voorhees PW (2013) Ostwald ripening in multicomponent alloys. Acta Mater. 61(11):4237–4244.
37. Argon A (2007) Strengthening mechanisms in crystal plasticity. Oxford University Press, Oxford, the United Kingdom
38. Spigarelli S, Cerri E, Bianchi P (1999) Interpretation of creep behavior of a 9Cr–Mo–Nb–V–N (T91) steel using threshold stress concept. Mater Sci Tech. 15:1433.
39. Voorhees PW, McFadden GB, Johnson WC (1992) On the morphological development of second-phase particles in elastically-stressed solids. Acta Metall. 40:2979–2992.
40. Su CH, Voorhees PW (1996) The dynamics of precipitate evolution in elastically stressed solids—II. Particle alignment. Acta Mater. 44:2001-2006.
41. Thompson M, Su C, Voorhees PW(1994) The equilibrium shape of a misfitting precipitate. Acta Metall. Mater. 42(6):2107–2122.
42. Wasilewski RJ (1996) Elastic constants and young's modulus of NiAl. Trans. Mct. Soc. AIME 236:455.
43. Rivera-Díaz-del-Castillo PEJ, Hayashi K, Galindo-Nava EI (2013) Computational design of nanostructured steels employing irreversible thermodynamics. Mater. Sci. Technol. 29:1206–1211.
44. Baik SI, Wang SY, Liaw PK, Dunand DC (2018) Increasing the creep resistance of Fe-Ni-Al-Cr superalloys via Ti additions by optimizing the $B2/L2_1$ ratio in composite nanoprecipitates. Acta Mater. 157:142–154.
45. Ardell AJ (1985) Precipitation hardening. Metall. Trans. A 16A:1985–2135.
46. Brown LM, Ham RK, Dislocation-particle interactions, in: A. Kelly, R.B Nicholson, Strengthening methods in crystals. Elsevier, Amsterdam
47. Jansson B, Melander A (1978) On the critical resolved shear stress from misfitting particles. Scripta Metall. 12:497–498.
48. Nembach E (1983) Precipitation hardening caused by a difference in shear modulus between particle and matrix. Phys. Stat. Sol. 78:571–581.
49. Pavlina EJ, Van Tyne CJ (2008) Correlation of yield strength and tensile strength with hardness for steels. J. Mater. Eng. Perform. 17:888–893.
50. Zhao YX, Fang QH, Liu YW, Wen PH, Liu Y (2015) Creep behavior as dislocation climb over NiAl nanoprecipitates in ferritic alloy: The effects of interface stresses and temperature. Int. J. Plast. 69:89–101.
51. G. Ghosh, G.B. Olson, The isotropic shear modulus of multicomponent Fe-base solid solutions. Acta Mater. 50 (2002) 2655–2675.

52. Rawlings MJS, Liebscher CH, Asta M, Dunand DC (2017) Effect of titanium additions upon microstructure and properties of precipitation-strengthened Fe-Ni-Al-Cr ferritic alloys. Acta Mater. 128:103–112.
53. Krug ME, Seidman DN, Dunand DC (2012) Creep properties and precipitate evolution in Al-Li alloys microalloyed with Sc and Yb. Mater. Sci. Eng. A 550:300–311.
54. Polvani RS, Tzeng WS, Strutt PR (1973) High temperature creep in a semi coherent NiAl-Ni_2AlTi alloy. Metall. Trans. A 7A:33–40.

Part VI
Additive: Powder and Processing

Surface Roughness of Additively Manufactured IN718 and H282 Superalloys from Multi-size and Multi-laser Machines

R. Subramanian, K. Cwiok, and A. Kulkarni

Abstract Laser Powder Bed Fusion (LPBF) of metallic components is unlocking new design options for high-efficiency gas turbine component designs not possible by conventional manufacturing technologies. Surface roughness is a key characteristic of LPBF components that impacts heat transfer correlations and crack initiation from co-located surface defects—both are critical for gas turbine component durability and performance. However, even for a single material, there is an increasing diversity in laser machines (single vs multi-laser), layer thicknesses (~20–80μ) and orientations to the build plate (upskin, vertical, and downskin) that result in significant variability in surface roughness. Build direction effects are particularly important when considering three-dimensional gas turbine components each having unique cooling features. This study systematically compares the external and internal surface roughness of two gas turbine superalloys—Inconel 718 and Haynes 282—from multi-laser and multi-size machines. This presented data will be discussed in detail, to show potential applicability of a 3D process signature surface across machines and substrate orientations for additively manufactured superalloys.

Keywords Surface roughness · Laser incidence angle · Surface orientation · Additive manufacturing · Laser powder bed fusion

Introduction

Additive manufacturing (AM), more specifically LPBF (Laser Powder Bed Fusion), can reduce design to manufacturing cycle time by almost 50% and contribute to faster time-to-market targets. The need for accelerated, multi-disciplinary iteration across design, materials, and manufacturing tasks is critical to develop and release a reliable process, minimizing component risks from inherent manufacturing process defects. This is indeed a demanding challenge, with the diversity in machine chamber size (for

R. Subramanian (✉) · K. Cwiok · A. Kulkarni
Orlando, FL, USA
e-mail: ramesh.subramanian@siemens-energy.com

E. A. Ott et al. (eds.), *Proceedings of the 10th International Symposium on Superalloy 718 and Derivatives*, The Minerals, Metals & Materials Series,
https://doi.org/10.1007/978-3-031-27447-3_31

example-EOSM290, EOSM400-1/4), number of lasers (single versus four) and materials (combustor alloys such as IN625, H282 to γ' strengthened turbine alloys IN939, Mar-M-247). With demands on higher LPBF productivity, the number of components in a chamber can vary from 1 (for large turbine components) to ~20 (for combustor burner components in a chamber)—this additionally increases the demands on manufacturing repeatability build-to-build, as well as for different locations within a build plate. Despite the widespread popularity and promising applications of LPBF, it is associated with challenges for the fabrication of fully dense parts with superior surface finish compared to manufacturing processes such as machining [1] and even conventional investment casting. LPBF components are prone to develop surface flaws due to the several concurrently active mechanisms during the fast-melting and solidification process. There are various effects contributing to surface roughness for LPBF surfaces—(a) the staircase effect (the roughness resulting from layer wise stacked 2D sections approximating of smooth 3D surface), (b) waviness created by the partially overlapping individual melt beads, (c) the balling effect (separation of the melt tracks in individual beads due melt pool instability), (d) spatter formation (the ejection of particles from melt pools), and (e) loose or partially melted powders that unavoidably occur when the melt pool is in contact with nearby loose powder. The above factors significantly impair the resulting surface integrity, which is detrimental to fatigue life [2–6]. The severity of the adverse phenomena during LPBF processes can be mitigated by tuning the several processing parameters involved, such as layer thickness, laser power, scan speed, and hatch spacing that will profoundly affect the physical and mechanical properties of the LPBF product [7–10]. Other build related factors affecting the gas flow and thermal history of the material also contribute to roughness such as exposure order, number of parts in the build, fume/laser interaction, and packing of parts in a layer but those influences are not the focus of this paper. In this paper, the significance of the surface inclination angle, face orientation towards or away from the laser, and position of the printed part on the build chamber are evaluated. Fig. 1 shows previous systematic studies [5, 11–13] of the effect of surface orientation to surface roughness. Recently, Rott et al. [14] further elaborated on the work done of Kleszczynski et al. [15] in demonstrating the application of surface laser relation angle (or laser incidence angle relative to the surface normal) to quantify the interdependency between the surface orientation to the laser (facing the laser or away from the laser) and location in the build plate and act as a parameter to predict surface roughness. This is shown in Fig. 2, with the surface roughness calculated from the surface laser relation angle. The surface laser relation angle can be utilized as a machine independent parameter. The surface laser relation angle can then improve comparability of studies across machines and materials (Figs. 3, 4, 5, 6, 7, 8).

The predicted surface roughness based on the CAD model orientation and placemen could enable a comprehensive roughness prediction for the full part in preprocessing. This early insight into the distribution of surface roughness of critical surfaces can then possibly be minimized by redesign, re-orientation, or repositioning. This could result in an accurate application of heat transfer correlations, possibly an improvement in fatigue or even reduced postprocessing afterwards.

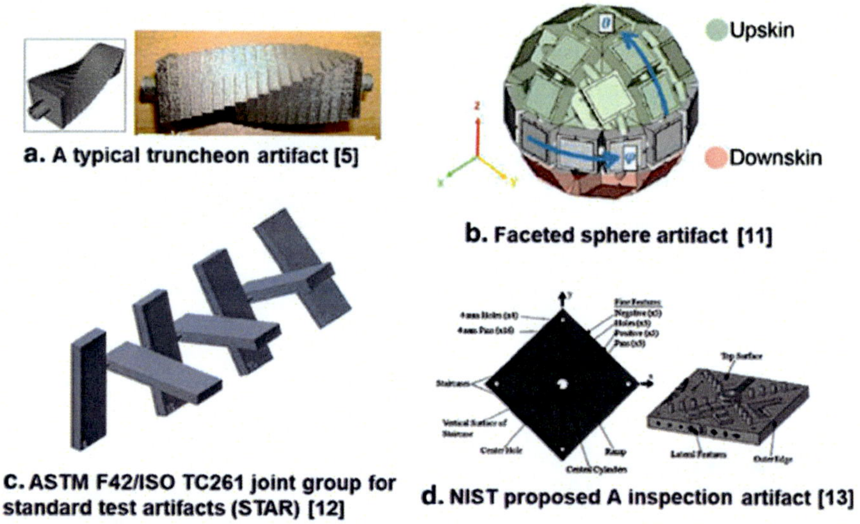

Fig. 1 Different coupon geometries/standards used for surface roughness characterization

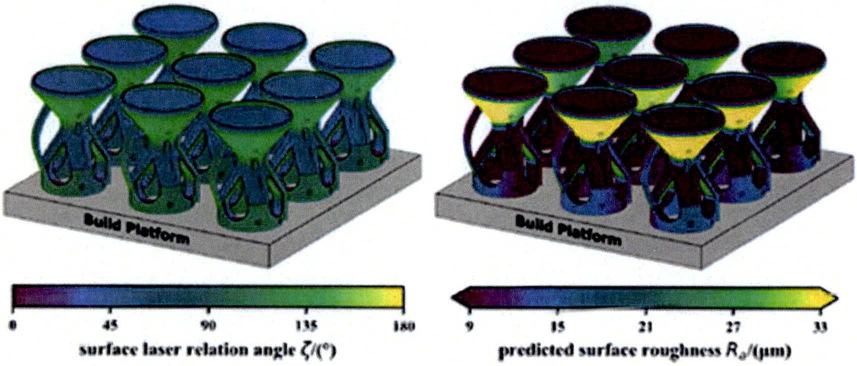

Fig. 2 Correlation of the surface roughness as function of the part location in the build plate [14]

In this paper, Siemens Energy and Siemens Technology have further evaluated the relationship between the laser incidence angle and surface roughness using a simple "chess piece" geometry that enables laser incidence angles to cover a range from 40 to 140°. This angle range covers almost all the possible orientations—upskin, vertical and downskin—for many LPBF component build set ups. The surface roughness coupon, described in this paper, can be used as a stand-alone geometry or along with other component geometries due to the compact size. As discussed in the paper below, the results show a repeatable relationship or "machine signature"

Fig. 3 Void area analysis of 40 μm parameter

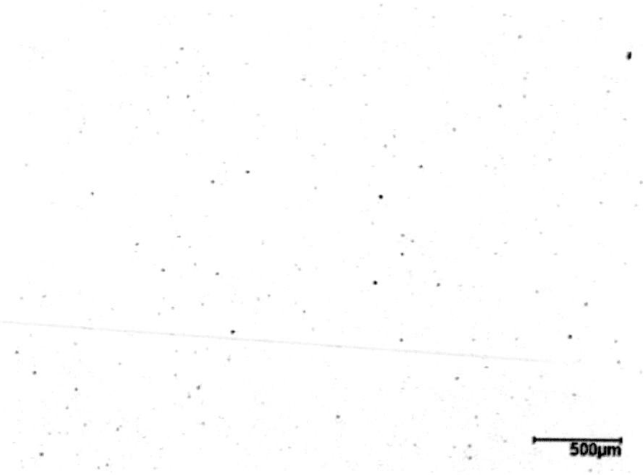

Fig. 4 Void area analysis of 80 μm parameter

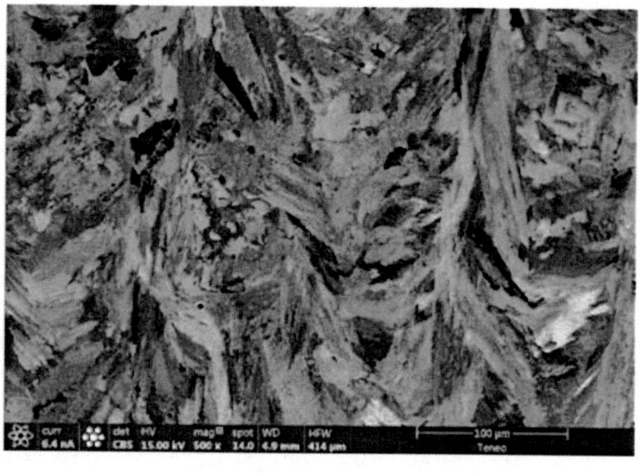

Fig. 5 SEM BSE image of 40 μm parameter grain structure

Fig. 6 SEM BSE image of 80 µm parameter grain structure

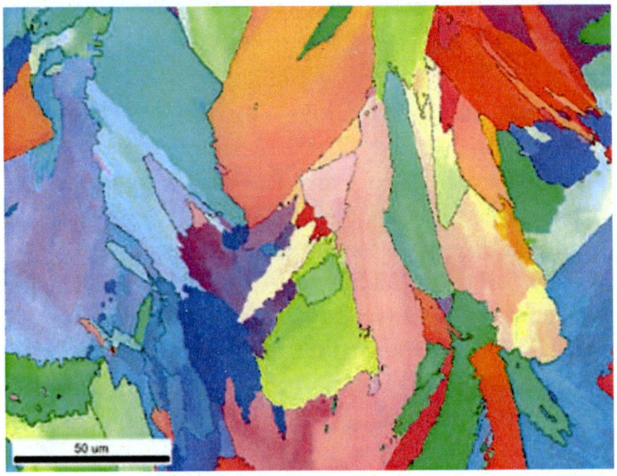

Fig. 7 EBSD image of 40 µm parameter

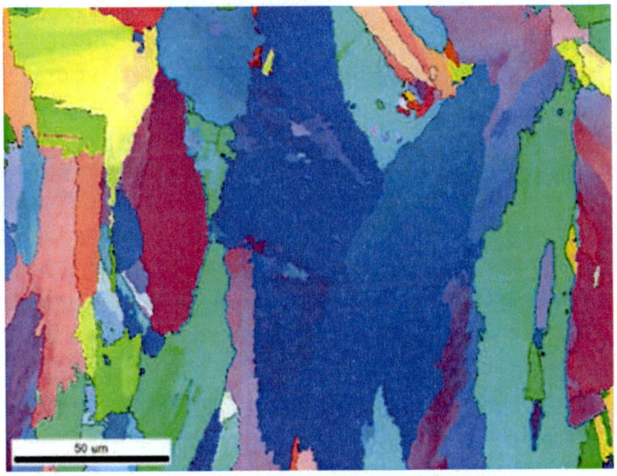

Fig. 8 EBSD image of 80 µm parameter

across single and multi-laser systems, layer thickness and materials. This methodology attempts to account for the laser angle effect's contribution to roughness to further isolate and understand the build space position influence on roughness. Additionally, the data creates a kind of signature of a process and machine combination useful checking for differences between machine serial numbers or for benchmarking different machine manufacturers and models. This signature will enable comparison of capabilities across machines and establish a benchmark for process stability, after process qualification of a LPBF machine, using a non-destructive property such as surface roughness, measurable on both coupons and components.

Materials and Methods

Materials

Two gas turbine superalloys—Inconel 718 (IN718) and Haynes 282 (H282)—were evaluated in this study. These two alloys are not only common for this application but are also common alloys for LPBF printing. As a result, many standards and specifications from ASTM and SAE AMS have been created to assist with the standardization of these alloys with this printing process.

The following tables detail the feedstock specification and printed mechanical properties of the two alloys. Tables 1 and 2 detail the typical chemistries of IN718 and H282 powders used for additive manufacturing, respectively. While chemistry is important for overall part quality, the powder size distribution of the feedstock is important in ensuring process stability and consistency. Following the test method in ASTM B822, Table 3 outlines the common D50 powder size distributions used for the process.

Mechanical Properties

Bulk printed mechanical properties of IN718 and H282 are seen below. These results are minimum values and observed at room temperature. Table 4 shows the IN718 printed material properties from AMS 7000 for LPBF processing. The results are post-heat treatment also detailed within AMS 7000. For H282, Table 5 is derived from an EOS material specification for their standardized H282 process parameters. These samples were also heat treated using a solution and aging step outlined in the specification. Both results are from test specimens printed in the Z-direction.

Table 1 Chemical composition of IN718 following AMS 7006

Element	Min	Max
Carbon	–	0.045
Manganese	–	0.35
Silicon	–	0.35
Phosphorus	–	0.010
Sulfur	–	0.010
Chromium	17.00	21.00
Nickel	50.0	55.0
Molybdunum	2.80	3.30
Columbium (Niobium)	4.75	5.50
Titanium	0.8	1.15
Aluminum	0.40	0.60
Cobalt	–	1.00
Tantulum	–	0.05
Columbium (Niobium) + Tantulum	4.87	5.20
Boron	–	0.0060 (60 ppm)
Copper	–	0.23
Lead	–	0.0010 (10 ppm)
Bismuth	–	0.00005 (0.5 ppm)
Selenium	–	0.0005 (5 ppm)
Magnesium[b]	–	0.0060 (60 ppm)
Calcium[b]	–	0.0030 (30 ppm)
Nitrogen		0.0300 (300 ppm)
Iron[a]	Balance	

Table 2 Chemical composition of H282 following Haynes International data sheet

Element	Nominal
Carbon	0.06
Manganese	0.3 max
Silicon	0.15 max
Chromium	20.0
Nickel	57.0 Balance
Molybdunum	8.5
Titanium	2.1
Aluminum	1.5
Cobalt	10.0
Boron	0.005
Iron	1.5 max

Table 3 Typical D50 particle size distribution for LPBF powder ranges

Size range (μm)	Size	Testing method	Min. [μm]	Max. [μm]
15-45	D50	ASTM B822	28	38
15-53	D50	ASTM B822	34	44
15-63	D50	ASTM B822	40	50

Table 4 Minimum room temperature mechanical data for IN718 LPBF material from AMS7000

Ultimate strength	827 MPa (120 ksi)
Yield Strength	345 MPa (50.0 ksi)
% Elongation	30

Table 5 Room temperature mechanical data for H282 LPBF material from the EOS H282 Datasheet

Ultimate strength	1200 MPa (174 ksi)
Yield Strength	751 MPa (109 ksi)
% Elongation	26.2

Experimental Setup

Surface roughness coupon: This is a 50 mm tall and 25 mm diameter "chess piece", with octagonal surfaces at different heights as shown in Fig. 9a. Each surface roughness coupon has 33 distinct surfaces—1 top flat surface, 8 20° upskin surfaces, 8 60° upskin surfaces 8 90° vertical surfaces, and 8 60° downskin surfaces. A laser incidence angle is defined as angle between the substrate normal and laser incidence direction. The surface roughness coupons are typically distributed on a M290-1 build plate as shown in Fig. 9b. The coupons alignment in the build plate is identified by a notch oriented towards the gas flow and the letters aligned from left to right, opposite to the recoater direction. This orientation is important to also evaluate the effects of gas flow and recoater direction on the surface roughness.

Surface roughness measurements: Surface roughness measurements were done by non-contact profilometry using Sensofar S-Neox90, with the focus variation setup. The focus variation setting takes several bright field images while scanning in the Z-axis. For each of the images, at a specific z-axis value, a focus operator algorithm records the axial position of the pixels in focus (at highest contrast) at each focal plane. A focus operator algorithm then records an axial displacement for pixels, at its maximum contrast. A stacked image, with the axial position information is stored for all the images with highest contrast, and this will result in a 3D surface topology measurement. After scanning the entire Z-range, a 3D image composed of stacked images is compiled and a 3D image is reconstructed. Images are taken with an objective lens 10X and the surface roughness measurement requirements of 0.8 mm cut-off length, with a L-Gaussian correction is evaluated to meet ISO 4287 standards, after application of a 5 × 5 median denoising filter.

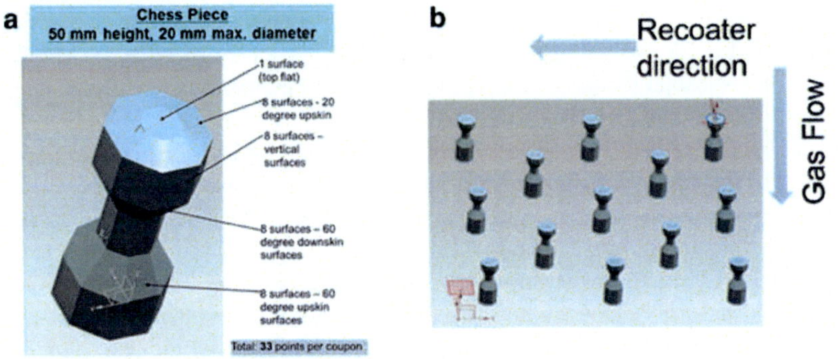

Fig. 9 **a** CAD model of "chess piece" **b** Typical layout of the "chess pieces" in the build plate

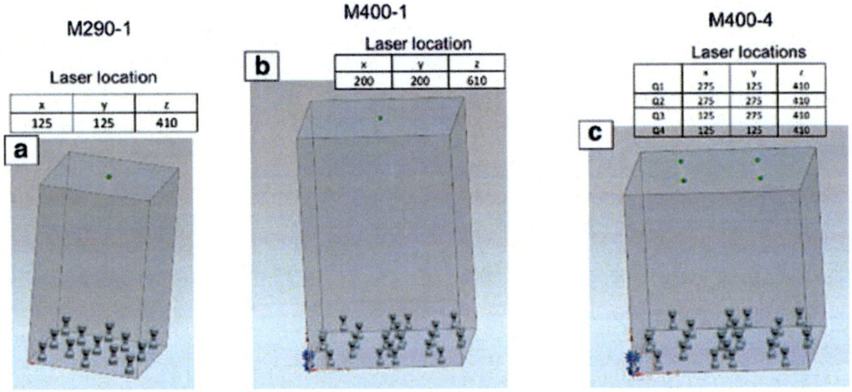

Fig. 10 Laser coordinates for EOS machines a M290-1, b M400-1, and c M400-4

In addition to M290-1, a single laser system, Fig. 10 shows the coordinates of the laser head for each of the other machines M400-1 (a taller single laser system) and M400-4 (a four laser system). Depending on the type of machine, the laser incidence angle can be calculated. The measured surface roughness can be plotted against the laser incidence angle for any machine.

Results and Discussion

Build Plate Location on Laser Incidence Angle

Location in the build plate of this octagonal chess piece will result in a laser incidence angle that traverses from horizontal (flat), upskin, vertical, and downskin surfaces, that face towards and away from the laser. Figure 11 shows an example of the laser incidence angles for the surface roughness coupon geometry in an EOS M290-1 machine. For the center location under the laser, the laser incidence angle is equivalent to the surface orientation relative to the build plate (as illustrated in Fig. 3b). However, for the corner piece, the laser incidence angle has a larger range from acute to obtuse angles due to the surfaces facing towards and away from the laser.

The variation in the laser incidence angle is expected to influence the surface roughness as shown in Fig. 12. The surfaces that have an acute laser incidence angle (mostly upskin) have the heat flow and consequently melting of a solid surface. The surfaces that are heated by an obtuse laser incidence, result in more of heat transfer into loose powder near the surface. These phenomena are the same for internal cooling channels. Therefore, the surface roughness trends from external surface measurements can be used to predict internal surface roughness. This is a very useful relationship, especially for internal cooling channels, where surface roughness cannot be easily measured.

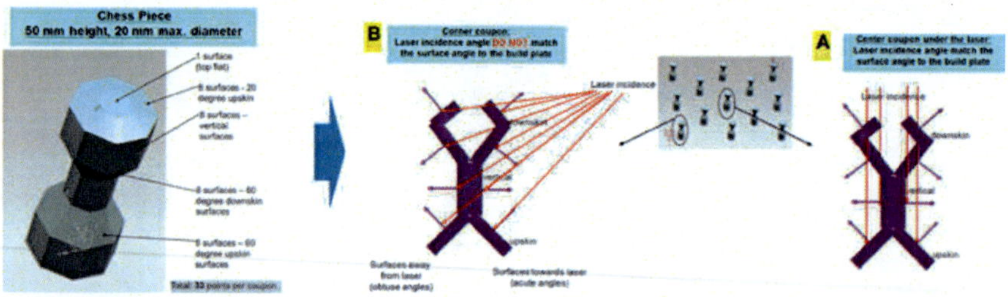

Fig. 11 **a** Schematic of laser incidence angles on center coupons under the laser and **b** Schematic of laser incidence angles on corner

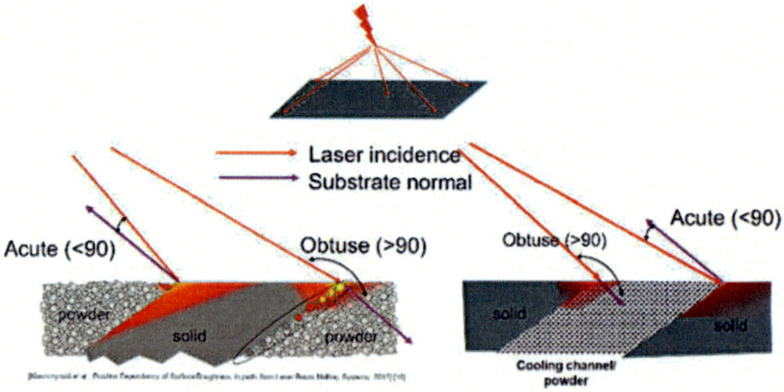

Fig. 12 Schematic of laser incidence angles on external surface roughness and the internal cooling channels

Surface Roughness Relative to Laser Incidence Angle

The measured surface roughness data for the center and corner coupon for IN718 is shown in Fig. 13. The lower curve (yellow color) shows the roughness variation in horizontal (0°), upskin (60°), vertical (90°), and downskin (120°) for the center coupon. For the same geometry, when the coupon is moved to the corner, the laser incidence angle range now increases from 40° to 135°, due to the differences in surfaces that face the laser and those that face away from the laser. This change in laser angle results in higher roughness values across all the incidence angles, but with a very similar parabolic trend.

Figure. 14 shows the comparison of the surface roughness pictures for the two coupons across the different surface orientations. Differences in the contributions to surface roughness are very clear—high upskin surface roughness at low laser incidence angles (due to the staircase effect as proposed by Strano [5]). The staircase effect is a description of the roughness resulting from layer-wise stacked 2D sections that approximate a smooth 3D surface and analyzed in detail by Strano [5]. This is followed by minimum surface roughness for vertical surfaces and finally, the increase in downskin roughness at high laser incidence angles—resulting from particle adhesion and unstable melt pools. The horizontal surfaces are smooth when directly under the laser in the center position but have a higher roughness for the corner coupons. From Fig. 13 although the trends are similar across the coupons in

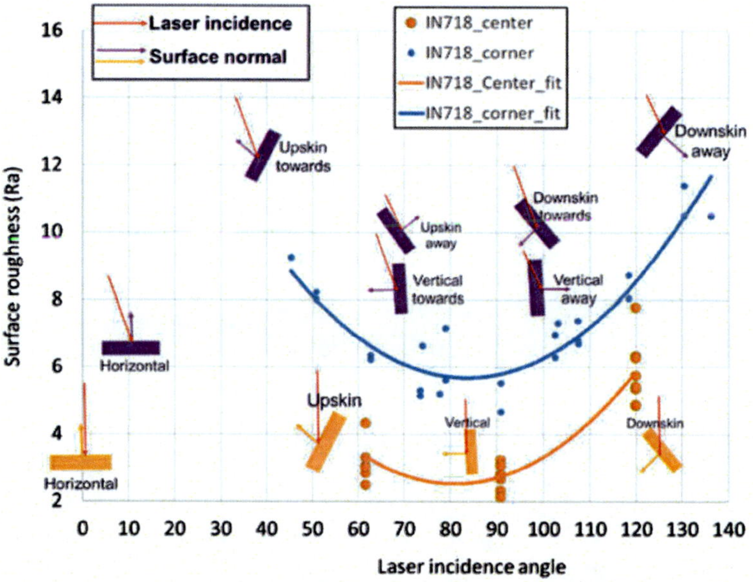

Fig. 13 Overview of laser incidence angles of upskin, vertical, and downskin surfaces facing and away from the laser

different locations of the build plate, the roughness values are higher for the corner coupons compared to the center coupons, for the same incidence angles between 60° and 120°. Two possible reasons that are likely are the effects of gas flow or recoater direction. To evaluate these differences, all surfaces (upskin, vertical, and downskin) facing towards and away from gas flow and recoater direction were measured.

The results are shown in Fig. 15 below (using average trendlines for clarity, without data points). Standard deviations in surface roughness measurements are typically within 1.5–2.5 Ra. Within this standard deviation and data scatter, Fig. 15 clearly shows that a surface roughness variation is not detected. Therefore, other reasons such as the melt pool size and the consequent decrease in local energy density can result in a higher roughness or variation in powder spreading near the build plate corners and edges. Understanding of these differences requires further investigation.

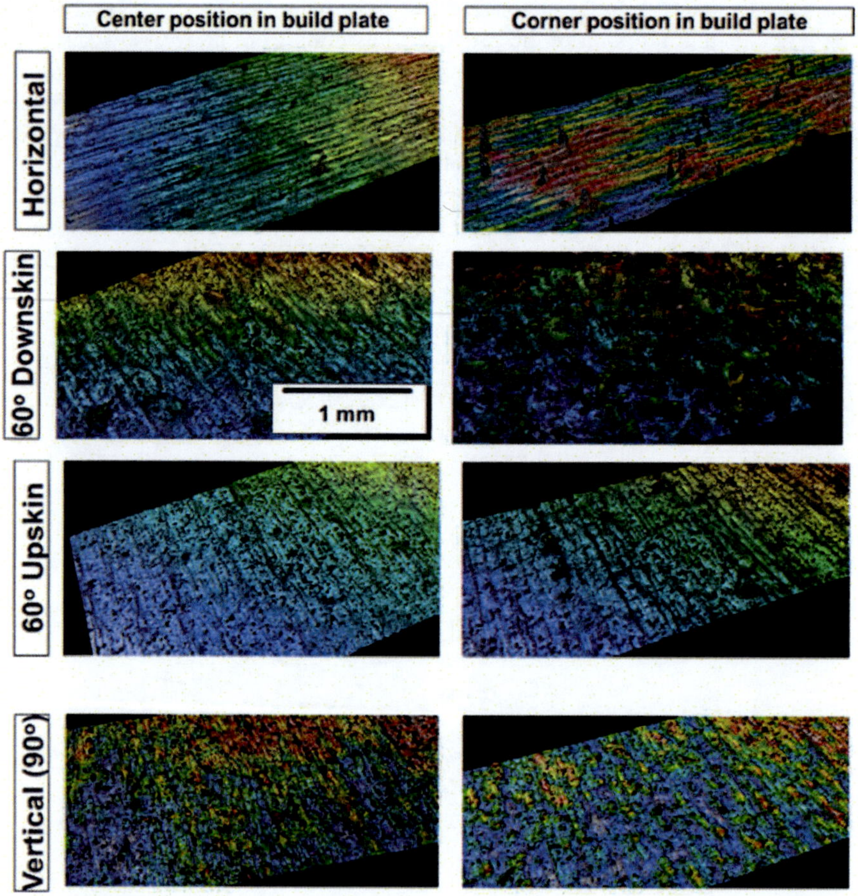

Fig. 14 Surface roughness images for IN718 center and corner coupons as function of orientation to the build plate

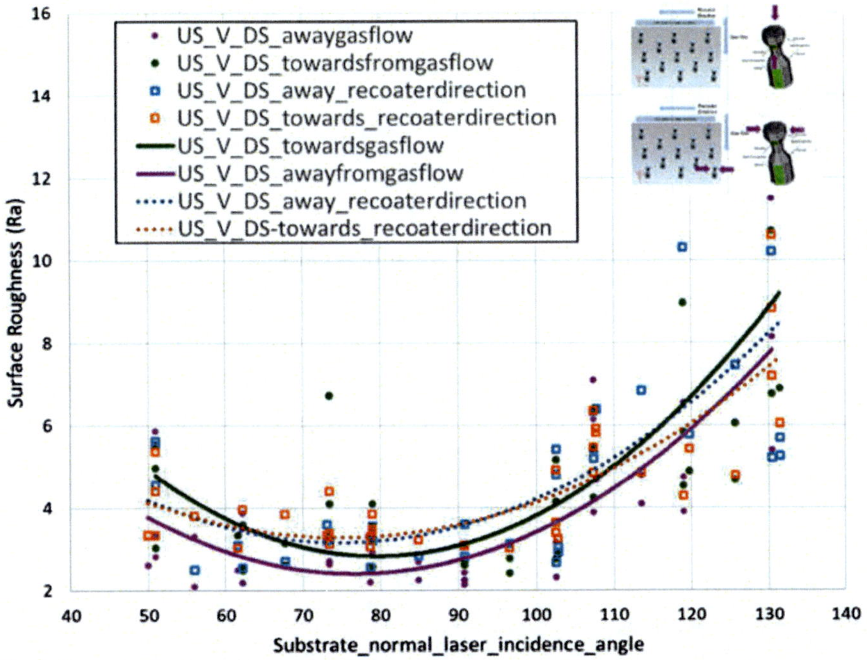

Fig. 15 IN718 surface roughness data towards and away from gas flow and recoater direction

Similar to layer thickness trends in literature [14], the surface roughness is greater at a 80μ layer thickness relative to a 40-μ layer thickness as seen in the external surfaces of the chess piece samples in Fig. 16. On average, the arithmetic mean roughness for the 80μ layer thickness is 1–13μ higher compared to the 40μ layer thickness for all surface orientations, except the downskin 60°, of the chess piece created using the M400-1 machine.

For all the surface orientations evaluated across the chess piece samples that were made using the different machines, surface roughness was typically the highest on the 60° downskin followed by the 20° upskin, 60° upskin, 90° vertical, then 0° horizontal surface orientations. The only machine that did not follow this trend was for the samples created using the M400-1 in a 80μ layer thickness. The surface roughness of the M400-1 80μ layer thickness is higher for the 20° upskin surface compared to the 60° downskin surface. After further investigation, the cause is a result of the stair stepping effect, which was exacerbated due to the 80μ layer thickness relative to a 40-μ layer thickness, that lead to an increase in roughness for the 20° upskin surface compared to the 60° downskin surface.

It is clear from the results in Fig. 16, that the surface roughness across all three machines (M290-1, M400-1, M400-4) for a 40-μ layer thickness are very similar, less than a 5μ difference. Some surface orientations such as the upskin 20°, upskin

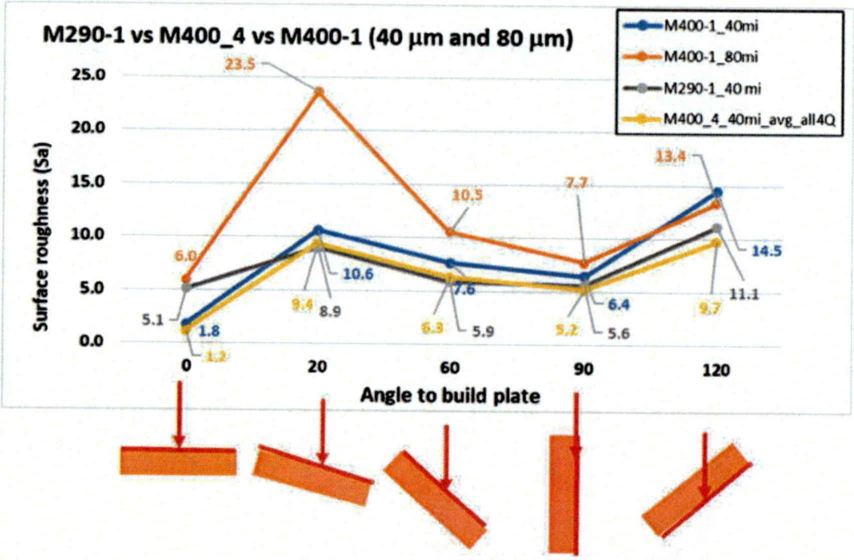

Fig. 16 Areal average surface roughness results (Sa) for each chess piece fabricated in 718 using two different single laser machines, two layer thicknesses, and a four laser machine. Average of 88 measurements for each point

60°, and vertical 90° are less than a 2µ difference between the 40µ layer thickness machines. The small difference in roughness level highlights that the machine-induced surface roughness variation is minimal on flat surfaces. In comparing the roughness of the single laser machines at a 40-µ layer thickness (M290-1 and M400-1), the roughness is 13-27% lower for the M290-1 compared to the M400-1 for all surface orientations except for the horizontal 0° surface orientation. The surface roughness across all build orientations was 12–40% higher for the single laser M400-1 chess piece sample compared to the four samples created at the different quadrants using the multi-laser M400-4.

Each chess piece that was fabricated at each of the four quadrants (Q1, Q2, Q3, Q4) in the four laser M400-4 are colored coded in Fig. 17. Results from Fig. 17 indicate that there is a minimal difference (less than 1µ) in surface roughness at every surface orientation for the four samples printed using the different lasers in the M400-4 machine. It is important to note that the samples printed at each of the four corners in the M400-4 shared the same radial distance from the laser source.

The next summary focuses on material differences between Ni-based H282 printed on an M290 and M400-1. The trend observed for IN718 surface roughness is also observed for H282 (see Fig. 18), although with a slightly higher surface roughness. The contributions to H282 surface roughness also have a similar trend as shown for IN718 in Fig. 17. For all surface roughness data with H282, the M290 and M400-4 are used to print with a 40 µm layer thickness only. Figure 18 details the surface roughness data obtained from the coupons. Like the 718 printing, results

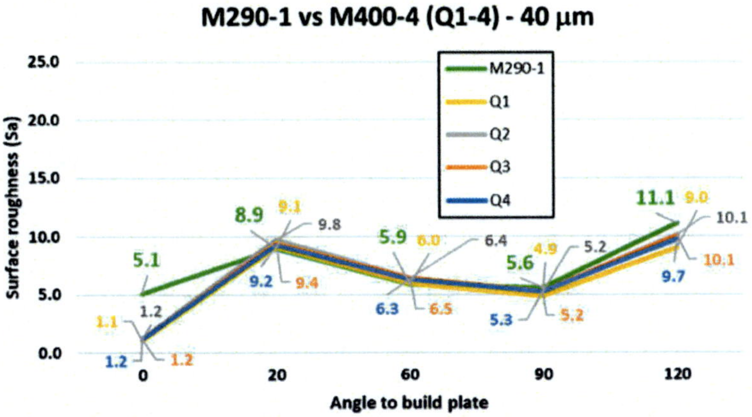

Fig. 17 M290 and M400-4 surface roughness results for IN718. Surface roughness are measured from all coupons in the M290 and M400-4 (Q1-Q4) build plate

between center coupons of each quadrant on the M400-4 have minimal differences. The maximum difference is 3 μm between Q2 and Q1/Q4. When comparing the differences between the M290 and M400-4, the differences are higher than 718, with a maximum difference of 7.7 μm, for the 0° and 60° surfaces. Overall, the surface roughness on the M290-1 is 30–141% lower than the M400-4 data for the 4 quadrants. The reason for this difference is not yet clear.

When comparing the two materials, IN718 and H282, the observed surface roughness for H282 is higher except for one data point at 20° on the M400-4. Figures 19 and 20 compare the results for each material in 40 μm layer thicknesses with the EOS M290-1 and the M400-4 machines, respectively.

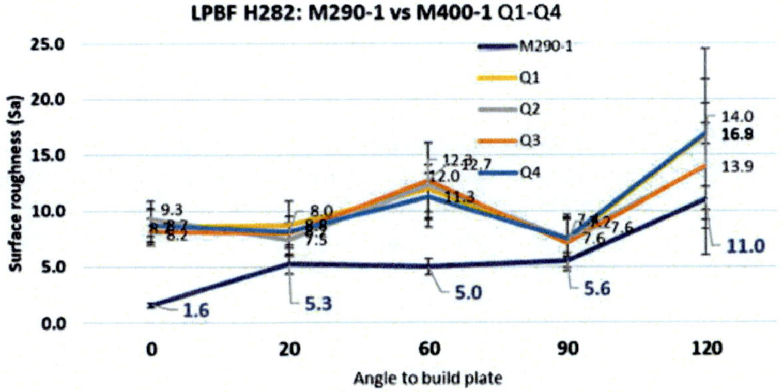

Fig. 18 M290 and M400-4 surface roughness results for H282. Surface roughness are measured from all coupons in the M290 and M400-4 (Q1-Q4) build plate

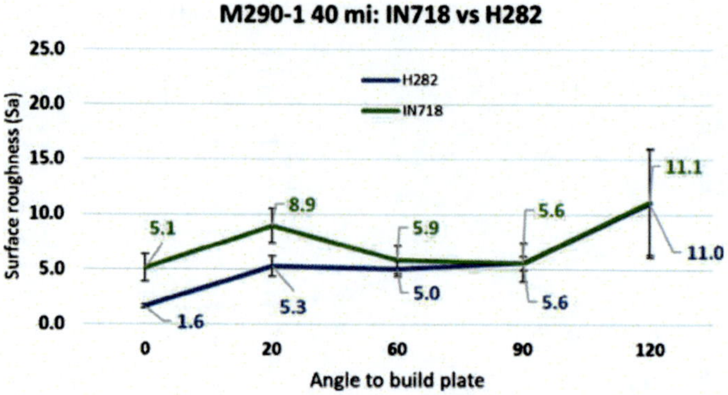

Fig. 19 Comparison of surface roughness results for IN718 and H282 printed on the M290-1

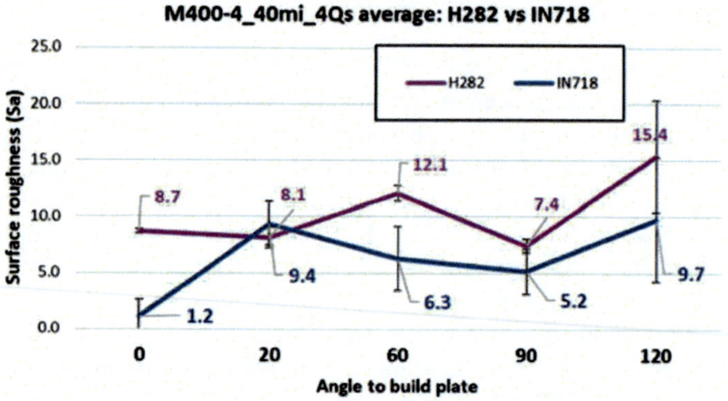

Fig. 20 Comparison of surface roughness results for IN718 and H282 printed on the M400-4 (average of all coupons in all 4 quadrants)

Surface roughness on the M290-1 for these two materials are very similar; however, the upskin regions have a maximum roughness difference of 3.6 μm on the 20° surface. The M400-4 results are not as similar, and do not appear to follow any trend. For this machine, the maximum difference in surface roughness measurements between these IN718 and H282 are 7.5 μm at the 0° surface. The only data point in which H282 surface roughness is lower than 718 is at the 20° surface with a difference of 1.3 μm (Figs. 19 and 20).

Conclusions

In summary, the laser incidence angle relative to the surface normal is a universal descriptor of surface orientation to the build plate, independent of location in the build plate, for any LPBF machine. This laser incidence angle has a significant effect on surface roughness with a consistent parabolic effect with high upskin and downskin roughness with a minimum surface roughness value for vertical faces. Upskin surfaces facing the laser have a higher roughness than those facing away from the laser—a trend understood by the staircase effect. The trend is reversed for downskin roughness, predominantly because of laser angle on particle adhesion and material extensions. The same roughness trends are observed across different materials—IN718 and H282 as well as in different LPBF machines with single and multi-laser systems. The data reported in this paper can be used for establishing fundamental relationships of laser incidence angles to surface roughness across LPBF machines for future process stability evaluations and process transfer check across machines.

Acknowledgements We would like to acknowledge Ross Henrys from the Siemens Innovation Center for LPBF prints in M290-1 and John Rapp for LPBF prints in M400-1. We also would like to acknowledge Michael Pepperman from Siemens Energy, Casselberry Labs for his diligence in systematic surface roughness measurements using non-contact profilometry.

References

1. F. Cabanettes, A. Joubert, G. Chardon, V. Dumas, J. Rech, C. Grosjean, and Z. Dimkovski, "Topography of as built surfaces generated in metal additive manufacturing: A multi scale analysis from form to roughness," Precision Engineering, vol. 52, pp. 249–265, 2018/04/01/ 2018, https://doi.org/10.1016/j.precisioneng.2018.01.002.
2. C. Emmelmann, J. Kranz, D. Herzog, and E. Wycisk, "Laser Additive Manufacturing of Metals," in Laser Technology in Biomimetics: Basics and Applications, V. Schmidt and M. R. Belegratis Eds. Berlin, Heidelberg: Springer Berlin Heidelberg, 2013, pp. 143–162.
3. Y. Liu, Y. Yang, S. Mai, D. Wang, and C. Song, "Investigation into spatter behavior during selective laser melting of AISI 316L stainless steel powder," Materials & Design, vol. 87, pp. 797–806, 2015/12/15/ 2015, https://doi.org/10.1016/j.matdes.2015.08.086.
4. M. Simonelli, C. Tuck, N. T. Aboulkhair, I. Maskery, I. Ashcroft, R. D. Wildman, and R. Hague, "A Study on the Laser Spatter and the Oxidation Reactions During Selective Laser Melting of 316L Stainless Steel, Al-Si10-Mg, and Ti-6Al-4V," Metallurgical and Materials Transactions A, vol. 46, no. 9, pp. 3842–3851, 2015/09/01 2015, https://doi.org/10.1007/s11661-015-2882-8.
5. G. Strano, L. Hao, R. M. Everson, and K. E. Evans, "Surface roughness analysis, modelling and prediction in selective laser melting," Journal of Materials Processing Technology, vol. 213, no. 4, pp. 589–597, 2013/04/01/ 2013, https://doi.org/10.1016/j.jmatprotec.2012.11.011.
6. S. Sun, M. Brandt, and M. Easton, "2 - Powder bed fusion processes: An overview," in Laser Additive Manufacturing, M. Brandt Ed.: Woodhead Publishing, 2017, pp. 55–77.
7. L. Carter, M. Attallah, and R. Reed, "Laser Powder Bed Fabrication of Nickel-Base Superalloys: Influence of Parameters; Characterisation, Quantification and Mitigation of Cracking," 2012, pp. 577–586.
8. Q. Jia and D. Gu, "Selective laser melting additive manufacturing of Inconel 718 superalloy parts: Densification, microstructure and properties," Journal of Alloys and Compounds, vol. 585, pp. 713–721, 2014/02/05/ 2014, https://doi.org/10.1016/j.jallcom.2013.09.171.
9. I. Koutiri, E. Pessard, P. Peyre, O. Amlou, and T. De Terris, "Influence of LPBF process parameters on the surface finish, porosity rate and fatigue behavior of as-built Inconel 625

parts," Journal of Materials Processing Technology, vol. 255, pp. 536–546, 2018/05/01/ 2018, https://doi.org/10.1016/j.jmatprotec.2017.12.043.
10. S. Pal, D. Igor, and T. Brajlih, "Physical behaviors of materials in selective laser melting process," 2018.
11. Grimm T, Wior G, Witt G. Characterization of typical surface effects in additive manufacturing with confocal microscopy. Surf Topogr: Metrol Prop 2015;3:014001.
12. Moylan S. Progress toward standardized additive manufacturing test artifacts. Achieving precision tolerances in additive manufacturing. In: ASPE2015 spring topical meeting. 2015. ASPE.
13. Moylan S, Slotwinski J, Cooke A, Jurrens K, Donmez MA. Proposal for a standardized test artifact for additive manufacturing machines and processes. In: Proceedings of the 2012 annual international solid freeform fabrication symposium. 2012. p. 6–8.
14. S. Rott, A. Ladewig, K. Friedberger, J. Casper, M. Full, J.H. Schleifenbaum, "Surface roughness in laser powder bed fusion–Interdependency of surface orientation and laser incidence", Additive Manufacturing 36 (2020) 101437.
15. S. Kleszczynski1, A. Ladewig, K. Friedberger, J. zur Jacobsmühlen, D. Merhof , G. Witt, "Position dependency of surface roughness in parts from laser beam melting systems", Conference: Proceedings of the 26th Internation Solid Free Form Fabrication (SFF) Symposium, January 2015.
16. O. Nazik, "Process development of the Merl72 material for an L-PBF multi-laser machine", M. Eng Thesis, Technical University, Berlin.
17. J. Snyder, K. Thole, "Tailoring Surface Roughness Using Additive Manufacturing to Improve Internal Cooling", Journal of Turbomachinery, Vol. 142 / 071004–1, July 2020.

Influence of Morphology and Size Distribution of Haynes 230 Particles on the Powder Spreading Behavior and Performance on Selective Laser Melting

Peng Zhang, Rui Wang, Shaoming Zhang, Zhongnan Bi, Xizhen Chen, Hailong Qin, and Guangbao Sun

Abstract There is increasing interest in the use of additive manufacturing (AM) for superalloys due to their broad applications in the aerospace industry. As the raw material, high-quality metal powder is very important for successful powder bed fusion in AM. In this work, Haynes 230 powders manufactured by Vacuum Induction Melting Inert Gas Atomization (VIGA) and Plasma Rotating Electrode Process (PREP) were characterized and compared. Results demonstrated P-230 powder is superior to V-230 powders. P-230 powder exhibits a lower Hausner coefficient and better flowability. Meanwhile, attributed to superior sphericity and fewer satellite particles, lower dynamic angle of repose and cohesive index were achieved by P-230 powder, which means better dynamic flow and spreading of the powder during the recoating process of selective laser melting (SLM). In terms of their performance on SLM, the powder bed density of the P-230 powder is higher, and samples prepared with P-230 powder exhibited higher relative density. Although both V-230 and P-230 samples were all HIPed at 1200 °C for 4 h, P-230 samples revealed higher yield strength at room temperature.

Keywords Ni-based superalloy · Additive manufacturing · Selective laser melting · Powder morphology · Powder size distribution · Powder spreading

Zhang and Wang: These authors contributed equally to the paper.

P. Zhang · S. Zhang
Central Iron and Steel Research Institute, Beijing 100081, China

R. Wang (✉) · Z. Bi · X. Chen · H. Qin · G. Sun
Beijing Key Laboratory of Advanced High Temperature Materials, Central Iron and Steel Research Institute, Beijing 100081, China
e-mail: wangrui@cisri-gaona.com.cn

P. Zhang · R. Wang · Z. Bi · X. Chen · H. Qin · G. Sun
Gaona Aero Material Co, Ltd, Beijing 100081, China

E. A. Ott et al. (eds.), *Proceedings of the 10th International Symposium on Superalloy 718 and Derivatives*, The Minerals, Metals & Materials Series,
https://doi.org/10.1007/978-3-031-27447-3_32

Introduction

In recent years, there have been many studies on nickel-based superalloys formed by the selective laser melting (SLM) process [1–4]. For solid solution strengthened superalloys, the most studied alloys include Hastelloy X [5–8] and IN625 [9]. The research mainly focuses on the influence of process parameters on the formation of pores and cracks and the cracking mechanism. For the precipitation hardening superalloys, the alloys currently studied mainly include IN718 [10, 11], IN738LC [12–15], CM247LC [16, 17], etc. The research focus is mainly on the distribution characteristics of the γ' and γ'' phases and their effects on mechanical properties, and the impact of element content on crack suppression. Haynes 230 alloy is a solid solution strengthened Ni-based superalloy used for combustors and flame tubes. Due to the demand in turbine engines for aviation and power generation, much research has been carried out on SLM of Haynes 230 alloy. Bauerd et al. studied the influence of different process parameters on apparent density, porosity, and cracks of SLMed Haynes 230 alloy, and compared the mechanical properties of SLMed samples with cast and forged materials [18]. Moneni studied the effects of powder layer thickness, laser spot size, and laser power on the width and depth of the molten pool in SLMed Haynes 230 alloy [19]. Yang et al. focused on the evolution of the microstructure of Haynes 230 alloy in the as-formed and heat-treated states [20].

Powder is the raw material of the SLM process, and the characteristics of the powder are one of the important factors affecting the powder packing density, sample defects, and mechanical performance [21–23]. By comparing the pores and cracks of samples prepared by 8 different batches of IN738LC powder, Engeli et al. found that under the same process parameters, the porosity of the samples prepared by gas atomization was lower than 1.2%, while the porosity of the samples prepared by water atomization is 3.5%. In addition, the powder with lower Hausner value has better flowability, and the formed sample has lower porosity and fewer micro cracks [24]. Smith et al. analyzed the effect of 16 different powders on microstructure and properties of samples built using SLM process. The result revealed a wide range of building qualities and microstructures despite all powder lots falling within the 718 chemical specification and being processed using the same prescribed build and heat treatment parameters. In particular, a strong correlation of the amount of carbon in the powder with the final grain size and strength of the heat-treated 718 parts [25]. In the Haynes 230 alloy, due to the high content of Cr and W elements and the high viscosity of the alloy, it is vulnerable to defects during the SLM process, The influence of powder characteristics on density, microstructure, and mechanical properties of SLMed specimen is not clear yet. Therefore, this paper will analyze the characteristics of Haynes 230 powders prepared by two different processes and their influence on samples formed by SLM process.

Experiment Method

In this paper, As-cast pre-alloyed ingots are obtained through the vacuum melting process, and then processed into rods of specified dimensions. The rods are remelted to prepare metal powder through Plasma Rotating Electrode Process (PREP) and Vacuum Induction Melting Inert Gas Atomization (VIGA). The powders prepared by the two processes were sieved to a particle size of 15–53 μm for SLM processing. The chemical compositions (Table 1) of the two powders were analyzed by Inductively Coupled Plasma Optical Emission Spectroscopy (ICP-OES). The particle size distribution of the powder was tested in the liquid dispersion state using a laser particle size distribution analyzer at 20°C ambient temperature and 40% humidity. The Span value of the powder particle size distribution is given by D10, D50, and D90, and the calculation formula is as follows:

$$Span = \frac{D90 - D10}{D50}$$

The morphology of the powder was observed by a Hitachi S4800 scanning electron microscope. The aspect ratio of the powder particles was calculated by ImagePro. The sphericity of the powder was calculated by the image method. The flowability of the powders was measured by a Hall flow meter.

In order to characterize the impact of internal friction between powder particles and the three forces (van der Waals force, electrostatic force, and capillary force) on the dynamic flowability of the two powders and the spreadability on the powder bed under dynamic recoating conditions, three powder rheology characterization methods were chosen: (1) GranuHeap instrument was used to measure powder static packing angle, cohesion index, and packing height; (2) GranuDrum instrument was used to measure powder dynamic flow angle and dynamic cohesion index; (3) GranuPack instrument was used to measure the bulk density, tap density. Hausner ratio of the powder is obtained by the following formula:

$Hausner = \frac{\rho_{Tap}}{\rho_{Bulk}}$, where ρ_{Tap} is tapped density and ρ_{Bulk} is bulk density.

The density of the powder bed is evaluated in situ by designing an enclosed hollow model that can seal powders on the powder bed inside during the printing process [23], which is shown in Fig. 1a. The enclosed model consists of an inner cavity and an outer wall. The inner diameter of the cavity is 20 mm. The wall thickness is 2 mm.

Table 1 Chemical composition of Haynes 230 Powders (wt.%)

Elements	C	Cr	B	Al	Fe	Mn	Mo
P-230	0.056	20.82	0.0052	0.42	1.79	0.58	1.95
V-230	0.053	20.60	0.0050	0.42	1.81	0.49	2.09
Element	Ti	Co	Si	La	W	Ni	/
P-230	0.004	0.27	0.41	0.005	13.35	Bal	/
V-230	0.013	0.31	0.42	0.0005	13.38	Bal	/

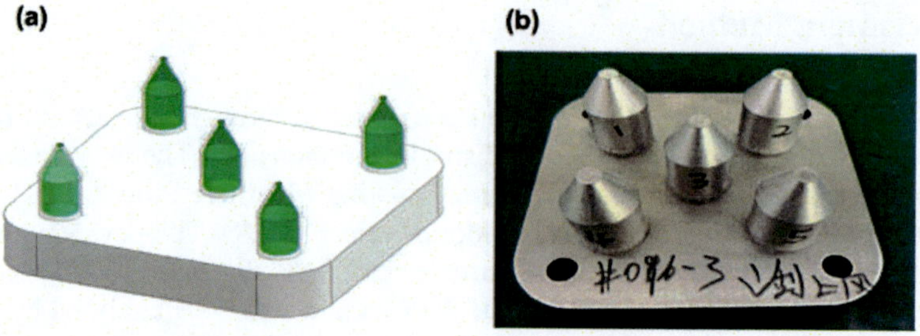

Fig. 1 Powder bed density analysis model (**a**) and SLMed sample (**b**)

The height is 38 mm. The upper taper angle is 60°, and the top port of the model is sealed with a 0.2 mm sheet. The model is distributed on the 4 corners and the center of the forming substrate as shown in Fig. 1a. The forming parameter: power 180W, scanning speed 700 mm/s, hatching spacing 0.09 mm, layer thickness 0.03 mm. The model after forming is shown in Fig. 1b. After formation, open the upper sheet of the model, weigh the powder weight M_c of the inner cavity, and then measure the volume V_c of the model cavity by injecting water into the cavity. The density of the powder bed (PBD) is calculated by the following formula:

$$PBD = \frac{M_c}{V_c}$$

SLM processing was conducted on an SLM 125HL machine (SLM Solutions GmbH, Germany) equipped with a 400 W fiber laser (IPG) under an argon atmosphere. 29 different processing parameters were used to build samples: laser power of 100~260 W, scanning speed of 300~1900 mm/s, hatching space of 90 μm, layer thickness of 30 μm, scanning rotation angle of 67°, strip scanning strategy with strip size 5 mm, and baseplate temperature of 200 °C. Following fabrication, samples were Hot Isostatically Pressed (HIP) at 1200 °C/140 MPa for four hours to remove any microcracks, large voids, and porosity.

The relative density of both as-built and HIPed samples was measured using the Archimedes method. For metallographic studies, both as-built and HIPed samples were mounted, mechanically grinded successively from 500 to 4000 Grit, and polished with diamond suspension from 3 to 0.5 μm and finally with OPS colloidal silica suspension. To reveal the grain boundary, the polished samples were etched for a few seconds using a 15 g copper sulfate +50 ml hydrochloric acid +3.5 ml sulfuric acid etchant. A JEOL JSM-7200F FEG scanning electron microscope (SEM) equipped with energy dispersive X-ray spectroscopy (EDS) and electron backscatter diffraction (EBSD) system from Oxford Instrument, was operated at 25 kV to examine the microstructure and texture. Note that all the SEM images and EBSD mappings are taken perpendicular to the build direction.

The tensile testing was performed using a Zwick/Roell Z100 machine with a maximum load of 100 KN at room temperature. The cross-head displacement speed was 1 mm/min before yielding and 5 mm/min after yielding. Round tensile bars with a gauge length of 25 mm and a gauge diameter of 5 mm were prepared by machining from $\Phi 13 \times 70$ mm SLMed cylinder bars. An extensometer was used to measure elastic strains and yield stress accurately.

Results and Discussion

Physical Properties of Powders

It can be seen from Table 2 that the span value of P-230 is lower than that of V-230, indicating that the particle size distribution of P-230 powder is more concentrated. Besides, the D10, D50, and D90 values of P-230 powder are larger than that of V-230, which means the average particle size of P-230 powder is bigger.

The morphology and sphericity of the two powders are demonstrated in Figs. 2–3. It can be seen that the P-230 powder is a smooth spherical particle with no satellite particle on the surface, while most of the V-230 powder particles are spherical, there are a certain number of satellite particles and irregular-shaped powder, which is related to the gas atomization process. In the process of gas atomization, the gas and the molten metal have a strong interaction, and the metal film is destabilized and broken to form droplets. In the process of flight, droplets collide and combine under the impact of turbulent airflow to form irregular particles. In addition, due to the different solidification and spheroidization rates of large and small droplets, small particles that solidify first tend to adhere to the surface of large particles that solidify later and form satellite powder. According to statistical analysis, the average sphericity of the P-230 powder is 0.97, which is higher than the average sphericity of the V-230 powder of 0.89. In addition, the sphericity distribution of P-230 powder is concentrated in a narrow range of 0.95–1.0, while the sphericity of V-230 powder fluctuates in a wide range of 0.5–1.0, which is mainly related to the irregular-shaped powders and satellite particles existing.

It was found that the surface morphology of the two powder particles showed a typical rapid solidification structure that obvious solidified dendrites and cellular structures can be observed (Fig. 4). The surface of the P-230 powder particle presents almost uniform cellular structure, while the surface of the V-230 powder particle exhibits coarse dendrites in addition to cellular structure. The secondary dendrite arms are more pronounced.

Table 2 Powder particle size distribution

Samples	D10 (μm)	D50 (μm)	D90 (μm)	Span
P-230	35.25	48.02	66.59	0.63
V-230	22.39	36.22	58.18	0.99

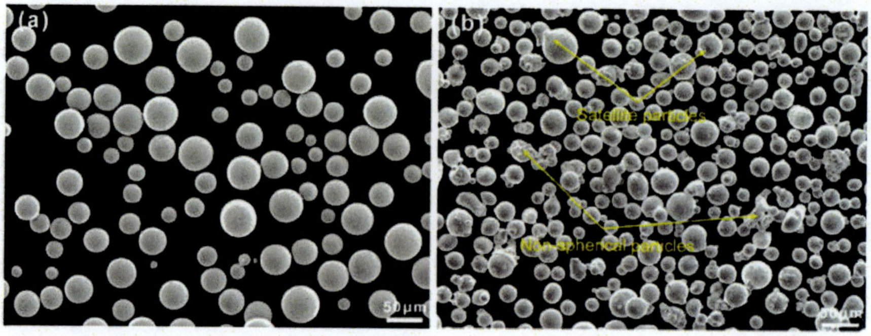

Fig. 2 Powder Morphology **a** P-230; **b** V-230

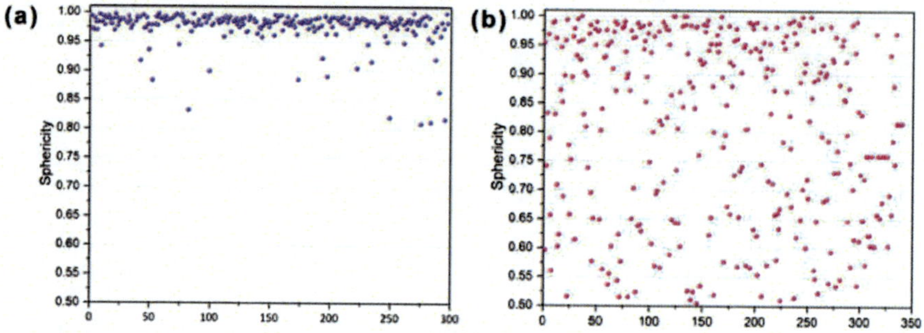

Fig. 3 Sphericity of powders **a** P-230; **b** V-230

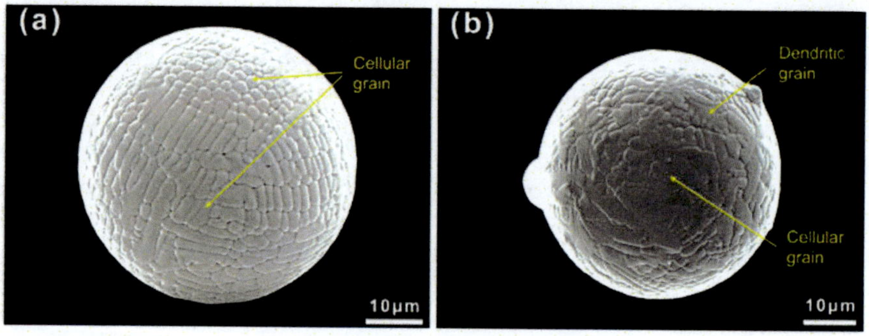

Fig. 4 Surface morphology of powder particles **a** P-230; **b** V-230

Figure 5 compares the static packing behavior of the two powders. On the one hand, the consistency of the static angle of repose of P-230 powder is higher than that of V-230, where the static angle of repose of P-230 powder fluctuates between 26–27° and that of V-230 powder fluctuates in the range of 42–44°. On the other hand, the average angle of repose of the P-230 powder (26.77°) is significantly lower than that

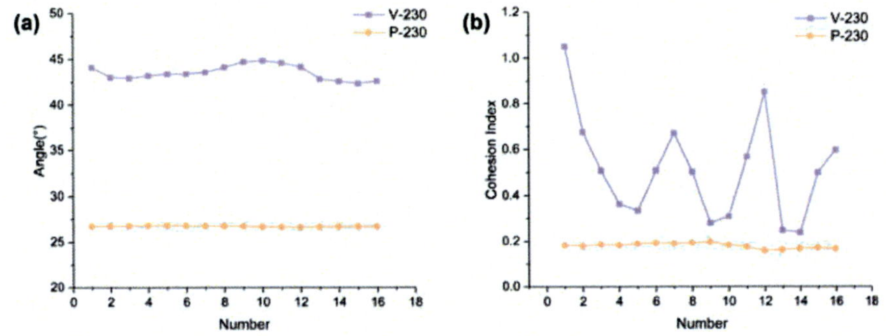

Fig. 5 **a** Static angle of repose **b** Static cohesion index

of the V-230 powder (43.54°), indicating that the inter-particle friction and adhesion of the P-230 powder are lower than those of the V-230 powder. From Fig. 5b, the average static cohesion index of P-230 powder, namely 0.18, is significantly lower than that of V-230 powder, which is 0.51. The excellent static packing behavior of P-230 powder is attributed to the smoother particle surface and regular and spherical particles. The existence of the irregular particles and satellite powder in V-230 powder leads to the increase of the roughness of the particle surface and the friction between the particles, which eventually causes the increase of the static stacking angle and cohesion index.

To further characterize the flow behavior of the two powders in the state of recoating, the dynamic flow behavior of the two powders was measured, and the results of the dynamic flow angle and cohesion index are shown in Fig. 6. It can be seen from Fig. 6a that the dynamic flow angles of both P-230 and V-230 powders increase with the increase of the drum speed, indicating that with the increase of the shear stress between particles, the flowability of the powders decreases, and both powder particles exhibit shear thickening behavior. It is obvious that the dynamic flow angles of the P-230 powder at different rotational speeds are smaller than those of the V-230 powder, indicating that the dynamic flowability of the P-230 powder is better than that of the V-230 powder, which is consistent with the angle of repose results. The dynamic cohesion index of P-230 is significantly lower than that of V-230 powder (Fig. 6b) as well, indicating that the spreadability of P-230 powder on the powder bed should be better than that of V-230 powder, which could lead to a uniform and dense powder layer on the powder bed, and is eventually beneficial to achieve high relative density of the printed specimen.

The results shown in Table 3 and Fig. 7 further reveal the influence of powder morphology, flowability, and cohesion index on powder bulk density, tap density, and powder bed density. Compared with V-230, P-230 powder has higher bulk density and lower Hausner ratio, meaning that P-230 powder has better flowability, which is also consistent with the results obtained by the dynamic flow angle of the powders. Through the Hall flow rate test, the flowability of P-230 is 11.8 s/50 g, and the flowability of V-230 is 14.3 s/50 g. The powder bed densities of P-230 and V-230

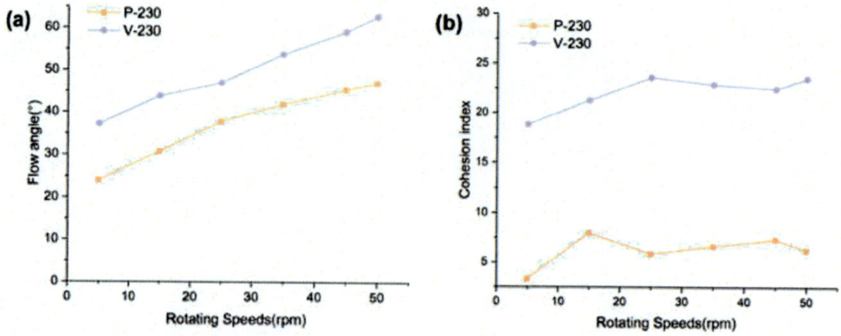

Fig. 6 **a** Dynamic flow angle **b** Dynamic cohesion index of two powders

Table 3 Comparation of various densities of two powders

Samples	Bulk density (g/cm^3)	Tap density (g/cm^3)	Hausner ratio	Powder bed density (g/cm^3)
P-230	5.18	5.59	1.08	5.68
V-230	4.38	5.19	1.18	5.26

are 5.68 g/cm^3 and 5.26 g/cm^3 respectively (Fig. 7b). Higher powder bed density of P-230 indicates that during the powder stacking process, the proportion of voids between particles is less, and under the condition of the same laser energy input, the molten pool can fill the voids better, which is easier to form a uniform and dense melt channel, and eventually higher relative density of specimen [26, 27]. In addition, it is found that the powder bed densities of P-230 and V-230 powders are both higher than the tap densities. This can be explained that during the powder recoating process when the scraper pushes the powder to move, the powder is subjected to the shearing force of the scraper, which generates compressive stress on the powder. Under the action of the compressive stress, the powder bed is further densified, resulting in the powder bed density exceeding the tap density of the powder.

Microstructures of Haynes 230 SLMed with Two Powders

The XY-plane microstructures of samples SLMed with P-230 and V-230 powders are shown in Figs. 8, 9, 10. Optical images reveal that after HIP, there are almost no visible pores and defects in both samples. HIP at 1200°C can act as a solution treatment as well. After HIP, carbides are formed along grain boundaries and inside grains. The average size of carbides in the P-230 sample is larger than that of V-230 sample. Besides, fewer carbides are formed inside grains in the P-230 sample. According to EBSD analysis, the average grain size of the P-230 sample is larger than the V-230 sample in as-build state. The inverse pole figure of samples shows

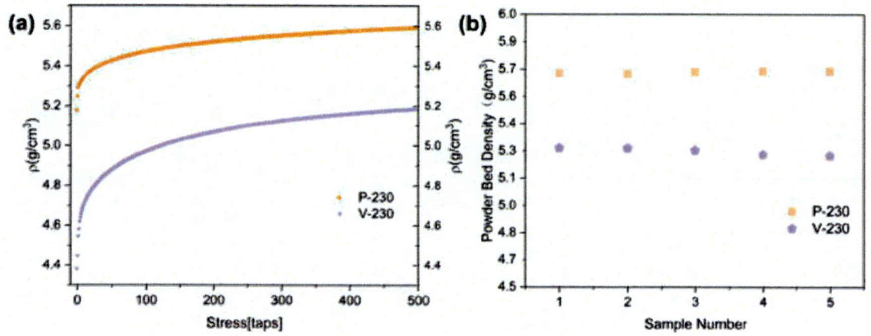

Fig. 7 **a** Tap density curves **b** Powder bed density of two powders

that the grains of P-230 has a strong tendency to align along (001) plane while the grains of the V-230 has a slight tendency to align along (101) plane. Recrystallization occurred after HIP, indicated by reduced grain sizes and randomized grain orientation. However, the recrystallization behavior of samples from two powders are diverse. Full recrystallization occurred for the V-230 sample, as the grain size is much smaller, grain morphology is more equiaxed, and a nearly random grain orientation distribution is observed. Besides, a large quantity of twin crystals is formed. While partial recrystallization occurred for the P-230 sample, although the grain size is smaller, the grain morphology is still elongated, and a slight tendency to align along (101) plane is observed .

As-built samples prepared by SLM typically exhibit large numbers of dislocations that are uniformly distributed across the sample. And large thermal gradient of the SLM process easily generates high residual stress and local strain. These are all requirements to trigger recrystallization. Recrystallization also requires a minimum temperature for the necessary atomic mechanisms to occur. In this case, both P-230 and V-230 samples experienced the same 1200°C HIP process, therefore, the thermal condition for recrystallization is the same. It is reasonable to suspect that

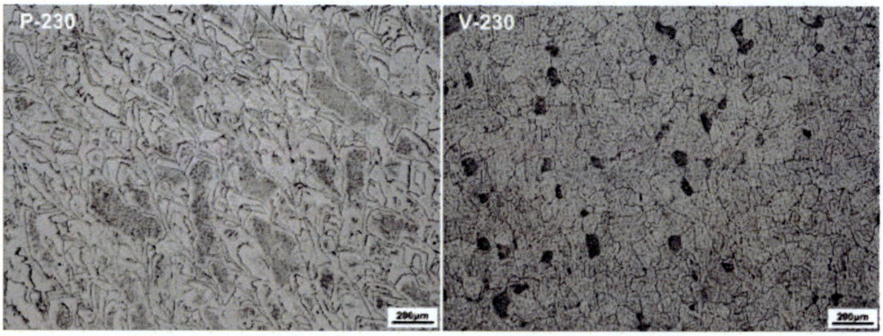

Fig. 8 Optical images of HIPed samples of XY-plane

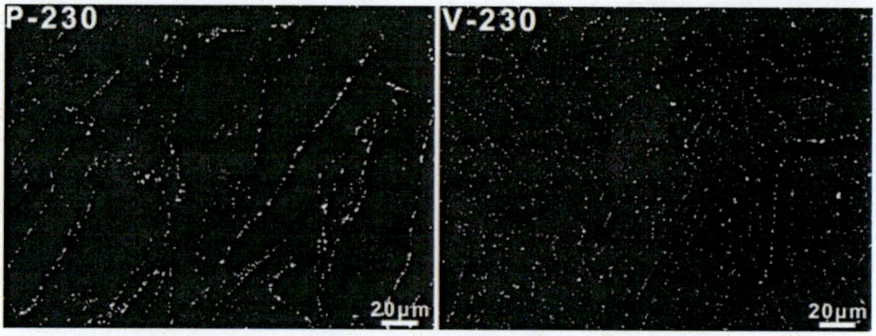

Fig. 9 Back scattered images of HIPed samples of XY-plane

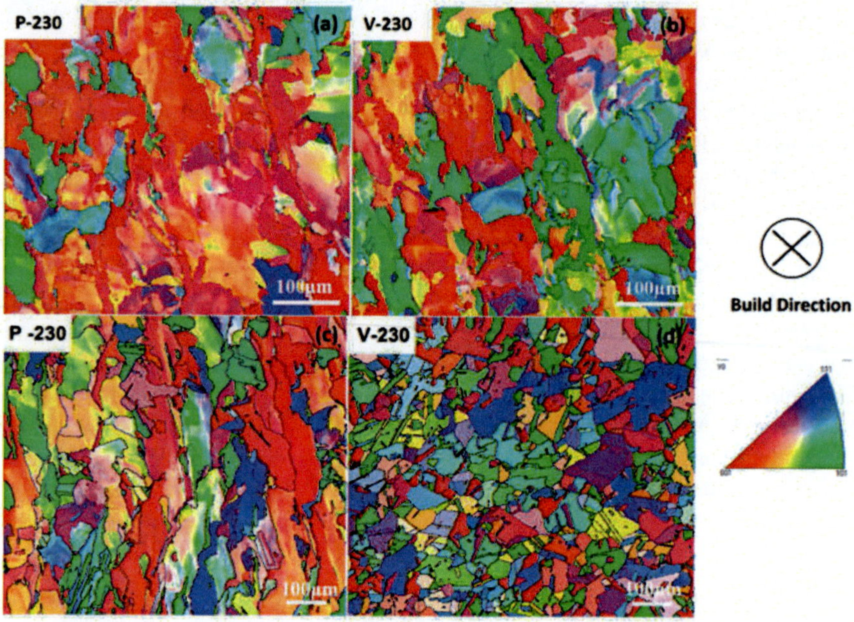

Fig. 10 EBSD crystallographic orientation map of the XY-plane, using a Z-based (build direction) projected IPF: **a b** As-built samples **c d** HIPed samples

the dislocations, residual stress, or local strain conditions are different for as-built P-230 and V-230 samples, which leads to distinct recrystallization behaviors. The exact reason needs further investigation.

Relative Density

The relative densities of samples SLMed with different process parameters using V-230 and P-230 powders respectively are illustrated in Fig. 11. It is clear that P-230 samples exhibit higher relative density than V-230 samples in both as-built state and HIPed state. In terms of as-built state, the relative densities of most V-230 samples fluctuate between 98 and 98.6%, while the relative densities of most P-230 samples are near or above 99%. For the #5 sample (100W, 900 mm/s), both two powders showed significantly low relative density, which is due to a large lack of fusion voids caused by insufficient energy density (41.2 J/mm^3). After HIP, the relative densities of samples are all increased and most P-230 samples achieve near 99.9%, while all V-230 samples achieve above 99%. The differences between samples from various process parameters are also reduced and the curves become flattened for both powders.

The relative density results of the P-230 and V-230 samples are closely related to powder properties. As mentioned above, the physical characteristics difference between P-230 and V-230 powders contribute to their dissimilar static and dynamic stacking behaviors, which eventually reflect on the powder bed densities. The higher relative densities of P-230 samples are a direct impact of the higher powder bed density of P-230 powder. With higher powder bed density, the energy of laser beam can be absorbed more efficiently which results in less lack of fusion. More molten metal and fewer voids between powders contribute to fewer defects. HIP is a common process used to reduce the micro-cracks and porosity of AM materials. The effect of HIP is proved by the improvement of relative densities compared to as-built state. However, the relative density gap between V-230 and P-230 powder are not closed after HIP, which means not all defects are eliminated in V-230 samples. Normally a certain amount of hollow powder is present in the VIGA powder [28], and the gas trapped inside could results in porosity during SLM process, but these pores are impossible to be fully closed by HIP. Parameter 13# (180W, 700 mm/s, 95.24 J/mm^3)

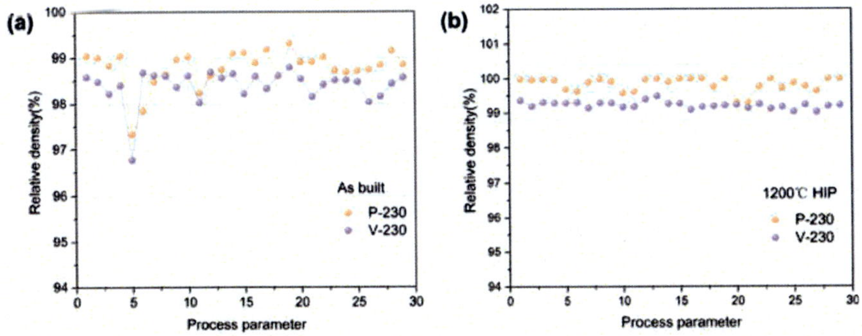

Fig. 11 Relative densities of samples SLMed by different process parameters using two powders: **a** As-built samples **b** HIP samples

was chosen, which corresponds to relative densities of 99.49% for VIGA powder and 99.95% for PREP powder, to build mechanical specimens.

Mechanical Properties

Figure 12 shows the quasi-static room temperature tensile true stress–strain curves of the samples SLMed with two powders. It can be found that the mechanical properties of the samples prepared with the two powders using the same forming parameters and HIP process are different. Among them, the yield stress of P-230 is 448 MPa, which is higher than that of V-230, namely 351 MPa, but the tensile stress and plastic elongation of the P-230 sample are lower than those of V-230. The tensile stress of P-230 is 860 MPa, and the elongation is 39%, while the tensile stress of V-230 is 892 MPa and the elongation is 46%.

The difference of mechanical properties is mainly caused by two factors: on the one hand, better physical characteristics of the P-230 powder results in higher powder bed density of the P-230 powder than that of the V-230 during the powder recoating process. And higher powder bed density eventually results in higher relative densities of samples printed. During the stretching process, the fully dense sample has more load bearing capacity by having more load bearing area, resulting in higher yield stress of the P-230 sample; on the other hand, the P-230 sample partially maintains the as-printed grain morphology, while the V-230 sample has obvious recrystallization after the same HIP process, and a large number of twin crystals are regenerated. After recrystallization, the residual stress in the sample is fully reduced, and the number of linear defects such as dislocations is eliminated, resulting in the material being easily deformed plastically, thus the yield strength of the V-230 sample is decreased. However, after full recrystallization, grain size is greatly reduced and a large number

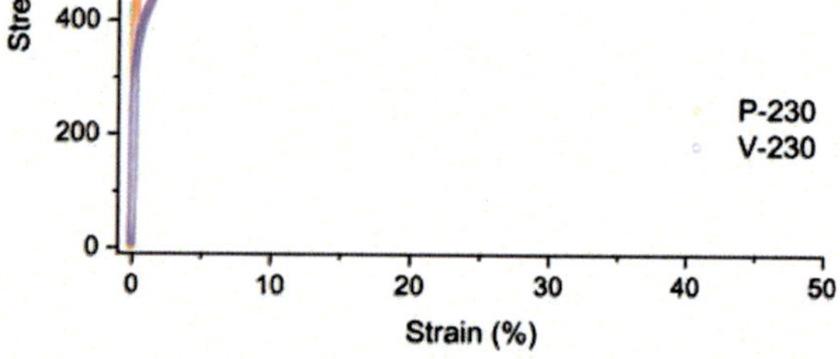

Fig. 12 Tensile true stress–strain curves of samples SLMed with two powders

of twin boundaries are formed. The increased grain boundaries and twin boundaries can accommodate more dislocation slip and plastic deformation, which contributes to the improvement of the plasticity of the V-230 sample.

Conclusion

(1) The characteristics of the powder have an important influence on the powder packing density and powder bed density. Compared with VIGA, the Haynes 230 powder prepared by PREP has a narrow particle size distribution, high sphericity, smooth particle surface, and lower Hausner ratio, which leads to decrease of cohesion and friction between the powder particles. These factors improve the dynamic flowability and spreadability of the powder during the powder spreading process, and further increase the powder bed density of the P-230.

(2) Under the same process condition and HIP process, the sample formed by P-230 powder has higher relative density and less recrystallization, which leads to higher yield strength of the material. Meanwhile, the sample from V-230 has obvious recrystallization, and grains are refined and twinning microstructure appears, which results in a decrease of the yield strength and an increase in plasticity.

Acknowledgements This work was supported by the National Key Research and Development Program of China (Grant No. 2021YFB3702500) and National Natural Science Foundation of China (Grant No. 12205055).

References

1. Sanchez S., Smith P., Xu Z., et al. (2021). Powder Bed Fusion of nickel-based superalloys: A review. International Journal of Machine Tools and Manufacture, 165: 103729. https://doi.org/10.1016/j.ijmachtools.2021.103729
2. Shahwaz M., Nath P., Sen, I. (2022). A critical review on the microstructure and mechanical properties correlation of additively manufactured nickel-based superalloys. Journal of Alloys and Compounds, 907: 164530. https://doi.org/10.1016/j.jallcom.2022.164530
3. Cobbinah P. V., Nzeukou R. A., Onawale O. T., et al. (2020). Laser powder bed fusion of potential superalloys: a review. Metals, 11(1): 58. https://doi.org/10.3390/met11010058
4. Tian Y., Muñiz–Lerma J. A., et al. (2017). Nickel-based superalloy microstructure obtained by pulsed laser powder bed fusion. Materials Characterization, 131: 306–315. https://doi.org/10.1016/j.matchar.2017.07.024
5. Han Q., Gu Y., Setchi R., et al. (2019). Additive manufacturing of high-strength crack-free Ni-based Hastelloy X superalloy. Additive Manufacturing, 30: 100919. https://doi.org/10.1016/j.addma.2019.100919

6. Calignano F., Minetola P. (2019). Influence of process parameters on the porosity, accuracy, roughness, and support structures of hastelloy X produced by laser powder bed fusion. Materials, 12(19): 3178. https://doi.org/10.3390/ma12193178
7. Marchese G., Basile G., Bassini E., et al. (2018). Study of the microstructure and cracking mechanisms of hastelloy X produced by laser powder bed fusion. Materials, 11(1): 106. https://doi.org/10.3390/ma11010106
8. Han Q., Mertens R., Montero–Sistiaga, et al. (2018). Laser powder bed fusion of Hastelloy X: effects of hot isostatic pressing and the hot cracking mechanism. Materials Science and Engineering: A, 732: 228–239. https://doi.org/10.1016/j.msea.2018.07.008
9. Tian Z., Zhang C., Wang D., et al. (2019). A review on laser powder bed fusion of inconel 625 nickel-based alloy. Applied Sciences, 10(1): 81. https://doi.org/10.3390/app10010081
10. Mostafa A., Picazo Rubio I., Brailovski V., et al. (2017). Structure, texture and phases in 3D printed IN718 alloy subjected to homogenization and HIP treatments. Metals, 7(6): 196. https://doi.org/10.3390/met7060196
11. Huynh T., Mehta A., Graydon K., et al. (2022). Microstructural Development in Inconel 718 Nickel-Based Superalloy Additively Manufactured by Laser Powder Bed Fusion. Metallography, Microstructure, and Analysis, 11(1): 88–107. https://doi.org/https://doi.org/10.1007/s13632-021-00811-0
12. Harrison N. J., Todd I., Mumtaz K. (2015). Reduction of micro-cracking in nickel superalloys processed by Selective Laser Melting: A fundamental alloy design approach. Acta Materialia, 94: 59–68. https://doi.org/10.1016/j.actamat.2015.04.035
13. Vilanova M., Garciandia F., Sainz S., et al. (2022). The limit of hot isostatic pressing for healing cracks present in an additively manufactured nickel superalloy. Journal of Materials Processing Technology, 300: 117398. https://doi.org/10.1016/j.jmatprotec.2021.117398
14. Zhou W., Tian Y., Tan Q., et al. (2022). Effect of carbon content on the microstructure, tensile properties and cracking susceptibility of IN738 superalloy processed by laser powder bed fusion. Additive Manufacturing, 58: 103016. https://doi.org/10.1016/j.addma.2022.103016
15. Raza M. M., Lo Y. L. (2021). Experimental investigation into microstructure, mechanical properties, and cracking mechanism of IN713LC processed by laser powder bed fusion. Materials Science and Engineering: A, 819: 141527. https://doi.org/10.1016/j.msea.2021.141527
16. Boswell J. H., Clark D., Li W., et al. (2019). Cracking during thermal post-processing of laser powder bed fabricated CM247LC Ni-superalloy. Materials & Design, 174: 107793. https://doi.org/10.1016/j.matdes.2019.107793
17. Basak A., Das S. (2017). Microstructure of nickel-base superalloy MAR-M247 additively manufactured through scanning laser epitaxy (SLE). Journal of Alloys and Compounds, 705: 806–816. https://doi.org/10.1016/j.jallcom.2017.02.013
18. Bauer T., Dawson K., Spierings A. B., et al. (2015). Microstructure and mechanical characterisation of SLM processed Haynes® 230®. In Proceedings of the 26th annual international solid freeform fabrication symposium (pp. 813–822). Laboratory for Freeform Fabrication and University of Texas at Austin. https://hdl.handle.net/2152/89380
19. Momeni K. (2021). Sensitivity of laser powder bed fusion additive manufactured HAYNES 230 to composition and print parameters. Journal of Materials Research and Technology, 15: 6453–6463. https://doi.org/10.1016/j.jmrt.2021.11.080
20. Yang B., Shang Z., Ding J., et al. (2022). Investigation of strengthening mechanisms in an additively manufactured Haynes 230 alloy. Acta Materialia, 222: 117404. https://doi.org/10.1016/j.actamat.2021.117404
21. Li R., Shi Y., Wang Z., et al. (2010). Densification behavior of gas and water atomized 316L stainless steel powder during selective laser melting. Applied surface science, 256(13): 4350–4356. https://doi.org/10.1016/j.apsusc.2010.02.030
22. Riener K., Albrecht N., Ziegelmeier S., et al. (2020). Influence of particle size distribution and morphology on the properties of the powder feedstock as well as of AlSi10Mg parts produced by laser powder bed fusion (LPBF). Additive Manufacturing, 34: 101286. https://doi.org/10.1016/j.addma.2020.101286

23. Farzadfar S. A., Murtagh M. J., Venugopal N. (2020). Impact of IN718 bimodal powder size distribution on the performance and productivity of laser powder bed fusion additive manufacturing process. Powder Technology, 375: 60–80. https://doi.org/10.1016/j.powtec.2020.07.092
24. Engeli R., Etter T., Hövel S., et al. (2016). Processability of different IN738LC powder batches by selective laser melting. Journal of Materials Processing Technology, 229:484–491. https://doi.org/10.1016/j.jmatprotec.2015.09.046
25. Smith, T. M., Gabb, T. P., Kantzos, C. A., et al. (2021). The effect of composition on microstructure and properties for additively manufactured superalloy 718. Journal of Alloys and Compounds, 873, 159789. https://doi.org/10.1016/j.jallcom.2021.159789
26. Abu-Lebdeh, T., Damptey, R., Lamberti, V., et al. (2019). Powder packing density and its impact on SLM-based additive manufacturing. In TMS 2019 148th Annual Meeting & Exhibition Supplemental Proceedings (pp. 355–367). Springer, Cham. https://doi.org/10.1007/978-3-030-05861-6_33
27. German, R. M. (1992). Prediction of sintered density for bimodal powder mixtures. Metallurgical Transactions A, 23(5), 1455-1465. https://doi.org/10.1007/BF02647329
28. Guo, R. P., Xu, L., Zong, B. Y. P., et al. (2017). Characterization of prealloyed Ti–6Al–4V powders from EIGA and PREP process and mechanical properties of HIPed powder compacts. Acta Metallurgica Sinica (English Letters), 30(8), 735-744. https://doi.org/10.1007/s40195-017-0540-4

Tensile Performance of Direct Energy Deposited IN718 and Oxide-Dispersed Strengthened IN718

2023 Superalloy 718 & Derivatives

Kyle Rozman, Bruce Kang, and Ömer N. Doğan

Abstract With the need to increase electric power generation efficiencies in order to reduce carbon emissions, operational temperatures of turbine components must increase. Oxide-dispersion strengthening (ODS) has been proven to be useful in increasing the operational temperature of steels. In this study, yttria oxides (Y_2O_3) particles were mechano-chemical bonded (MCB) to IN718 powder to investigate the ability of IN718 to be strengthened by ODS particles at temperatures above 1000 °C. The Y_2O_3 infused IN718 powder was direct energy deposited using two parameter sets for testing. Plates of standard IN718 power were built using the same deposition parameters for reference. Specimen blanks were cut from plates and tested as built, with 1080 and 1200 °C homogenization heat treatments. The initial microstructure for all four conditions was dendric with significant Mo, Nb, and Ti segregation. Homogenization was observed to break up the initial large Mo and Nb precipitates and form Mo–, Nb–, and Ti–rich spherical precipitates. The 1080 °C homogenization heat treatment retained the dendritic structure, while the 1200 °C was observed to partially recrystallize the grain structure. Minor changes in the tensile properties were observed at 1050 °C by additions of Y_2O_3. This suggests that with further optimization, IN718 may be utilized at temperatures above 1000 °C with ODS methods.

Keywords IN718 · Oxide-dispersed strengthened · Microstructure · Tensile properties

---

K. Rozman (✉) · Ö. N. Doğan
National Energy Technology Laboratory, 1450 Queen Avenue SW, Albany, OR 97321, USA
e-mail: kyle.rozman@netl.doe.gov

K. Rozman
NETL Support Contractor, 1450 Queen Avenue SW, Albany, OR 97321, USA

B. Kang
Mechanican and Aerospace Engineering Department, West Virginia University, Morgantown, WV 26506, USA

E. A. Ott et al. (eds.), *Proceedings of the 10th International Symposium on Superalloy 718 and Derivatives*, The Minerals, Metals & Materials Series,
https://doi.org/10.1007/978-3-031-27447-3_33

Introduction

Additive manufacturing involves fusing metal powder layer-by-layer to build up the desired geometry. These parts typically have complex microstructures and high-residual stresses. Researchers have shown coating the powder with yttria oxides (Y_2O_3) prior to fusing can improve the strength of alloys. Utilization of oxide dispersion strengthening (ODS) is a relatively recent development. Much research has been done in steels, however less research has been published in Nickel base alloys. Smith et al. [1] used an ultrasonic methodology to impregnate NiCoCr powder with 1.0 wt% Y_2O_3. Smith et al. [1] showed a > 20% improvement in ultimate tensile strength (UTS) and > 50% improvement in the ductility of the alloy at 1093 °C. This significant increase in mechanical properties was found by Smith et al. [1], is encouraging for further research in ODS of nickel-base alloys. Other researchers have shown mild tensile improvements by additions of Y_2O_3 ODS to commercially available alloys. Song et al. [2] used ball milling to impregnate Inconel 718 (IN718) powder with 1.0 wt% Y_2O_3. Song et al. [2] showed a 10% improvement in yield and UTS, and a −24% loss of ductility for the alloy at 650 °C. Guo et al. [3] showed a 2.6% increase in UTS of IN738LC at 850 °C with the addition of 0.05 wt% Y_2O_3 also utilizing ball milling. Luu et al. [11] studied IN718 with 0.4−1.5 wt% of Y_2O_3 finding peak strength at a composition of 1.0 wt% Y_2O_3 at room temperature.

In addition to the strength improvement offered by Y_2O_3 additions, significant changes to the microstructure of IN718 and other superalloys have been reported. The Y_2O_3 has been shown to combine with Al, Ti, Nb, and Zr to form complex precipitates [3, 4], potentially robbing the matrix of γ' and γ" formers [4] or coarsening the Y_2O_3 particles [4], reducing the Orowan strengthening provided by the small dispersoids. This may have a significant impact on the elevated temperature strength of commercial alloys as the microstructure evolves away from base microstructure of the alloy which has been well characterized for certifications.

IN718 contains gamma double prime (γ"−Ni_3Nb) with rapid strength degradation above 650 °C. Other strengthening phase of IN718 is gamma prime phase (γ'−Ni_3Al or Ni_3Ti) with significant coarsening after 700 °C. Thus, IN718 is typically limited to service temperatures below 650°C. If the ODS IN718 alloy can demonstrate an increase in the temperature limit of IN718 to 850 °C or above, this would be a major research accomplishment.

Materials and Methods

Inconel 718 (IN718) powder (size: 44–106 μm) was acquired from Carpenter Additive. Half of the powder was left as received, while the other half of the powder was mechanically alloyed with Y_2O_3 to 0.5% by weight. A 0.5 wt% of Y_2O_3 was selected as a compromise between alloy degradation at high temperature and anticipated strength. Mechanical alloying was done by a mechano-chemical bonding

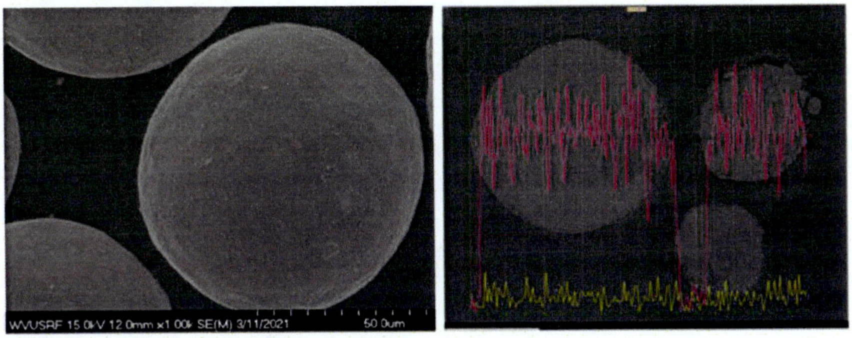

Fig. 1 SEM images of the MCB IN718 powder. Left) featuring powder surface. Right) X-ray line trace of Ni (purple) and Y (yellow)

(MCB) method recently developed at West Virginia University, where the powders are rotated in a drum with the Y_2O_3 under high rpm (2000–4000 rpm). This novel MCB method enables to breakdown the strengthening phase (i.e. Y_2O_3) to nanosize (<10 nm) and infuse the ultrafine Y_2O_3 into as-received gas atomized powder. The advantage of MCB is the uniform dispersion of the strengthening phases.

The MCB process shows strong uniform bonding of the Y_2O_3 to the IN718 powder. A scanning electron microscope (SEM) micrograph of the MCB powder is provided in Fig. 1, where the left image shows uniform distribution of the Y_2O_3 powder, and the right image shows the X-ray trace of a line profile of the powder with Ni content in purple and Y content in yellow.

The powder was subsequently direct energy deposited (DED) by Synergy Additive into 100 × 75 × 10 mm plates using a 90° cross hatching pattern between layers. A total of four plates were manufactured. Two using IN718 powder (IN718-1 and IN718-2) and two using the MCB Y_2O_3 718 (ODS718-1 and ODS718-2) powder. Each of the two IN718 and ODS718 plates were built with different parameters. An example plate deposited onto a 1018 steel substrate is shown in Fig. 2.

Microstructure coupons and tensile specimens were machined from the plates. Tensile specimens had a gage diameter of 6.33 mm and gage length of 31 mm. Tensile testing was conducted to ASTM E8 [5] and E21 [6] standards. The tensile strain rate was $1 \cdot 10^{-4}$ s^{-1} until 3% extension and $1 \cdot 10^{-3}$ s^{-1} until failure. Tensile tests were conducted on an Instron 8852 tensile frame with a clamshell furnace and high-temperature extensometer. Instron Bluehill 3 software was used to analyze the tensile data.

An FEI inspect F scanning electron microscope (SEM) was used to gather initial microstructure images. ImageJ was used to analyze grain size and porosity. Grain size was measured by the circular intercept method outlined in ATSM E112 [7]. However, only 7 mm of perimeter was measured, which is less than 25 mm recommended by ASTM E112 [7]. Porosity was measured from a single field at 250X and 4000X. ImageJ was used to apply a bandpass and thresholding filters, where

Fig. 2 DED manufactured IN718-2 plate on a 1018 substrate. Scale in inches

particle size analysis was automatically conducted by ImageJ. Energy-dispersive X-ray spectroscopy (EDXS) mapping was conducted for a minimum of 45 min.

Specimens were tested as printed (AP), with 1080 °C for 1 h, and 1200 °C for 1 h homogenization heat treatment. Raghavan et al. [8] has shown the ductility for selective laser melted IN718 has improved after a 1200 °C homogenization treatment. Standard aging treatment of 720 °C for 8 h, furnace cooling to 620 °C, and holding for 8 h and air cooling, was given to all specimens. At ThermoCalc simulation of IN718 shows homogenization heat treatments at 1080 and 1200 °C is sufficient to dissolve all microstructural phases except for a face centered cubic L_{12} carbide phase at minor concentration, Fig. 3.

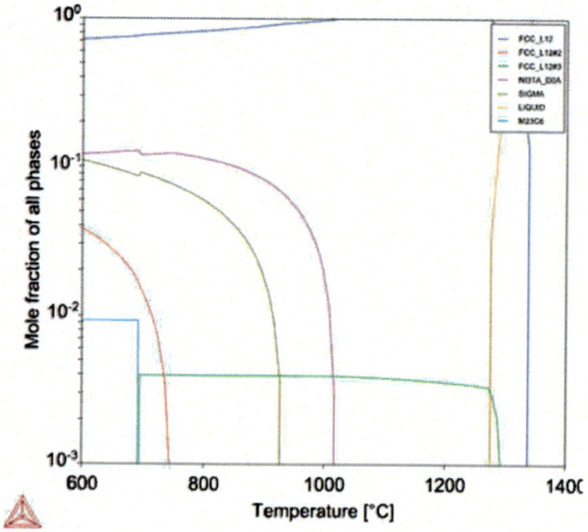

Fig. 3 ThermoCalc simulation of IN718

Results

Microstructure

The initial microstructures are shown in Fig. 4 for build parameter 1. The initial AP microstructure (Fig. 4 top and bottom left) appeared dendritic with Nb and Mo rich phases interspersed. X-ray energy dispersive spectroscopy (XEDS) spot analysis in the SEM was used to determine the constituents of phases. A homogenization heat treatment of 1080 °C (Fig. 4, top and bottom middle) dissolved the Nb & Mo rich phases but retained a dendritic grain structure. At a homogenization heat treatment of 1200 °C (Fig. 4 top and bottom right), the Nb & Mo rich phases also dissolved, and additionally the grains have re-crystallized into a more equi-axial structure.

Due to the dendritic nature of the microstructure, grain size was only measured for the specimens given a 1200 °C. For IN718-1 the grain size measured G-2.06 (157 μm average diameter), IN718-2 measured G-0.44 (275 μm average diameter), ODS718-1 measured G-1.15 (215 μm average diameter) and ODS718-2 measured G-2.48 (136 μm average diameter).

At low magnifications there are some 10–20 μm diameter pores present, accounting for ~ 0.1−0.5% porosity, Fig. 5. The horizontal axis in Fig. 5 is the homogenization heat treatment, with AP being in the as printed state. For IN718-1 and ODS718-1 the porosity was relatively constant with respect to homogenization

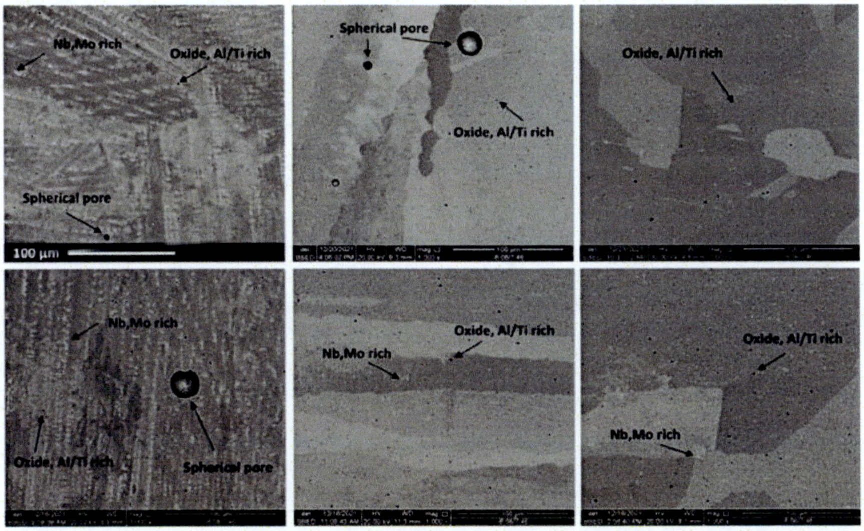

Fig. 4 Initial microstructure of ODS/718. Top left) IN718-1 AP; Top middle) IN718-1 with 1080 °C homogenization; Top Right) IN718-1 with 1200 °C homogenization; Bottom left) ODS718-1 AP; Bottom middle) ODS718-1 with 1080 °C homogenization; and Bottom right) ODS718-1 with 1200 °C homogenization

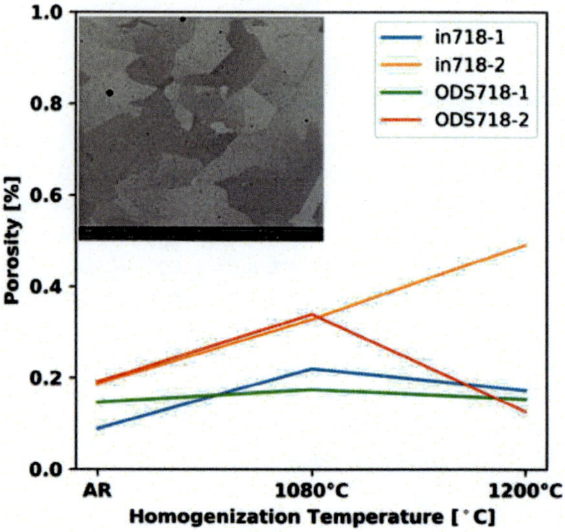

Fig. 5 Porosity measurements in % area against homogenization temperature. Inset from IN718-1 alloy given the 1200 °C homogenization heat treatment, representative of the specimen porosity. Measured at 250X

heat treatment. While for IN718-2 the porosity increased with increasing the applied homogenization heat treatment. For ODS718-2 porosity peaked for the 1080 °C heat treatment. It is suspected if more fields of view were measured the average porosity would remain constant with respect to the applied homogenization temperature.

XEDS maps were taken of the initial microstructure, shown in Fig. 6, where the left section is IN718-2 and the right section is ODS718-1. In Fig. 6, the columns are representative of the heat treatment, with the left column as built, middle column given 1080 °C for 1 h homogenization, and right column given 1200 °C for 1 h homogenization. The rows represent the element measured, with the top row representative of the backscatter electron image (BSE) and Cr, Mo, Nb, O, Ti, Al, and Y in descending order. Nickel was omitted for brevity. Elements Nb and Mo overlapped and showed the same patterns for all samples.

For the IN718, Fig. 6 left, the homogenization heat treatments removed the segregation of Mo and Nb in the dendritic structure. The Mo and Nb were observed to aggregate into discrete laves particles for both homogenization temperatures. Titanium appears to aggregate with Mo and Nb in the laves phases at 1080 °C heat treatment. At the higher 1200 °C heat treatment, titanium was also incorporated into the laves phases, however, also seemed to be segregated in the matrix. At 1080 °C the aluminum remained as discrete particles coincident with titanium; however, at 1200 °C homogenization temperature, the aluminum distribution appeared uniform, absent a small Al-rich oxide. Only minor traces of oxygen were present in IN718, likely coming from surface oxidation of the powder.

For the ODS718, Fig. 6 right, the homogenization heat treatments removed the dendritic distribution of Mo and Nb. The Nb and Mo were also observed to aggregate

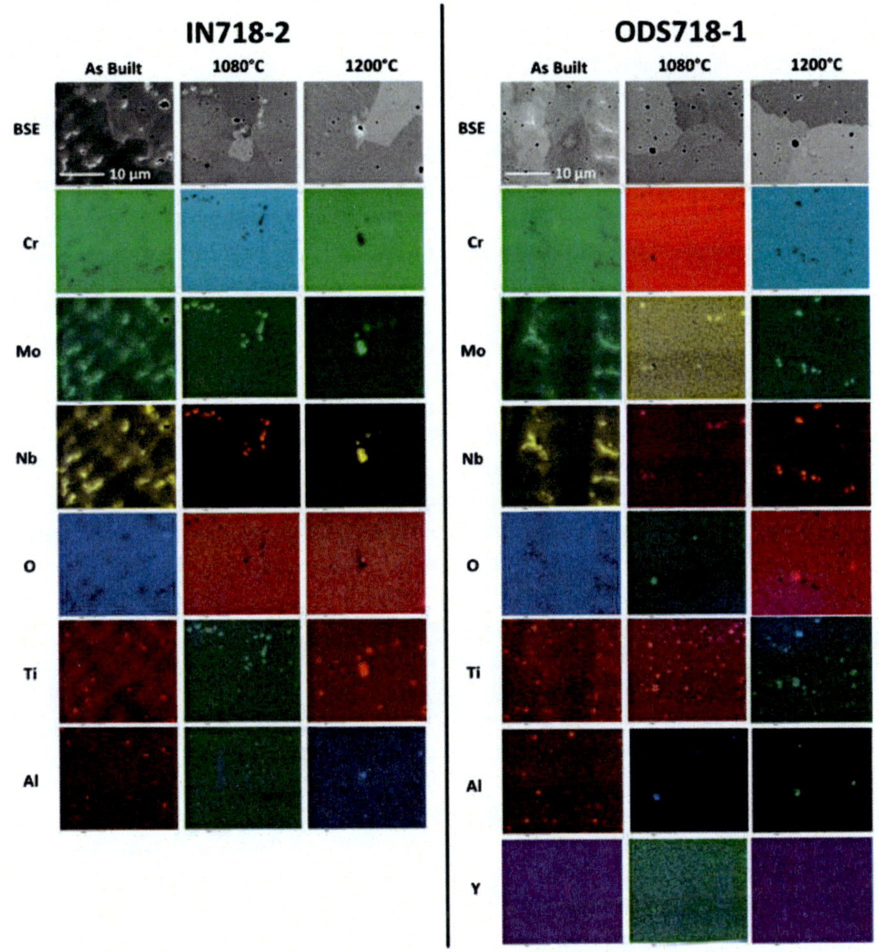

Fig. 6 XEDS maps of initial microstructure of IN718-2. Left column as printed; Middle column given 1080 °C homogenization for 1 h; Right column given 1200 °C homogenization for 1 h. Rows are labeled by element

into laves phases. Titanium appears to have no change in distribution after homogenization. Al and O appear to aggregate together after both homogenization temperatures, suggesting the formation of Al-rich oxides. Faint contrast is showing some yttrium discretization; however, it appears that the Y does not diffuse but remains clustered in the dendritic Mo/Nb rich areas AP.

Tensile Properties

Tensile tests were performed on the IN/ODS718 material. A temperature of 850 °C was chosen as the tensile performance of IN718 is significantly reduced at this temperature. A temperature of 1050 °C was chosen as it is above the γ' and γ" solvus temperature, leaving only the ODS and $M_{23}C_6$ carbides (Fig. 3) to strengthen the matrix.

Tensile results are plotted in bar format in Fig. 7. In Fig. 7, the top row is the yield strength in MPa, the middle row is UTS and the bottom row is the elongation. In the left column of Fig. 7 are the tensile properties at 850 °C, while the right column is at 1050 °C. Each homogenization heat treatment was grouped together, while the individual alloys are plotted in different colors. Overall, yield and UTS are largely unchanged between the alloys. Both at 850 and 1050 °C there are minor variations in yield and UTS between IN718 and ODS718 for both build parameter set 1 and 2.

At 850 °C the initial homogenization temperature had significant effects on the tensile properties. The yield and UTS were lowest for the AP specimens, while the specimens given a homogenization heat treatment at 1200 °C for 1 h had the highest yield and UTS. The opposite is true for the elongation, where the AP specimens had the highest tensile elongation, and the specimens given a homogenization heat treatment at 1200 °C for 1 h had the least elongation.

At 1050 °C the initial homogenization temperature had significant effects on the yield and elongation tensile properties, but not the UTS. The yield strength for specimens given a 1080 °C homogenization for 1 h had slightly higher yield strength and approximately double the elongation than the specimens given at 1200 °C homogenization for 1 h. The specimens given a 1200 °C homogenization for 1 h were tested both at the ASTM standard tensile strain rate of $1 \cdot 10^{-4}$ s^{-1} until 3% extension, and $1 \cdot 10^{-3}$ s^{-1} until failure and a single strain rate (SST) of $1 \cdot 10^{-4}$ s^{-1}. The SST tests had a higher yield strength and greater elongation than the dual strain rate test. The ODS718 with build parameter set 1 had consistently less elongation than the other alloys tested at 1050 °C.

Fracture Surfaces

Low magnification overview of the fracture surfaces from the specimens tensile tested at 1050 °C are provided in Fig. 8. Where the top row are specimens given a 1080 °C homogenization heat treatment and the bottom row are specimens given a 1200 °C homogenization heat treatment. The left two columns are standard IN718, while the right two columns are ODS718. It is apparent that the specimens given a 1080 °C homogenization heat treatment failed at dendrite boundaries. The specimens given a 1200 °C homogenization heat treatment still failed at apparent dendrite boundaries, however, they were broken up somewhat by recrystallization suggesting intergranular failure modes. The ductility of the specimens given a 1080 °C homogenization heat

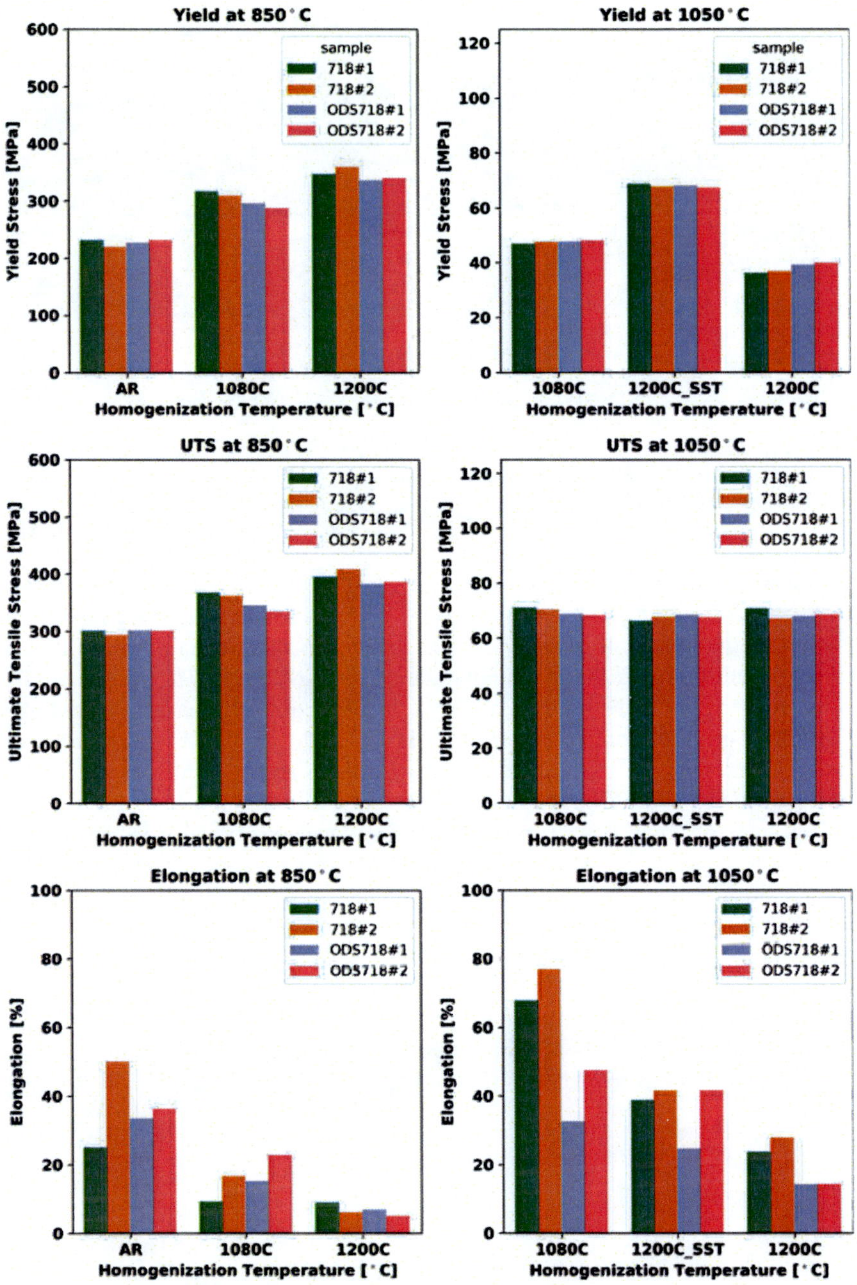

Fig. 7 Bar chart of mechanical properties for IN/ODS718. Top left: Yield strength at 850 °C; Top right: Yield strength at 1050 °C; Middle left: Ultimate tensile strength at 850 °C; Middle right: Ultimate tensile strength at 1050 °C. Bottom left: Elongation at 850 °C, Bottom right: Elongation at 1050 °C. SST = single strain rate, AP = as printed

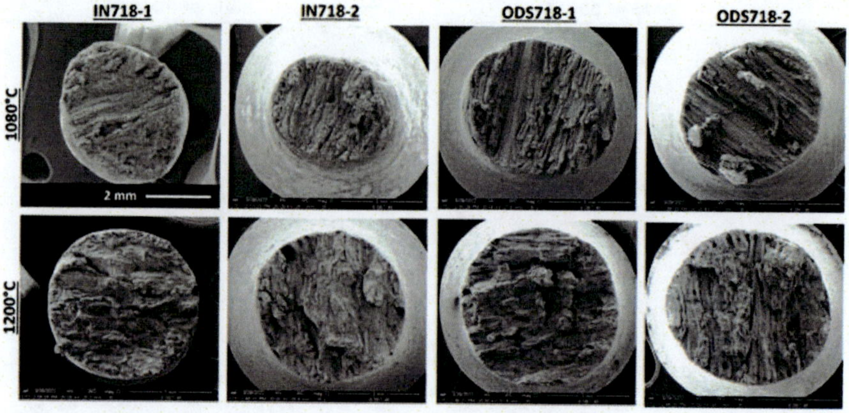

Fig. 8 Fracture surface overview for specimens tensile tested at 1050 °C. Top row: given 1080 °C homogenization for 1 h; Bottom row: column given 1200 °C homogenization for 1 h; Columns refer to alloy as labeled

treatment was much larger than the specimens given a 1200 °C homogenization heat treatment. Comparing the top and bottom rows of Fig. 8, it is apparent that a significant amount of the reduction in area occurred parallel to the growth direction of the primary dendrite arms in the specimens homogenized at 1080 °C. The fracture profile for each specimen in the top row of Fig. 8 appears contracted along the dendritic direction. The most contracted specimen is IN718-2, which also corresponds to the specimen with the most ductility (Fig. 7 bottom right). While the specimens homogenized at 1200 °C have a rounder fracture profile.

High-magnification images of the fracture surfaces are provided in Fig. 9. The reader is referred to Fig. 8 for detailed sub-figure descriptions. For the specimens given 1080 °C homogenization heat treatment, the specimens failed at the dendritic boundaries without any signs of ductility. Secondary cracking between dendrites is also apparent. For the specimens given a 1200 °C homogenization heat treatment, the apparent failure location is at grain boundaries also without any signs of ductility. Some nodule-like features are present on all specimens at high-magnification, Fig. 9. With the powder size ~ 50–60 μm in diameter (Fig. 1 left), these nodule features are much too small to be related to fusion defects from the powder.

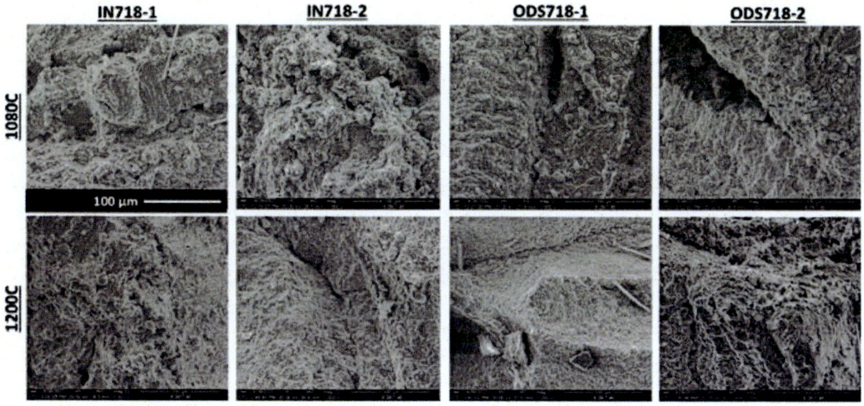

Fig. 9 High-magnification fracture surface for specimens tensile tested at 1050 °C. Top row: given 1080 °C homogenization for 1 h; Bottom row: column given 1200 °C homogenization for 1 h; Columns refer to alloy as labeled

Discussion

Microstructure Evolution

Application of a homogenization step prior to peak aging successfully reduced the elemental distribution as expected for both IN718 and ODS718. The primary differences between the 1080 °C and 1200 °C 1 h heat homogenizations were at the higher homogenization temperature the laves phases were larger and grains structure changed from dendritic to somewhat equi-axed. This is also as expected as grain growth is reported to occur above 1020 °C in IN718 [9]. For the ODS718, some additional complexities involving the Y_2O_3 are discussed below.

Guo et al. [3] studied laser powder bed fusion (LPBF) Inconel 738 LC (IN738LC) infused with 0.05 wt% Y_2O_3 nano-particles. Guo et al. used a homogenization heat treatment of 1120 °C for 2 h followed by aging treatment of 850 °C for 24 h. Their results showed an increase in UTS for the alloy when tested both at room temperature and 850 °C. Additionally, Guo et al. showed various Y-Al-Nb-Zr-O compounds. With no Zr in IN718, the Y-Al-Nb–O compounds are particularly relevant to this research. Guo et al. showed $Y_4Al_2O_9$ with formation energy of −300.4 eV, $Y_3NbAl_2O_9$ with formation energy of −514.75 eV, $YNb_3Al_2O_9$ with formation energy of −211.41 eV, and $Nb_4Al_2O_9$ with formation energy of −190.6 eV. Yu et al. [4] showed evidence of $Y_2Ti_2O_7$ formation in a high-Al and Ti superalloy. Reviewing the AP and homogenized microstructures in Fig. 6, no oxygen compounds were observed in IN718. Aluminum compounds formed with oxygen when given homogenization heat treatments were given to ODS718 (Fig. 6). While Nb and Ti were in the vicinity of these Y and O rich precipitates, EDS spot analysis of ODS718-1 with the 1080 °C homogenization shows oxygen rich areas of 51.1 at% O, 30.5 at% Al, 7.3 at% Ni, 3.9 at%

Cr, 3 at% Fe, 1.8 at% Ti, and 1.4 at% Y. For ODS718-1 with an applied 1200 °C homogenization heat treatment, the oxygen rich areas shows 65.7 at% O, 29.7 at% Al, 1.4 at% Y, 1.35 at% Ti, 0.9 at% Ni, 0.6 at% Cr, and 0.4 at% Fe. This suggests that the Y_2O_3 may be reacting with the matrix to form Al-Y–O compounds [3, 10]. Luu et al. [11] has reported complex carbo-nitrides may envelope Y_2O_3 cores and coarsen, which may have also occurred in this research. Additional TEM characterization would confirm this speculation. Additionally, the Y_2O_3 may be robbing the local microstructure of γ' and γ" formers (Al and Ti), which may lead to reduction of strength by denuding the local microstructure of the strengthening phase.

Tensile Properties

There were no significant changes in tensile properties between IN718 and ODS718 alloys for the same homogenization heat treatment. This suggests the Y_2O_3 additions at 0.5% wt had a negligible effect on the tensile properties of the alloy. With the observation of Al-rich oxide precipitate in the matrix, its likely Y-Al-O compounds were formed [3, 10]. This could have robbed the microstructure of γ' forming elements, weakening the alloy. This likely explains the small, but repeatable drop in yield strength at 850 °C for the ODS718 homogenized alloys. The AP specimens did not show a difference in yield strength at 850 °C between the IN718 and ODS718 alloys. This confirms the heat treatment made small changes in the microstructure, reducing the strength of the ODS718 slightly at 850 °C.

It has been reported that laves phases will dissolve at 1150 °C [12, 13]. The Nb and Mo rich laves phases were observed in both alloys after homogenization at 1080 °C and 1200 °C. The laves phases appear to have precipitated intra-granularly. Many authors have reported detrimental mechanical properties where laves phases have precipitated in IN718 [14–16]. Liu et al. [17] have shown that delta aging (890 °C for 12 h) prior to homogenization breaks apart the laves phases into smaller particles. This was not performed in this study. When comparing the size of the laves phases in Fig. 6, they are smaller with the lower homogenization temperature of 1080 °C. The relatively larger laves phases from the 1200 °C heat treatment likely explains the reduction in elongation at the higher applied homogenization temperature.

Luu et al. [11] studied selective laser sintered IN718 with 0, 0.4, 1.0, and 1.5 wt% Y_2O_3 infused by ball milling. While their research was conducted at room temperature, Luu et al. [11] did show 1 wt% Y_2O_3 to have the best combination of elongation, yield, and UTS of the powder weights considered. Luu et al. [11] contributed this to complex carbo-nitrides enveloping the Y_2O_3 particles ruining their ability to strengthen the material through Orowan mechanisms. Furthermore, Luu et al. [11] found the laves phases were minimized at 1% wt Y_2O_3. This may suggest increasing the Y_2O_3 content from 0.5 to 1.0 wt% may increase the mechanical performance at higher temperatures.

Conclusion

As-received gas atomized Inconel 718 powder was infused with 0.5 wt% Y_2O_3 by a mechano-chemical bonding processing. The powder was then direct energy deposited onto a substrate in plate form. Both the microstructure and mechanical properties were investigated for the virgin and Y_2O_3 infused IN718 powders. The as printed microstructure was dendritic and severely segregated. A minimal amount of porosity was measured, between 0.1 and 0.5%. A homogenization heat treatment of 1080 °C for 1 h mitigated elemental segregation and increasing the homogenization temperature to 1200 °C for 1 h induced some grain growth. For the Y_2O_3 infused plates formation of Y-Al-rich oxides was observed after the homogenization heat treatments.

The yield and ultimate tensile strength of the virgin powder exceeded that of the Y_2O_3 infused powders for tensile tests at 850 °C. This was true regardless of the homogenization temperature. At 1050 °C, the 1200 °C homogenization heat treatment showed a very minor increase in yield strength. It is suspected that formation of Y-Al-O compounds robbed the microstructure of aluminum, which forms the strengthener particle γ', explaining the slight reduction in yield strength for the ODS718 alloy. At the lower tensile test temperature of 850 °C, yield and ultimate tensile strength improved equally for all IN718/ODS718 alloys with increasing homogenization temperature. However, at the higher tensile test temperature of 1050 °C, the applied homogenization temperature had no influence on these mechanical properties. At both tensile temperatures, the elongation was reduced with increasing applied homogenization temperature.

Further investigations are planned to increase the content of Y_2O_3 which is anticipated to result in better tensile properties at 1050 °C.

Acknowledgements This research effort was sponsored by the U.S. DOE's Energy Efficiency and Renewable Energy and Advanced Manufacturing Office. We are grateful for Dr. Bob Gemmer's technical guidance, support, and encouragement throughout this effort. We also acknowledge Matthew Thom for assistance for helping with the powder production.

References

1. Smith TM, Thompson AC, Gabb TP, Bowman CL, Kantzos CA. Efficient production of a high-performance dispersion strengthened, multi-principal element alloy. Scientific Reports. 2020;10:9663.
2. Song Q-s, Zhang Y, Wei Y-f, Zhou X-y, Shen Y-f, Zhou Y-m, et al. Microstructure and mechanical performance of ODS superalloys manufactured by selective laser melting. Optics & Laser Technology. 2021;144:107423
3. Guo C, Yu Z, Hu X, Li G, Zhou F, Xu Z, et al. Y2O3 nanoparticles decorated IN738LC superalloy manufactured by laser powder bed fusion: Cracking inhibition, microstructures and mechanical properties. Composites Part B: Engineering. 2022;230:109555.
4. Yu L, Lu Z, Peng S, Li X. Effect of Al/Ti ratio on γ' and oxide dispersion strengthening in Ni-based ODS superalloys. Materials Science and Engineering: A. 2022;845:143240.
5. ASTM. ASTM E8/E8M-16a, Standard Test Methods for Tension Testing of Metallic Materials. West Conshohocken, PA: ASTM International; 2016
6. ASTM. E21–20 Standard Test Methods for Elevated Temperature Tension Tests of Metallic Materials. ASTM International; 2020
7. ASTM. E112–13(2021) Standard Test Methods for Determining Average Grain Size. 2021.
8. Raghavan S, Zhang B, Wang P, Sun C-N, Nai MLS, Li T, et al. Effect of different heat treatments on the microstructure and mechanical properties in selective laser melted INCONEL 718 alloy. Materials and Manufacturing Processes. 2017;32:1588-95.
9. Liu F, Lin X, Song M, Zhao W, Chen J, Huang W. Effect of intermediate heat treatment temperature on microstructure and notch sensitivity of laser solid formed Inconel 718 superalloy. Journal of Wuhan University of Technology-Mater Sci Ed. 2011;26:908-13.
10. Xia T, Yang C, Zeng W, Xie Y, Zhang Y, Zhang D, et al. Dispersoids and γ' precipitates in an ultrafine grained René 88DT – 5vol.%Y2O3 alloy with outstanding thermal stability. Materials Characterization. 2018;141:139–47
11. Luu DN, Zhou W, Nai SML. Influence of nano-Y2O3 addition on the mechanical properties of selective laser melted Inconel 718. Materials Science and Engineering: A. 2022;845:143233.
12. Changshen XDT. Research on Laves Phase Re-dissolution Behaviors of GH4169Alloy Steel During Heat Treatment Process Based on Mathematical Model. Foundry Technology. 2014:2570–2
13. Chen WL, CC; Li, H.;. Effects of heat treatment on microstructure and mechanical properties of modified inconel 718C alloy. Heat treatment of Metals. 2007;6:81–7
14. Sui S, Chen J, Zhang R, Ming X, Liu F, Lin X. The tensile deformation behavior of laser repaired Inconel 718 with a non-uniform microstructure. Materials Science and Engineering: A. 2017;688:480-7.
15. Ming XC, J.; Tan, H.;. Study on the long-lasting fracture mechanism of laser repairing GH4169 superalloy. Chinese Journal of Lasers. 2015;42:403005
16. Tabernero I, Lamikiz A, Martínez S, Ukar E, Figueras J. Evaluation of the mechanical properties of Inconel 718 components built by laser cladding. International Journal of Machine Tools and Manufacture. 2011;51:465-70.
17. Liu F, Lyu F, Liu F, Lin X, Huang C. Laves phase control of inconel 718 superalloy fabricated by laser direct energy deposition via δ aging and solution treatment. Journal of Materials Research and Technology. 2020;9:9753-65.
18. Sun D, Liang C, Shang J, Yin J, Song Y, Li W, et al. Effect of Y2O3 contents on oxidation resistance at 1150 °C and mechanical properties at room temperature of ODS Ni-20Cr-5Al alloy. Applied Surface Science. 2016;385:587-96

Effects of Scan Strategy Induced Microstructural Differences on Thin-Wall SLM IN718 Fatigue Performance

Tracy Connor Varney, Md. Imran Noor, and Paul F. Rottmann

Abstract The as-printed microstructure of additively manufactured parts is a function of many variables that span from scan strategy to part geometry. This is particularly relevant in precipitation strengthened alloys (e.g. IN718), as thermal history—which itself varies across a build—dictates the distribution of precipitates in the microstructure. Elucidation of the complex relationship between geometry, scan strategy, and resultant microstructure is necessary to optimize future scan strategies. In this study, a series of IN718 dogbone samples were printed via selective laser melting at 0.8 mm in thickness using a contour + hatching scan strategy with both rotating (67°) and static beam directions. To quantify the effect of subsurface porosity on fatigue crack initiation, the as-built surface was left intact on the thin-wall samples. As a reference, a 10 mm thick "bulk" sample was also printed and had 0.8 mm thick dogbones removed from it. To investigate the influence of scan strategy on the low-cycle fatigue (LCF) resistance of AM IN718, these samples were tested to failure using a custom micromechanical test setup equipped with a linear actuator and digital image correlation (DIC) to identify the onset of plasticity and the initiation and propagation of fatigue cracks across the sample surfaces. The driving hypothesis for this study is that the rotated scan strategy increases the fatigue resistance due to a more uniform microstructure with smaller grains. As-printed defects and fracture surfaces were characterized utilizing SEM imaging and DIC analysis and compared to observations from mechanical testing.

Keywords Additive manufacturing · Fatigue performance · IN718

Introduction

Additive manufacturing (AM) enables the fabrication of complex structures that are either impossible or impractical using conventional processing techniques. The incorporation of additive parts into aerospace structures greatly aids lightweighting

T. C. Varney (✉) · Md. I. Noor · P. F. Rottmann
Department of Chemical and Materials Engineering, University of Kentucky, Lexington, KY, USA
e-mail: connor.varney@uky.edu

E. A. Ott et al. (eds.), *Proceedings of the 10th International Symposium on Superalloy 718 and Derivatives*, The Minerals, Metals & Materials Series,
https://doi.org/10.1007/978-3-031-27447-3_34

efforts due to this geometrical freedom in part design. The benefit of AM processing brings with it substantial challenges, however. AM alloys, in general, differ substantially from their conventionally manufactured counterparts [1]. Largely resulting from multiple rapid thermal excursions required to locally melt and solidify the part followed by extremely fast, uncontrolled cooling rates [1, 2]. This generates suboptimal microstructures that are far from equilibrium, often possessing porosity and atomic segregation [3–6]. This is of particular concern for alloys that depend on a controlled microstructure for optimal properties, as is the case for Ni-based superalloys in which suboptimal thermal processing can result in poor homogenization of elements and the formation of undesirable phases [2, 7, 8].

Another important factor to consider is the extent to which undesirable phases will form at various locations across a part with complex geometry arising from the variable thermal history [9, 10]. Yang et.al. demonstrated a significant departure from 'bulk' properties and microstructure in parts with thin walls (single track up to 10× melt pool width); thin wall sections exhibited drastically different precipitate formation and microstructures based on the local beam conditions [2, 11]. The physical rationale for this is that, unlike the bulk of the part, thin walls are mostly surrounded by un-melted powder which conducts heat much slower than the solidified material causing the cooling rate of thin walls to be orders of magnitude slower than the rest of the part [2, 11, 12]. Unlike the long aging times of IN718 demonstrated by Theska et al., these slower cooling rates have been shown by several groups (Mantri et al., Yang et al., Xiao et al.) to allow for the nucleation of brittle Laves and Delta phases due to segregation of Nb and Mo [2, 3, 10, 12, 13]. While the formation of these phases, to some degree, can be controlled by tuning process parameters the presence of these phases in as-deposited AM structures to some extent is likely unavoidable.

Salzbrenner et al. showed in 17-4PH stainless steel that even when ignoring the extremes seen in thin walls, the AM process is still very non-equilibrium when compared to the standard wrought material, and as such, the properties can be highly stochastic [14–16]. Gribbin et al. showed that in IN718, interestingly, the fatigue strength of as-built samples was greater than the wrought samples at lower stress amplitudes, but worse as the stress amplitude increases [17, 18]. They attributed this to part porosity, inherit to AM, not contributing at lower stresses, but acting as stress concentrators at the higher amplitudes resulting in premature failure [17]. Generally, as-built parts tested in fatigue have their surfaces machined away as the normal contour-hatching based scan strategies are prone to generate porosity at the interface between these regions, as well as the roughness leading to premature crack initiation and failure [18, 19]. Gradl et al., among others, have shown that to utilize the expanded geometrical freedom of AM, there will be areas where this type of treatment is not possible [9, 20, 21]. The mechanical properties of these regions are poorly understood even with research like that by Campione et al., who studied fatigue in as-built samples but like previously mentioned, the surface has been machined and the actual initiation of the fatigue cracks was not considered given the use of compact tension samples including a starter notch [22].

This paper will focus on the fatigue properties of thin wall Inconel 718 dogbone samples printed using differing build strategies (aligned and rotated) to those of a "bulk" AM part. These tests were monitored using a digital image correlation system allowing for the study of the crack initiation and evolution as the test progressed. We hypothesize that the bulk samples will outperform both of the thin wall scan strategies due to a reduction in the ability to initiate a crack, with the rotated samples performing closer to bulk than the aligned samples. These mechanical results will be backed up by representative SEM imaging to further understand what mechanisms are contributing to the material's overall performance. This will all go towards furthering the understanding of the interconnection between geometry, processing conditions, and final part performance.

Methods

SLM Fabrication

The as-built tensile specimens were printed using an EOS M 290 SLM system using their IN718 gas-atomized powder. The specimens were printed using two different build strategies; "aligned" with no scan rotation and "rotated" which has a prescribed 67° hatching beam path rotation between layers. The aligned samples had the beam direction roughly parallel to the width direction of each sample and both strategies had 2 contour passes to improve the overall surface quality. Otherwise, the print parameters between the different strategies are consistent with one another with both using a beam power of 285W with a speed of 960 $\frac{mm}{s}$. Additionally, the samples share a hatch spacing of 110 μm, a layer thickness of 40 μm, and a platform temperature of 80 °C. The geometry and scan strategy differences are shown below in Fig. 1. The test samples were printed in the same machine, during the same print run to eliminate machine and build-to-build variability. Along with the printed thin walls, 4 bulk specimens were printed, two vertical and two horizontal with respect to the build direction. The bulk samples were printed with the aligned build strategy.

Specimen Design and Preparation

All samples were printed vertically with respect to the build plate and to help with the printability of this orientation, the samples had "shields" printed around them to take up any stress caused by the recoater blade which are visible in the upper left of Fig. 2. The as-built samples were small enough and the angle for the upper grip section was shallow enough that support structures were unnecessary. The samples themselves were designed to echo the ASTM E8 and E345 standards but in a miniature form. Overall, the samples are 10 mm in length and 6 mm wide, with a gauge section that is 1 mm in length, 1.25 mm wide, and a thickness of either 0.8 or 0.6 mm. Eight of these dogbones were attached at the grips together to form a single row, with 3

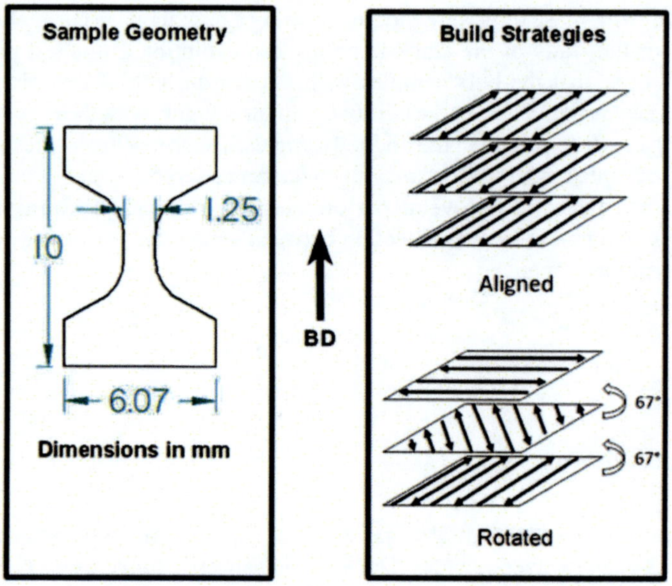

Fig. 1 Shows the sample geometry used on the left, and the different build strategies on the right. The complete part geometry that was printed is visible in the upper left of Fig. 2

sets of 3 rows per "shield" for a total of 72 samples per thickness. To serve as a point of comparison several dogbones of a similar size were removed from the bulk samples by means of an electrical discharge machining setup (EDM). To remove the EDM surface finished, the samples were low-stress ground and polished, and then speckled using a white base paint and fine carbon powder to get good contrast for digital image correlation described in section "Fatigue Testing". Otherwise, no post-processing was carried out, and the samples were tested in the as-built condition.

Fatigue Testing

The samples were tested using a custom micromechanical setup. The whole setup is controlled by a customizable python script that allows a large amount of testing flexibility and is shown in Fig. 2.

The data in this research was collected using a Futek LSB302 load cell and testing was controlled using a Transmotec DLA linear actuator powered by a JRK G2 controller utilizing speed control. A linear air bearing was utilized to assist with alignment and minimize friction. Vic-Gauge software, developed by Correlated Solutions, utilizes a Basler Ace acA2440 camera to collect 5MP monochrome images at up to 75fps; for this study, images were acquired spanning the entire sample surface initially every 30 cycles for the lower stress amplitude tests and 10 cycles

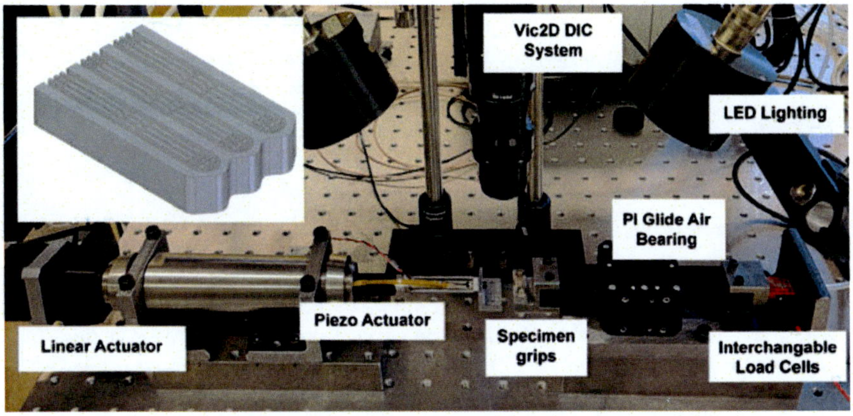

Fig. 2 Labels the major components of the testing setup used in this research, in the inset in the upper left is the "shield" geometry used during the fabrication of the as-built samples

for the higher stress samples. After a minimum number of cycles had passed, the program switched to taking an image every cycle to better capture the localization and propagation of failure in each sample. These images were taken by maintaining displacement for 0.4 s at the peak of a given cycle to allow for the setup to settle at which point an image was taken. The samples in this study were tested using a load control scheme targeting the max load and the required minimum load to achieve an R ratio between 0 and 0.1. The samples were initially loaded to the max stress at a strain rate of 10^{-4}/s after which they were cycled at a rate between 1 and 2 Hz depending on the max load. Bulk samples were tested at 650 and 750 MPa while as-built samples were tested at 650, 750, and 850 MPa.

Digital Image Correlation

The DIC system works by periodically taking an image of the sample surface, these images can then be correlated with one another allowing for the extraction of displacements to be calculated. To carry out this correlation, the sample surface must contain a fine random pattern to have the proper contrast for the DIC system to correctly track. This was simple for the as-built samples as the inherent surface roughness worked well, however, scattering alumina powder on the polished bulk samples was required. The software uses this surface contrast to project a grid of a given size (subset size) onto the sample that contains several points (determined by the step size or how far apart the data points are in pixels) and then as the sample deforms, so to do the subsets and their data points. Utilizing continuum mechanics and running this process across the entire surface allows for the displacements and strains to be visualized as a map throughout a test. In these experiments, subset sizes from ~ 25 up to~ 60px were used depending on the quality of the surface contrast, and a step size of 2px was used as a balance between accuracy and analysis speed. The final maps shown in this paper are representative of the local Lagrangian strain at each point on the sample surface.

SEM Characterization

Representative samples from both scan strategies were selected utilizing the DIC maps generated during testing, for fracture surface imaging. All imaging steps took place in an FEI Helios Nanolab 660 SEM. The bulk of the imaging was carried out at a beam voltage of 2 kV and a current of 0.20 nA utilizing both the Everhart–Thornley Detector (ETD) and an Ion Conversion and Electron detector (ICE). Images were taken at varying zoom levels and tilts to elucidate the mechanisms that dictated each sample's fatigue performance. The samples selected for SEM analysis were one of each type (bulk, aligned, and rotated) at the 650 MPa stress level, as well as one of each as-built sample (aligned and rotated) at the 850 MPa stress level.

Results

Fatigue Testing

As stated in section "Specimen Design and Preparation", three dogbones of each type (Aligned, Rotated, and Bulk) were tested at 650 and 750 MPa, with an additional three of the as-built dogbones, also being tested at 850 MPa. This was carried out using a load control scheme, with R ~ 0.1 at a frequency of between 1 and 2 Hz. Figure 3 shows this data in terms of a standard SN-plot with arrows denoting runout samples which were broken in uniaxial tension.

The average results for the fatigue testing can be found in Table 1. Exploring these, at the higher stress amplitude of 850 MPa, the rotated samples while having a similar average number of cycles (7513 versus 7773) were more consistent than the aligned group having a standard deviation 34% smaller compared to the aligned samples (1963 versus 4599). This trend generally continues for the other stress levels, with the rotated and aligned samples having a similar number of cycles to failure but with the rotated samples having lower standard deviation compared to the aligned samples. This is confirmed statistically by considering the P values for the comparison of each stress amplitude, which all end up much higher than the $\alpha = 0.05$ level, indicating no observed statistical difference from among the test populations. The bulk samples, however, sustained more cycles to failure than both as-built groups as well as exhibited reduced standard deviation of tested samples. This again is evident

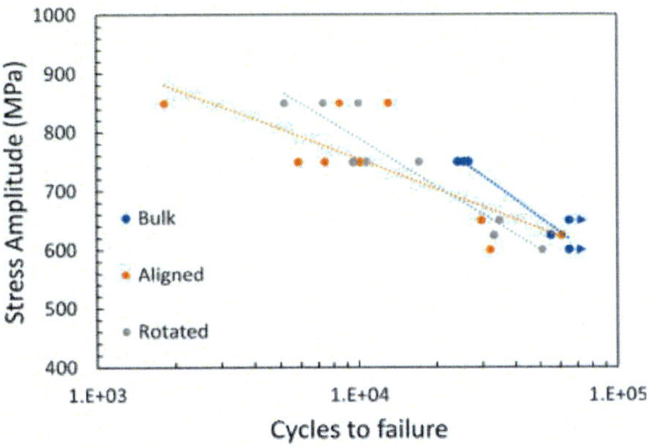

Fig. 3 Fatigue results of the as-built and bulk dogbone samples. Tests were carried out at an R = 0.1 using a load control scheme, at a frequency between 1 and 2 Hz. The bulk samples that exceeded 65,000 cycles and were considered an outlier is marked with an arrow

Table 1 Lists the average cycles to failure of each sample group as well as the standard deviation of each as a measure of the scatter

	Bulk		Aligned as-built		Rotated as-built	
	cycles	StdDev	cycles	StdDev	cycles	StdDev
650 MPa	61,713	4648	40,870	14,155	39,660	8215
750 MPa	25,630	981	7810	1747	12,460	3336
850 MPa	–	–	7773	4599	7513	1963

in the T-test results with the bulk samples having an average P value of 0.038 when compared to the aligned and rotated samples.

DIC Analysis

Significantly more information beyond the overall fatigue results described in section "Fatigue Testing" can be obtained by analyzing the local strain of the samples throughout the duration of the tests. This was done using the DIC images taken during testing, with which full strain maps were created. From these maps, virtual extensometers could be placed on the samples and strain vs. image plots were created. Using the Vic-2D software, these full field strain maps were generated for every frame allowing for the playback of each test while being able to observe strain localization and failure of the sample. Figure 4 highlights distinct differences in two tests that were performed by plotting the average strain across the sample as a function of image number. Figure 4a, from the rotated 650 MPa sample, shows a typical plot that exhibits expected behavior for such a plot: a gradual increase to a baseline level of strain, followed by a long period of relatively constant strain, and lastly a short portion of increasing strain until eventual failure. While these features are all present in Fig. 4b, which is of the aligned 650 Mpa sample, the area in which the strain is

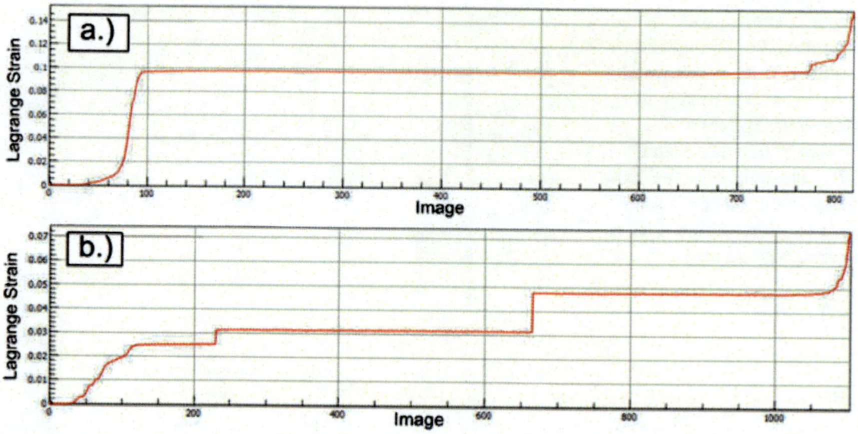

Fig. 4 Shows the virtual extensometer strain as a function of image during the testing of the as-built samples. Part **a.** shows the expected slow rise in strain until complete failure, of the 850 Mpa rotated sample while part **b.** shows the strain discontinuities observed in certain samples, this specifically the 650 Mpa aligned sample

constant is interrupted by two discontinuous increases in strain at the 225th image and 675th image. Further investigation into the DIC maps at these points confirms that these discontinuous increases of strain were not setup related (e.g. slippage, camera movements, etc.) and indeed reflect an increase in specimen strain.

Significantly more information can be obtained by analyzing the full 2D strain maps to characterize the localization and propagation of plasticity and crack propagation in these samples. For the majority of samples, localized deformation appeared upon initial loading and continued to slowly increase until it accelerates as the material begins to fail in tensile overloading mirroring the overall displacement behavior showing in Fig. 4a. This behavior, shown via two frames of the DIC strain video of the Rotated 850 Mpa, is summarized in Fig. 5a–c. Some samples, however, showed more than one area of strain localization as the testing progressed which is best shown by the 650 Mpa rotated sample in Fig. 5d–f where the two areas are circled in yellow.

SEM Imaging

These results warranted further study of the sample fracture surfaces to explore the differences seen in the cycles to failure and DIC. Overall, the fracture surfaces of each sample appear similar; there is generally a distinct fatigue region that gives way to the overloading region of the sample. The bulk sample as well as the two as-built samples at the 650 Mpa stress amplitude had a larger fatigue region than the samples studied at the 850 Mpa stress range which can be seen in Fig. 6. Additionally, these samples exhibited a larger transition zone between pure fatigue striations and pure

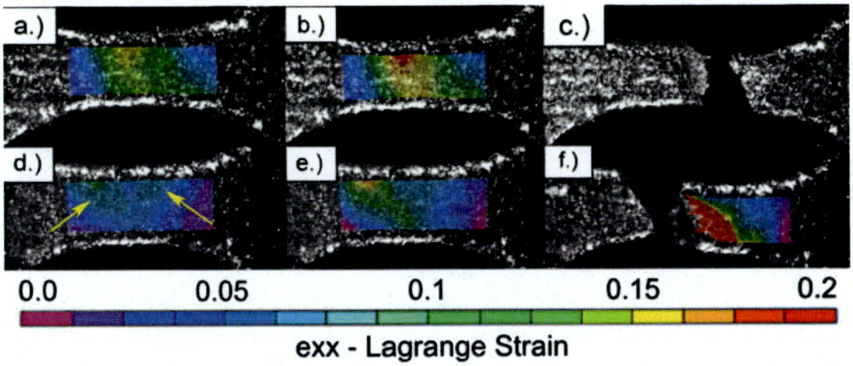

Fig. 5 **a**, **b**, and **c** showing the 850 Mpa aligned sample which only localized in one area with **d**, **e**, and **f** showing the two strain localization areas present during its testing of the 650 Mpa rotated sample, each highlighted with a yellow arrow

dimple fracture with these areas consisting of increasingly stochastic fatigue dimples similarly to those detailed by Griebel [23].

Overall, when comparing both flat and tilted images of the fracture surfaces, at a given stress level none are subjectively more tortuous than the others. Comparing across stress levels, however, differences start to arise with the 650 MPa sample fracture surfaces being relatively flat (constrained to 1 or 2 build layers) compared to the surfaces from the 850 MPa sample which span several build layers and can be seen in Fig. 7. In both cases at 850 MPa, the main fracture appears to have joined up with other smaller surface cracks (several build layers away) during the final failure of the sample. These surface cracks seem to nucleate on either a pore, or along a layer boundary shown best in Fig. 7a.

Calculating the crack propagation rates using the fatigue striations results in small differences between scan strategies and the bulk. The crack growth rates were calculated near the assumed initiation site, as well as within the transition zone between fatigue and overloading. These results are presented in Table 2.

Based on the values in Table 2, it can be seen that the bulk sample had the slowest crack growth of all the samples. The as-built samples show similar results to each other but exhibit slightly faster crack growth than the bulk sample. Interestingly, when calculating this in the transition zone, the opposite is true with the bulk sample having the highest growth rate followed by the aligned samples and finally the rotated samples having the slowest crack growth within the transition zone. The 850 MPa aligned sample was omitted from this calculation as its transition zone was hard to distinguish. An example of the striations can be seen in Fig. 8.

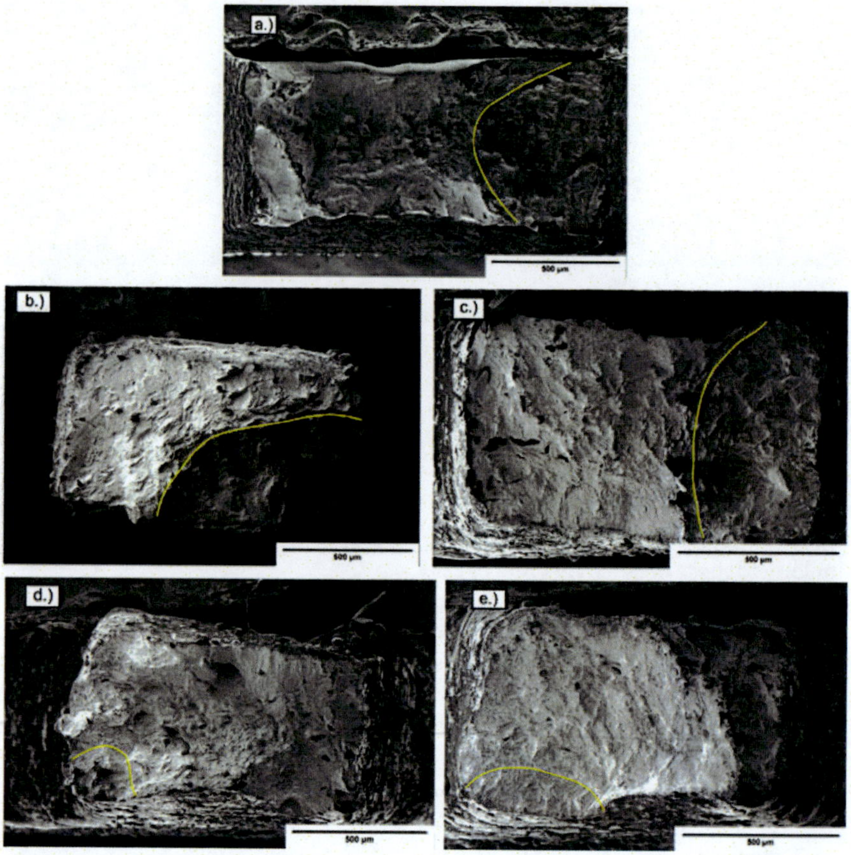

Fig. 6 Presents SEM micrographs of the entire fracture surfaces, each containing a yellow curve which denotes the approximate transition point between fatigue and overloading regions. **a** represents the bulk samples at 650 MPa, **b** and **c** are respectively aligned and rotated at 650 MPa, and **d** and **e** are respectively Aligned and Rotated strategies at 850 MPa. These micrographs also depict the reduction in fatigue area at 850 MPa when compared to the same samples at 650 MPa

Discussion

Given in Table 3 are the average yield and ultimate strengths for the samples tested in this research which are from a paper currently in preparation for publishing [24]. This clearly shows that in uniaxial tension, the as-built samples, in the same surface condition as this research, have a significantly higher yield and ultimate strength than the bulk samples. This was attributed mostly to the repeating columnar microstructure present in the bulk samples similar to that observed by Wan et al. that was generally absent in the as-built samples [25]. That is why, when compared to the fatigue results in Fig. 3 and Table 1, the outcome is surprising that the bulk samples were able

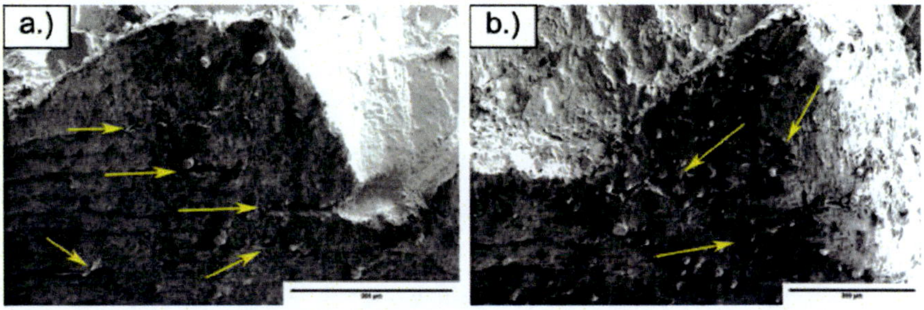

Fig. 7 Highlights with yellow arrows the smaller surface cracks that formed during fatigue testing which the main crack joined up with in certain cases jumping several layers to do so. **a** shows a micrograph of the aligned 850 MPa sample while **b** shows the rotated 850 MPa sample

Table 2 Details the measured average crack growth rate in units of μm/cycle for each sample near the assumed initiation site as well as within the transition zone. The aligned 850 MPa sample was omitted as the transition zone was hard to distinguish on the fracture surface

Units are in μm/cycle	Bulk	Aligned 650 MPa	Aligned 850 MPa	Rotated 650 Mpa	Rotated 850 Mpa
Initiation	0.0563	0.1302	0.1423	0.1583	0.1538
Transition	0.5927	0.5515	_	0.4558	0.4240

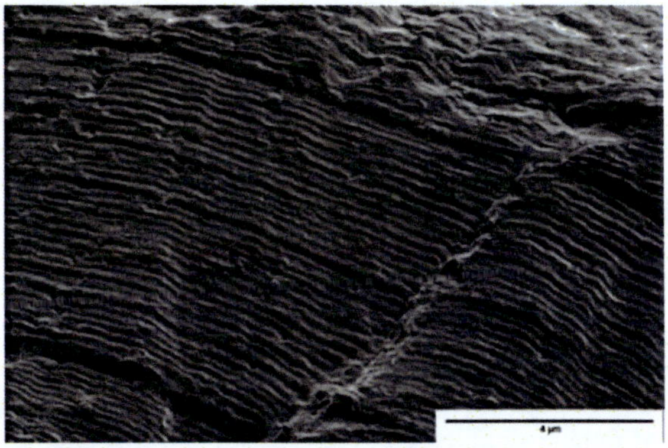

Fig. 8 Depicts the fatigue striations present in the fatigue portion of the fracture surface, in particular, this micrograph is from the rotated 850 MPa sample

to sustain on average 68.3% more cycles before failure than either of the as-built samples even when they were tested at a higher percentage of the UTS. On top of that, the bulk samples also managed to be more consistent with a 37.5% smaller standard deviation across the two stress levels at which the bulk samples were tested.

This large increase in fatigue resistance, as well as a reduction in scatter is likely due to the polished surface finish of the bulk samples. Removing the rough as-built/EDM surface causes the energy required to initiate a crack to increase. Similar to what Gribbin et al. saw in their testing, at the higher stress levels the as-built samples

Table 3 Reports the average yield and ultimate strengths of the materials used in this study collected over the course of several uniaxial tension experiments from a manuscript still being prepared for publishing [24]

	Bulk	Aligned	Rotated
Average yield strength (MPa)	571.3	657.7	709.1
Average ultimate strength (MPa)	799.3	992.8	1008.1

lost their lead over the wrought material which they attributed to the porosity in their samples having a greater effect on the failure of the material at the higher stress levels by means of stress concentration [17]. It is likely that a similar process is occurring here, as the bulk samples are mostly devoid of printing defects and possess a polished surface, while the as-built samples have been shown to contain subsurface porosity as well as a rough surface from the SLM process. The difference in the ability to initiate a crack allows for the weaker material in uniaxial tension (bulk samples) to sustain a higher number of cycles before failing as it is harder for a crack to form and propagate across the material.

Starting with the higher stress samples (850 MPa), this is most easily seen in Figs. 6d-e and 7 where the surface roughness/porosity is evident. This high stress level combined with a rough surface allows for the initiation of multiple cracks along the surface. These cracks seem to have nucleated along layer boundaries, or in some cases subsurface defects, and likely caused the premature failure of these samples as multiple cracks were able to propagate into the material, sometimes joining the main crack. Comparing the different build strategies at this stress level, while they have similar numbers of cycles to failure, the rotated samples are much more consistent having a 34% decrease in the scatter of samples. Further study is needed to confirm, but as shown by Arisoy et al. using a 67° rotating scan strategy can result in a finer grain structure and other researchers such as Nezhadfar et al. have shown a moderate increase in fatigue performance with a refined, more equiaxed grain structure even though a significant increase was not observed in this study [26, 27].

Again, shown in Fig. 6, the amount of area on the fracture surface occupied by the fatigue crack varies with the stress amplitude that the sample was subjected to; with the higher stress (850 MPa) samples having a reduced fatigue area compared to the lower (650 MPa) samples. There is also a noticeable reduction in the amount of fatigue dimples that are present, showing that the 850 MPa transitioned faster between fatigue and overloading failure. This makes sense as the overall stress will be higher as the area shrinks and will hit the material's ultimate strength sooner as the load bearing area decreases especially if the material has initiated multiple cracks. Interestingly, however, at the 850 MPa stress level, the crack growth rate remains similar to the 650 MPa samples, but not much can be learned from this as the growth rate was only calculated for 2 samples, and this may not hold true for all samples.

Focusing on just the 650 MPa samples, the fracture surfaces are much less tortuous, and there is little evidence that more than the main crack was active in the material. The fracture surfaces were flatter and generally stayed within one to two build layers.

These samples were likely at a stress level low enough that multiple crack initiations were unfavorable with only a main crack forming and continuing to propagate. Surprisingly, this is backed up by an increase in the scatter of the samples. At this stress level, if there only exists a single crack it is likely to be very dependent on the defect composition of the sample. Take the aligned samples for example, of the three tested the cycles to failure were 29,670, 32,100, and 60,840. This trend is also seen in the rotated samples but to a lesser extent, with the worse performing samples likely containing a similarly sized defect that the fatigue crack could readily initiate on, while the higher performing sample happened to not include as large of a defect and so it took longer for a crack to form and propagate.

This is visible in the DIC imaging as it is possible to pick out when a crack starts to propagate across the material by the localization of the strain. For the aligned samples, the worse performing samples quickly showed a localization of strain which proceeded to increase until the sample failed. The sample with more cycles to failure took longer to start localizing strain and it seemed to accumulate slower. This show that even at the low spatial resolution of these tests, it is possible to see areas of interest/smaller fatigue cracks as there will be multiple areas of localization on the surface. This can help determine where further analysis of samples is needed, and which samples may hold interesting areas of study. It was unfortunately not possible to pinpoint exactly where the crack tip was in these samples like similar studies have done likely due to the short distances involved and likely not having a high enough spatial resolution [28].

Continuing with DIC, the strain discontinuities seen in Fig. 4b are likely the result of these additional cracks forming in the material as the overall strain of the sample increases. An example of this can be seen in d. of Fig. 5 where there are 2 areas that initially start to localize strain before the larger area takes over for the final failure. While it's also possible that these discontinuities are a result of the main crack running into and opening a large void, expanding the crack quickly, this was not observed to be true on the fracture surface of the samples studied. The fracture surfaces are relatively void free, with occasional small (~10-micron diameter) voids present other than those connected to the surface.

Conclusions

Altogether, these results show that the thin wall samples fall behind the bulk material in terms of fatigue performance, but by adjusting the scan strategy, and consequently the thermal history of the parts, the resulting properties can be improved. By combining this testing with DIC analysis and SEM imaging of the fracture surfaces, this work moves towards creating a better understanding of the interconnection between geometry, processing conditions, and part performance. Key findings are summarized below.

- While the bulk samples were weaker in uniaxial tension, they exhibited a greater fatigue resistance than either of the as-built scan strategies having 68.3% higher average cycles to failure across 650 and 750 MPa when compared to the highest performing as-built samples likely due to a higher threshold for crack initiation and reduced defect population.
- While having similar average cycles to failure, the rotated samples showed a reduction in the scatter of the results having a standard deviation 34% smaller across an average of all 3 stress amplitudes compared to the aligned samples.
- Fracture surface imaging showed that along with the main crack, several smaller cracks were able to initiate on the surface of each sample, in some cases the main crack went through several build layers to connect with these smaller cracks likely leading to the sample's premature failure. This also highlighted that the 850 MPa samples had a reduced fatigue area and transitioned faster from fatigue to overloading failure.
- Digital image correlation (DIC) analysis allows for the determination of the approximate crack initiation/evolution and highlights samples of further study by showing multiple areas of strain localization as well as strain discontinuities as the material generates new cracks.

References

1. Y.-H. Li, B. Wang, C.-P. Ma et al. (2019) Material Characterization, Thermal Analysis, and Mechanical Performance of a Laser-Polished Ti Alloy Prepared by Selective Laser Melting. Metals vol. 9, no. 2, p. 112. https://doi.org/10.3390/met9020112
2. H. Yang, J. Yang, W. Huang, Z. Wang, and X. Zeng (2018) The printability, microstructure, crystallographic features and microhardness of selective laser melted Inconel 718 thin wall. Materials & Design vol. 156, pp. 407–418. https://doi.org/10.1016/j.matdes.2018.07.007
3. S. A. Mantri, S. Dasari, A. Sharma et al. (2021) Effect of micro-segregation of alloying elements on the precipitation behaviour in laser surface engineered Alloy 718. Acta Materialia vol. 210, p. 116844. https://doi.org/10.1016/j.actamat.2021.116844
4. P. Tao, H. Li, B. Huang, Q. Hu, S. Gong, and Q. Xu (2019) The crystal growth, intercellular spacing and microsegregation of selective laser melted Inconel 718 superalloy. Vacuum vol. 159, pp. 382–390. https://doi.org/10.1016/j.vacuum.2018.10.074
5. P. Promoppatum and S.-C. Yao (2019) Analytical evaluation of defect generation for selective laser melting of metals. International Journal of Advanced Manufacturing Technology.
6. G. Kasperovich, J. Haubrich, J. Gussone, and G. Requena (2016) Correlation between porosity and processing parameters in TiAl6V4 produced by selective laser melting. Materials and Design vol. 105.
7. C. K. Yong, G. J. Gibbons, C. C. Wong, and G. West (2020) A Critical Review of the Material Characteristics of Additive Manufactured IN718 for High-Temperature Application. Metals vol. 10, no. 12, p. 1576. https://doi.org/10.3390/met10121576
8. X. Liu, K. Wang, P. Hu, X. He, B. Yan, and X. Zhao (2021) Formability, Microstructure and Properties of Inconel 718 Superalloy Fabricated by Selective Laser Melting Additive Manufacture Technology. Materials vol. 14, no. 4, p. 991. https://doi.org/10.3390/ma14040991
9. G. Kasperovich, R. Becker, K. Artzt, P. Barriobero-Vila, G. Requena, and J. Haubrich (2021) The effect of build direction and geometric optimization in laser powder bed fusion of Inconel

718 structures with internal channels. Materials & Design vol. 207, p. 109858. https://doi.org/10.1016/j.matdes.2021.109858
10. F. Theska, K. Nomoto, F. Godor et al. (2020) On the early stages of precipitation during direct ageing of Alloy 718. Acta Materialia vol. 188, pp. 492–503. https://doi.org/10.1016/j.actamat.2020.02.034
11. C.-H. Yu, R. L. Peng, V. Luzin et al. (2020) Thin-wall effects and anisotropic deformation mechanisms of an additively manufactured Ni-based superalloy. Additive Manufacturing vol. 36, p. 101672. https://doi.org/10.1016/j.addma.2020.101672
12. A. Alhuzaim, S. Imbrogno, and M. M. Attallah (2021) Controlling microstructural and mechanical properties of direct laser deposited Inconel 718 via laser power. Journal of Alloys and Compounds vol. 872, p. 159588. https://doi.org/10.1016/j.jallcom.2021.159588
13. H. Xiao, X. Liu, W. Xiao et al. (2022) Influence of molten-pool cooling parameters on solidification structure and mechanical property of laser additive manufactured Inconel 718. Journal of Materials Research and Technology. https://doi.org/10.1016/j.jmrt.2022.06.162
14. H. Y. Wan, W. K. Yang, L. Y. Wang et al. (2021) Toward Qualification of Additively Manufactured Metal Parts: Tensile and Fatigue Properties of Selective Laser Melted Inconel 718 Evaluated using Miniature Specimens. Journal of Materials Science & Technology. https://doi.org/10.1016/j.jmst.2021.04.049
15. B. L. Boyce, B. C. Salzbrenner, J. M. Rodelas et al. (2017) Extreme-Value Statistics Reveal Rare Failure-Critical Defects in Additive Manufacturing Advanced Engineering Materials vol. 19, no. 8, p. 1700102. https://doi.org/10.1002/adem.201700102
16. B. C. Salzbrenner, J. M. Rodelas, J. D. Madison et al. (2017) High-throughput stochastic tensile performance of additively manufactured stainless steel. Journal of Materials Processing Technology vol. 241, pp. 1–12. https://doi.org/10.1016/j.jmatprotec.2016.10.023
17. S. Gribbin, J. Bicknell, L. Jorgensen, I. Tsukrov, and M. Knezevic (2016) Low cycle fatigue behavior of direct metal laser sintered Inconel alloy 718. International Journal of Fatigue vol. 93, pp. 156–167. https://doi.org/10.1016/j.ijfatigue.2016.08.019
18. M. Muhammad, P. Frye, J. Simsiriwong, S. Shao, and N. Shamsaei (2021) An investigation into the effects of cyclic strain rate on the high cycle and very high cycle fatigue behaviors of wrought and additively manufactured Inconel 718. International Journal of Fatigue vol. 144, p. 106038. https://doi.org/10.1016/j.ijfatigue.2020.106038
19. T. C. Varney, R. N. Quammen, N. Telesz, T. J. Balk, A. Wessman, and P. F. Rottmann (2021) Effects of Pore Geometry on the Fatigue Properties of Electron Beam Melted Titanium-6Al-4V. Metallurgical and Materials Transactions A vol. 52, no. 5, pp. 1836–1849. https://doi.org/10.1007/s11661-021-06194-9
20. P. Gradl, D. C. Tinker, A. Park et al. (2022) Robust Metal Additive Manufacturing Process Selection and Development for Aerospace Components. Journal of Materials Engineering and Performance. https://doi.org/10.1007/s11665-022-06850-0
21. P. Wang, H. Zhou, L. Zhang et al. (2020) In situ X-ray micro-computed tomography study of the damage evolution of prefabricated through-holes in SLM-Printed AlSi10Mg alloy under tension. Journal of Alloys and Compounds vol. 821, p. 153576. https://doi.org/10.1016/j.jallcom.2019.153576
22. I. Campione, T. M. Brugo, G. Minak, J. J. Tomić, N. Bogojević, S. C. Kostić (2020) Investigation by digital image correlation of mixed mode I and II fracture behavior of metallic IASCB specimens with additive manufactured crack-like notch. Metals 10(3). https://doi.org/10.3390/met10030400
23. A. Griebel (2009) Technical Brief: Fatigue Dimples. Journal of Failure Analysis and Prevention vol. 9, no. 3, pp. 193–196. https://doi.org/10.1007/s11668-009-9228-z
24. T. C. Varney, T. Oldham, I. Noor, and P. F. Rottmann (2023) Investigating the Influence of Scan Strategy and Small-scale Geometrical Complexity on the Microstructure and Mechanical Properties of Thin Wall SLM IN718. Manuscript in preparation.
25. H. Y. Wan, W. K. Yang, L. Y. Wang et al. (2022) Toward qualification of additively manufactured metal parts: Tensile and fatigue properties of selective laser melted Inconel 718 evaluated using miniature specimens. Journal of Materials Science & Technology 97, pp. 239–253. https://doi.org/10.1016/j.jmst.2021.04.049

26. Y. M. Arısoy, L. E. Criales, T. Özel, B. Lane, S. Moylan, and A. Donmez (2017) Influence of scan strategy and process parameters on microstructure and its optimization in additively manufactured nickel alloy 625 via laser powder bed fusion. The International Journal of Advanced Manufacturing Technology Vol. 90 No. 5, 1393–1417. https://doi.org/10.1007/s00170-016-9429-z
27. P. D. Nezhadfar, K. Anderson-Wedge, S. R. Daniewicz, N. Phan, S. Shao, and N. Shamsaei (2020) Improved high cycle fatigue performance of additively manufactured 17-4 PH stainless steel via in-process refining micro-/defect-structure. Additive Manufacturing Vol. 36, p. 101604. https://doi.org/10.1016/j.addma.2020.101604
28. J. M. Vasco-Olmo, F. A. Díaz, F. V. Antunes, and M. N. James (2019) Characterisation of fatigue crack growth using digital image correlation measurements of plastic CTOD. Theoretical and Applied Fracture Mechanics Vol. 101, pp. 332–341. https://doi.org/10.1016/j.tafmec.2019.03.009

Characterization of Laser Powder Bed Fusion of Nickel-Based Superalloy Haynes 282

Kameshwaran Swaminathan, Jonas Olsson, Tahira Raza, Peter Harlin, and Joel Andersson

Abstract Nickel-based superalloy Haynes 282 specimens were manufactured using the Laser Powder Bed Fusion process with a powder layer thickness of 60 and 90 microns to study the effect of laser power, laser scan speed, and hatch distance on the melt pool dimensions and porosity. The melt pool dimensions and porosity were measured at the center of the cubes parallel to the build direction. Variation of melt pool depth and overlap exist within the same sample signifying the scatter present in the process. Laser scan speed was found to be the most significant parameter for porosity and hatch distance was found to be the most significant parameter affecting the average melt pool overlap depth in the cubes built with 60 microns layer thickness. Interaction of speed and hatch distance was found to be the most significant parameter for porosity and Laser scan speed was the most significant parameter for average melt pool overlap depth in cubes built with 90 microns layer thickness. Comparison of measured responses with individual parameters provides partial trends of melt pool dimensions and porosity. As the heat input is captured better in line energy and area energy density, a better trend of the melt pool dimensions data and marginal trend of porosity in comparison with energy densities is discussed. The ratio of maximum length to minimum length of defects such as porosity and lack of fusion is measured to determine the shape of the defects and averaged to provide insight into the dominant shape of defect for a given set of parameters.

Keywords Haynes 282 · Laser powder bed fusion · Melt pool dimensions · High layer thickness · Process parameter study

---

K. Swaminathan (✉) · J. Olsson · T. Raza · P. Harlin · J. Andersson
Department of Engineering Science, University West, 461 86 Trollhättan, Sweden
e-mail: kameshwaran.swaminathan@hv.se

P. Harlin
Sandvik Additive Manufacturing, Sandviken, Sweden

E. A. Ott et al. (eds.), *Proceedings of the 10th International Symposium on Superalloy 718 and Derivatives*, The Minerals, Metals & Materials Series,
https://doi.org/10.1007/978-3-031-27447-3_35

Introduction

Haynes 282 is a Ni-based superalloy developed in 2005, targeting applications at temperatures higher than 650°C combined with ease of manufacture (formability and weldability). Haynes 282 is a γ' strengthening alloy with sluggish precipitation kinetics enabling the reduced loss of strength at high temperatures. The specific applications include transition sections and exhaust sections of gas turbines. A relatively low volume fraction (~19%) of γ' for the given strength and creep resistance provides good formability to the alloy, compared to Waspaloy and Rene 41 [1].

Powder Bed Fusion (PBF) additive manufacturing (AM) process is building components layer by layer [2, 3]. PBF is advantageous over conventional manufacturing in producing components with design complexity. The heat source used for the PBF of metals is usually a laser or electron beam. Powder Bed Fusion-Laser Beam (PBF-LB) and Powder Bed Fusion-Electron Beam (PBF-EB) are the nomenclatures used in this manuscript, based on the heat source. PBF-LB is usually used for components with finer features and smoother surfaces, but the process imparts lesser energy into the powder particles due to reflection losses and a lower build rate compared to PBF-EB. PBF-EB imparts more energy into the powder particles and can build at a faster rate but with a lower resolution of surface features and roughness. Pre-heating before the start of the build is also widely varied and affects the quality of the build. PBF-EB can pre-heat the build plate and powder particles up to 1000°C or even higher as necessary, by using the same energy source used for melting the powder, whereas PBF-LB machines typically have powder bed heating from the base plate to around 200 °C. Recently, newer PBF-LB machines are developed that can heat the build plate up to 1200 °C [2–4].

As the AM research of Nickel-based superalloys has matured over the years, multiple research groups have worked on PBF-LB and PBF-EB characterization of Alloy 718 [5–12]. Other superalloys have recently been explored and Haynes 282 is one such alloy.

In a study by K. A. Unocic et al., PBF-EB of Haynes 282 was explored using process parameters such as focus offset, hatch spacing, line order, and speed function (speed function controls the power and velocity of the beam), no cracks were observed, with density ranging from 98.5 to 99.5%, suggesting low amount of porosity [13]. Strong [001] texture and epitaxial grain growth was found parallel to the build direction, caused by the direction of heat removal, although the scan strategy can be varied to modify the texture [14].

PBF-LB of Haynes 282 was explored by multiple research groups based on process parameters and post-treatment parameters with layer thickness ranging from 20 to 40 μm [15–22]. Generally, no processability issues were found, given the good weldability of the alloy [1], although Boswell et al., and Hariharan et al. have reported one or few microcracks. R. Otto et al. has reported multiple cracks in samples built without powder bed heating, however, no cracks were reported in the samples built with heated powder bed. Texture along [001] direction was found in the build direction as expected in the PBF-LB process. This reflects in the anisotropic grain size and mechanical properties in as-built conditions. Various types of post treatments like hot isostatic pressing (HIP), conventional heat treatment (HT) used for wrought alloys, and low-temperature aging have been shown to reduce the anisotropy and comparable mechanical properties in horizontal (perpendicular to build direction) and vertical (parallel to build direction) orientations [16–20].

There is more scope for the fundamental understanding of the PBF AM process, and the effect of process parameters on Haynes 282 material. The heat distribution during the PBF-LB process and the resultant melt pool behaviour and dimensions have shown to be critical for process understanding through in-situ measurements and simulation of Ti-6Al-4V, SS316, and 718 alloys [23–25]. The current work intends to study the melt pool dimensions and 2D porosity parallel to build direction for samples built with 60 microns and 90 microns layer thickness for various combinations of process parameters for Haynes 282 alloy. This understanding of the effect of process parameters on melt pool behaviour is expected to provide insights into the shape and amount of porosity in the built part. This knowledge will be useful to characterize the static and dynamic behaviour of components.

Methods and Manufacture

$10 \times 10 \times 10$ mm samples were manufactured using SLM 280 2.0 Twin-Laser 400W machine. The design of experiments was planned with the centre point at Alloy 718 optimized parameters (theme) for a PBF-LB machine manufactured by SLM Solutions Group AG. The theme uses a laser power of 350 W, laser beam scan speed of 800 mm/s, and a hatch distance of 0.15 mm. Alloy 718 theme was set as the centre point for 90 micron layer thickness as well, except 750 mm/s as the centre point for speed, to increase the heat input. The alloy 718 theme was chosen as the centre point of the experiment since the reflectance of laser incidence is similar for the Haynes 282 and 718 powders with comparable size distribution [15]. The extent of variation of various parameters for each sample is shown in Table 1. The process was designed using MODDE software, with three replicates for the centre point, similar to Adegoke et al. [26]. The focus parameter was maintained at +4 mm (laser focal plane was approximately 4 mm above the build plate) for all the samples. The nominal spot size was 80 microns and a focus offset of +4 mm results in a nominal spot size calculated as 105 microns. The chemical composition of the HAYNES® 282® powder used in this study is given in Table 2. The powder was produced through vacuum induction melting followed by inert gas atomisation process using argon gas in the same powder production equipment by Sandvik under special license from Haynes International, Inc. The reported particle size distribution was D_{10} – 18.8μm, D_{50} – 29.8 μm, and D_{90} – 46.3μm.

The samples were sectioned using Struers Secotom 10 cutting machine, parallel to the build direction for characterization as shown in Fig. 1a. The typical metallographic procedures for mounting, grinding, and polishing were followed with the final polishing step of Mastermet 0.2 μm colloidal silica. Optical microscopy was

Table 1 Process parameters used for samples manufactured with layer thickness of 60 microns (A1–A17) and 90 microns (B1–B17)

ID	Power (W)	Speed (mm/s)	Hatch distance (mm)	ID	Power (W)	Speed (mm/s)	Hatch distance (mm)
A1	323.0	720.0	0.135	B1	323.0	550.0	0.132
A2	377.0	720.0	0.135	B2	377.0	550.0	0.132
A3	323.0	880.0	0.135	B3	323.0	950.0	0.132
A4	377.0	880.0	0.135	B4	377.0	950.0	0.132
A5	323.0	720.0	0.165	B5	323.0	550.0	0.168
A6	377.0	720.0	0.165	B6	377.0	550.0	0.168
A7	323.0	880.0	0.165	B7	323.0	950.0	0.168
A8	377.0	880.0	0.165	B8	377.0	950.0	0.168
A9	304.6	800.0	0.150	B9	304.6	750.0	0.150
A10	395.4	800.0	0.150	B10	395.4	750.0	0.150
A11	350.0	665.4	0.150	B11	350.0	413.6	0.150
A12	350.0	934.6	0.150	B12	350.0	1086.4	0.150
A13	350.0	800.0	0.125	B13	350.0	750.0	0.120
A14	350.0	800.0	0.175	B14	350.0	750.0	0.180
A15	350.0	800.0	0.150	B15	350.0	750.0	0.150
A16	350.0	800.0	0.150	B16	350.0	750.0	0.150
A17	350.0	800.0	0.150	B17	350.0	750.0	0.150

Table 2 Chemical composition of Haynes 282 powder used in this study

Element	Cr	Co	Mo	Ti	Al	Fe	Mn	Nb	Si	C	B	Ni	O	N
Wt%	18.94	9.5	8.3	2.04	1.44	0.2	0.17	0.16	0.12	0.07	0.0005	Bal.	0.011	0.004

performed using a Zeiss AXIO microscope and electron microscopy was performed using a Zeiss EVO 50 Scanning Electron Microscope (SEM). The porosity measurements were performed in the as-polished condition. The samples were electrolytically etched with 10% oxalic acid and 3V for ~5–10 s for melt pool imaging and measurements.

A 7 × 7 mm square area was drawn centred on each sample, as shown in Fig. 1b and the porosities were identified using a set threshold value in the Zeiss imaging software. This area and the porosities within the 7 × 7 mm square area on the sample were utilized to calculate the defect area and other porosity-related measurements.

The top layer of the sample was imaged and used for measuring the depth and overlap depth of the melt pool. At least ten measurements of depth and ten measurements of overlap depth were performed for each sample as shown in Fig. 2.

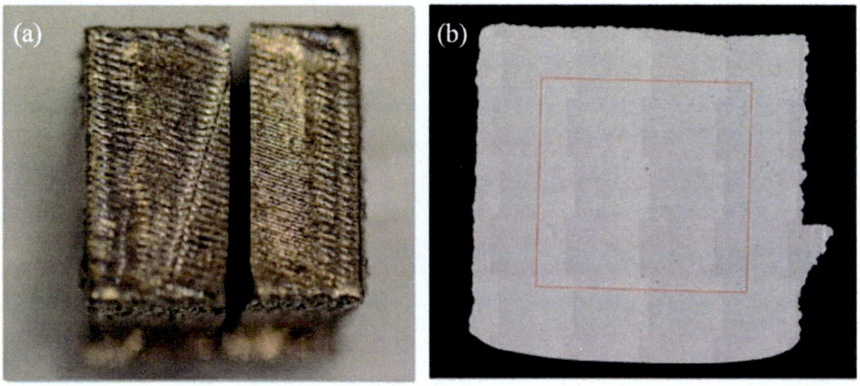

Fig. 1 **a** 10 × 10 mm top surface of the cube sample is shown after sectioning, the blurred portion at the bottom of the image is a marker for cube orientation in the machine. **b** The sectioned face is mounted, ground, and polished for further analysis, including a 7 × 7 mm square (shown in red) drawn on the centre of the cube for porosity analysis. The marker for cube orientation is seen on the right side of the cube. The build direction is from the bottom to the top of the image on the right

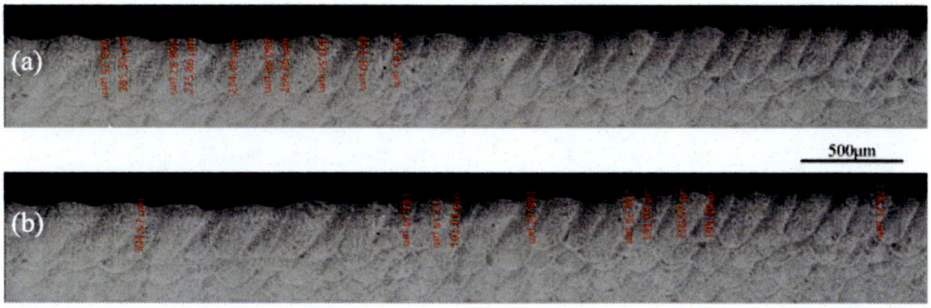

Fig. 2 Melt pool depth and melt pool overlap depth measurements were performed for all the samples. The example shown here is 60 micron A10 sample. **a** The melt pool depth was measured until the bottom of the melt pool was in the topmost layer. **b** Melt pool overlap was measured to the intersection of two melt pools from the top

Results and Discussion

Melt pool depth measurements varied between 100 and 350 microns and overlap depth measurements varied between 50 and 300 microns as shown in Fig. 3 for both 60 micron and 90 micron layer thickness. The extent of variation signifies the inherent scatter on melt pool behaviour even with the same set of parameters [23, 24]. The variation of average melt pool dimensions with process parameters such as power, speed, and hatch distance are presented for 60 micron and 90 micron samples in Figs. 4, 5 and 6.

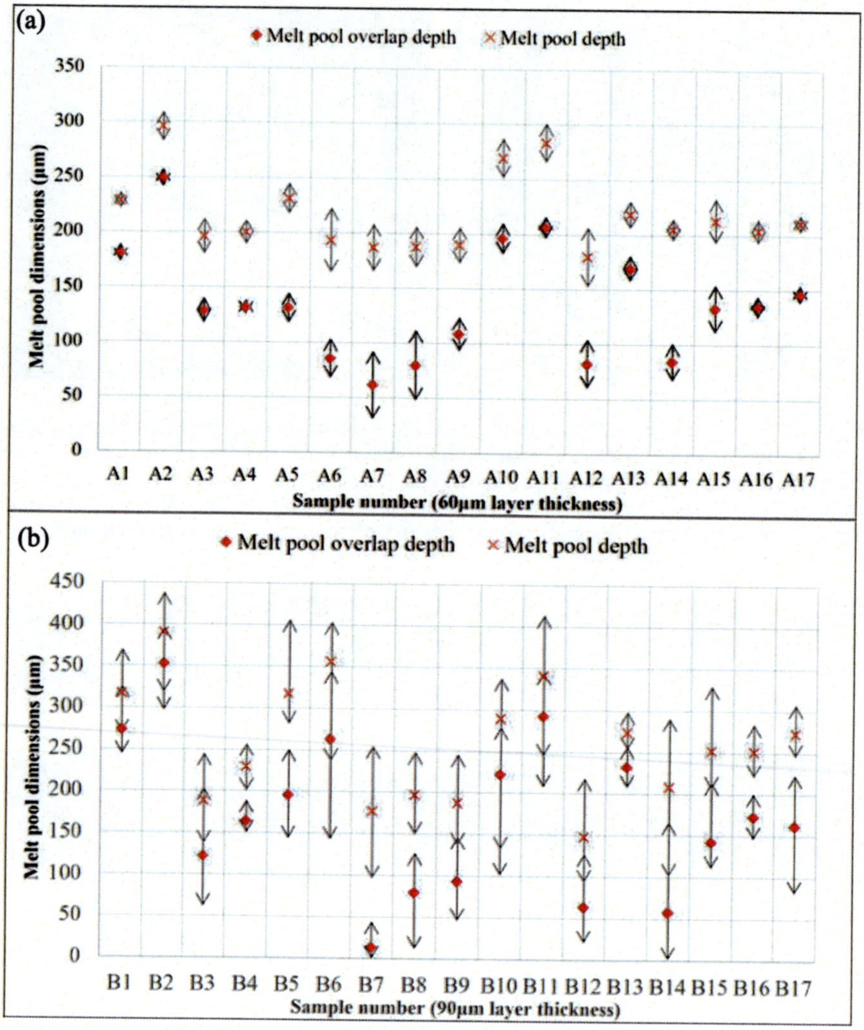

Fig. 3 Variation of melt pool depth and melt pool overlap depth (**a**) 60 microns and (**b**) 90 microns. Error bars signify the standard deviation of ten measurements in each sample

In the 60 micron and 90 micron samples, the influence of variation of power on melt pool depth and overlap does not seem to follow any specific trend. The effect of speed and hatch distance on the melt pool dimensions seems to have an overall inverse relation as expected. The increase in speed reduces the dwell time of the laser to interface with the melt pool resulting in lower values for melt pool depth and melt pool overlap depth. The increase in hatch distance increases the distance between two laser scan lines reducing the depth and overlap of the melt pool. The increase in hatch distance and speed affects the cumulative heat along and across the

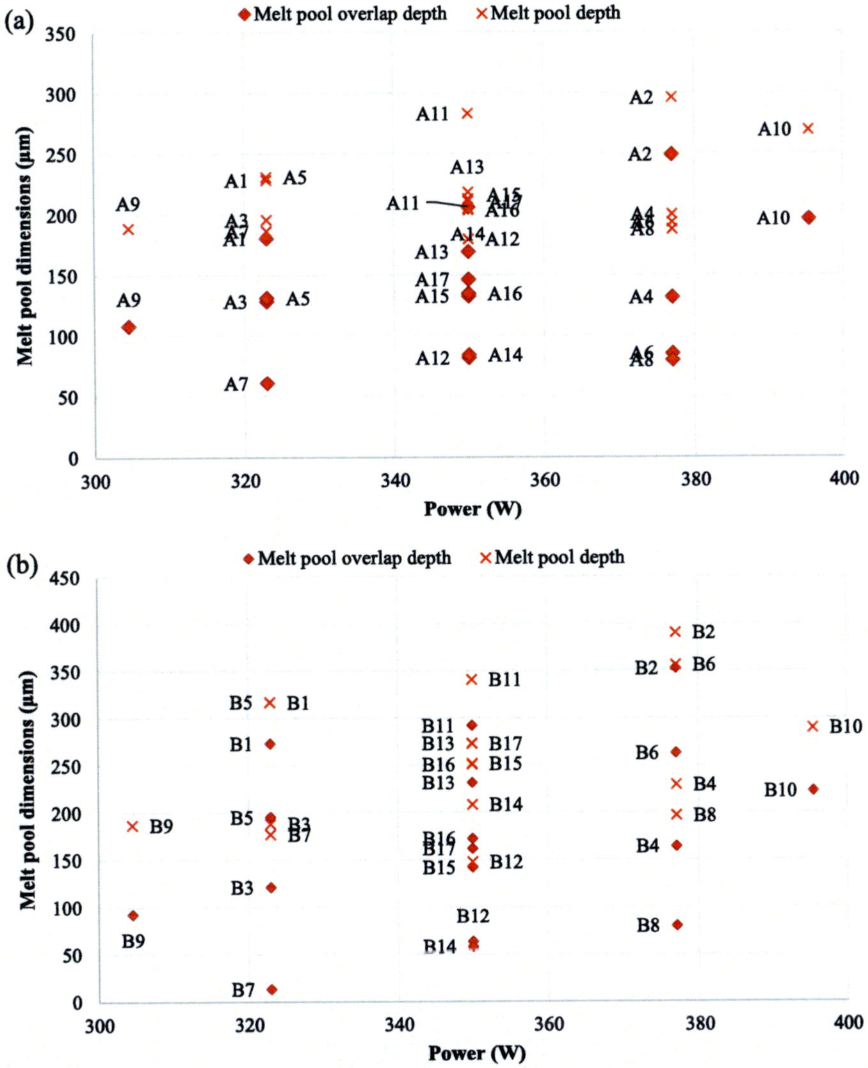

Fig. 4 Variation of average melt pool dimensions with power for **(a)** 60 microns and **(b)** 90 microns samples

laser scan lines resulting in reduced depth and overlap of melt pools. The variation of average melt pool dimensions with a single value of power or speed or hatch distance provides an insight into the possibilities of process modifications. If the overall set-up of manufacturing is either limited by the power of the laser, or reduced build time, the process variations allow completely remelted layers to be produced in the final samples or components to enhance part integrity.

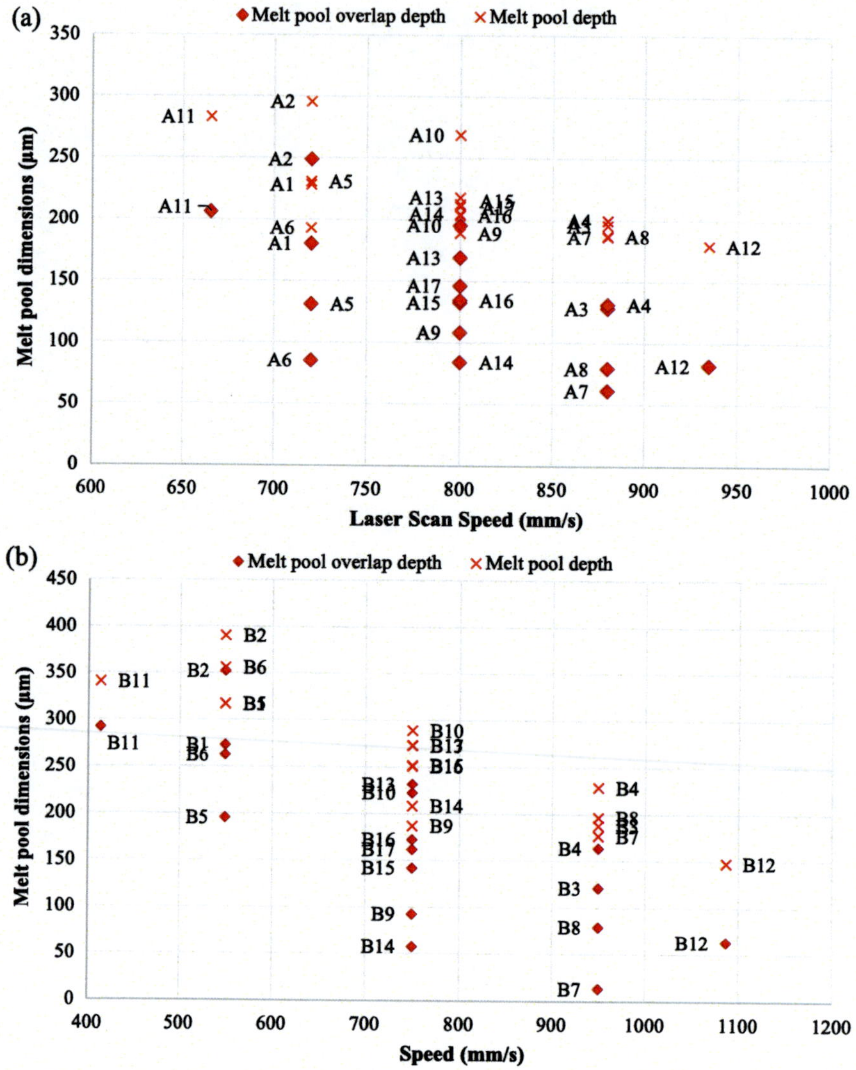

Fig. 5 Variation of average melt pool dimensions with Laser Scan Speed for (**a**) 60 microns and (**b**) 90 microns samples

The effect of individual and combined parameters on porosity and melt pool overlap depth studied using MODDE software similar to the 247LC study by Adegoke et al. [26], is shown in Fig. 7. Laser scan speed was the most significant parameter for porosity and hatch distance was the most significant parameter affecting the average melt pool overlap depth in the cubes built with 60 microns layer

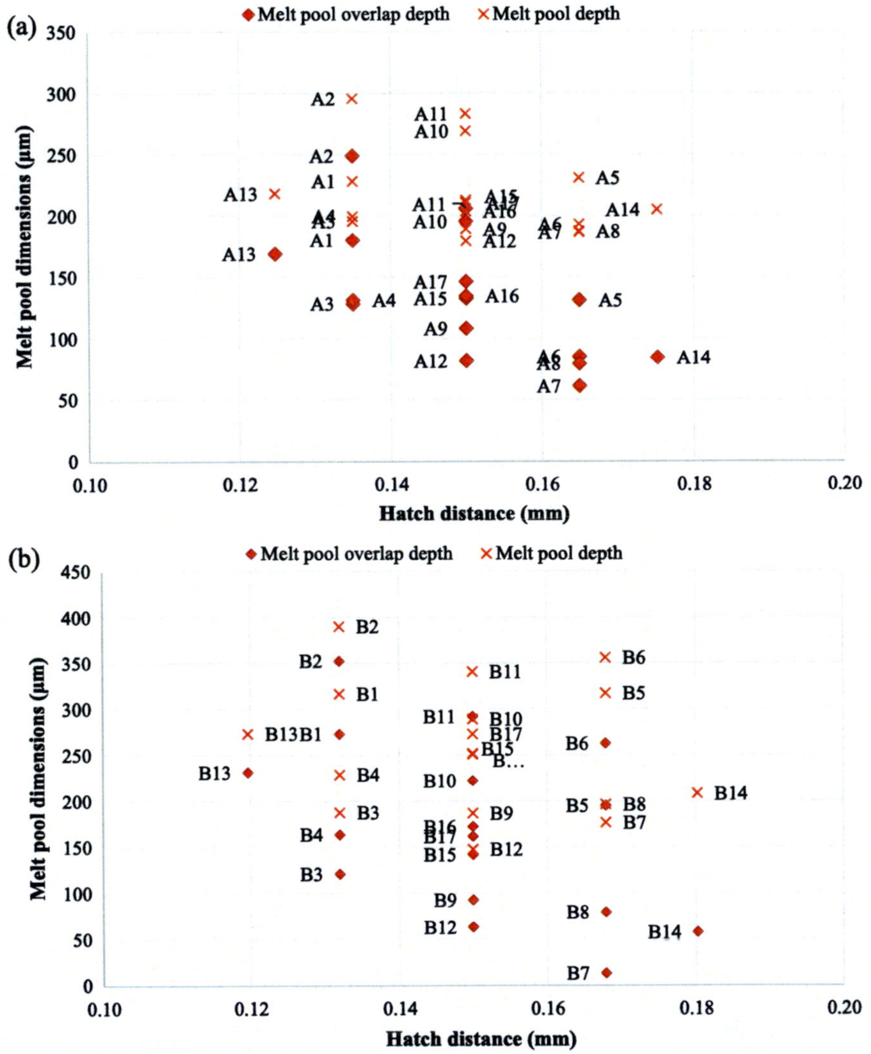

Fig. 6 Variation of average melt pool dimensions with Hatch distance for (**a**) 60 micron and (**b**) 90 micron samples

thickness. Interaction of speed and hatch distance was the most significant parameter for porosity and Laser scan speed was the most significant parameter for average melt pool overlap depth in cubes built with 90 microns layer thickness. Although the individual effect and combined effect of speed and hatch distance are apparent in the graphs, there are some overlapping values with mixed effects of the parameters, that can be separated further through line energy and area energy density, which are discussed later. It is also noted that the laser power classified as an insignificant parameter, may be valid within the tested parameter range of speed and hatch distance, but the Laser power is imminent and important in the PBF-LB process. For example, samples A9 and A10 and B9 and B10 have only the power parameter at the extremes with the same values for all other parameters. The melt pool depth is almost 50% higher and the melt pool overlap is almost 100% higher for samples A10 and B10, with higher power compared to A9 and B9, respectively.

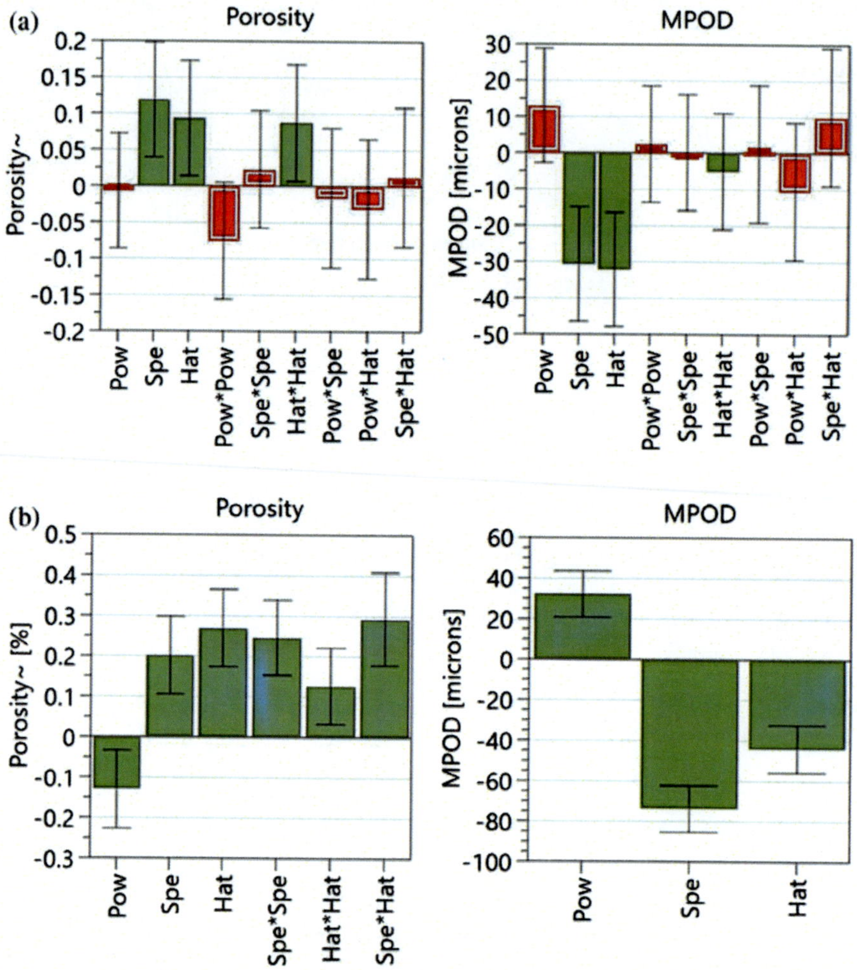

Fig. 7 Coefficient of each factor with respect to porosity and melt pool overlap depth (MPOD) analysed through MODDE software for (**a**) 60 microns and (**b**) 90 microns samples. The significant factors are shown in green and insignificant factors are shown in red. All the factors were considered for both plots, but only significant factors are shown for 90 micron plot

Line energy (Laser Power/Laser scan speed) considers the amount of energy available in each laser scan line. Melt pool dimensions increased with increasing line energy in general, although multiple values for the same line energy due to the variation in hatch distance can be seen in Figs. 8 and 9, with A6 being a clear outlier. To further include the effect of power, speed, and hatch distance in a single parameter, the area energy (power divided by the product of speed and hatch distance) is plotted against the melt pool dimensions in Fig. 10. Although there is not a distinct trend, there is an overall increase in melt pool depth and the melt pool overlap depth starting from 2 to 4 J/mm^2 in 60 micron samples. For the 90 micron samples, the trend remains the same, with 2 to 5 J/mm^2, but the highest energy density sample (B11) has marginally lesser melt pool overlap than the previous sample (B2). This could be due to the onset of transition from conduction to keyhole mode of melting due to higher energy input, or orientation of the previous laser scan layer or both. The keyhole manifests in the microstructure as voids close to the bottom of the melt pool due to the slow nature of the gas entrapment to exit the gas–liquid interface before the collapse of the keyhole. There are a few outliers from the linear trend, especially A6, which exhibited uneven top layer, either due to the scan strategy or inherent nature of scatter present in the laser melting process as shown by Cunningham et al. [23], resulting in measurements lesser than expected level for the given area energy. But the overall effect of area energy in increasing the melt pool dimensions is understandable given that the area energy captures the overall heat available in the given area of laser scan.

All the melt pools had a conduction mode melting due to the focus offset parameter set to +4 for all the cubes. The focus plane of the laser was approximately 4 mm above the build plate during the manufacture of the cubes. The defocussed beam provides conduction mode melting even at high energy inputs [27].

The ferret ratio is a measure of the proportion of the largest line (between parallel planes) that can be drawn inside a defect to that of the smallest line that can be drawn inside the same defect. The intent behind this measurement is to provide an idea about the shape of the defect. A ferret ratio of 1 signifies a circular defect and the extent of the ratio less than 1 signifies the amount of linearity or non-circularity of the defect. The shape of the defect is usually a critical factor during static or cyclical loading. The ferret ratio varied between 0.7 and 1 for 60 micron samples and 0.5 to 1 for 90 micron samples. The variation shows the extent of the variation of process parameters and the resultant area energy of the 90 micron samples was wider than that of 60 micron samples. If the shape of the defect is close to linear than circular (designated by the difference in ferret ratio average from 1) the mechanical behaviour, especially fatigue properties are expected to be lower depending on the loading direction. The ferret ratio was averaged for each sample based on the defects present in the sample, that are greater than 100 μm^2 in area, to reduce the skewing of data based on small spherical pores or particles present.

The porosity measurement and ferret ratio for porosities with an area greater than 100 μm^2 against area energy is plotted in Fig. 11. All the porosity values were below 1% for 60 micron layer thickness samples. All the porosity values were below 4% for 90 micron layer thickness samples. The increase in layer thickness could be the

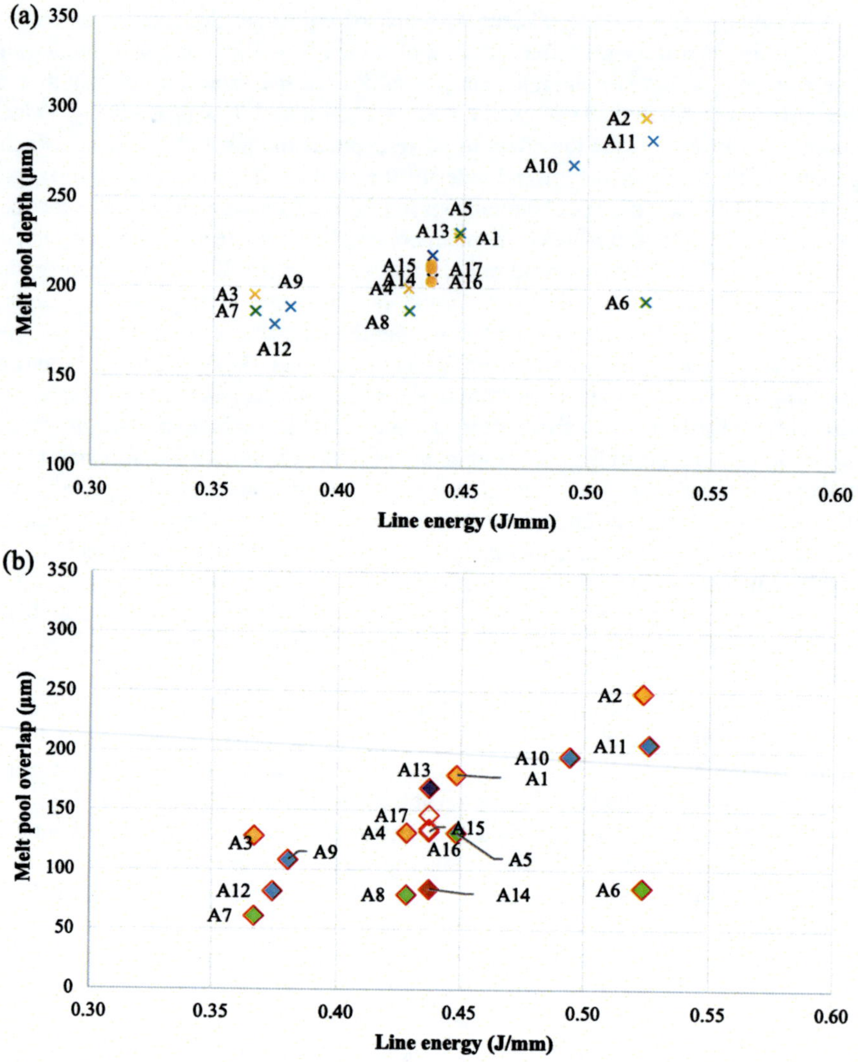

Fig. 8 Variation of (**a**) average melt pool depth and (**b**) melt pool overlap with line energy. The colours represent various hatch distance settings. A15, A16, and A17 are samples with the same process parameter settings

reason for the increase in porosity levels. The general PBF-LB processability for structurally integral parts using Haynes 282 is good within the process parameter range chosen and the type of build tilized in this study. Although the porosity is low for 60 micron samples and varied for 90 micron samples, it is a measure using a 2D section of the cube, so further volumetric analysis can provide more insights into the defect distribution. In general, the "porosity seem" to reduce with increased area

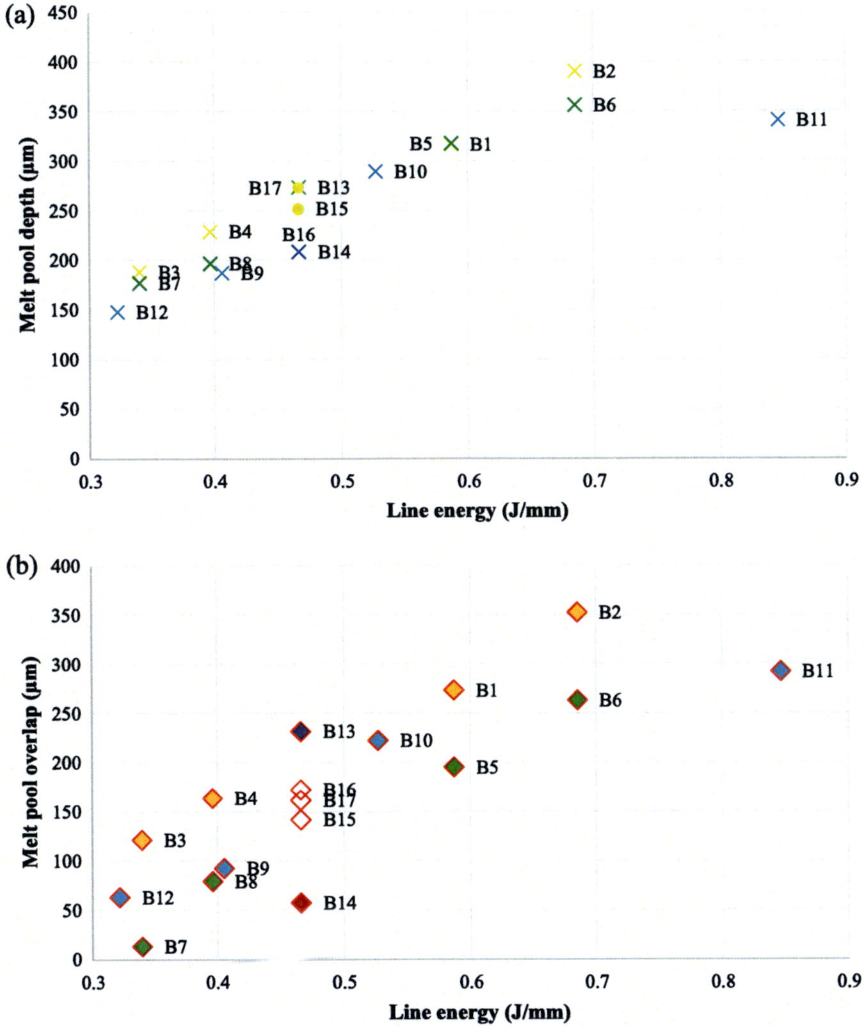

Fig. 9 Variation of (**a**) average melt pool depth and (**b**) melt pool overlap with line energy. The colours represent various hatch distance settings. B15, B16, and B17 are samples with the same process parameter settings

energy density and melt pool overlap depth, possibly due to the increased remelting of underlying layers.

From Figs. 10 and 11, the melt pool overlap depth can be correlated to the porosity level and to the ferret ratio average. The lack of fusion porosity is expected to be minimised when the overlap depth is at least equal or greater than the layer thickness, due to the repeated remelting of the underlying layer and filling in the lack of fusion if present [28]. In this study, only one sample (A7) in 60 micron sample study and

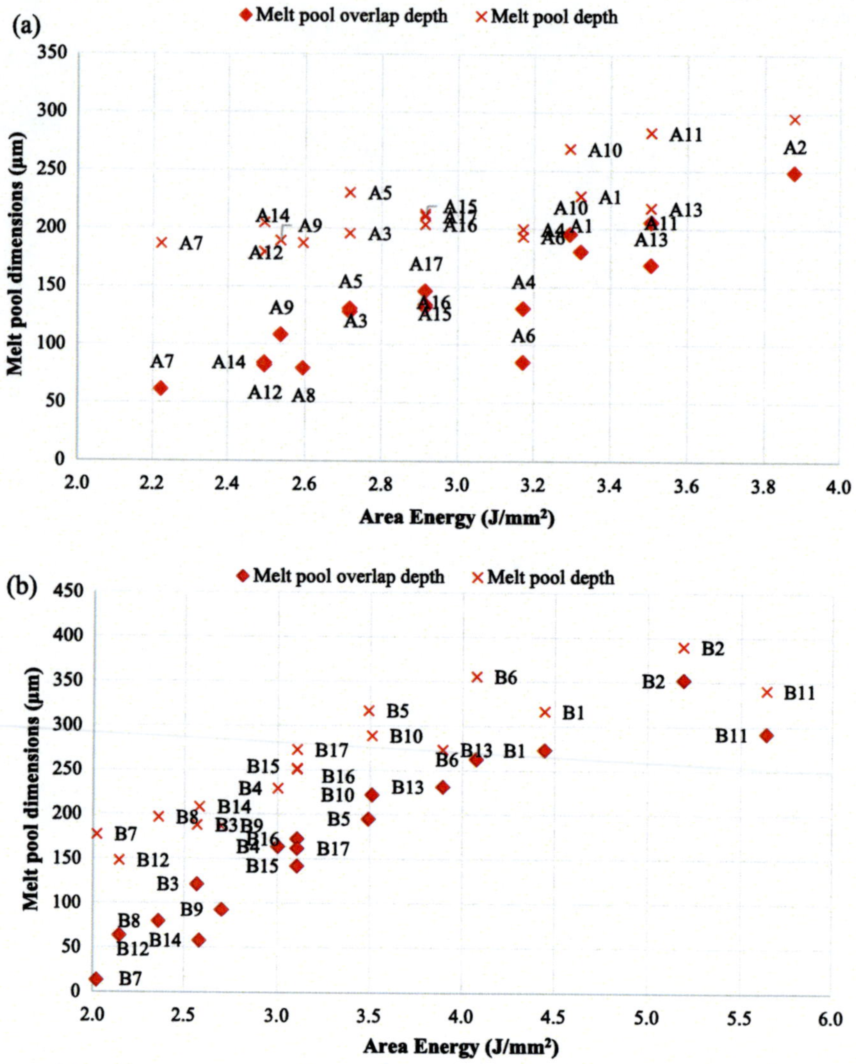

Fig. 10 Variation of average melt pool depth and melt pool overlap depth with area energy for **(a)** 60 micron and **(b)** 90 micron samples

about 4–5 samples in 90 micron study seem to have melt pool overlap less than the layer thickness. In 60 micron samples, A7 seems to have the lowest melt pool overlap and highest porosity. The same is true for 90 micron sample B7. The SEM images shown in Fig. 12 show the lack of fusion prevalent in those samples, more so in B7 due to the higher layer thickness. It is also noticeable that the samples having melt pool overlap depth lesser than or equal to layer thickness also have a ferret ratio of

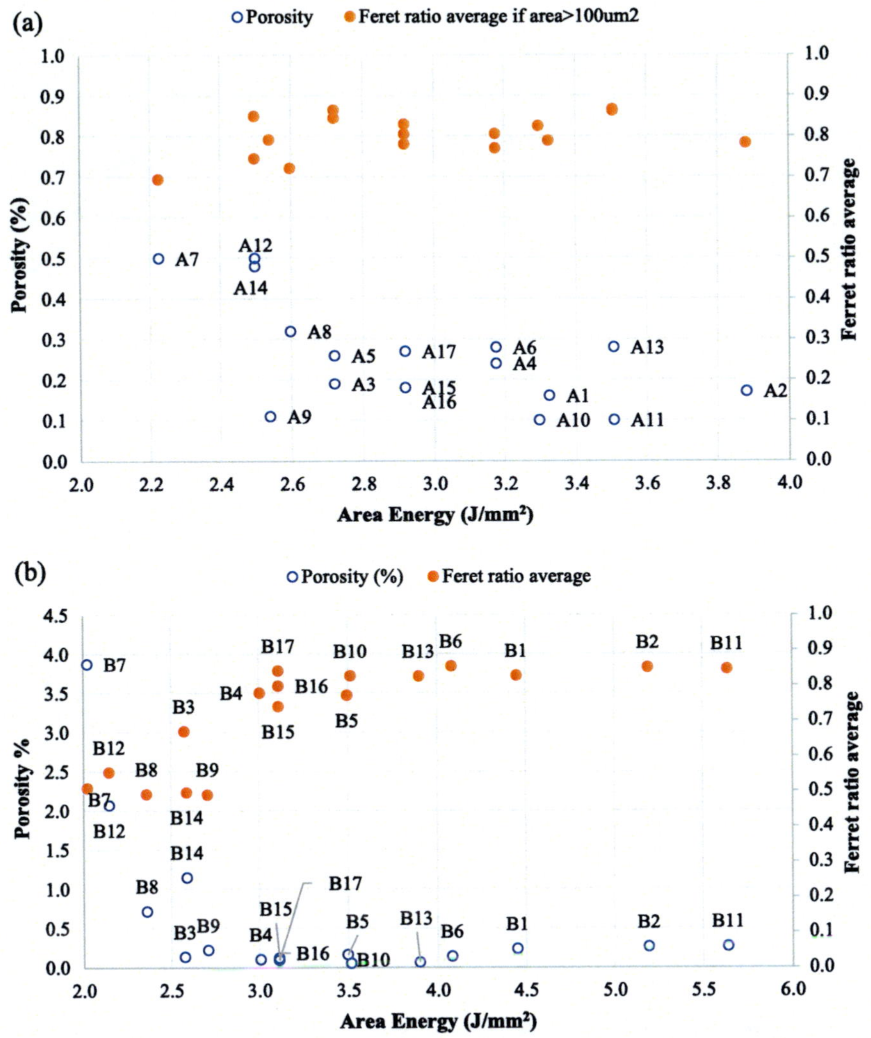

Fig. 11 Variation of porosity and ferret ratio average of defects with area >100 μm^2 with area energy density for (**a**) 60 micron and (**b**) 90 micron samples

0.7 or lower, hinting at more linear defects than round defects predominantly due to lack of fusion. This correlation of melt pool overlap depth with porosity level and dominant shape of the defects present can be utilized in planning different process parameters as per the limitations of the machine.

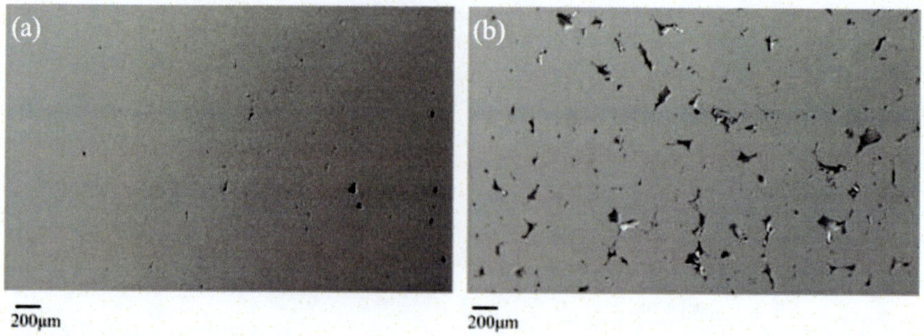

Fig. 12 SEM image of samples (**a**) A7 and (**b**) B7 showing lack of fusion defects due to the lowest energy input in the respective set of experiments. The build direction is from the bottom to the top of the image

Conclusion

In this study, the effect of individual process parameters on melt pool dimensions and porosity was analysed. Although individual or combined parameters of speed and hatch distance were found to be the most significant parameter for different layer thicknesses, line energy and area energy density were shown to better correlate to the trends of the melt pool dimensions, porosity level, and average shape of defects. The area energy density captures the overall heat incident on the powder particles and hence provides the best parameter to determine the melt pool behaviour and the resultant integrity of the built components. The variation of melt pool dimensions can be used to determine the layer thickness and building time trade off, between layer thickness and hatch distance. If higher laser power is available, the higher layer thickness can be used at a lower hatch distance and still have a reasonable building time. If the laser power is limited, then to optimize build time, hatch distance and layer thickness can be varied without compromising on quality of the build. We have addressed the first part of the trade off, showing even if the hatch distance is lower, higher layer thickness can be used to improve the build rate. The melt pool overlap depth in comparison to layer thickness can be correlated to the porosity level and the dominant shape of the defects present.

References

1. Haynes International (2020) HAYNES ® 282 ® alloy Material Brochure. https://haynesintl.com/docs/default-source/pdfs/new-alloy-brochures/high-temperature-alloys/brochures/282-brochure.pdf?sfvrsn=20
2. Sames, W. J., List, F. A., Pannala, S., Dehoff, R. R., & Babu, S. S. (2016) The metallurgy and processing science of metal additive manufacturing. International Materials Reviews, 61(5), 315–360. https://doi.org/10.1080/09506608.2015.1116649

3. Gibson, I., Rosen, D., & Stucker, B. (2015) Additive manufacturing technologies: 3D printing, rapid prototyping, and direct digital manufacturing, second edition. https://doi.org/10.1007/978-1-4939-2113-3
4. Aconity MIDI | Aconity3D. Retrieved June 28, 2022, https://aconity3d.com/products/aconity-midi/
5. Deng, D. (2019) On the microstructures and anisotropic mechanical behaviours of additively manufactured in718, Doctoral thesis, Linköping University.
6. Kirka, M. M., Lee, Y., Greeley, D. A., Okello, A., Goin, M. J., Pearce, M. T., & Dehoff, R. R. (2017) Strategy for Texture Management in Metals Additive Manufacturing. The Journal of The Minerals, Metals & Materials Society 69, 523–531. https://doi.org/10.1007/s11837-017-2264-3
7. Raghavan, N. (2017) Understanding Process-Structure Relationship for Site-Specific Microstructure Control in Electron Beam Powder Bed Additive Manufacturing Process Using Numerical Modeling, Doctoral Dissertations, The University of Tennessee, Knoxville.
8. Popovich, V. A., Borisov, E. v., Popovich, A. A., Sufiiarov, V. S., Masaylo, D. v., & Alzina, L. (2017) Impact of heat treatment on mechanical behaviour of Inconel 718 processed with tailored microstructure by selective laser melting. Materials & Design, 131, 12–22. https://doi.org/10.1016/J.MATDES.2017.05.065
9. Tillmann, W., Schaak, C., Nellesen, J., Schaper, M., Aydinöz, M. E., & Hoyer, K. P. (2017) Hot isostatic pressing of IN718 components manufactured by selective laser melting. Additive Manufacturing, 13, 93–102. https://doi.org/10.1016/J.ADDMA.2016.11.006
10. Raza, T., Andersson, J., & Svensson, L. E. (2018) Microstructure of selective laser melted alloy 718 in as manufactured and post heat treated condition. Procedia Manufacturing, 25, 450–458. https://doi.org/10.1016/J.PROMFG.2018.06.100
11. Goel, S., Olsson, J., Ahlfors, M., Klement, U., & Joshi, S. (2018) The effect of location and post-treatment on the microstructure of EBM-built alloy 718. Minerals, Metals and Materials Series, 2018-June, 115–129. https://doi.org/10.1007/978-3-319-89480-5_6/COVER/
12. Balachandramurthi, A. R., Moverare, J., Mahade, S., & Pederson, R. (2018) Additive manufacturing of alloy 718 via electron beam melting: Effect of post-treatment on the microstructure and the mechanical properties. Materials, 12(1) https://doi.org/10.3390/ma12010068
13. Unocic, K. A., Kirka, M. M., Cakmak, E., Greeley, D., Okello, A. O., & Dryepondt, S. (2020) Evaluation of additive electron beam melting of Haynes 282 alloy. Materials Science and Engineering A, 772. https://doi.org/10.1016/j.msea.2019.138607.
14. Fernandez-Zelaia, P., Kirka, M. M., Dryepondt, S. N., & Gussev, M. N. (2020) Crystallographic texture control in electron beam additive manufacturing via conductive manipulation. Materials and Design, 195. https://doi.org/10.1016/j.matdes.2020.109010.
15. Otto, R., Brøtan, V., Carvalho, P. A., Reiersen, M., Graff, J. S., Sunding, M. F., Berg, O. Å., Diplas, S., & Azar, A. S. (2021) Roadmap for additive manufacturing of HAYNES® 282® superalloy by laser beam powder bed fusion (PBF-LB) technology. Materials and Design, 204. https://doi.org/10.1016/j.matdes.2021.109656.
16. Deshpande, A., Nath, S. D., Atre, S., & Hsu, K. (2020) Effect of post processing heat treatment routes on microstructure and mechanical property evolution of Haynes 282 Ni-based superalloy fabricated with selective laser melting (SLM). *Metals*, *10*(5) https://doi.org/10.3390/met10050629.
17. Boswell, J., Jones, J., Barnard, N., Clark, D., Whittaker, M., & Lancaster, R. (2021) The effects of energy density and heat treatment on the microstructure and mechanical properties of laser additive manufactured Haynes 282. Materials and Design, 205. https://doi.org/10.1016/j.matdes.2021.109725.
18. Shaikh, A. S., Schulz, F., Minet-Lallemand, K., & Hryha, E. (2021) Microstructure and mechanical properties of Haynes 282 superalloy produced by laser powder bed fusion. Materials Today Communications, 26. https://doi.org/10.1016/j.mtcomm.2021.102038.
19. Ghiaasiaan, R., Ahmad, N., Gradl, P. R., Shao, S., & Shamsaei, N. (2022) Additively manufactured Haynes 282: effect of unimodal vs. bimodal γ'- microstructure on mechanical properties. Materials Science and Engineering A, 831. https://doi.org/10.1016/j.msea.2021.142234

20. Christofidou, K. A., Pang, H. T., Li, W., Pardhi, Y., Jones, C. N., Jones, N. G., & Stone, H. J. (2020) Microstructural Control and Optimization of Haynes 282 Manufactured Through Laser Powder. Superalloys 2020 Minerals, Metals and Materials Series, 1014–1023. https://doi.org/10.1007/978-3-030-51834-9_99
21. Islam, Z., Agrawal, A. K., Rankouhi, B., Magnin, C., Anderson, M. H., Pfefferkorn, F. E., & Thoma, D. J. (2022). A High-Throughput Method to Define Additive Manufacturing Process Parameters: Application to Haynes 282. Metallurgical and Materials Transactions A: Physical Metallurgy and Materials Science, 53(1), 250–263. https://doi.org/10.1007/s11661-021-06517-w.
22. Hariharan, V. S., Pramod, S., Kesavan, D., Murty, B. S., & Phanikumar, G. (2022). ICME framework to simulate microstructure evolution during laser powder bed fusion of Haynes 282 nickel-based superalloy. Journal of Materials Science, 57(21), 9693–9713. https://doi.org/10.1007/s10853-022-07170-3.
23. Cunningham, R., Zhao, C., Parab, N., Kantzos, C., Pauza, J., Fezzaa, K., Sun, T., & Rollett, A. D. (2019a). Keyhole threshold and morphology in laser melting revealed by ultrahigh-speed x-ray imaging. In Science (Vol. 363). https://www.science.org
24. Khairallah, S. A., Martin, A. A., Lee, J. R. I., Guss, G., Calta, N. P., Hammons, J. A., Nielsen, M. H., Chaput, K., Schwalbach, E., Shah, M. N., Chapman, M. G., Willey, T. M., Rubenchik, A. M., Anderson, A. T., Wang, Y. M., Matthews, M. J., & King, W. E. (2020a). Controlling interdependent meso-nanosecond dynamics and defect generation in metal 3D printing. In Science (Vol. 368). https://www.science.org
25. Vallabh, C. K. P., Sridar, S., Xiong, W., & Zhao, X. (2022). Predicting melt pool depth and grain length using multiple signatures from in-situ single camera two-wavelength imaging pyrometry for laser powder bed fusion. Journal of Materials Processing Technology, 308, 117724. https://doi.org/10.1016/j.jmatprotec.2022.117724.
26. Adegoke, O., Andersson, J., Brodin, H., & Pederson, R. (2020). Influence of laser powder bed fusion process parameters on voids, cracks, and microhardness of nickel-based superalloy alloy 247LC. Materials, 13(17). https://doi.org/10.3390/MA13173770
27. Metelkova, J., Kinds, Y., Kempen, K., de Formanoir, C., Witvrouw, A., & van Hooreweder, B. (2018). On the influence of laser defocusing in Selective Laser Melting of 316L. Additive Manufacturing, 23, 161–169. https://doi.org/10.1016/j.addma.2018.08.006.
28. Tang, M., Pistorius, P. C., & Beuth, J. L. (2017). Prediction of lack-of-fusion porosity for powder bed fusion. Additive Manufacturing, 14, 39–48. https://doi.org/10.1016/j.addma.2016.12.001.

Investigating the Influence of Build Parameters and Porosity on Fatigue of AM IN718

Alexander Caputo, Richard W. Neu, Chaitanya Vallabh, Xiayun Zhao, and Haolin Zhang

Abstract Using laser powder bed fusion additive manufacturing (L-PBF AM), a series of 10 sample walls were made, sectioned into fatigue specimens, and tested using high cycle fatigue (HCF) testing with a stress ratio of 0.1 at 538 °C. Each wall was built with a different process pedigree, or set of process parameters, both to explore the effects of different process regimes (conduction, transition, keyhole) on the porosity and microstructure of AM IN718 and to serve as a database of AM process conditions and their resultant defects. The internal porosity in the full gage regions of all fatigue specimens was characterized prior to HCF testing using X-ray computed tomography. Following fatigue testing, SEM fractography was used to identify the locations of fatigue critical flaws that led to failure. Using the data acquired in this work, the relationships between fatigue performance of IN718 at high temperature and the process conditions and associated porosity will be interpreted. These relationships could be adapted and used in a quality assurance model for L-PBF AM IN718.

Keywords Additive manufacturing · Alloy 718 · High-cycle fatigue · XCT

A. Caputo (✉) · R. W. Neu
School of Materials Science and Engineering, Georgia Institute of Technology, Atlanta, GA, USA
e-mail: acaputo7@gatech.edu

R. W. Neu
Woodruff School of Mechanical Engineering, Georgia Institute of Technology, Atlanta, GA, USA

C. Vallabh · X. Zhao · H. Zhang
Mechanical Engineering & Materials Science, University of Pittsburgh, Pittsburgh, PA, USA

C. Vallabh
Department of Mechanical Engineering, Stevens Institute of Technology, Hoboken, NJ, USA

E. A. Ott et al. (eds.), *Proceedings of the 10th International Symposium on Superalloy 718 and Derivatives*, The Minerals, Metals & Materials Series,
https://doi.org/10.1007/978-3-031-27447-3_36

Introduction

Additive manufacturing provides an innovative solution for modern engineering challenges, but with all the potential gains, this technology can provide some challenges need to be overcome. The microstructure and properties of additively manufactured materials are dependent on the local process parameters experienced by any specific region of the build, the thermal history and fluctuations caused by subsequent layers, and the conditions of the previous layer surface they are printed on. The localized microstructure dependence is due to the variance of process parameters around the build plate and build volume. Whether it be preferential spatter downwind of the inert gas flow, poor powder spreading along the recoater blade path, laser intensity variance caused by vapor clouds created during the melting step, or localized increases or decreases in volumetric energy density caused by part geometries, there are many types of nonuniformities introduced during the additive process that cause localized variance in process conditions [1–5]. While we can conduct studies to determine the optimum process parameters for a certain material, we cannot trust that all regions of the build will experience these optimized parameters as the AM process currently stands [6, 7]. It is because of this understanding that this work is being conducted. We are developing an effective means for quality assurance for AM parts so that even with all the localized variance in process parameters and the defects in the microstructure created by them, we will still be able to qualify the fatigue properties of AM IN718 components. This work is also contributing to a longer-term goal of creating a system capable of recognizing defects in real time and correcting them.

Due to the prevalence of IN718 in aerospace and energy turbine manufacturing, AM of this alloy yields a specific advantage compared to conventional methods. Not only does AM enable the manufacturing of complex geometries impossible through conventional IN718 forming methods, but also the increased design freedom and the speed of prototyping new components provided by AM help to foster more innovation and efficiency. In turbine-based applications where fatigue becomes the critical property, in addition to new design freedoms, AM introduces microstructural anomalies in the form of porosity and unconventional grain structures which can have a dramatic effect on the spread of the fatigue lives of components.

In tandem with this work, multiple in-process monitoring systems are employed during the L-PBF AM process to track the laser shape and temperature as well as its position and any spatter it might generate [8, 9]. These in-process data will be associated with post-process X-ray computed tomography (XCT) and scanning electron microscope (SEM)-based electron backscatter diffraction (EBSD) characterization conducted on a set of L-PBF AM IN718 components with varying combinations of optimal and non-optimal process parameters. Using these data and the correlations between in-situ monitoring and ex-situ characterization, the relationships between the in-situ monitoring and fatigue properties can be determined with the goal of creating a system capable of certifying part quality from in-situ monitoring. It is by studying the relationships between the in-process monitoring data, the post-process data, and the fatigue properties that we will be able to develop a system capable of efficiently assuring the quality of the part as soon as it is taken off the build plate [8, 9].

Due to the nature of AM-induced anomalies in the IN718 microstructure, studies in the literature have shown that AM IN718 can have fatigue performance significantly less than that of its wrought counterpart [10–14], though other studies show that optimum processing can result in fatigue performance directly comparable to that of wrought [12, 14]. This large variance is expected due to the stochastic nature of fatigue, but the inconsistencies in the manufacturing and testing of AM alloys in general, IN718 included, further inflate this variance. This work aims to gauge a statistically significant amount of AM IN718 specimens from varying process conditions using the same surface condition and testing apparatus to understand the distribution of fatigue behavior without many of the confounding factors found when compiling results from other studies.

Previous works conducted by Scime and Beuth have shown that in-process monitoring data can be directly correlated to post-process characterization in L-PBF AM IN718 [15, 16]. Li et al. has used similar methods to derive relationships between process-induced defects and fatigue properties of Ti-6Al-4 V [17]. Through these works in tandem, it has been shown that linkages between process monitoring data and process-induced defects can be correlated directly to the fatigue behavior of metallic AM alloys. In work by Luo et al. they have shown that porosity in AM IN718 can be used to understand and predict fatigue properties [18]. Luo has solidified that porosity specifically can influence the fatigue behavior of AM IN718. Through these studies, the framework applied in this work has proven successful for other properties and material systems.

Motivated by these other studies in the field, this work aims to understand how AM material microstructure is related to fatigue properties in IN718 through various in-process and post-process techniques [19].

The goal of this work is to understand the relationship between the post-process characterization data and the fatigue properties of the AM IN718. To achieve this end, x-ray computed tomography characterization of AM IN718 fatigue specimens is analyzed in relation to their fracture surfaces and their respective fatigue lives. The findings of this work are directly contributing to the larger goal of establishing an effective quality assurance method for AM materials using the in-process data.

Methods

Build Plate Layout

The L-PBF AM IN718 fatigue specimens were extracted from a single build in an EOS M290 DMLS printer running EOSPRINT 2.6. The AM machine runs using a 1 Yb single mode fiber laser with a max power of 400 W running continuously.

The fatigue specimens were extracted from 16 sample walls of dimension 77.07 × 8 × 80 mm and with cross section in shape of fatigue specimens manufactured each using different process pedigrees, or build parameter combinations. Their geometry and locations on the build plate are shown in Fig. 1. For this work specifically, only process pedigrees 6 through 15 were used because the other process pedigree samples were unfortunately lost in shipping. Some contextual information relevant to Fig. 1 is that inert Ar gas flow is from the top to the bottom in the image, or from the back to the front in the machine, and the recoater blade direction is from the right to the left. The relevant process parameters and their respective volumetric energy densities are calculated by the equation

$$VED = \frac{P}{t * v * h}$$

where P is laser power in W, t is layer thickness in μm, which is 40 μm for all pedigrees, v is scanning speed in mm/s, h is hatch spacing in μm, and VED is the volumetric energy density given in J/mm^3; shown in Table 1. The physical meaning of most process parameters is simple, but to be complete: laser power is the target power set for the laser as it scans the bulk of the sample, this parameter is altered by machine subroutines for contour scans, Layer thickness is the distance the build plate moves down between subsequent layers. Scanning speed is the absolute velocity of the continuous laser spot as it rasters through the 2D sample geometries. Hatch spacing is the distance between the centers of adjacent laser paths. It is relevant to note here that while the hatch spacing varies from 80 to 120 μm, the laser spot size was 100 μm on average. The final relevant build parameters that are consistent through each process pedigree are the scanning strategy which involved stripes with

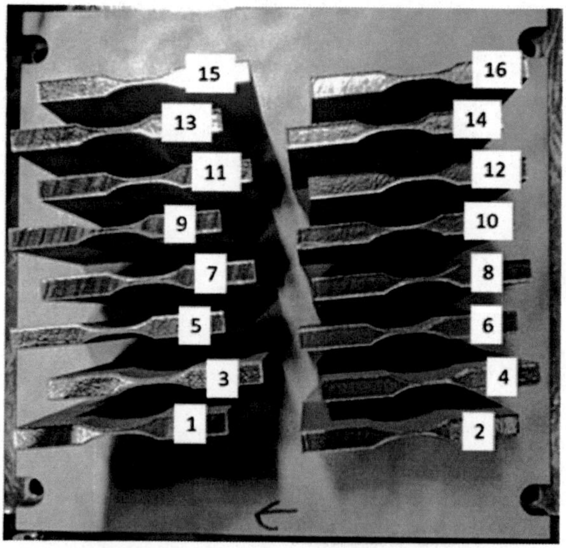

Fig. 1 Image of sample walls before removal from build plate with front of machine at the bottom

Table 1 Process pedigrees listed with their corresponding AM process parameters and calculated volumetric energy densities

Process pedigree	Process regime	Power (W)	Scan speed (mm/s)	Hatch spacing (μm)	Volumetric energy density (J/mm³)
6	Transition	300	1000	110	68.18
7	Keyhole	350	1000	110	79.55
8	Keyhole	200	500	110	90.91
9	Keyhole	250	500	110	113.64
10	Keyhole	300	500	110	136.36
11	Conduction	200	1000	80	62.50
12	Conduction	200	1000	120	41.67
13	Keyhole	250	500	80	156.25
14	Keyhole	250	500	120	104.17
15	Conduction	200	1500	110	30.30

a 67 ° rotation between layers. The build plate was C45 carbon steel and preheated to 80 °C. The metal powder was IN718 with a size range of 20–63 μm created by vacuum induction melting and inert gas atomization by VDM (Werdohl, Germany) but was not virgin powder in this build and had a reuse fraction of roughly 80%. The combinations of the process parameters used across these different pedigrees amount to three main process regimes: the conduction, transition, and keyhole regime. The differentiating factor demarking these regimes is the way the laser interacts with the metal powder. In the conduction regime, shallow melt pools are created due to the lesser heat input. In the keyhole regime, the melt pools created are deep and turbulent due to high heat input that even causes metal vaporization. The heat input of the conduction regime often manifests in the creation of a lack of fusion porosity, while the heat input of the transition regime often manifests in keyhole pores of trapped gas. The transition regime between these extremes can contain defects from both, and its melt pool morphology reflects aspects of both as well. The defects common in these regimes are displayed in Fig. 2 adapted from Mostafaei et al. [20].

In-process Monitoring

Collaborators on this work developed and implemented an in-process monitoring system composed of two cameras: an on-axis camera set up with a beam splitter to track the melt pool size, shape, and temperature, and an off-axis camera to track laser position and any occurrences of spatter as shown in Fig. 3. These systems in tandem are used to register the in-process monitoring data to the sample specific coordinate axes. It is planned to later investigate the relationships between the data acquired in-process and the post-process data described in this work.

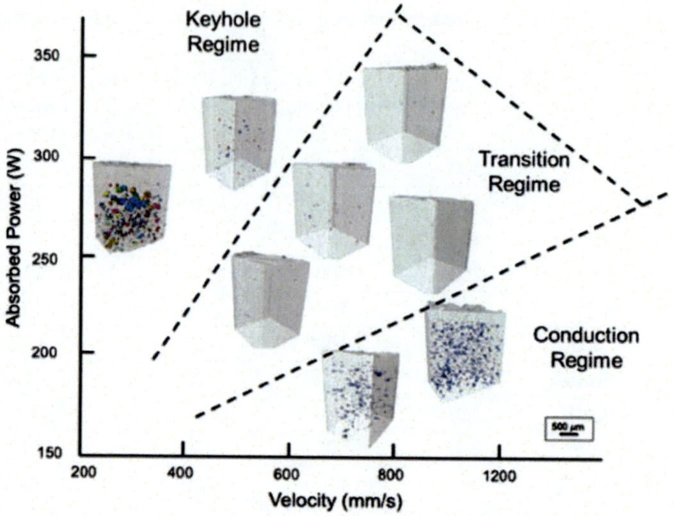

Fig. 2 Process window diagram showing the different process regimes for IN718 with corresponding characteristic porosities adapted from Mostafaei et al. [20]

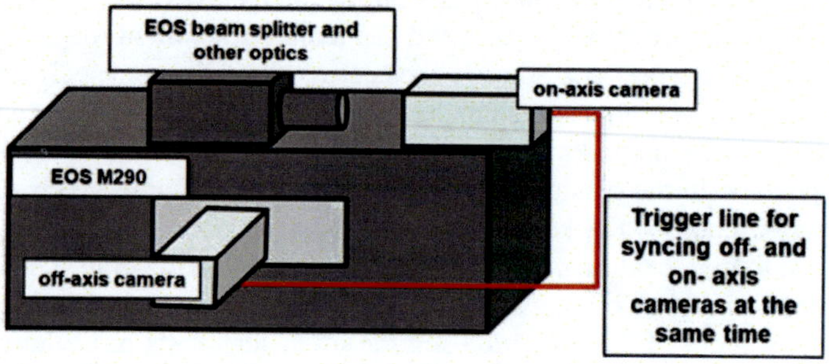

Fig. 3 Schematic of the on-axis and off-axis in-process monitoring cameras

Sample Heat Treatment Procedure

In order to fit into the tube furnace used for heat treatments, each sample wall was first sectioned at half the build height, which was the 80 mm dimension. To preserve the microstructure resulting directly from the AM build conditions, the heat treatment procedure chosen for this work was a direct age procedure involving no homogenization or solution heat treatment steps prior to aging. This direct age treatment promotes the formation of strengthening γ' and γ'' precipitates while preserving the as-built microstructure. The protocol consisted of heating all sectioned sample walls in a tube furnace in an Ar atmosphere at 720 °C for 8 h with a furnace cool to 620 °C and held for 8 h before air cooling to room temperature [21–23]. No additional processing, such as hot isostatically pressing, was performed prior to sectioning into fatigue and microstructure specimens.

Sample Sectioning Procedure and Surface Preparation

All sample walls were sectioned into fatigue specimens after being heat treated. Each nominally 2 mm thick specimen was cut from each build wall using wire electrical discharge machining. This results in a fatigue specimen with gage section volume of 2 × 2 × 6 mm. Each of this dogbone fatigue is loaded along the 6 mm direction with the build direction being in one of the 2 mm transverse directions. Hence, the fatigue properties are measured perpendicular to the build direction.

The exact location where each specimen was chosen specifically corresponds with the in-process monitoring data taken during the AM process. The nominal build heights for each letter are A starting at 78 mm, B at 75.98 mm, C at 72.3 mm, D at 49.7 mm, E at 47.46 mm, F at 43.5 mm, G at 35.7 mm, H at 32.3 mm, I at 28.5 mm, J at 4.1 mm, and K at 2.14 mm. An example sample wall with fatigue specimen geometries marked is illustrated in Fig. 4. The heights where each alphabetical fatigue specimens were sectioned from were chosen specifically to coincide with layers that were monitored by in-process systems during the build.

After sectioning, each fatigue specimen was prepared for XCT and fatigue testing by polishing the gage region surface to a near mirror finish. After the nominal thickness and width of the fatigue specimen, the gage region is measured and recorded using a Dremel head mounted in a drill press; all specimen surfaces were polished

Fig. 4 Square cross section fatigue specimen geometry in relation to the sample walls with all dimensions given in mm

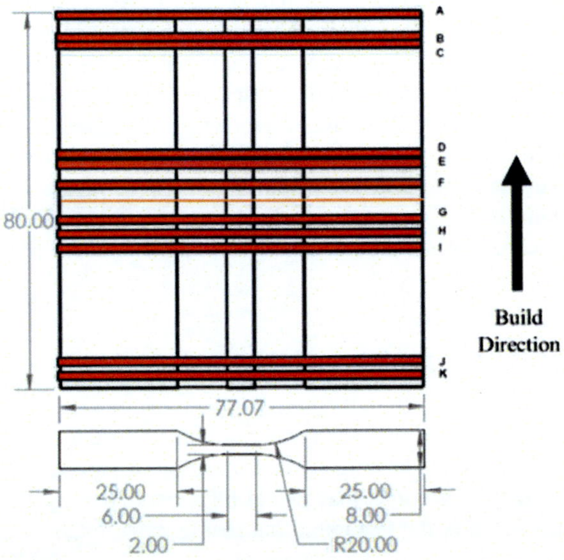

using Struers SiC polishing pads from P60 to P4000 grit (European) to eliminate surface roughness and wire EDM induced microstructural changes from impacting the fatigue performance. After polishing, the gage regions are measured again to track the thickness of the material being removed. The goal of the study was not to focus on how the surface condition of AM samples affects fatigue performance, but rather to focus on how internal defects observable through in-process and post-process characterization affect fatigue.

XCT Machine Parameters and Data Processing

Prior to fatigue testing, every specimen is characterized by XCT to map all porosity within the gage region. The machine used in this study was a Zeiss Metrotom 800 130 kV (Oberkochen, Germany) running Zeiss Metrotom OS. All relevant machine parameters are given in Table 2. The Zeiss Metrotom OS handled the XCT reconstruction by utilizing the Feldkamp-David-Kress reconstruction algorithm and a Shepp-Logan digital filter. The exported reconstruction was processed using the Dragonfly software (Version 2021.3.0.1069) by Object Research Systems (Montreal, QC, Canada). In Dragonfly, the XCT reconstruction was registered with the build plate coordinate system, and the centroid of the gage region was translated to be centered at the origin. Using the manual segmentation tools in Dragonfly, all porosity within the $6 \times 2 \times 2$ mm gage region was identified and quantified. The pore segmentation was conducted manually using the intensity histograms of each XCT reconstruction, all porosity was verified by visually inspecting all XCT cross-sectional slices. All porosity smaller than three voxels was filtered out to eliminate incorrectly segmented local perturbations in the reconstruction voxel intensity. For all segmented porosity, the data detailing their volume, number of voxels, aspect ratio, XYZ coordinates, max/mean/min Feret diameter, and relative distance from specimen surface (created using a distance map illustrated in Fig. 5) are exported for use in later machine learning training. In this distance map, the intensity of each pixel increases linearly with respect to the minimum distance from the outer surface, demarcated by the red outline near the edges of the figure. Using distance maps like this, the porosity in fatigue specimens was labeled by the minimum distance to the surface and colorized accordingly with purple being the closest and red being the furthest.

High Temperature High Cycle Fatigue Testing Procedure

After XCT porosity characterization, all specimens were subjected to high temperature high cycle fatigue (HCF) testing to simulate the operating environment and loading conditions that would be experienced by IN718 in turbine-based applications and is the test condition that is most notch sensitive. All testing in this work was conducted in force control on an MTS model 370 servohydraulic test system

Table 2 XCT machine specific parameters

Parameter	Value
Voxel size	8 μm
Focal spot size	8 μm
Voltage	130 kV
Current	61 μA
No. projections	800–900
Physical filter	0.25 mm, Cu

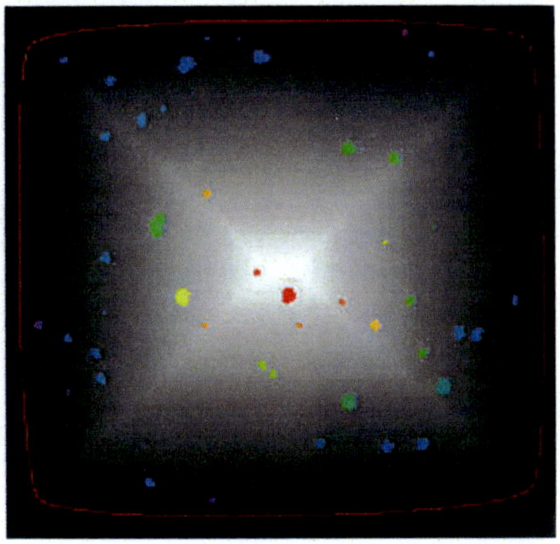

Fig. 5 XCT slice section of specimen 8 K showing the specimen surface and a distance map with intensity increasing linearly with distance from the surface

(Eden Prairie, MN, USA). Specimens were gripped using hydraulic wedge grips with serrated steel inserts using 1.6 mm annealed Cu shims to prevent fatigue specimens from a fracture in the grip section. Guides were used to ensure the alignment of fatigue specimens between the wedge grips. The guides were installed using a flat calibration specimen and a level to insure vertical alignment of fatigue specimens. The HCF test conditions comprised of a stress ratio (*R*) of 0.1, a maximum stress (σ_{max}) of 690 MPa, and a loading frequency of 20 Hz with a sinusoidal wave form at 538 °C facilitated by induction heating in lab air. The test stress was chosen to ensure fatigue failure which occurred in under 2 million cycles per MMPDS data [24]. Failure was defined by specimen separation. The unique induction coil used for these sub-sized specimens, shown in Fig. 6, provided a uniform temperature distribution in the gage section while allowing access for measurement of strain. It was powered by an Ambrell HOTSHOT 3.5 kW power supply unit (Rochester, NY, USA) with its voltage input moderated by a temperature controller with feedback from a thermocouple spot welded near the top and bottom of the gage region shown in Fig. 6.

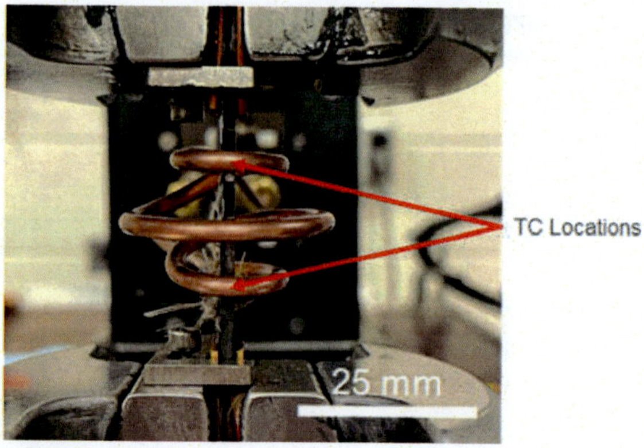

Fig. 6 Fatigue specimen gripped in test system detailing the induction coil design with thermocouple locations

SEM Fractography Procedure

All post-mortem fractography of fatigue samples was conducted on a TESCAN Mira FE-SEM (Brno, Czech Republic) using a secondary electron detector and an acceleration voltage of 10 kV. Specimens were cleaned with an acetone solution in an ultrasonic cleaner for 5 min immediately prior to imaging the fracture surface in the SEM to eliminate organic matter contamination that may have been deposited during specimen testing and storage.

Results and Discussion

XCT Analysis

The segmented porosity data acquired through XCT analysis yielded direct volume measures of every pore using the number of voxels present in each tied to the calculated size of each voxel determined by the respective position of the specimen and the detector in the XCT. Using all the measured volume of porosity within each sample, the volume fraction of porosity was calculated by dividing the cumulative volume of all porosity within the gage region of a given specimen by the idealized volume of the $2 \times 2 \times 6$ mm gage region. The corresponding data is plotted with respect to the nominal VED of each specimen in Fig. 7. It is readily apparent in this plot that there is a range of volumetric energy densities within which porosity is minimized. Specimens from samples 6, 7, 11, and 12 which all lie within this range have noticeably less porosity than others, but as shown in the fatigue section later, the amount of porosity is not only the sole contributor to the mechanical properties of each specimen, but the location and shape of the porosity, as well as the underlying grain structure of the specimens, also influence the fatigue lives.

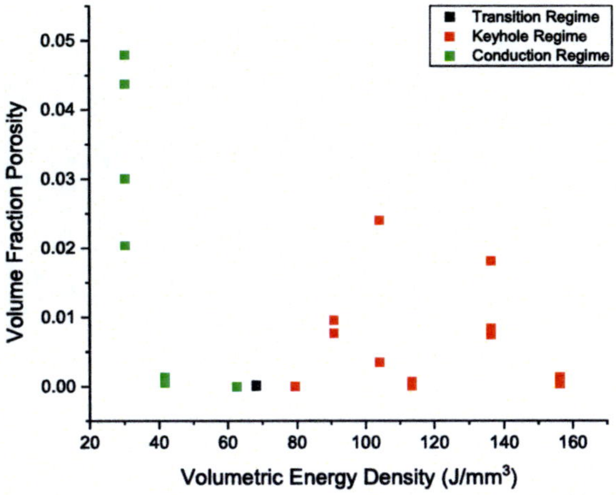

Fig. 7 Plot depicting the relationship between volumetric energy density and volume fraction of porosity inside the gage region of each specimen

The transition between the conduction regime and its associated anomalies, and the keyhole regime and its associated anomalies is depicted in this VED range where porosity is minimized as shown in Fig. 8. In process pedigree 15, pores are generally large, irregularly shaped, and often interconnected. As the VED increases, in pedigree 12, some evidence of localized lack of fusion porosity is observed, but it is infrequent. Over the transition between conduction and keyhole regime through pedigrees 11, 6, and 7, porosity is minimized and almost entirely relegated to the surfaces where local thermal perturbations from contouring scans and turns in the laser path rastering created pores. In pedigree 8, spherical keyhole porosity is apparent throughout the microstructure. Figure 8 illustrates the change in the amount of porosity and morphology of porosity observed between these process pedigrees.

Using the distance maps like the one shown in Fig. 5, the porosity of each fatigue specimen was ascribed an intensity value directly related to the minimum distance between the pore and the nearest surface. These intensity values were generated using the distance map by assigning the minimum intensity value of all distance map voxels within the pore volume to the pore itself. These intensity values scale linearly with respect to distance from the surface, but since the cross-sectional area of each specimen is slightly different due to the machining and surface polishing procedure, the quantitative distances of these intensity values vary slightly from sample to sample. To expand upon this, Fig. 9 shows a plot of the number of pores with respect to their relative distance from the surface for multiple process pedigrees. It is pertinent to note here that the radial volume decreases as you move from the surface of the sample towards the center of the sample, and thus the number of pores observed decreases along with the decreasing volume as the distance from the surface increases. This aside, the data show that between the different process regimes, there is an observable difference in how the porosity manifests in the specimens. Many of the keyhole regime and the transition regime specimens have more pores near the surface and less pores as you move away from the surface, seen in Fig. 8 as well,

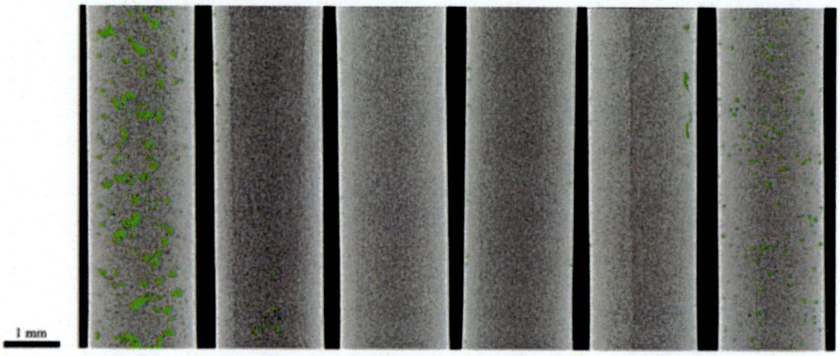

Pedigree	15	12	11	6	7	8
P (W)	200	200	200	300	350	200
SS (mm/s)	1500	1000	1000	1000	1000	500
HS (μm)	110	120	80	110	110	110
VED (J/mm³)	30.30	41.67	62.50	68.18	79.55	90.91

Fig. 8 XCT cross sections of the gage regions of fatigue specimens from a lower range of VEDs presented with pertinent process parameters. The build direction for all specimens is on the page

more than the general trend of decreasing porosity. This contrasts directly with the conduction regime specimens which have significantly less porosity directly at the surface and more porosity at a moderate distance from the surface. This difference in cross-sectional area-wise distribution of porosity can have a direct effect on the fatigue behavior due to the creation of stress concentrations near the surface acting as crack initiation sites for eventual fatigue failure.

The difference in porosity distribution can likely be ascribed to how the corners of the rastering path of the laser affect the locally observed process parameters. In conduction regime samples where the local areal/volumetric energy density is less then optimal, having the laser spend more time in a local spot as it turns around causes a localized increase in energy density which helps to better melt the powder.

Fig. 9 Plot depicting the relationship between the number of pores and the relative distance from the polished as-built surfaces in the specimen gage section

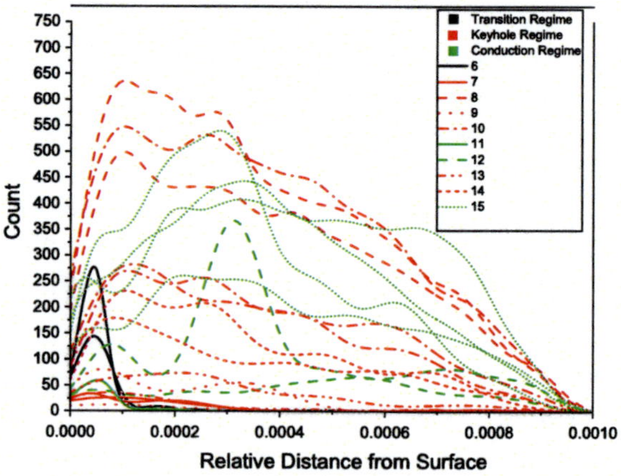

This effect combines with the contouring scan to further improve the local process conditions near the edges of specimens which causes many low VED samples, especially those from sample 15, to have little to no porosity near the as-built edges of the build wall as seen in Fig. 9. This helpful increase in energy density observed in conduction regime samples is directly contrasted by the increase in porosity in the keyhole and transition regime specimens. When the local energy density is at or a bit above the optimal process conditions to minimize porosity, as it is in keyhole regime samples, a local increase caused by the ends of the raster path in addition to the contouring scans can serve to locally induce more keyhole porosity. The number of keyhole pores near the as-built surfaces of these specimens is further reinforced when operating the laser in the keyhole mode since the turns at the end of the raster paths can lead to the collapse of the keyhole melt pool and leave behind additional porosity.

Fatigue

With respect to the fatigue life of the AM IN718 specimens, there are many contributing factors. Fatigue properties are especially sensitive to surface defects due to the nature of how fatigue cracks nucleate and grow, so the location of pores, along with their size and shape, are especially pertinent to understand their effect on fatigue. The general shape of porosity in additively manufactured IN718 can range from highly irregular and often sharp lack of fusion defects, produced in regions where VED is too low to properly fuse adjacent layers in the conduction regime, to near-spherical keyhole porosity, that can be generated while operating in the keyhole regime. Figure 10 contains cropped lack of fusion and keyhole porosity. Irregular lack of fusion defects are more deleterious to fatigue performance due to their often sharp geometry which acts as an easy site for crack nucleation while the keyhole porosity, which serves as a site for stress concentration, must first sharpen before it can serve as a site for crack nucleation. The amount of porosity appears to be highly correlated to fatigue life for specimens with higher degrees of porosity (>0.01) with only a few outliers as shown in Fig. 11. These outliers (from pedigrees 8, 10, and 14) with moderate porosity but high fatigue lives are specimens that while having many pores, lack a large, sharp critical defect to significantly decrement their fatigue life. The fatigue lives of specimens with more porosity in the specimen gage region are lower; however, this does not tell the whole story. It is not simply the amount of porosity that decreases the fatigue life, but the increased amount of porosity creates a higher probability of a sharp critical defect appearing near the surface of a specimen. For this reason, in specimens with minimal porosity (<0.01), there is scatter in fatigue lives ranging from 100 K to nearly 2 M cycles. Specimens with only a small volume fraction of porosity still have a finite chance of having a critical pore in a location to cause early fatigue failure. It is for this reason that efficient quality assurance methods are so necessary. If a specimen has minimal porosity but there exists a sizeable pore in a location that experiences extreme stress states during operation, that component should not be cleared for use. On the other hand, if porosity exists in an AM component, but the regions that experience the most extreme stresses are devoid of critical porosity, the component should be cleared for use.

The fatigue data is plotted on a stress-life plot in Fig. 12. Plotted alongside the test specimen data are fatigue life predictions from the NASGRO fatigue crack growth

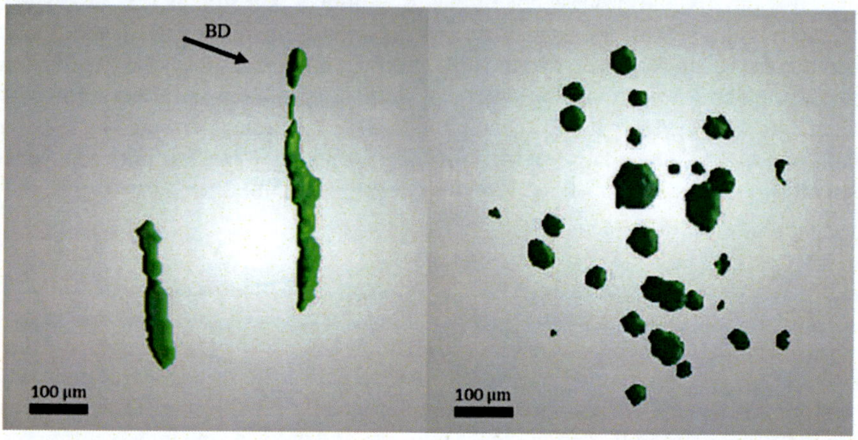

Fig. 10 Segmented porosity showing of lack of fusion pores (left) and keyhole pores (right) found in specimens from pedigree 7 and 8, respectively. The build direction is to the right and out of the page, and the loading direction is vertical

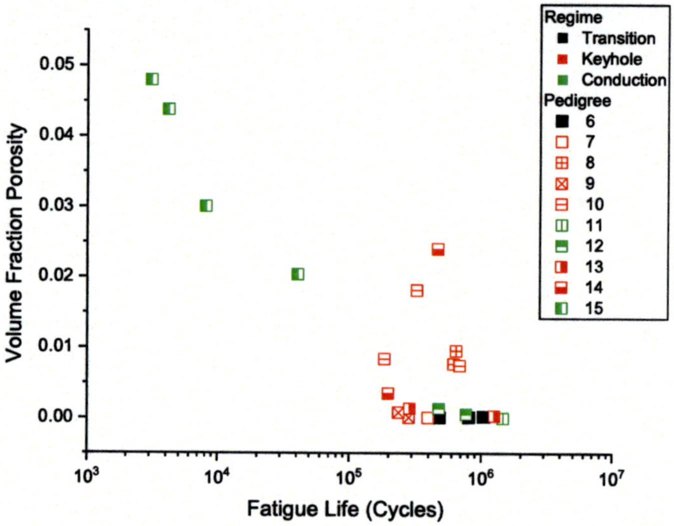

Fig. 11 Plot depicting the relationship between volume fraction of porosity and fatigue life

software (Southwest Research Institute, San Antonio, TX, USA). For the predictions, semi-circular cracks with radii varying from 0.01 to 0.5 mm were placed along one of the sides of the specimen and grown through cyclic fatigue until either the stress intensity of the crack reached the fracture toughness value or net section yielding occurred. The radii used correspond to different initial sizes of perfectly sharp cracks and were considered to generate a family of stress-life curves that reflect the range of expected AM defect sizes. The fatigue crack growth data used in the model was a representative wrought IN718 available in the NASGRO software database (Q3LBB3AA18) that underwent a similar heat treatment (with a solution treat step) tested at the same temperature, 538 °C. Since in the NASGRO model, the cracks are perfectly sharp, they required no crack incubation and sharpening prior to growth. Therefore, the predictions from NASGRO should be conservative, especially when the porosity in AM samples has spherical morphologies and crack incubation and nucleation are required prior to growth and failure. By comparing the NASGRO predictions with the actual lives, a measure of the incubation life can be obtained since real cracks require sharpening prior to growth but NASGRO flaws grow immediately upon loading. On the other extreme, similarly, heat treated (had solution treatment), unnotched wrought IN718 tested at 538 °C extracted from MMPDS [24] is shown which provides an upper bound on the fatigue lives of the AM samples. The curve extracted from the MMPDS data was collected by interpolating the data from specimens tested at adjacent mean stresses with stress ratios of 0 and 0.2. Shown in Fig. 12, the fatigue lives of the tested specimens span a range between the NASGRO predictions and the MMPDS data with some specimens performing nearly identically to wrought IN718. The range in fatigue performance is assuredly due to the various levels and locations of porosity throughout all specimens, further analysis of porosity and its effects on the fatigue lives is ongoing. The fatigue life data are deconvoluted in Fig. 13 where the fatigue lives of all specimens in the S–N plot are plotted with respect to their volumetric energy density. While again, volumetric energy density cannot be used as a sole parameter to describe fatigue life, it does show that there exists a threshold of roughly 40 J/mm^3 below which fatigue lives are drastically decremented. While above this threshold, fatigue lives are longer due to the reduced porosity of the gage regions. This difference in porosity observed is shown in Fig. 8 where specimens from pedigrees 6, 7, 11, and 12 have drastically reduced amounts of porosity when compared to pedigrees 8 and 15. It can also be observed that there appears to be an additional threshold of approximately 90 J/mm^3 above which fatigue lives are slightly decremented. The cause of this threshold is not clear. Specimens from sample 8 (90 J/mm^3) lie below it, while specimens from samples 14 (104 J/mm^3) and 9 (114 J/mm^3) lie above it. Between these three specimens, as shown in Fig. 14, the porosity within the gage region is dramatically different in frequency, size, and location.

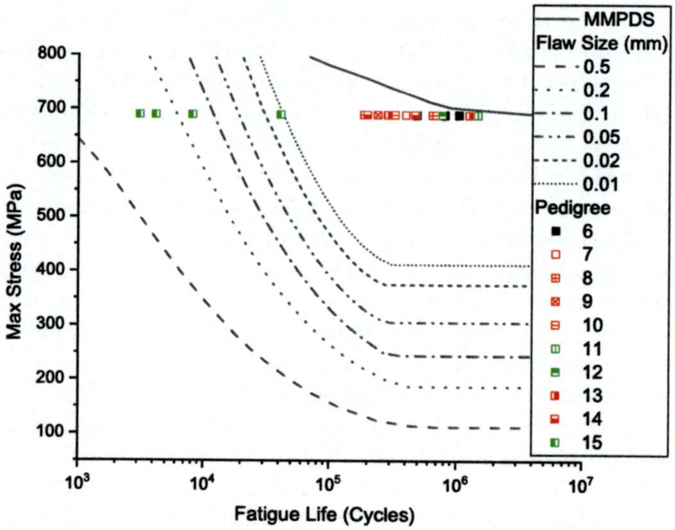

Fig. 12 S–N curve showing NASGRO life predictions, MMPDS fatigue data, and the fatigue lives of specimens of various process pedigrees [24]

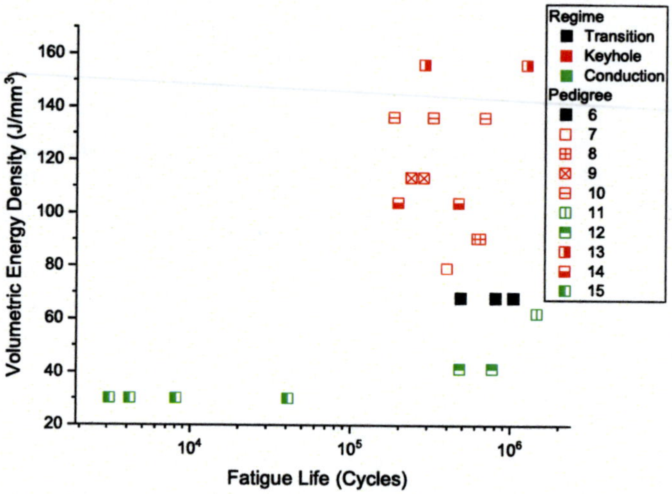

Fig. 13 Plot showing the effect of volumetric energy density on fatigue life of AM specimens

Fractography

Fractography shows that fatigue cracks can initiate from both porous and crystallographic microstructural features. While some results in this work indicate that the

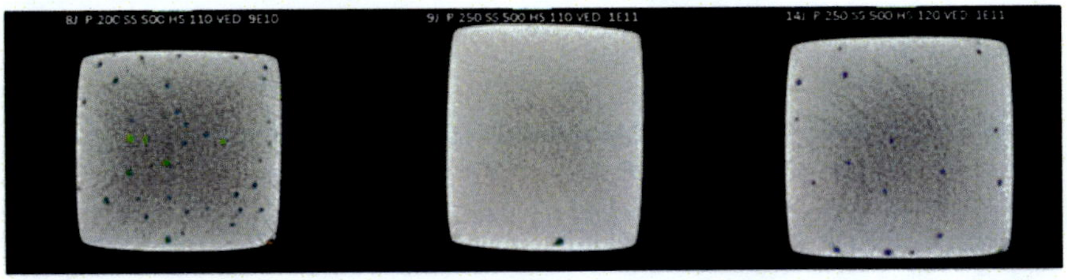

Fig. 14 XCT cross sections of specimens 8 J (680 K cycles), 9 J (285 K cycles), and 14 J (470 K cycles), with porosity colorized by mean Feret diameter

fatigue life of AM IN718 can be highly correlated to the amount of porosity within the gage region of a dogbone specimen, the fractographic analysis further reinforces the fact that this is not the entire picture. Higher volume fractions of porosity lead to a higher probability that a pore of critical size or shape will be located close to the surface, but the probability of these defects appearing in specimens with low volume fractions of porosity is not zero. In the fracture surfaces of specimens 6 J in Figs. 15 and 11K in Fig. 16, the angular faceted nature around the fatigue crack initiation site indicates that there was a locally vulnerable crystallographic slip system near the surface that led to eventual failure, not an AM induced pore near the surface. This type of fatigue failure is comparable to the mechanisms seen in wrought IN718 [25, 26]. The angular nature of the fracture surface in and around where fatigue cracks appear to initiate in these specimens is evidence that the cracks originated from cyclic slip along certain crystallographic planes within grains with slip systems oriented most favorably with respect to the loading direction near the surface. In-depth analysis of the microstructure of each process pedigree using EBSD is ongoing. Alternatively, the fracture surfaces of specimens 8 J (0.8% porosity), 13 J (0.04% porosity), and 9 J (0.01%) in Figs. 15 and 16 show evidence that fatigue cracks initiated from locations that had both a large pore and a locally vulnerable slip system due to the faceted surfaces surrounding one or more pores. Conversely to other specimens, the large, sharp, irregular nature of the porosity present in the microstructure of specimen 15 J quickly led to failure from crack initiation from multiple sites simultaneously.

Conclusions

In this work, IN718 fatigue specimens created by L-PBF AM with various process pedigrees were heat treated by direct aging then characterized by XCT and tested in HCF at high temperature. The effects of the location, size, and shape of porosity within the AM microstructure on the fatigue life were explored. The following conclusions were drawn from this work:

(1) For AM IN718, the process pedigree has a direct effect on the geometry of porosity and how porosity tends to be distributed in the cross section of fatigue

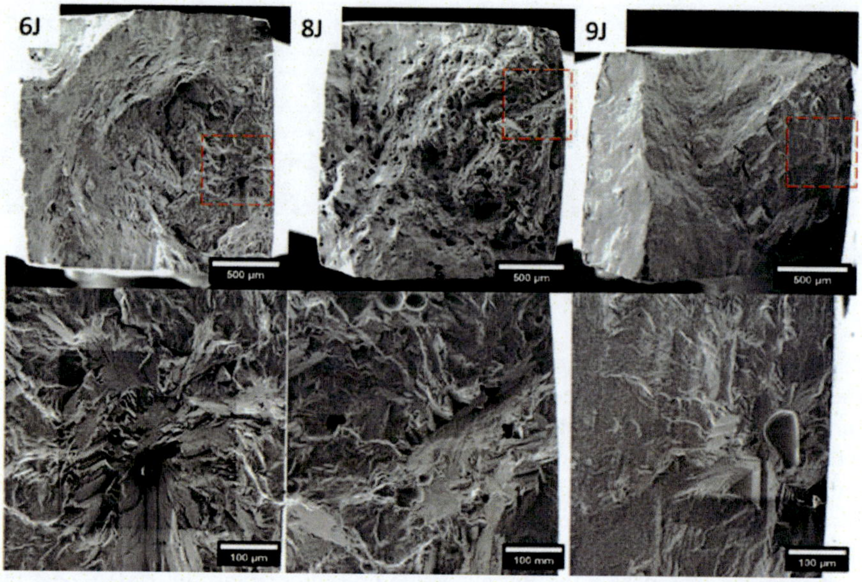

Fig. 15 SE-SEM images of the fracture surfaces at low and high magnification for specimens 6 J (800 K cycles), 8 J (680 K cycles), and 9 J (285 K cycles) oriented with the build direction pointed down

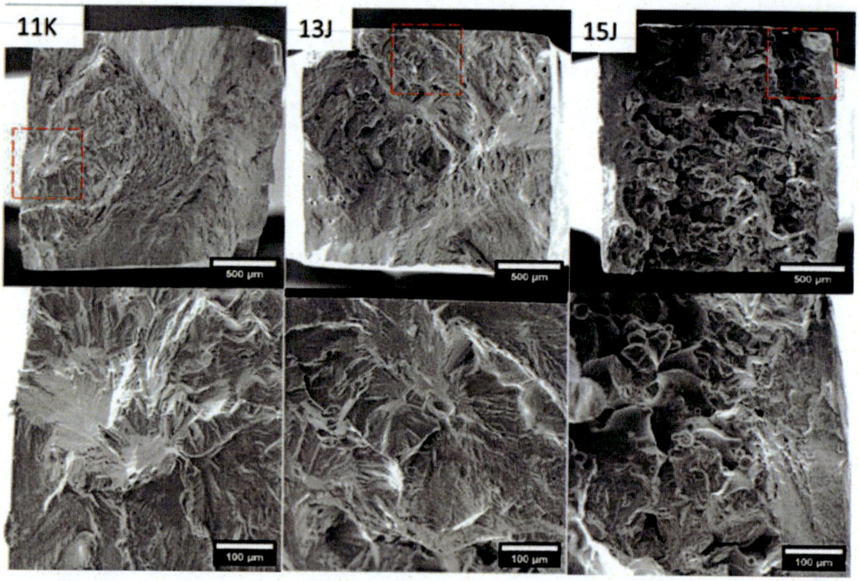

Fig. 16 SE-SEM images of the fracture surfaces at low and high magnification for specimens 11 K (1.46 M cycles), 13 J (1.26 M cycles), and 15 J (3.1 K cycles) oriented with the build direction pointed down

specimens. Spherical porosity generated in the keyhole regime is less detrimental to fatigue than the irregular shaped lack of fusion defects prevalent in the conduction regime.

(2) Local increases in VED near the as-built surface of specimens caused by contouring scans and the turns in the laser rastering path can serve to reduce the presence of porosity near the surface in conduction regime samples while increasing the amount of porosity near the surface in transition and keyhole regime samples.

(3) There appears to be a range of VEDs across various process regimes where porosity can be minimized within the microstructure. In this transition regime, the fatigue strengths are approaching those of the wrought material with similar age treatment.

(4) While there is a general inverse relationship between the volume fraction of porosity within the gage region and the fatigue life of the AM IN718 specimen, the fatigue lives cannot be correlated simply with the volume fraction of porosity alone. The morphology, location, and size of porosity are also important. Pores near the surface with irregular or sharp morphologies are more detrimental.

(5) Fatigue crack initiation in AM IN718 occurs due to either transgranular cyclic shear in particularly vulnerable grains, due to stress concentrations created by sharp pores near the surface of the fatigue specimens, or some combination of both mechanisms.

References

1. Z. A. Young *et al.*, "Types of spatter and their features and formation mechanisms in laser powder bed fusion additive manufacturing process," *Addit Manuf*, vol. 36, Jul. 2020, https://doi.org/10.1016/J.ADDMA.2020.101438.
2. C. Wang, X. P. Tan, S. B. Tor, and C. S. Lim, "Machine learning in additive manufacturing: State-of-the-art and perspectives," Addit Manuf, vol. 36, p. 101538, Dec. 2020, https://doi.org/10.1016/J.ADDMA.2020.101538.
3. J. Berez, L. Sheridan, and C. Saldaña, "Extreme variation in fatigue: Fatigue life prediction and dependence on build volume location in laser powder bed fusion of 17-4 stainless steel," Int J Fatigue, vol. 158, p. 106737, May 2022, https://doi.org/10.1016/j.ijfatigue.2022.106737.
4. M. Ahmed, M. Pasha, W. Nan, and M. Ghadiri, "A simple method for assessing powder spreadability for additive manufacturing," Powder Technol, vol. 367, pp. 671–679, May 2020, https://doi.org/10.1016/J.POWTEC.2020.04.033.
5. X. Zhao, J. Chen, X. Lin, and W. Huang, "Study on microstructure and mechanical properties of laser rapid forming Inconel 718," Materials Science and Engineering: A, vol. 478, no. 1–2, pp. 119–124, Apr. 2008, https://doi.org/10.1016/J.MSEA.2007.05.079.
6. W. Abd-Elaziem et al., "On the current research progress of metallic materials fabricated by laser powder bed fusion process: a review," Journal of Materials Research and Technology, vol. 20, pp. 681–707, Sep. 2022, https://doi.org/10.1016/J.JMRT.2022.07.085.
7. D. L. Bourell, D. W. Rosen, and M. C. Leu, "The roadmap for additive manufacturing and its impact," *3D Print Addit Manuf*, vol. 1, no. 1, pp. 6–9, Mar. 2014, https://doi.org/10.1089/3DP.2013.0002/FORMAT/EPUB.

8. C. Krishna, P. Vallabh, and X. Zhao, "Continuous Comprehensive Monitoring of Melt Pool Morphology Under Realistic Printing Scenarios with Laser Powder Bed Fusion," https://home.liebertpub.com/3dp, Sep. 2021, https://doi.org/10.1089/3DP.2021.0060.
9. C. K. Prasad Vallabh, Y. Xiong, and X. Zhao, "In-situ monitoring of laser powder bed fusion process anomalies via a comprehensive analysis of off-axis camera data," *ASME 2020 15th International Manufacturing Science and Engineering Conference, MSEC 2020*, vol. 1, 2020, https://doi.org/10.1115/MSEC2020-8300.
10. C. Pei, D. Shi, H. Yuan, and H. Li, "Assessment of mechanical properties and fatigue performance of a selective laser melted nickel-base superalloy Inconel 718," Materials Science and Engineering: A, vol. 759, pp. 278–287, Jun. 2019, https://doi.org/10.1016/J.MSEA.2019.05.007.
11. Y. Yamashita, T. Murakami, R. Mihara, M. Okada, and Y. Murakami, "Defect analysis and fatigue design basis for Ni-based superalloy 718 manufactured by selective laser melting," Int J Fatigue, vol. 117, pp. 485–495, Dec. 2018, https://doi.org/10.1016/J.IJFATIGUE.2018.08.002.
12. A. S. Johnson, S. Shao, N. Shamsaei, S. M. Thompson, and L. Bian, "Microstructure, Fatigue Behavior, and Failure Mechanisms of Direct Laser-Deposited Inconel 718," JOM, vol. 69, no. 3, pp. 597–603, Mar. 2017, https://doi.org/10.1007/S11837-016-2225-2/FIGURES/4.
13. F. Sausto et al., "Anisotropic mechanical and fatigue behaviour of Inconel718 produced by SLM in LCF and high-temperature conditions," Fatigue Fract Eng Mater Struct, vol. 44, no. 1, pp. 271–292, Jan. 2021, https://doi.org/10.1111/FFE.13373.
14. P. D. Nezhadfar, A. S. Johnson, and N. Shamsaei, "Fatigue behavior and microstructural evolution of additively manufactured Inconel 718 under cyclic loading at elevated temperature," Int J Fatigue, vol. 136, p. 105598, Jul. 2020, https://doi.org/10.1016/J.IJFATIGUE.2020.105598.
15. L. Scime and J. Beuth, "Melt pool geometry and morphology variability for the Inconel 718 alloy in a laser powder bed fusion additive manufacturing process," Addit Manuf, vol. 29, p. 100830, Oct. 2019, https://doi.org/10.1016/J.ADDMA.2019.100830.
16. L. Scime and J. Beuth, "Using machine learning to identify in-situ melt pool signatures indicative of flaw formation in a laser powder bed fusion additive manufacturing process," Addit Manuf, vol. 25, pp. 151–165, Jan. 2019, https://doi.org/10.1016/J.ADDMA.2018.11.010.
17. J. Li, Z. Yang, G. Qian, and F. Berto, "Machine learning based very-high-cycle fatigue life prediction of Ti-6Al-4V alloy fabricated by selective laser melting," Int J Fatigue, vol. 158, p. 106764, May 2022, https://doi.org/10.1016/J.IJFATIGUE.2022.106764.
18. Y. W. Luo et al., "Pore-affected fatigue life scattering and prediction of additively manufactured Inconel 718: An investigation based on miniature specimen testing and machine learning approach," Materials Science and Engineering: A, vol. 802, p. 140693, Jan. 2021, https://doi.org/10.1016/J.MSEA.2020.140693.
19. EOS GmbH, "EOS Nickel Alloy IN718 Material Data Sheet Metal Solutions," 2021.
20. A. Mostafaei et al., "Defects and anomalies in powder bed fusion metal additive manufacturing," Curr Opin Solid State Mater Sci, vol. 26, no. 2, p. 100974, Apr. 2022, https://doi.org/10.1016/J.COSSMS.2021.100974.
21. T. G. Gallmeyer, "The effects of heat treatment on structure-property relationships of additively manufactured Inconel 718," Doctoral Dissertation, Colorado School of Mines, 2020.
22. D. Zhang, W. Niu, X. Cao, and Z. Liu, "Effect of standard heat treatment on the microstructure and mechanical properties of selective laser melting manufactured Inconel 718 superalloy," Materials Science and Engineering: A, vol. 644, pp. 32–40, 2015, https://doi.org/10.1016/j.msea.2015.06.021.
23. M. Ni, S. Liu, C. Chen, R. Li, X. Zhang, and K. Zhou, "Effect of heat treatment on the microstructural evolution of a precipitation-hardened superalloy produced by selective laser melting," Materials Science and Engineering: A, vol. 748, pp. 275–285, 2019, https://doi.org/10.1016/j.msea.2019.01.109.
24. *Metallic materials properties development and standardization (MMPDS) ; MMPDS-01.* Columbus, OH: Federal Aviation Administration, 2002.

25. C. Mercer and W. O. Soboyejo, "Fatigue Crack Growth Mechanisms in a Forged IN 718 Nickel-Based Superalloy," in *Superalloys 718, 625, 706 and Various Derivatives (1997)*, 1997, pp. 577–586. https://doi.org/10.7449/1997/Superalloys_1997_577_586.
26. D. G. Leo Prakash, M. J. Walsh, D. Maclachlan, and A. M. Korsunsky, "Crack growth micromechanisms in the IN718 alloy under the combined influence of fatigue, creep and oxidation," *Int J Fatigue*, vol. 31, no. 11–12, pp. 1966–1977, Nov. 2009, https://doi.org/10.1016/J.IJFATIGUE.2009.01.023.

Part VII
Additive: Microstructure and Properties

Correlating Alloy Inconel 718 Solidification Microstructure to Local Thermal History Using Laser Powder Bed Fusion Process Monitoring

Yi Zhang, Nazmul Hasan, John Middendorf, Thomas Spears, Timothy Smith, Fan Zhang, Mohammed Shafae, and Andrew Wessman

Abstract Additive manufacturing processes such as laser powder bed fusion produce material by localized melting of a powder feedstock layer by layer. The small melt pools and high energy density generate very different microstructures in nickel superalloys when compared to more traditional cast or wrought processing, including features such as cellular structures and epitaxial grain growth. The features of these microstructures vary depending on local thermal history, alloy chemistry, and processing parameters. There is a need to develop a systematic understanding of the influence the local thermal conditions during solidification have on the resulting microstructure. Such understanding will be useful in predicting and ultimately avoiding microstructural defects such as undesirable phases or non-optimal grain structures. In this work, in-situ Longwave Infrared imaging of a laser powder bed fusion process is used to characterize the local thermal conditions throughout additively manufactured builds for alloy IN718 processed using systematically varied process parameters. This information is then correlated to observations of the microstructural features of these alloys in the as-built condition. This correlation analysis shows clear influence of the local thermal conditions during solidification on the dimensions of the dendritic microstructures formed during the build process for IN718. These dendritic structures arise due to segregation of elements such as niobium during solidification, an observation which can be predicted using a Scheil modeling approach.

Keywords Laser powder bed fusion · Alloy IN718 · Nickel-based superalloys

Y. Zhang · N. Hasan · M. Shafae · A. Wessman (✉)
University of Arizona, Tucson, AZ, USA
e-mail: wessman@arizona.edu

J. Middendorf · T. Spears
Open Additive LLC, Beavercreek, OH, USA

T. Smith
NASA Glenn Research Center, Cleveland, OH, USA

F. Zhang
Computherm LLC, Madison, WI, USA

E. A. Ott et al. (eds.), *Proceedings of the 10th International Symposium on Superalloy 718 and Derivatives*, The Minerals, Metals & Materials Series,
https://doi.org/10.1007/978-3-031-27447-3_37

Introduction

Nickel-based superalloys are known for maintaining strength, toughness, fatigue performance and resistance to corrosion and oxidation at elevated temperature and pressure. These properties make nickel-based superalloys key materials in turbine engines for aircraft and power generation. Some components of gas turbines require complex geometries for which nickel-based superalloys can be difficult to machine and fabricate. Additive manufacturing (AM), a method that builds parts in layer-wise process, enables the production of parts with complex geometries where traditional manufacturing methods have limited ability. Thus, AM, especially Laser Powder Bed Fusion (LPBF), has found growing applications in fabricating intricate part geometries with hard-to-machine metals used in aviation and aerospace [1, 2]. In a LPBF system, as the focused laser beam scans each layer of powder, it melts the powder into the shape of the cross section of the part to form the designed 3D model. This building process generates microstructures with features including cellular structures and epitaxial grain growth which is very different from those formed in traditional processes such as forging and casting. The local thermal history, alloy chemistry, and processing parameters all contribute to the features of the microstructures. In this work, the local thermal conditions during the LPBF building process of alloy IN718 are recorded using in-situ process monitoring system for different parts built using varying process parameters. The monitoring data is correlated to the microstructural features of these alloys in the as-built condition characterized by electron microscopy and optical microscopy. This information is then compared to computational modeling of the segregation and dendrite structures formed during the solidification. This together moves towards developing a deeper systematic understanding of the interplay between local thermal condition variation during solidification and the resulting microstructure. This will allow for better process designs to mitigate not only macroscopic defects such as porosity and cracks but also microstructural defects such as undesirable phases or non-optimal grain structures.

General Background on LPBF of 718

IN718 is a common nickel-based superalloy. IN718 is a precipitation hardening alloy, its excellent mechanical properties benefit from the precipitation phase γ'' (Ni_3Nb). For the last few decades, IN718 has been used in components such as casings, shafts, disks and compressor blades and vanes of jet engines [3, 4]

The deposition of layers in LPBF involves highly localized laser energy input, high laser scanning velocity and short interaction time with the melt pool. These conditions result in large thermal gradients and high cooling rates during the solidification and thus cause directional grain growth accompanied by micro-segregation and precipitation of metastable phases. Studies on IN718 fabricated by LPBF have shown epitaxial columnar dendrites approximately along the build direction [4–6]. In the as-built material, carbides and Laves precipitates were found in the interdendritic spaces with segregation of niobium, titanium, and molybdenum [4, 7–9], and the presence of γ' was indicated by XRD in [10]. After heat treatment, partial recrystallization and homogenization may occur depending on the temperature and γ', γ'' and δ precipitates were found at the grain boundaries [4, 7, 11, 13–15]

Observations of Solidification Structures of AM Superalloys-Rene 65, IN625, Mar-M-247, CM247LC, Rene 108, IN738

Similar solidification structures have been found in some nickel-based superalloys whose chemistries are similar to IN718. For example, Wessman et al. [12] studied the microstructure of Rene 65, a γ' strengthened alloy, processed by LPBF. Cellular dendritic structures rich in γ' forming elements were found in the as-built material. The microstructures developed in heat treatment at temperature below and above γ' solvus were examined. The subsolvus grains still showed some elongation (aspect ratio about 1.5) in the build direction. The supersolvus grains were close to equiaxed. Both heat treated microstructures showed a multimodal distribution of γ' precipitates, and the supersolvus material had a generally finer distribution than subsolvus material. IN625 is a solid solution strengthened alloy. The as-built LPBF IN625 consists of cellular dendrites and elongated grains in the build direction [13]. Amato et al. observed columnar arrays of fine γ'' nanoparticles along the boundaries of the melt pool [14]. C247LC is a nickel-based superalloy considered hard to weld due to the high Ti and Al content and is susceptible to strain-age cracking. Boswell et al. observed finγ' cellular dendrites interspersed with carbides at cell and grain boundaries in LPBF fabricated CM247LC heat treated below 700 °C [15]. After heat treatment above 750 °C up to 975 °C, precipitation of intra-cellular γ', γ' films at cell and grain boundaries and $M_{23}C_6$ carbides at grain boundaries were observed. Alloy IN738LC has poor weldability and is susceptible to hot cracking as well. In the study of IN738LC by Rickenbacher et al. [16], dendrite structures with no distinct γ' precipitates were found in the as-built condition. A unimodal γ' distribution was developed after solution and aging heat treatment. When the hot isostatic pressing (HIP) was applied before solution and aging heat treatments a bi-modal γ' distribution with fine secondary γ' between coarser γ' precipitates was observed. Microcracks formed during LPBF process can be reduced by HIP.

Modeling of Solidification Structures in AM, Emphasis on Superalloys

Although the essentials of AM process make the grains prone to grow into columnar dendrites, the transition of columnar to fine equiaxed grains is found at the top of each layer in some experiments. In the publications of Bermingham et al. [17] and Prasad et al. [18], the mechanisms of nucleation and grain growth in AM process have been studied by modeling the microstructures formed during solidification. Their research concluded that the growth of equiaxed grains requires: a low temperature gradient G, a high cooling rate and an alloy containing grain refining solute and potent nucleant particles. These conclusions are in accordance with what is demonstrated in the experiments. For instance, Zhang et al. [19] reported the columnar to equiaxed transition (CET) at the top of each layer where the temperature gradient is low and the cooling rate is high during laser melting deposition (LMD) of Ti-2Al-7Mo alloy. Hadadzadeh et al. [20] reported that the CET in cylindrical samples of AlSi10Mg alloy produced by direct metal laser sintering (another name for LPBF) is dependent on the build direction. Bermingham et al. [21] reported a mix of columnar and equiaxed grains in titanium alloys containing La_2O_3 particles fabricated by a wire arc AM process. While these models describe the mechanisms that drive grain structure formation, so far there is limited work available on modeling the formation of solidification structures such as cellular dendrites in LPBF manufactured nickel superalloys.

General Background on Process Monitoring in LPBF

The development of in-situ process monitoring and control will remarkably improve the robustness of the AM process and the quality of additively manufactured parts. The basic idea of in-situ monitoring in the LPBF process is to collect the information of "process signatures" associated with the melt pool and surrounding heat affected zone (HAZ). Process signatures can include electromagnetic signatures such as plasma emission/absorption, reflected/scattered light, and radiated light [22]. Prior research has also examined acoustic signatures, and the frequency response of acoustic signals has shown correlation with several weld quality metrics and process phenomena, such as keyhole formation, plasma formation, and crack propagation [22, 23]. Due to practical system integration limitations, there has been limited research on the adoption of acoustic monitoring in LPBF system. The electromagnetic signals can be used to determine the melt pool geometry, temperature, cooling rate and thermal history in the HAZ. Many in-process monitoring systems based on electromagnetic sensing have been developed for LPBF. The electromagnetic sensors could include photodiodes [24, 25], pyrometry [25], and visible light or infrared cameras [26]. In the work of Berumen et al. [24], a high-speed camera and a photodiode has been utilized to measure the dimension of the melt pool and mean radiation emitted respectively. With this method, only the active area in LPBF is monitored. Single channel detectors such as photodiodes and pyrometers have small field of view but high sensitivity, fast data collection speed and low cost [22]. The cameras using CCD or COMS detectors, either visible or IR, enable the spatial resolution of imaging and thus allow the mapping of the build area in LPBF. However, the cameras have slower response time than single-channel detectors and post processing of data is a great challenge [22, 24].

In this work, we will utilize in-situ process monitoring data streams to describe the local thermal history and its effect on microstructure formation in PBF-LB builds of IN718. In particular, we will use Longwave Infrared data to construct the time–temperature history at locations throughout a designed experiment and correlate these histories with features of solidification structures such as dendrite spacing that are determined by cooling rates during the solidification process. In this study we will use a designed experiment with intentional variation of energy density, but in contrast to most prior studies that have taken this approach we will allow for relatively modest variation in process parameters such that the majority of the material studied will be free from defects such as cracks and porosity and our analysis will focus on the variation in microstructure seen in as-built materials that would be generally deemed of high quality by end users.

Experimental Methods

All parts in this study were built with gas atomized powder. The chemical composition of the IN718 powder provided by the manufacturer is given in Table 1. The IN718 samples were built using a custom designed and fabricated PBF-LB test bed by Open Additive. This machine has a notional build volume of 6″ by 6″ by 6″ delivering the 1070 nm fiber laser light via a SCANLAB varioSCAN and intelliSCAN which enables coaxial meltpool thermal imaging. Each IN718 sample is 10 mm tall–the bottom 7 mm is inverted pyramid shaped, and the top 3 mm is square cuboid shaped with 10 $\times$ 10 mm square cross-section. The samples were built with varied laser power and laser scanning velocity. All samples were built with 30 μm build layers using a 70 μm hatch spacing within a 5 mm tile size in a hatch strip approach. The 7 mm tall, inverted pyramid structured base of each sample was built with a common parameter using a 175W laser power and 1100 mm/s scan speed to reduce the effects of proximity to the build plate heat sink in the samples. The top square cuboid portion of each sample was built with varying processing condition in the experiment by varying laser power and scan speed across the experimental layout. The layout of the samples is illustrated in Fig. 1, samples with varied parameters are assigned ID from 11 to 88.

Table 1 Composition of IN718 powder

Element	Al	C	Co	Cr	Fe	Mo	N	Nb	Ni	O	S	Si	Ti
wt.%	0.44	0.03	0.03	19.23	18.04	3.06	0.01	5.12	52.85	0.02	0.002	0.02	0.94

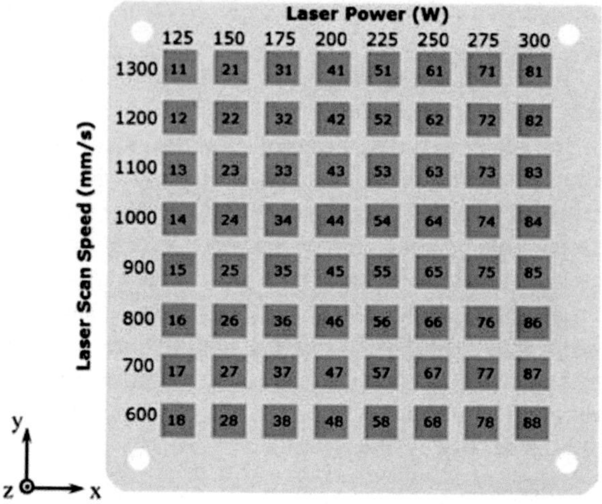

Fig. 1 Layout and building parameters for IN718

Process Monitoring Methods

The machine used to create the metallurgical samples was equipped with Open Additive's AMSENSE sensing and analytics platform outfitted with both commercially available and R&D sensors. There was Longwave IR, thermal tomography, spatter, recoat imaging, galvo position data, and coaxial melt pool thermal imaging recorded for each build.

Recoat imaging takes an image before and after recoat using a standard silicon based CMOS sensor. White light LEDs are used for illumination and the optical filtration is just to prevent the passage of laser light. These images reveal insights into the quality of the recoat process including short feeds, damaged recoater blades, part peel up, or other process anomalies that can disturb the layer of powder. Thermal tomography is also a silicon based CMOS camera that images the entire powder bed while being filtered to being sensitive only in the Near IR (NIR) with a long integration time of 250 ms and a frame rate of 4 fps. Individual images are stitched together to create a composite image on a layer-wise basis. Thermal tomography is sensitive to many process anomalies such as spatter events, uneven gas flow, and part warping among many others. The spatter camera is also a silicon CMOS filtered for the same band in the NIR but has a shorter integration time and operates at 150 fps. Typical exposure times are on the order of 500 μs but for extremely bright materials such as tungsten can be as low as 25 μs. A GPU accelerated analytic works to detect the presence of slow-moving hot objects, referred to as 'welded spatter' that tend to get incorporated into the resultant microstructure and given their mass much larger than individual powder particles result in Lack of Fusion (LOF). Other metadata from the images is also measured and logged in an HDF5 file for post process analysis as desired.

Longwave IR imaging is a valuable tool for measuring the temporal dynamics of the solid state cooling in the LPBF process. By its nature, LWIR imaging is done off axis typically viewing the process through a germanium window. The Optris PI640 is a VGA resolution microbolometer that can run at 32 Hz, is sensitive from 7 to 13 μm, and has four temperature ranges that will measure as low as −20–1500 °C. The temperatures returned from this camera are useful for relative temperature changes but absolute temperatures can vary by 50 °C or more due to changes or ambiguity in the material's emissivity. For the builds, the camera was run at full resolution at 32 Hz in the 150–900 °C range for its full duration. The field of view captured the whole build area allowing different parts to be segmented from the data set.

Coaxial melt pool thermal imaging is accomplished using a NIR filtered CMOS camera that has been blackbody calibrated to return temperature measurements. It is windowed down to run between 1000 and 1500 Hz depending on application and for these builds was set to 1000 Hz. The camera is situated so that it images down the beamline and before the scanner such that the field of view of the camera is always centered on the melt pool. To enable registration of the images to their location on the build an FPGA was developed that records the commanded scanner position and is written to disk in a CSV file. Using the time stamps of the galvo positions and the melt pool images enable registration of one to the other facilitating more detailed analysis.

Metallurgical Analysis

The samples were removed from the build plate and cut along center x–z cross section by a wire EDM machine. The sectioned samples were mounted in phenolic resin by compression mounting process. Samples for microstructure analysis were prepared using metallographic methods including successive grinding using 240, 320, 400, 600, 800 and 1200 grit silicon carbide papers. Final polishing was performed progressively with 9 micron diamond slurry, 1 micron diamond slurry and 0.05 micron Alumina abrasive paste. The porosity in the as-polished x–z cross section was examined by KEYENCE VHX-700 optical microscope. IN718 samples were electrolytically etched in 10% phosphoric acid solution at 2.5 V. A TESCAN MIRA3 Scanning Electron Microscope (SEM) with Schottky field emission gun (FEG) was used to characterize the microstructures including epitaxial grains, cellular dendrite spacing, and micro-segregation.

Image Analysis Methods

The porosity was measured by the optical microscope based on the contrast difference between the holes and surrounding the surface. The measurement of the dendrite spacing was performed on the SEM micrographs of the x–z cross sections at 10kx magnification using the image analysis software ImageJ. On the 10kx SEM images, clusters of cellular dendrites growing over a couple of melt pools along z axis (build direction) were targeted. The width of each cluster was measured and the number of dendrites within each cluster was counted. The dendrite spacing of a sample was calculated by averaging the total width of the targeted clusters over the total number of dendrites.

Results

Porosity

The porosity of IN718 samples 11, 22, 44, 66, 77 and 88 are measured to evaluate the defect density of samples across the broadest range of energy densities in the experimental matrix. The examined area of each sample in the center x–z plane is 10 × 3 mm. The optical microscope images of sample 11, 44 and 88 are listed in Fig. 2, the porosity of samples are shown in Fig. 2(d). Sample 11 built with parameter 125W laser power and 1300 mm/s scan speed has the lowest energy density input. The formation of holes in it could be attributed to the lack of fusion while printing. The irregular shape of holes shown in Fig. 2(a) also indicates the lack of fusion. Sample 88 is built with 300 W and 600 mm/s. It has the highest energy density input among samples. The formation of holes in this sample are more spherical in nature and likely due to keyhole formation at higher energy density.

Local Thermal History Data Collection and Processing

Longwave IR imaging data was collected throughout the build and analyzed first at layers of interest and then locations of interest for extracting local thermal history data. Figure 3 shows an adjusted and color-scaled LWIR image. It is important to note that the temperatures captured by the LWIR camera undergo spatial and temporal averaging. Therefore, the local thermal history recorded by the camera is not absolute but rather relative. Hence, though the complete melting of IN718 takes place at a temperature range of 1370–1430 °C in the LPBF process, the maximum temperature visible in the color-scaled image of Fig. 3 is 430 °C.

Layer-wise local thermal history is obtained by collecting the calibrated temperature values from all the LWIR images captured during the printing of a single

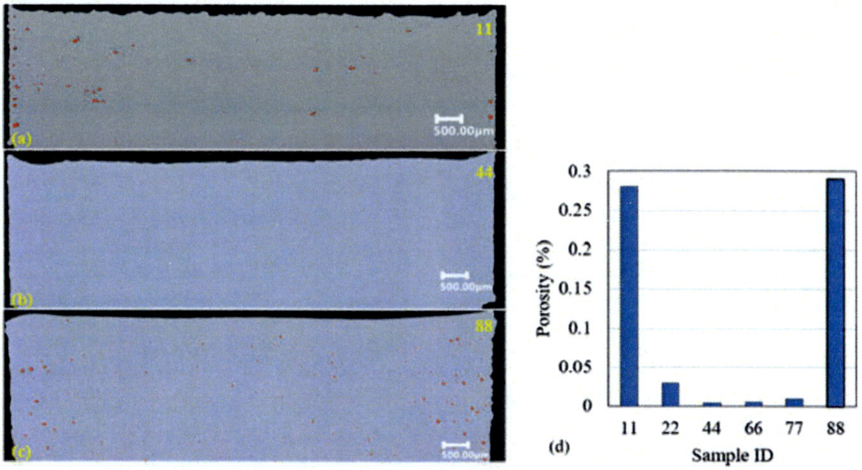

Fig. 2 **a**, **b** and **c** Optical microscope images of IN718 samples 11, 44 and 88, holes are filled by red color. **d** IN718 porosity of IN718 diagonal samples

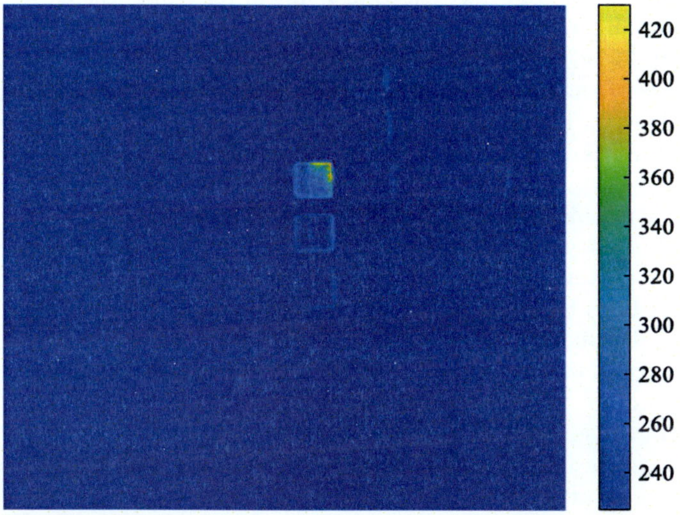

Fig. 3 Sample color-scaled LWIR image

layer. Figure 4(a) shows a sample local thermal history of a pixel point for the time of printing one layer of the build. In this work, our primary focus is on the cooling profile starting from the highest peak temperature and until the temperature reaches a steady ambient temperature which was around 225 °C for the experiments conducted in this study. A pixel-level cooling profile is illustrated in Fig. 4(b) Assuming that the cooling rate itself is not significantly affected by the aforementioned spatial

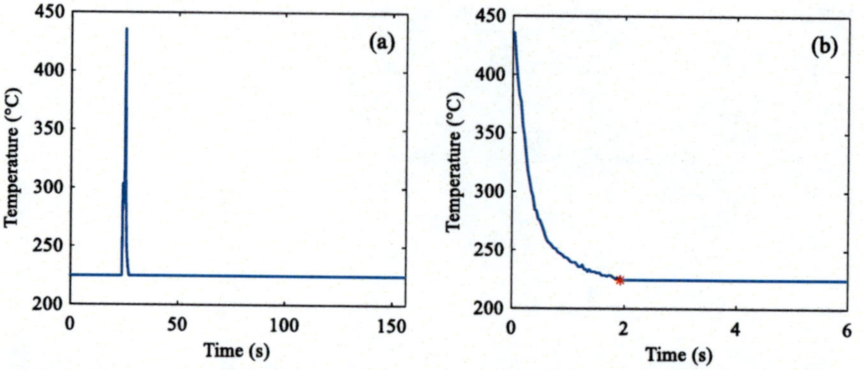

Fig. 4 Sample layer-based local thermal history–**a** thermal history at a pixel point for the time of printing one layer; **b** corresponding cooling profile starting from the highest peak temperature. *Note* The asterisk (*) symbol indicates the time when the temperature reaches steady ambient temperature starting from the highest peak temperature

and temporal averaging, the cooling profile can be backward extrapolated to obtain a profile starting from the melting temperature of the build material in the LPBF process. In doing so, the measured pixel-level cooling profile is used to estimate the temperature decay model for a specific pixel.

The cooling data is best approximated by the second-order exponential (Exp2) decay function of the form $f_{Exp2}(t) = a \times e^{b \times t} + c \times e^{d \times t}$ since it can closely model the physics of cooling in the LPBF process with the superposition of its two components. In the LPBF process, after the laser passes by a point and melting happens, the material temperature starts cooling exponentially towards the ambient temperature at a rapid decay rate—this effect is captured by the first component of the $f_{Exp2}(t)$. However, this decay in temperature changes its rate after a short period of time thanks to heat conduction from the neighboring laser-traversed points—this effect is captured by the second component of the $f_{Exp2}(t)$. Figure 5(a) shows the Exp2 curve fitting to the cooling profile shown in Fig. 4(b) along with the two components of the fitted curve. The equation of the Exp2 fitted curve in Fig. 5(a) is $f_{Exp2}(t) = 221.5 \times e^{-2.85t} + 225.1 \times e^{-3.56\times10^{-6}t}$ and the R^2-value of the fit is 0.9962. Using the fitted second-order exponential model, the local cooling profile can be backward extrapolated to the melting point of the build material in the LPBF process. Figure 5(b) illustrates the extrapolation of the Exp2 fitted curve shown in Fig. 5(a) up to 1500 °C.

Figure 6 shows the cooling profiles at a selected pixel point in the IN718 samples 11, 22, 44, 77, and 88 where differences in cooling rate and cooling time are clearly visible. The selected points are located at the centroid of the top square cuboid portion of the samples. Recall that the samples are 10 mm tall–the bottom 7 mm is inverted pyramid shaped, and the top 3 mm is square cuboid shaped with 10 × 10 mm square cross-section. Therefore, the selected points are located at a height of 8.5 mm (7 mm + 3/2 mm) and at the center of the 10 × 10 mm square cross-section

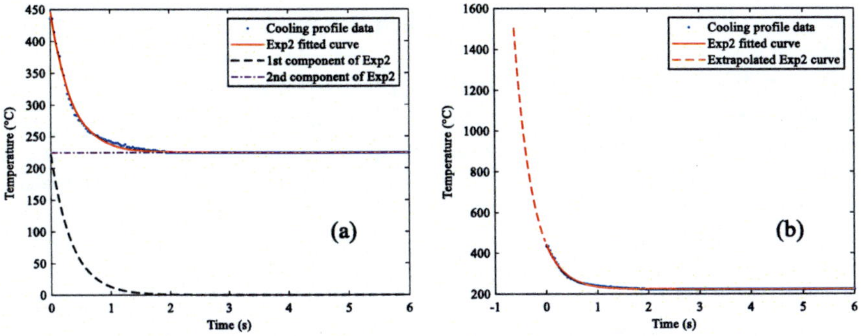

Fig. 5 **a** Second-order exponential (Exp2) fitted curve to the cooling profile and its two components; **b** extrapolation of the 2nd order exponential (Exp2) fitted curve up to 1500 °C

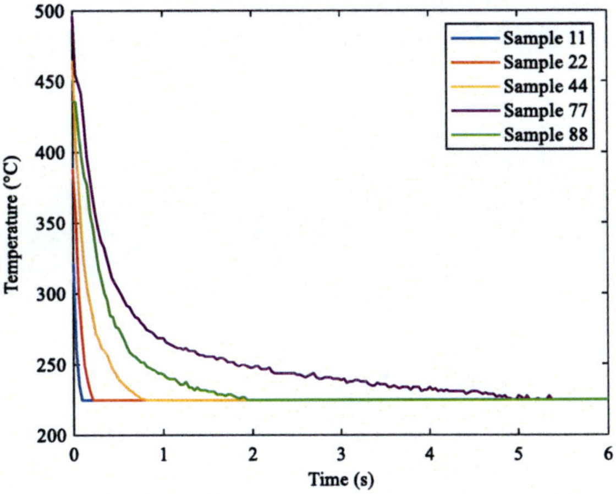

Fig. 6 Cooling profiles at the centroid of the IN718 samples 11, 22, 44, 77 and 88

of the square cuboid. Note that the thermal history and cooling profile presented in Figs. 4 and 5 correspond to the selected point in Sample 88. The differences in the local thermal history can be correlated to the microstructural features of the samples under investigation—which will be discussed in the next section.

Solidification Structure

The epitaxial grain structure is found in all of the IN718 diagonal samples. Figure 7(a) and (b) show the clusters of cellular dendrites along build direction found in the samples. The dendrite cell spacing of 11, 22, 44, 66, 77 and 88 are measured and given in Fig. 7(c). As the volumetric energy density input increases from sample 11 to sample 88, the dendrite spacing increases. The higher energy input leads to a higher thermal gradient and longer cooling time which allows more time for dendrite growth and micro-segregation. The trend of increasing dendrite cell spacing increasing with corresponding increases in power density is consistent across the entire range of the experiment, however there is an observed deviation from the trend of increasing cooling time with increasing power density for the highest power density sample 88. In Fig. 6 the cooling time is less for sample 88 than for sample 77, despite a higher power density parameter used for sample 88. While the exact cause of this deviation is not yet clear, analysis of alternate locations in samples 77 and 88 shows significant variability in cooling time for sample 88 relative to other samples. This suggests that at very high power densities that result in the generation of significant keyhole porosity defects, which are shown in Fig. 2c for sample 88, there may be phenomena such as vaporization and condensation in turbulent melt pools that cause additional complexities in the thermal history of the material. Ongoing work will utilize additional in-situ process monitoring methods to better understand this.

These dendritic structures are formed during solidification due to the preferential segregation of elements such as niobium to the liquid phase. This segregation can be

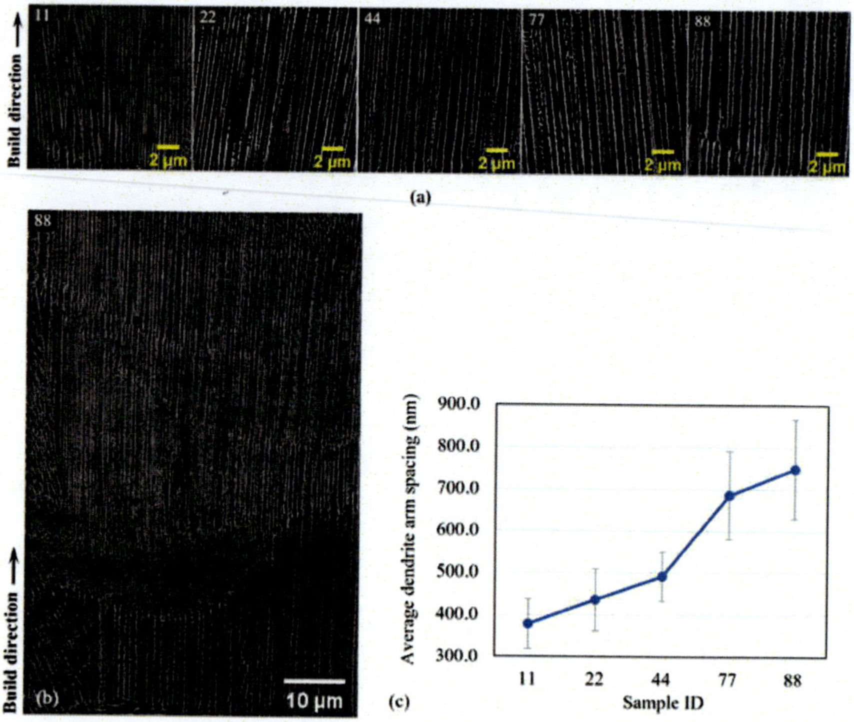

Fig. 7 **a** Cellular dendrites along z direction found in IN718 samples 11, 22, 44, 77 and 88. Images were taken by SEM at 10 k magnification. **b** A long range of dendrites over a couple of melt pools in sample 88. The image was taken by SEM at 2 k magnification

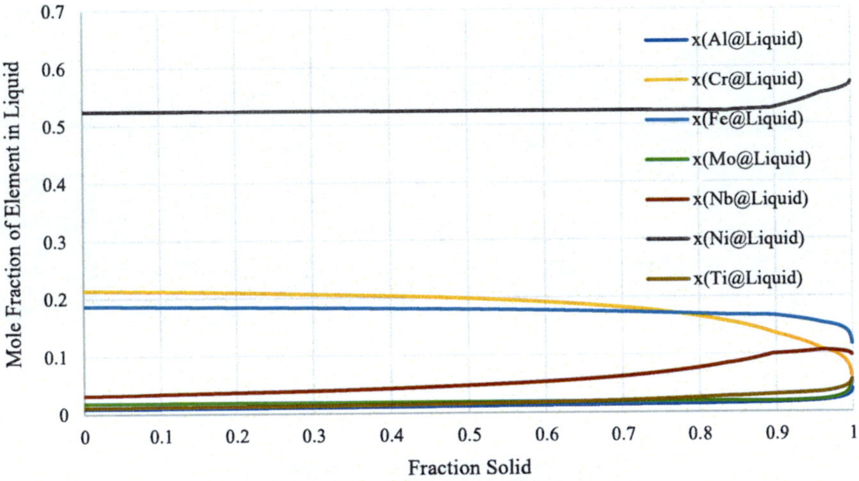

Fig. 8 Scheil model elemental segregation results for IN718 solidification

predicted using CALPHAD simulation tools such as Pandat. Utilizing a Scheil model in the Pandat software (PanPhaseDiagram module), the solidification segregation predictions are shown in Fig. 8. As IN718 solidifies, elements such as chromium and iron are expected to preferentially incorporate in the solid dendrite cores, while elements such as niobium, molybdenum, titanium and aluminum are pushed into the interdendritic regions where they form the dendrite cell boundaries upon complete solidification.

EDS maps of IN718 sample 88 are shown in Fig. 9. Niobium and titanium are found rich in the interdendritic regions as the CALPHAD model predicts. Figure 10 shows the EDS line scan data on another site of the same sample. The titanium peak indicates a Ti–rich particle on the cell wall, presumably a titanium containing carbide or carbonitride particle due to the presence of carbon also noted in the cell walls.

Discussion

The results presented in this work show that the dendrite spacing in as-built IN718 vary as a function of processing parameters and energy density, and that this variation can be related to time–temperature information that can be obtained using Longwave IR in-situ process monitoring. This variation is observed even within a relatively narrow range of processing parameters that can be used to build material free of significant numbers of defects such as cracks and porosity. Material evaluated in this work would be generally considered acceptable if analyzed via non destructive evaluation or metallographic analysis of witness coupons.

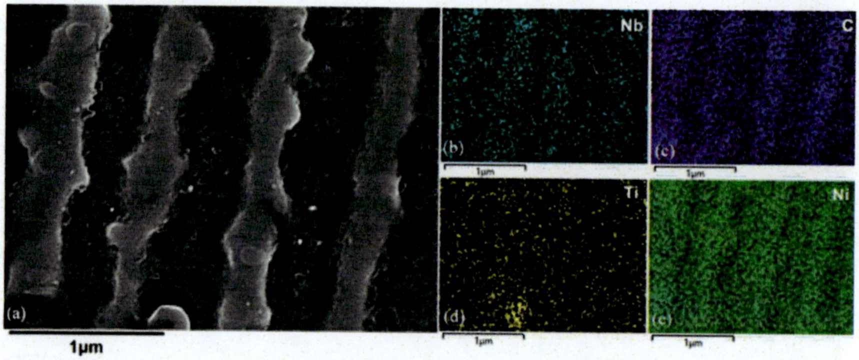

Fig. 9 EDS maps of IN718 sample 88. **a** Secondary electron image at 50kx. **b–e** EDS signal of Nb, C, Ti and Ni

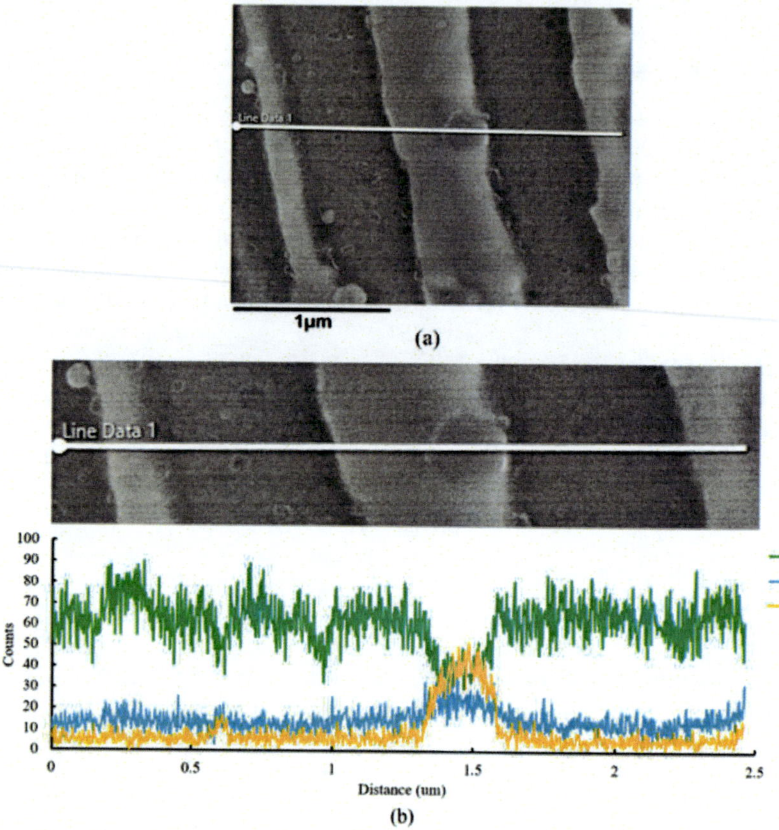

Fig. 10 EDS line scan data of IN718 sample 88

Significant variation could also be expected to be present in material built with a single fixed process parameters but having complex variations in thermal history due to geometric complexity in the component design. This variation in thermal history and dendritic structure leads to variations in the spatial distribution of elements that are important in the formation of phases in these alloys, for example the segregation of niobium in IN718 can lead to regions of Laves phase formation. This size and distribution of these phases can have a significant impact on mechanical properties of the material. Control of the microstructure of these alloys via closed loop process control enabled by in-situ process monitoring could produce materials with more consistent mechanical behavior for the end user.

While this work provides a useful initial correlation of process monitoring data to microstructure formation, significant additional work is underway to improve both the in-situ monitoring data collection and microstructure predictions. Longwave IR data provides a useful measure of temporal variation of solid state cooling throughout the build, by nature of the wide viewing area and relatively low data collection rate this method captures heat flow in a lower temperature regime across the build plate. This analysis has shown that this can be correlated to trends in microstructure formations and the data can be extrapolated to higher temperatures to construct location specific cooling curves. However, addition of additional process monitoring data streams such as thermal tomography and melt pool thermal imaging can capture complementary higher temperature data which can be used to improve the accuracy of thermal history reconstruction at specific locations in the build. These reconstructed cooling curves could then be used in more sophisticated solidification models such as Computherm's Pandat software (PanSolidification module) that utilize mobility information to make predictions of solidification structures that include spatial information such as dendrite arm spacing. In the long term, this coupling of in-situ monitoring of local thermal conditions and prediction of microstructure can be used to control LPBF processes and validate component quality for AM parts that can be difficult to inspect adequately using available NDE methods.

Conclusions

In this work we have shown that in-situ process monitoring can provide useful information towards understanding the variation observed in solidification structures of nickel superalloys such as IN718. The following conclusions can be drawn from this work:

- Dendrite spacing varies in IN718 as a function of energy density in LPBF builds
- This dendrite spacing variation is driven by variations in local cooling rates in the material
- The time–temperature history of the material throughout the build can be characterized and understood by utilizing in-situ process monitoring methods including Longwave IR data collection and processing

- Dendritic solidification leads to micro-segregation of elements such as niobium in IN718 that could have important implications for phase formation during subsequent processing
- Further work is needed to develop in-situ monitoring data collection and analysis methods that can provide input into CALPHAD modeling tools to predict phase formation during the build process and subsequent heat treatments to ensure consistent microstructural response in geometrically complex components

Acknowledgements Support of this work by the NASA Transformational Tools and Technologies Project is gratefully acknowledged.

References

1. B. Blakey-Milner, P. Gradl, G. Snedden, M. Brooks, J. Pitot, E. Lopez, M. Leary, F. Berto, A. du Plessis, Metal additive manufacturing in aerospace: A review, Mater Des. 209 (2021) 110008. https://doi.org/10.1016/J.MATDES.2021.110008.
2. M. Gorelik, Additive manufacturing in the context of structural integrity, Int J Fatigue. 94 (2017) 168–177. https://doi.org/10.1016/j.ijfatigue.2016.07.005.
3. J.F. Barker, The Initial Years of Alloy 718 - A G.E. Perspective, in: E.A. Loria (Ed.), The 9th International Symposium on Superalloy 718 & Derivatives, 1989: pp. 269–277.
4. K.N. Amato, S.M. Gaytan, L.E. Murr, E. Martinez, P.W. Shindo, J. Hernandez, S. Collins, F. Medina, Microstructures and mechanical behavior of Inconel 718 fabricated by selective laser melting, Acta Mater. 60 (2012) 2229–2239. https://doi.org/10.1016/J.ACTAMAT.2011.12.032.
5. Z. Wang, K. Guan, M. Gao, X. Li, X. Chen, X. Zeng, The microstructure and mechanical properties of deposited-IN718 by selective laser melting, J Alloys Compd. 513 (2012) 518–523. https://doi.org/10.1016/J.JALLCOM.2011.10.107.
6. J. Strößner, M. Terock, U. Glatzel, Mechanical and Microstructural Investigation of Nickel-Based Superalloy IN718 Manufactured by Selective Laser Melting (SLM), AdvEng Mater. 17 (2015) 1099–1105. https://doi.org/10.1002/ADEM.201500158.
7. E. Chlebus, K. Gruber, B. Kuźnicka, J. Kurzac, T. Kurzynowski, Effect of heat treatment on the microstructure and mechanical properties of Inconel 718 processed by selective laser melting, Materials Science and Engineering: A. 639 (2015) 647–655. https://doi.org/10.1016/J.MSEA.2015.05.035.
8. W.M. Tucho, P. Cuvillier, A. Sjolyst-Kverneland, V. Hansen, Microstructure and hardness studies of Inconel 718 manufactured by selective laser melting before and after solution heat treatment, Materials Science and Engineering: A. 689 (2017) 220–232. https://doi.org/10.1016/J.MSEA.2017.02.062.
9. D.H. Smith, J. Bicknell, L. Jorgensen, B.M. Patterson, N.L. Cordes, I. Tsukrov, M. Knezevic, Microstructure and mechanical behavior of direct metal laser sintered Inconel alloy 718, Mater Charact. 113 (2016) 1–9. https://doi.org/10.1016/J.MATCHAR.2016.01.003.
10. Q. Jia, D. Gu, Selective laser melting additive manufacturing of Inconel 718 superalloy parts: Densification, microstructure and properties, J Alloys Compd. 585 (2014) 713–721. https://doi.org/10.1016/J.JALLCOM.2013.09.171.
11. G.H. Cao, T.Y. Sun, C.H. Wang, X. Li, M. Liu, Z.X. Zhang, P.F. Hu, A.M. Russell, R. Schneider, D. Gerthsen, Z.J. Zhou, C.P. Li, G.F. Chen, Investigations of γ' γ'' and δ precipitates in heat-treated Inconel 718 alloy fabricated by selective laser melting, Mater Charact. 136 (2018) 398–406. https://doi.org/10.1016/j.matchar.2018.01.006.

12. A. Wessman, J. Cormier, F. Hamon, K. Rainey, S. Tin, D. Tiparti, L. Dial, Microstructure and Mechanical Properties of Additively Manufactured Rene 65, in: Superalloys 2020. The Minerals, Metals and Materials Series, Springer, Cham, 2020: pp. 961–971. https://doi.org/10.1007/978-3-030-51834-9_94.
13. Z. Tian, C. Zhang, D. Wang, W. Liu, X. Fang, D. Wellmann, Y. Zhao, Y. Tian, A Review on Laser Powder Bed Fusion of Inconel 625 Nickel-Based Alloy, Applied Sciences. 10 (2019) 81. https://doi.org/10.3390/app10010081.
14. K. Amato, J. Hernandez, L.E. Murr, E. Martinez, S.M. Gaytan, P.W. Shindo, S. Collins, Comparison of Microstructures and Properties for a Ni-Base Superalloy (Alloy 625) Fabricated by Electron Beam Melting, Journal of Materials Science Research. 1 (2012) p3. https://doi.org/10.5539/JMSR.V1N2P3.
15. J.H. Boswell, D. Clark, W. Li, M.M. Attallah, Cracking during thermal post-processing of laser powder bed fabricated CM247LC Ni-superalloy, Mater Des. 174 (2019) 107793. https://doi.org/10.1016/J.MATDES.2019.107793.
16. L. Rickenbacher, T. Etter, S. Hövel, K. Wegener, High temperature material properties of IN738LC processed by selective laser melting (SLM) technology, Rapid Prototyp J. 19 (2013) 282–290. https://doi.org/10.1108/13552541311323281.
17. M. Bermingham, D. StJohn, M. Easton, L. Yuan, M. Dargusch, Revealing the Mechanisms of Grain Nucleation and Formation During Additive Manufacturing, in: JOM, Springer, 2020: pp. 1065–1073. https://doi.org/10.1007/S11837-020-04019-5.
18. A. Prasad, L. Yuan, P. Lee, M. Patel, D. Qiu, M. Easton, D. StJohn, Towards understanding grain nucleation under Additive Manufacturing solidification conditions, Acta Mater. 195 (2020) 392–403. https://doi.org/10.1016/J.ACTAMAT.2020.05.012.
19. F. Zhang, M. Yang, A.T. Clare, X. Lin, H. Tan, Y. Chen, Microstructure and mechanical properties of Ti-2Al alloyed with Mo formed in laser additive manufacture, J Alloys Compd. 727 (2017) 821–831. https://doi.org/10.1016/J.JALLCOM.2017.07.324.
20. A. Hadadzadeh, B.S. Amirkhiz, J. Li, M. Mohammadi, Columnar to equiaxed transition during direct metal laser sintering of AlSi10Mg alloy: Effect of building direction, Addit Manuf. 23 (2018) 121–131. https://doi.org/10.1016/J.ADDMA.2018.08.001.
21. M.J. Bermingham, D.H. StJohn, J. Krynen, S. Tedman-Jones, M.S. Dargusch, Promoting the columnar to equiaxed transition and grain refinement of titanium alloys during additive manufacturing, Acta Mater. 168 (2019) 261–274. https://doi.org/10.1016/J.ACTAMAT.2019.02.020.
22. T.G. Spears, S.A. Gold, In-process sensing in selective laser melting (SLM) additive manufacturing, Integr Mater Manuf Innov. 5 (2016) 16–40. https://doi.org/10.1186/s40192-016-0045-4.
23. T. Purtonen, A. Kalliosaari, A. Salminen, Monitoring and Adaptive Control of Laser Processes, Phys Procedia. 56 (2014) 1218–1231. 1016/J.PHPRO.2014.08.038.
24. S. Berumen, F. Bechmann, S. Lindner, J.P. Kruth, T. Craeghs, Quality control of laser- and powder bed-based Additive Manufacturing (AM) technologies, Phys Procedia. 5 (2010) 617–622. https://doi.org/10.1016/J.PHPRO.2010.08.089.
25. M. Pavlov, M. Doubenskaia, I. Smurov, Pyrometric analysis of thermal processes in SLM technology, Phys Procedia. 5 (2010) 523–531. https://doi.org/10.1016/J.PHPRO.2010.08.080.
26. H. Krauss, C. Eschey, M.F. Zaeh, Thermography for Monitoring the Selective Laser Melting Process, in: 2012 International Solid Freeform Fabrication Symposium, Austin, TX, 2012: pp. 999–1014. https://repositories.lib.utexas.edu/handle/2152/88469 (accessed July 27, 2022).

Understanding Annealing Behavior During Post-Built Heat Treatment of Ni-Based Alloys Across Additive Manufacturing Processes

Juan Gonzalez, Yi Zhang, Andrew Wessman, and Jonah Klemm-Toole

Abstract Ni-based alloys are used for high temperature structural components that span from small, highly complex, with fine feature resolution to large, simple shapes with low dimensional tolerances, necessitating the use of processes spanning from laser powder bed fusion (LPBF) to wire arc additive manufacturing (WAAM). However, there is very little understanding about how annealing behavior during post-build heat treatments varies between additive manufacturing processes. In this work, we explore the annealing behavior of IN625 and Haynes 282, manufactured with WAAM and LPBF, under the same annealing conditions. The results of hardness measurements after annealing indicate that for both IN625 and Haynes 282, the LPBF samples show larger decreases in hardness between the as-built condition and after annealing at 1200 °C for 1 h compared to the WAAM samples. LPBF IN625 and Haynes 282 samples annealed at 1200 °C for 1 h, all show complete and partial recrystallization, respectively, whereas none of the WAAM samples annealed at this temperature show recrystallization. For a given alloy, both LPBF and WAAM samples annealed at 1200 °C show particles with compositions consistent with MC carbides that are predicted from thermodynamic simulations. The MC particles present are of similar size and distribution in both LPBF and WAAM samples indicating a similar capacity for these particles to pin moving boundaries during recrystallization. In concert, these results suggest that LPBF samples have more stored energy in the as-built condition compared to their WAAM counterparts, and therefore have a higher driving force for recovery and recrystallization.

Keywords Post-build heat treatment · Laser powder bed fusion · Wire arc additive manufacturing

J. Gonzalez · J. Klemm-Toole (✉)
Colorado School of Mines, 1500 Illinois Street, Golden, CO 80401, USA
e-mail: jklemmto@mines.edu

Y. Zhang · A. Wessman
University of Arizona, Tucson, AZ 85721, USA
e-mail: wessman@arizona.edu

E. A. Ott et al. (eds.), *Proceedings of the 10th International Symposium on Superalloy 718 and Derivatives*, The Minerals, Metals & Materials Series,
https://doi.org/10.1007/978-3-031-27447-3_38

Introduction

Ni-based alloys are used for high temperature structural components that span from small, highly complex, with fine feature resolution to large, simple shapes with low dimensional tolerances. Accordingly, a range of additive manufacturing (AM) processes from high precision laser powder bed fusion (LPBF) to high deposition rate wire arc additive manufacturing (WAAM) can be used to produce components. The majority of Nickel based alloys used in additive manufacturing were designed to be used in cast or wrought forms requiring several heat treatment processes to control secondary phases such as γ', carbides, and topologically close-packed (TCP) phases to optimize mechanical performance. As a result, post-build heat treatments are needed after additive manufacturing. However, the microstructure evolution during post processing can vary considerably between LPBF and WAAM.

There have been several works published regarding the microstructure evolution and annealing behavior of LPBF processed Ni-based alloys during post-build heat treatment. Christofidou et al. [1], Shaikh et al. [2] Deshpande et al. [3] found no recrystallization in LPBF Haynes 282 after heat treatment below 1150 °C for 1 h. Christofidou et al. found partial recrystallization after heat treating at 1190 °C for 1 h and full recrystallization at 1240 °C for 1 h and suggested that heat treatments must be performed above the primary MC carbide solvus in order to obtain recrystallization [1]. However, Deshpande et al. observed recrystallization after a hot isostatic pressing (HIP) treatment at 1185 °C for 3.5 h, which is below the MC carbide solvus for Haynes 282 [3]. Kreitcberg et al. [4] and Marchese et al. [5] did not observe complete recrystallization after heat treating LPBF IN625 below 1080 °C for 1 h. Kreitcberg et al. showed that HIP heat treatments at 1120–1175 °C for 2 h resulted in nearly complete recrystallization even though MC carbides were present in the microstructure [4]. Marchese et al. reported nearly complete recrystallization after a heat treatment at 1150 °C for 2 h and also noted that MC carbides were present [5]. The literature on recrystallization of LPBF Haynes 282 and IN625 appears to generally suggest that recrystallization can occur even with MC carbides present in the microstructure, although there is considerable variation in the heat treatment parameters with which complete recrystallization is observed.

Unlike LPBF, there is considerably less information available about the annealing of WAAM or other directed energy deposition (DED) AM processed Ni alloys. Zhang et al. evaluated carbide precipitation after heat treating WAAM Haynes 282 at 1150 °C for 2 h but provided no information regarding recrystallization [6]. To the authors' knowledge, this is the only report of microstructure development in WAAM Haynes 282 during post-build heat treatment. Hu et al. observed partial recrystallization in laser blown powder DED after heat treating at 1200 °C for 0.5 h and HIP at 1150 °C for 3 h [7]. Hu et al. also found nearly complete recrystallization of laser blown powder DED IN625 after heat treating at 1200 °C for 90 min [8]. Tanvir et al. showed that after heat treating WAAM IN625 at 980 °C for 1 h, no recrystallization was observed, and MC carbides were present [9]. Overall, there is significantly less information in the literature regarding the annealing phenomena of

WAAM IN625 and Haynes 282 compared to LPBF, and considerable differences in heat treatment parameters associated with recrystallization are reported.

In light of the gaps in the literature we have described, we seek to explore and understand the annealing behavior of IN625 and Haynes 282, manufactured with WAAM and LPBF, under the same annealing conditions. These two disparate AM processes are expected to encompass the breadth of annealing behaviors exhibited by additively manufactured forms of these alloys and shed light on routes to control microstructure and mechanical properties for demanding structural applications.

Methods and Materials

Materials and Additive Manufacturing Parameters

Two Ni-based alloys were investigated in this work: a solid solution strengthened alloy, IN625, and a γ'-strengthened alloy, Haynes 282. The nominal compositions of these alloys representative of wire and powder feedstocks used in this study are shown in Table 1. The composition of IN625 reported in Table 1 corresponds to a gas metal arc welding (GMAW) wire feedstock, and the composition of Haynes 282 is the typical one for the wrought form.

LPBF IN625 samples were produced on an SLM Solutions SLM280 machine under flowing Argon gas. A 30 μm powder layer, 60 μm hatch spacing, 350 W laser power, and 2.8 m/s raster speed were used to produce 12.7 mm diameter and 50 mm long cylindrical specimens. LPBF Haynes 282 samples were produced on an Open Additive PANDA under flowing Argon gas. A 30 μm powder layer, 70 μm hatch spacing, 175 W laser power, and 0.8 m/s raster speed were used to produce

Table 1 Nominal compositions of IN625 and Haynes 282

Element	IN625 [10]	Haynes 282 [11]
Ni	Balance	Balance
Cr	22	20
Fe	5	1.5
Mo	9	8.5
Nb	3.65	–
Mn	0.5	0.3
Si	0.5	0.12
Al	0.4	1.5
Ti	0.4	2.1
Co	1	10
C	0.1	0.06
B	–	0.005

to make 9.5 mm cuboidal specimens. WAAM IN625 samples were produced with a Fronius Cold Metal Transfer (CMT) GMAW heat source using 1.14 mm (0.045 in) diameter IN625 GMAW wire fed at 106 mm/s (250 in/min), Ar + 5% CO_2 shielding gas, an average potential of 17.3 V and average current of 136 A, with a travel speed of 4.2 mm/s (10 in/min). Single pass wide wall-shaped builds, measuring approximately 150 mm long, 38 mm tall, and 6 mm wide were produced. 4 pass wide builds were also produced for IN625 with similar dimensions except the thickness was approximately 16 mm thick. WAAM Haynes 282 samples were produced with a Lincoln Power Wave S500 GMAW heat source using 1.14 mm (0.045 in) diameter Haynes 282 GMAW wire fed an 85 mm/s (200 in/min), Ar + 25% He shielding gas, an average potential of 17.5 V and average current of 115 A, with a travel speed of 6.4 mm/s (15 in/min). Only single pass wide wall-shaped builds were produced of Haynes 282 with similar dimensions to the single pass wide builds of WAAM IN625.

Thermodynamic Predictions

Thermo-Calc© version 2022a using the TCNI11 database was used to predict phase amounts and phase compositions as a function of temperature. Single axis equilibrium simulations were performed using the compositions shown in Table 1. No restrictions were placed on the phases included in the simulations.

Heat Treatment

Samples from each alloy and AM process were taken in the as-built condition and subjected to heat treatment for 1 h each at 1000, 1100, and 1200 °C. Samples were enclosed in an Ar filled stainless steel bag and heat treated in a Carbolite air furnace. Samples were allowed to air cool after heat treatment.

Hardness Testing

Hardness testing in the as-built condition and after annealing was performed on mounted and metallographically polished samples using a LECO AMH55 Automatic Hardness Testing System with a LM110AT indenter unit with a 500 g load. At least 20 hardness measurements were taken for each condition.

Microstructure Characterization

Samples were mounted and metallographically prepared prior to microstructure characterization. In preparation for electron backscatter diffraction (EBSD), samples were polished down to 1 μm diamond and then vibratory polished with 0.05 μm diameter colloidal silica for 12 h. EBSD was performed on a JEOL JSM-7000F field emission scanning electron microscope (SEM) with an accelerating voltage of 20 kV. Nearest neighbor pattern average reindexing (NPAR) was used to improve the indexing on each map. Samples for energy dispersive spectroscopy (EDS) were electro-etched using 2% nitric acid in ethanol with 3.5–V for 10–40 s at room temperature. EDS was performed using an FEI Teneo Field Emission SEM with an accelerating voltage of 10 kV.

Results and Discussion

Thermodynamic Predictions of Phases Present During Post-Build Heat Treatment

Figure 1 shows the results of thermodynamic predictions for IN625 and Haynes 282 as a function of temperature. MC carbide with a composition of approximately 80% Nb–7% Ti–11% C–balance other elements in the alloy (wt %) is predicted to be present below approximately 1290 °C in IN625 as shown in Fig. 1a. The predicted σ phase solvus in IN625 is approximately 1040 °C. Similarly, within the temperature range experimentally evaluated in this work, Haynes 282 is expected to have MC carbide and TiB2 present below 1200 °C as shown in Fig. 1b. The composition of MC carbide in Haynes 282 is predicted to be approximately 73% Ti–8% Mo–17% C–balance other elements in the alloy (wt%). The solvus temperatures of $M_{23}C_6$ and γ' are predicted to be below 1000 °C and should not be present during the heat treatments performed in this work. MC carbides are predicted to be present during annealing heat treatments between 1000 and 1200 °C for IN625 and Haynes 282 evaluated in this work.

Hardness After Post-Build Heat Treatment

Table 2 summarizes the hardness of LPBF and WAAM IN625 and Haynes 282 samples in the as-built condition and after annealing at 1000, 1100, or 1200 °C for 1 h. As expected, the hardness decreases from the as-built condition with increasing annealing temperature. Figure 2 shows the changes in hardness of the LPBF and WAAM samples as a function of annealing temperature, and relevant solvus temperatures of secondary phases are also shown. In general, the hardness values of IN625

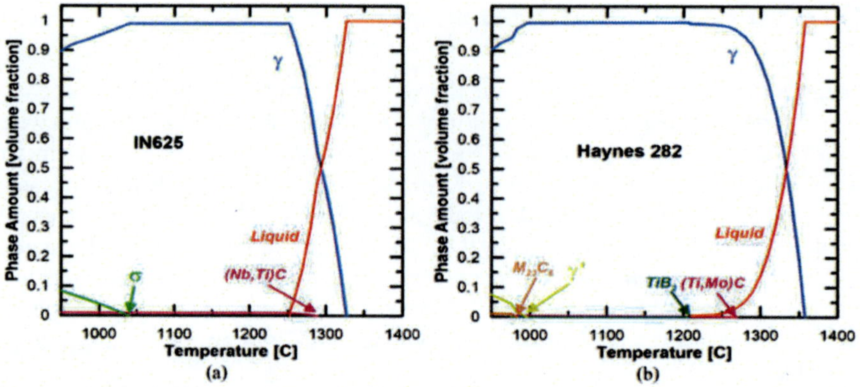

Fig. 1 Thermo-Calc © predictions of phase amounts present as a function of temperature for **a** IN625 and **b** Haynes 282

conditions (Fig. 2a) are lower than that of Haynes 282 (Fig. 2b). LPBF samples of both materials show higher hardness in the as-built condition compared to WAAM counter parts. After annealing at 1200 °C for 1 h, there is not a significant difference between the hardness of IN625 LPBF and WAAM samples. However, for Haynes 282, the LPBF samples maintain a higher hardness than WAAM counterparts after annealing at 1200 °C. Alternatively stated, LPBF samples exhibit much greater reductions in hardness compared to WAAM samples for the same annealing treatments. These reductions in hardness are likely related to recovery, recrystallization, and possibly grain growth. Negligible differences in hardness as a function of annealing temperature are observed between the 1 pass wide and 4 pass wide IN625 WAAM conditions.

Table 2 Average hardness values for LPBF and WAAM IN625 and Haynes 282 in the as-built and annealed conditions

Alloy	Process	Average hardness ($HV_{0.5}$)			
		As-built	1000 °C	1100 °C	1200 °C
IN625	LPBF	316	285	266	212
	WAAM 1 Pass	243	239	224	208
	WAAM 4 Pass	236	228	235	214
Haynes 282	LPBF	394	298	310	286
	WAAM 1 Pass	297	316	292	255

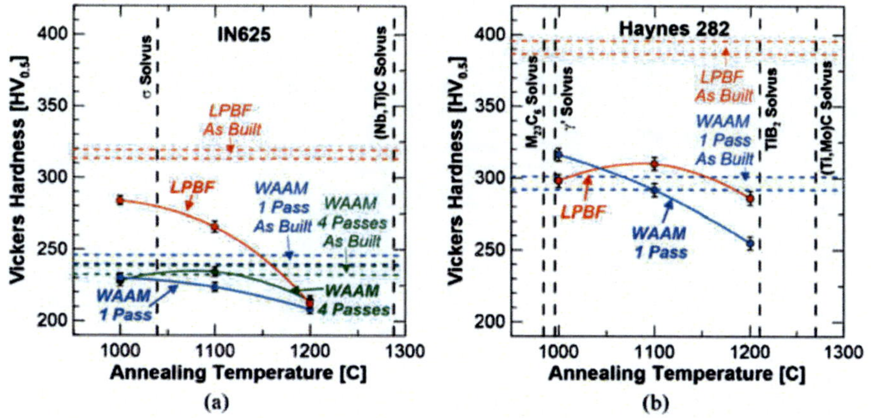

Fig. 2 Hardness measured at room temperature as a function of annealing temperature for LPBF and WAAM **a** IN625 and **b** Haynes 282. Horizontal dashed lines in each plot indicate the hardness of each condition in the as-built condition. Vertical dashed lines show solvus temperatures predicted by Thermo-Calc©

Recrystallization from Post-Build Heat Treatment

Electron backscatter diffraction (EBSD) was performed to evaluate the extent of recrystallization after annealing. Figure 3 shows results for LPBF IN625. Figure 3a shows a superimposed inverse pole figure (IPF) and image quality (IQ) map of the as-built condition of LPBF IN625, where grains elongated parallel to the build direction can be observed with grey scale values within grains indicating low image quality likely due to dislocation substructure. Figure 3b shows LPBF IN625 after annealing at 1100 °C for 1 h. The grain structure is still elongated parallel to the build direction, but many regions show higher image quality indicating that recovery has occurred. After annealing at 1200 °C for 1 h, complete recrystallization is observed as shown in Fig. 3c. Several grains contain annealing twins, the average grain size appears to be greater than the as-built condition, and the grain morphology appears to be more equiaxed, indicating some grain growth may have occurred in addition to recrystallization.

The 1 pass and 4 pass WAAM IN625 in the as-built condition also show elongated grains parallel to the build direction, Fig. 4a and b. Figures 4c and d show the 1 pass and 4 pass WAAM IN625 builds after annealing at 1200 °C for 1 h in which no recrystallization is observed. Generally, higher image quality is observed after annealing at 1200 °C indicating recovery is likely responsible for the reductions in hardness observed in Fig. 2a.

Figure 5 shows EBSD results of LPBF Haynes 282 samples after annealing. EBSD scan from the as-built condition (Fig. 5a) exhibited large columnar grains elongated parallel to the build direction. Regions of higher image quality are observed after annealing at 1100 °C for 1 h (Fig. 5b) indicating recovery is responsible for hardness

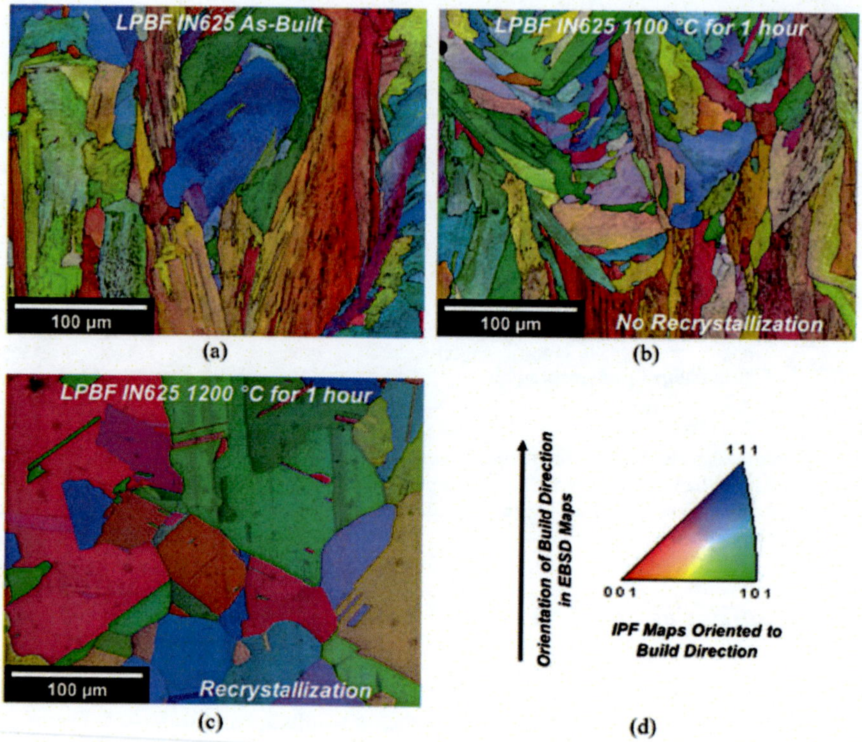

Fig. 3 EBSD IPF + IQ maps of LPBF IN625 **a** as-built, **b** after annealing at 1100 °C for 1 h, **c** after annealing at 1200 C for 1 h, **d** orientation of the build direction and coloration key of the IPF maps

reductions observed. Partial recrystallization has occurred after annealing at 1200 °C for 1 h as evidenced by the presence of grains with annealing twins in addition to elongated towards the build direction, Fig. 5c. In general, the annealing behavior of LPBF Haynes 282 is similar to LPBF IN625, except only partial recrystallization occurs in Haynes 282 after annealing at 1200 °C, whereas IN625 is fully recrystallized.

Similar to WAAM IN625 samples, the 1 pass wide WAAM Haynes 282 sample in the as-built condition exhibits large columnar grains elongated parallel to the build direction as shown in Fig. 6a. After annealing at 1200 °C, increases in image quality indicating recovery are observable, but recrystallization has not occurred, as shown in Fig. 6b.

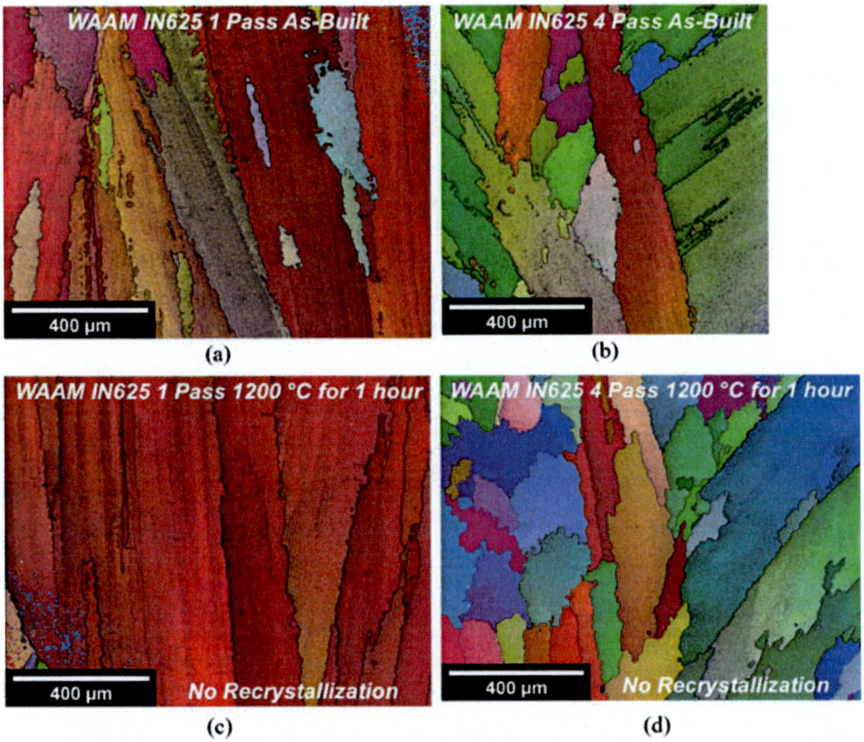

Fig. 4 EBSD IPF + IQ maps of WAAM IN625 **a** 1 pass wide build in the as-built condition, **b** 4 passes wide in the as-built condition, **c** 1 pass wide after annealing at 1200 °C for 1 h, and **d** 4 passes wide after annealing at 1200 °C for 1 h. The orientation of the build direction and IPF key is the same as shown in Fig. 3d

Secondary Phases Present After Post-Build Heat Treatment

Samples LPBF and WAAM, IN625, and Haynes 282, annealed at 1200 °C for 1 h were evaluated using EDS spot measurements to determine whether the MC carbides predicted to be present from thermodynamic simulations are observed. Particles enriched in Nb, Ti, and C are observed in both LPBF and 1 pass wide WAAM IN625 specimens annealed at 1200 °C, indicating the presence of (Nb,Ti)C predicted by a Thermo-Calc© in Fig. 1a. Note that particles in the LPBF (Fig. 7a) and WAAM (Fig. 7b) IN625 specimens annealed at 1200 °C are observed both at grain boundaries as well as grain interiors.

Similar to IN625 specimens, EDS spot measurements identified particles enriched with Mo, Ti, and C in Haynes 282 specimens annealed at 1200 °C for both LPBF and WAAM, indicating the presence of (Ti,Mo)C predicted by Thermo-Calc© in Fig. 1b. Again, note that both the LPBF (Fig. 7c) and WAAM (Fig. 7d) specimens annealed at 1200 °C exhibit particles at grain boundaries and grain interiors. In all of the EDS

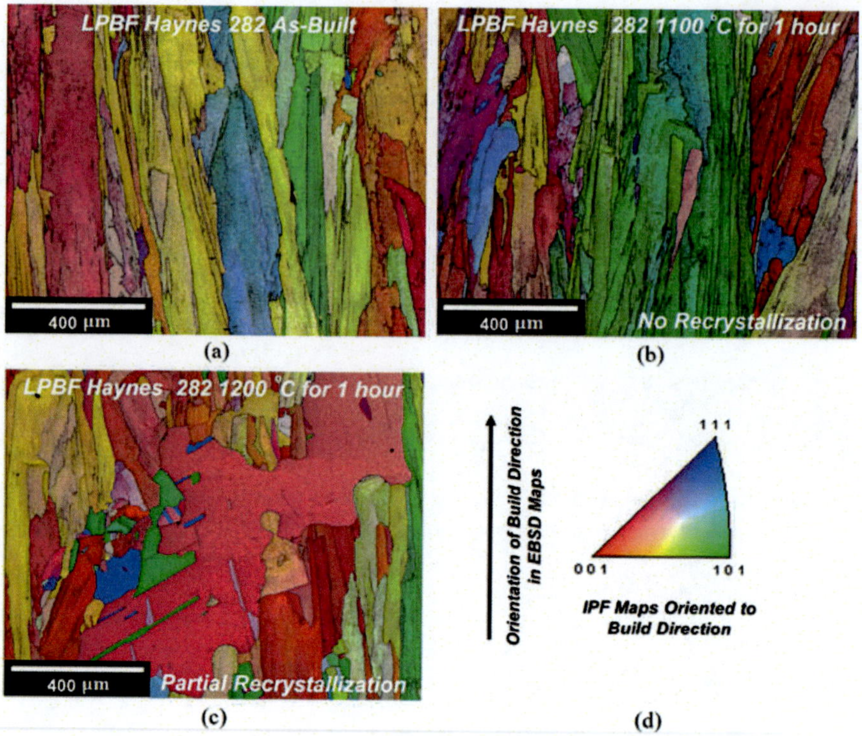

Fig. 5 EBSD IPF + IQ maps of LPBF Haynes 282 **a** as-built, **b** after annealing at 1100 °C for 1 h, **c** after annealing at 1200 °C for 1 h, **d** orientation of the build direction and coloration key of the IPF maps

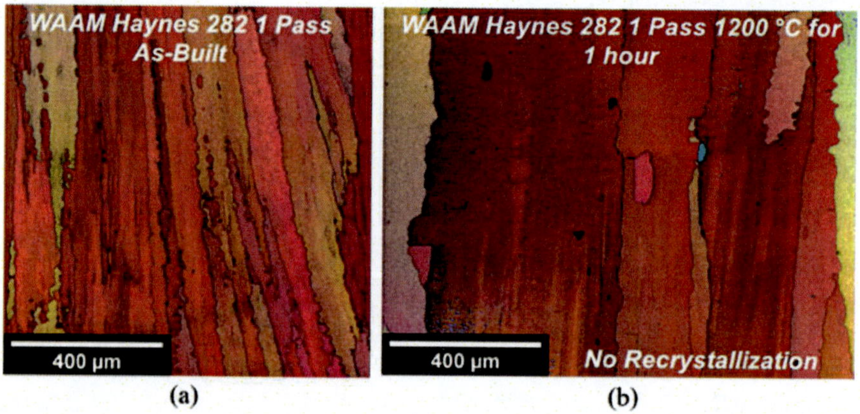

Fig. 6 EBSD IPF + IQ maps of WAAM Haynes 282 **a** 1 pass wide build in the as-built condition **b** 1 pass wide build after annealing at 1200 °C for 1 h. The orientation of the build direction and IPF key is the same as shown in Fig. 3d

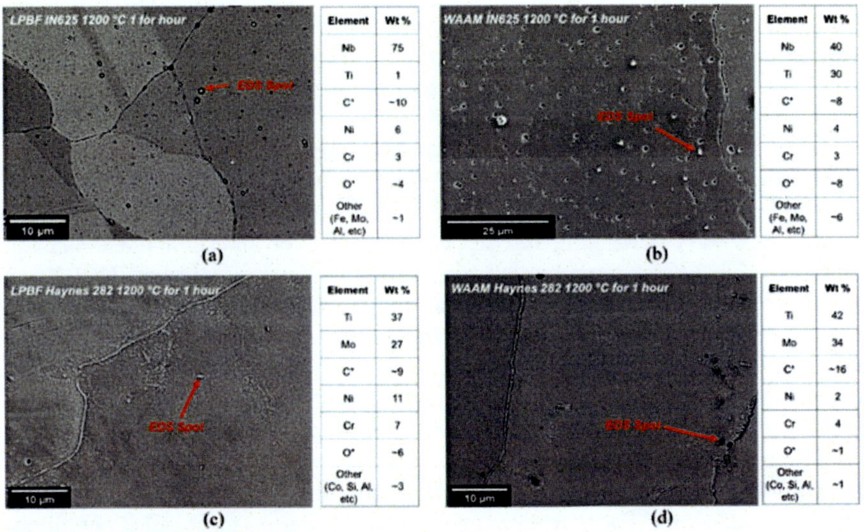

(a)

Element	Wt %
Nb	75
Ti	1
C*	~10
Ni	6
Cr	3
O*	~4
Other (Fe, Mo, Al, etc)	~1

(b)

Element	Wt %
Nb	40
Ti	30
C*	~8
Ni	4
Cr	3
O*	~8
Other (Fe, Mo, Al, etc)	~6

(c)

Element	Wt %
Ti	37
Mo	27
C*	~9
Ni	11
Cr	7
O*	~6
Other (Co, Si, Al, etc)	~3

(d)

Element	Wt %
Ti	42
Mo	34
C*	~16
Ni	2
Cr	4
O*	~1
Other (Co, Si, Al, etc)	~1

Fig. 7 SEM SE micrographs with EDS point measurement after annealing at 1200 C for 1 h **a** LPBF IN625, **b** WAAM IN625, **c** LPBF Haynes 282, and **d** WAAM Haynes 282. Measurements of C and O have an asterisk to indicate that the accuracy of these measurements is lower than that of the other elements detected by EDS point measurements

spot measurements shown in Fig. 7, some amount of oxygen is detected. This is likely an artifact of using an oxidizing acid (nitric acid) to etch the microstructure to reveal carbide particles. It is acknowledged that oxide particles are present in the microstructure, but the particles shown in Fig. 7 are likely the MC carbides discussed. Finally, elements such as Ni and Cr are also detected which is likely due to the interaction volume of the electron beam being big enough to sample the matrix surrounding the MC carbide particles.

Alloy and Process Influences on Annealing Phenomena During Post-Build Heat Treatment

The results of hardness measurements after annealing indicate that for both IN625 and Haynes 282, the LPBF samples show larger decreases in hardness between the as-built condition and after annealing at 1200 °C for 1 h compared to the WAAM samples. Furthermore, the IN625 LPBF and WAAM samples show similar hardness after annealing at 1200 °C, whereas the LPBF Haynes 282 samples still show a higher hardness than WAAM Haynes 282 samples after annealing at the sample temperature. IN625 and Haynes 282 LPBF samples annealed at 1200 °C for 1 h exhibited complete recrystallization and partial recrystallization respectively, whereas none of the WAAM samples annealed at this temperature show recrystallization. For a given alloy, both LPBF and WAAM samples annealed at 1200 °C show particles with compositions consistent with MC carbides that are predicted from thermodynamic simulations. The MC particles present are of similar size and distribution in both LPBF and WAAM samples indicating a similar capacity for these particles to pin moving boundaries during recrystallization. In concert, these results suggest that LPBF samples generally have more stored energy in the as-built condition compared to their WAAM counterparts, and therefore they have a higher driving force for recovery and recrystallization. Terris et al. interpreted the annealing phenomena of LPBF IN625 in the context of the equivalent stored energy from cold work [12]. X-ray diffraction experiments revealed that the as-built condition had dislocation densities on the order of 10^{15} m^2, which is comparable to extensive plastic strain. Depending on the exact processing parameters, LPBF samples showed equivalent cold work amounts of 13–20%, and the conditions with greater equivalent cold work recrystallized at lower temperatures, consistent with fundamental annealing concepts. Leicht et al. suggested that sample geometry and built orientation strongly affect the microstructure of LPBF 316L [13], which may also influence the stored energy that drives annealing. These results suggest that the thermo-mechanical conditions during AM processing directly impacts the stored energy available to drive annealing during post-build heat treatment. It could be that the differences in processing conditions and build geometry resulted in the slightly different recrystallization kinetics between the LPBF IN625 and Haynes 282 samples in addition to the larger differences between LPBF and WAAM.

A valid question that remains is: why do LPBF samples have a greater amount of stored energy compared to WAAM samples? LPBF melt pools are generally on the order of 100–150 μm wide and 100 m deep, and the samples analyzed in this work are on the order of 10 mm cubes. Such small melt pools relative to sample dimensions result in highly constrained conditions in the vicinity of the melt pool. The resulting thermal strains from the LPBF melt passes likely result in significant plastic deformation. WAAM melt pools are approximately 6 mm wide by comparison, which is much closer in dimension to the sample thickness. Samples that are 1 pass wide have melt pools and sample dimensions that are the same, which allows the thermal expansion around the melt pool to occur unconstrained resulting in relatively less plastic deformation. Even 4 pass wide samples have thickness dimensions only 2–2.5 X the dimension of a melt pool. In order to get a degree of constraint in a WAAM to build equivalent to that of LPBF, the dimensions of the sample would have to be on the order of 600 mm, which is experimentally feasible, but very little information about annealing of such sized builds are in the literature. In short, we hypothesize that the highly constrained nature of LPBF melts pools results in significantly more plastic strain accumulation during the build process compared to the WAAM samples analyzed here, and the greater constraint results in a greater amount of stored energy to drive annealing during post-build heat treatment.

Implications for Alloy and Process Design for Additive Manufacturing

The results of this work suggest that for a given alloy, samples produced with different AM processes have different amounts of stored energy in the as-built condition, which leads to systematic differences in annealing behavior. Thermodynamic simulations of second phase particles do not appear to be sufficient to inform annealing behavior. Not only thermal, but thermo-mechanical simulations capable of predicting the extent of plastic strain during the build process are needed to design AM processing parameters to tune the desired annealing behavior during post-build heat treatment. Furthermore, parallel alloy design efforts could enable the tailoring of secondary phases and/or microstructural features such that the desired annealing behavior could be achieved for a given degree of plastic strain from the additive manufacturing process.

Conclusions

In this work, the annealing behavior during post-build heat treatment between 1000 and 1200 °C for 1 h was investigated for IN625 and Haynes 282 processed with LPBF and WAAM using hardness testing, EBSD, and EDS. Based on the results, the following conclusions can be made:

- LPBF samples of IN625 and Haynes 282 show complete and partial recrystallization respectively, after annealing at 1200 °C, but WAAM samples of the same alloys heat treated at the same temperature do not show recrystallization.
- Despite large differences in hardness in the as-built condition, IN625 LPBF and WAAM samples show very similar hardness values after annealing at 1200 °C, despite significant differences in grain size and shape.
- LPBF Haynes 282 samples exhibit higher hardness values than WAAM Haynes 282 samples annealed at 1200 °C, likely due to only partial recrystallization of the LPBF samples.
- MC carbides are observed in LPBF and WAAM processed IN625 and Haynes 282 after annealing at 1200 °C for 1 h, which are predicted to be present using Thermo-Calc simulations.
- Because MC carbides were present in LPBF samples that recrystallized and WAAM samples that did not recrystallize, these particles likely do not control recrystallization in additively manufactured IN625 and Haynes 282.
- LPBF samples of both alloys show much larger decreases in hardness during annealing suggesting that there is more stored strain energy in the as-built condition compared to WAAM samples. The amount of stored energy in the as-built condition appears to have a greater effect on recrystallization compared to high temperature carbides in the alloys and processing conditions investigated here.

- In order to predict annealing behavior across vastly disparate processes, thermomechanical simulations capable of estimating plastic strain accumulation during additive manufacturing, and thus the driving force for annealing, are needed.

Acknowledgements The authors would like to acknowledge support from the Metallurgical and Materials Engineering Department at Colorado School of Mines as well as the Materials Science and Engineering Department at the University of Arizona. JG and JKT would also like to acknowledge the Center for Advanced Non-Ferrous Structural Alloys (CANFSA), USA, a National Science Foundation Industry/University Cooperative Research Center (I/UCRC), USA, [Award No. 2137243] at the Colorado School of Mines (Mines).

Credit author statement Conceptualization: AW, JKT; Methodology: JG, YZ, AW, JKT; Formal Analysis: JG, JKT; Investigation: JG, JKT; Resources: AW, JKT; Writing - Original Draft: JG, JKT; Writing – Review and Editing: JG, YZ, AW, JKT; Supervision: AW, JKT; Funding Acquisition: AW, JKT.

References

1. K. Christofidou, H. Pang, W. Li, C. Jones, N. Jones and H. Stone., "Microstructure control and optimization of Haynes 282 manufactured through laser powder bed fusion," in *Superalloys 2020*, Virtual, 2020.
2. A. S. Shaikh, F. Schulz, K. Minet-Lallemand and E. Hryha, "Microstructure and mechanical properties of Haynes 282 superalloy produced by laser powder bed fusion," *Materials Today Communications,* vol. 26, no. 102038, 2021.
3. Deshpande, S. D. Nath, S. Atre and K. Hsu., "Effect of post processing heat treatment routes on the microstructure and mechanical property evolution of Haynes 282 Ni-based superalloy fabricated with selective laser melting (SLM)," *Metals,* vol. 10, no. 629, pp. 1–13, 2020.
4. Kreitcberg, V. Brailovski and S. Turenne., "Effect of heat treatment and hot isostatic pressing on the microstructure and mechanical properties of Inconel 625 alloy processed by laser poder bed fusion," *Materials Science and Engineering A,* vol. 689, pp. 1–10, 2017.
5. G. Marchese, S. Parizia, M. Rashidi, A. Saboori, D. Manfredi, D. Ugues, M. Lombardi, E. Hryha and S. Biamino, "The role of texturing and microstructure evolution on the tensile behavior of heat-treated Inconel 625 produced via laser powder bed fusion," *Materials Science and Engineering A,* vol. 769, no. 138500, 2020.
6. H. Zhang, Y. Wang, R. R. D. Vecchis and W. Xiong., "Evolution of carbide precipitates in Haynes 282 superalloy processed by wire arc additive manufacturing," *Journal of Materials Processing Technology,* vol. 305, no. 117597, 2022.
7. Y. Hu, X. Lin, Y. Li, S. Zhang, Q. Zhang, W. Chen, W. Li and W. Huang., "Influence of heat treatments on the microstructure and mechanical properties of Inconel 625 fabricated by directed energy deposition," *Materials Science and Engineering A,* vol. 817, no. 141309, 2021.
8. Y. Hu, Y. Li, S. Zhang, X. Lin, Z. Wang and W. Wang, "Effect of solution temperature on static recrystallization and ductility of Inconel 625 superalloy fabricated by directed energy deposition," *Materials Science and Engineering A,* vol. 772, no. 138711, 2020.
9. Tanvir, M. Ahsan, C. Ji, W. Hawkins, B. Bates and D. Kim., "Heat treatment effects on Inconel 625 components fabricated by wire + arc additive manufacturing (WAAM) - part 1: microstructure characterization," *The International Journal of Advanced Manufacturing Technology,* vol. 103, pp. 3785 - 3798, 2019.
10. Lincoln Electric, "Techalloy 625," 2016. [Online]. Available: https://www.lincolnelectric.com/assets/global/Products/Consumable_NickelAlloys-Techalloy-Techalloy625/c141010.pdf. [Accessed 2022].

11. Haynes International, "Haynes 282 Alloy," 2021. [Online]. Available: https://haynesintl.com/docs/default-source/pdfs/new-alloy-brochures/high-temperature-alloys/brochures/282-brochure.pdf?sfvrsn=20. [Accessed 2022].
12. T. D. Terris, O. Castelnau, Z. Hadjem-Hamouche, H. Haddadi, V. Michel and P. Peyre., "Analysis of as-built microstructures and recrystallization phenomena on Inconel 625 alloy obtained via laser powder bed fusion (L-PBF)," *Metals,* vol. 11, no. 619, pp. 1–22, 2021.
13. Leicht, U Klement, E. Hryha., "Effect of build geometry on the microstructural development of 316L parts produced by additive manufacturing," *Materials Characterization,* 143 (2018) 137–143

High-Temperature Properties of Alloy 718 Made by Laser Powder-Bed Fusion

David Witkin, Tait McLouth, Glenn Bean, Julian Lohser, and Robert W. Hayes

Abstract The properties of alloy 718 made by laser powder-bed fusion (LPBF) have been widely reported, and while their room-temperature static properties are often similar to wrought material, elevated temperature properties have proven inferior, especially at slow strain rates. Stress rupture tests performed on LBPF 718 material after Hot isostatic Pressing (HIP) and heat treating in conformance with AMS 5662 have persistently led to brittle notch failures with limited elongation. Creep testing at 650 °C and 690 MPa shows a similar tendency, with samples showing capability of sustaining a load but with limited elongation and abrupt, intergranular failure. Alternative heat treatments performed to enhance high-temperature ductility have shown success in adjusting delta phase population but without benefit for stress rupture. Alternative HIP schedules have also been performed that led to changes in room-temperature and elevated temperature static strength but no benefit in stress rupture behavior. The root cause of this behavior is attributed to the dispersion of NbC that is a consequence of the LPBF process that results in fine-scale segregation of Nb and C during rapid solidification. Deliberately lowering the carbon content of the powder feedstock led to a greater number of smaller Y" particles and smaller size NbC particles which resulted in an increase in static strength at room and elevated temperatures but no improvement in stress rupture. The presence of a large number of NbC particles leads to environmental sensitivity of LPBF 718 that is most apparent at elevated temperatures and slow strain rates.

Keywords Inconel 718 · Laser powder bed fusion · Additive manufacturing · Elevated temperature properties · Creep rupture

D. Witkin (✉) · T. McLouth · G. Bean · J. Lohser
The Aerospace Corporation, El Segundo, CA, USA
e-mail: david.b.witkin@aero.org

R. W. Hayes
Metals Technology Inc., Northridge, CA, USA

E. A. Ott et al. (eds.), *Proceedings of the 10th International Symposium on Superalloy 718 and Derivatives*, The Minerals, Metals & Materials Series,
https://doi.org/10.1007/978-3-031-27447-3_39

Introduction

Alloy 718 processed by additive manufacturing (AM) techniques has continued to attract the attention of researchers. As a weldable, heat-treatable nickel-base superalloy with widespread use in high-value applications, alloy 718 is well-suited to AM processing and the potential benefits of fabrication of complex parts and assemblies using AM are applicable to high-value parts made with 718. A review of mechanical properties of AM 718 [1] concluded that AM processing parameters, heat treatment, and processing defects play an important role in determining properties. For example, tensile strength was said to be intermediate to expected values for cast or wrought product forms. While there was significant variation in creep properties, that could be attributed to lack of consistency between different published accounts based on creep mode (i.e., tensile versus compressive) and heat treatment of the test materials. One report claimed superior performance for laser powder-bed fusion (LPBF) processed material in comparison to wrought material but this was based on compression creep testing to only 1% plastic strain [2]. Subsequent publications on creep have tested material that retained a significant as-built microstructural character due to heat treatment at temperatures insufficient to cause recrystallization of the γ matrix phase prior to solution treatment and aging [3, 4].

For high-reliability applications such as aerospace systems, it is expected that AM 718 would be subjected to hot isostatic pressing (HIP) to remove defects to the greatest extent possible followed by heat treatment to take advantage of the alloy's intrinsic strengthening capability via the precipitation of γ''. In this light, it is instructive to understand the mechanical properties of alloy 718 in comparison to traditional requirements for wrought material as indicated by minimum acceptance values in aerospace standards AMS 5662, AMS 5663, AMS 5664, and AMS 5596. In the case of creep, LPBF 718 has been generally shown to be inferior to wrought material when time-to-rupture and strain-to-failure even after HIP and solution and age. Investigations that focused on manipulation of microstructure via laser scanning strategies found that elongated grains parallel to the loading axis improved creep life in LPBF [5–7]. At the same time, even in particular cases where time-to-rupture was similar to or even exceeded wrought material, the elongation of the creep samples prior to failure was only approximately one quarter of the wrought material which reached strains on the order of 30% engineering strain prior to failure.

The occurrence of resistance to deformation seen as long times to failure accompanied by brittle fracture was also documented for the specific case of notched stress rupture that is part of the material acceptance testing under AMS 5662 [8]. In this test, a combination specimen consisting of a notched section with Kt ~ 3.9 and a second uniform gauge section is subjected to a 690 MPa tensile load at 650 °C and is required to (1) sustain the load for at least 23 h; (2) exhibit a minimum of 4% elongation in the uniform section; and (3) rupture in the uniform section, not at the notch. For samples reaching 23 h under this condition, the load can be increased by 34 MPa (5 ksi) at successive 8-h intervals, but the other two requirements are still in force. All LPBF samples, whether heat treated to AMS 5662 or 5664 schedules, failed the test due to brittle failure at the notch, although several greatly exceeded 23 h [8]. This result is analogous to reports of creep tests of LPBF 718 where the rupture time was comparable to wrought material while the strain at failure was much lower than wrought. Although there was a range of failure times in combination stress rupture from failure within minutes to failure beyond 40 h of loading time, the lack of elongation in the uniform section accompanied by notch failure indicates that the material may exhibit strong but brittle behavior at elevated temperatures. Such behavior is not apparent when performing quasi-static tensile tests at room-temperature.

The question of what conditions (temperature and strain rate) lead to degraded ductility of LPBF 718 compared to wrought material was addressed through a series of slow strain-rate tensile tests conducted from 550 to 750 °C and strain rates from 10^{-3} to 10^{-6} s^{-1} [9] for materials heat treated per AMS 5662. The lower end of the strain rate range is higher than published steady-state creep rates in the range of 10^{-8} to 10^{-7} s^{-1} at 650 °C where lower ductility had been established in creep tests. At 650 °C, the yield and ultimate strength of both LPBF and wrought material decreased from 10^{-3} to 10^{-6} s^{-1} strain rates, but wrought material had similar elongation and reduction of area (RA) regardless of strain rate. In LPBF 718, the elongation and reduction in area (RA) declined significantly from 10^{-3} to 10^{-4} s^{-1}. When temperature was the test variable at 10^{-4} s^{-1} strain rate, wrought material exhibited a characteristic ductility dip at 650 °C compared to 550 and 750 °C. In the LPBF material, the elongation was much lower and the ductility dip, while present, was much less apparent. The strength of the wrought material was higher at 550 and 650 °C, but at 750 °C both types had similar strength. These trends are illustrated in Fig. 1.

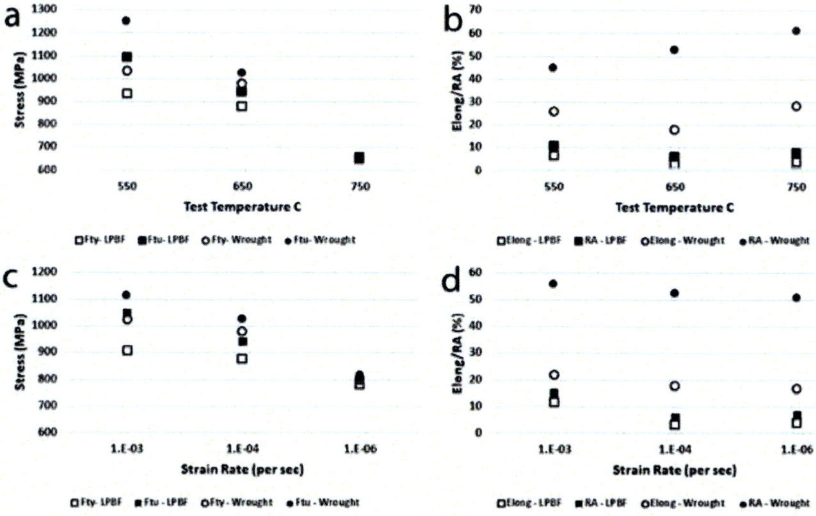

Fig. 1 Variation of tensile yield strength (Fty), ultimate tensile strength (Ftu), elongation (Elong), reduction in area (RA) for LPBF and wrought alloy 718 as a function of temperature at strain rate 10^{-4} s^{-1} **a** and **b** and strain rate at 650 °C **c** and **d**

This survey of published elevated temperature property data shows that LPBF 718 has generally lower ductility than wrought material at elevated temperatures, but the differences are exacerbated at lower strain rates characteristic of creep. Tensile yield and ultimate strength values also tend to be lower than wrought material, even after solution treatment and aging using standard schedules. Interpretation of these results in light of the microstructural features and fracture surfaces led to a conclusion that dynamic embrittlement due to the presence of a dispersion of numerous small NbC particles, often on grain boundaries, is responsible for the lower ductility [9]. Even at similar carbon content, the NbC population of wrought material manifests many fewer but larger particles. The difference in NbC distribution is due to the small-scale rapid solidification that occurs in the LPBF, the segregation of Nb and C to the inter-dendritic regions of the fused material, and the absence of thermo-mechanical processing in net- and near net-shape fabrication enabled by AM. Several alternative heat treatments were attempted to improve these properties that did lead to alterations in the microstructure and static strength but did not improve the high-temperature ductility. More promising results were obtained by minimizing the total carbon content of the LPBF powder, but while these had a considerable effect on the γ" and NbC populations and improved static strength by freeing more Nb for precipitation of γ", the elevated temperature ductility did not improve as the carbide population was merely altered, and not eliminated.

Materials and Methods

All LPBF test materials were produced on a Concept Laser M2 printer using gas atomized alloy 718 powder purchased from Praxair Surface Technologies (Indianapolis). Details of powder composition, bulk material composition, and processing conditions have been previously reported [5, 8–10]. In brief, test materials were made using a Concept Laser M2 under an argon protective environment with a laser power of 180 W, scanning speed of 600 mm/s, a hatch spacing of 0.105 mm, and a layer thickness of 30 μm. These process parameters equate to a volumetric laser energy density of 95 J/mm^3. Specimens were prepared using an island scanning strategy with 5 $\times$ 5 mm squares.

Most materials had a carbon content of 0.02–0.04 wt %, similar to wrought material content but one study used a customized heat of powder with carbon reduced to 0.01 wt % [11]. Wrought material was purchased in the form of ½" diameter (12.7 mm) solution annealed bars per AMS 5662. LPBF samples were hot isostatically pressed (HIP), and LPBF were given a variety of pre- and post-HIP thermal treatments as described below. Mechanical testing was performed in general conformance with applicable ASTM standards, including E8 for room-temperature tension, E21 for elevated temperature tension, E139 for creep and creep rupture, and E292 for notched rupture. Tests were performed at one of three facilities: The Aerospace Corporation (El Segundo, CA), Metals Technology Inc. (Northridge, CA), or Westmoreland Mechanical Testing and Research (Youngstown, PA).

HIP and heat treatments were performed in a variety of combinations. A baseline heat treatment consisted of HIP at 1163 °C and 100 MPa for three hours, followed by solution treatment at 954 °C for one hour and water quench, and then dual aging at 718 °C for eight hours, ramp down to 621 °C with hold for additional eight hours. This is HIP followed by solution treatment and aging per AMS 5662.

Two modified HIP schedules were performed:

- MHIP1: 1140 °C for 1 h, ramp to 1200 °C, hold for 3 h, both at 15ksi
- MHIP2: 1065° C for 3 h at 30 ksi

Samples using these two modified HIP schedules and the baseline schedule were HIP'ed at Quintus Technologies and then subjected to either the baseline AMS 5663 solution and age schedule or two alternative heat treatments.

- MHT1: Solution/stabilize 920 °C 2 h, ramp to 980 °C 1 h, water quench, Age 718 °C, 8 h, FC, 625 °C, 8.25 h
- MHT2: Solution/stabilize 920 °C 6 h, ramp to 980 °C 1 h, water quench, Age 718 °C, 8 h, FC, 625 °C, 8.25 h

In total, there were nine possible combinations of three HIP schedules and three heat treatments. Samples that were in this experimental matrix were tensile tested and combination stress rupture tested at 650 °C following AMS 5662. One exception was that the same strain rate was used throughout the tensile test, instead of increasing strain rate after yield strength (as 0.2% offset) has been established, which is allowed per the standard.

For the three HIP schedules, two lots of material were subjected to one of two heat treatments after HIP:

- Solution 1038 °C 1 h oil quench; Age 760 °C 10 h furnace cool to 650 °C to a total age cycle 20 h;
- Solution 1038 °C 1 h oil quench; Stabilize/age 960 °C 1 h, cool to 760 °C 10 h furnace cool to 650 °C to a total age cycle 20 h

The aging cycle for these two heat treatments is identical to AMS 5664, while the solution treatment is 28 °C lower than the 1066 °C temperature in AMS 5664. The combination of solution temperature above the delta solvus followed by the stabilization step at 960 °C was intended to remove any delta that formed during cooling from the HIP cycle with the addition of a delta phase precipitation that would normally be absent in AMS 5664 treated materials. These samples were subjected to stress rupture testing at 758 MPa and 650 °C.

An additional set of samples were HIP'ed at Kittyhawk Products (Garden Grove, CA) using the baseline HIP schedule, then subjected to either the baseline AMS 5662 solution and age schedule or one of three alternative heat treatments:

- HTX: Solution/stabilize 900 °C, 2 h, Ramp to 954 °C, 1 h, water quench; 718 °C, 8 h, furnace cool to 625 °C, 8.25 h
- HTY: 1200 °C, 3 h, air cool; solution/stabilize 900 °C, 2 h, Ramp to 954 °C, 1 h, water quench; 718 °C, 8 h, furnace cool to 625 °C, 8.25 h

- HTZ: 1200 °C, 3 h, air cool; 954 °C, 1 h, water quench; 718 °C, 8 h, FC, 625 °C, 8.25 h (DA)

The 1200 °C soaks were intended primarily to reduce the carbide population and dissolve any Laves phase present that might have formed during HIP or during slow cool down in the HIP chamber after the cycle. These samples were subjected to tensile testing at and combination stress rupture testing at 650 °C following AMS 5663 requirements.

The primary difference between HIP cycles at the two vendors is the use of rapid, forced-air quenching from the HIP temperature at Quintus Technologies.

A sub-set of materials were heat treated using the AMS 5664 solution treatment temperature and time (1066 °C for one hour), then given a lower-temperature stabilization at 843 °C for 6 h, followed by two-step aging at the standard AMS 5664 conditions. These samples were tensile tested at room-temperature and 650 °C but no stress rupture testing was performed.

Results

Tensile Properties–Baseline HIP and Modified AMS 5664 Heat Treatment

The delta stabilization treatment between solution treatment and aging cycles was intended to generate a population of delta phase particles within grain interiors where it has been speculated they can serve as oxygen sinks that prevent excessive oxidation of grain boundaries that promotes intergranular brittle fracture [12]. At the same time, it can be expected to consume Nb that would otherwise be available for γ'' precipitation, thus decreasing its strength. Tensile properties of LPBF material for ½" diameter vertical oriented rods with standard AMS 5664 heat treatment after HIP were previously reported [8], and are compared to material that received the delta stabilization treatment in Table 1.

Although these represent samples taken from different build and heat treatment lots, it does demonstrate that delta stabilization will reduce static properties.

Table 1 Tensile properties of materials after baseline HIP and AMS 5664-based solution and aging treatments. Delta stabilization was performed after solution treatment

Condition	Test temperature (°C)	Fty (MPa)	Ftu (MPa)	Elong (%)	R.A. (%)
AMS 5664	23	1094	1362	23	36
	650	924	1110	18	22
AMS 5664 with delta stabilization	23	986	1303	23	34
	650	862	1055	19	23

Table 2 Summary of previously published tensile data for samples receiving Baseline HIP and AMS 5662 solution and aging treatments

Condition	Test temperature (°C)	Fty (MPa)	Ftu (MPa)	Elong (%)	R.A. (%)
AMS 5662 (Baseline)	23	1020–1055	1365–1385	22–28	30–37
	650	896–945	1110–1151	16–26	17–31

Tensile Properties–Baseline HIP and AMS 5662 Heat Treatment

Room-temperature and elevated-temperature tensile properties for vertically oriented specimens with baseline HIP and AMS 5662 heat treatments from the same or similar powder lots and the same powder-bed fusion apparatus have been previously published [5, 8, 13] and are cited here as ranges in values for subsequent comparison to materials with alternative heat treatments where the focus was on tensile properties at 650 °C and stress- or creep-rupture performance. All specimens were HIP'ed at Kittyhawk Products, implying furnace cooling rates after completion of the HIP cycle (Table 2).

Except for the low end of the range of room-temperature yield values, where the AMS 5662 minimum acceptance value is 1034 MPa, all tested samples meet specification minimum requirements.

Tensile Properties–Baseline HIP with Modified Heat Treatments

Table 3 shows results of tensile testing at 650 °C for materials that were HIP'ed by one of the three schedules at Quintus Technologies and heat treated by either the baseline heat treatment or one of the two alternative heat treatment schedules. A set of additional coupons that had been HIP'ed at Kittyhawk Technologies (marked by an asterisk) and wrought material that were heat treated in the same heat treat lot as the other baseline heat treatment samples are included for comparison. The results are mean values along with standard deviation for three samples at each condition.

Samples that were HIP'ed using baseline conditions and then heat treated using one of the modified heat treatments HTX, HTY, or HTZ are summarized in Table 4 [14], which includes results for both room-temperature and 650 °C.

Table 3 Elevated temperature (650 °C) tensile properties of samples after modified HIP and heat treatment schedules. "Baseline" HIP refers to a single 1163 °C soak under pressure, while Baseline heat treatment refers to solution and age per AMS 5662. Samples were HIP'ed using rapid cooling except where marked by an asterisk

Condition (HIP–heat treat)	Yield strength (MPa)	Ultimate tensile strength (MPa)	Elongation (%)	Reduction of area (%)
Baseline–Baseline	918.8 ± 2.4	935.8 ± 1.0	2.2 ± 1.0	8.3 ± 1.6
Baseline–MHT1	937.7 ± 16.6	974.0 ± 14.0	3.3 ± 2.4	11.7 ± 0.1
Baseline–MHT2	900.1 ± 48.3	926.0 ± 63.4	4.9 ± 2.5	8.2 ± 0.5
MHIP1–Baseline	988.7 ± xx	1019.7 ± 17.4	4.8 ± 0.2	15.9 ± 0.5
MHIP1–MHT1	977.1 ± 6.7	994.6 ± 13.2	6.2 ± 2.4	12.5 ± 4.2
MHIP1–MHT2	923.6 ± 10.2	998.4 ± 59.8	7.4 ± 3.0	15.0 ± 0.4
MHIP2–Baseline	946.7 ± 24.4	985.8 ± 4.5	6.5 ± 0.6	21.8 ± 1.1
MHIP2–MHT1	957.0 ± 55.6	976.8 ± 34.4	6.6 ± 0.5	18.8 ± 4.4
MHIP2–MHT2	959.8 ± 39.0	968.2 ± 28.0	6.8 ± 0.7	18.0 ± 0.5
Baseline–Baseline*	875.6 ± 17.1	940.4 ± 11.7	2.9 ± 0.4	5.9 ± 0.9
Wrought–Baseline	977.7 ± 15.9	1023.9 ± 19.3	17.8 ± 1.0	52.4 ± 0.9

Table 4 Summary of room- and elevated-temperature tensile properties for samples HIP'ed at baseline conditions with chamber cooling from HIP cycle and alternative heat treatments

Condition (HIP–Heat Treat)	Temperature (°C)	Yield strength (MPa)	Ultimate tensile strength (MPa)	Elongation (%)	Reduction of area (%)
Baseline–HTX	23	1050.7 ± 17.3	1359.5 ± 18.3	24.6 ± 0.6	Not reported
	650	875.4 ± 4.5	974.2 ± 4.5	7.4 ± 1.1	13.6 ± 1.1
Baseline–HTY	23	1077.6 ± 12.0	1374.4 ± 23.7	23.4 ± 0.8	Not reported
	650	868.7 ± 19.8	953.8 ± 27.3	3.3 ± 2.4	11.7 ± 0.1
Baseline–HTZ	23	1149.5 ± 4.2	1375.1 ± 1.4	23.6 ± 1.8	Not reported
	650	900.1 ± 48.3	926.0 ± 63.4	4.9 ± 2.5	8.2 ± 0.5
Baseline–Baseline	23	1042.7 ± 5.1	1373.8 ± 2.5	22.6 ± 0.6	Not reported
Wrought–Baseline	23	1194.9 ± 10.7	1424.6 ± 15.1	25.5 ± 2.0	Not reported

Table 5 Combination stress rupture results for samples with various combinations of HIP and heat treatment

Condition (HIP–Heat treat)	Time to failure (hrs)	Elongation at failure (%)	Failure location	Final load (MPa)
Baseline–Baseline	1.0	0	Notch	690
Baseline–MHT1	1.7	0	Notch	690
Baseline–MHT2	0.4	0	Notch	690
MHIP1–Baseline	5.7	0	Notch	690
MHIP1–MHT1	1.1	0	Notch	690
MHIP1–MHT2	2.9	0	Notch	690
MHIP2–Baseline	18.0	0	Notch	690
MHIP2–MHT1	9.5	0	Notch	690
MHIP2–MHT2	2.0	0	Notch	690
Baseline–Baseline*	3	0	Notch	690
Wrought–Baseline	61	27	Smooth	896

Combination Stress Rupture–Baseline and Modified HIP and Modified Heat Treatments

For the samples that were HIP'ed using one of three schedules at Quintus Technologies and then heat treated using the Baseline, MHT1, or MHT2, a single sample was tested as a combination stress rupture specimen at initial load of 690 MPa and 650 °C. These results are shown in Table 5. The values for the wrought material and LPBF material HIP'ed at Kittyhawk (marked by an asterisk in Table 5) are the average of two samples.

Combination Stress Rupture–Baseline HIP with Modified Heat Treatments

Samples that received a baseline HIP treatment at Kittyhawk Products and then either HTX, HTY, or HTZ and then tested using a combination are summarized in Table 6 [14]. The values are the average of two samples. Additional samples from HTY and HTZ lots were tested in a backfilled argon atmosphere. Two samples of each heat lot were tested in this condition. Baseline heat treatment per AMS 5662 and wrought material comparisons can be made to values in Table 5.

The results in argon show a significant difference from those tested in air. In the case of HTY samples, both samples reached the 23-h minimum and were loaded with an additional load equivalent to ~ 50 MPa before failing, although each failed at the notch with zero elongation of the smooth section of the CSR specimen. One of the HTZ samples tested in argon met the elongation and fracture requirements of the test by exceeding 4% smooth section failure and fracture in the smooth section, but neither met the 23-h minimum requirement.

Table 6 Combination stress rupture results for samples receiving Baseline HIP and alternative heat treatments

Condition (HIP–Heat treat)	Time to failure (hrs)	Elongation at failure (%)	Failure location	Final load (MPa)
Baseline–HTX	21	0	Notch	690
Baseline–HTY	8.6	0	Notch	690
Baseline–HYZ	13.3	0	Notch	690
Samples tested in a backfilled Ar atmosphere				
MHIP1–HTY	29.8	0	Notch	741
MHIP1–HYZ	14.5	0/5.9	Notch/Smooth	690

Stress Rupture–Baseline and Modified HIP with Modified AMS 5664 Heat Treatments

One set of samples was tested with a uniform gauge stress rupture specimen at 650 °C and 758 MPa. These samples were HIP'ed at Quintus Technologies and then heat treated as described above beginning with a solution treatment at 1038 °C followed by oil quench and a double aging treatment per AMS 5664 with or without a preceding delta stabilization soak at 958 °C. Because this specimen design did not contain a notch and testing was performed at a 10% higher load, the results cannot be directly compared to the CSR results except to note under what circumstances they exceeded the 23-h loading time and 4% elongation minimums. Three samples were tested for each condition and shown in Table 6 with time to failure and elongation at failure listed in corresponding order for each condition.

These results indicate that delta stabilization treatment may lower the time-to-rupture, but increase the strain at failure, but only for Baseline HIP and MHIP1 samples. Given the intrinsic variability of time-to-rupture, it is not apparent that one heat treatment is clearly better than another, although the lower-temperature, higher pressure MHIP2 condition appears to reduce stress rupture time at these test conditions.

A summary of stress rupture results at two stress levels is shown in Fig. 2. Data are taken from Table 7 and published [5, 14] creep data. Also plotted as the letter "W" are approximate values for wrought alloy 718 from MMPDS [15] based on the estimated time-to-rupture derived from data in Fig. 6.7(e) according to the best-fit equation

$$logt = c + b_1T + b_2X + b_3X^2 + b_4X^3$$

where t is time in hours, T is temperature on the Rankine scale (650 °C is 1660 R), X is applied stress as ksi, and c and b_n are constants determined by regression of the MMPDS data (values of these constants are given in MMPDS). This comparison shows that wrought material can be expected to outperform LPBF at 690 MPa but at

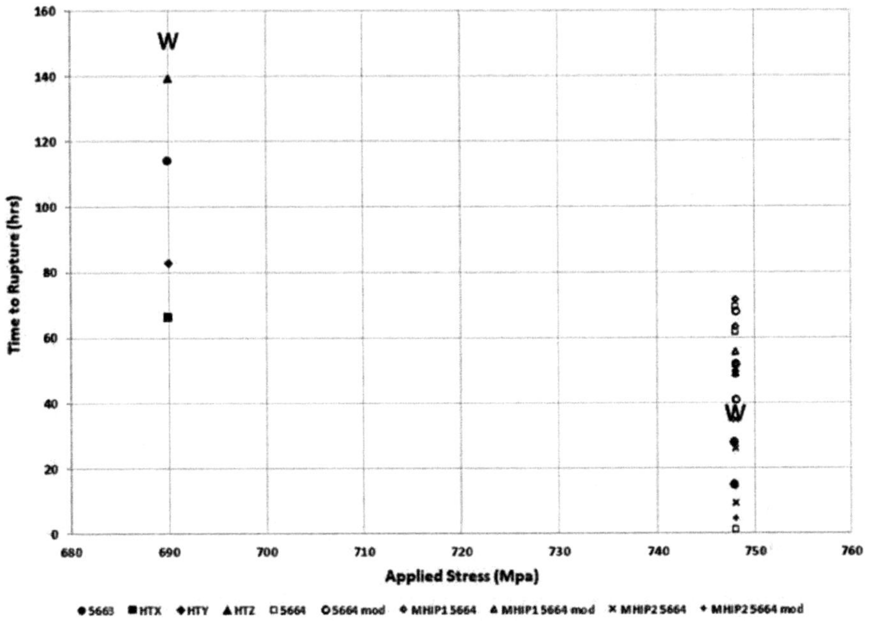

Fig. 2 Time to rupture for stress rupture tests and creep tests performed at 650 °C and either 690 or 748 MPa. Values marked as "W" are taken from MMPDS as explained in the text

Table 7 Results of stress rupture testing performed at 650 °C and 748 MPa. There were three samples per condition, with all results shown

HIP condition	Time to failure (hrs)	Elongation (%)	Time to failure (hrs)	Elongation (%)
	1038 °C solution, oil quench, age at 760 °C/650 °C		1038 °C solution, oil quench, soak at 958 °C 1 h, age at 760 °C/650 °C	
Baseline	69.4, 61.6, 1.3	2.3, 2.1, 7.0	26.5, 41.0, 68.1	3.6, 3.0, 3.0
MHIP1	71.6, 49.5, 63.2	3.0, 2.8, 2,6	51.9, 49.0, 55.5	4.0, 4.0, 4.0
MHIP2	9.2, 25.9, 34.9	5.0, 3.0, 7.0	48.3, 14.2, 4.5	8.0, 9.0, 3.0

the 10% higher stress the wrought material is within the overall scatter in the LPBF results. Strain before failure is not accounted for in this comparison.

Discussion

Alternative heat treatments were investigated specifically to determine if heat treatment of an as-printed or as-HIP'ed LPBF 718 material could be tailored to address issues of low ductility compared to wrought material at elevated temperatures. A summary of observations from elevated temperature tensile testing, stress rupture, creep, and combination stress rupture includes:

- Static strength of the LPBF version of the alloy is lower than wrought material but still meets traditional minimum requirements for acceptance testing;
- Ductility of LPBF is significantly lower than wrought, even though room-temperature ductility is sufficient to meet required standard minimum levels;
- Slower strain rates, such as those encountered during creep and stress rupture conditions exacerbate the differences in ductility between LPBF and wrought;
- Fracture at slow strain rates is transgranular and ductile in wrought but intergranular and brittle in LPBF;
- Despite lower ductility and propensity to brittle fracture, LPBF does exhibit the ability to sustain a load at high temperatures, but generally exhibits a wide variation in time-to-rupture in creep and combination stress rupture.

The tentative conclusion to these observations was that dynamic embrittlement traceable to NbC was responsible for lower ductility. The notch sensitivity seen in combination stress rupture testing means that a standardized means for evaluating the quality of wrought 718 alloy is either inappropriate or too demanding of the LPBF version of the alloy. Given the high solvus temperature of NbC, initial efforts to address ductility and notch sensitivity were focused on manipulating the δ phase content because δ phase is known to influence each of these [12, 16, 17], although HTY, HTZ, and MHIP1 each contained a soak at 1200 °C that was intended to reduce the NbC population. Examples of the difference in δ phase in samples given HTX and HTZ heat treatments after baseline HIP are shown in Fig. 3. In Fig. 3, it is notable that despite the high-temperature exposure of HTZ, the γ phase grain size is similar for both.

Quasi-static tensile tests for MHIP1 samples that were exposed to 1200 °C during the HIP cycle showed a small increase in yield strength relative to baseline HIP or MHIP2 (Table 3) and samples exposed to HTZ show a significant increase in strength after baseline HIP treatment (Table 4). The difference in strength between samples with HTY and HTZ heat treatment is due to the depletion of Nb in the former during the 900 °C soak preceding solution and aging.

Nevertheless, although there are differences in time-to-rupture (Table 6) between the different heat treatments, there was no general improvement in terms of notch failure or reaching the 23-h minimum load time requirement. Analysis of the phase fractions of NbC and delta phase based on scanning electron microscopy shows that the 1200 °C soak had a limited impact on the carbide fraction (Fig. 4, Table 8), and the delta stabilization treatments dramatically increased the δ phase fraction. The result was a reduction in strength but no apparent benefit for stress rupture from the increase in δ, which has been attributed to, among other mechanisms, improved grain-boundary compliance and providing an oxygen sink that prevents grain boundary embrittlement via oxidation.

Figure 4 shows an example of a delta-rich HTX sample after heat treatment where carbide and delta fractions have been separated by image analysis based on particle aspect ratio and size. The identity of carbide particles was made based on

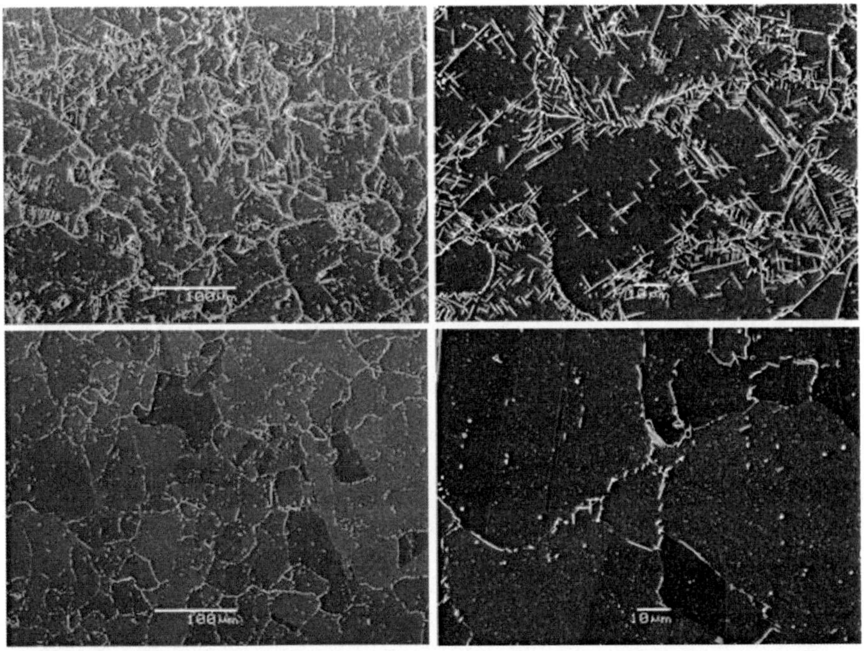

Fig. 3 Electron micrographs of HTX (top) and HTZ (bottom) samples illustrating the differences in δ phase abundance and distribution

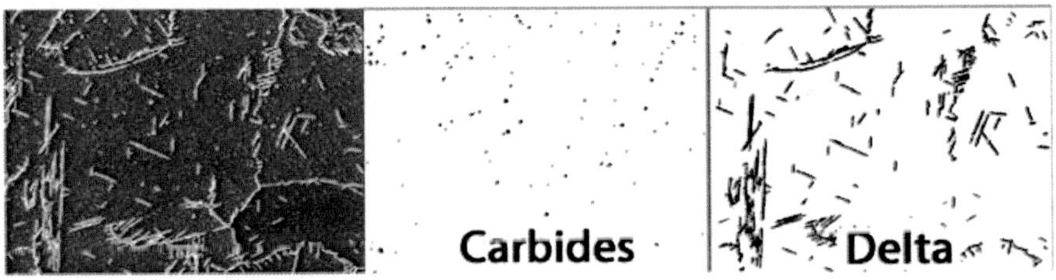

Fig. 4 Example microstructure and image analysis to separately measure area fraction of carbide and δ particles

Table 8 Estimated NbC and δ content of LPBF and wrought 718 after different heat treatments

Heat treatment	Area % NbC	Area % delta
As HIP (Baseline HIP)	0.56	0.06
As HIP + 1200 °C 3 h	0.48	0.0
Baseline (AMS 5662)	0.64	0.07
HTX	0.36	9.88
HTY	0.45	6.45
HTZ	0.49	1.52
Wrought (Baseline)	0.36	5.20

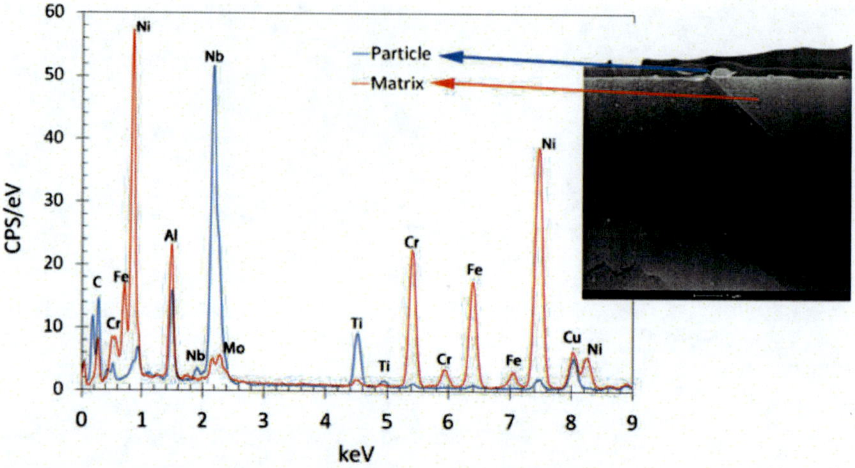

Fig. 5 EDS spectra from a carbide particle and fcc matrix showing relative enrichment (Nb, C, Ti) and depletion of matrix elements (Ni, Cr) in particles relative to γ phase)

energy dispersive spectroscopy (EDS); an example derived from a sample prepared for Transmission Kikuchi diffraction (TKD; [5]) is shown in Fig. 5. The roughly equi-axed carbide particle on the surface of the sample is seen to be enriched in carbide-forming elements Nb and Ti and depleted (relative to the fcc matrix phase) in Ni, Cr, and Fe. Phase identification of the particle and matrix were assigned based on Kikuchi diffraction [5].

Analysis of samples after full HTX, HTY, and HTZ treatments are compared with samples after baseline HIP and HIP, and 1200 °C homogenization prior to any additional heat treatment is shown in Table 8.

The small δ phase content of samples that did not receive a 900 °C delta stabilization treatment nevertheless were solution treated at 954 °C, a temperature at which δ is expected to be stable as a separate phase [18]. Furthermore, the wrought material undergoing a similar solution treatment has much greater delta content than LPBF following the same post-HIP heat treatment. An example of wrought microstructure is shown in Fig. 6, where arrows highlight the much larger, more widely dispersed carbide phase compared to the LPBF version in Fig. 3. These results show that microstructural manipulation of the LPBF 718 alloy is not sufficient to ameliorate the influence of a dispersion of numerous fine carbide particles that have their origin in segregation during solidification of the original melt pools and play an outsize influence in environmental susceptibility and notch sensitivity of the LPBF heat treated microstructures. The influence of elemental segregation during the solidification of LPBF alloys has been noted previously for Inconel 625 [19] and an iron-based superalloy [20]. Table 8 suggests that the carbide population develops during the HIP cycle and remains relatively stable throughout subsequent thermal treatments. The elongation and within-gauge fracture of only a single CSR sample tested in an inert atmosphere indicates that even reducing the oxidation source is not sufficient

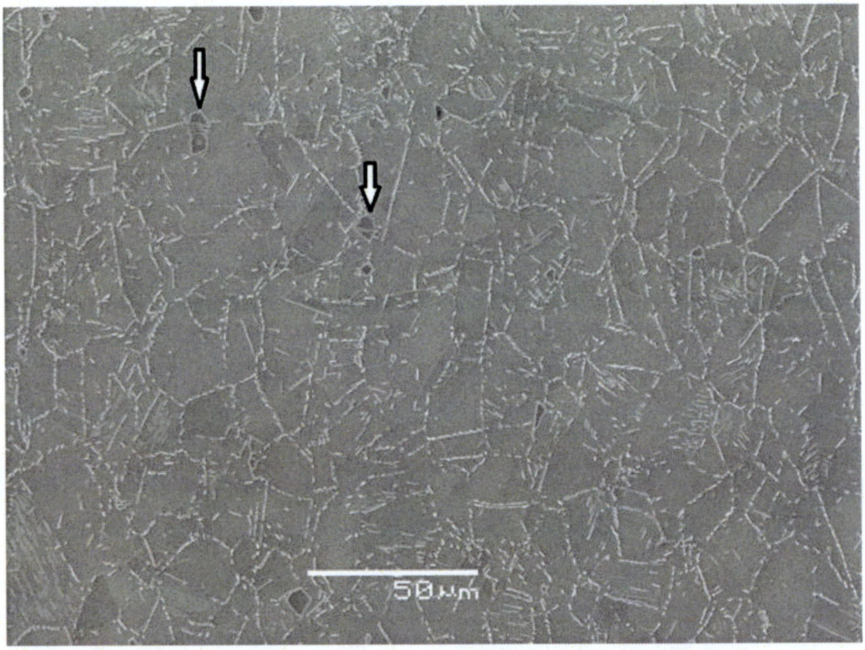

Fig. 6 Microstructure of wrought 718 highlighting occurrence of NbC particles

to improve the performance, suggesting a general lack of plasticity at high temperatures. Contradicting this is the observations that some samples have sustained loads for time periods similar to wrought material prior to brittle, intergranular failure.

Because the high-temperature ductility and notch sensitivity could not be improved via microstructural mechanisms, chemical composition was targeted as a possible solution. A custom blend of 718 powder with the minimum practical amount of carbon was tested, leading to a bulk carbon content of 0.01 wt%, compared to a typical bulk carbon content of 0.03–0.04 wt% for other lots of LPBF represented herein [11]. The reduction in carbon content was correlated with a similar scale reduction in carbide percentage (1.4 versus 0.4 measured area percent), an increase in grain size for the low-carbon variety (50 µm average grain size in standard material versus 77 µm in low-carbon material) and a decrease in γ" particle size (66 nm in standard material versus 35 nm in low-carbon material) [11]. For material subjected to the AMS 5662 solution and age schedules, this led to a significant increase in the yield strength of the low-carbon material at room-temperature, and an increase in both yield and ultimate tensile strength at 650 °C. In the high-temperature case, the tensile strength was comparable to wrought material in the same condition, but ductility remained much lower. The increase in strength was attributed to having less Nb partitioned into NbC after carbide formation during HIP, because less carbon on both a mass and atomic basis would have been available in inter-dendritic regions of the as-solidified material.

Table 9 Comparison of Creep and Combination Stress Rupture results for LPBF 718 with standard carbon load, low-carbon content, and wrought material

Sample	Creep			CSR		
	Time to failure (hrs)	Strain at failure (%)	Minimum creep rate (sec^{-1})	Time to failure (hrs)	Failure location	Elongation (%)
LPBF	114.2	1.84	$1.1 \cdot 10^{-8}$	3.0	Notch	0
LPBF Low-C	102.5	0.75	$5.2 \cdot 10^{-9}$	2.30	Notch	0
Wrought	184.6	27.25	$2.9 \cdot 10^{-9}$	61.0	Smooth	27

Despite the gains in static strength [11], creep and CSR behavior (reported here for the first time) did not show appreciable improvement over LPBF material with a higher carbon content, exhibiting low ductility and brittle failure. A summary of creep and CSR observations for low-carbon and typical 718 powder heats and wrought material, all given the same solution and age treatments is shown in Table 9. These results also reinforce the importance of considering both times to rupture and elongation, as the LPBF is evidently strengthened to some extent by a different mechanism allowing it to sustain a load without plastically deforming. It is expected that other important testing conditions for this alloy, especially low-cycle fatigue, would reinforce the shortcomings of LPBF relative to wrought, further limiting its adoption in high-reliability applications.

Conclusions

The low ductility and environmental and notch sensitivity in alloy 718 produced by laser powder-bed fusion has been attributed to a population of NbC that is not found in either wrought or cast forms. Acceptance tests for wrought material enshrined in AMS standards, especially combination stress rupture or notched rupture testing, are not likely to be applicable to LPBF 718, raising questions as to whether the quality and performance of LPBF 718 can be assessed using traditional approaches, and whether there are thus limitations of the operating conditions for this version of the alloy. Attempts to improve high-temperature ductility based on heat treatments alone were not successful in addressing these shortfalls. Reduction in carbon content of the material was successful in reducing the carbide population, altering the γ" and improving static strength, but still did not improve ductility or creep and creep rupture behavior. Further reductions in carbon content might be beneficial, but the expense of removing carbon below the 0.01 wt % level were prohibitive in this investigation.

The dispersion of carbide particles derives from the unique processing conditions of LPBF, including small-scale, rapid solidification. Subsequent thermal processing leads to a large population of NbC at grain boundaries where they increase the propensity to grain boundary oxidation and weakening. The delta stabilization was intended to introduce features that might balance this condition, through a combination of depleting grain boundaries of γ" and thus making them more compliant and adding possible sinks for diffusing oxygen, but they were clearly not sufficient. Grain boundary engineering [21] reduced dynamic embrittlement but required alternating thermo-mechanical processing and annealing, which is not an option for AM materials. Other elemental constituents that have been shown to improve the environmental sensitivity of wrought 718 material include B and Cr [22, 23]. Successful strategies to improve high-temperature mechanical behavior may need to rely on creating grain boundary conditions with lower NbC content and the particles that are present on grain boundaries are oxidized less favorably than other constituents.

Acknowledgements Ali Guzel performed image analysis of microstructures during a virtual summer internship in 2020. The assistance of Tyler Slinger of Praxair Surface Technologies in arranging for low-carbon alloy 718 powder is gratefully acknowledged. Magnus Ahlfors and Chad Beamer of Quintus Technologies provided HIP services at no cost. The work was supported in part by the Independent Research and Development program at The Aerospace Corporation. All service marks, trademarks, and trade names are the property of their respective owners.

References

1. Hosseini, E. and V.A. Popovich, *A review of mechanical properties of additively manufactured Inconel 718*. Additive Manufacturing, 2019. **30**: p. 100877.
2. Probstle, M., et al., *Superior creep strength of a nickel-based superalloy produced by selective laser melting*. Materials Science and Engineering A, 2016. **674**: p. 299-307.
3. Shi, J.J., et al., *Microstructure and creep anisotropy of Inconel 718 alloy processed by selective laser melting*. Materials Science and Engineering A, 2021. **805**: p. 140583.
4. Xu, Z., et al., *Creep behaviour of inconel 718 processed by laser powder bed fusion*. Journal of Materials Processing Technology, 2018. **256**: p. 13-24.
5. McLouth, T.D., et al., *Variations in ambient and elevated temperature mechanical behavior of IN718 manufactured by selective laser melting via process parameter control*. Materials Science and Engineering A, 2020. **780**: p. 139184.
6. Sanchez, S., et al., *The creep behaviour of nickel alloy 718 manufactured by laser powder bed fusion*. Materials and Design, 2021. **204**: p. 109647.
7. Sanchez, S., et al., *Multi-laser scan strategies for enhancing creep performance in LPBF*. Additive Manufacturing, 2021. **41**: p. 101948.
8. Witkin, D.B., et al., *Anomalous Notch Rupture Behavior of Nickel-Based Superalloy Inconel 718 Due to Fabrication by Additive Manufacturing*. Metallurgical and Materials Transactions A, 2019. **50**(8): p. 3458-3465.
9. McLouth, T.D., et al., *Temperature and strain-rate dependence of elevated temperature ductility of Inconel 718 prepared by selective laser melting*. Materials Science and Engineering A, 2021. **824**: p. 141814.
10. Bean, G.E., et al., *Effect of laser focus shift on surface quality and density of Inconel 718 parts produced via selective laser melting*. Additive Manufacturing, 2018. **22**: p. 207-215.
11. McLouth, T.D., et al., *Influence of reduced carbon content on microstructure and mechanical behavior of Inconel 718 prepared by laser powder bed fusion*. Additive Manufacturing Letters, 2022. **3**: p. 100037.
12. Hayes, R.W., et al., *Effect of environment on the rupture behavior of alloys 909 and 718*. Materials Science and Engineering A, 1994. **177**: p. 43-53.

13. Bean, G.B., et al., *Build Orientation Effects on Texture and Mechanical Properties of Selective Laser Melting Inconel 718.* Journal of Material Engineering and Performance, 2019. **28**(4): p. 1942-1949.
14. McLouth, T.D., *Microstructural Control and Elevated Temperature Mechanical Properties of Inconel 718 Manufactured by Selective Laser Melting*, in *Materials Science and Engineering.* 2020, University of California Los Angeles: Los Angeles. p. 228.
15. Government, U.S., *Metallic Materials and Elements for Aerospace Vehicle Structures*, D.o. Defense, Editor. 2003, U.S. Government: Washington, D.C.
16. Sjoberg, G., N.-G. Ingesten, and R.G. Carlson. *Grain Boundary d-phase Morphologies, Carbides and Notch Rupture Sensitivity of Cast Alloy 718.* in *Superalloys 718, 625 and Various Derivatives.* 1991. The Minerals, Metals and Materials Society.
17. Cai, D., et al., *Dissolution kinetics of delta phase and its influence on the notch sensitivity of Inconel 718.* Materials Characterization, 2007. **58**: p. 220-225.
18. Oradei-Basile, A. and J.F. Radavich, *A Current T-T-T Diagram for Wrought Alloy 718*, in *Superalloys 718, 625 and Various Derivatives*, E.A. Loria, Editor. 1991, TMS: Pittsburgh. p. 325-335.
19. Zhang, F., et al., *Effect of heat treatment on the microstrucutral evolution of nickel-based superalloy additive-manufactured by laser powder bed fusion.* Acta Materialia, 2018. **152**: p. 200-214.
20. Gradl, P.R., et al., *Process development and hot-fire testing of additively manufactured NASA HR-1 for liquid rocket engine applications*, in *AIAA Propulsion and Energy Forum.* 2021, AIAA: VIRTUAL. p. 23.
21. Krupp, U., W.M. Kane, X. Liu, O. Dueber, C. Lair, and C.J. McMahon, *The effect of grain-boundary-engineering-type processing on oxygen induced cracking of IN718.* Materials Science and Engineering A, 2003. **349**: p. 213-217.
22. Woodford, D.A., *Gas phase embrittlement and time dependent crack of nickel based superalloys.* Energy Materials, 2006. **1**(1): p. 59-79.
23. Pineau, A. and S.D. Antolovich, *High temperature fatigue of nickel-base superalloys - A review with special emphasis on deformation modes and oxidation.* Engineering Failure Analysis, 2009. **16**: p. 2668-2697.

Microstructure and Mechanical Properties of Selective Laser Melting Processed TiC/GTD222 Nickel-Based Composite

Rui Wang, Zhe Zhang, Peng Zhang, Hailong Qin, and Zhongnan Bi

Abstract In this study, the microstructure and deformation mechanisms of selective laser melted GTD222 and TiC/GTD222 composite were studied. The results show that the TiC/GTD222 composite has finer grains and more precipitation phases. Meanwhile, TiC/GTD222 composite has higher yield strength both at room and high temperatures, which can be mainly attributed to the synergistic effect of the TiC strengthening and γ' strengthening. The deformation mechanisms of TiC/GTD222 composite at 800 °C were identified as isolated stacking faults shearing the γ' phase, continuous stacking faults shearing the γ and γ' phases, dislocations cutting the γ' phase, and dislocation slip within the γ matrix. This study provides insights for understanding the influence of TiC particles on the deformation mechanisms of additive manufactured nickel-based alloys.

Keywords Nickel-based composite · Microstructures · Mechanical properties · Strengthening mechanisms · Plastic deformation mechanisms

---

R. Wang (✉) · P. Zhang · H. Qin · Z. Bi
Gaona Aero Material Co., Ltd., Beijing, China
e-mail: corhs@foxmail.com

Beijing Key Laboratory of Advanced High Temperature Materials, Central Iron & Steel Research Institute, Beijing, China

R. Wang
State Key Laboratory of Metal Matrix Composites, Shanghai Jiao Tong University, 800 Dongchuan Road, Shanghai 200240, PR China

Z. Zhang
College of Mechanical and Electronic Engineering, Shandong University of Science and Technology, Qingdao 266590, Shandong, China

E. A. Ott et al. (eds.), *Proceedings of the 10th International Symposium on Superalloy 718 and Derivatives*, The Minerals, Metals & Materials Series,
https://doi.org/10.1007/978-3-031-27447-3_40

Introduction

Nickel-based superalloys are widely used in aero-engine hot-end parts due to their excellent high temperature strength, corrosion resistance, and high creep properties. However, those hot-end parts are relatively complex in structure and require high precision, their processing requirements preclude traditional processing methods like casting or forging in practice [1–4]. Selective laser melting (SLM), an advanced additive manufacturing (AM) technology that provides great flexibility in manufacturing parts with complex geometrical shapes, has become the important development direction of hot-end parts [5–10].

In recent years, high temperature resistance requirements of hot-end parts are increasing quickly, which needs nickel-based superalloys to possess higher mechanical performance. Metal matrix composites (MMCs), which provide great potential for improving the physical and mechanical properties of alloys, offer an opportunity to meet the requirements of new generation hot-end parts [11, 12]. By incorporating ceramic particles with high hardness, high modulus, and high temperature resistance, Yang et al. [13] successfully prepared Hastelloy X (HX) composite containing 2wt.%TiB_2 particles by SLM technology. The results show that compared with HX alloy, the hardness of TiB_2/HX composite at room temperature and high temperature (850 °C) increases by 36.8 and 44.5%, respectively. In addition, the pitting potential of TiB_2/HX composite is higher than that of HX, indicating that the pitting corrosion resistance has been improved. Zhou et al. [14] used SLM to prepare nano-TiC reinforced IN738 composite. Compared with IN738 alloy, the ultimate tensile strength and yield strength of 2.5wt.%TiC/IN738 are increased by 574 and 429 MPa, respectively, and the mechanical properties are significantly improved.

In this work, TiC/GTD222 powder was prepared by in-situ synthesis method, and TiC/GTD222 composite samples were prepared by SLM technology. The microstructures and mechanical properties of GTD222 and TiC/GTD222 composite after heat treatment were studied, and the strengthening mechanisms and high temperature deformation behavior of TiC/GTD222 composite after heat treatment were discussed.

Materials and Methods

In this work, GTD222 and TiC/GTD222 ingots were obtained using a medium frequency self-induction resistance melting furnace. The GTD222 and TiC/GTD222 powders were vacuum induced gas atomized, and then, sieved to obtain the powder with the particle size distribution of 15–53 μm. The chemical composition of powder and deposited samples are shown in Table 1.

GTD222 and TiC/GTD222 composite samples were prepared using EOS M 290 intelligent 3D printing system. The process parameters were set as follows: laser power 280 W, scanning speed 960 mm/s, scanning pitch 80 μm, and thickness of each layer 20 μm. The heat treatment process for both GTD222 and TiC/GTD222

Table 1 Chemical composition of powders and deposited samples (wt.%)

Element	Al	Co	Ta	Nb	Cr	C	W	Ti	Ni
GTD222 powder	1.2	19.1	1.0	0.8	22.2	0.1	2.0	2.3	Bal.
TiC/GTD222 powder	2.2	19.0	1.0	0.8	22.1	0.3	2.0	3.5	Bal.
Deposited GTD222	1.2	19.1	0.98	0.8	22.2	0.1	1.9	2.27	Bal.
Deposited TiC/GTD222	2.2	18.9	0.99	0.8	22.0	0.29	1.9	3.49	Bal.

composite was 1250 °C for 2 h, air cooling, and then the aging treatment was at 800 °C for 8 h, followed by air cooling.

For the heat treated (HTed) GTD222 and TiC/GTD222 composite, the molten pool morphology was obtained by electrolytic etching after grinding and polishing. The samples were etched for 5 s in 12 ml $H_3PO_4 + 48$ ml $H_2SO_4 + 40$ ml HNO_3 solution, 13 V. The microstructure of GTD222 and TiC/GTD222 composite were investigated using scanning electron microscopy (SEM, JEF7800), equipped with an electron backscattered diffraction (EBSD) detector. The dislocation structure of TiC/GTD222 composite was observed using transmission electron microscopy (TEM, FEI Tecnai G2 F20).

The mechanical properties were tested in Zwick Amsler 100 HFT 5100 tensile testers with a tensile rate of 10^{-3} s^{-1}. The loading axis was perpendicular to the construction direction of the sample manufactured with SLM. Three replicate experiments were performed on each alloy to ensure the accuracy of the data.

Results

The microstructures of HTed GTD222 and TiC/GTD222 composite are shown in Fig. 1. Figure 1a shows the distribution of γ' and carbides in GTD222, Fig. 1b shows an amplified image of blue rectangle area marked in Fig. 1a. Carbides precipitated in GTD222 are indicated using white arrows and nanoscale spherical γ' were highlighted using yellow arrows, as shown in Fig. 1b. Compared with GTD222 alloy, TiC/GTD222 composite had more carbides after heat treatment, and the size of carbides was slightly larger than that of GTD222 alloy, as displayed in Fig. 1c–d. The carbide precipitation in TiC/GTD222 composite was promoted by the adjustment of alloy elements.

The grain morphologies and grain size distribution of HTed GTD222 and TiC/GTD222 composite were shown in Fig. 2. According to the inverse pole figures (IPF) displayed in Fig. 2a, b, both GTD222 and TiC/GTD222 composite samples consisted of columnar grains. The selective orientation of the grains was obvious, and most of them were elongated axially along the building direction. The average grain sizes of GTD222 and TiC/GTD222 composite are 87 and 75 μm, respectively,

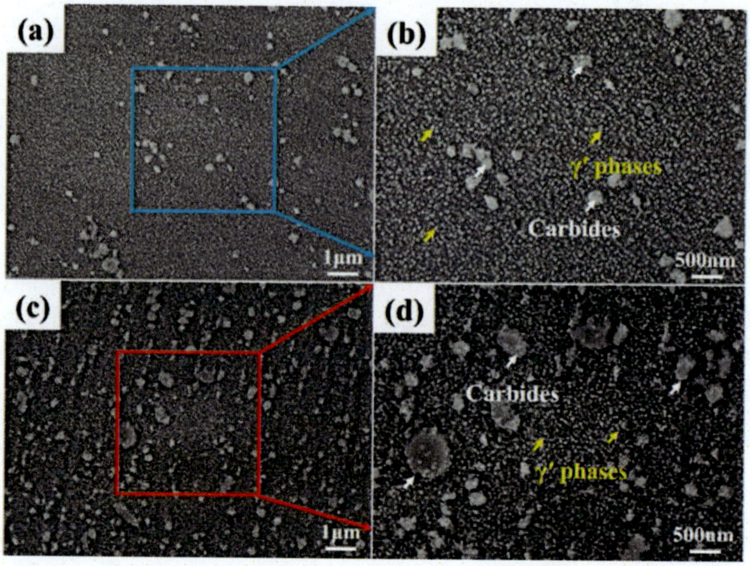

Fig. 1 SEM images of GTD222 and TiC/GTD222 composite after heat treatment: **a** GTD222, **b** Enlarged image of square area in **c** TiC/GTD222 composite, **d** Enlarged image of square area in **c**

as shown in Fig. 2c, d. TiC/GTD222 composite had finer grains than its counterpart GTD222. In addition, no cracks were observed in IPF images, suggesting the excellent processability of GTD222 and TiC/GTD222.

The TEM images of GTD222 and TiC/GTD222 composite after heat treatment were shown in Fig. 3. As shown in Fig. 3a, b, the grains of GTD222 mainly consisted of columnar and cellular structures and TiC/GTD222 composite had similar microstructure, as shown in Fig. 3d, e. The dislocations at the edges of columnar and cellular structures in the composite were significantly reduced after heat treatment, as displayed in the high-magnification image of GTD222 and TiC/GTD222 composite, Fig. 3c, f. The distribution of carbides in GTD222 was the same as that of the TiC/GTD222 composite, distributed along the boundary of the columnar structure. TiC particles were also identified according to the EDS mapping results, as shown in Fig. 4.

Figure 5 was the typical tensile stress–strain curves of HTed GTD222 and TiC/GTD222 composite at different temperatures. The yield strength of the TiC/GTD222 composite at room temperature was 1270 ± 13 MPa, which was about 13.4% higher than that of GTD222. When tested at 800–900 °C, yield strengths of GTD222 were decreased, and the TiC/GTD222 composites had the same trend. According to the stress–strain curves, the yield strength of TiC/GTD222 composite was higher than that of GTD222 alloy, and the elongation was lower than that of GTD222 alloy. The specific values were shown in Table 2.

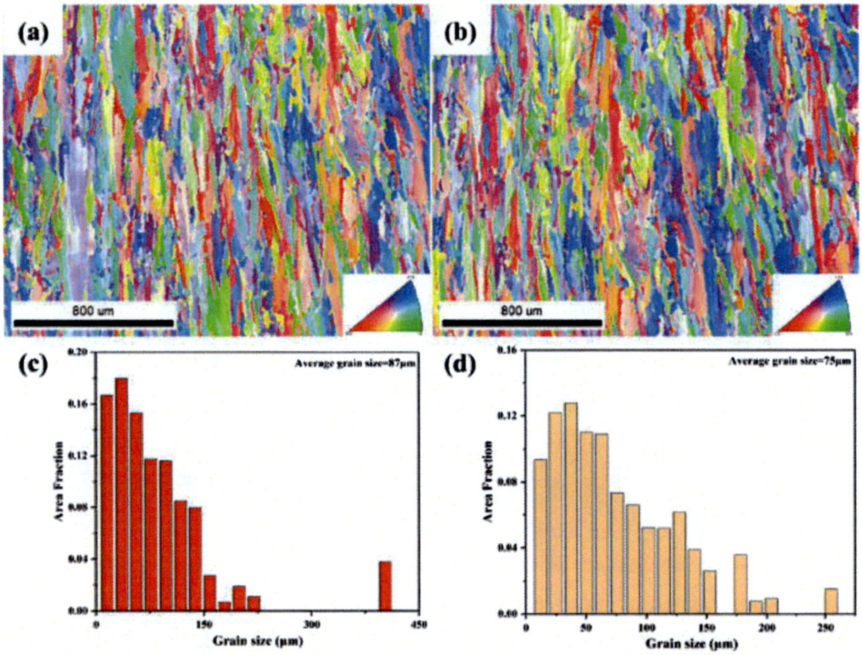

Fig. 2 EBSD images of GTD222 and TiC/GTD222 composite after heat treatment: **a** IPF map of GTD222, **b** IPF map of TiC/GTD222 composite, **c** Grain size distributions of GTD222, **d** Grain size distributions of TiC/GTD222 composite

Discussion

Strengthening Mechanisms of Composite

Compared with GTD222, the yield strength of TiC/GTD222 composite increased significantly after heat treatment. The EBSD results indicated that the TiC/GTD222 composite had finer grains after heat treatment, and their contribution to yield strength was attributed to the grain boundary fraction that hinders dislocation movement, which can be described by Hall–Petch relationship [15–18].

$$\Delta\sigma_{Hall-Petch} = kd^{-1/2}$$

where k was the strengthening coefficient. Therefore, finer grains potentially made a certain contribution to the improvement of the yield strength of the TiC/GTD222 composite.

The TEM image showing the interface between the carbide and the matrix is shown in Fig. 6. A Fourier transform image of the TiC was displayed in the inset image in Fig. 6a. Figure 6a indicates that the interface between the carbide and the

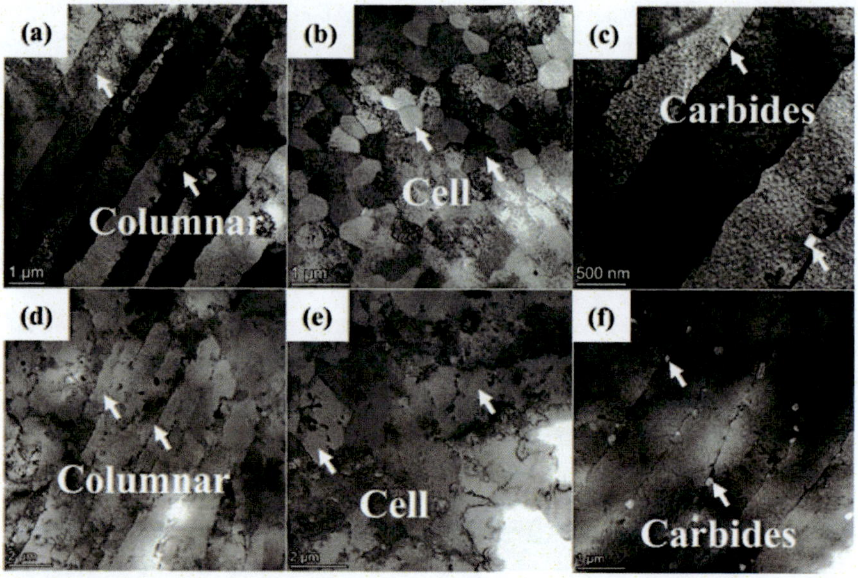

Fig. 3 TEM images of microstructures of GTD222 and TiC/GTD222 composite after heat treatment: **a** GTD222 columnar structures, **b** GTD222 cellular structures, **c** GTD222 carbide distribution, **d** TiC/GTD222 columnar structures, **e** TiC/GTD222 cellular structures, **f** TiC/GTD222 carbide distribution

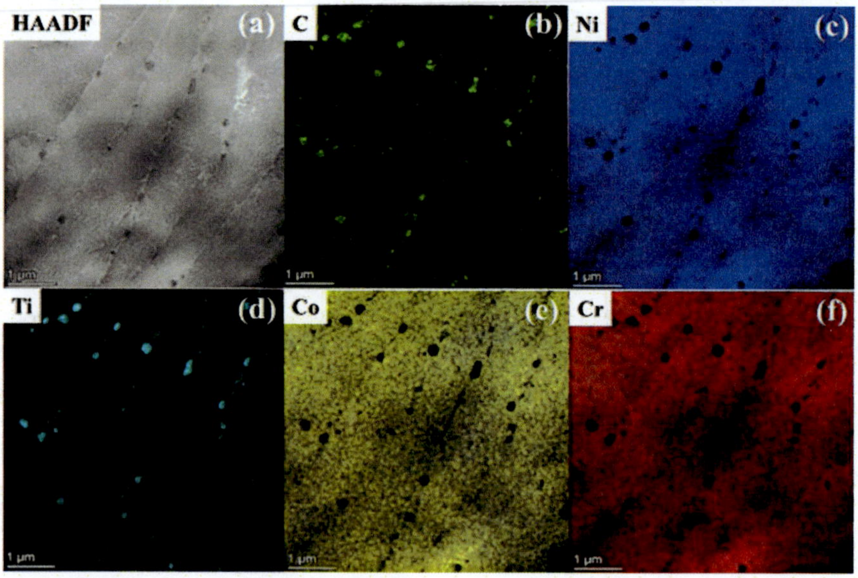

Fig. 4 EDS mapping of heat treated composite

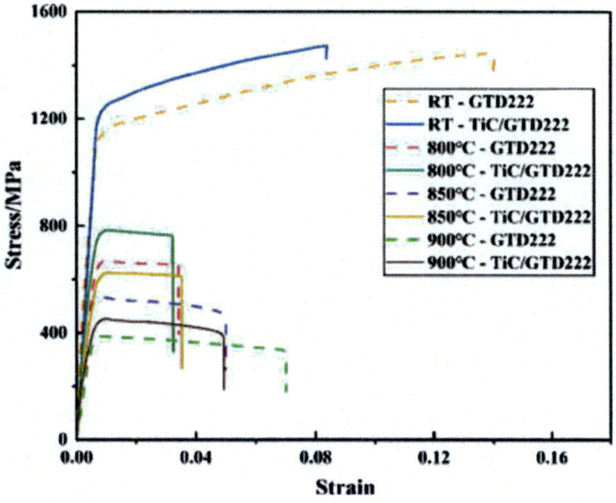

Fig. 5 Typical tensile curves of heat treated GTD222 and TiC/GTD222 composite at different temperatures

Table 2 Mechanical properties of heat treated GTD222 and TiC/GTD222 composite at different temperatures

Test temperature/°C	Materials	YS/MPa	UTS/MPa	El/%
Room temperature	GTD222	1120 ± 11	1380 ± 12	15 ± 1.3
	TiC/GTD222	1270 ± 13	1470 ± 11	8.1 ± 1.1
800	GTD222	660 ± 10	683 ± 9	3.3 ± 0.2
	TiC/GTD222	789 ± 11	797 ± 10	3 ± 0.1
850	GTD222	525 ± 8	533 ± 7	4.7 ± 0.1
	TiC/GTD222	620 ± 10	636 ± 9	3.4 ± 0.2
900	GTD222	340 ± 7	368 ± 9	6.5 ± 0.2
	TiC/GTD222	426 ± 9	432 ± 8	4.4 ± 0.2

matrix was well bonded and there were no reactants observed. This interface could effectively transfer the load from the matrix to the reinforcement, suggesting the load-bearing strengthening of particles was an important strengthening mechanism of the TiC/GTD222 composite. And because the lattice constant of TiC was different from that of nickel matrix, many dislocations emerged at the interface, as highlighted by the yellow T symbol in Fig. 6b.

Owing to the thermal expansion coefficient difference between TiC particles and nickel matrix, many dislocations will be formed during cooling. The specific values were calculated by the model proposed by Arsenault et al. [19].

$$\rho = 12\frac{\Delta\alpha\Delta T V_{Carbide}}{br(1 - V_{Carbide})}$$

Here, $\Delta\alpha$ is the difference in the thermal expansion coefficient between the carbide particles and the Ni matrix, ΔT is the temperature difference between the SLM

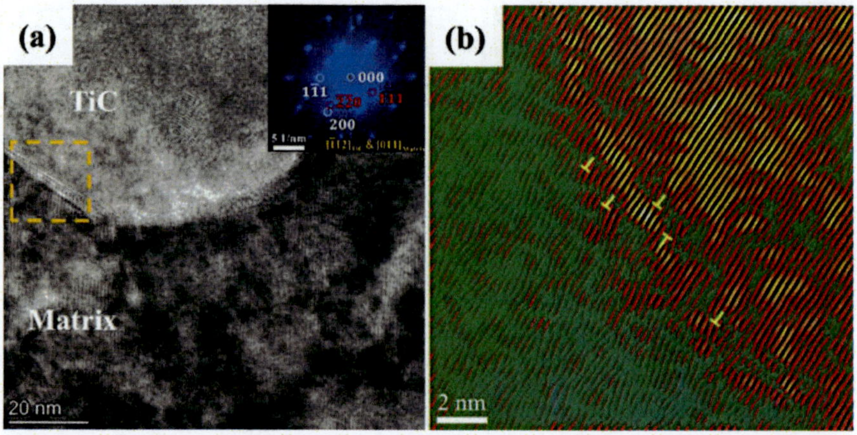

Fig. 6 Bright-field TEM images of TiC/GTD222 composite: **a** The TEM image at the interface between TiC and nickel matrix, and the embedded image is the selected area electron diffraction pattern in the yellow box, **b** The inverse Fourier transform image corresponding to the yellow box area

process and room temperature, b is the Burgers vector, and r is the size of carbides. The calculated dislocation density of GTD222 and TiC/GTD222 composite are $9.832 \times 10^{14}/\text{m}^2$ and $1.466 \times 10^{15}/\text{m}^2$, respectively. The dislocation density of the composite increased by about 49.1%. Therefore, we believe that dislocation strengthening was a factor in the improvement of yield strength.

According to Fig. 3f, the TiC particles were mainly distributed along the boundaries of the columnar structures. The TiC particles with high strength and good bonding with the matrix effectively impeded the dislocation movement inside the grain, resulting in elevated yield strength of the TiC/GTD222 composite through Orowan strengthening. In addition, the high proportion of spherical γ' phase in the composite will hinder the dislocation movement, increasing the stress needed for dislocation slip and further strengthening the matrix.

Deformation Mechanisms of the Composite at High Temperature

To understand the high temperature tensile plastic deformation mechanism of the TiC/GTD222 composite, the microstructure near the fracture surface was investigated using a focused ion beam (FIB) and TEM. Detailed microstructures inside the TiC/GTD222 sample were displayed in Fig. 7a–d. The macroscopic morphology of the FIBed slice is shown in the inset image of Fig. 7a. As shown in Fig. 7a, many slip lines inside the sample were present in matrix, as highlighted by the yellow arrow, which was caused by the start of the slip system. In addition, a large number of planar defects were observed in Fig. 7a, b, as shown in the yellow circle mark, which isolated stacking faults according to the shape and size of the plane defects [20]. As shown in Fig. 7c, a stacking fault with a smaller width and longer length was also observed in the microstructure, which was a continuous stacking fault morphology [21]. Dislocation pairs were also observed in the TiC/GTD222 composite, as shown in Fig. 7d, indicating antiphase boundary shearing occurred during plastic deformation [22].

From the above observations, the main deformation mechanisms of TiC/GTD222 composite were speculated as isolated stacking faults shearing γ' phases, continuous stacking faults shearing γ and γ' phases, dislocation pairs shearing γ' phases to form antiphase boundary and dislocation slip within the matrix.

According to existing research, the reduction of stacking fault energy of matrix γ phase or the narrowing of matrix γ channel promoted the formation of continuous stacking faults [23]. In this study, Al elements (1wt.%) were added to TiC/GTD222 composite. These Al enhanced the formation of Ni_3Al and increased the proportion of γ'-Ni_3Al, resulting in a narrower matrix channel between γ' particles and more continuous stacking faults. Regarding the formation of isolated stacking faults, it was caused by the dislocation reaction in a/2 <101> , and the reaction equation can be expressed as [24]:

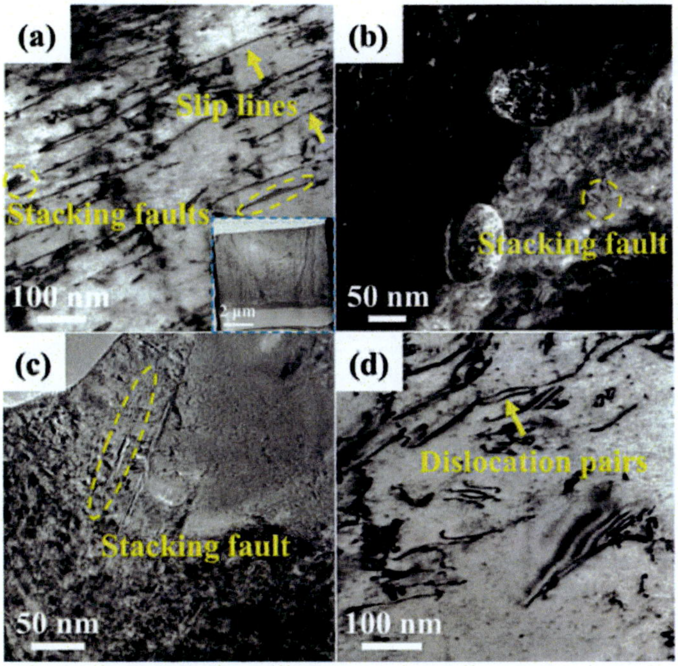

Fig. 7 TEM images of TiC/GTD222 composite near the tensile fracture after stretching at 800 °C: **a** Slip lines and isolated stacking faults, the inset is the FIB cut sample for TEM observation, **b** Isolated stacking faults, **c** Continuous stacking faults, **d** Dislocation pairs

$$a/2\langle 101\rangle \rightarrow a/6\langle 112\rangle + a/6\langle 2\bar{1}1\rangle \text{at } \gamma/\gamma' \text{ interface}$$

$$a/6\langle 112\rangle + a/6\langle 112\rangle \rightarrow a/3\langle 112\rangle \text{in} \gamma'$$

According to Fig. 7d, dislocation pairs were also observed in the deformed sample. During deforming, a large number of dislocations accumulated at the γ/γ' interface owing to the hindrance of γ' ordered structure, resulting in stress concentration. With deformation carrying on, dislocations with the same Burgers vector emerged as dislocation pairs [22], and when the local stress increased to a certain extent, dislocation pairs cut γ' phases to form antiphase boundaries [22].

As shown in Fig. 5, the plasticity of TiC/GTD222 composite was lower than that of GTD222 at 800 °C, and the slip plane separation caused by local deformation may be the main reason for the decrease in plasticity of Ni-base superalloys. In the process of stacking faults shearing γ', the cross-slip of the dislocation a/3 <112> cannot take place, resulting in local plastic deformation [25]. Once the local plastic deformation occurred, the slip surface will separate and the alloy will fracture under a low stress condition of finite strain, which may be the main reason for the decrease in the plasticity of TiC/GTD222 composite.

Conclusion

In this study, the microstructure and mechanical properties of HTed GTD222 and TiC/GTD222 composite were investigated. The main conclusions could be drawn as follows:

1. In contrast to GTD222, finer grains, more γ' and carbides, and higher dislocation density were observed in the HTed TiC/GTD222 composite. The carbides were mainly distributed along the boundaries of the TiC/GTD222 composite.
2. The yield strengths of TiC/GTD222 composite were higher than GTD222 at both room temperature and high-temperature. The superior yield strength of TiC/GTD222 composite can be attributed to the synergistic effect of TiC particles and γ' phase.
3. The main deformation mechanisms of TiC/GTD222 composite at 800 °C were isolated stacking faults shearing γ' phases, continuous stacking faults shearing γ and γ' phases, dislocation pairs shearing γ' phases forming antiphase boundary, and dislocation slip in the matrix.
4. The decrease in the plasticity of TiC/GTD222 composite at 800 °C was that the stacking fault in the γ' phases hinders the shear action of the dislocation, which leads to the early separation fracture of the slip surface.

Acknowledgements The authors acknowledge the support from the National Key R&D Program of China (Grant No. 2021YFB3702500) and the National Natural Science Foundation of China (Grant No. 12205055).

References

1. Q. Han, H. Gu, S. Soe, R. Setchi, F. Lacan, J. Hill, Manufacturability of $AlSi_{10}Mg$ overhang structures fabricated by laser powder bed fusion, Mater. Des. 160 (2018) 1080–1095, https://doi.org/10.1016/j.matdes.2018.10.043.
2. L. Yang, Z.C. Pang, M. Li, Y.G. Wang, D. Wang, C.H. Song, S.X. Li, Investigation into the dynamic mechanical properties of selective laser melted Ti-6Al-4V alloy at high strain rate tensile loading, Mater Sci Eng., A 745 (2019) 440–449, https://doi.org/10.1016/j.msea.2019.01.010.
3. M. Wang, B. Song, Q. Wei, Y. Zhang, Y. Shi, Effects of annealing on the microstructure and mechanical properties of selective laser melted AlSi7Mg alloy, Mater Sci Eng., A 739 (2019) 463–472, https://doi.org/10.1016/j.msea.2018.10.047.
4. Q. Han, R. Liu, Theoretical model for CNC whirling of screw shafts using standard cutters, Int. J. Adv. Manuf. Technol. 69 (2013) 2437–2444, https://doi.org/10.1007/s00170-013-5214-4.
5. Z. Sun, X. Tan, S.B. Tor, W.Y. Yeong, Selective laser melting of stainless steel 316L with low porosity and high build rates, Mater. Des. 104 (2016) 197–204, https://doi.org/10.1016/j.matdes.2016.05.035.
6. N.J. Harrison, I. Todd, K. Mumtaz, Reduction of micro-cracking in nickel super-alloys processed by Selective Laser Melting: a fundamental alloy design approach, Acta Mater. 94 (2015) 59–68, https://doi.org/10.1016/j.actamat.2015.04.035.
7. Q. Han, Y. Gu, S. Soe, F. Lacan, R. Setchi, Effect of hot cracking on the mechanical properties of Hastelloy X superalloy fabricated by laser powder bed fusion additive manufacturing, Optic Laser. Technol. 124 (2020) 105–984, https://doi.org/10.1016/j.optlastec.2019.105984.
8. N. Guo, M.C. Leu, Additive manufacturing: Technology, applications and research needs, Front. Mech. Eng. 8 (2013) 215–243, https://doi.org/10.1007/s11465-013-0248-8.
9. Q. Han, R. Setchi, F. Lacan, D. Gu, S.L. Evans, Selective laser melting of advanced Al-Al_2O_3 nanocomposites: simulation, microstructure and mechanical properties, Mater. Sci. Eng., A. 698 (2017) 162–173, https://doi.org/10.1016/j.msea.2017.05.061.
10. M. Rombouts, J.P. Kruth, L. Froyen, P. Mercelis, Fundamentals of selective laser melting of alloyed steel powders, CIRP Ann. Manuf. Technol. (2006), https://doi.org/10.1016/S0007-8506(07)60395-3.
11. L.J. Qian, X.M. Pang, J.Q. Zhou, J.X. Yang, S.S. Lin, D. Hui, Theoretical model and finite element simulation on the effective thermal conductivity of particulate composite materials, Compos. B Eng. 116 (2017) 291–297, https://doi.org/10.1016/j.compositesb.2016.10.067.
12. V.A. Popov, M. Burghammer, M. Rosenthal, A. Kotov, In situ synthesis of TiC nano-reinforcements in aluminum matrix composites during mechanical alloying, Compos. B Eng. 145 (2018) 57–61, https://doi.org/10.1016/j.compositesb.2018.02.023.
13. S.Z. Yang, Q.Q. Han, Y.Y. Yin, J. Gao, Z.H. Zhang, Y.C. Gu, K.W.Q. Low, Effects of micrometer-sized TiB2 on crack mitigation, mechanical and electrochemical performance of a Ni-based alloy fabricated by selective laser melting, Optic Laser. Technol. 142 (2021) 107–240, https://doi.org/10.1016/j.optlastec.2021.107240.
14. W.Z. Zhou, G.L. Zhu, R. Wang, C. Yang, Y.S. Tian, L. Zhang, A.P. Dong, D.H. Wang, D. Shu, B.D. Sun, Inhibition of cracking by grain boundary modification in a non-weldable nickel-based superalloy processed by laser powder bed fusion, Mater Sci Eng., A 791 (2020) 139–745, https://doi.org/10.1016/j.msea.2020.139745.
15. R. Wang, G.L. Zhu, C. Yang, W.Z. Zhou, D.H. Wang, A.P. Dong, D. Shu, B.D. Sun, Novel selective laser melting processed in-situ TiC particle-reinforced Ni matrix composite with excellent processability and mechanical properties, Mater. Sci. Eng., A 797 (2020) 140–145, https://doi.org/10.1016/j.msea.2020.140145.
16. M. Habibnejad-Korayem, R. Mahmudi, W.J. Poole, Enhanced properties of Mg-based nano-composites reinforced with Al2O3 nano-particles, Mater. Sci. Eng., A 519 (2009) 198–203, https://doi.org/10.1016/j.msea.2009.05.001.

17. J.B. Ferguson, F. Sheykh-Jaberi, C.S. Kim, P.K. Rohatgi, K. Cho, On the strength and strain to failure in particle-reinforced magnesium metal-matrix nanocomposites (Mg MMNCs), Mater. Sci. Eng., A 558 (2012) 193–204, https://doi.org/10.1016/j.msea.2012.07.111.
18. R.S.G. Ji, P.S. Karthik, K.B.S. Rao, K.V. Rajulapati, Strengthening mechanisms in equiatomic ultrafine grained AlCoCrCuFeNi high-entropy alloy studied by micro- and nanoindentation methods, Acta Mater. 125 (2017) 58–68, https://doi.org/10.1016/j.actamat.2016.11.046.
19. R.J. Arsenault, N. Shi, Dislocation generation due to differences between the coefficients of thermal expansion, Mater. Sci. Eng., A 81 (1986) 175–187, https://doi.org/10.1016/0025-5416(86)90261-2.
20. J.S. Huo, J.T. Gou, L.Z. Zhou, X.Z. Qin, G.S. Li, High temperature creep deformation mechanisms of a hot corrosion-resistant nickel-based superalloy, J. Mater. Eng. Perform. 16 (2007) 55–62, https://doi.org/10.1007/s11665-006-9008-9.
21. Y. Yuan, Y.F. Gu, T. Osada, Z.H. Zhong, T. Yokokawa, H. Harada, Deformation mechanisms in a new disc superalloy at low and intermediate temperatures, Scripta Mater. 67 (2012) 137–140, https://doi.org/10.1016/j.scriptamat.2012.03.042.
22. P. Zhang, Y. Yuan, S.C. Shen, B. Li, R.H. Zhu, G.X. Yang, X.L. Song, Tensile deformation mechanisms at various temperatures in a new directionally solidified Ni-base superalloy, J. Alloys Compd. 694 (2017) 502–509, https://doi.org/10.1016/j.jallcom.2016.09.303.
23. P. Caron, T. Khan, P. Veyssi'ere, On precipitate shearing by superlattice stacking faults in superalloys, Philos. Mag. A 57 (1988) 859–875, https://doi.org/10.1080/01418618808204522.
24. G.B. Viswanathan, P.M. Sarosi, M.F. Henry, D.D. Whitis, W.W. Milligan, M.J. Mills, Investigation of creep deformation mechanisms at intermediate temperatures in Rene 88 DT, Acta Mater. 53 (2005) 3041–3057, https://doi.org/10.1016/j.actamat.2005.03.017.
25. P. Duan, P. Zhang, J. Li, B. Li, X.F. Gong, G.X. Yang, X.L. Song, Intermediate temperature brittleness in a directionally solidified nickel-based superalloy M4706, Mater. Sci. Eng., A 759 (2019) 530–536, https://doi.org/10.1016/j.msea.2019.05.037.

Fabrication and Weldability Aspects of Ni- and Ni–Fe Based Superalloys—A Review

Joel Andersson

Abstract Superalloys are commonly used in structural components of aero-engines. Superalloys in general, Ni- and Ni–Fe-based superalloys, belong to an important group of materials used in aerospace applications. Fabrication and associated weldability aspects of structural components for the hot section of aero-engine gas turbines continue to be of high importance to the manufacturing industry within this discipline. Cracking and specifically hot cracking as well as strain age cracking is a serious concern during the welding and additive manufacturing (AM) of these structural components. The cracking phenomena can occur during welding, AM or subsequent heat treatment of precipitation-hardening superalloys. The cracking behaviour can be influenced by several factors, i.e., chemical composition in terms of hardening elements and impurities, the microstructure of base material, and weld zone, together with corresponding welding, AM and post-treatment process parameters. This paper provides a review of Ni- and Ni–Fe-based superalloys concerning fabrication and weldability aspects within the context of structural components of aero-engines. Also, the paper offers insight and analyses to research publication data of welding and AM of superalloys in the context of annual publication developed over the years as well as specific contributions from countries, affiliations, and specific researchers.

Keywords Hot cracking · Strain age cracking · Superalloys · Welding · Additive manufacturing

Introduction

Superalloys, in particular, the precipitation hardening Ni- and Ni–Fe-based superalloys, belong to the most critical class of materials in realizing the aero-engines seen today and on long-term future horizon. The trend in the past decades, regarding

J. Andersson (✉)
Department of Engineering Science, University West, 46186 Trollhättan, Sweden
e-mail: joel.andersson@hv.se

E. A. Ott et al. (eds.), *Proceedings of the 10th International Symposium on Superalloy 718 and Derivatives*, The Minerals, Metals & Materials Series,
https://doi.org/10.1007/978-3-031-27447-3_41

large hot structural aero-engine components, has been through fabrication. Fabrication of hot structural components such as a turbine exhaust casing requires extensive knowledge of welding of the aforementioned class of materials. Fabrication is, apart from welding, also dependent on know-how of metallurgical aspects of different material forms, i.e., sheet, forging, and casting as well as their implications on welding, where additive manufacturing (AM) is a recent tool that potentially can further enhance the fabrication concept of hot structural components. The fabrication concept, as such, benefits from the fact that one can utilize the best out of each type of material form, i.e., castings can be utilized where complex geometrical shapes are needed, and sheet/wrought material form can be used when high strength and less complex geometries are to be used. It also means that it is possible to utilize different types of alloys within the same component, hence, expensive alloys can be minimized and used where it is really needed. This rationale provides a path of tailoring complex structural components which in turn can reduce the overall weight and cost, depending on the circumstances. However, the journey to fully comprehend, master, and understand the fabrication concept requires a comprehensive understanding of material aspects in relation to material form and processes such as welding and AM. There are several obstacles such as cracking aspects in welding and AM which need to be addressed properly to avoid costly mistakes.

Research of Superalloys, Welding, and Additive Manufacturing and Its Global Presence

Research within AM technology has sky-rocketed since 2010. In Fig. 1, the amount of research (publications) in connection with "welding", "additive manufacturing", and "superalloys" are presented. Publications on "welding" and "superalloys" date back to ~1950s whereas "additive manufacturing" appeared for the first time in ~1985. Both "welding" and "additive manufacturing" have increased significantly in 2000 and 2010, respectively, whereas research on "superalloys" have levelled up from a quite low level from its start in the 1950s to a more moderate level since 1990s until today.

In Fig. 2, it can be seen that when combining the search strings of "welding" and "additive manufacturing" with "superalloys", the amount of research within the field of superalloys in combination with welding or AM is about ~60 and ~40 times less, respectively, than the overall amount of research carried out within the research fields of welding or AM.

Primary production as well as research of superalloys have traditionally had a strong presence in G8 countries (France, Germany, Italy, Japan, United Kingdom, United States, Canada, and Russia). Looking into the number of publications of "welding AND superalloys" and "additive manufacturing AND superalloys", Figs. 3and4, the top-ten countries regarding number of publications comprises the following countries (full list in Appendices 1 and 2):

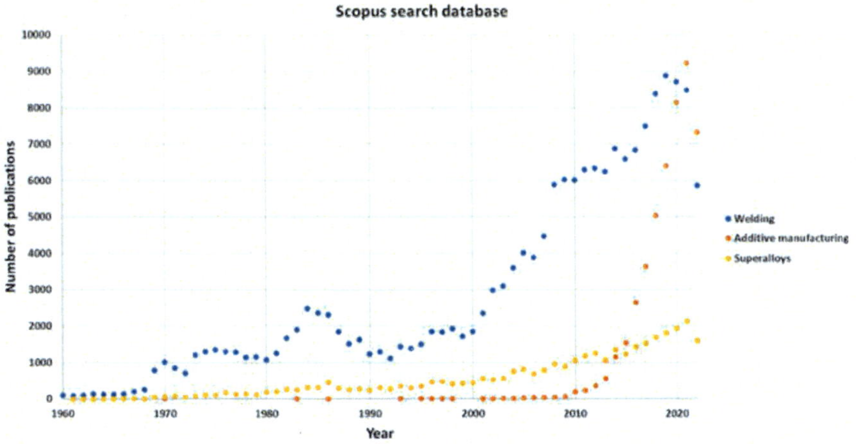

Fig. 1 Number of publications versus years for "welding", "additive manufacturing", and "superalloys" [1–3]

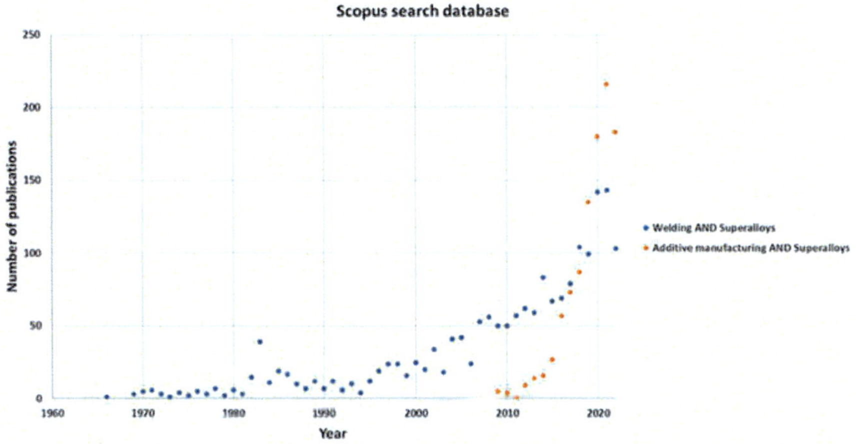

Fig. 2 Number of publications versus years for "welding AND superalloys" and "additive manufacturing AND superalloys" [4, 5]

- "welding AND superalloys"
  1. China
  2. United States
  3. India
  4. Canada
  5. Japan
  6. United Kingdom
  7. Iran

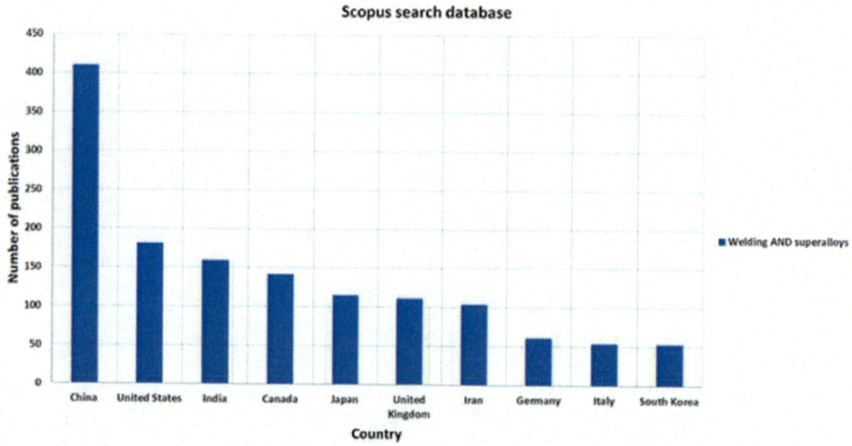

Fig. 3 Number of publications versus country for "welding AND superalloys" [6]

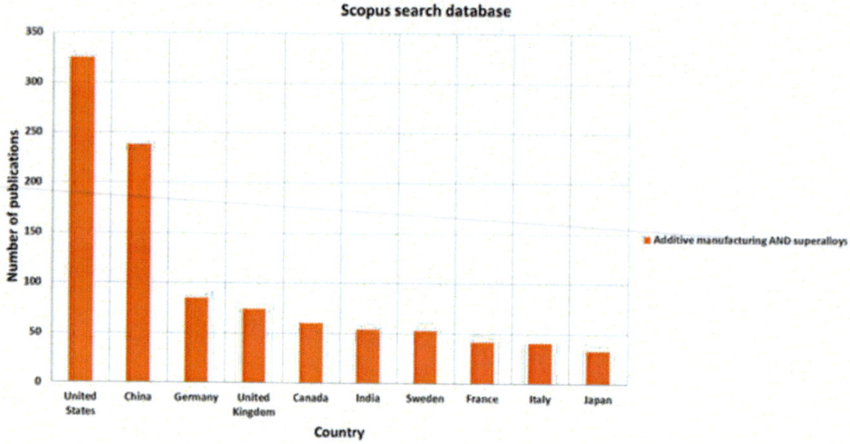

Fig. 4 Number of publications versus country for "additive manufacturing AND superalloys"

8. Germany
9. Italy
10. South Korea

- "additive manufacturing AND superalloys"

1. United States
2. China
3. Germany
4. United Kingdom
5. Canada

6. India
7. Sweden
8. France
9. Italy
10. Japan

Most of these countries holds a "leading" position within both fields of research ("welding AND superalloys" and "additive manufacturing AND superalloys"), except Iran and South Korea which do not enter the top-ten list of "additive manufacturing AND superalloys" as well as Sweden and France who do not enter the top-ten list of "welding AND superalloys". China is by far producing most of publications within the research field of "welding AND superalloys" where the remaining publications are evenly distributed among the other countries. However, United States followed by China is producing most of the publications in the field of "additive manufacturing AND superalloys" where the remaining countries are far behind regarding publications.

Looking further into research affiliations, Figs. 5and6, for the fields of interest the top-ten affiliations are as follows (full list in Appendices 3 and 4):

- "welding AND superalloys"

  1. University of Manitoba, Canada
  2. Northwestern Polytechnical University, China
  3. Chinese Academy of Sciences, China
  4. Harbin Institute of Technology, China
  5. Vellore Institute of Technology, India
  6. Osaka University, Japan
  7. Ministry of Education China, China

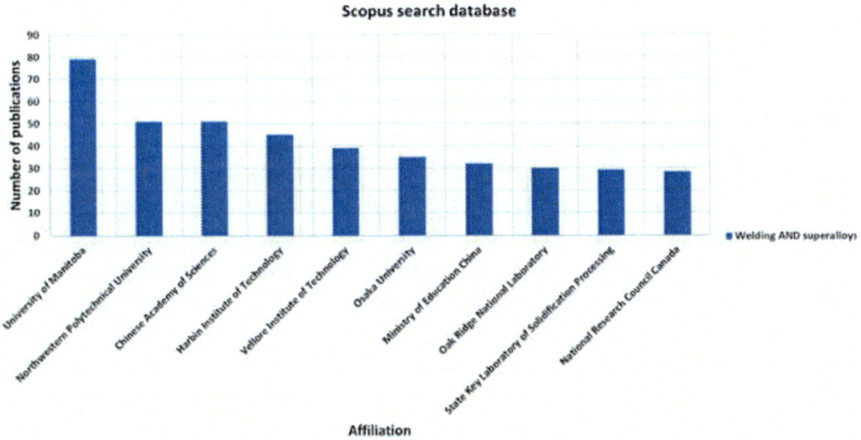

Fig. 5 Number of publications versus affiliation for "welding AND superalloys" [8]

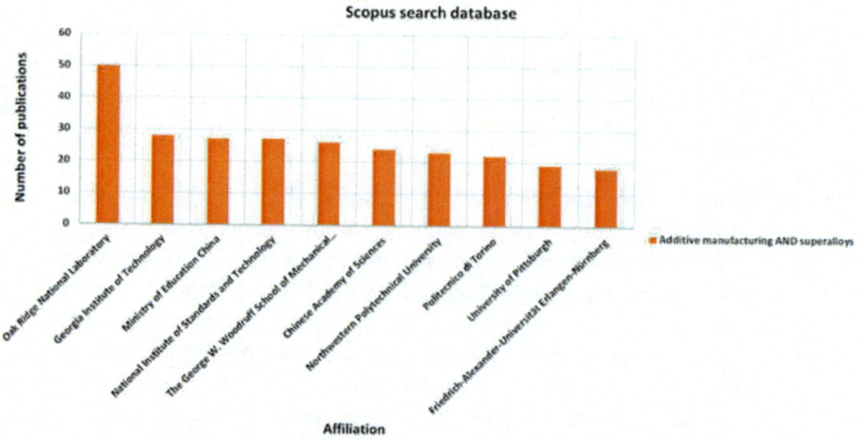

Fig. 6 Number of publications versus affiliation for "additive manufacturing AND superalloys" [9]

8. Oak Ridge National Laboratory, USA
9. State Key Laboratory of Solidification Processing, China
10. National Research Council Canada, Canada

- "additive manufacturing AND superalloys"

1. Oak Ridge National Laboratory, USA
2. Georgia Institute of Technology, USA
3. Ministry of Education China, China
4. National Institute of Standards and Technology, USA
5. The George W. Woodruff School of Mechanical Engineering, USA
6. Chinese Academy of Sciences, China
7. Northwestern Polytechnical University, China
8. Politecnico di Torino, Italy
9. University of Pittsburgh, USA
10. Friedrich-Alexander-Universität Erlangen-Nürnberg, Germany

There is a mix of institutes and universities among the two fields of research, however, Oak Ridge National Laboratory, Chinese Academy of Sciences, Northwestern Polytechnical University, and Ministry of Education China holds a top-ten position in both fields where University of Manitoba and Oak Ridge National Laboratory outrun the others in the research fields of "welding AND superalloys" and "additive manufacturing AND superalloys", respectively.

On an individual author basis, the number of publications of "welding AND superalloys" and "additive manufacturing AND superalloys", Figs. 7and8, disclose the following top-ten list of authors (full list in Appendices 5 and 6):

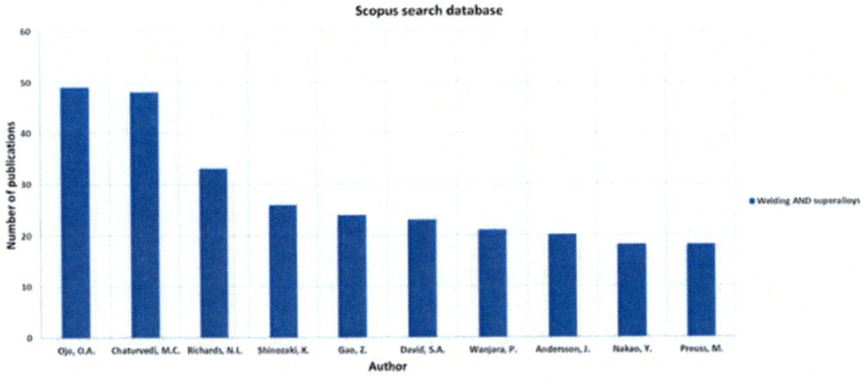

Fig. 7 Number of publications versus author for "welding AND superalloys" [10]

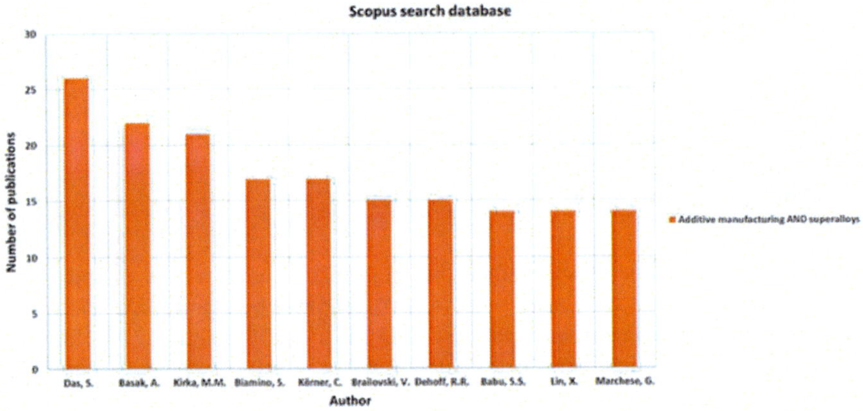

Fig. 8 Number of publications versus author for "additive manufacturing AND superalloys" [11]

- "welding AND superalloys"
    1. Ojo, O.A.
    2. Chaturvedi, M.C.
    3. Richards, N.L.
    4. Shinozaki, K.
    5. Gao, Z.
    6. David, S.A.
    7. Wanjara, P.
    8. Andersson, J.
    9. Nakao, Y.
    10. Preuss, M.

- "additive manufacturing AND superalloys"

    1. Das, S.
    2. Basak, A.
    3. Kirka, M.M.
    4. Biamino, S.
    5. Körner, C.
    6. Brailovski, V.
    7. Dehoff, R.R.
    8. Babu, S.S.
    9. Lin, X.
    10. Marchese, G.

Opposite to what has been seen regarding countries and research affiliations, there is no individual author that "qualifies" in the top-ten list of both fields of research.

The above search in the number of publications versus year, country, affiliation, and author do give some insight to the spread and presence of research in the fields of interest, however, a more in-depth analyses considering patents and research funds would add further insights.

Welding of Superalloys

There is a fair number of superalloys being used in a wide range of industrial applications where the most used ones in aerospace can be classified as either Ni and Ni–Fe-based solid solution hardening (i.e., Inconel 625, Hastelloy X, Nimonic 75, Incoloy 800, Haynes 556, Multimet N-155) or precipitation hardening (i.e., Waspaloy, Haynes 282, Inconel 939, Inconel 718, Alloy A-286, Incoloy 925) type of superalloys. The aerospace sector alone accounts for about 70% of the total market share with respect to superalloy usage [12]. However, a comprehensive classification of superalloys takes a stand on the base elements (Ni, Ni–Fe, and Co), primary hardening mechanism (solid solution hardening, precipitation hardening, and oxide dispersion strengthening), and material form (i.e., cast, wrought, and powder). The solid solution hardening superalloys most often find their usage in applications where resistance towards thermal loads and environmental resistance are of importance whereas the precipitation hardening superalloys are frequently used when higher demands for strength are required.

In welding with both arc and high-energy beam processes, welding results in high temperatures and large temperature gradients at the location of the heat source, which leads to thermal stresses [13–15] and microstructural inhomogeneities [16] being formed at the region of the weld and these can in turn lead to weld cracking [17, 18]. The magnitude of stress and inhomogeneities vary between the different processes and there are many pros and cons to the selection of a specific process. Gas tungsten arc welding process (GTAW) is frequently used [19–21] since it provides means for clean and high-quality welds as well as being in-expensive in comparison to high-energy beam processes, however, problems related to high heat input, lower penetration, and more limited ability regarding automation tend to favor laser beam welding (LBW) [22, 23]. There has been significant development on GTAW process enabling deeper penetration and increased robustness, which in some way makes it rival other high-energy beam processes like LBW. LBW, despite being significantly more capital intensive [24] in comparison to GTAW, still possesses advantages regarding automation and control, particularly for directed energy deposition (DED) applications. However, recent techniques such as pulsed mode of gas metal arc welding (GMAW), cold metal transfer (CMT), and various modes of GTAW as Force GTAW, K-TIG, and high frequency (HF) GTAW show very promising results in producing high-quality welds not at least including weld thicknesses larger than 3 mm, which historically has been a limit thickness for GTAW [25–29].

Still, issues regarding weld cracking persist to various degrees depending on the specific alloy of interest and irrespective of welding process. In general, the driving force for weld cracking originates from both internal and external sources. The internal sources have already been implicitly mentioned above, i.e., solidification shrinkage stresses, which vary significantly depending on welding process and process parameters. Welding process control has developed tremendously in the last decades, which has resulted in better control and the ability to use very low heat input for preventing weld cracking.

Weld Cracking Mechanisms of Superalloys

Weldability can be defined as "a measure of the ease with which a metal or an alloy can be welded or joined without degradation to the weldment microstructure or properties during or after welding and for the duration of its intended service" [30]. In this context, susceptibility towards cracking during welding and post-weld heat treatment (PWHT) is of major importance. There are different types of weld cracking mechanisms that come to play during welding and AM of superalloys, where hot cracking and solid-state type of cracking, primarily strain age cracking (SAC), are of high importance especially for Ni- and Ni–Fe-based precipitation hardening superalloys utilized in hot structural component for aero-engines [31, 32]. Many types of crack criteria exist to explain why and how cracking occurs, where the first criteria dates back to ~70 years [33–36]. The relationships between metallurgical and mechanical factors are complex, making it hard to include them in a crack criterion, while most criteria do not account for the physical phenomena linked to cracking, but instead are more related to macro-, and micro-conditions than can result in cracking [36]. In general, most cracking criteria can be divided into mechanical and non-mechanical [36] where the non-mechanical ones commonly are based on aspects such as thermal history, chemical composition, and brittle temperature range (BTR) [33, 34]. The mechanical-based criteria, on the other hand, are normally based on critical strain rate, critical strain, or critical stress [33–36]. It has been found that the mechanical criterion is of special use in predicting cracking based on numerical simulations of various kinds, whereas the non-mechanical criterion has been successful in predicting compositional dependencies of cracking [36].

Hot Cracking

The weld cracking that takes place during welding is referred to as hot cracking and is further divided into solidification cracking and heat affected zone (HAZ) liquation cracking, respectively.

Solidification Cracking

Solidification cracking is found in the fusion zone (FZ), whereas HAZ liquation cracks occur in the HAZ. There are several theories that have been developed over the years to explain these types of cracking mechanisms [37–39]. One concept referred to as "Borland's generalized theory", Fig. 9, separates the solidification range into four different stages [39]:

- Stage 1: Primary dendrites form as the temperature drops below the liquidus temperature.
- Stage 2: The dendrites interlock by forming a solid network. In this stage, only the liquid can move between the dendrites.
- Stage 3: The grain development occurs. Here, the liquid is not capable of free movement as it is hindered by the continuous network.
- Stage 4: All the liquid solidifies.

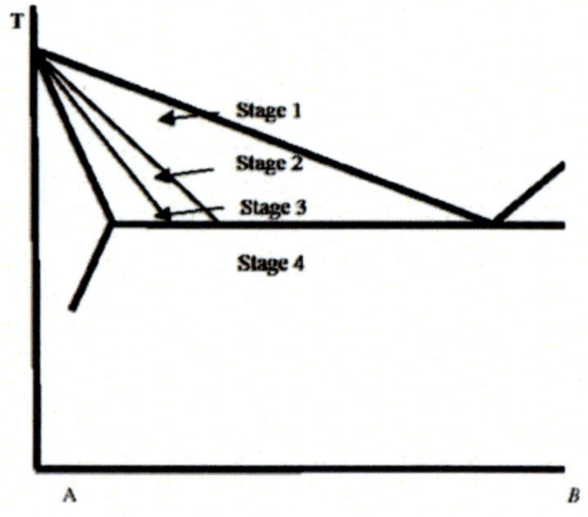

Fig. 9 Solidification stages according to Borland´s "generalized theory"; adapted from [39]

Borland postulated that cracks could form in stage 2 if a continuous dendrite network is formed, however, healing through backfilling is also possible at this stage. Borland defined a "critical solidification range" (CSR) associated with stage 3. In stage 3, substantial cracking can take place since the liquid backfilling is no longer possible. Borland's generalized concept has been further developed by Matsuda et al. [40]. Nevertheless, most existing theories concur on the fact that susceptibility towards solidification cracking is dependent on the combination of metallurgical factors together with local strain at the final stage of solidification. The solidification range is one important factor regarding solidification cracking where a wider range enhances cracking susceptibility [41, 42]. However, liquid distribution and grain boundary wetting are also of high importance, where a low surface tension tends to enable the liquid to better wet the grain boundaries and increase the risk of cracking.

Ni- and Ni–Fe-based superalloys will always be at some risk to solidification cracking since they solidify fully austenitic, however, welding procedure and parameters combined with the chemical composition of the specific alloy determine the susceptibility of an alloy. High heat input is, for instance, something which in general aggravates cracking since it decreases the temperature gradient which in turn increases the solid–liquid region where cracking can take place. There are several aspects to look out for regarding the effects of chemical composition of an alloy, where certain elements, such as, B, C, Si, Mo, and Nb are more important than others when it comes to the influence of solidification temperature range as well as the terminal eutectic type that form at the end of solidification [42–44]. So, solidification cracking is strongly related to the solidification process within the FZ; the amount of solute redistribution and solidification path, solidification temperature range and amount as well as distribution of solute rich interdendritic liquid at the end of solidification [45].

Heat Affected Zone Liquation Cracking

The region located just outside the FZ refers to the partially melted zone of the HAZ. When the applied strain can no longer sustain the locally induced strain, a crack appears. The susceptibility towards this type of cracking normally worsens with increasing heat input, however, the parent metal grain size and alloying content are the most important factors in determining the susceptibility towards HAZ liquation cracking. The nature of grain boundary liquation is of central importance, where trace elements such as B, P, and S are important aside from major elements like Nb. The actual liquation mechanism is triggered by two different types: penetration mechanism and segregation mechanism. The aforementioned trace elements are of crucial importance in the segregation mechanism where B segregates via non-equilibrium or equilibrium mechanisms. Equilibrium segregation decreases with increasing temperature due to increased diffusivity whereas non-equilibrium segregation increases with increasing temperature and takes place during cooling from high to low temperature in terms of diffusion of vacancy-solute complexes to the grain boundaries [46–48].

In a study on cast Haynes 282 where the influence on pseudo hot isostatic pressing temperature on susceptibility towards HAZ liquation cracking was investigated using Varestraint weldability testing [49] it was concluded, by performing nanoSIMS analysis, that free B was not present at 1120 and 1160 °C for a dwell time of four hours, meaning that non-equilibrium segregation was not effective at these temperatures. However, at 1190 °C for a dwell time of 4 h the decomposition of C-B precipitates allowed the B to diffuse and segregate at the boundaries which were believed to be the cause of the aggravated extent of cracking, Fig. 10 and 11.

Apart from segregation driven liquation, the following phenomena (see bullet list below) regarding penetration liquation mechanism have been reported in the literature [50–52], where the penetration mechanism primarily is claimed to be governed by eutectic melting, and constitutional liquation [42, 53]:

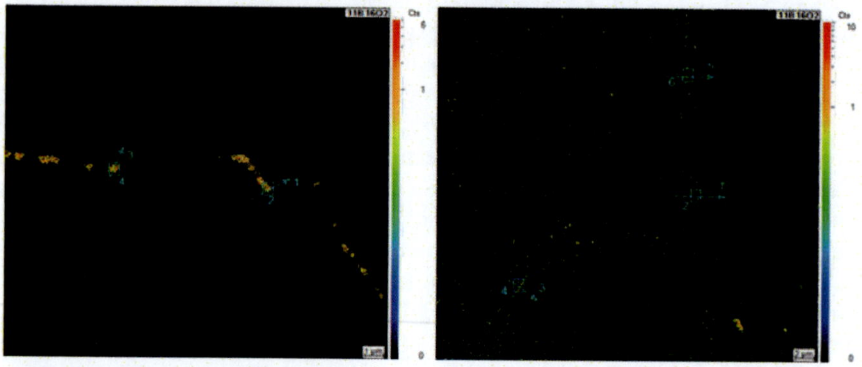

Fig. 10 NanoSIMS elemental map showing the particle segregation of B (left image) in the grain boundary in the 1120 °C /4 h condition and enrichment of B along the grain boundaries in the 1190 °C /4 condition, respectively adapted from Singh and Andersson [49]

Fig. 11 Average total crack length in the HAZ of cast Haynes 282 [49]

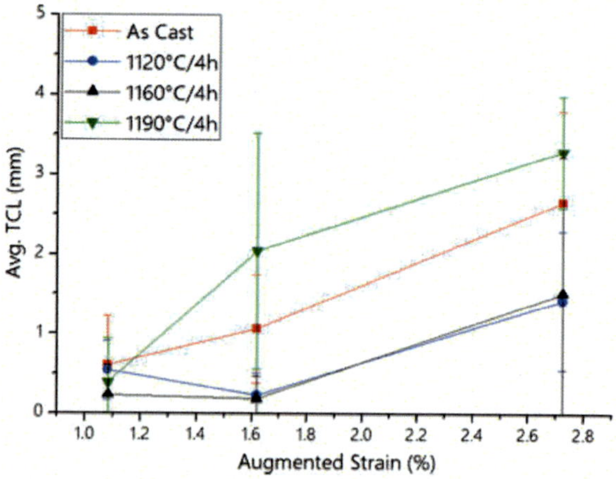

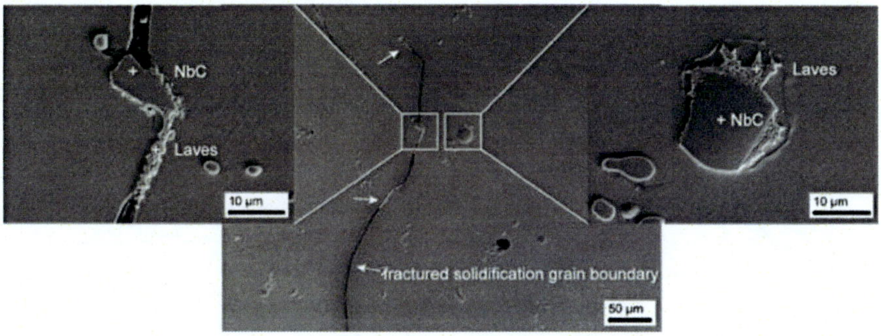

Fig. 12 Grain boundary cracking within a Gleeble weld-simulated HAZ microstructure with constitutional liquation of Nb(Ti) MC-type carbides located on and in close vicinity to the cracked grain boundary. The γ-Laves eutectic surrounding the particles indicates the formation of liquid where the presence of further gamma-Laves eutectic is indicated by right facing arrows [56]

- Constitutional liquation of secondary phases.
- Melting of the matrix.
- Liquation of precipitation hardening phases in high-volume fraction γ' alloys.
- Melting of residual eutectic in cast material.

Constitutional liquation was originally proposed by Pepe and Savage for "maraging steel" weldments [51], where the main aspect of this mechanism refers to a secondary particle being exposed to fast heating rate during welding. This implicates insufficient time for the particle to completely dissolve through solid state diffusion and as a result it encounters liquation since a compositional gradient is generated at the local particle–matrix interface, leading to partial dissolution of the precipitate and a solute-enriched area. A liquid develops consequently as soon as the local eutectic composition is reached and it is most common in Nb-bearing superalloys [49, 54, 55]. An example of a constitutional liquation of NbC in ATI 718Plus can be seen in Fig. 12 [56].

The liquation mechanism in cast ATI 718Plus during weld thermal cycling can be seen as a combination of several contributing factors, as discussed in the previous sections, and is summarized in Fig. 13 [55, 57].

The HAZ thermal cycle in Fig. 13 is divided into three stages, based on temperature ranges between the eutectic temperature (Te), solidus temperature (Ts), and peak temperature (Tp), which are highlighted in the pseudo-binary phase diagram. Stage 1 (up to Te) is characterized by the dissolution of Nb-rich phases, with a higher diffusion rate along the grain boundaries as compared with the grain interior. In this stage, liquation occurs from the solute segregation of minor elements, which lowers the effective solidus of the alloy. In Stage 2 (between Te and Ts), liquation is possible from the melting of the Laves phase or by constitutional liquation of MC carbides, depending on the chemistry after the homogenization heat treatments. In Stage 3 (between Ts and Tp), bulk melting occurs, with the extent of melting based on the amount of Nb in solid solution. The overall contribution will depend on the

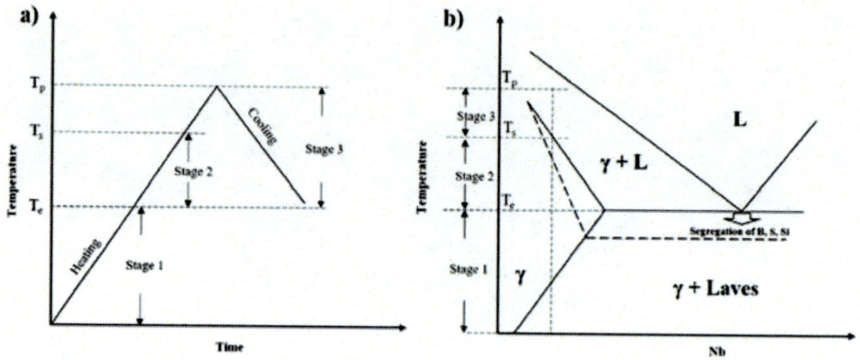

Fig. 13 **a** HAZ thermal cycle divided into three stages based on the temperatures and **b** corresponding pseudo-binary phase diagram for cast ATI 718Plus [55]. Adapted from Radhakrishnan and Thompson [57]

contribution from solute segregation, Laves melting, constitutional liquation of MC carbides and supersolidus melting of the matrix [49, 54, 55].

Strain Age Cracking

Ductility dip cracking (DDC) is a solid-state type of cracking mechanism commonly observed in solid solution hardening Ni-based superalloys and it occurs during cooling in the temperature range of Ts (solidus) and ~0.5Ts, which is concomitant with a significant drop in ductility [45]. When significant contraction stresses take place simultaneously with the drop in ductility, DDC could be observed and have been significantly researched for alloys such as 617 and 740 [45]. However, another solid-state type of cracking phenomena being more predominant in comparison to DDC when dealing with aero-engine structural components is referred to as strain age cracking (SAC). SAC is a type of cracking that occurs in the solid state, without any prerequisites of liquid phase, when hardening takes place in the alloy. As it is closely related to the kinetics of hardening in superalloys, this type of cracking phenomena is primarily of concern for γ' hardening type of alloys and was one of the original drivers for the development of Alloy 718 [58], an alloy that is immune against SAC, however, still suffers from hot cracking. Strain age cracking, as a cracking phenomenon, dates back to the early 1960s. In general, SAC occurs during the post-weld heat treatment (PWHT) and is consequently sometimes referred to as "PWHT cracking" or "Reheat cracking" because of high weld stresses at the same time that hardening occurs in γ' precipitation hardening Ni-based superalloys [59, 60]. The actual cracking mechanism is related to low ductility in the weld HAZ and the material's inability to accommodate stress relaxation. So, stress localization in the HAZ occurs because of reduced strength in comparison to the base material. It is therefore reasonable to assume that a softer material decreases the risk of encountering problems with SAC, whereas a harder material tends to be less resistant [61–67]. Over-aging the material is another approach to mitigate the risk of SAC if a normal solution heat treatment is not feasible, since it reduces the material's strength in favor of stress relaxation in the base material [63].

Chemical Composition and Microstructure

One of the first correlations on susceptibility towards SAC regarding the influence of alloying elements was suggested by Prager and Shira [61], who proposed a significant influence of Al and Ti concentrations on precipitation behavior of the γ' phase. The more of these alloying elements, the higher the volume fraction and precipitation kinetics of the γ′ phase. The more of these alloying elements, the higher the volume fraction and precipitation kinetics of the γ′ phase, which in turn adversely affects the risk of SAC. In Fig. 14, which is based on what Prager and Shira postulated [61], it can be seen that a combined level of Al + Ti content above 6at-% leads to severe problems with regard to SAC, which is also evident by comparing the high susceptibility of René 41 [68] with that of Alloy 718. The sluggish strengthening reaction of (γ" in Alloy 718 compared) enables stress relaxation to occur in advance of any hardening reactions, which puts them in a highly immune state regarding SAC [69].

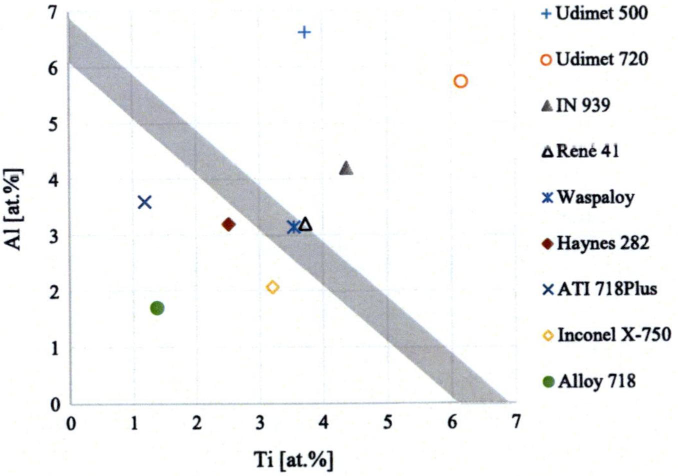

Fig. 14 Adapted Prager-Shira diagram showing [61] Al and Ti concentrations (at.%) of commercially available superalloys. Reduced weldability above the grey line [60]

Other alloying elements, such as C and B, have been claimed to be beneficial in enhancing the resistance towards SAC, due to grain size control [70–72] as well as grain boundary cohesion [71, 73], respectively.

From a microstructural perspective, finer grain size as well as homogenous microstructure have been found to be of importance in preventing SAC as compared to coarse grain size and segregated microstructure, since the distribution of strain can be accommodated over a larger grain boundary area [62, 66, 67, 74]. The aforementioned strain can, apart from residual stresses and external restraint, originate from the mismatch between the austenitic matrix and the γ' phase and lead to the development of contraction stresses, that in turn, induce tensile stresses on the grain boundaries and consequently raise the susceptibility towards SAC [61, 75–77]. However, current research work presumes a significant correlation between lattice misfit and susceptibility towards SAC even though no available quantifiable data provide the actual time–temperature range where SAC takes place [66, 67, 78–81]. So, in summary it can be concluded that the main factors that influence susceptibility towards SAC are related to [60]:

- Strain
  - External weld restraint as well as solidification shrinkage strain
  - Precipitation induced stress
- Stress localization at grain boundaries
  - Grain size
  - Carbides
- Precipitation kinetics
  - Chemical composition
    - γ' phase volume fraction
  - Deformation hardening
- Stress relaxation
  - Young's modulus
  - Time-temperature regime.

Weldability Testing

Weldability as such is quite a broad concept covering everything from actual welding to service performance and should therefore be treated with care to avoid misunderstanding of its meaning. When it comes to material's inherent weldability and actual fabrication of hot structural components and specifically susceptibility towards cracking during welding and PWHT one needs to be careful in assessing the overall cracking susceptibility to avoid costly mistakes. So, based on experience [31, 49,

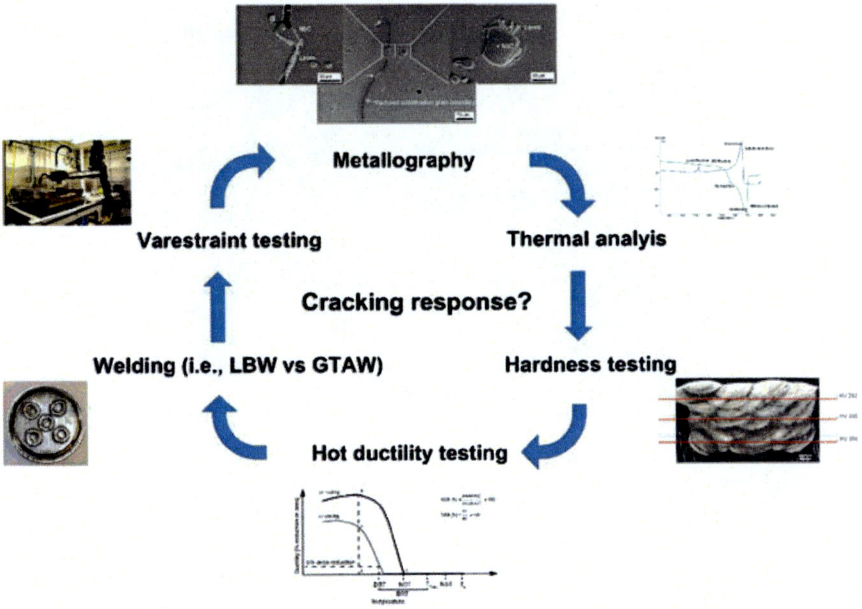

Fig. 15 Proposed procedure in assessing susceptibility toward cracking during welding and PWHT of superalloys. Adapted from Andersson [77]

82–86] in the evaluation of cracking susceptibility and in utilizing different types of assessment methods, an approach [77] is proposed as disclosed in Fig. 15. The proposed procedure in assessing susceptibility toward cracking is utilizing a variety of tools where each one of them provides insightful knowledge on the specific material's behavior.

Utilizing characterization tools such as optical microscopy or scanning electron microscopy (SEM) is something that nowadays is easily accessible and that provides good insight to what phases and constituents are present in the actual material. In Fig. 16, a detailed high-resolution image of the Laves eutectic can be seen in cast Alloy 718 [87].

Information gained from microstructural characterization can be used to enhance understanding of cracking mechanisms [88]. Various kinds of thermal analyses, such as differential scanning calorimetry (DSC) in Fig. 17, can be used to investigate possible phase reactions during heating and cooling in welding, even though these methods are significantly slower in heating and cooling rate in comparison.

Information from thermal analyses and microstructure characterization can be coupled to increase understanding and develop pseudo-binary phase diagrams, Fig. 18. Pseudo-binary phase diagrams can be used to explain possible liquation mechanisms such as constitutional liquation, Fig. 18.

Hardness testing is a valuable tool to quickly gain insight into susceptibility toward SAC, in terms of precipitation kinetics especially when combined with repair welding

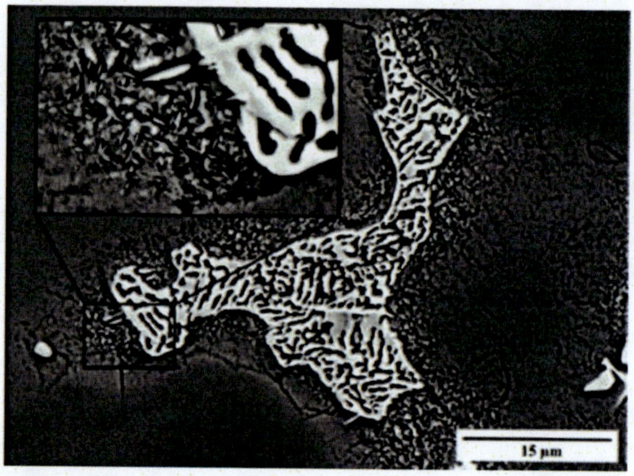

Fig. 16 Backscattered SEM image of Laves phase with eutectic morphology in cast Alloy 718 [87]

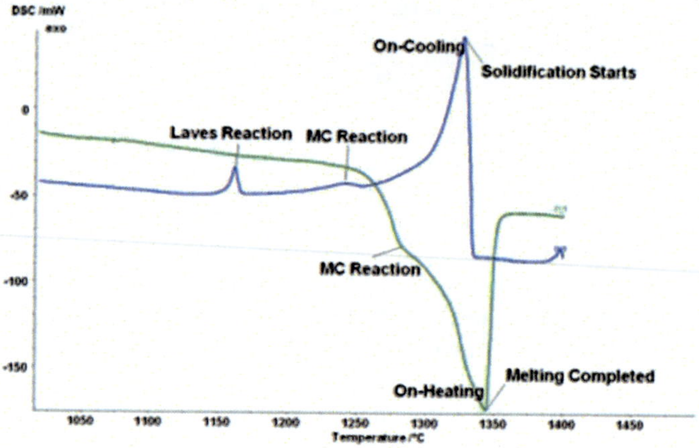

Fig. 17 A differential scanning calorimetric thermograph of Alloy 718 disclosing different phase reactions upon heating (green curve) and cooling (blue curve) [50]

or heat treatments of various kinds as illustrated in the time–temperature-hardness (TTH) diagrams in Fig. 19.

Hot ductility testing through Gleeble simulation can be used to tailor specific thermal–mechanical cycles to study the underlying mechanism to liquation and provide quantitative information on the weldability of materials. The traditional hot ductility type of test, Fig. 20 [82], can be used to gain insight into the HAZ and corresponding liquation mechanisms and to derive criteria for ranking materials against one another. The ductility signature measured in Fig. 20 shows a rapid ductility loss during on-heating curve, which is related to the onset of liquation in the material [42]. The on-cooling tests are executed by first heating a test specimen

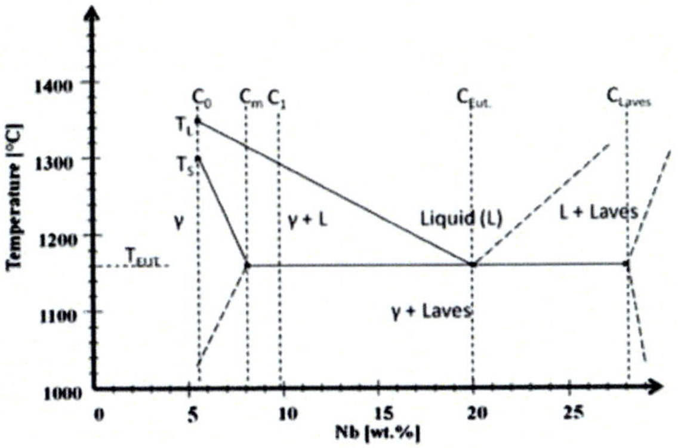

Fig. 18 Pseudo-binary phase diagram of ATI®718Plus™ [89]

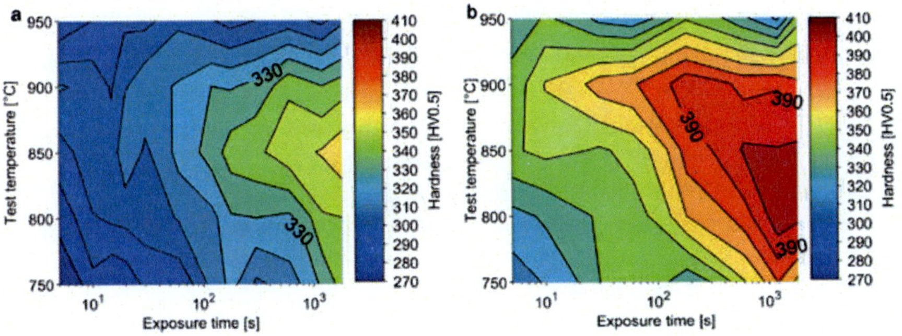

Fig. 19 TTH diagrams for Haynes® 282® (**a**) and Waspaloy (**b**) for exposure times up to 1800s [67]

to a peak temperature (T_{Peak}), derived from the nil-strength temperature (NST), typically 50 °C below. Other parameters and criteria such as the nil-ductility temperature (NDT), ductility recovery temperature (DRT), the ductility recovery rate (DRR), the ratio of ductility recovery (RDR), and the brittle temperature range can be used as measures for how susceptible a material would be towards HAZ liquation cracking. The method allows studying the liquation behavior in the HAZ during welding and can be used to compare different materials [50].

There is also a great potential in utilizing the tool to gain insight to SAC [60] where tests such as constant load rupture tests, stress relaxation tests, stress to fracture tests, and tests measuring ductility have been tried in the past [60]. Another frequently used approach follows the rationale of simulating the PWHT, and to evaluate how the ductility is affected in the temperature range where SAC occurs. This type of test approach is referred to as constant heating rate test (CHRT) [74]. Since the CHRT uses a constant heating rate to the different test temperatures, the effect of hardening reactions cannot be investigated. This has led to the development of a new approach

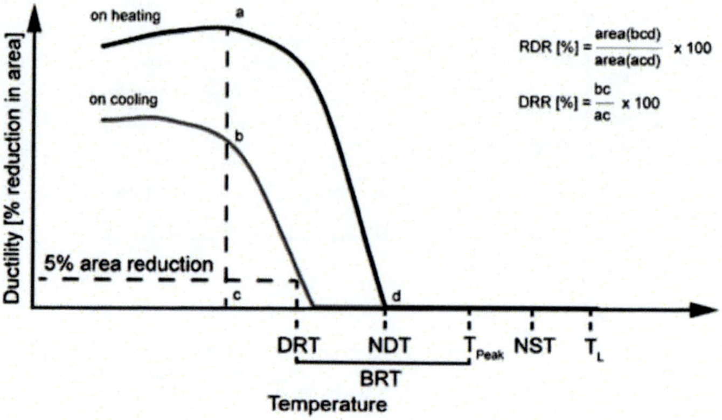

Fig. 20 Schematic hot ductility signature curve from the hot ductility test with corresponding test temperatures indicated in the graph [82]

that utilizes fast heating of 1000 °C/s and subsequent isothermal exposure to obtain microstructures with varying precipitation structure [66, 67, 82].

Variable restraint (Varestraint) weldability testing is a method that can be used to investigate susceptibility towards hot cracking [42, 85]. The test is carried out by applying an external load to a specimen at the same time as it is being welded. Varestraint testing provides a way to rank materials, conditions, welding processes, and parameters against one another. In Fig. 21, an example is presented regarding Varestraint testing of cast Alloy 718 in three different conditions, as cast, hot isostatic pressing (HIP)-1120 °C-4 h and HIP-1190 °C-4 h. [87]. The scatter in cracking response is unfortunately normally large when dealing with cast materials, however, the average data indicate the condition of HIP-1190 °C-4 h being the worst of the investigated conditions, which, despite the large scatter, have been also observed for other alloys [49, 56]. So, scattering in test data and other types of factors that add uncertainties to the understanding of cracking behavior is just another proof of the importance of not relying on single type of test method, one should utilize a wide portfolio of methods to successively build a comprehensive and reliable understanding of the underlying mechanisms.

Conclusions

Research within AM technology has increased exponentially since 2010, dating back to ~1985, however, the number of publications for "welding" and "superalloys" has had a more modest increase dating back to ~1950s. The top-ten countries as well as affiliations regarding the greatest number of publications within the field of "welding AND superalloys" coincide to a large extent to the top-ten countries in the field of "additive manufacturing AND superalloys" with few exceptions. Fabrication can

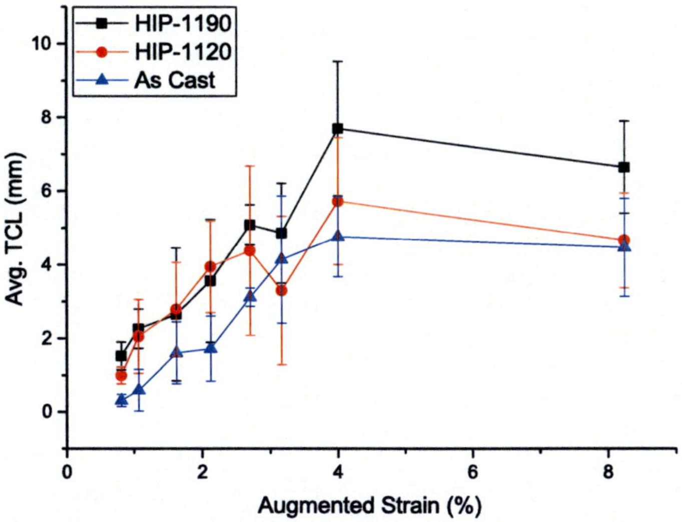

Fig. 21 HAZ liquation cracking response with standard deviations for Varestraint testing of cast Alloy 718 in the as cast, HIP-1120 °C-4 h and HIP-1190 °C-4 h conditions [87]

be seen as an enabler for more efficient aero-engines by reducing the weight of structural components. It is dependent on know-how of metallurgical aspects of different material forms, i.e., sheet, forging, and casting as well as their implications on welding. Additive manufacturing is an additional recent tool that can potentially further enhance the fabrication concept of hot structural components. The aerospace sector dominates the total market share (70%) with respect to superalloy usage. Weldability as a concept can be defined as a measure of the ease with which a metal or an alloy can be welded or joined without degradation to the weldment microstructure or properties during or after welding and for the duration of its intended service. In this context, susceptibility towards cracking during welding and post-weld heat treatment (PWHT) is of major importance. Weld cracking in Ni- and Ni–Fe-based precipitation hardening superalloys is predominantly related to hot cracking and strain age cracking (SAC), which are of high importance especially for hot structural applications in aero-engines. There has been significant development on gas metal arc welding processes such as pulsed mode of GMAW, cold metal transfer, keyhole, and high frequency, which provide deeper penetration and a more robustness, however, laser beam welding despite being significantly more capital intensive still possesses advantages regarding automation and control, not at least of high importance in directed energy deposition type of processes.

Ni- and Ni–Fe-based superalloys will always be at some risk of solidification cracking where welding procedure and parameters combined with the chemical composition determine the susceptibility of an alloy. High heat input, in general, aggravates cracking since it decreases the temperature gradient which in turn increases the region of solid–liquid where cracking can take place. Certain elements such as B, C, Si, Mo, and Nb are more important than others when it comes to the influence of solidification temperature range and formation of terminal eutectic at the end of solidification.

Heat-affected zone liquation cracking on the other hand is very much related to grain boundary liquation, where trace elements such as B, P, S, and Nb are important. The actual liquation mechanism is triggered by two different types: penetration mechanism and segregation mechanism. Trace elements such as B are of crucial importance in the segregation mechanism, where it segregates via non-equilibrium or equilibrium mechanisms, whereas the penetration mechanism primarily is claimed to be governed by eutectic melting, and constitutional liquation.

Strain age cracking (SAC) occurs in the solid state, without any prerequisite of liquid phase, and occurs when hardening takes place in the alloy. It is closely related to the kinetics of hardening in superalloys, and it is primarily of concern for γ' hardening superalloys. SAC occurs during the post-weld heat treatment (PWHT) and is consequently sometimes referred to as "PWHT cracking" or "Reheat cracking" because of the high welding stresses that develop at the same time that hardening occurs in γ' precipitation hardening Ni-based superalloys. The main factors that influence susceptibility towards SAC are related to induced strain, stress localization at grain boundaries, precipitation kinetics, and the ability for stress relaxation.

Weldability testing is necessary to account for the aforementioned concerns related to cracking, however, weldability, as such, is quite a broad concept that covers everything from actual welding to service performance and should therefore be treated with care to avoid misunderstanding of its meaning. A proposed procedure in assessing susceptibility toward cracking is to utilize a variety of tools, where each one of them provides insightful knowledge on the specific material's behavior. The proposed procedure involves systematic work of metallography, thermal analysis, hardness testing in combination with thermal treatments, hot ductility testing, welding using different processes, as well as Varestraint testing.

Acknowledgements I sincerely thank Prof. Olanrewaju Ojo at the University of Manitoba for reviewing and providing fruitful input to the paper.

Appendix 1

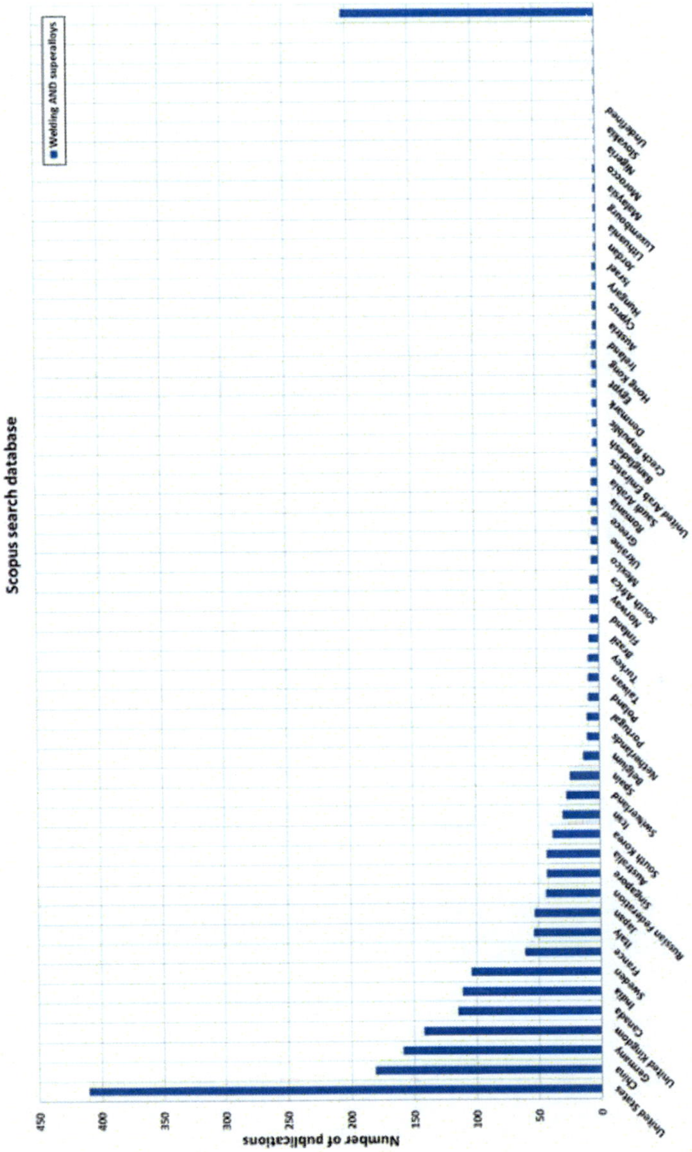
Scopus search database
Welding AND superalloys
Number of publications
Country

Appendix 2

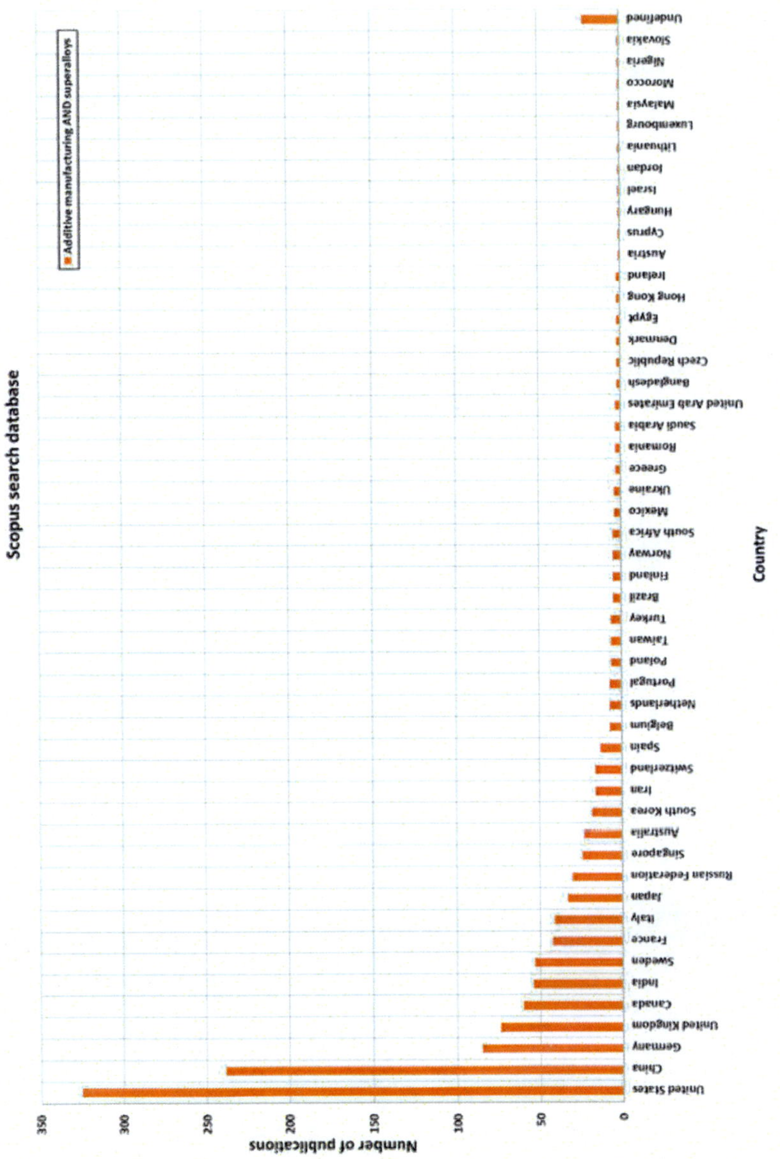
Scopus search database
Additive manufacturing AND superalloys
Number of publications
0
50
100
150
200
250
300
350
Country
United States
China
Germany
United Kingdom
Canada
India
Sweden
France
Italy
Japan
Russian Federation
Singapore
Australia
South Korea
Iran
Switzerland
Spain
Belgium
Netherlands
Portugal
Poland
Taiwan
Turkey
Brazil
Finland
Norway
South Africa
Mexico
Ukraine
Greece
Romania
Saudi Arabia
United Arab Emirates
Bangladesh
Czech Republic
Denmark
Egypt
Hong Kong
Ireland
Austria
Cyprus
Hungary
Israel
Jordan
Lithuania
Luxembourg
Malaysia
Morocco
Nigeria
Slovakia
Undefined

Appendix 3

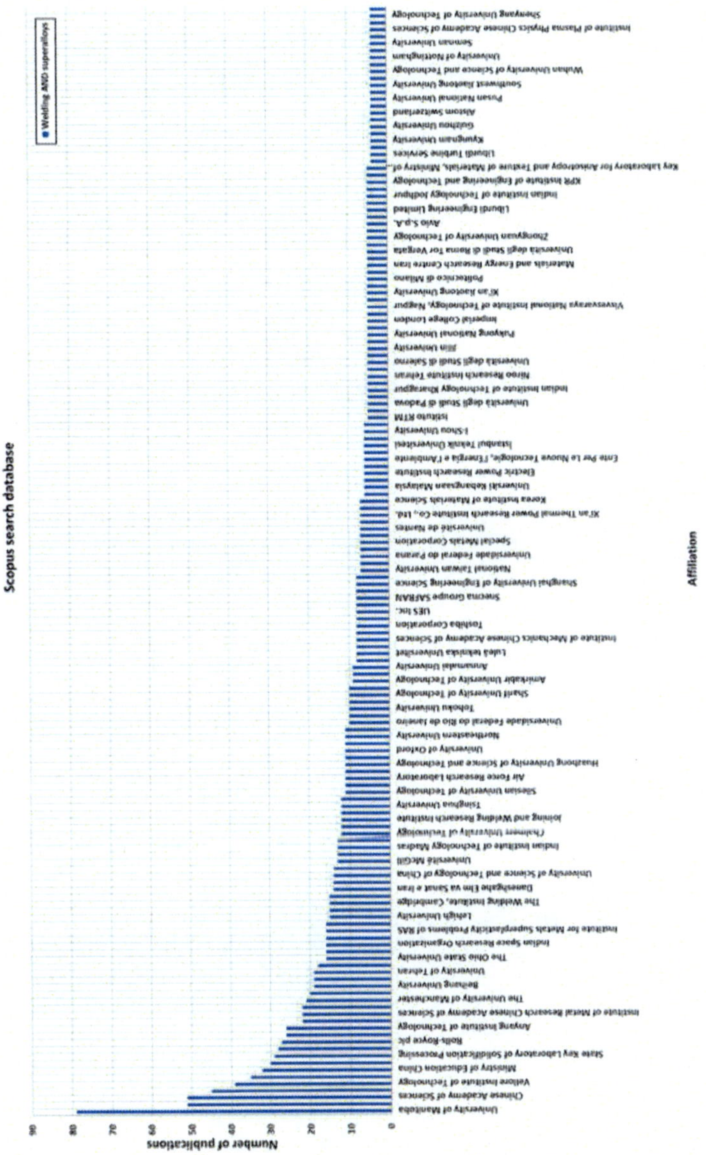
Scopus search database
Welding AND superalloys
Number of publications
Affiliation

Appendix 4

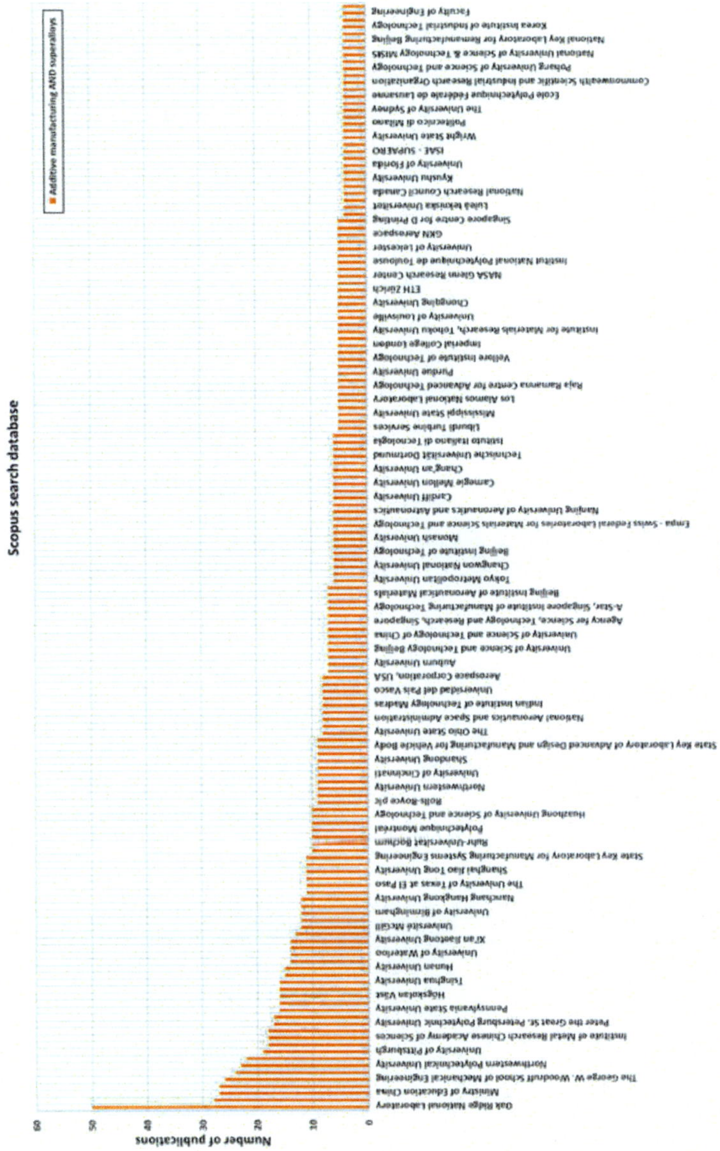
Scopus search database
Additive manufacturing AND superalloys
Number of publications
0
10
20
30
40
50
60
Affiliation
Oak Ridge National Laboratory
Ministry of Education China
The George W. Woodruff School of Mechanical Engineering
Northwestern Polytechnical University
University of Pittsburgh
Institute of Metal Research Chinese Academy of Sciences
Peter the Great St. Petersburg Polytechnic University
Pennsylvania State University
Högskolan Väst
Tsinghua University
Hunan University
University of Waterloo
Xi'an Jiaotong University
Université McGill
University of Birmingham
Nanchang Hangkong University
The University of Texas at El Paso
Shanghai Jiao Tong University
State Key Laboratory for Manufacturing Systems Engineering
Ruhr-Universitat Bochum
Polytechnique Montreal
Huazhong University of Science and Technology
Rolls-Royce plc
Northwestern University
University of Cincinnati
Shandong University
State Key Laboratory of Advanced Design and Manufacturing for Vehicle Body
The Ohio State University
National Aeronautics and Space Administration
Indian Institute of Technology Madras
Universidad del Pais Vasco
Aerospace Corporation, USA
Auburn University
University of Science and Technology Beijing
University of Science and Technology of China
Agency for Science, Technology and Research, Singapore
A-Star, Singapore Institute of Manufacturing Technology
Beijing Institute of Aeronautical Materials
Tokyo Metropolitan University
Changwon National University
Beijing Institute of Technology
Monash University
Empa - Swiss Federal Laboratories for Materials Science and Technology
Nanjing University of Aeronautics and Astronautics
Cardiff University
Carnegie Mellon University
Chang'an University
Technische Universität Dortmund
Istituto Italiano di Tecnologia
Ubundi Turbine Services
Mississippi State University
Los Alamos National Laboratory
Raja Ramanna Centre for Advanced Technology
Purdue University
Vellore Institute of Technology
Imperial College London
Institute for Materials Research, Tohoku University
University of Louisville
Chongqing University
ETH Zürich
NASA Glenn Research Center
Institut National Polytechnique de Toulouse
University of Leicester
GKN Aerospace
Singapore Centre for 3D Printing
Luleå tekniska Universitet
National Research Council Canada
Kyushu University
University of Florida
ISAE - SUPAERO
Wright State University
Politecnico di Milano
The University of Sydney
Ecole Polytechnique Fédérale de Lausanne
Commonwealth Scientific and Industrial Research Organization
Pohang University of Science and Technology
National University of Science & Technology MISIS
National Key Laboratory for Remanufacturing Beijing
Korea Institute of Industrial Technology
Faculty of Engineering

Appendix 5

Scopus search database
Author
Number of publications

Appendix 6

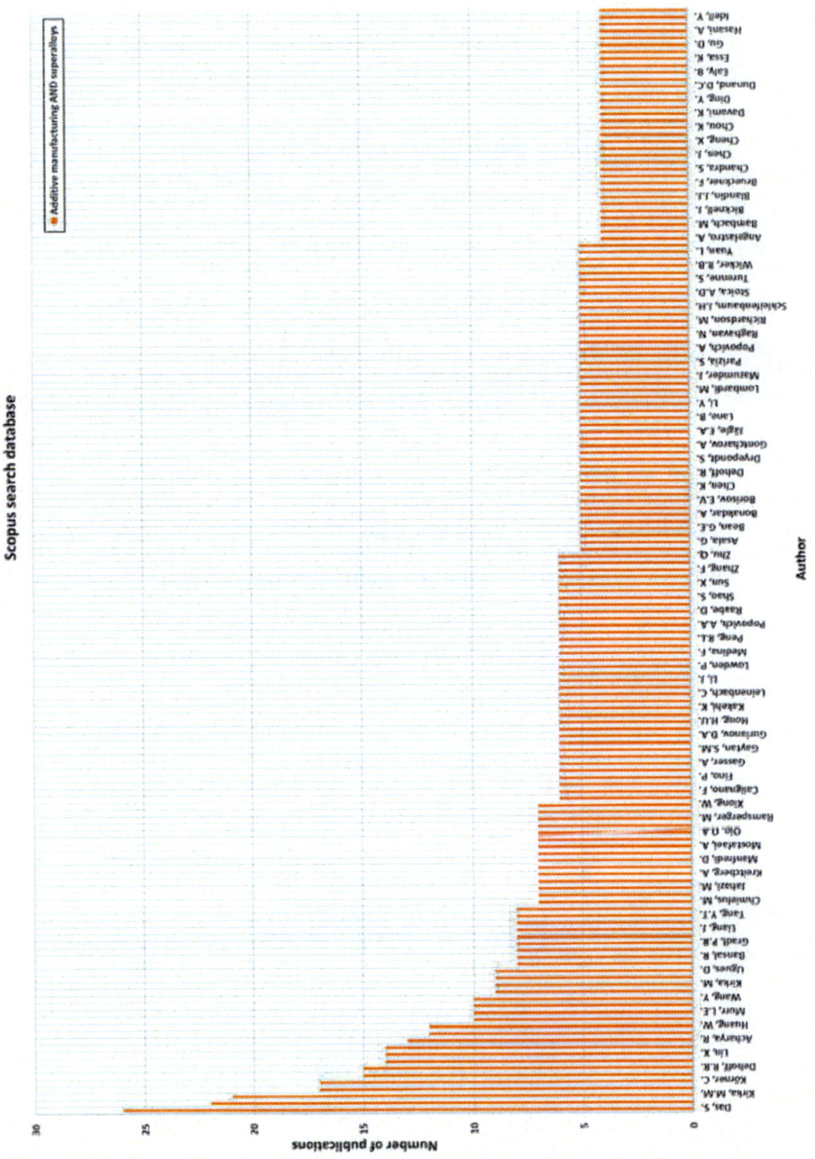
Scopus search database
Additive manufacturing AND superalloys
Number of publications
Author

References

1. Scopus.com, (2022). Analyze Results - Number of publications versus years "welding". Available at: http://www.scopus.com [Accessed: 2022–09–17].
2. Scopus.com, (2022). Analyze Results - Number of publications versus years "additive manufacturing". Available at: http://www.scopus.com [Accessed: 2022–09–17].
3. Scopus.com, (2022). Analyze Results - Number of publications versus years "superalloys". Available at: http://www.scopus.com [Accessed: 2022–09–17].
4. Scopus.com, (2022). Analyze Results - Number of publications versus years "welding AND superalloys". Available at: http://www.scopus.com [Accessed: 20e22–09–17].
5. Scopus.com, (2022). Analyze Results - Number of publications versus years "additive manufacturing AND superalloys". Available at: http://www.scopus.com [Accessed: 2022–09–17].
6. Scopus.com, (2022). Analyze Results - Number of publications versus country "welding AND superalloys". Available at: http://www.scopus.com' [Accessed: 2022–09–17].
7. Scopus.com, (2022). Analyze Results - Number of publications versus country "additive manufacturing AND superalloys". Available at: http://www.scopus.com [Accessed: 2022–09–17].
8. Scopus.com, (2022). Analyze Results - Number of publications versus affiliation "welding AND superalloys". Available at: http://www.scopus.com [Accessed: 2022–09–17].
9. Scopus.com, (2022). Analyze Results - Number of publications versus affiliation "additive manufacturing AND superalloys". Available at: http://www.scopus.com [Accessed: 2022–09–17].
10. Scopus.com, (2022). Analyze Results - Number of publications versus author for "welding AND superalloys". Available at: http://www.scopus.com [Accessed: 2022–09–17].
11. Scopus.com, (2022). Analyze Results - Number of publications versus author for "additive manufacturing AND superalloys". Available at: http://www.scopus.com [Accessed: 2022–09–17].
12. S. Sravan Sashank, S. Rajakumar, R. Karthikeyan, and D. S. Nagaraju, Weldability, Mechanical Properties and Microstructure of Nickel Based Super Alloys: a review, E3S Web of Conferences 184, 01040 (2020), https://doi.org/10.1051/e3sconf/202018401040.
13. J.-M. Drezet and D. Allehaux, "Application of the rappaz-drezet-gremaud hot tearing criterion to welding of aluminium alloys," in Hot cracking phenomena in welds II, pp. 27–45, Springer, 2008.
14. M. Rappaz, J.-M. Drezet, and M. Gremaud, "A new hot-tearing criterion," Metallurgical and materials transactions A, vol. 30, no. 2, pp. 449–455, 1999.
15. C. Bordreuil and A. Niel, "Modelling of hot cracking in welding with a cellular automaton combined with an intergranular fluid flow model," Computational Materials Science, vol. 82, pp. 442–450, 2014.
16. H. Z. Rajani and A. Phillion, "3d multi-scale multi-physics modelling of hot cracking in welding," Materials & Design, vol. 144, pp. 45–54, 2018.
17. T. Sonar, V. Balasubramanian, S. Malarvizhi, T. Venkateswaran, D. Sivakumar, An overview on welding of Inconel 718 alloy - Effect of welding processes on microstructural evolution and mechanical properties of joints, Materials Characterization 174 (2021), https://doi.org/10.1016/j.matchar.2021.110997.
18. A. Mashhuriazar, H. Omidvar, C. H. Gur, and Z. Sajuri, Effect of Welding Parameters on the Liquation Cracking Behavior of High-Chromium Ni-Based Superalloy, Journal of Materials Engineering and Performance (2020) 29:7843–7852, https://doi.org/10.1007/s11665-020-05292-w.
19. T. Sonar, V. Balasubramanian, S. Malarvizhi, T. Venkateswaran, D. Sivakumar, Microstructural characteristics and tensile properties of gas tungsten constricted arc (GTCA) welded Inconel 718 superalloy sheets for gas turbine engine components, Mater. Test 67 (2020) 1099–1108, https://doi.org/10.3139/120.111576.
20. G.D. Janaki Ram, A. Venugopal Reddy, K. Prasad Rao, G. Madhusudhan Reddy, Control of laves phase in Inconel 718 GTA welds with current pulsing, Sci. Technol.Weld. Join. 9 (5) (2004) 390–398, https://doi.org/10.1179/136217104225021788.

21. K. Sivaprasad, S.G.S. Raman, P. Mastanaiah, G. Madhusudhan Reddy, Influence of magnetic arc oscillation and current pulsing on microstructure and high temperature tensile strength of alloy 718 TIG weldments, Mater. Sci. Eng. A 428 (2006) 327–331, https://doi.org/10.1016/j.msea.2006.05.046.
22. G.D. Janaki Ram, A. Venugopal Reddy, K. Prasad Rao, G. Madhusudhan Reddy, J. K. Sarin Sundar, Microstructure and tensile properties of Inconel 718 pulsed Nd-YAG laser welds, J. Mater. Proc. Technol. 167 (2005) 73–82, https://doi.org/10.1016/j.jmatprotec.2004.09.081.
23. G.D. Janaki Ram, A. Venugopal Reddy, K. Prasad Rao, G. Madhusudhan Reddy, High temperature mechanical properties of Inconel 718 pulsed Nd–YAG laser welds, Mater. High Temp. 23 (2006) 29–37, https://doi.org/10.1179/mht.2006.003.
24. L. Kuzmikova, N. Larkin, Z. Pan, M. Callaghan, H. Li, J. Norrish, Investigation into feasibility of hybrid laser-GMAW process for welding high strength quenched and tempered steel, (2012).
25. A. Ariaseta, D. Pick, J. Andersson, and O. Ojo, Keyhole TIG welding of new Co-lean nickel-based superalloy G27, Proceedings of the 10th International Symposium on Superalloy 718 and Derivatives, Springer, 2023.
26. S. Cui, Z. Liu, Y. Fang, Z. Luo, S.M. Manladan, S. Yi, Keyhole process in K-TIG welding on 4 mm thick 304 stainless steel, Journal of Materials Processing Technology 243 (2017) 217–228, 0924-0136.
27. Z. Liu, Y. Fang, S. Cui, Z. Luo, W. Liu, Z. Liu, Q. Jiang, S. Yi, Stable keyhole welding process with K-TIG, Journal of Materials Processing Technology 238 (2016) 65–72, 0924-0136.
28. Z. Fei, Z. Pan, D. Cuiuri, H. Li, B. Wu, D. Ding, L. Su, A.A. Gazder, Investigation into the viability of K-TIG for joining armour grade quenched and tempered steel, Journal of Manufacturing Processes 32 (2018) 482–493 1526-6125.
29. A. Ariaseta, N. Sadeghinia, and J. Andersson, Influence of keyhole TIG welding parameters on weld geometry of newly-developed superalloy VDM Alloy 780, Proceedings of the 33rd congress of the international council of the aeronautical sciences, Stockholm Sweden, 4–9 September 2022.
30. S. A. David, J. A. Siefert and Z. Feng, Welding and weldability of candidate ferritic alloys for future advanced ultrasupercritical fossil power plants, Science of Technology of Welding and Joining, 2013, 18 (8), pp. 631–651
31. J. Andersson, Review of Weldability of Precipitation Hardening Ni-and Fe-Ni-Based Superalloys, Proceedings of the 9th International Symposium on Superalloy 718 & Derivatives: Energy, Aerospace, and Industrial Applications, Springer, 2018, pp. 899–916.
32. Welding of Metallic Materials: Methods, Metallurgy, and Performance, edited by F. Khoshnaw, ISBN: 9780323905527.
33. N Coniglio and CE Cross. Initiation and growth mechanisms for weld solidification cracking. International Materials Reviews 58.7 (2013), pp. 375–397.
34. DG Eskin, L Katgerman, et al. Mechanical properties in the semi-solid state and hot tearing of aluminium alloys. Progress in materials science 49.5 (2004), pp. 629–711.
35. Jiangfeng Song, Fusheng Pan, Bin Jiang, Andrej Atrens, Ming-Xing Zhang, and Yun Lu. A review on hot tearing of magnesium alloys. Journal of Magnesium and Alloys 4.3 (2016), pp. 151–172.
36. DG Eskin and L Katgerman. A quest for a new hot tearing criterion. Metallurgical and Materials Transactions A 38.7 (2007), pp. 1511–1519.
37. W.I. Pumphrey, P.H. Jennings, A consideration of the nature of brittleness at temperatures above the solidus in castings and welds in aluminium alloys, Journal of the Institute of Metals, 1948, 75(4), 235–256.
38. W.S. Pellini, Strain theory of hot tearing, Foundry, 1952, 80 (11), 125–133.
39. J. Borland, Generalized theory of super-solidus cracking in welds (and castings), British Welding Journal 1960, 7(8), 508–512.
40. F. Matsuda, H. Nakagawa, K. Sorada, Dynamic Observation of Solidification and Solidification Cracking during Welding with Optical Microscope (I): Solidification Front and Behavior of Cracking, Materials, Metallurgy & Weldability, Transactions of JWRI, 1982, 11(2), 67–77.
41. S. Kou, Welding metallurgy, 2nd ed. Hoboken, N.J: Wiley-Interscience, 2003.

42. J.N. Dupont, J.C. Lippold, S.D. Kiser, Welding metallurgy and weldability of nickel-base alloys, Hoboken, N.J: John Wiley & Sons, 2009.
43. N. Saito, N. Komai and Y. Takei: 'Fabrication trials of Ni-based alloys for advanced USC boiler application' 'Proc. 7th Int. Conf. on 'Advances in materials technology for fossil power plants', Waikoloa, HI, USA, October 2013, ASM International', Paper 156 190–201; 2013a.
44. S. C. Ernst: 'Weldability studies of Haynes 230 alloy', Weld. J., 1994, 73, (4), 80-s–89-s.
45. J. A. Siefert, J. P. Shingledecker, J. N. DuPont & S. A. David (2016) Weldability and weld performance of candidate nickel based superalloys for advanced ultrasupercritical fossil power plants Part II: weldability and cross-weld creep performance, Science and Technology of Welding and Joining, 21:5, 397–427, https://doi.org/10.1080/13621718.2016.1143708.
46. L. Karlsson, H. Nordén, H. Odelius, Overview no. 63 Non-equilibrium grain boundary segregation of boron in austenitic stainless steel—I. Large scale segregation behaviour, Acta Metallurgica, 1988, 36(1), 1–12.
47. L. Karlsson, H. Norden, Grain boundary segregation of boron, An experimental and theoretical study, Le Journal de Physique Colloques, 1986, 47(C7), C7-257.
48. L. Karlsson and H. Nordén, Overview no. 63, Non-equilibrium grain boundary segregation of boron in austenitic stainless steel-II, Fine scale segregation behaviour, Acta Metallurgica, 1988, 36(1), 13–24.
49. S. Singh, and J. Andersson, Heat-Affected-Zone Liquation Cracking in Welded Cast Haynes® 282®, Metals, 10, 29, https://doi.org/10.3390/met10010029, 2020.
50. J. Andersson, Weldability of precipitation hardening superalloys–influence of microstructure. PhD Thesis, Chalmers University of Technology, Gothenburg, Sweden, 2011.
51. J.J. Pepe, W.F. Savage, Effects of constitutional liquation in 18-Ni maraging steel weldments (Microsegregation and grain boundary liquation in heat affected zone of 18-Ni maraging steel welds), Welding Journal, Research Supplement, 1967, vol. 46.
52. M. C. Chaturvedi, Liquation Cracking in Heat Affected Zone in Ni Superalloy Welds, Materials Science Forum, 2007, vol. 546–549, 1163–1170.
53. J.C. Lippold, Welding metallurgy and weldability. Hoboken, New Jersey: John Wiley & Sons Inc, 2015.
54. S. Singh, F. Hanning, J. Andersson, Influence of hot isostatic pressing on hot ductility of cast Alloy 718: effect of Niobium and minor elements on liquation mechanisms, Metallurgical and Materials Transactions A, 2020, https://doi.org/10.1007/s11661-020-06004-8.
55. S. Singh, F. Hanning, J. Andersson, Influence of homogenisation treatments on the hot ductility of Cast ATI® 718Plus®: Effect of Niobium and minor elements on liquation characteristics, Materials Science and Engineering A, 2021, Volume 799, 140151.
56. F. Hanning, A. K. Khan, J. Andersson, and O. Ojo, Advanced microstructural characterisation of cast ATI 718Plus®—effect of homogenisation heat treatments on secondary phases and repair welding behaviour, Welding in the World, https://doi.org/10.1007/s40194-020-00851-0.
57. B. Radhakrishnan and R.G. Thompson, A model for the formation and solidification of grain boundary liquid in the heat-affected zone (HAZ) of welds, Metall. Trans. A, vol. 23, no. 6, 1992, pp. 1783–1799.
58. H. L. Eiselstein. Metallurgy of a Columbium-Hardened Nickel-Chromium-Iron Alloy. In: Committee A-10, editor. Adv. Technol. Stainl. Steels Relat. Alloys, 100 Barr Harbor Drive, PO Box C700, West Conshohocken, PA 19428–2959: ASTM International; 1965, pp. 62–79.
59. C. T. Sims, N. S. Stoloff, and W. C. Hagel, Superalloys II. New York: Wiley, 1987.
60. F. Hanning and J. Andersson, A Review of Strain Age Cracking in Nickel Based Superalloys, in Conference Proceedings of the 7th International Swedish Production Symposium, Lund, SE, 2016.
61. M. Prager and C. S. Shira. Welding of Precipitation-Hardening Nickel-Base Alloys. WRC Bull, 128, pp. 1–55, 1968.
62. J. E. Franklin and W. F. Savage. Stress Relaxation and Strain-Age Cracking in Rene 41 Weldments. Weld J, 53, pp. 380–387, 1974.
63. T. F. Berry and W. P. Hughes. A Study of the Strain-Age Cracking Characteristics in Welded René 41 - Phase II. Weld J, 48, pp. 505–513, 1969.

64. E. G. Thompson, S. Nunez, and M. Prager. Practical Solutions to Strain-Age Cracking of René 41. Weld J, 47, pp. 299–313, 1968.
65. D. S. Duvall and W. A. Owczarski. Heat Treatments for Improving the Weldability and Formability of Udimet 700. Weld J,50, pp. 401–409, 1971.
66. F. Hanning, A. K. Khan, J. Steffenburg-Nordenström, O. Ojo, and J. Andersson, Investigation of the Effect of Short Exposure in the Temperature Range of 750–950°C on the Ductility of Haynes® 282® by Advanced Microstructural Characterization, Metals 2019, 9, 1357, https://doi.org/10.3390/met9121357.
67. F. Hanning, A. K. Khan, O. Ojo, and J. Andersson, Effect of short-term isothermal exposure on the ductility signature of Waspaloy in the temperature range of 750–950°C – a comparison with Haynes® 282®, Proceedings of the 10th International Symposium on Superalloy 718 and Derivatives, Springer, 2023.
68. W. P. Hughes and T. F. Berry. A Study of the Strain-Age Cracking Characteristics in Welded René 41 - Phase I. Weld J 1967;46:361s–370s.
69. D. S. Duvall and W. A. Owczarski. Studies of Postweld Heat-Treatment Cracking in Nickel-Base Alloys. Weld J, 48, pp. 10–22, 1969.
70. Dix AW, Savage WF (1971) Factors influencing strain-age cracking in Inconel X-750. Weld J 50:247s–252s.
71. Thamburaj R, Goldak JA, Wallace W (1979) The Influence of Chemical Composition in Post-Weld Heat Treatment Cracking in René 41. SAMPE Q 4:6–12.
72. Prager M, Sines G (1970) A Mechanism for Cracking During Postwelding Heat Treatment of Nickel-Base Alloys. WRC Bull 150:24–32.
73. Cao W-D, Kennedy RL (1997) Effect and Mechanism of Phosphorous and Boron on Creep Deformation of Alloy 718. In: Superalloys 718, 625, 706 and Various Derivatives. pp 511–520.
74. Carlton JB, Prager M (1970) Variables Influencing the Strain-Age Cracking and Mechanical Properties of René 41 and Related Alloys. WRC Bull 150:13–23.
75. Wu KC, Herfert RE (1967) Microstructural Studies of René 41 Simulated Weld Heat-Affected Zones. Weld J 46:32s–38s.
76. W. Schwenk, Trabold AF (1963) Weldability of René 41. Weld J 42:460s–645s.
77. J. Andersson (2014) Weldability of Ni-Based Superalloys. In: Proceedings of the 8th International Symposium on Superalloy 718 and Derivatives. The Minerals, Metals & Materials Society, pp 249–262.
78. Liu WC, Xiao FR, Yao M, et al (1997) Relationship between the lattice constant of ϒ phase and the content of δ phase, γ″ and γ′ phases in inconel 718. Scr Mater 37:59–64, https://doi.org/10.1016/S1359-6462(97)00064-X.
79. Tiley J, Srinivasan R, Banerjee R, et al (2009) Application of X-ray and neutron diffraction to determine lattice parameters and precipitate volume fractions in low misfit nickel base superalloys. Mater Sci Technol 25:1369–1374, https://doi.org/10.1179/174328409X399010.
80. Whitmore L, Ahmadi MR, Stockinger M, et al (2014) Microstructural investigation of thermally aged nickel-based superalloy 718Plus. Mater Sci Eng A 594:253–259, https://doi.org/10.1016/j.msea.2013.11.037.
81. Dirand L, Cormier J, Jacques A, et al (2013) Measurement of the effective γ/γ′ lattice mismatch during high temperature creep of Ni-based single crystal superalloy. Mater Charact 77:32–46, https://doi.org/10.1016/j.matchar.2012.12.003.
82. F. Hanning, 'Weld Cracking of Precipitation Hardening Ni-based Superalloys - Investigation of repair welding characteristics and susceptibility towards strain age cracking', PhD thesis, Chalmers University of Technology, Gothenburg, Sweden, 2020.
83. S. Singh, 'Weldability of Cast Superalloys - Effect of homogenization heat treatments on hot cracking susceptibility of cast Alloy 718, ATI® 718Plus®, and Haynes® 282®', PhD thesis, Chalmers University of Technology, Gothenburg, Sweden, 2020.
84. T. Raza, 'Process Understanding and Weldability of Laser-Powder Bed Fusion Manufactured Alloy 718', PhD thesis, University West, Trollhättan, Sweden, 2020.
85. J. Andersson, J. Jacobsson, C. Lundin, A Historical Perspective on Varestraint Testing and the Importance of Testing Parameters, Cracking Phenomena in Welds IV. Springer International Publishing, 2016, 3–23.

86. J. Jacobsson, Weldability and Testing Methodology in Precipitation Hardening Superalloys, Licentiate thesis, Chalmers University of Technology, Gothenburg, Sweden, 2017.
87. S. Singh, J. Andersson, Hot cracking in cast alloy 718, Science and Technology of Welding and Joining, 23:7, 568–574, https://doi.org/10.1080/13621718.2018.1429238, 2018.
88. L. Karlsson, E. L. Bergquist, S. Rigdal, and N. Thalberg, 'Evaluating Hot Cracking Susceptibility of Ni-Base SAW Consumables for Welding of 9% Ni Steel', Hot Cracking Phenomena in Welds II, ed. T. Bollinghaus, H. Herold, C. E. Cross, and J. C. Lippold, Springer-Verlag Berlin Heidelberg, 2008, pp. 329–347.
89. J. Andersson, S. Raza, A. Eliasson, and K. B. Surredi; "Solidification of Alloy 718, ATI 718Plus and Waspaloy", Proceeding of the 8th International Symposium on Superalloy 718 and Derivatives, TMS (The Minerals, Metals & Materials Society); September 28–October 1, 2014, Pittsburgh, Pennsylvania, USA, pp 181–195.

Part VIII
Welding, Deposition, Manufacturing, and Repair

Tensile Properties of Inconel 718 Produced by LMD-Wire

J. Cormier, S. Cabeza, G. Burlot, R. Bordas, M. Bordas-Czaplicki, F. Machado Alves da Fonseca, S. Polenz, F. Marquardt, E. Lopez, and P. Villechaise

Abstract The anisotropy in tensile properties of Wire Laser Metal Deposited Inconel 718 (LMD-w) has been investigated from room temperature up to 750 °C at a strain rate of 5.0 10^{-4} s^{-1}. These properties have been investigated along, at 45° and perpendicular to the building direction. Moreover, different heat treatments have been used: as-built, solution heat treated to dissolve Laves phases, solution treated + aged to trigger γ'/γ'' precipitation and direct-aged. According to this extensive characterization of tensile properties, complemented by SEM and EBSD characterizations, it is shown that, whatever the temperature, Yield stress and tensile resistance have a very weak anisotropy and that tensile properties are mostly dependent to the prior heat treatment state. The anisotropy is mostly observed on elastic properties, due to a pronounced crystallographic texture inherited from the directional thermal gradient during the building process. Moreover, Laves phases do not seem to have a strong impact on tensile properties for this coarse grain material. Tensile strength in such an LMD-processed Inconel 718 is mostly controlled by the γ'/γ'' precipitation and stored "processing" dislocations. A loss of tensile ductility has been evidenced at 750 °C, due to grain boundary oxidation.

Keywords Inconel 718 · LMD-w · Microstructure · Tensile properties · Laves phases · Anisotropy

---

J. Cormier (✉) · G. Burlot · R. Bordas · M. Bordas-Czaplicki · F. Machado Alves da Fonseca · P. Villechaise
Institut Pprime, UPR CNRS 3346, Physics and Mechanics of Materials Department, ISAE-ENSMA, 1 Avenue Clément Ader, BP 40109, 86961 Futuroscope-Chasseneuil, France
e-mail: jonathan.cormier@ensma.fr

S. Cabeza
Institut Laüe Langevin (ILL), Diffraction Unit, 71 Avenue des Martyrs, 38042 Grenoble, France

S. Polenz · F. Marquardt · E. Lopez
Additive Manufacturing Division, Fraunhofer IWS, Winterbergstrasse 28, 01277 Dresden, Germany

E. A. Ott et al. (eds.), *Proceedings of the 10th International Symposium on Superalloy 718 and Derivatives*, The Minerals, Metals & Materials Series,
https://doi.org/10.1007/978-3-031-27447-3_42

Introduction

Inconel 718 is the most widely used superalloy thanks to an attractive cost in addition to both high mechanical strength at high temperature (of up to 650–680 °C) and good corrosion/oxidation resistance [1–3]. This alloy finds applications in nuclear power plants [4], oil & gas [5], turbochargers for formula 1 racing cars, industrial gas turbines for power generation and aero-engines [6–11]. It is typically used in this last field for the manufacturing of high-pressure turbine and compressor disks, low-pressure turbine rings, casings, seal rings, fasteners as well as some airfoils and other static parts. Another reason for the extensive use of Inconel 718 is its good processability and weldability allowing for repair operations to be performed more easily compared to γ'-strengthened alloys [12, 13], like, e.g., Waspaloy, Rene 65, Udimet 720Li, Inconel 738LC, Mar-M247.

As original engine makers are now selling their gas turbines more and more with Rate-Per-Flight-Hour contracts [14], including maintenance at fixed (extended) period of time, the repair and refurbishment of components have become major issues. Maintenance, Repair and Overhaul (MRO) services have a main goal in repairing more engines, more components, faster and better than before, aiming to extend engine life and to reduce the cost associated with the entire replacement of the damaged part. Within this context, Metal Additive Manufacturing (AM) is a very promising processing route, especially using Direct Energy Deposition (DED) techniques [15, 16], to repair components such as blade tips, seal rings, shrouds and shroud fences and many other precise operations, more specifically considering Inconel 718 alloy thanks to its wide range of applications and its good weldability.

In this article, a special focus is paid to the tensile properties of LMD-w processed alloy 718 in a wide temperature range (room temperature up to 750 °C) and with different post-processing heat treatments to better understand how such a process can affect at tensile properties. In fact, if tensile properties of Inconel 718 processed by different AM techniques have already been widely studied over the past five years [17, 18], very few of them have focused on LMD-w [18, 19], a technique known to generate quite pronounced dendritic segregations in the as-processed state as well as a strong texture. The work presented in this article is part of an on-going effort between Faunhofer IWS in Dresden (Germany), Institut Laüe Langevin in Grenoble (France) and Institut Pprime in Poitiers (France) aimed at better characterizing the development of residual stress during the LMD-w process of Inconel 718 and during post-processing heat treatment and how building conditions and residual stresses affect at the subsequent mechanical properties such as tensile and relaxation ones.

Experimental Procedures

Material

An Inconel 718 (IN718) feedstock wire material with a diameter of 1 mm has been used in this study. The wire has been provided by VDM Germany (Nicrofer 5219 Nb) and its chemical composition before printing is presented in Table 1. The elemental composition of IN718 feedstock wire was measured via inductively coupled plasma atomic emission spectroscopy (ICP-OES) technique using a SPECTRO ARCOS ICP-OES spectrometer. As a substrate for the wire-DED process, non-textured forged Inconel 718 baseplates were used. No specific consideration to the baseplates will be given in this article as no specific focus will be placed on the interface between the printed material and the substrate.

LMD-w manufacturing technique has been employed to fabricate IN718 walls at Fraunhofer institute for this study with a typical dimension of 15 × 100 × 50 mm^3 (see Fig. 1). Laserline LDF 4000–30 diode laser with coaxial laser wire cladding head (COAXwire) [20] has been used with a laser power of 900 W for the 5 first layers, and then an 800 W power for further layers. Such a decrease in laser power has been used to avoid wire stuck to the nozzle due to the high thermal input. A wire feeding rate of 1250 mm/min and a 25 L/min Argon protective flow rate have been used to print these walls. As a printing strategy, meander scanning strategy with ±45° rotation on each layer have been selected. Building Direction (also called "Vertical" hereafter), Wall Axis (also called "Horizontal" hereafter) and Through Thickness directions are defined in Fig. 1.

Heat Treatment

In an aim to dissolve Laves phases and to further trigger the precipitation of $\gamma' + \gamma''$ strengthening particles, a Nabertherm 1800 furnace has been used. Temperature has been controlled with an S-type thermocouple attached to each specimen, with a ±2 °C accuracy. Solution heat treatments from 1000 °C up to 1250 °C have been performed to dissolve as much as possible Laves phases and to homogenize the chemical composition across the dendritic structure of the alloy. Water quench has been applied to specimens at the end of each solution treatment trial. Once a given optimal solution treatment has been selected, further "classical" aeronautic aging heat treatment (i.e., 8 h at 720 °C + 8 h at 620 °C, cooling at 50 °C/h in between

Table 1 Chemical composition of the Inconel 718 wire (wt %) used in this study

Fe	Ni	Cr	Nb	Mo	Ti	Al	Co	Si	Mn	Cu
17.5	53.5	18.6	5.1	2.9	1.0	0.50	Max 1.0	Max 0.35	Max 0.35	Max 0.30

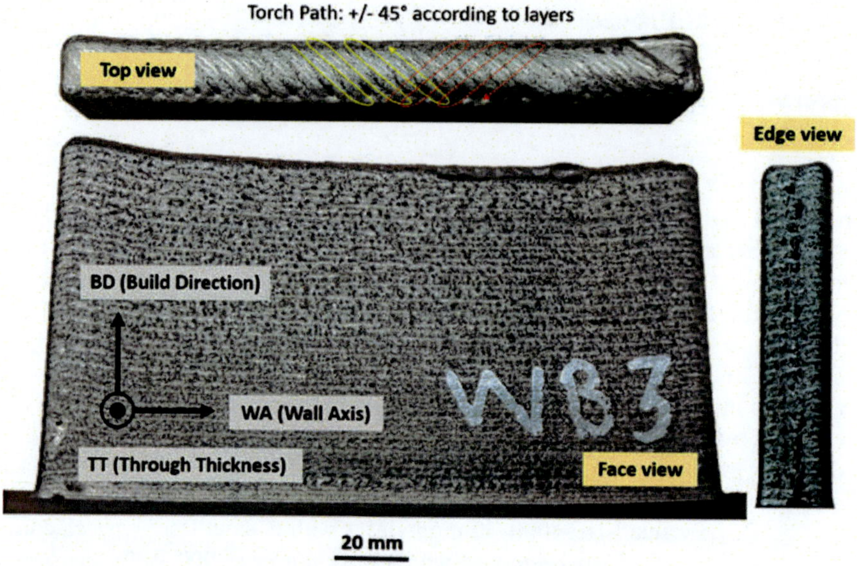

Fig. 1 Photograph of an IN718 LMD-w process wall labelled with axis directions

both steps) has been applied to the specimens to provide a $\gamma' + \gamma''$ precipitation. It is worth noticing that such heat treatments have been performed on cylinders before final machining of tensile specimens.

Metallographic Examination

Observations of Laves phases has been performed on specimens mechanically polished up to SiC 4000 paper, and then polished to 1 μm with diamond suspension. For a better observation of the microstructure, the samples received electrolytic polishing using A3 electrolyte from Struers for 8 s at 45 V and 4 °C.

Scanning Electron Microscope (SEM) observations were all performed using a field emission gun (FEG) SEM JEOL 7000F with an accelerating voltage of 25 kV. Some fractographic observations after tensile tests were also performed using the same microscope. The area fraction and size of Laves phases have been quantified using the Visilog software and in-house developed algorithms [21].

Finally, large electron back-scattered diffraction (EBSD) maps (340 × 1000 μm^2) have been performed using a JEOL 6100 SEM operating at 20 kV. EBSD data have all been post-processed and analyzed using the OIM software. A scanning step of 0.5 μm with hexagonal grid pattern has been used for such EBSD characterizations.

Mechanical Testing

Tensile samples were extracted from the walls using electric discharge machining, both vertically (loading direction BD) and horizontally (loading direction TD). Additional specimens were extracted at 45° of both directions. Before tensile testing, specimens were mechanically polished up to 4000 grade SiC abrasive paper to remove the surface affected by work hardening due to the turning operations. The final polishing step was performed along the specimens' loading axis.

Tensile tests were conducted from room temperature (RT) up to 750 °C at a strain rate of 5.0 10^{-4} s^{-1} using an Instron 8562 testing system. Temperature was controlled with a two-zone resistive furnace with a ±1 °C accuracy and a ±3 °C temperature heterogeneity along the gauge length. A 15 °C/min heating rate was applied up to 30 °C below the target temperature and then a final heating at 3 °C/min up to the target temperature. Before tensile straining, a 15 min soak time was applied before each experiment, except for those performed at 750 °C. The tensile strain was measured using in-house water cooled extensometer with an initial basis of 12.5 mm. Experiments were performed using cylindrical specimens with a 42 mm total length, 14 mm gauge length and a 4 mm gauge diameter. These specimens have M6 × 100 threaded heads. An additional test at 750 °C has also been performed in high vacuum ($P < 10^{-6}$ mbar) [22–24] to assess at the role of environment on ductility.

Results and Discussion

Solution Heat Treatment

In an aim to optimize tensile properties of LMD-w processed IN718, different solution heat treatments from RT up to 1250 °C have been performed so as to dissolve the Laves phases as much as possible. In fact, Laves particles, located in the interdendritic spacings and rich in Nb and Mo (Nb content higher than 9 wt. %, according to EDS characterizations not shown here) results from the dendritic solidification during casting/welding/AM processes [25]. Figure 2 presents the evolution of Laves precipitation state as a function of temperature for 1 h solution treatments (ST). A decrease in area fraction of white particles is observed by increasing the temperature. A Laves surface fraction of <0.5% has been reached after a 1 h ST at 1200 °C, starting from a 6–8% area fraction in as-built (AB) conditions (see Fig. 3a). It is also worth mentioning that the Laves area fraction can be as high as ~10% locally, close to wire interlayers. Figure 3b also shows that the hotter and the longer the ST, the greater the Laves phases dissolution, as expected. Contrary to other studies showing a total dissolution of Laves phases for ST in the 1050–1150 °C temperature range [17, 18, 26, 27], it has never been able to reach a full dissolution of Laves phases, even by

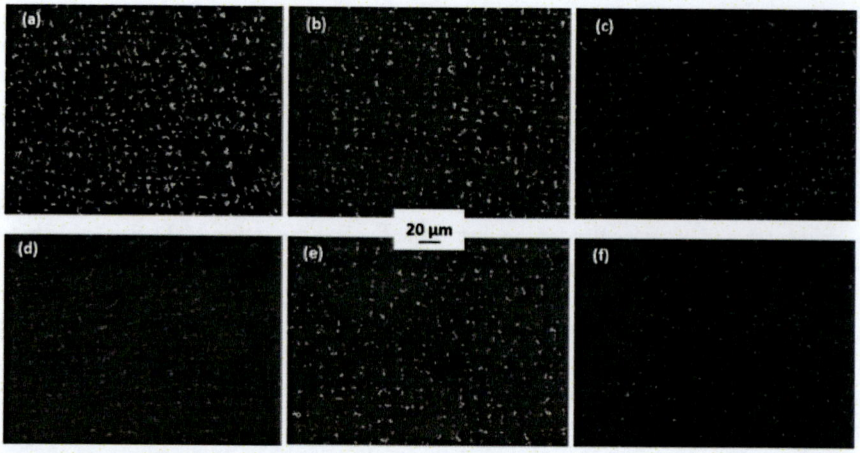

Fig. 2 Laves phases precipitation structure in as-built state (**a**), and after 1 h solution heat treatment at 1000 °C (**b**), 1050 °C (**c**), 1100 °C (**d**), 1150 °C (**e**) and 1200 °C (**f**)

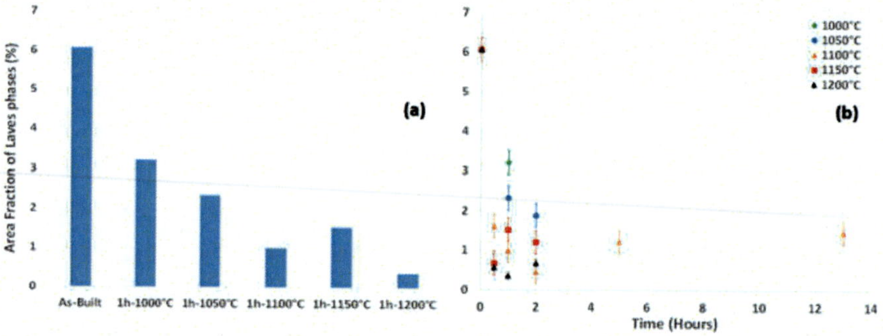

Fig. 3 Evolution of Laves phases area fraction as a function of temperature for 1 h solution heat treatment (**a**) and as a function of time for different temperatures (**b**)

extending the duration of ST. This is probably due to a very high dendritic segregation compared to other studies. Further ST at 1225 and 1250 °C led to a similar amount of residual Laves particles (~0.5% area fraction) but with some evidence of local incipient melting at the hottest. Hence, given the nature of the process and possible heterogeneities from bottom top of the walls [18], it has been decided to only apply a 1 h/1200 °C/Air Quench solution heat treatment for specimens tested in tension. These specimens will be denoted as "ST" in the following of this article. Moreover, solution treated specimen using this solution heat treatment and further aged using the aeronautic heat treatment will be denoted as "ST + Aged". Finally, some specimens will be direct-aged (DA), by applying the aeronautic aging sequence on the AB state.

Crystallographic Structure

Figures 4 and 5 present the crystallographic structure in the AB and ST states, respectively. These figures are presenting Index quality maps (Figs. 4a and 5a), inverse pole figures (IPF) along building direction (Figs. 4b and 5b) and along wall axis (Figs. 4c and 5c). It is seen in Fig. 4b and c some noise, corresponding to each Laves particles, as they do not diffract like the austenitic FCC matrix. Such a "noise" corresponds to all dark spots in Fig. 4a and is almost absent in the ST state. As a very striking result, these EBSD maps are presenting a very columnar type microstructure, with multi-millimetric grains along BD. Some recrystallized grains can be observed between each layer in the AB state (Fig. 4), grains transforming to columnar ones after solution treatment (see left part of EBSD maps in Fig. 5b and c)

As a most striking result from these EBSD characterizations, both states present a very textured crystallographic structure. Texture maps coded along BD and WA are presented in Fig. 6. Both states are showing a marked ~<012> crystallographic texture along BD and a ~<112> texture along WA (texture index >20 for both states), which is hardly affected by the ST. Indeed, the small variations in main crystallographic texture between both states can results from small differences in the exact location where specimens have been machined out from the walls as well as to the growth of smaller columnar from interlayers. All in all, these walls are presenting a crystallographic structure very similar to directionally-solidified (DS) type structure [23, 28, 29], with a collection of ~<012> oriented grains along building direction.

Tensile Properties at Room Temperature, 450 and 650 °C

True stress versus true strain tensile curves are presented in Fig. 7. Thirty seven tests are included in this figure, for specimens tested along BD (Vertical), WA (Horizontal), at 45° of the building direction, and in four different microstructural states: as-built, solution treated, solution treated and aged and direct-aged. The DA state has only been characterized along horizontal direction. According to Fig. 7, LMD-w IN718 performs better in terms of yield stress (defined at 0.2% plastic offset—YS) and ultimate tensile stress (UTS) in DA state, followed by ST + aged, AB and then ST states at RT, 450 and 650 °C (Fig. 7a–c respectively). An opposite trend is obtained for ductility, ST specimens presenting the highest strain at failure. Such a trend in tensile resistance is not a surprise given that $\gamma' + \gamma''$ is the main source of strengthening in alloy 718 [30–32]. A similar trend was already observed in cast & wrought (C&W) alloy 625 regarding γ'' precipitation [33, 34]. The effect of post-processing heat treatment appears to affect much more tensile properties of LMD-w IN718 in comparison to the loading direction. Tensile results at 750 °C (Fig. 7a) will commented later in this article.

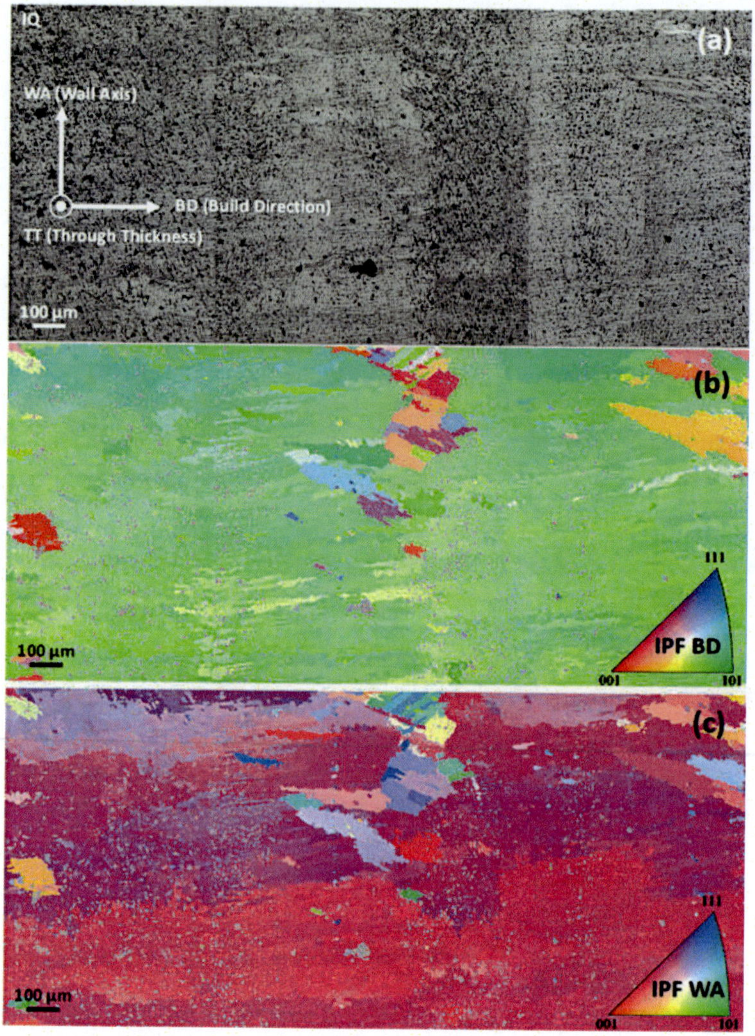

Fig. 4 EBSD characterizations of the as-built microstructure: Index Quality (**a**) and inverse pole figures coded along BD/Vertical (**b**) and along WA/Horizontal (**c**) directions

Figure 8 summarizes the evolution of Young's modulus, YS, UTS and ductility as a function of temperature in the RT-650 °C temperature range. According to Fig. 8a, Young's modulus is far higher for specimens machined at 45° of the loading direction in the AB state, and more generally, for most of the specimens machined at 45° of the BD. The fact that 45° specimen in AB state have a very high Young's modulus (values typical of ~<111> oriented single crystals) compared to all other direction while other 45° specimens ST or ST + Aged have Young's modulus more comparable to other orientations may arise from the fact that machining at +45° or −45° with

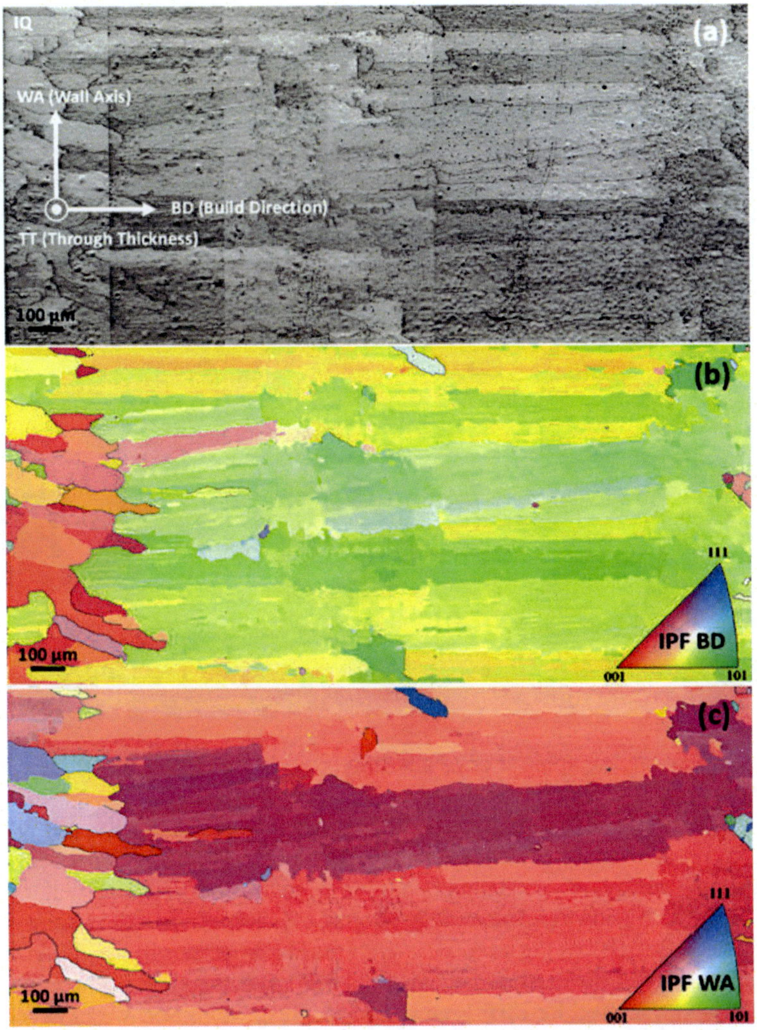

Fig. 5 EBSD characterizations after a 1 h/1200 °C solution treatment: Index Quality (**a**) and inverse pole figures coded along BD/Vertical (**b**) and along WA/Horizontal (**c**) directions

respect to the BD does not lead to the same crystallographic grain distribution in these specimens, given the pronounced texture illustrated in Figs. 4 and 5. Also, it is seen that horizontal and vertical specimens are Young's moduli not so different, what could be a surprise according to the very pronounced texture illustrated in Figs. 4 and 5. Such a result can be understood in the sense that vertical specimens have a ~<012> orientation while horizontal ones have a ~<112> texture, two orientations being very close in terms of Young's modulus given the cubic anisotropy of Ni-based fcc materials [35, 36]. It is also worth recalling that elasticity in alloy 718 is mostly

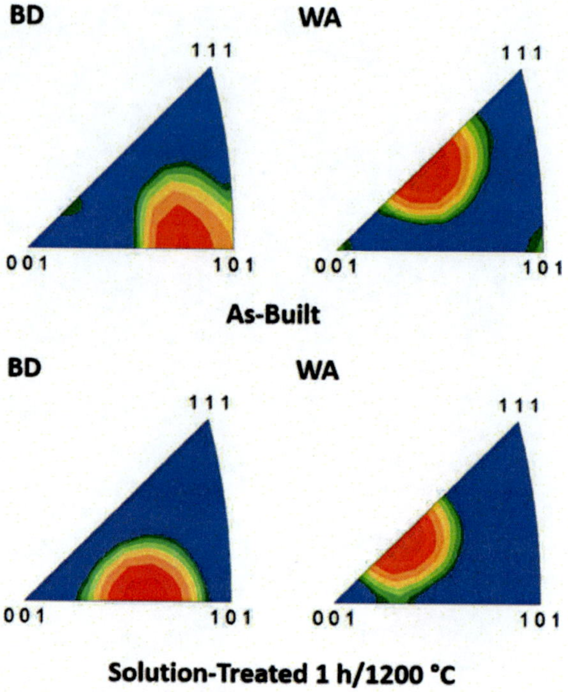

Fig. 6 Texture maps in the As-Built and 1 h/1200 °C Solution-Treated states coded along BD and WA directions

affected by crystallographic texture and only weakly by the precipitation state, as already proven recently by Cormier et al. for C&W IN718 [37].

The evolution of YS, UTS and ductility as a function of temperature (Fig. 8b–d respectively) are showing that these tensile properties are mostly dependent to the intragranular precipitation state, and possible stored dislocation density during the LMD process, in good agreement with prior studies [17]. In fact, it is known from the literature that IN718 presents quite high tensile properties in the AB state due to a high density of dislocations organized in cell structures (see, e.g., some results about IN718 processed by PBF-L or DED [38, 39]). Moreover, we have recently demonstrated that high levels of residual stresses are obtained after printing IN718 cylinders by LMD-w [40]. So as to get a better idea of the level of tensile properties of these LMD-w printed IN718 specimens, a comparison has been made in Fig. 8b–d with C&W specimens machined out from a forged component submitted to a DA heat treatment. These specimens had a very fine grain (ASTM 10-11), and, more importantly, have been tested using the same set-up, specimen's geometry and strain rate [7, 41]. One can see that C&W DA specimens have ~300 MPa higher YS and ~100 to 200 MPa higher UTS compared to its LMD-w counterpart, in all the temperature range. Such a decrease in yield and tensile strength mainly results from the very large grain size of LMD printed specimens, leading to a lower Hall & Petch strengthening. This larger grain size, however, always provide a better ductility in all the temperature range compared to the C&W fine grain DA alloy.

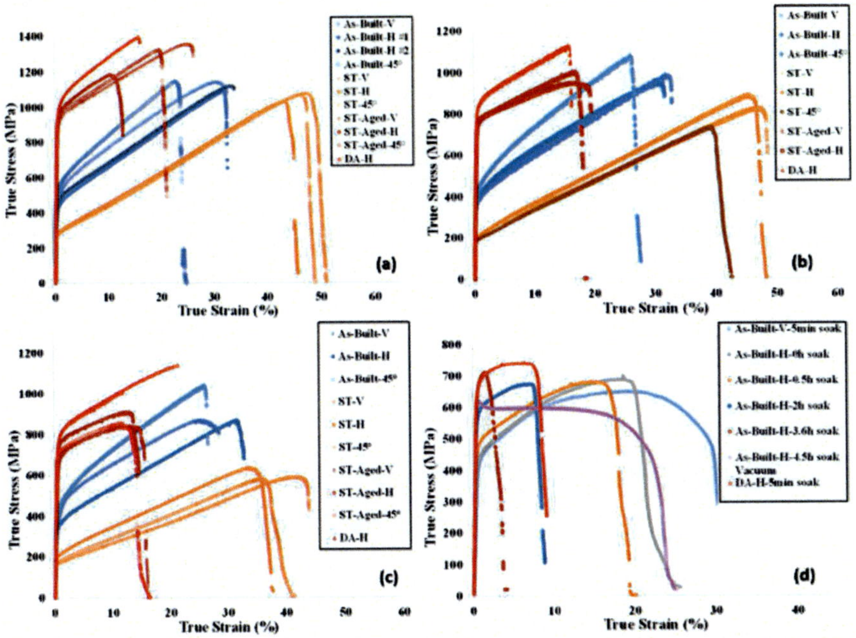

Fig. 7 Tensile curves at 5.0 10^{-4} s^{-1} of LMD-w IN718 at room temperature (**a**), 450 °C (**b**), 650 °C (**c**) and 750 °C (**d**). A variation in soak time at maximum temperature before testing has been used for tests performed at 750 °C, this soak time being indicated in the legend of (**d**). These plots are representing the True Stress versus True Strain tensile data

As YS is one of the most critical design criteria for the design of aero-engine components, the YS anisotropy has been evaluated for all microstructural state, taken the properties along vertical direction as a reference (i.e., the YS anisotropy coefficient is defined as the ratio between YS along one direction and YS along vertical—BD—direction). Figure 9 shows the evolution of the coefficient of anisotropy as a function of temperature. Variation in tensile properties of, at most, 10% are observed at RT and 450 °C. This anisotropy seems to increase at 650 °C, peaking up to ~25% for the ST state. This could be a result of an unstable microstructure at this temperature (i.e., onset of intermetallic precipitation) but such a result deserves to be further investigated. All in all, given the results presented in Fig. 9, a rather limited anisotropy in tensile properties is observed in our LMD-w specimens, a results pretty similar to those obtained by Moverare et al. using PBF-L specimen [42] and Seow et al. using WAAM processed specimens [26]. A larger anisotropy in tensile properties has, however, been obtained by other teams using PBF-L processed specimens [17, 43, 44], probably due to different grain sizes and/or initial density of dislocations. It is also worth mentioning that a very similar level of tensile anisotropy has been obtained for cold metal transfer processed Waspaloy by Sazerat et al., a process also leading to a very pronounced columnar <001> texture along BD [45].

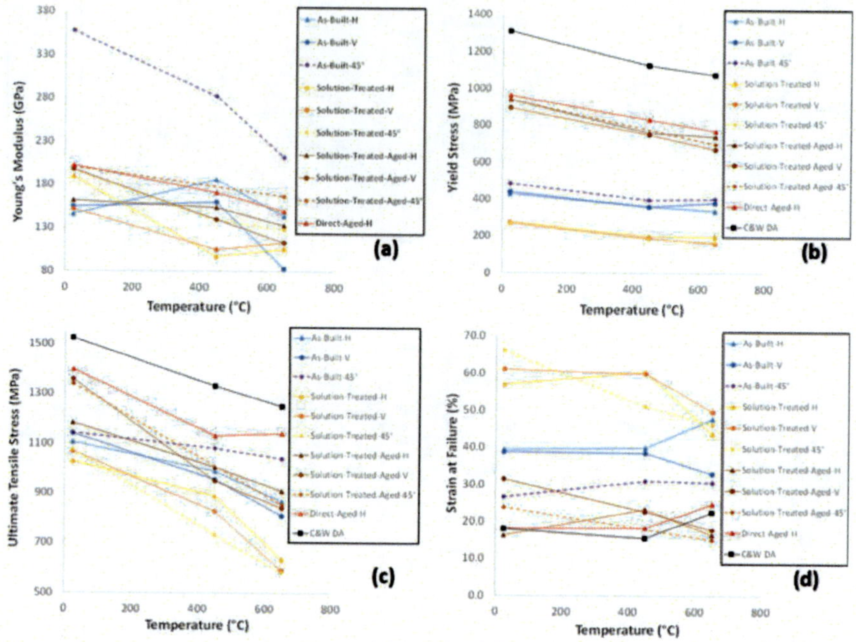

Fig. 8 Evolution in tensile properties at 5.0 10^{-4} s^{-1} of LMD-w IN718: Young's modulus (**a**), Yield Stress at 0.2% plastic offset (**b**), Ultimate Tensile Stress (**c**) and Ductility (**d**). Data for fine grain (ASTM 10-11) C&W direct-aged material have been extracted from Nicolaÿ et al. [7, 41]

Fig. 9 Evolution of the yield stress anisotropy coefficient taking data along the Building/Vertical direction as a reference

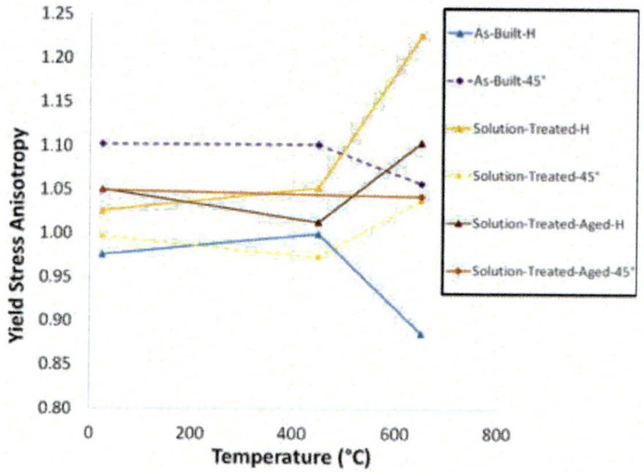

Figure 10 presents some typical fracture surfaces after tensile tests at RT. It has been chosen to only show results after RT testing as such fracture surfaces are representative of those obtained at 450 and 650 °C. Despite a very limited anisotropy in tensile properties as discussed before, such ductile fracture surfaces are all presenting a very anisotropic failure, with an elliptical fracture surface for all specimens. Such an

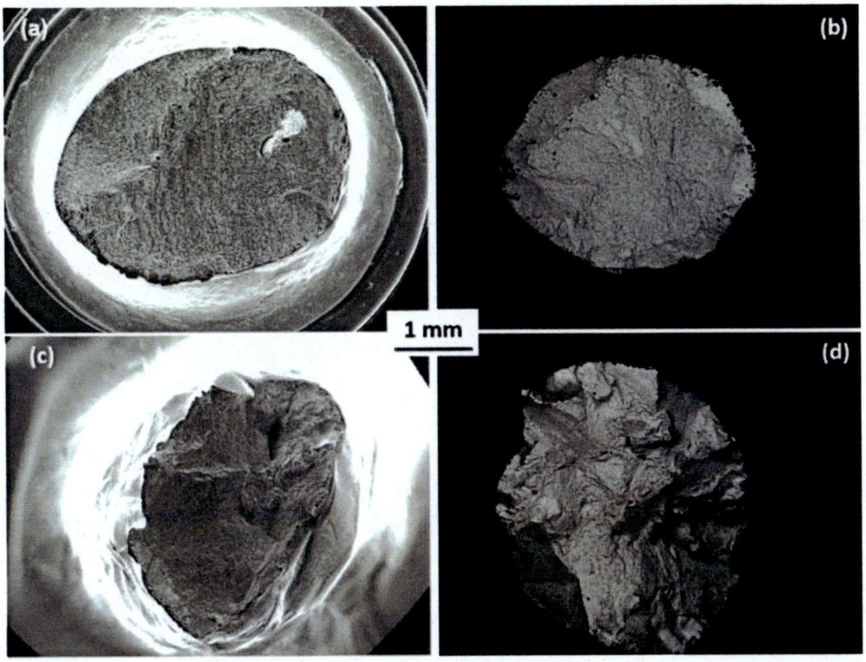

Fig. 10 Fracture surface after tensile tests at room temperature in as-built horizontal (**a**), as-built vertical (**b**), Solution-treated horizontal (**c**) and Solution-Treated + Aged Horizontal (**d**) specimens

elliptical fracture surface is characteristic of a plastic deformation mostly dominated by single slip activity at the grain scale, as expected for <012> oriented Ni-based superalloys [46–48]. Such spectacular fracture surfaces also result from the very large grain size of our specimens, leading to a mechanical behavior very similar to those that can be obtained in Ni-based single crystalline specimens.

Tensile Properties at 750 °C

The tensile behavior at 750 °C of LMD-w IN718 specimens is presented in Fig. 7a. Only AB specimens have been tested at this temperature as the microstructure is known to evolve rapidly at this temperature in the form of γ' and γ'' coarsening, but also through the $\gamma'' \rightarrow \delta$ transformation [49, 50]. For very short soak time (5 min and less) horizontal and vertical specimen present a very similar YS. However, the soak time before testing at 750 °C has been observed to have a spectacular impact on the tensile behavior. In fact, the longer the soak time, the higher is the YS lower is the ductility. Figure 11 shows that a YS increase of ~250 MPa and a decrease of ductility by a factor of 6 have been obtained by increasing the soak time at 750 °C

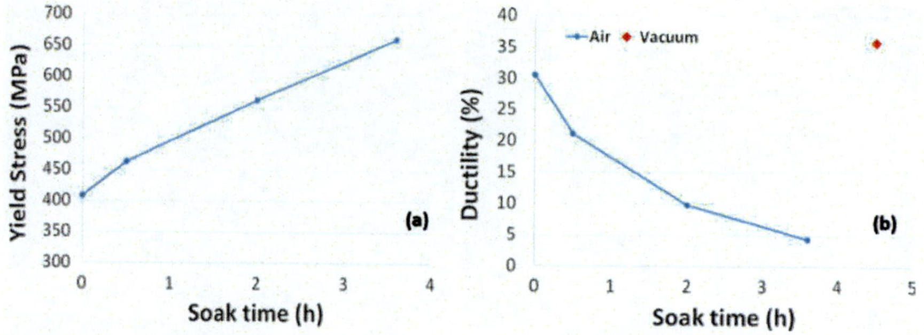

Fig. 11 Evolution of the yield stress at 0.2% plastic offset (**a**) and of the ductility (**b**) as a function of the soak time at the testing temperature for tensile tests at 750 °C/5.0 10^{-4} s^{-1}

before testing up to ~4 h. To better understand such a result, and, more specifically, the loss of ductility, a tensile test in high vacuum using the same tensile parameters and specimen's geometry has been performed. It is observed a ductility very similar and even better than the specimen without soak time at 750 °C. Such a result clearly highlight that rapid γ' and γ'' precipitation is occurring at 750 °C before and during the tests, a result in good agreement with Slama et al. and Balan et al. [51, 52]. Moreover, a grain boundary embrittlement due to oxidation occurs at such a temperature [53, 54], what could be a serious limitation for repaired components should they be exposed to such very high temperature. It can even be inferred that testing at tensile strain rates lower than 5.0 10^{-4} s^{-1} would lead to an even more severe drop in tensile ductility [55, 56].

On the Impact of Laves Phases

One of the striking results from our study is the excellent ductility of As-Built specimens, whatever the orientation and temperature (see Fig. 8d). This ductility is always larger than 25% of plastic deformation at failure, despite a Laves phase content approaching 10%. While Laves phases are often considered to be detrimental to mechanical properties by inducing a debit in ductility, fracture toughness and fatigue life [25, 57, 58], it seems that such a trend is not observed in our present case. It is speculated by the authors that this non-effect of Laves phases on ductility mostly result from the extremely large grain size in our specimens. The possible effect of the high amount of Laves phases in our specimens could be a lack of γ'' precipitation upon further direct aging of the material, as a high fraction of Nb element is trapped in these intermetallic phases. Such a lack of γ'' precipitation does not seem to be observed as our DA specimens have always exhibited the highest YS and UTS values among all our LMD-w processed specimens (see Fig. 8b and c), especially compared to ST + Aged specimens. The respective contributions of the initial stored dislocation density after printing and $\gamma' + \gamma''$ precipitation to the tensile strength have to be objectively established (like what has been done on C&W IN718 [7, 59, 60]) to get a full picture of the impact of Laves phases.

Conclusions

From the present study devoted to the characterization of tensile properties of LMD-w processed IN718 alloy in the room temperature −750 °C temperature range, the following conclusions can be established:

- A very textured material is obtained after LMD processing, with a strong ~<012> texture along building direction and a ~<112> texture along wall axis. This textured material is weakly affected by the application of a very high temperature solution heat treatment. Such a directionally-solidified like granular structure leads to an anisotropic deformation of tensile specimens, whatever the loading direction and testing temperature.
- A high content of Laves phases is obtained in the as-built state (up to 9–10% area fraction). A solution heat treatment of 1 h at 1200 °C has been determined to be optimal to reduce this Laves phase content to nearly 0.4–0.5% but it has been found impossible to fully remove Laves particles within the investigated time/temperature conditions.
- Direct aging after LMD processing provides the best tensile properties at room temperature, 450 °C and 650 °C, followed by the solution treated + aged state, then As-Built state and finally, the solution treated state. Tensile strength is mostly controlled by the precipitation of $\gamma' + \gamma''$ particles and then, by the initial dislocation density after LMD processing. Tensile strength of all our LMD-processed specimens is, however, far lower than a fine grain C&W direct-aged IN718 tested in the same conditions.
- Despite a very textured material, a weak overall tensile anisotropy has been observed at all tested temperatures.
- Laves phases do not induce a debit in tensile properties and, more specifically, a debit in tensile ductility. This is probably due to the very large grain size of our material.
- Tensile properties at 750 °C of As-Built material are highly sensitive to the testing procedure. The longer is the soak time at 750 °C before testing, the higher is the Yield stress and the lower is the tensile ductility, due to grain boundary embrittlement.

Acknowledgements Pprime Institute gratefully acknowledges "Contrat de Plan Etat - Région Nouvelle-Aquitaine" (CPER) as well as the "Fonds Européen de Développement Régional (FEDER)" for their financial support to the reported work in the context of the OPERA program (convention P-2020-BAFE-3). This work was partially funded by the French Government program "Investissements d'Avenir" (EQUIPEX GAP, reference ANR-11-EQPX-0018).

References

1. Loria, E.A., *Recent developments in the progress of superalloy 718.* JOM, 1992. **44**(6): p. 33–36.
2. Longuet, A., C. Dumont, and E. Georges, *Advanced modeling tools for processing and lifing of aeroengine components*, in *Superalloys 2020.* 2020, Springer. p. 3–15.
3. Alizadeh-Sh, M., et al., *Prediction of solidification cracking by an empirical-statistical analysis for laser cladding of Inconel 718 powder on a non-weldable substrate.* Optics & Laser Technology, 2020. **128**: p. 106244.
4. Galliano, F., et al., *Effect of temperature on hydrogen embrittlement susceptibility of alloy 718 in Light Water Reactor environment.* International Journal of Hydrogen Energy, 2017. **42**(33): p. 21371–21378.
5. Valle, L., et al., *The influence of heat treatments on the corrosion behaviour of nickel-based alloy 718.* Journal of Alloys and Compounds, 2019. **809**: p. 151781.
6. Donachie, M.J. and S.J. Donachie, *Superalloys: a technical guide.* 2002: ASM international.
7. Nicolaÿ, A., et al. *Metallurgical Analysis of Direct Aging Effect on Tensile and Creep Properties In Inconel 718 Forgings.* in *Superalloys 2020.* 2020. Seven Springs, PA, USA: TMS.
8. Texier, D., et al., *Microstructural features controlling the variability in low cycle fatigue properties of alloy Inconel 718DA at intermediate temperature.* Metallurgical and Materials Transactions A, 2016. **47A**: p. 1096–1109.
9. Texier, D., et al., *Crack initiation sensitivity of wrought direct aged alloy 718 in the very high cycle fatigue regime: the role of non-metallic inclusions.* Material Science and Engineering A, 2016. **678**: p. 122–136.
10. Texier, D., et al. *Role of Non-metallic Inclusions and Twins on the Variability in Fatigue Life in Alloy 718 Nickel Base Superalloy.* in *Superalloys 2020.* 2020. Seven Springs, PA, USA: TMS.
11. Texier, D., et al., *Short crack propagation from cracked non-metallic inclusions in a Ni-based polycrystalline superalloy.* Acta Materialia, 2019. **165**: p. 241–258.
12. Lippold, J.C., S.D. Kiser, and J.N. DuPont, *Welding metallurgy and weldability of nickel-base alloys.* 2011: John Wiley & Sons.
13. Andersson, J., *Weldability of precipitation hardening superalloys: influence of microstructure*, PhD Thesis, 2011, Chalmers University (Sweden).
14. CFM International, https://www.cfmaeroengines.com/services/maintenance/. 2022.
15. Onuike, B. and A. Bandyopadhyay, *Additive manufacturing in repair: Influence of processing parameters on properties of Inconel 718.* Materials Letters, 2019. **252**: p. 256-259.
16. Saboori, A., et al., *Application of directed energy deposition-based additive manufacturing in repair.* Applied Sciences, 2019. **9**(16): p. 3316.
17. Hosseini, E. and V. Popovich,*A review of mechanical properties of additively manufactured Inconel 718.* Additive Manufacturing, 2019. **30**: p. 100877.
18. Adomako, N.K., N. Haghdadi, and S. Primig, *Electron and laser-based additive manufacturing of Ni-based superalloys: a review of heterogeneities in microstructure and mechanical properties.* Materials & Design, 2022: p. 111245.
19. Kumar, S.P., et al.,*A review on properties of Inconel 625 and Inconel 718 fabricated using direct energy deposition.* Materials Today: Proceedings, 2021. **46**: p. 7892–7906.
20. Heckert, S. and B. Wedel. *Laser processing heads for cladding and heat treatment applications.* in *the 2nd Annual Laser Additive Manufacturing Workshop (LAM 2010), Houston, TX, USA (Laser Institute of America, Orlando, FL, USA, 2010).* 2010.
21. Vaunois, J.R., et al. *Influence of both γ' distribution and grain size on the tensile properties of UDIMET 720Li at room temperature.* in *7th International Symposium on Superalloy 718 and Derivatives.* 2010. Pittsburgh, PA, USA: TMS.
22. Mataveli Suave, L., et al., *Anisotropy in creep properties of DS200 + Hf alloy.* Materials at High Temperatures, 2016. **33**(4–5): p. 361–371.
23. Mataveli Suave, L., et al. *High temperature creep damage mechanisms in a directionally solidified alloy: impact of crystallography and environment.* in *Superalloys 2016.* 2016. Seven Springs, Champion, PA, USA: TMS.

24. Mataveli Suave, L., et al., *Thin-wall debit in creep of DS200+Hf alloy.* Metallurgical and Materials Transactions A, 2018. **49A**: p. 4012–4028.
25. Radhakrishna, C. and K. Prasad Rao, *The formation and control of Laves phase in superalloy 718 welds.* Journal of Materials Science, 1997. **32**(8): p. 1977–1984.
26. Seow, C.E., et al., *Wire+ Arc Additively Manufactured Inconel 718: Effect of post-deposition heat treatments on microstructure and tensile properties.* Materials & Design, 2019. **183**: p. 108157.
27. You, X., et al., *Effect of solution heat treatment on the precipitation behavior and strengthening mechanisms of electron beam smelted Inconel 718 superalloy.* Materials Science and Engineering: A, 2017. **689**: p. 257–268.
28. Torfeh, M. et al., *Intermediate Temperature Creep Damage Mechanisms of a Directionally Solidified Ni-Based Superalloy.* Materials at High Temperatures, 2022. **39**(3): p. 193–205.
29. Stinville, J.C., et al., *Damage Nucleation During Transverse Creep of a Directionally Solidified Ni-based Superalloy.* Material Science & Engineering A, 2022. **858**: p. 144089.
30. Chaturvedi, M. and Y.-f. Han, *Strengthening mechanisms in Inconel 718 superalloy.* Metal science, 1983. **17**(3): p. 145–149.
31. Cozar, R. and A. Pineau, Morphology of γ' and γ'' *precipitates and thermal stability of inconel 718 type alloys.* Metallurgical Transactions, 1973. **4**(1): p. 47–59.
32. Paulonis, D., J. Oblak, and D. Duvall, *Precipitation in Nickel-base Alloy 718.* 1969, Pratt and Whitney Aircraft, Middletown, Conn.
33. Mataveli Suave, L., et al., *High temperature low cycle fatigue properties of Alloy 625.* Materials Science and Engineering A, 2016. **650**: p. 161–170.
34. Mataveli Suave, L., et al., *Microstructural evolutions during thermal aging of Alloy 625: impact of temperature and forming process.* Metallurgical and Materials Transactions A, 2014. **45A**(7): p. 2963–2982.
35. Fedelich, B., *Crystal orientation and elastic properties*, in *Nickel base single Crystals across length scales.* 2022, Elsevier. p. 41–67.
36. Gadaud, P., *Crystal Elasticity.* 2022: John Wiley & Sons.
37. Cormier, J., et al., *In-Situ Determination of Precipitation Kinetics during Heat Treatment of Superalloy 718.* Metallurgical and Materials Transactions A, 2021. **52A**(2): p. 500–511.
38. Ho, I.-T., et al.,*Effects of CoAl2O4 inoculants on microstructure and mechanical properties of IN718 processed by selective laser melting.* Additive Manufacturing, 2020. **35**: p. 101328.
39. Hsu, T.-H., et al., *Effect of Carbide Inoculants Additions in IN718 Fabricated by Selective Laser Melting Process*, in *Superalloys 2020.* 2020, Springer. p. 982–989.
40. Cabeza, S., et al. *Strain Monitoring of Laser Metal Deposition of Inconel 718 by Neutron Diffraction.* in *Superalloys 2020.* 2020. Seven Springs, PA, USA: TMS.
41. Nicolaÿ, A., *Microstructure et propriétés de l'Inconel 718 DA forgé en presse à vis dans le domaine subsolvus δ*, PhD thesis, 2019, Mines ParisTech: Sophia Antipolis, France.
42. Deng, D., et al., *Microstructure and mechanical properties of Inconel 718 produced by selective laser melting: Sample orientation dependence and effects of post heat treatments.* Materials Science and Engineering: A, 2018. **713**: p. 294–306.
43. Liu, S., et al.,*The effect of energy density on texture and mechanical anisotropy in selective laser melted Inconel 718.* Materials & Design, 2020. **191**: p. 108642.
44. Bean, G., et al., *Build orientation effects on texture and mechanical properties of selective laser melting Inconel 718.* Journal of Materials Engineering and Performance, 2019. **28**(4): p. 1942–1949.
45. Sazerat, M., et al. *Microstructural and tensile properties evolutions of direct-aged Waspaloy produced by Wire Arc Additive Manufacturing* in *10th International Symposium on Superalloy 718 and Derivatives.* 2023. Pittsburgh, PA, USA: TMS.
46. Kubin, L., B. Lisiecki, and P. Caron, *Octahedral slip instabilities in* γ/γ' *superalloy single crystals CMSX-2 and AM3.* Philosophical Magazine A, 1995. **71**(5): p. 991–1009.
47. MacKay, R.A. and R.D. Maier, *The influence of orientation on the stress rupture properties of nickel-base superalloy single crystals.* Metallurgical Transactions A, 1982. **13**(10): p. 1747–1754.

48. Stinville, J., K. Gallup, and T. Pollock, *Transverse creep of nickel-base superalloy bicrystals.* Metallurgical and Materials Transactions A, 2015. **46**(6): p. 2516–2529.
49. Agnoli, A., et al., *Mechanical properties* evolution of γ'/γ'' *nickel-base superalloys during long term thermal over-aging.* Metallurgical and Materials Transactions A, 2018. **49A**: p. 4290–4300.
50. Jouiad, M., et al., *Microstructure and mechanical properties evolutions of Alloy 718 during isothermal and thermal cycling over-aging.* Materials & Design, 2016. **102**: p. 284–296.
51. Slama, C., C. Servant, and G. Cizeron, *Aging of the Inconel 718 alloy between 500 and 750 C.* Journal of materials research, 1997. **12**(9): p. 2298–2316.
52. Balan, A., et al., *Precipitation of* γ*" in Inconel 718 alloy from microstructure to mechanical properties.* Materialia, 2021. **20**: p. 101187.
53. Rezende, M.C., et al., *Oxidation assisted intergranular cracking under loading at dynamic strain aging temperatures in Inconel 718 superalloy.* Journal of alloys and compounds, 2015. **643**: p. S256–S259.
54. Wagenhuber, E., V. Trindade, and U. Krupp. *The role of oxygen-grain-boundary diffusion during intercrystalline oxidation and intergranular fatigue crack propagation in alloy 718.* in *Proc. Int. Symp. Superalloys Var. Deriv. Pittsburgh, PA.* 2005.
55. Thébaud, L., et al. *Relationships between microstructural paramaters and time-dependent mechanical properties of a new nickel based superalloy AD730*^{TM}. in *Superalloys 2016*. 2016. Seven Springs, Champion, PA, USA: TMS.
56. Gabb, T.P., et al. *Use of Slow Strain Rate Tensile Testing to Assess the Ability of Several Superalloys to Resist Environmentally-Assisted Intergranular Cracking.* in *8th International Symposium on Superalloy 718 and Derivatives.* 2014. Wiley Online Library.
57. Janaki Ram, G., et al., *Control of Laves phase in Inconel 718 GTA welds with current pulsing.* Science and technology of welding and joining, 2004. **9**(5): p. 390–398.
58. Sui, S., et al., *The influence of Laves phases on the high-cycle fatigue behavior of laser additive manufactured Inconel 718.* Materials Science and Engineering: A, 2017. **695**: p. 6–13.
59. Theska, F., et al., *On conventional versus direct ageing of Alloy 718.* Acta Materialia, 2018. **156**: p. 116–124.
60. Theska, F., et al., *Microstructure-property relationships in directly aged Alloy 718 turbine disks.* Materials Science and Engineering: A, 2020. **776**: p. 138967.

Microstructural and Tensile Properties Evolutions of Direct-Aged Waspaloy Produced by Wire Arc Additive Manufacturing

Marjolaine Sazerat, Azdine Nait-Ali, Lucie Barot, Alice Cervellon, Inmaculada Lopez-Galilea, Dominique Eyidi, Anne Joulain, Patrick Villechaise, Jonathan Cormier, Sebastian Weber, and Roland Fortunier

Abstract The microstructure and tensile properties of direct-aged Waspaloy manufactured using wire arc-based Cold Metal Transfer (CMT) have been investigated. Samples were exposed to temperatures ranging from 700 to 900 °C, for up to 96 h. In the as-deposited condition, pronounced chemical segregation is inherited from the process, leading to heterogeneous γ' precipitation between dendrite cores and interdendritic spacings. γ' size and distribution were measured in both areas for each heat treatment, and a diffusion-controlled coarsening behavior following the Lifshitz–Slyozov–Wagner theory was observed for temperatures above 760 °C. Activation energies were calculated. Tensile tests at room temperature were carried out not only on the additively processed alloy before and after aging but also on wrought sub-solvus and super-solvus treated material for reference. Results showed that heat treatment significantly increased the yield strength and ultimate tensile strength of the CMT samples, of up to +340 MPa compared to the as-built conditions. Elongation, however, decreased from 40–45% to 16–28%. Direct-aged CMT Waspaloy exhibited similar behavior to that of wrought super-solvus Waspaloy, due to their large grains (~200–250 μm). Anisotropy in tensile properties was estimated by calculating

M. Sazerat (✉) · A. Nait-Ali · P. Villechaise · J. Cormier
Institut Pprime, Physics and Mechanics of Materials Department, UPR CNRS 3346, ISAE-ENSMA, 1 Avenue Clément Ader, BP 40109, 86961 Futuroscope-Chasseneuil, France
e-mail: marjolaine.sazerat@ensma.fr

L. Barot · A. Cervellon
Engineering Department, Processes Methods Team, Safran Aircraft Engines, 1 Rue Maryse Bastié, 86100 Châtellerault, France

I. Lopez-Galilea · S. Weber
Institute for Materials, Ruhr University Bochum, Universitätsstraße 150, 44801 Bochum, Germany

D. Eyidi · A. Joulain
Institut Pprime, Physics and Mechanics of Materials Department, UPR CNRS 3346, Université de Poitiers, 11 Boulevard Marie Et Pierre Curie, TSA 41123, Poitiers, France

R. Fortunier
UMR CNRS 5513, LTDS, ENISE, 58 Rue Jean Parot, 42100 Saint-Etienne, France

E. A. Ott et al. (eds.), *Proceedings of the 10th International Symposium on Superalloy 718 and Derivatives*, The Minerals, Metals & Materials Series,
https://doi.org/10.1007/978-3-031-27447-3_43

the ratio of properties for horizontal and vertical specimens. Finally, the formation of intermetallic phases was assessed. Thermodynamic calculations predicted the formation of $M_{23}C_6$, η, and σ phases in interdendritic spacings at thermodynamic equilibrium in the range 700–900 °C. Using electron diffraction patterns and energy-dispersive X-ray spectrometry, intergranular $(Cr, Mo)_{23}C_6$ secondary carbides decorating grain boundaries and located near (Ti, Mo)C primary carbides in the interdendritic spacings were observed to nucleate and grow.

Keywords Waspaloy · WAAM-CMT · Aging · Microstructure · Tensile properties · Secondary phase

Introduction

Waspaloy is a nickel-based superalloy first developed in the 1950s and patented in 1960 [1]. Widely used in its wrought form for high-temperature components in aircraft engines, its mechanical properties are excellent for service temperatures up to 750 °C, i.e., 100 °C higher compared to Inconel 718, thanks to the high stability of the γ' phase [2–4]. With original engine makers now selling engines with fly-by-the-hour contracts, including a fixed maintenance cost over an extended period of time, repair and remanufacturing have become major issues. Maintenance, Repair, and Overhaul (MRO) services are challenged in repairing more engines, faster and better than before, aiming to extend engine life and reduce the cost associated with the entire replacement of the damaged part. Moreover, Metal Additive Manufacturing (AM) has gained interest over the past decade and provides a wide range of new refurbishing methods. Welding-based Wire Arc Additive Manufacturing (WAAM) processes offer great advantages for the repair of large components considering their high deposition rate, high efficiency, and lower cost compared to powder technologies [5]. In particular, Cold Metal Transfer (CMT), a variant of Gas Metal Arc Welding (GMAW), allows for a lower heat input, hence a smaller heat affected zone (HAZ), as a result of the short-circuiting droplet transfer [6]. A significant reduction in deformation and residual stress is therefore accomplished [7].

While Inconel 718 produced by WAAM [8–10], and more specifically by CMT [7, 11, 12], has been a frequent subject of published research, only a few articles were issued on TIG Waspaloy [9, 13, 14]. Data on this material using CMT is lacking in the literature. Due to its higher Al/Ti ratio, Waspaloy is more susceptible to Strain Age Cracking (SAC) and is therefore often considered as marginally weldable [2, 15]. However, Inconel 718 has a larger solidification range (170 °C) in comparison with Waspaloy (70 °C), meaning the latter is less likely to experience hot cracking, whether it would be solidification or liquation cracking, during welding [16].

Post-weld heat treatments (PWHT), especially aging, are frequently performed on age-hardenable superalloys to enhance microstructural stability and increase mechanical properties by triggering/optimizing γ' precipitation through coarsening. During aging, the formation of secondary carbides (M_6C and $M_{23}C_6$) can be observed due to the decomposition of primary carbides of MC type [17, 18]. Long-term thermal exposure can also allow the precipitation of intermetallic compounds such as Topologically Close-Packed (TCP) phases, the most common ones being μ, σ, P, and Laves [19, 20]. Typically composed of Cr, Co, Mo, and W, their presence has been proven to be deleterious to mechanical properties [21], especially creep [18, 22, 23]. Such phases can also cause a significant decrease in oxidation resistance [24]. Moreover, residual stresses are a critical issue in AM due to the steep temperature gradient associated with the thermal cycle. Highly localized heating followed by cooling generates thermal expansion and contraction within the material [25, 26]. PWHT can be a way to relieve the residual stresses inherited from the process [2]. Stress relief, however, tends to happen simultaneously with γ' precipitation hardening during PWHT which can lead to SAC. For this reason, PWHT should be a compromise allowing for improved mechanical properties and residual stress relief, but avoiding the formation of unfavorable compounds, TCP phases for instance, and preventing SAC.

Within this context, the purpose of the present article is to study microstructural evolutions of CMT Waspaloy in the 700–900 °C temperature range, as well as the impact on tensile properties at room temperature. The as-deposited material was first investigated in order to assess the changes after direct aging, and the γ' coarsening kinetics were evaluated.

Experimental Procedures

Material

A single-pass multi-layer wall (Fig. 1) of Waspaloy filler wire was deposited on a wrought Hastelloy X substrate using a Fronius welding system operating in CMT mode. The wire had a diameter of 0.9 mm, and its chemical composition is given in Table 1. Argon was used for shielding, with a flow rate of 15 $L \cdot min^{-1}$ at the torch gas nozzle. The Wire Feed Speed (WFS) was set to 7 $m \cdot min^{-1}$ and the Travel Speed (TS) to 250 $mm \cdot min^{-1}$. A few experiments were also conducted on the cast and wrought (C&W) Waspaloy supplied by the company Aubert & Duval, serving as a reference to draw the comparison with the CMT material.

Heat Treatment

Rectangular samples and tensile specimens were extracted by electron discharge machining (EDM) from the thin wall in both longitudinal and transverse directions and placed in a Nabertherm P330 furnace to receive direct aging heat treatments. The samples were positioned in a crucible in contact with an S-type thermocouple,

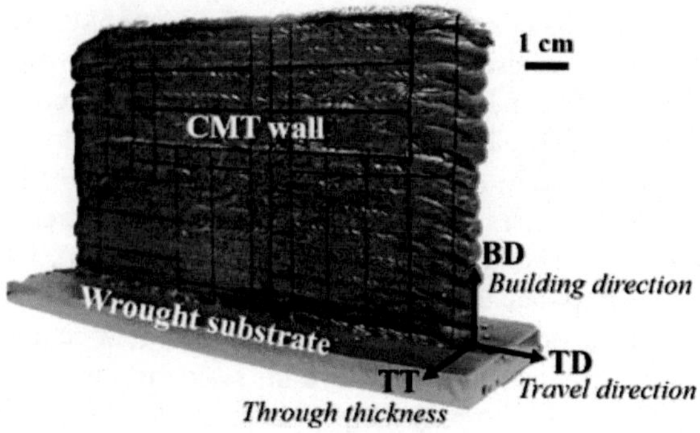

Fig. 1 Photograph of a single-pass multi-layer CMT wall (130 × 80 × 6.5 mm) labelled with axis directions

Table 1 Chemical composition of the Waspaloy wire (at %), provided by Safran Aircraft Engines

Ni	Cr	Co	Mo	Ti	Al	Fe	C	Si	Mn	Cu	B	S	P
55.48	21.15	12.70	2.54	3.43	2.94	1.22	0.38	0.12	0.03	0.02	0.016	<0.01	<0.01

leading to a temperature accuracy of ±1 °C. Exposure times ranged from 4 to 96 h and temperatures ranged from 700 to 900 °C. The samples were air quenched after aging.

The wrought Waspaloy samples received two different heat treatments to examine the effect of grain size on microstructure and mechanical properties. A sub-solvus solution heat treatment, 1020 °C for 4 h/Air Quench (AQ), was carried out on the first batch to obtain small grains. Double aging (850 °C for 4 h/AQ +760 °C for 16 h/AQ) was then performed. The second batch received a super-solvus solution heat treatment, 1135 °C for 2 h, to trigger larger grains before double aging (1040 °C for 4 h/AQ + 760 °C for 16 h/AQ).

Metallographic Examination

Aged samples were polished using SiC abrasive papers up to grade 4000, and then polished to 1 μm with diamond suspension. For a better observation of the microstructure, the samples received electrolytic polishing using A3 electrolyte from Struers for 8 s at 45 V and 4 °C, followed by chemical etching using aqua regia (75% HCl + 25% HNO_3).

Two Scanning Electron Microscopes (SEM) were used in this study. The general microstructure as well as γ' precipitation was characterized using a field emission gun (FEG) SEM JEOL 7000F with an accelerating voltage of 25 kV. Chemical segregation was evaluated using Energy-Dispersive X-ray Spectrometry (EDS) on a JEOL 6100 SEM at 15 kV.

A Thermo Scientific Talos F200S G2 Transmission Electron Microscope (TEM) was used at 200 kV for Selected Area Electron Diffraction (SAED) to determine the crystallographic nature of an unknown intermetallic phase precipitating after aging heat treatment. [001] zone axis observations were performed on a thin sample prepared using Focused Ion Beam (FIB), while [110] and [111] observations were done on a second thin sample prepared manually by grinding down to a 40 μm thickness, followed by thinning and milling with a precision ion polishing system (PIPS). EDS was also performed on the same equipment to study the chemical composition of secondary phases.

The average γ' size was measured for each aging condition in both primary dendrite arms and interdendritic regions using the image processing software ImageJ. SEM images containing at least 200 precipitates were analyzed.

Mechanical Testing

Tensile tests were conducted at room temperature at a strain rate of 10^{-3} s^{-1} using an Instron 8862 testing system. The displacement was measured using an Instron extensometer with a modified basis of 8.78 mm. Experiments were performed on flat dog-bone-shaped micro specimens with a 10 mm gauge length, 2 mm width, 1 mm thickness, and 32 mm total length. Tensile samples were extracted from the wall using EDM, both vertically (loading direction BD) and horizontally (loading direction TD). Each aging condition, as well as the as-deposited material, has therefore been tested twice to assess tensile anisotropy. Tensile samples for the wrought material were cylindrical (M6 type) with a 4 mm diameter, 14 mm gauge length, and 42 mm total length. A total of 34 tensile tests were achieved. Prior to testing, the samples were mechanically polished up to 4000-grade SiC abrasive paper to remove the surface affected by work hardening. The final polishing step was performed along the specimens' loading axis.

Results and Discussion

As-Deposited Microstructure

To study the effect of direct aging, the as-deposited microstructure was analyzed first. SEM images revealed a dendritic structure (Fig. 2a–b). As shown in Fig. 2c, MC primary carbides with a blocky shape are present in interdendritic spacings. Ultra-fine spherical γ' precipitation was observed, with a pronounced size difference between dendritic cores and interdendritic spacings. γ' precipitates in interdendritic spacings have an average diameter of 16.1 ± 3.2 nm, while those in the dendritic cores displayed an average diameter of 6.0 ± 0.8 nm, suggesting that there is a substantial chemical segregation inherited from the CMT process and its rapid solidification. Figure 2d shows an EBSD map of two CMT layers deposited on a wrought substrate. While the latter has small equiaxed grains (22.78 μm on average), the additive

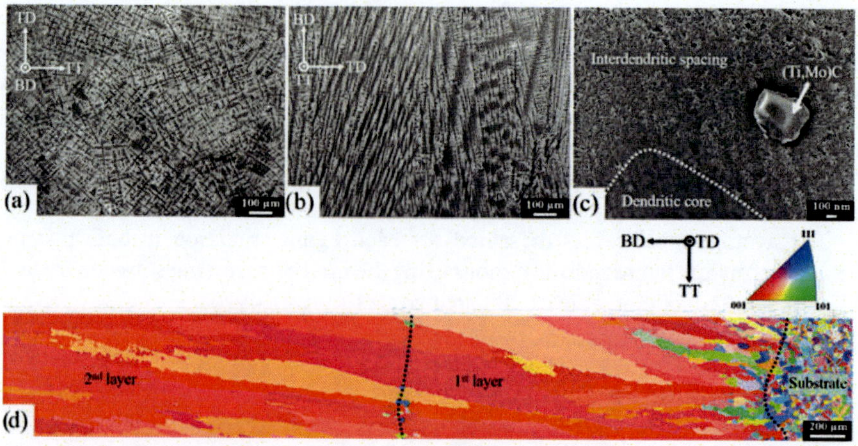

Fig. 2 As-deposited material (**a**, **b**) dendritic structure perpendicular and parallel to the building direction respectively, (**c**) inhomogeneous γ' precipitation and **d** EBSD inverse pole figure map of the interface and the first two layers, with respect to the building direction (BD)

Table 2 Chemical composition of the different areas in the as-deposited material (at %) obtained by SEM–EDS

	Ni	Cr	Co	Mo	Ti
Bulk	56.24 ± 0.18	20.18 ± 0.07	13.58 ± 0.05	2.26 ± 0.02	3.55 ± 0.12
Dendritic cores	56.75 ± 0.13	20.36 ± 0.08	14.19 ± 0.15	2.06 ± 0.03	2.48 ± 0.13
Interdendritic regions	55.53 ± 0.75	19.84 ± 0.18	12.91 ± 0.25	2.50 ± 0.24	4.97 ± 1.04

	Fe	Al	Mn	Si
Bulk	1.40 ± 0.02	2.70 ± 0.01	0.05 ± 0.01	0.05 ± 0.00
Dendritic cores	1.49 ± 0.03	2.62 ± 0.04	0.03 ± 0.01	0.03 ± 0.02
Interdendritic regions	1.31 ± 0.04	2.78 ± 0.26	0.04 ± 0.02	0.12 ± 0.06

material displays a strong <001> texture with columnar grains oriented in the building direction.

This phenomenon was confirmed with SEM–EDS measurements of the average chemical composition in both regions, as shown in Table 2. As expected from the previous observations, interdendritic regions are enriched in γ' formers Ti and Al, leading to a larger precipitate size. The analysis also demonstrated that MC carbides were (Ti, Mo)C.

Effect of Thermal Exposure on γ' Evolution

Figure 3 shows an example of γ' precipitation evolution in both primary dendrite arms and interdendritic regions after exposure at 800 °C, from 4 to 96 h. Using similar SEM observations, γ' average diameter was measured and the distribution was plotted for each direct aging condition. Figure 4 presents the effect of aging time and temperature on the average γ' diameter, in comparison with the as-deposited reference. For temperatures above 760 °C, γ' coarsened with increasing aging times and almost homogenized at 96 h, reaching about the same size in both regions. The data for 700 °C, however, show almost no coarsening within the investigated time range, with even a small decrease in size after 4 h. This could be explained by γ' nucleating at this temperature, instead of coarsening. Indeed, the size distribution graphs in Fig. 5a–b show very few differences for the data at 700 °C and various exposure times, especially in the dendritic cores where the maximum peaks remain centered at 10–15 nm. On the contrary, at 845 °C (Fig. 5c-d), the values flattened and shifted right towards greater diameters when aging time was increased, as a clear sign of particle growth.

Since γ' precipitates in CMT Waspaloy are spherical, the coarsening behavior can be expected to follow the cubic law of Lifshitz, Slyozov, and Wagner (LSW) [27–29]:

$$\langle r(t)\rangle^3 - \langle r_0\rangle^3 = k \cdot t \tag{1}$$

$$\text{with } k = \frac{8\gamma \cdot C_\infty \cdot {V_m}^2 \cdot D}{9R \cdot T} \tag{2}$$

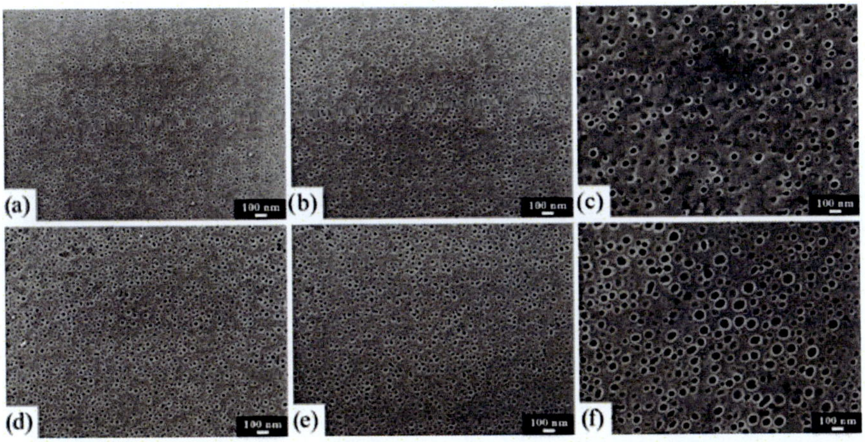

Fig. 3 Secondary electron images of γ' precipitation **a–c** in dendritic cores after 4, 16, and 96 h at 800 °C, respectively, **d–f** in the interdendritic spacings after 4, 16, and 96 h at 800 °C, respectively

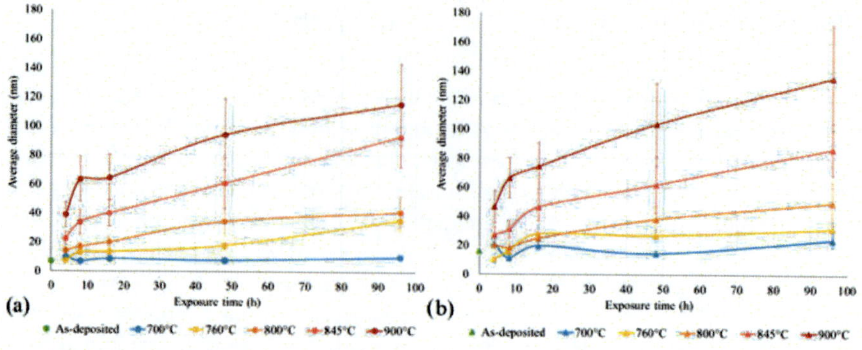

Fig. 4 Average diameter of γ′ precipitates as a function of aging time at different temperatures in **a** dendrite cores and **b** interdendritic spacings

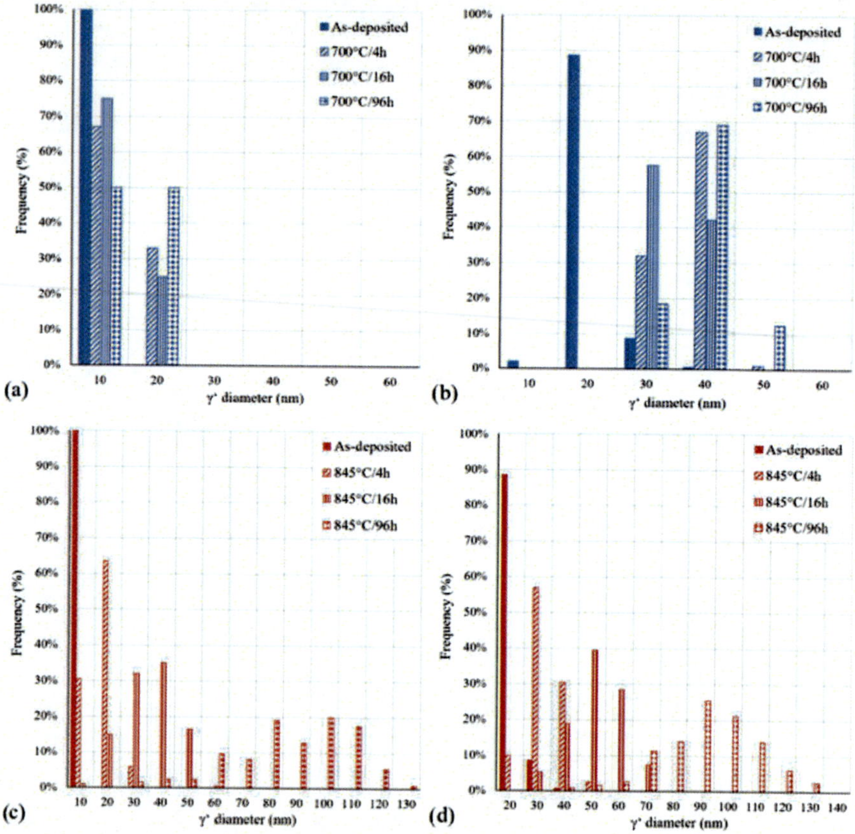

Fig. 5 γ′ size distribution for different aging times in **a** dendritic cores at 700 °C, **b** interdendritic spacings at 700 °C, **c** dendritic cores at 845 °C, and **d** interdendritic spacings at 845 °C

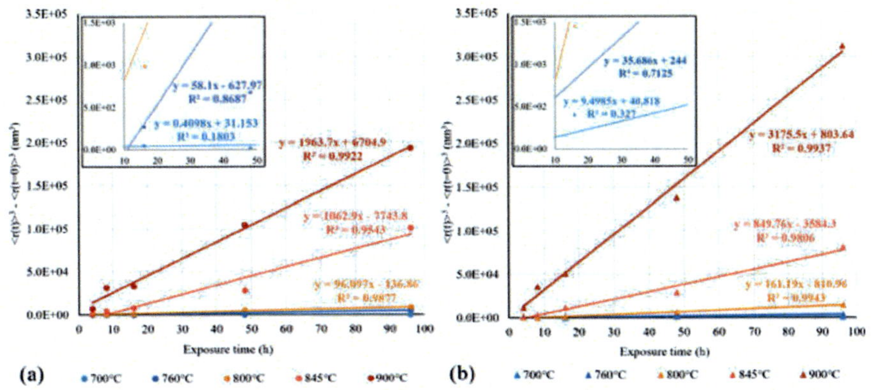

Fig. 6 Coarsening plots of γ' precipitates in **a** dendritic cores and **b** interdendritic spacings, with linear fitting curves and associated equations. The insets show magnifications of the data for 700 and 760 °C

where t is the aging time, r_0 is the initial radius in the as-deposited condition, k is the kinetics coefficient, γ is the surface free energy of the precipitate–matrix interface, C_∞ the solute concentration in the matrix at equilibrium, V_m is the molar volume of the precipitate, R is the universal gas constant, T is the absolute temperature, and D is the diffusion coefficient of solute atoms in the matrix. The plot of $<r(t)>^3 - <r_0>^3$ as a function of exposure time t is presented in Fig. 6. A linear relationship between the two variables is obvious in the 800–900°C temperature range. The LSW theory is therefore validated for this temperature range, with very satisfactory coefficients of determination ($R^2 > 0.98$). Exposure at 760 °C was also consistent but with slightly lower R^2. On the other hand, 700 °C did not display a linear behavior at all, presumably because of γ'-nucleation taking place at this temperature, as explained previously.

Moreover, the slopes of each linear fitting curve in Fig. 6 represent the temperature-dependent kinetics coefficient k. The diffusion coefficient D has an Arrhenius type behavior and can be defined as $D = D_0 \cdot e^{-\frac{Q_{ac}}{R \cdot T}}$, with D_0 being a constant [29]. Therefore, Eq. (2) can be expressed as

$$k = \frac{A}{T} \cdot e^{-\frac{Q_{ac}}{R \cdot T}} \tag{3}$$

where A is a constant, Q_{ac} is the activation energy for diffusion-controlled coarsening, R is the universal gas constant, and T is the absolute temperature. By plotting $ln(k \cdot T)$ as a function of $1/T$ (Fig. 7), the activation energies for γ' precipitate coarsening in both dendritic cores and interdendritic spacings can be determined, based on experimental data. Data obtained in the 760–900 °C temperature range were considered since their coarsening correlated well with the LSW theory, while data at 700 °C were excluded from this coarsening analysis because of the aforementioned reason. Linear fitting curves with reasonable determination coefficients ($R^2 > 0.92$) were obtained. Using the slope of the linear curves, the activation energy for diffusion-controlled coarsening was estimated at 289.1 $kJ \cdot mol^{-1}$ for dendritic cores and 335.9 $kJ \cdot mol^{-1}$

for interdendritic regions. Despite a lower γ' volume fraction, a lower energy is thus required for precipitate growth in dendritic cores, probably due to the absence of any other intermetallic phase precipitating in these areas (see subsequent section). Moreover, the value obtained in primary dendrite arms is close to the activation energy of Al and Ti diffusion in pure Ni (270 $kJ \cdot mol^{-1}$ and 257 $kJ \cdot mol^{-1}$, respectively [30, 31]), suggesting they are the main elements whose diffusion in the matrix is controlling the coarsening process. On the other hand, interdendritic spacings require about 15% more energy for precipitate growth, meaning the reaction is slower.

Wang et al. studied the coarsening kinetics of wrought Waspaloy at 960 °C and 1000 °C; activation energies were estimated to be 255.4 and 256.5 $kJ \cdot mol^{-1}$, respectively [32]. The values found in the present work for CMT Waspaloy are therefore a bit higher than expected. However, most nickel-based superalloys have activation energy in the range 250–290 $kJ \cdot mol^{-1}$ [29, 33–35], so the value calculated for dendritic cores is in good agreement with prior literature. The one obtained for interdendritic spacings could be due to the distribution broadening and the presence of very different γ' sizes with various coarsening kinetics. For larger sizes, the growth rate slows as a consequence of the adjacent matrix being depleted of the solute elements [36]. In addition, the formation of intermetallic phases preferentially in these regions has been observed (see subsequent section), and there could be a competitive relationship between their precipitation and the γ' coarsening process. The value is, however, still consistent when compared to N18 (335 $kJ \cdot mol^{-1}$ [37]) and René 65 (350–358 $kJ \cdot mol^{-1}$ [38, 39]).

From the results presented previously, it seems that thermal aging allows mitigating the chemical segregation and leads to a better chemical homogeneity between dendritic cores and interdendritic spacings, thanks to the diffusion of solute elements in the matrix. To verify this, SEM–EDS measurements were carried out again after 4 different aging treatments. The segregation coefficient k' was calculated for the main alloying elements i based on their atomic fractions x_i:

$$k' = \frac{x_{i,\ \text{dendrite core}}}{x_{i,\ \text{interdendritic spacing}}} \tag{4}$$

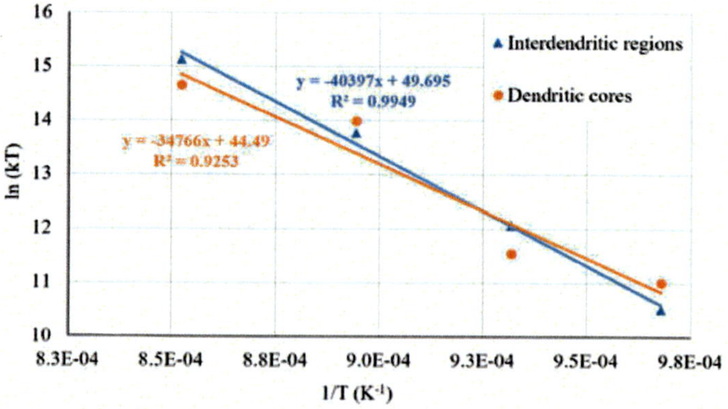

Fig. 7 Arrhenius plot for γ' coarsening in CMT Waspaloy

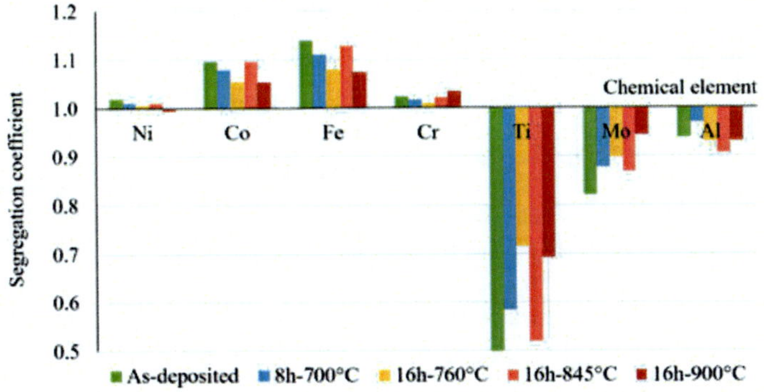

Fig. 8 Segregation coefficient evolution with aging heat treatments

Figure 8 shows k' in the as-deposited condition and its evolution with various aging temperatures and times. While $k' = 1$ implies the alloying element is perfectly distributed, $k' > 1$ means the element is preferentially located in the dendrite cores and $k' < 1$ in the interdendritic spacings. Ni and Cr are uniformly spread out unlike Co and Fe which are mostly in the dendritic cores. Ti, Mo, and Al are concentrated in the interdendritic spacings. With the exception of the data at 845 °C, there is a general trend for k' to approach 1 when increasing the aging temperature, especially when comparing the as-deposited state with the results after a 16 h exposure at 900 °C. Co and Fe segregation coefficients decrease lightly towards a value of 1.0, while Ti and Mo coefficients are getting closer to 1.0 in a more obvious way. The major deviation for 845 °C could be explained by the formation of intermetallic phases in rather large amounts at this temperature. It is also worth recalling that since the additive material is very heterogeneous by nature, variations could also be due to the sample extraction location in the wall or the position of the analyzed surface within a layer.

Room Temperature Tensile Properties After Aging

Figure 9a shows the tensile properties at room temperature for the as-deposited material, along loading directions TD and BD, with a comparison to the C&W materials (sub- and super-solvus). It can be seen that additive samples, prior to aging, have a remarkable ductility. The vertical specimen has an elongation of 44.4% due to the columnar grain growth orientation being parallel to the loading direction. Classical values for strain at failure in wrought Waspaloy optimized with complete heat treatment (solution annealing and double aging) are about 24.5% [40]. The yield strength (YS) and ultimate tensile strength (UTS) prior to aging, however, are significantly lower than for the wrought Waspaloy: 557 and 850 MPa respectively, on average. This is in agreement with the works of *Xu* et al. on Plasma Arc Welding (PAW)-WAAM Inconel 718; an average YS of 515 MPa and UTS of 793 MPa were found in this as-deposited material [8]. The curves for the horizontal and the

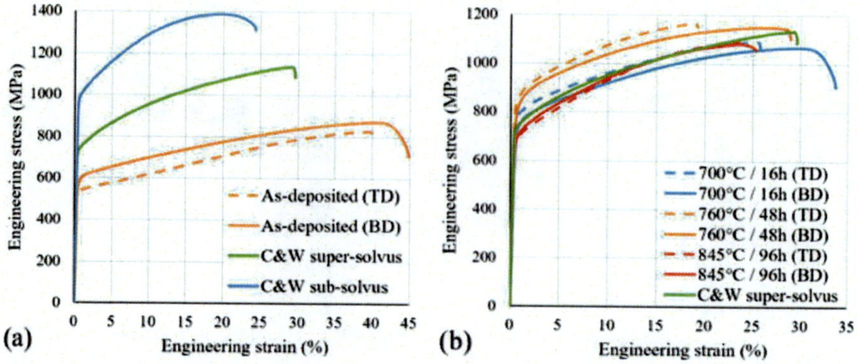

Fig. 9 Engineering stress-strain curves at room temperature for CMT Waspaloy **a** as-deposited and **b** direct-aged, including tests on C&W material for comparison

vertical CMT are different, indicating that there is anisotropy inherited from the additive process. As expected, the sub-solvus wrought sample exhibited better tensile behavior than the super-solvus, thanks to the much smaller grain size (21 vs. 270 μm for sub-solvus/super-solvus, respectively).

As can be seen in Fig. 9b, direct aging allows for a significant increase in YS and UTS of CMT Waspaloy. Because such a type of direct aging does not affect grain size, this is achieved mainly through the nucleation and coarsening of γ', as demonstrated previously. Elongation, on the other hand, decreases but remains within a reasonable 16–28% range. Within the ranges of temperature and exposure times studied in this work, aging results in such a major improvement of tensile properties that the additive samples exhibit a behavior close to the one of the super-solvus wrought material. Columnar grains in CMT Waspaloy have an average width of about 200 μm (see Fig. 2d, similar to WAAM Inconel 718 [41]), meaning that the debit in tensile strength observed in Fig. 9b is mainly a result of the out-of-equilibrium γ' precipitation due to the fast cooling during the CMT process.

The evolution of tensile properties (YS, UTS, and elongation) at room temperature as a function of exposure time for each aging temperature is presented in Fig. 10. At 700 °C (Fig. 10a), YS and UTS improved with increasing aging time for both TD and BD loading directions. Moreover, it is important to note the major difference in properties after only 4 h at this lower temperature. As it has been shown γ' size does not considerably change at this temperature (see Fig. 5a–b), this improvement must be due to an increase in precipitate volume fraction. At 760 °C (Fig. 10b), vertical and horizontal samples showed different behaviors. Although YS and UTS strongly increased at first for both, a decrease was observed after 48 h for loading in TD while BD steadily escalated. At 845 °C (Fig. 10c), despite an increase of properties after 4 h of annealing, a decrease in YS and UTS is then observed when extending the exposure time.

A general decrease in ductility (−26% on average) occurs after direct aging, and for a majority of the tests carried out, the minimal elongation is reached after

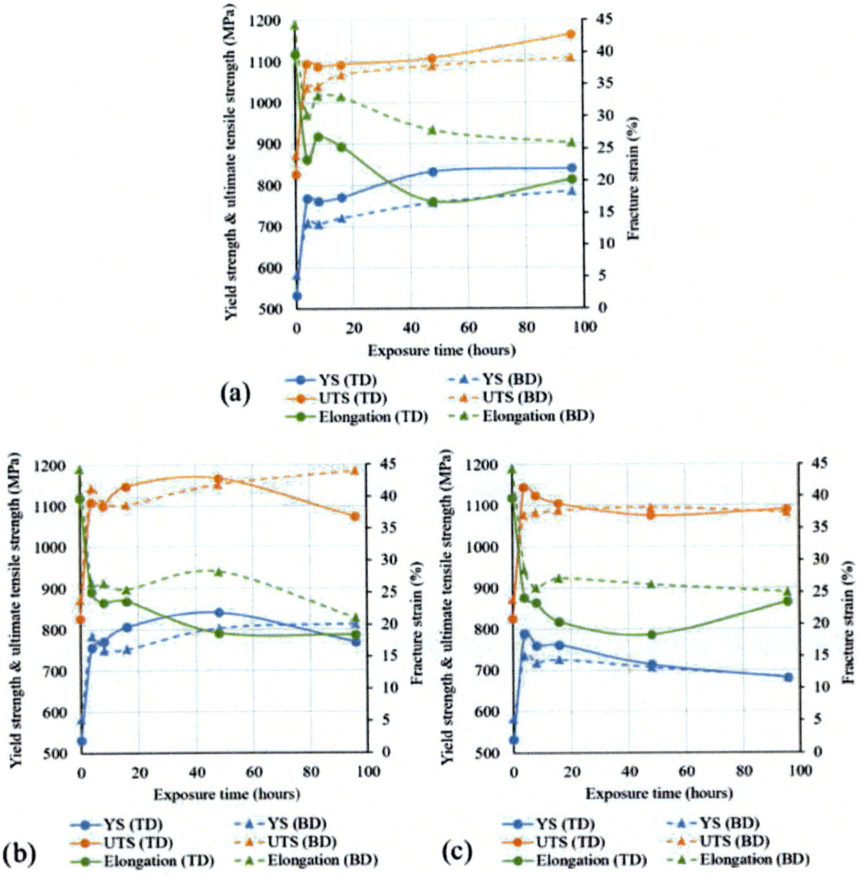

Fig. 10 YS, UTS, and fracture strain evolutions as a function of exposure time at **a** 700 °C, **b** 760 °C, and **c** 845 °C

48 h. Performing an aging heat treatment on CMT Waspaloy is, nevertheless, quickly advantageous for YS and UTS, especially at lower temperatures (700–760 °C) and short times (4 h). Indeed, enhancements of +308 and +338 MPa were obtained respectively, on average.

In order to estimate a potential evolution in tensile anisotropy, property ratios $\frac{YS_H}{YS_V}$ and $\frac{UTS_H}{UTS_V}$ can be calculated. The closer the ratio is to 1, the more isotropic the material is. Figure 11 shows the extent of anisotropy at room temperature in the as-deposited condition, as well as a few examples of direct aging heat treatments that greatly reduced anisotropy.

All the data presented in Figs. 10 and 11 suggest that 4 h at 760 °C may be the most interesting compromise with such printing parameters to achieve close-to-isotropic CMT Waspaloy in tension. In fact, it allows to significantly increase YS (+182 MPa) and UTS (+217 MPa), maintain a fair ductility (25%), and reduce tensile

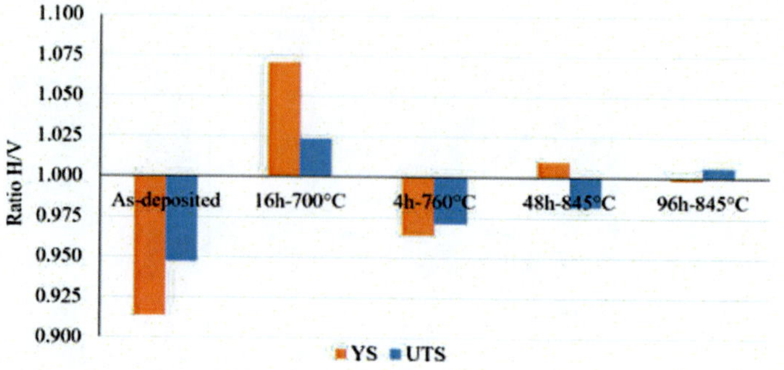

Fig. 11 Property ratios of yield strength (YS) and ultimate tensile strength (UTS) for CMT Waspaloy at room temperature

anisotropy by half compared to the as-deposited state. Moreover, this heat treatment is quick and at a reasonably low temperature, meaning that it could be applicable for refurbishment practices as it may not further degrade the bulk microstructure of the repaired substrate.

Effect of Thermal Exposure on Phase Formation

While aging allows for γ' coarsening and improvement of tensile properties, precipitation of intermetallic phases also occurs. Grain boundary decoration is observed for as soon as 4 h at 700 °C, as well as the formation of bright plate-like precipitates in the interdendritic spacings, as shown in Fig. 12a. While the grain boundary particles are typical for this material, the plates are unusual. Preferentially located around primary (Ti, Mo)C carbides, this phase seems to have a favorable precipitation orientation since all plates exhibit a parallel and perpendicular relationship to each other (Fig. 12b). Due to this specific morphology and orientation, the first hypothesis on the nature of such white phases leaned towards TCP phases.

Figure 13 displays the TEM-EDS element maps, showing the grain boundary is an alternation of large γ' (150–200 nm) with secondary carbides enriched in Cr and Mo. Quantitative analysis revealed they are $(Cr, Mo)_{23}C_6$, as reported in the literature [3, 42]. The plate-like precipitates also exhibited a similar elemental distribution, but since they are thinner and surrounded by more γ matrix, they do not show up as well on the distribution maps.

Nevertheless, chemical composition and precipitate morphology are not sufficient to identify an unknown phase. TCP phases σ and μ, for example, are known for having needle and plate-like morphologies, and the following chemical compositions: $(Cr, Mo)_6(Ni, Co)_4$ [43, 44] and $(Co, Fe, Ni)_7(Mo, W, Cr)_6$ [45], respectively. For this reason, the crystal structure of these white phases was studied using SAED. Figure 14 presents electron diffraction patterns in zone axes [001] and [111] of three different phases at once: γ, γ', and the plate-like particles. The latter appears to have a cubic lattice, with a d-spacing (10.7 Å) almost three times larger than that of γ/γ' (3.6 Å).

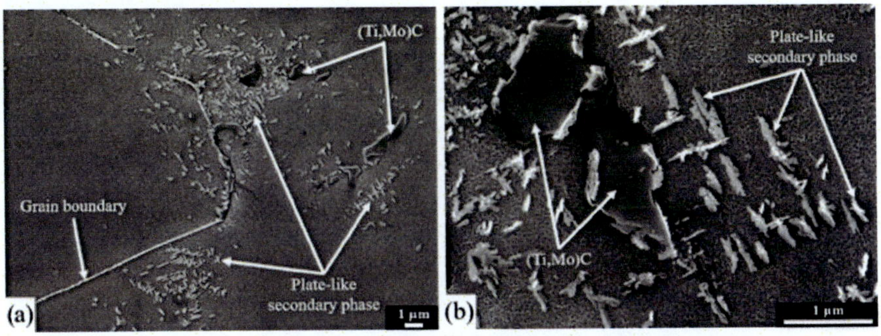

Fig. 12 Secondary electron images of CMT Waspaloy microstructure after exposure at 800 °C for 16 h **a** general overview and **b** interdendritic plate-like precipitates at higher magnification

[110] zone axis was also analyzed to verify that the crystallographic structure of these white phases is indeed cubic, and not tetragonal. Identical patterns were found for grain boundary secondary carbides. All these features are typical for $M_{23}C_6$ in nickel-based superalloys [46, 47].

Moreover, quantitative TEM-EDS analyses in extremely thin areas of TEM foils allowed for an estimation of the chemical composition of the interdendritic precipitates: 46.55 Cr, 23.07 C, 14.61 Ni, 9.84 Mo, 4.69 Co, 0.81 Ti, 0.34 Al, and 0.08 Fe (at %). Therefore, it can be concluded the plate-like precipitates are also secondary carbides of the $M_{23}C_6$ type. Even so, they relate to the matrix according to the well-known cube-on-cube orientation relationship: $(111)_{M23C6}$ // $(111)_{\gamma/\gamma'}$ and $[110]_{M23C6}$ // $[110]_{\gamma/\gamma'}$.

Finally, thermodynamic calculations were carried out using Thermo-Calc and the TCNi10 database. From SEM observations, it was assumed that CMT Waspaloy had a ratio of 25/75 between the volumetric fraction of interdendritic regions and that of dendritic regions, meaning the solidification of the last 25% of liquid generates the former. Based on the chemical composition of the wire used to manufacture the samples (see Table 1), the Scheil–Gulliver model was applied to determine the chemical composition of the last 25% of liquid. Equilibrium calculations were then carried out using the previously calculated composition as input data. This allowed to simulate more precisely which phases are present in the interdendritic spacings at equilibrium and what their compositions are. Figure 15 shows the evolution of phase molar fraction with temperature. As observed in this study, the software predicted the formation of TiC and $M_{23}C_6$. Although not detected experimentally in the 700–900 °C temperature range, the precipitation of TCP phases σ and η is also indicated, in equilibrium conditions. An explanation for this difference is that the CMT process induces a non-equilibrium solidification, and the material therefore requires much longer exposure times to approach equilibrium. The average chemical composition predicted by Thermo-Calc for $M_{23}C_6$, η, and σ between 700 and 900 °C is shown in Table 3.

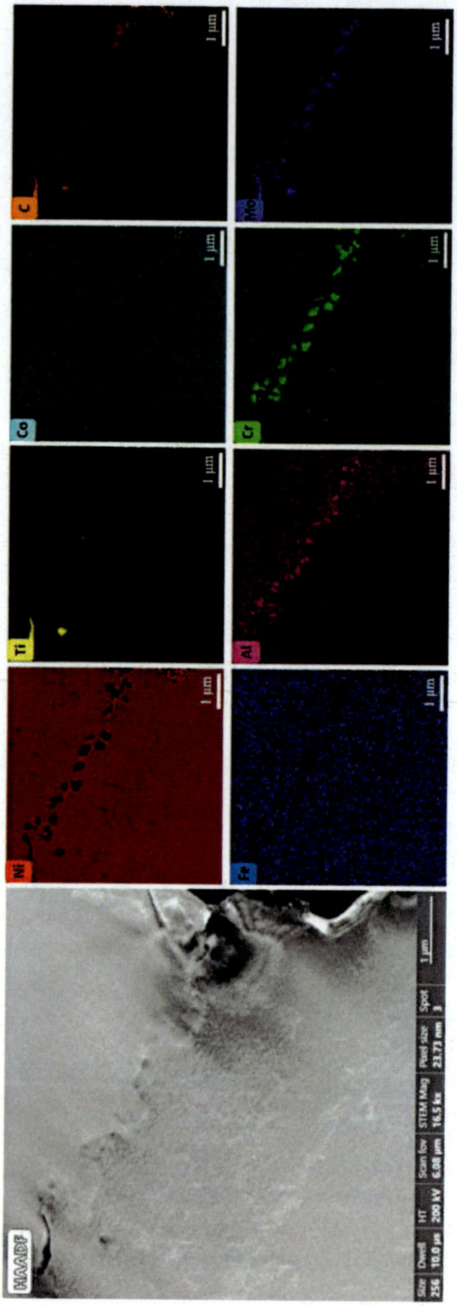

Fig. 13 TEM-EDS element distribution maps of a grain boundary and plate-like precipitates in a sample aged 48 h at 760 °C

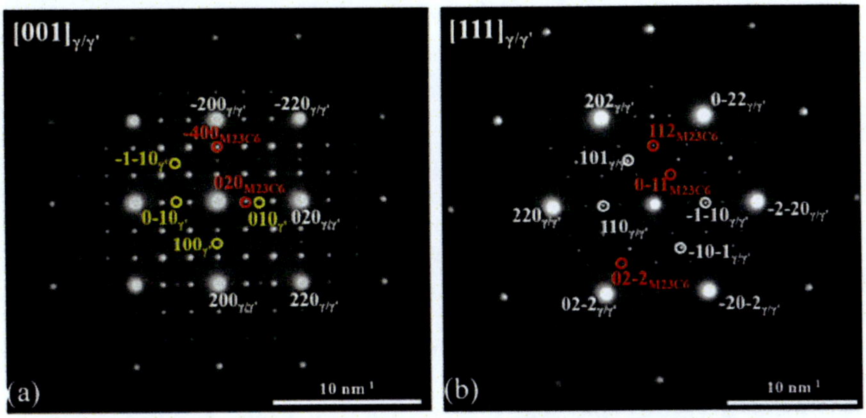

Fig. 14 SAED patterns for γ, γ′, and $M_{23}C_6$ in **a** [001] zone axis and **b** [111] zone axis

Table 3 Chemical composition (at %) of the intermetallic phases in interdendritic regions of CMT Waspaloy at 700–900 °C, as predicted by Thermo-Calc

	Cr	Co	Mo	Ti	Fe	Al	Mn	Si	Ni	C
$M_{23}C_6$	66.31 ±0.49	0.92 ±0.18	9.43 ±0.16	0	0.20 ±0.02	0	~0	0	~2.45 ±0.45	20.70 ±0.01
η	42.25 ±0.55	12.16 ±0.27	21.21 ±0.43	~0	1.23 ±0.15	~0	~0	~0	~23.17 ±0.53	0
σ	0.61 ±0.04	3.17 ±0.41	0.11 ±0.05	22.34 ±0.33	0.49 ±0.04	1.97 ±0.32	0	~0	~71.33 ± 0.52	0

Fig. 15 Equilibrium molar fraction of phases as a function of temperature in the interdendritic regions of CMT Waspaloy, as predicted by Thermo-Calc from the composition of the last 25% of liquid

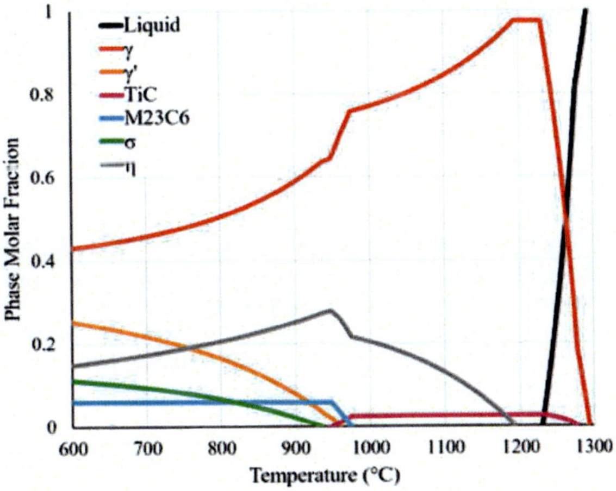

Conclusions

The effect of thermal exposure in the 700–900 °C temperature range on the microstructure and tensile properties of CMT Waspaloy has been investigated. The as-deposited microstructure consists of columnar grains with a strong <001> texture and dendritic structure within, leading to major chemical segregation and therefore heterogeneous γ' precipitation. While mostly nucleation occurred at 700 °C, the growth kinetics of γ' in both dendritic cores and interdendritic spacings in the 760–900 °C temperature range are in good agreement with the LSW theory. Activation energy for the diffusion-controlled coarsening was determined to be 279.9 $kJ \cdot mol^{-1}$ for dendritic cores and 326.7 $kJ \cdot mol^{-1}$ for interdendritic regions. Moreover, aging allowed for the reduction of chemical segregation through the diffusion of solutes and resulted in γ' size homogenizing between the two regions.

Thanks to aging and its resulting γ' nucleation/coarsening, tensile properties (YS and UTS) at room temperature were increased compared to the as-deposited material up to levels similar to a super-solvus C&W material. The tensile anisotropy was also reduced after aging compared to as-built conditions.

Finally, the formation of intermetallic phases was assessed. $(Cr, Mo)_{23}C_6$ secondary carbides decorate the grain boundaries, while plate-like precipitates appear near (Ti, Mo)C primary carbides in the interdendritic spacings. SAED revealed the plates have a cubic lattice with a d-spacing estimated at 10.7 Å. TEM-EDS highlighted that they are enriched in Cr (46.55 at %), C (23.07 at %), and Mo (9.84 at %), therefore they were identified as $M_{23}C_6$ carbides as well. Thermodynamic calculations were in agreement with the experimental observations, predicting the formation of both MC and $M_{23}C_6$ in interdendritic spacings at equilibrium conditions. Although η and σ phases were expected from the simulations, they were not found in the materials under the aging conditions studied in the present work.

Acknowledgements This study is part of the ongoing project on Process Optimization in Additive Repair with Safran Aircraft Engines, more specifically the MRO center of Safran Aircraft Engines of Châtellerault. Fruitful discussions with A. Malié, F. Corpace, G. Seure (all at Safran Aircraft Engines), and with S. Rix (formerly at Safran Aircraft Engines) are gratefully acknowledged. Pprime Institute gratefully acknowledges "Contrat de Plan Etat—Région Nouvelle-Aquitaine" (CPER) as well as the "Fonds Européen de Développement Régional (FEDER)" for their financial support to the reported work in the context of the OPERA program (convention P-2020-BAFE-3). This work was partially funded by the French Government program "Investissements d'Avenir" (EQUIPEX GAP, reference ANR-11-EQPX-0018). I. Lopez-Galilea acknowledges funding by the "Deutsche Forschungsgemeinschaft" (DFG) in the framework of the collaborative research center SFB/TR 103 through project T4.

References

1. Thielemann RH (1960) High Temperature Nickel Base Alloy. US. Patent 2,948,606. 9 August 1960
2. DuPont JN, Lippold JC, Kiser SD (2009) Welding Metallurgy and Weldability of Nickel-base alloys. Wiley
3. Andersson J (2011) Weldability of precipitation hardening superalloys: influence of microstructure. PhD thesis, Chalmers University of Technology
4. Agnoli A, Le Gall C, Thebault J, et al (2018) Mechanical Properties Evolution of γ'/γ'' Nickel-Base Superalloys During Long-Term Thermal Over-Aging. Metallurgical and Materials Transactions A 49:4290–4300. https://doi.org/10.1007/s11661-018-4778-x
5. Cong B, Ding J, Williams S (2015) Effect of arc mode in cold metal transfer process on porosity of additively manufactured Al-6.3%Cu alloy. The International Journal of Advanced Manufacturing Technology 76:1593–1606. https://doi.org/10.1007/s00170-014-6346-x.
6. Fronius (2017) Cold Metal Transfer
7. Benoit A, Jobez S, Paillard P, et al (2011) Study of Inconel 718 weldability using MIG CMT process. Science and Technology of Welding and Joining 16:477–482. https://doi.org/10.1179/1362171811Y.0000000031
8. Xu X, Ganguly S, Ding J, et al (2018) Enhancing mechanical properties of wire + arc additively manufactured INCONEL 718 superalloy through in-process thermomechanical processing. Materials & Design 160:1042–1051. https://doi.org/10.1016/j.matdes.2018.10.038
9. Andersson J, Sjöberg GP (2012) Repair welding of wrought superalloys: Alloy 718, Allvac 718Plus and Waspaloy. Science and Technology of Welding and Joining 17:49–59. https://doi.org/10.1179/1362171811Y.0000000077
10. Seow CE, Coules HE, Wu G, et al (2019) Wire + Arc Additively Manufactured Inconel 718: Effect of post-deposition heat treatments on microstructure and tensile properties. Materials & Design 183:108157. https://doi.org/10.1016/j.matdes.2019.108157
11. Kindermann RM, Roy MJ, Morana R, Prangnell PB (2020) Process response of Inconel 718 to wire + arc additive manufacturing with cold metal transfer. Materials & Design 195:109031. https://doi.org/10.1016/j.matdes.2020.109031
12. Ola OT, Doern FE (2014) A study of cold metal transfer clads in nickel-base INCONEL 718 superalloy. Materials & Design 57:51–59. https://doi.org/10.1016/j.matdes.2013.12.060
13. Kaldellis A, Alexandratou A, Kladis A, et al (2021) Comparative Study of Microstructural Evolution and Mechanical Properties of Inconel® 718 and Waspaloy® Welds. MATEC Web of Conferences 349:02004. https://doi.org/10.1051/matecconf/202134902004
14. Gregori A, Bertaso D (2007) Welding and Deposition of Nickel Superalloys 718, Waspaloy and Single Crystal Alloy CMSX-10. Welding in the World 51:34–47. https://doi.org/10.1007/BF03266607
15. Prager M, Shira CS (1968) Welding of Precipitation Hardening Nickel-Base Alloys. Weld Research Council Bulletin 6:128–155.
16. Andersson J, Raza S, Eliasson A, Surreddi KB (2014) Solidification of Alloy 718, ATI 718Plus and Waspaloy. John Wiley & Sons, Inc., pp 145–156
17. Sims CT, Stoloff NS, Hagel WC (1987) Superalloys II: High-Temperature Materials for Aerospace and Industrial Power. John Wiley & Sons, Inc.
18. Reed RC (2008) The Superalloys: Fundamentals and Applications. Cambridge University Press
19. Darolia R, Lahrman DF, Field RD (1988) Formation of Topologically Closed Packed Phases in Nickel Base Single Crystal Superalloys. Superalloys 1988 255–264
20. Rideout S, Manly WD, Kamen EL, et al (1951) Intermediate Phases in Ternary Alloy Systems Of Transition Elements. Journal of Metals 3:872–876. https://doi.org/10.1007/BF03397394
21. Long F, Yoo YS, Jo CY, et al (2009) Formation of η and σ phase in three polycrystalline superalloys and their impact on tensile properties. Materials Science and Engineering: A 527:361–369. https://doi.org/10.1016/j.msea.2009.09.016
22. Simonetti M, Caron P (1998) Role and behaviour of μ phase during deformation of a nickel-based single crystal superalloy. Materials Science and Engineering: A 254:1–12

23. Acharya MV, Fuchs GE (2004) The effect of long-term thermal exposures on the microstructure and properties of CMSX-10 single crystal Ni-base superalloys. Materials Science and Engineering: A 381:143–153. https://doi.org/10.1016/j.msea.2004.04.001
24. Zhao K, YH M, ZQ H (2005) μ phase in a nickel base directionally solidified alloy. Materials transactions 46:54–58
25. DebRoy T, Wei HL, Zuback JS, et al (2018) Additive manufacturing of metallic components – Process, structure and properties. Progress in Materials Science 92:112–224. https://doi.org/10.1016/j.pmatsci.2017.10.001.
26. Cabeza S, Özcan B, Cormier J, et al (2020) Strain Monitoring During Laser Metal Deposition of Inconel 718 by Neutron Diffraction. In: Tin S, Hardy M, Clews J, et al (eds) Superalloys 2020. Springer International Publishing, Cham, pp 1033–1045
27. Wagner C (1961) Theorie der Alterung von Niederschlägen durch Umlösen (Ostwald-Reifung). Zeitschrift für Elektrochemie, Berichte der Bunsengesellschaft für physikalische Chemie 65:581–591. https://doi.org/10.1002/bbpc.19610650704
28. Lifshitz IM, Slyozov VV (1961) The kinetics of precipitation from supersaturated solid solutions. Journal of Physics and Chemistry of Solids 19:35–50. https://doi.org/10.1016/0022-3697(61)90054-3
29. Baldan A (2002) Review Progress in Ostwald ripening theories and their applications to nickel-base superalloys -Part II: Nickel-base superalloys. Journal of Materials Science 37:2379–2405. https://doi.org/10.1023/A:1015408116016
30. Swalin RA (1957) A model for solute diffusion in metals based on elasticity concepts. Acta Metallurgica 5:443–448. https://doi.org/10.1016/0001-6160(57)90062-7
31. Swalin RA, Martin A (1956) Solute Diffusion in Nickel-Base Substitutional Solid Solutions. Journal of Metals 8:567–571. https://doi.org/10.1007/BF03377730
32. Wang H, Liu D, Shi Y, et al (2019) Matrix-Diffusion-Controlled Coarsening of the γ′ Phase in Waspaloy. Metals and Materials International 25:1410–1419. https://doi.org/10.1007/s12540-019-00274-7
33. Ardell AJ, Nicholson RB (1966) On the modulated structure of aged Ni-Al alloys: with an Appendix On the elastic interaction between inclusions by J. D. Eshelby. Acta Metallurgica 14:1295–1309. https://doi.org/10.1016/0001-6160(66)90247-1
34. Ardell AJ (1970) The growth of gamma prime precipitates in aged Ni-Ti alloys. Metallurgical Transactions B 1:525–534
35. Muralidharan G, Chen H (2000) Coarsening kinetics of coherent γ′ precipitates in ternary Ni–based alloys: the Ni–Al–Si system. Science and Technology of Advanced Materials 1:51–62. https://doi.org/10.1016/S1468-6996(00)00005-X
36. Robson JD (2004) Modelling the evolution of particle size distribution during nucleation, growth and coarsening. Materials Science and Technology 20:441–448. https://doi.org/10.1179/026708304225016725
37. Flageolet B, Villechaise P, Jouiad M, Mendez J (2004) Ageing Characterization of the Power Metallurgy Superalloy N18. TMS, pp 371–379
38. Laurence A, Cormier J, Villechaise P, et al (2014) Impact of the solution cooling rate and of thermal aging on the creep properties of the new cast & wrought René 65 Ni-based superalloy. pp 333–348
39. Wessman AE, Laurence A, Cormier J, et al (2016) Thermal Stability of Cast and Wrought Alloy Rene 65. The Minerals, Metals & Materials Society, Seven Springs, PA, USA, pp 793–800
40. Haynes International (2017) HAYNES® Waspaloy alloy
41. Xu X, Ding J, Ganguly S, Williams S (2019) Investigation of process factors affecting mechanical properties of INCONEL 718 superalloy in wire + arc additive manufacture process. Journal of Materials Processing Technology 265:201–209. https://doi.org/10.1016/j.jmatprotec.2018.10.023
42. Liu G, Kong L, Xiao X, Birosca S (2022) Microstructure evolution and phase transformation in a nickel-based superalloy with varying Ti/Al ratios: Part 1 - Microstructure evolution. Materials Science and Engineering: A 831:142228. https://doi.org/10.1016/j.msea.2021.142228

43. Antonov S, Huo J, Feng Q, et al (2017) σ and η Phase Formation in Advanced Polycrystalline Ni-base Superalloys. Materials Science and Engineering: A 687:232–240. https://doi.org/10.1016/j.msea.2017.01.064
44. Jackson MP (1998) Modelling and Characterisation of Phase Transformation in Nickel-base Superalloys. PhD Thesis, University of Cambridge
45. Pessah-Simonetti M (1994) Effets des instabilités structurales sur les propriétés mécaniques du superalliage monocristallin MC2. PhD Thesis, Université de Paris-Sud
46. Carter CB, Williams DB (2016) Transmission Electron Microscopy: Diffraction, Imaging, and Spectrometry. Springer Cham
47. Wilson AS, Christofidou KA, Evans A, et al (2019) Comparison of Methods for Quantification of Topologically Close-Packed Phases in Ni-Based Superalloys. Metallurgical and Materials Transactions A 50:5925–5934. https://doi.org/10.1007/s11661-019-05442-3

IN718 Cold Gas Repair Spray of Large Cavities—Microstructure and Residual Stresses

Florian Lang, Johannes-Christian Schmitt, Sandra Cabeza, Thilo Pirling, Jochen Fiebig, Robert Vassen, and Jens Gibmeier

Abstract Cold gas spray is an established process for coating substrates with similar or dissimilar materials. By use of a high-pressure process gas stream, solid particles are accelerated onto a substrate at supersonic velocities. The method is particularly suited for repair applications since neither structural changes nor oxidation occur during the process. To investigate the suitability of the cold gas spray process for the repair of major defects with up to 4 mm depth in Inconel 718 components, sample geometries were manufactured, containing tapered cavities. The specimen cavities were filled with Inconel 718 particles by a cold gas spray process. Non-destructive high-resolution neutron diffraction experiments were performed by use of the SALSA instrument at the Institut Laue-Langevin (ILL) to evaluate the local residual stress state in the as-sprayed condition. 2D maps of the residual stress distribution over the cross-sectional area of the filled cavities were determined. The results indicate compressive residual stresses within the filled process zone. Metallographic

F. Lang (✉) · J. Gibmeier
Institute for Applied Materials (IAM-WK), Karlsruhe Institute of Technology (KIT), Engelbert-Arnold-Str. 4, Karlsruhe 76131, Germany
e-mail: Florian.Lang@kit.edu

J. Gibmeier
e-mail: Jens.Gibmeier@kit.edu

J.-C. Schmitt · J. Fiebig · R. Vassen
Institute of Energy and Climate Research (IEK-1), Forschungszentrum Jülich GmbH, Wilhelm-Johnen-Strasse, Jülich, 52425, Germany
e-mail: Jo.Schmitt@fz-juelich.de

J. Fiebig
e-mail: J.Fiebig@fz-juelich.de

R. Vassen
e-mail: R.Vassen@fz-juelich.de

S. Cabeza · T. Pirling
Institut Laue-Langevin (ILL), 71 Avenue des Martyrs, Grenoble, 38000, France
e-mail: Cabeza@ill.fr

T. Pirling
e-mail: Pirling@ill.eu

E. A. Ott et al. (eds.), *Proceedings of the 10th International Symposium on Superalloy 718 and Derivatives*, The Minerals, Metals & Materials Series,
https://doi.org/10.1007/978-3-031-27447-3_44

examinations show a good bonding between the repair filling and the substrate as well as strongly deformed particles within the repaired region. The latter indicates significant plastic deformation during cold gas spray, which is also in good agreement with increased diffraction line width from the neutron diffraction analyses in the filled process zone compared to the surrounding substrate.

Keywords Cold gas repair spray · Residual stress analysis · Neutron diffraction · Inconel 718

Introduction

Cold gas spray (CGS) is an established thermal spraying technique for depositing similar or dissimilar materials onto a substrate by combining two or more solid phases. By use of a high-pressure process gas stream, solid particles with a diameter of $1\,\mu m \leq d_{particle} \leq 100\,\mu m$ are accelerated to supersonic velocities (up to $1500\,ms^{-1}$ [1–4]) and directed onto a substrate, where they form a closed coating. To achieve this, particles are injected into a convergent-divergent de-Laval nozzle where they are mixed with a pre-heated main gas stream (T_{gas} ≤1500 K). While passing the nozzle, the gas-particle stream is accelerated to supersonic speed and cools down in the process. Pre-heating the main gas flow increases the sonic speed of the gas and allows for the acceleration of the particles to higher velocities. Usually, Helium or Nitrogen (or a mixture of both) are used as process gas, due to their thermodynamic properties [5]. Figure 1 shows a schematic of the CGS process, wherein a process gas is divided into a main and a by-pass flow. The main flow is heated in a gas heater to increase the sonic speed. The by-pass flow is used to transport particles from the powder feeder into a de-Laval nozzle where it is mixed with the main flow and then accelerated onto a substrate where a dense coating is formed. The bonding of particles to the substrate takes place due to the high kinetic energy of the particles, which leads to strong plastic deformations and adiabatic shear instabilities [1, 6, 7]. They in turn lead to a form-fit connection of particles to the substrate. In contrast to other thermal spray processes, CGS is considered as a low-temperature process, since the deposition takes place in the solid instead of the liquid or gaseous state. It is also more environmentally friendly, since the power requirement is comparatively low, and no toxic gases or chemicals are needed [8–10]. Applications of CGS include surface functionalisation, e.g. to improve wear and corrosion resistance [11, 12] or create an antimicrobial surface [13], additive manufacturing of parts (CSAM—Cold Spray Additive Manufacturing) [14–16], and the repair of damaged components (CGRS—Cold Gas Repair Spray). Recently, CGRS has been the subject of increasing research activities. Some examples from the literature where CGRS has been successfully used to repair a variety of components with superficial, corrosion, or wear-related damage in different applications are briefly mentioned below: Wiedener et al. used CGRS to repair the hard-to-reach internal sealing bore hole of a 6061 aluminium alloy hydraulic valve actuator for naval applications with similar materials [5]. In military

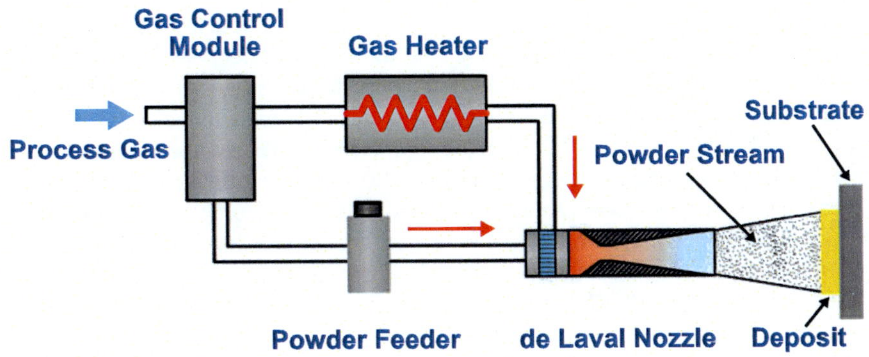

Fig. 1 Diagram of the cold gas spray process

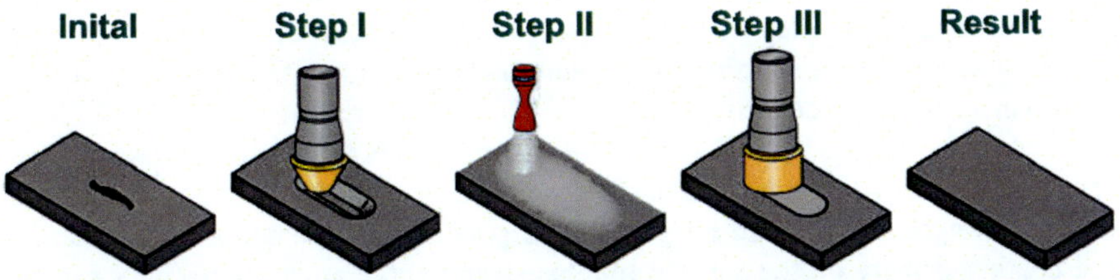

Fig. 2 Sequence scheme of cold gas repair spray process: a near-surface damage which occurred during operation (Initial) is milled out and the surface is prepared for the coating (Step I). The cavity is filled by means of cold gas spray (Step II). The specimen is then machined to restore the original geometry (Step III) in order to obtain a refurbished component for recommissioning (Result)

aircraft, chafing around fastener holes in aircraft skin leads to wear damages and results in the fit of the panel exceeding tolerances. Using CGRS, 2024-T6 aluminium skin panels were repaired using 6061 aluminium particles instead of being replaced saving money and conserving resources [17, 18]. MOOG Aircraft Group found that worn Inconel nose-wheel steering actuators of a Boeing 737, repaired by means of CGRS, showed better corrosion resistance when compared to replacement parts [19]. Figure 2 schematically illustrates the procedure to CGRS components: The initial state is a near-surface damage that occurred during operation, e.g. erosion of the surface (Initial). Similar to other repair techniques, the appropriate specimen preparation consists of carving out the defect and conditioning the surface by means of peening or sandblasting, as the complex surface topography, e.g. showing wear and cracks, as well as the unclean surface in the damaged area can affect the quality of the repair (Step I). The machined cavity is filled by means of CGS (Step II). The specimen is then reworked to the original geometry, i.e. to remove surplus material (Step III), in order to obtain a restored component for recommissioning (Result). CGS is particularly suited for repair applications, since neither structural changes nor oxidation occur during the deposition [20].

Usually, for CGS the substrate surface is pre-treated by sandblasting for cleaning and improved adhesion. In combination with the blasting effects of CGS, this leads to compressive residual stresses (RS) in the near-surface region of the component [3, 21–23]. This can be positive with regard to the mechanical integrity for coatings that cover the entire surface as long as the compressive RS is not high enough for the stress state to cause spallation of the coating. However, when filling confined cavities, the influence must be investigated as compressive RS are accompanied by balancing tensile RS in the surrounding area. Depending on the location, this can also have a negative influence on the fatigue strength. A number of studies were performed on CGS of Inconel 718 (IN718), especially with regard to a potential application in aerospace components. Wong et al., for example, compared microstructural and mechanical properties of 6mm thick IN718 coatings on Al-alloys for the as-sprayed condition with different subsequent heat treatments and found very low ductility in the as-sprayed condition. They observed improved tensile strength and ductility in heat-treated samples with the degree in improvement depending on the heat treatment and hypothesised sintering effects during heat treatment that improved the interconnection in the coating [24].

Singh et al. produced very dense IN718 coatings on IN718 substrates, using Nitrogen as process gas and investigated effects of surface roughness, stand-off distance, and spray angle. They found a significant dependence of the deposition efficiency and coating density on the surface roughness and spray angle. Especially for fine grit blasted substrates and coating thickness, they observed coating peel-off [25]. In a subsequent study, Singh et al. analysed the through-thickness RS distribution and adhesion strength for different coating thicknesses of IN718 particles deposited on IN718 substrates by means of CGS as a tool for aircraft engine component repair. They found axisymmetric compressive RS increasing and adhesion strength decreasing with increasing coating thickness and concluded that the adhesion strength is influenced by the coating thickness as well as the RS state [22].

Vassen et al. studied CGS IN718 coatings with different thicknesses on IN718 substrates. Their analyses revealed similar microstructure and axisymmetric RS depth profiles for all investigated thicknesses as well as decreasing adhesion strength for increasing thickness due to a higher energy release rate in thick coatings. A subsequent heat treatment leads to RS relaxation in the coating and an increase in adhesion strength [23]. Fiebig et al. identified high compressive RS as a possible reason for the coating thickness dependency of the adhesion strength [26].

In order to extend CGS to the repair of large defects, especially in IN718 components used in turbine engines, the interaction of the repair site with the component must be investigated. An important aspect is a good and, above all, complete bonding of the repair to the substrate. To be able to evaluate the repair also in terms of mechanical integrity in use, the microstructure and RS are important. This applies not only to the filling, but also especially to the interface and the component beyond. Considering the intended application for the repair of aerospace components, which complicates subsequent heat treatment, it should be investigated whether the properties of the repair can already be controlled by suitable process parameters so that a condition is achieved that is not susceptible to distortion and has adequate mechanical properties.

To investigate the suitability of CGRS for the repair of defects with up to 4 mm depth, large tapered cavities were milled in IN718 specimens, which were then filled with similar material by means of CGS using two sets of parameters. The bonding of the CGS filling to the substrate was evaluated using microscopy techniques. The filled process zone was characterised in terms of hardness distribution, microstructure, and RS using neutron diffraction, as this is one of the few non-destructive methods for in-depth stress analysis. The method allows the determination of residual stresses in the filled process zone, at the interface, and also in the surrounding substrate. It permits quantification not only of the compressive residual stresses in the CS filling, but also of the compensating tensile residual stresses in the surrounding material in the as-sprayed state, which are crucial for assessing the repair and distortion potential during mechanical or thermal post-treatment.

Experimental

Sample Preparation

CGRS was performed on two cuboid samples made of Inconel 718 in a solution annealed state with a length L_S =100 mm, width W_S = 50 mm, and height H_S =12 mm. The repair site was machined using an angular milling cutter tool which allows for a tapered cavity. The cavity has a length of L_C =50 mm, a width of W_C =20 mm, and a depth of D_C =4 mm. The angle of the tapered sidewalls is $\alpha = 60°$. Figure 3 illustrates the dimensions of the specimen geometry and the specimen coordinate system. Tapered sidewalls were chosen to facilitate gas flow during CGS, as pre-trials using perpendicular sidewalls showed insufficient bonding of the filling to the substrate.

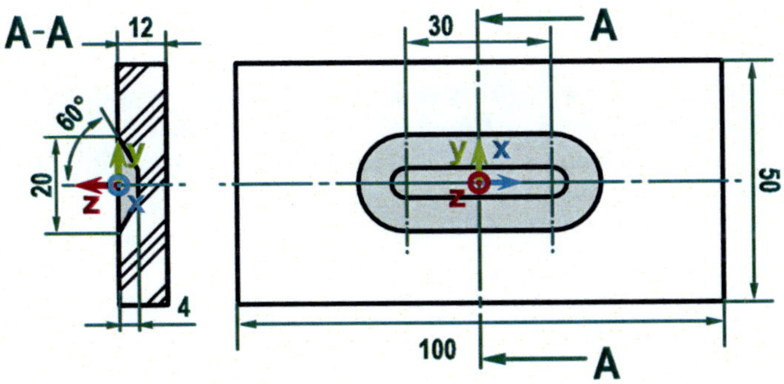

Fig. 3 Technical drawing of the sample geometry with a specimen coordinate system. All dimensions are given in mm

Table 1 Normalised cold gas spray process parameters of the specimens

Sample designation	Parameter I	Parameter II
Normalised gas pressure p^*_{gas}	1.0	1.125
Normalised gas temperature T^*_{gas}	1.0	1.11

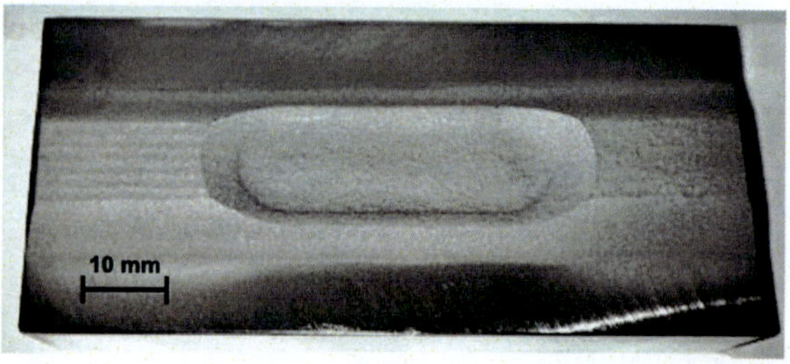

Fig. 4 Specimen in the as-sprayed condition. The image shows the sample Parameter II, according to Table 1. The surplus material is a result of the used spraying trajectory

After preparation, the repair site was filled by means of CGS with particles of the same chemical composition as IN718. The AE10718 powder from Oerlikon-Metco, Troy, MI, USA, has a spherical morphology and a mean particle size of $D_{50} = 14\,\mu\text{m}$. The IN718 powder particles are mixed with Al-rich Ni particles. This corresponds to an ongoing patent application by Oerlikon-Metco Inc. to improve the deposition of dense coatings [27]. More details about the powder are given in Vassen et al. [25] and Singh et al. [23]. An Impact Gun 5/11 from Impact Innovations GmbH, Rattenkirchen, Germany, was used as CGS system with Nitrogen as process gas. The 4 mm deep cavity is filled with 46 layers of a meander consisting of 20 line profiles running along the length of the cavity. To adapt the temperature of the substrate surface to the gas temperature and to reduce quenching RS, the sample surface is conditioned by three pre-heating cycles. The spray angle was set to $\alpha_{spray} = 90°$. CGS of the samples was performed at the Forschungszentrum Jülich GmbH, Jülich, Germany.

Two sample states were created using the CGS parameters in Table 1. This work is focused on the RS distributions in the as-sprayed condition, so after the second step (see Step 2 in Fig. 2) and before the original geometry is restored. Figure 4 shows an image of the sample as it was investigated in the as-sprayed condition with surplus material, before the original geometry is restored.

Metallographic Analyses

Metallographic examinations were performed to evaluate the capability of CGRS to successfully fill comparatively large cavities of IN718 with similar material. To that end, light optical microscopy (LOM) and scanning electron microscopy (SEM) as well as microhardness testing were performed on metallographically (ground and polished) prepared cross sections (A-A reference in Fig. 3) in the centre of the specimens for both parameters. The LOM images were taken using an inverted microscope Axio Vert 200 MAT from Carl Zeiss Microscopy Deutschland GmbH, Oberkochen, Germany. Images were taken at different magnifications. SEM microstructure observations were performed using a Zeiss LEO 1530, Carl Zeiss Microscopy Deutschland GmbH, Oberkochen, Deutschland, in the back-scattered electron imaging (BSE) mode with magnifications of up to 1000x. Vickers microhardness testing (HV 0.1) was performed according to DIN EN ISO 6507-1 [28], using a Q10 A+ from ATM Qness GmbH, Golling an der Salzach, Austria. A total of 1225 indentations were arranged in a $12 \times 6\text{mm}^2$ grid, spanning from the centre of the cavity into the surrounding substrate (c.f. A-A reference in Fig. 3). The distance between the indentations is $\Delta y = \Delta z = 250\,\mu\text{m}$.

Neutron Diffraction

Neutron diffraction analyses were carried out at the SALSA instrument [29] at Institut Laue Langevin, Grenoble, France, in the three principal directions x (longitudinal), y (transversal), and z (normal) inferred from the sample geometry (c.f. Fig. 3) according to Hutchings [30]. The data set is given in Gibmeier et al.[31]. A nominal gauge volume of $4 \times 0.6 \times 0.6\,\text{mm}^3$, defined by radial collimators (FWHM 0.6 mm), at the incident and diffracted beam paths was used. The gauge volume length of 4 mm is in the x-direction to improve the grain statistic, while assuming a homogeneous condition in this direction. This also results in a reduced acquisition time, due to the enlarged gauge volume. Figure 5 shows a schematic of the beam path. The displayed sample setup corresponds to a measurement in the z-direction (normal). Using the {422} reflection of a double-focusing bent Si-crystal monochromator, the wavelength was set to $\lambda \approx 1.5$ Å resulting in $2\theta_0 \approx 91°$ for the γ-Ni {311} reflection of IN718. Triaxial RS distributions were mapped at 50 positions arranged in a 10×5 grid over the mid-plane cross section of the filled cavity in the as-sprayed condition. The first row of the grid is positioned 1.5 mm below the specimen surface, to ensure that the gauge volume is entirely immersed into the material volume. The distance between the measurement positions is $\Delta y = 2$ mm and $\Delta z = 1$ mm in y- and z-directions (see A-A reference in Fig. 3), respectively. Additionally, d_0 reference samples were studied, using the same setup, to obtain the strain-free lattice parameter. The reference samples consisted of $2 \times 2 \times 4\text{mm}^3$ cuboids, which were electric discharge machine (EDM) wire cut from free-standing bar structures fabricated with the same coating parameters as the tested samples, as well as a $2 \times 2\times 2\text{mm}^3$ cuboid, EDM wire cut from the substrate material. Each d_0 value was then used for the calculation of strains in the respective material volume. After background subtraction, the peaks were fitted using a Gaussian function. All data treatment was carried out in LAMP software [32]. RS were calculated using the diffraction elastic constants $s_1^{hkl} = 1.523 \times 10^{-6}\,\text{MPa}^{-1}$

Fig. 5 Instrument setup of the SALSA instrument at the Institut Laue-Langevin in Grenoble, France. The displayed sample setup corresponds to a measurement in the z-direction (normal)

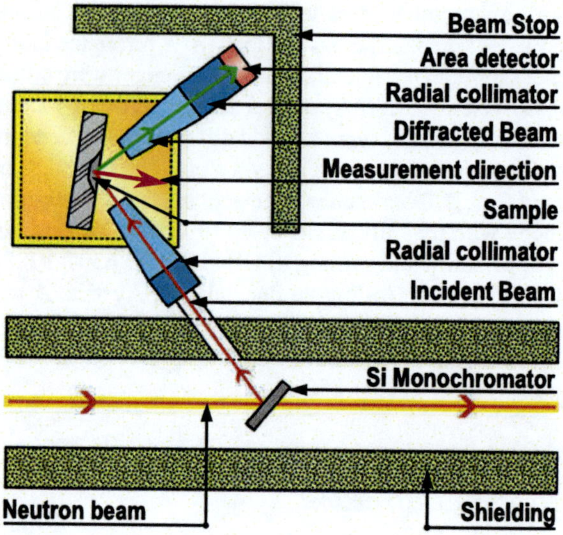

and $\frac{1}{2}s_2^{hkl}$ =6.445 × 10^{-6} MPa^{-1} according to Kröner [33] based on single crystal coefficients given in Dye et al. [34].

Results and Discussion

Metallographic Analyses

Figure 6 shows BSE-SEM micrographs taken in the substrate and the filler material in a reasonable distance from the interface for Parameter II (see A-A reference in Fig. 3). In the filled process zone, a typical microstructure for IN718 particles sprayed onto IN718 substrate is shown, as described in [25]. The repair fillings are macroscopically homogeneous for Parameter I and Parameter II. The orientation contrast allows for the identification of individual grains and even small angle grain boundaries in the substrate (Fig. 6 red). In the filled process zone (Fig. 6 orange), however, individual crystallites cannot be discerned by orientation contrast due to the strong plastic deformation the particles experience during deposition. Only the Al-rich Ni particles can be identified, as they appear darker due to the materials contrast (Z-contrast) of the BSE-SEM image. Metallographic analyses showed that there are no significant differences between Parameter I and Parameter II. To assess the bonding of the filler material to the substrate, the interface was investigated using conventional light optical and scanning electron microscopy at several different magnifications. Figure 7 shows four micrographs taken along the interface at several locations for different magnifications (see A-A reference in Fig. 3). The results show

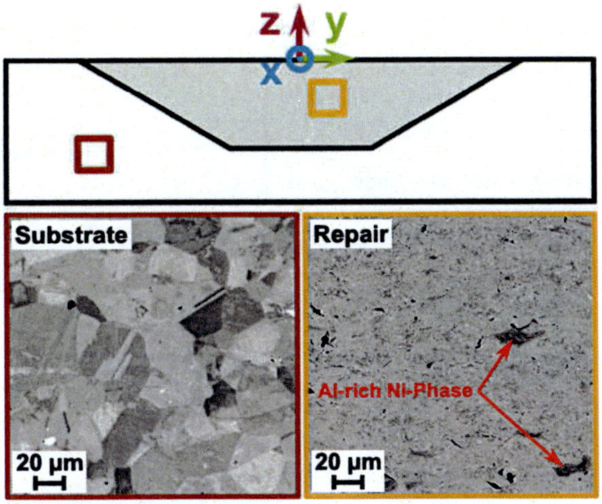

Fig. 6 Schematic representation of the specimen cross section (top) and BSE-SEM micrographs at a reasonable distance from the interface showing the microstructure of the substrate (red frame) and the filler material (orange frame)

good bonding of the filling to the surrounding substrate, even at locations critical for the gas flow, like the tapered sidewall (Fig. 7 blue and green) or the interior interface edge (see Fig. 7 yellow) where eddies could form and hinder deposition. The dark grey, deformed particles visible in the filled process zone are Al-rich Ni particles present in the powder. The triangle above the dashed blue line in Fig. 7 blue is surplus material, deposited as a result of the used spraying trajectory and has to be removed prior to recommissioning of the repaired component (c.f. section. "Experimental"). The result shown in Fig. 7 is exemplary given for Parameter I, as the analyses have shown that there are no significant differences between Parameter I and Parameter II. In contrast to results from the literature, e.g. [22, 23, 26] that indicate decreasing adhesion and even delamination for increasing coating thickness on flat surfaces, the results suggest the suitability of CGRS to repair large cavities in IN718 components with similar material.

Work Hardening

The contour plots in Fig. 8 show the normalised full width at half maximum (FWHM) of neutron diffraction lines and the Vickers hardness distribution (HV 0.1) for Parameter I and Parameter II in Fig. 8 top and bottom, respectively. The cross hatched areas represent the specimen. The dashed line represents the contour of the cavity (c.f. A-A reference in Fig. 3). The black dots represent measurement positions. The FWHM is normalised to the average FHWM value of the surrounding substrate material, away from the interface, for visualisation purposes. The graphs on the left (Fig. 8 left) reveal strongly broadened peaks within the cavity, which can be attributed to the strong plastic deformation (c.f. Fig. 6). The level of FWHM increase is similar

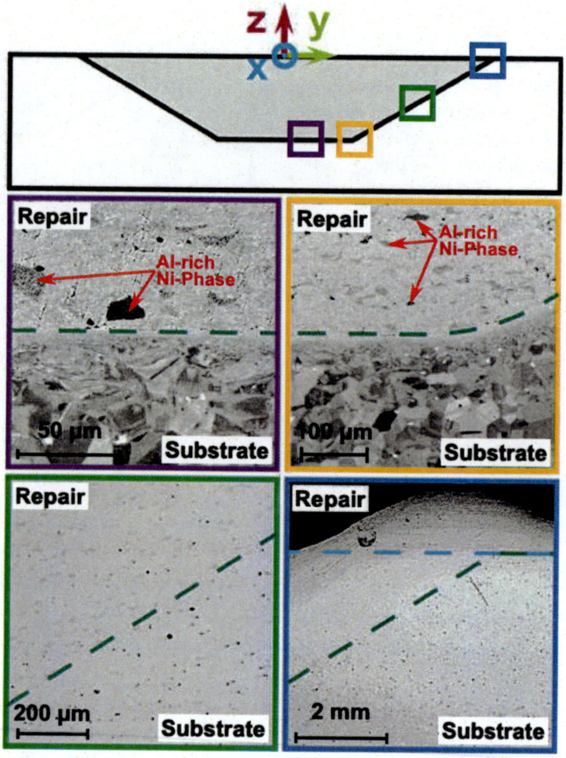

Fig. 7 Schematic representation of the specimen cross section (top) and BSE-SEM (mid) and LOM (bottom) micrographs taken at the cavity base (purple frame), the interior interface edge (yellow frame), the tapered sidewall (green frame), and the cavity top edge (blue frame) at different magnifications. The dashed green line represents the contour of the interface. The dashed blue line represents the original geometry with surplus material above the line

for both samples and almost homogeneous with an average value of $\approx 3.95 \pm 0.4$ for Parameter I and $\approx 3.82 \pm 0.66$ for Parameter II within the filled cavity, which suggests a homogeneous state throughout the repair zone. Figure 6 right shows contour plots of the respective Vickers hardness distribution (HV 0.1). The results show similar hardness values of on average $\approx 510 \pm 26$ HV 0.1 in the filled process zone and $\approx 237 \pm 15$ HV 0.1 in the surrounding substrate (c.f. A-A reference in Fig. 3). The increase in hardness within the filled process zone is due to work hardening as a result of strong plastic deformation. The powder particles used to fill the cavities by CGRS do not contain precipitates. The hardness distribution for both cases is almost homogeneous despite the small indentation size and the high spatial resolution, further corroborating a homogeneous condition throughout the filled cavities.

Residual Stresses

The results of the neutron diffraction RS mapping are shown in Fig. 9 left and right for Parameter I and Parameter II, respectively. The contour plots for the as-sprayed

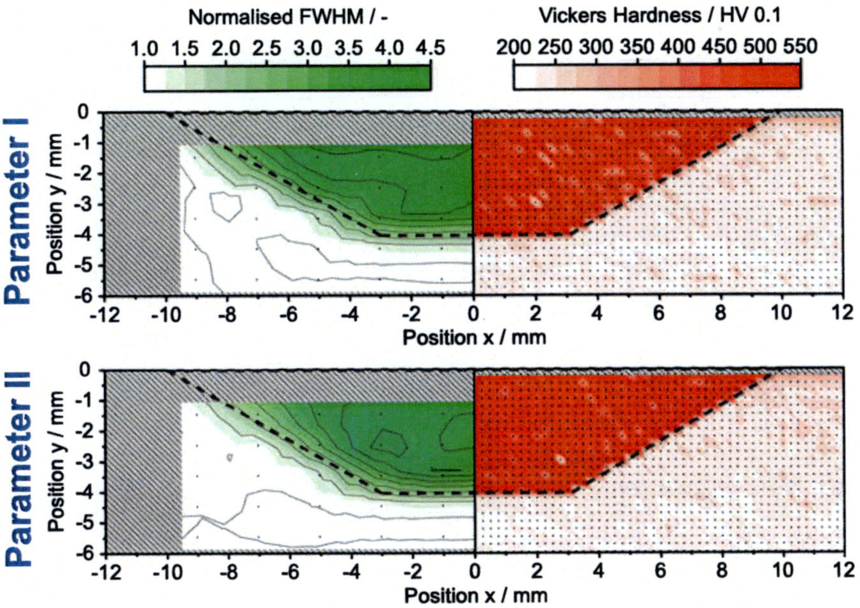

Fig. 8 Results for Parameter I (top) and Parameter II (bottom): Contour graphs of the normalised FWHM (left) and Vickers microhardness mapping (right). The cross hatched areas represent the specimen. The dashed line represents the contour of the cavity. The black dots represent measurement locations

state show compressive RS throughout the cavity and balancing tensile RS in the substrate near the interface (c.f. A-A reference in Fig. 3), indicated by the dashed grey line. The cross hatched areas denote the sample outside of the mapping grid. The black dots represent measurement locations. The results show that the highest compressive RS for both parameters are in the y-direction ($\sigma_{max} \approx 400$ MPa for Parameter I and $\sigma_{max} \approx 350$ MPa for Parameter II). In contrast to the results for thin coatings on planar surfaces in, e.g. [22, 23], the RS are direction dependent with lower compressive RS in x-direction. This might be attributed to stronger geometric constraints in the y-direction, i.e. the length of the cavity L_c compared to the width W_c, which could result in different temperature gradients in each direction. Although the RS distributions for both samples appear qualitatively similar, the compressive RS level in x, y, and z-direction is lower throughout the cavity for Parameter II (Fig. 9 right) when compared to Parameter I (Fig. 9 left). This may be traced back to the larger heat input in the sample repaired with Parameter II, as the gas temperature is higher (c.f. Table 1) and could be beneficial for the distortion behaviour with regard to a mechanical refurbishing and/or thermal post-treatment. The balancing tensile RS in x as well as in y-direction are similar at $\approx$300MPa for Parameter I and Parameter II (Fig. 9 top and mid). The results in z-direction for both parameters (Fig. 9 bottom) show RS levels close to zero in comparison to the other directions as well as a less

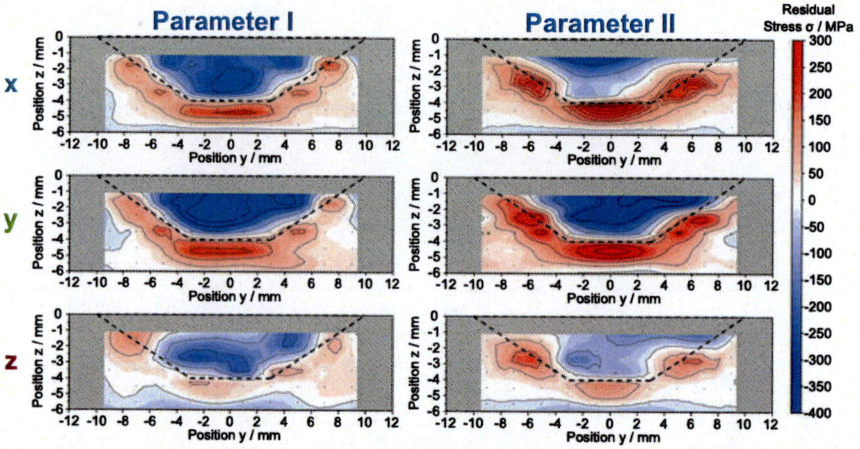

Fig. 9 Results of the triaxial residual stress mapping for Parameter I (left) and Parameter II (right). Contour graphs of the RS distribution in *x*-direction (top), *y*-direction (mid), and *z*-direction (bottom). The cross hatched areas represent the specimen. The dashed line represents the contour of the cavity. The black dots represent measurement positions

uniform distribution. This might be due to the small magnitude of strains in this direction and increased diffraction line widths that complicate the determination of an exact peak position. Further uncertainties in RS analyses by neutron diffraction experiments could be related to coarse grains in combination with the rather small gauge volume and the complexity in the determination of strain-free lattice parameter d_0. The error in stress analysis is comparable in all three directions and accumulates to $\sigma_{\text{substrate}}^{\text{err}} \approx \pm 2$ MPa and $\sigma_{\text{filling}}^{\text{err}} \approx \pm 54$ MPa in the substrate and filling, respectively. It can however be observed that in contrast to hardness and microstructure the RS in repair fillings can be well adjusted by the appropriate choice of spraying parameters.

Conclusion

RS analyses and metallographic investigations were carried out on flat IN718 specimens, containing 4 mm deep tapered cavities filled with IN718-type particles by means of CGRS. The main conclusions based on these analyses can be summarised as follows:

- CGRS is a suitable thermal spray process to fill large cavities in IN718 components with a depth of $D_c = 4$ mm with similar material for two combinations of process temperature and pressure (Parameter I and Parameter II).
- The connection of the filler to the surface of the large tapered cavity is good. Especially in flow-mechanically difficult areas, neither delamination nor poor quality of the bond could be observed in the investigated cross section. The CGS filling

shows a plastically deformed, uniform microstructure throughout the cavity. This is corroborated by an almost homogeneous FWHM increase as well as a uniform hardness distribution higher that the substrate material due to the work hardening induced through CGS.

- Using neutron diffraction characterisation in the filled process zone and the surrounding substrate, we found compressive RS throughout the filling and balancing tensile RS in the substrate near the interface of the as-sprayed condition. Contrary to results for thin flat coatings, the RS are direction dependent as a result of geometric constraints. The RS level introduced by CGRS can be adjusted through the appropriate choice of spraying parameters while maintaining almost identical metallographic results. In particular, higher gas temperature and pressure lead to fewer RS in the investigated volume, which could prove to be beneficial regarding the distortion of the component.

Acknowledgements This project was funded by the German Research Foundation (DFG) through the project 452606919. The authors are grateful to the Institut Laue-Langevin for granting beamtime [31] at the SALSA instrument and for their support. Special thanks are given to Dr. Daniel Schliephake from IAM-WK at KIT for his support and SEM imaging.

Authors' contributions Florian Lang: Conceptualisation, Formal analysis, Data curation, Writing—original draft, Writing—review & editing, Visualisation. Johannes-Christian Schmitt: Resources, Writing—review & editing. Sandra Cabeza: Investigation, Resources, Writing—review & editing. Thilo Pirling: Investigation, Resources, Writing—review & editing. Jochen Fiebig: Resources, Writing—review & editing. Robert Vassen: Resources, Project administration, Funding acquisition, Writing—review & editing. Jens Gibmeier: Resources, Supervision, Project administration, Funding acquisition, Writing—review & editing.

References

1. Villafuerte, J.: Modern Cold Spray, pp. 1–429. Springer, Cham (2015). https://doi.org/10.1007/978-3-319-16772-5
2. Murray, J.W., Zuccoli, M.V., Hussain, T.: Heat treatment of cold-sprayed C355 Al for repair: Microstructure and mechanical properties. Journal of Thermal Spray Technology **27**(1-2), 159–168 (2018). https://doi.org/10.1007/s11666-017-0665-z
3. Ghelichi, R., MacDonald, D., Bagherifard, S., Jahed, H., Guagliano, M., Jodoin, B.: Microstructure and fatigue behavior of cold spray coated Al5052. Acta Materialia **60**(19), 6555–6561 (2012). https://doi.org/10.1016/j.actamat.2012.08.020
4. Ziemian, C.W., Sharma, M.M., Bouffard, B.D., Nissley, T., Eden, T.J.: Effect of substrate surface roughening and cold spray coating on the fatigue life of AA2024 specimens. Materials and Design **54**, 212–221 (2014). https://doi.org/10.1016/j.matdes.2013.08.061
5. Widener, C.A., Carter, M.J., Ozdemir, O.C., Hrabe, R.H., Hoiland, B., Stamey, T.E., Champagne, V.K., Eden, T.J.: Application of high-pressure cold spray for an internal bore repair of a navy valve actuator. Journal of Thermal Spray Technology **25**(1–2), 193–201 (2016). https://doi.org/10.1007/s11666-015-0366-4
6. Champagne, V.K.: The Cold Spray Materials Deposition Process, pp. 1–362. Woodhead Publishing Limited, Cambridge, England (2007). https://doi.org/10.1533/9781845693787
7. Assadi, H., Gärtner, F., Stoltenhoff, T., Kreye, H.: Bonding mechanism in cold gas spraying. Acta Materialia **51**(15), 4379–4394 (2003). https://doi.org/10.1016/S1359-6454(03)00274-X

8. Petráčková, K., Kondás, J., Guagliano, M.: Fixing a hole (with cold spray). International Journal of Fatigue **110**, 144–152 (2018). https://doi.org/10.1016/j.ijfatigue.2018.01.014
9. Stier, O.: Fundamental cost analysis of cold spray. Journal of Thermal Spray Technology **23**(1-2), 131–139 (2014). https://doi.org/10.1007/s11666-013-9972-1
10. Faccoli, M., Cornacchia, G., Maestrini, D., Marconi, G.P., Roberti, R.: Cold spray repair of martensitic stainless steel components. Journal of Thermal Spray Technology **23**(8), 1270–1280 (2014). https://doi.org/10.1007/s11666-014-0129-7
11. Dosta, S., Couto, M., Guilemany, J.M.: Cold spray deposition of a WC-25Co cermet onto Al7075-T6 and carbon steel substrates. Acta Materialia **61**(2), 643–652 (2013). https://doi.org/10.1016/J.ACTAMAT.2012.10.011
12. Cinca, N., Guilemany, J.M.: Cold gas sprayed Stellite-6 coatings and their wear resistance. Journal of Material Science and Engineering **02**(02) (2013). https://doi.org/10.4172/2169-0022.1000122
13. Champagne, V.K., Helfritch, D.J.: A demonstration of the antimicrobial effectiveness of various copper surfaces. Journal of Biological Engineering **7**(1) (2013). https://doi.org/10.1186/1754-1611-7-8
14. Farjam, A., Cormier, Y., Dupuis, P., Jodoin, B., Corbeil, A.: Influence of alumina addition to aluminum fins for compact heat exchangers produced by cold spray additive manufacturing. Journal of Thermal Spray Technology **24**(7), 1256–1268 (2015). https://doi.org/10.1007/S11666-015-0305-4
15. Luzin, V., Kirstein, O., Zahiri, S.H., Fraser, D.: Residual stress buildup in Ti components produced by cold spray additive manufacturing (CSAM). Journal of Thermal Spray Technology **29**(6), 1498–1507 (2020). https://doi.org/10.1007/s11666-020-01048-z
16. Vargas-Uscategui, A., King, P.C., Styles, M.J., Saleh, M., Luzin, V., Thorogood, K.: Residual stresses in cold spray additively manufactured hollow titanium cylinders. Journal of Thermal Spray Technology **29**(6), 1508–1524 (2020). https://doi.org/10.1007/s11666-020-01028-3
17. Widener, C.A., Hrabe, R.H., James, B., Champagne, V.K.: B1 bomber-FEB panel repair by cold spray. In: Cold Spray Action Team (CSAT) Meeting 2013 (Worcester Polytechnic Institute, MA, 2013), Worcester, MA, USA (2013)
18. James, B.L.: B-1 cold spray initiative. In: Cold Spray Action Team Meeting 2016, Worcester, MA, USA (2016)
19. Champagne, V.K.J.: Practical Cold Spray. Springer, Cham (2021). https://doi.org/10.1007/978-3-030-70056-0
20. Assadi, H., Kreye, H., Gärtner, F., Klassen, T.: Cold spraying - a materials perspective. Acta Materialia **116**, 382–407 (2016). https://doi.org/10.1016/j.actamat.2016.06.034
21. Shayegan, G., Mahmoudi, H., Ghelichi, R., Villafuerte, J., Wang, J., Guagliano, M., Jahed, H.: Residual stress induced by cold spray coating of magnesium AZ31B extrusion. Materials and Design **60**, 72–84 (2014). https://doi.org/10.1016/j.matdes.2014.03.054
22. Singh, R., Schruefer, S., Wilson, S., Gibmeier, J., Vassen, R.: Influence of coating thickness on residual stress and adhesion-strength of cold-sprayed Inconel 718 coatings. Surface and Coatings Technology **350**, 64–73 (2018). https://doi.org/10.1016/j.surfcoat.2018.06.080
23. Vassen, R., Fiebig, J., Kalfhaus, T., Gibmeier, J., Kostka, A., Schrüfer, S.: Correlation of microstructure and properties of cold gas sprayed Inconel 718 coatings. Journal of Thermal Spray Technology **29**(6), 1455–1465 (2020). https://doi.org/10.1007/s11666-020-00988-w
24. Wong, W., Irissou, E., Vo, P., Sone, M., Bernier, F., Legoux, J.G., Fukanuma, H., Yue, S.: Cold spray forming of inconel 718. Journal of Thermal Spray Technology **22**(2–3), 413–421 (2013). https://doi.org/10.1007/s11666-012-9827-1
25. Singh, R., Rauwald, K.-H.H., Wessel, E., Mauer, G., Schruefer, S., Barth, A., Wilson, S., Vassen, R.: Effects of substrate roughness and spray-angle on deposition behavior of cold-sprayed Inconel 718. Surface and Coatings Technology **319**, 249–259 (2017). https://doi.org/10.1016/j.surfcoat.2017.03.072
26. Fiebig, J., Bakan, E., Kalfhaus, T., Mauer, G., Guillon, O., Vaßen, R.: Thermal spray processes for the repair of gas turbine components. Advanced Engineering Materials **22**(6), 1901237 (2020). https://doi.org/10.1002/adem.201901237

27. Wilson, S., Barth, A., Nestler, M., Kupada, S.: Cold gas spray coating methods and compositions, patent application, WO 2017/003427 A1 (2015)
28. DIN EN ISO 6507-1:2018-07, Metallische Werkstoffe - Härteprüfung Nach Vickers - Teil 1: Prüfverfahren, Deutsche Fassung EN ISO 6507-1:2018. Beuth-Verlag, Berlin (2018). https://doi.org/10.31030/2778746
29. Pirling, T., Bruno, G., Withers, P.J.: SALSA: Advances in residual stress measurement at ILL. Materials Science Forum **524–525**, 217–222 (2006). https://doi.org/10.4028/www.scientific.net/MSF.524-525.217
30. Hutchings, M.T.: Neutron diffraction measurement of residual stress fields-The answer to the engineers' prayer? Nondestructive Testing and Evaluation **5**(5–6), 395–413 (1990). https://doi.org/10.1080/02780899008952981
31. Gibmeier, J., Cabeza, S., Lang, F., Pirling, T.: Assessment of triaxial residual stress distributions after local repair of components by means of cold gas spraying [Data Set]. Institut Laue-Langevin (ILL), Grenoble (2021). https://doi.org/10.5291/ILL-DATA.1-02-345
32. Richard, D., Ferrand, M., Kearley, G.J.: Analysis and visualisation of neutron-scattering data. Journal of Neutron Research **4**(1), 33–39 (1996). https://doi.org/10.1080/10238169608200065
33. Kröner, E.: Berechnung der elastischen Konstanten des Vielkristalls aus den Konstanten des Einkristalls. Zeitschrift für Physik **151**(4), 504–518 (1958). https://doi.org/10.1007/BF01337948
34. Dye, D., Roberts, S.M., Withers, P.J., Reed, R.C.: The determination of the residual strains and stresses in a tungsten inert gas welded sheet of IN718 superalloy using neutron diffraction. The Journal of Strain Analysis for Engineering Design **35**(4), 247–259 (2000). https://doi.org/10.1243/0309324001514396

Design of Graded Transition Interlayer for Joining Inconel 740H Superalloy with P91 Steel Using Wire-Arc Additive Manufacturing

Soumya Sridar, Xin Wang, Mitra Shabani, Michael A. Klecka, and Wei Xiong

Abstract In this work, two graded transition interlayers were designed using a CALPHAD-based ICME framework (CALPHAD: Calculation of Phase Diagrams; ICME: Integrated Computational Materials Engineering) for joining Inconel 740H superalloy with P91 steel. Successful builds with the designed interlayers (content of P91 steel are 60 and 85 wt.%) sandwiched between the constituent materials were fabricated using wire-arc additive manufacturing. 60% P91 interlayer exhibited an FCC structure with low hardness, while the 85% P91 interlayer had a martensitic structure with high hardness. A two-step post-heat treatment consisting of homogenization at 1150°C and aging at 760°C was designed. 60% P91 interlayer showed no improvement in hardness after aging. It agrees with the CALPHAD modeling that predicts a lack of effective strengthening precipitates at 760°C, whereas the hardness of 85% P91 increased significantly after aging for 8 h due to the precipitation of the $M_{23}C_6$ phase. Mechanical tests equipped with digital image correlation were performed to determine the location of the failure and tensile properties. As-built and heat treated 60% P91 build failed in the graded alloy block, whereas the as-built 85% P91 alloy failed at the 85% P91/740H interface, and the aged alloy failed in the pure P91 region. This proves that post-heat treated 85% P91 is much stronger than pure P91, and the alloy design strategy used in this work proves successful for developing interlayers for dissimilar joining.

Keywords Wire-arc additive manufacturing · Graded interlayer · CALPHAD · ICME · Digital image correlation

---

S. Sridar · X. Wang · M. Shabani · W. Xiong (✉)
Physical Metallurgy and Materials Design Laboratory, Department of Mechanical Engineering and Materials Science, University of Pittsburgh, Pittsburgh, PA 15261, USA
e-mail: weixiong@pitt.edu

M. A. Klecka
Raytheon Technologies Research Center, 411, Silver Lane, East Hartford, CT 06108, USA

E. A. Ott et al. (eds.), *Proceedings of the 10th International Symposium on Superalloy 718 and Derivatives*, The Minerals, Metals & Materials Series,
https://doi.org/10.1007/978-3-031-27447-3_45

Introduction

In advanced ultra-supercritical (A-USC) powerplant, high operating temperatures (>700°C) and pressures (35 MPa) demands the use of materials with superior creep and oxidation resistance. This led to the development of Inconel 740H superalloy (referred to as 740H hereafter) with a high amount of Cr and Co to attain excellent creep and oxidation properties till 850°C [1], which is expensive. However, there are several regions that operate below 650°C, where the use of 740H superalloys is not required, whereas ferritic-martensitic P91 steel with good creep resistance till 650°C [2] and cheaper than 740H superalloy will be sufficient. The use of different materials in different regions of an A-USC powerplant necessitates the joining of dissimilar materials.

Dissimilar materials can be joined using traditional techniques such as welding [3, 4], brazing [5], or diffusion bonding [6]. In these processes, there are several disadvantages from a process viewpoint, such as long processing time, high production cost, and complex setup with multiple steps involved [7]. From a microstructure standpoint, porosity, crack formation, as well as structural distortion due to the formation of wider heat affected zone (HAZ), are the major drawbacks. In order to overcome these difficulties, additive manufacturing (AM) has become the most sought-after technique to accomplish the joining of dissimilar materials. In an AM process, the component is built layer-by-layer using 3D CAD models [8]. Hence, it has advantages such as freedom to produce complex parts, reduced material wastage, and high efficiency [9].

Several reports are available for multi-material deposition using directed energy deposition (DED)-based systems such as laser engineered net shaping (LENS) using powder as starting material [10–13]. Wire-arc additive manufacturing (WAAM) is another DED-based technique that uses wire as starting material which is melted and deposited using an electric/plasma arc. The high deposition rates in WAAM make it suitable for fabricating large parts that are valuable for real-time applications. Besides, the cost of powders is 2–5 times higher than wire and hence, the materials costs are much lesser for the WAAM process [14]. Owing to these advantages, the WAAM technique will be an appropriate technique for multi-material manufacturing and few reports are available in the literature for using WAAM to join dissimilar aluminum alloys [15, 16], dissimilar steels [17] as well as stainless steel with Ni-base superalloy [18].

Dissimilar joining can be accomplished either by direct bonding of the materials or by introducing an interlayer that leads to a smooth change in composition and properties between the constituent materials. The P91 steel has a martensitic structure, while the 740H superalloy has an γ matrix with FCC structure. Moreover, there is a drastic difference in the coefficient of thermal properties of P91 steel and 740H superalloy. As reported in our previous work [19], when P91 steel was directly bonded over 740H superalloy, cracks formed in the interfacial region and the thermal residual stresses developed due to the differences in thermal properties were relieved by the crack formation. Therefore, the introduction of an interlayer is imperative to avoid cracking while joining P91 steel with 740H superalloy, and an appropriate interlayer that can ensure a gradual transition between the constituent materials needs to be identified.

In this work, a CALPHAD-based ICME framework (CALPHAD: Calculation of Phase Diagrams; ICME: Integrated Computational Materials Engineering) has been applied to design potential graded alloy compositions that can act as an interlayer between P91 steel and 740H superalloy. Builds with the designed graded alloy interlayer sandwiched between the constituent materials were successfully fabricated using WAAM. The microstructure and mechanical properties of the WAAM builds have been investigated extensively and a suitable post-heat treatment has been designed to remove microstructure inhomogeneities and induce precipitation to improve the properties. This work will prove useful for identifying the graded alloy compositions that can be suitably used as transition interlayers while joining dissimilar materials using any manufacturing technique.

Materials and Methods

Computational Details

In order to identify the compositions that can potentially act as an interlayer, high-throughput thermodynamic calculations were performed using TC-API (Thermo-Calc Application Programming Interface) toolkit known as TC-Python (based on the Python programming language) which is included with the Thermo-Calc software. The content of P91 for the graded alloys was calculated in wt.% and will be expressed as %, hereafter. The fraction of phases at 760°C for different mixtures (101 entries) of P91 steel and 740H superalloy whose compositions were determined by calculating the weighted mean for each element for varying content of P91 steel (in steps of 1%) were calculated. TCNI8, the commercial multicomponent thermodynamic database developed specifically for Ni-base superalloys was used for these calculations. In addition, the Martensite start (Ms) temperatures were determined for these compositions using predictive machine learning models developed by Agrawal et al. [20] for the determination of Ms temperature.

WAAM Processing

The composition (in wt.%) as quantified by the manufacturers for the P91 steel (ER90S-B91, manufactured by Euroweld, Mooresville, NC, USA) and the 740H superalloy (Special Metals Welding Products Co., Newton, NC, USA) wires that were used for fabricating the graded alloy builds using WAAM are listed in Table 1. The nominal composition of the designed interlayers, namely, 60 and 85% P91 graded alloys determined using the computing the weighted average for each element, is also listed in the same table. The diameter of the P91 steel and 740H superalloy wires were 0.9 and 1 mm, respectively. The builds were fabricated using a robotic WAAM system (ABB robotic platform) in Raytheon Technologies Research Center (RTRC). The system consists of a six-axis robot connected to the twin-wire feeder as well as a plasma arc welding (PAW) torch and the material was deposited over a 12 mm thick mild steel substrate. Purified argon gas was employed for creating the plasma and also used as a shielding gas for the PAW torch with a flow rate of 1.1 L/min to avoid oxidation during deposition. A square wave pattern with a raster pattern width of 11 mm and step-over of 3 mm was used for depositing each layer (Fig. 1) with a constant deposition rate of 0.9 kg/hr. The distance from the torch to the deposition surface and the wire feeding angle were maintained at 15 mm and 30°, respectively.

The printing parameters used for each layer are listed in Table 2. The printing parameters were chosen based on design of experiments for deposition of the constituent materials. Several conditions were tested with various travel speeds, wire feeds, and power, until a stable melt pool was achieved. An average voltage of 20.9 V was maintained during the deposition. The layer height was set as 2 mm for the deposition of P91 steel layers while 1.8 mm for the 740H superalloy. A layer height of ~1.5 mm was maintained for the deposition of the graded alloy. A higher current and lower travel speed as well as wire feed rate were used for the first and

Table 1 Composition (in wt.%) of the P91 and 740H wires used for deposition in the WAAM process and the nominal composition (in wt.%) of 60 and 85% P91 alloy

Material	Composition (wt.%)											
	Al	C	Co	Cr	Cu	Fe	Mn	Mo	Nb	Ni	Ti	Si	V
P91	0.003	0.09	–	9.2	0.03	Bal.	0.45	0.91	0.052	0.4	–	0.26	0.21
740H	1.4	0.03	20.3	24.6	0.02	0.2	0.24	0.5	1.49	Bal	1.5	0.1	–
60% P91	0.56	0.066	8.12	15.4	0.026	53.1	0.37	0.75	0.63	20.1	0.6	0.2	0.13
85% P91	0.21	0.081	3.05	11.5	0.03	75.2	0.42	0.85	0.27	7.8	0.23	0.24	0.18

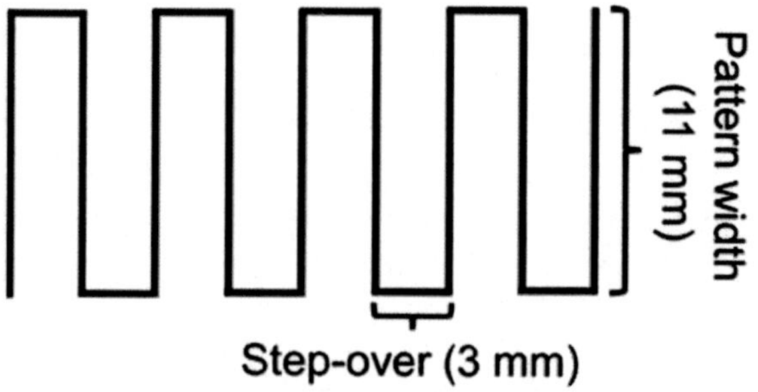

Fig. 1 Schematic of the tool path used for the fabrication of graded alloy builds using WAAM

Table 2 Processing parameters for the fabrication of 60 and 85% P91 graded alloy builds using WAAM

Layer no.	Average current (A)	Travel speed (mm/s)	Wire feed (m/min)	Material
60% P91 graded build				
Layer 29–44	166	5	2.44	740H
Layer 13–28	166	5	1.88 P91 0.97 740H	60% P91
Layer 3–12	166	5	3.10	P91
Layer 2	200	4	2.90	P91
Layer 1	220	3.5	2.90	P91
85% P91 graded build				
Layer 29–44	166	5	2.44	740H
Layer 13–28	166	5	2.67 P91 0.37 740H	85% P91
Layer 3–12	166	5	3.10	P91
Layer 2	200	4	2.90	P91
Layer 1	220	3.5	2.90	P91

second layers of P91 steel to ensure good adhesion with the mild steel substrate. The diameter of the P91 steel and 740H superalloy wires, along with their material densities, were used to calculate the relative weight percentage and resulting wire feed values to achieve the desired blends for the graded alloy blocks. The interpass temperature was fixed as 175°C during the deposition. The interpass time for each layer was not fixed and instead it was dictated by the time taken for the sample to cool below 175°C.

Experimental Details

The builds were sectioned along the build (Z) direction, i.e., the XZ plane of the build using electric discharge machining (EDM, Mitsubishi MV2400S, Japan). The sections were ground from 800 to 1200 grit SiC emery papers followed by polishing using suspensions of diamond with particle sizes of 3 and 1 μm as well as colloidal silica with a particle size of 0.04 μm. The as-polished surface was viewed under FEI Scios Dual Beam focused ion beam (FIB)—scanning electron microscope (SEM) attached with a field emission gun (FEG) source. The elemental composition was determined using an energy dispersive spectroscope (EDS, Octane Elite EDS system) attached to the SEM. Detailed phase and grain structure analysis was performed using electron backscattered diffraction (EBSD, EDAX Hickory EBSD system) attached to the SEM, and the data was analyzed further using TSL-OIM software (version 8).

Hardness measurements were carried out using an automated Vicker's micro-hardness tester (AMH55 with LM310AT Microindenter, LECO Corporation, USA) with a load of 300 g and dwell time of 10 s. Around 400 indentations were made along the build direction (Z direction) to determine the hardness distribution. Flat dog bone-shaped tensile bars (gauge length = 25 mm, total length = 66 mm, width = 5 mm, and thickness = 4 mm) were extracted from the XZ plane of the sample such that both the interfaces are included within the gauge section. Uniaxial tensile tests were carried out with a strain rate of 10^{-3} s^{-1} using MTS 880 universal testing machine with 100 kN capacity. The strain was measured using non-contact digital image correlation (DIC, VIC-2D 7, Correlated Solutions Inc., USA), in order the identify the failure location. The spatial distribution of strains in the gauge section was captured using DIC. The strain was averaged over the complete gauge section and extracted for plotting the stress–strain curves.

Results and Discussion

Computational Design and WAAM Processing

The phase fractions at 760°C predicted using CALPHAD method and the Ms temperatures estimated using machine learning model as a function of the P91 steel content are shown in Fig. 2a and b, respectively. The thermodynamic calculations were performed at 760°C since it is the optimized aging temperature for P91steel and 740H superalloy fabricated using WAAM, based on our previous works [21, 22]. According to the thermodynamic predictions, the matrix phase will have the FCC structure from 0 to 90% P91 and the BCC phase will start to form after 90% P91. Amongst the secondary phases, MX (M: Ti, Nb; X: C) is predicted to form in all the compositions, whereas $M_{23}C_6$ forms only between 83 and 100% P91. Since 740H is a γ' strengthened superalloy, it appears between 0 and 50% P91 and it is accompanied by σ phase formation from 8 to 45% P91. However, the formation of σ phase is not expected due to the slow diffusion kinetics involved.

The predicted Ms temperature shows a linear trend as a function of P91 content with a small discontinuity close to 60% P91 (Fig. 2b). Two graded alloy compositions, namely, 60 and 85% P91 were identified to test their potential to act as an interlayer between P91 steel and 740H superalloy. Both these compositions are expected to have an FCC matrix structure. However, the calculated Ms temperature of 85% P91 is close to 250°C thus, martensite is expected to form during the cooling process. Since the Ms temperature for the 60% P91 graded alloy is less than room temperature (~15°C), an FCC matrix is expected to form for this composition. In addition, based on the thermodynamic predictions, 85% P91 alloy is expected to be strengthened by the formation of $M_{23}C_6$ precipitates along with the MX phase while the 60% P91 graded alloy will not have any of the major strengthening precipitates except the MX phase.

The graded alloy builds with the designed interlayer composition block sandwiched between the P91 steel and 740H superalloy are shown in Figs. 2c and 2d.

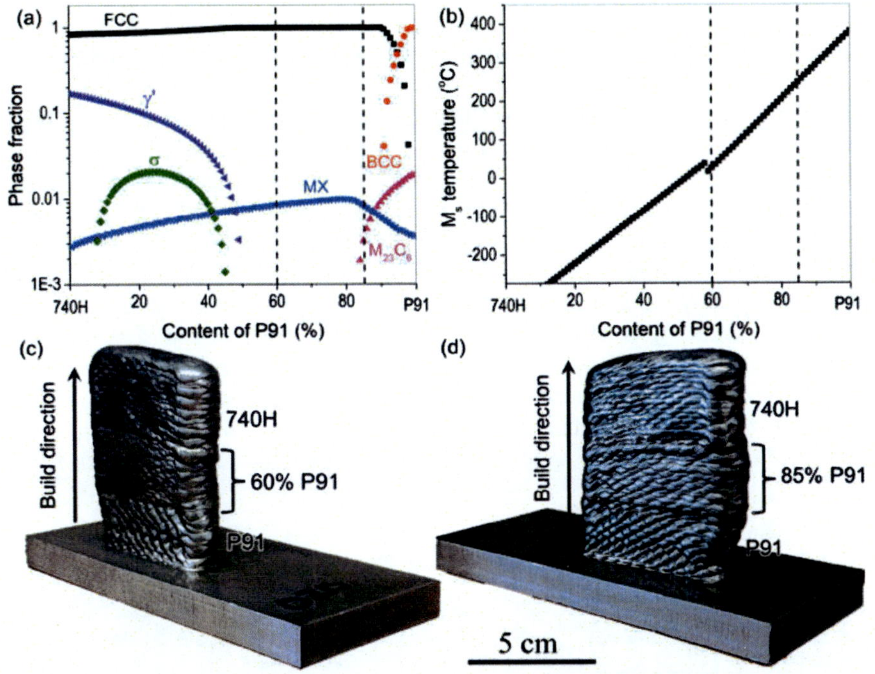

Fig. 2 **a** Phase fraction and **b** Ms temperatures predicted as a function of P91 content. Successful crack-free builds fabricated using WAAM for **c** 60% P91 and **d** 85% P91 graded alloys

The builds were 100 mm long, 10 mm wide, and the total height of the build including the substrate was 87 mm. Different regions of the build were identified based on the differences in the melt pool size as seen from the lateral sides of the build. These builds were intended to be deposited for a total ideal height of nearly 77 mm from the substrate with an ideal height of 24, 24, and 28.8 mm for P91 steel, graded alloy, and 740H superalloy, respectively. The measured heights of each block were ~21 mm for P91 steel, 24 mm for 60 and 85% P91 graded alloys, and ~29 mm for the 740H superalloy with a total height close to 74 mm which is very close to the ideal height of the build. Since the first two layers of P91 steel were mixed completely with the substrate for good adhesion, the height of the P91 steel block was slightly smaller than the ideal height. Hence, the interpass temperature (175°C) used for depositing these builds has been effective in reducing the dilution between the layers.

The composition of the 60 and 85% P91 graded alloy builds along the build direction of the sample measured using EDS is shown in Fig. 3a and b. It was observed that the measured composition of the graded alloy interlayer is close to the nominal composition listed in Table 1. This proves that WAAM is a robust AM technique to deposit a graded alloy using the twin-wire feeder. Moreover, it can be clearly seen that Fe and Ni concentration spikes downward throughout the graded alloy and 740H superalloy regions along with a simultaneous increase in the composition of Mo, Nb, or Ti. Further EDS mapping confirms the presence of secondary phases in the graded alloy blocks as shown in Fig. 3c and d. From Fig. 3(c1–c4) it is evident that the bright phase with irregular structure is rich in Nb, Mo, and Ti, which corresponds to the Laves phase. Figures 3(d1–d3) shows that the dark square-shaped phase has a high content of Ti and a moderate amount of Nb which can be correlated with the MC carbide.

The inverse pole figure (IPF) and phase maps obtained using EBSD for the 60 and 85% P91 alloys are shown in Fig. 4. From the phase maps shown in Fig. 4b and d, it can be found that the 60% P91 graded alloy has a matrix with FCC structure while the 85% P91 alloy has a martensitic structure. This proves that the Ms temperature prediction matches well with the experimental observation for the designed graded alloys. Moreover, a coarse-grained microstructure with no particular texture was observed in the 60% P91 graded alloy, as shown in Fig. 4a, which could be possibly due to the multiple heating and cooling cycles underwent during the deposition. In the 85% P91 graded alloy (Fig. 4b), a clear partition was observed, where one side

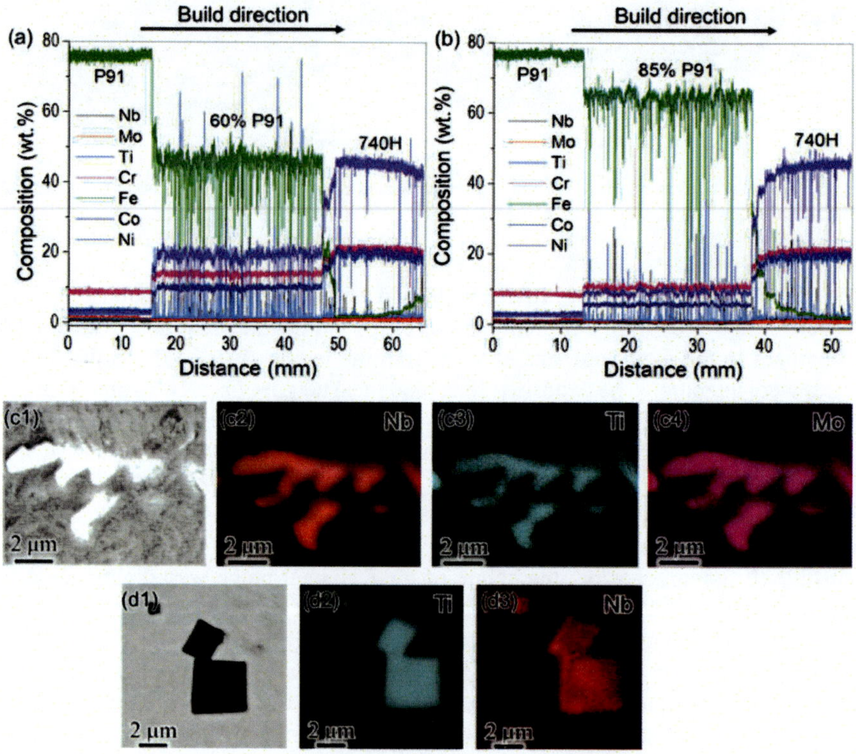

Fig. 3 Composition profile along the build direction determined using EDS for **a** 60 and **b** 85% P91 graded alloy blocks. Compositions profiles for **c1–c4** Laves phase and **d1–d3** MC carbide obtained using area EDS mapping showing that the Laves phase is rich in Mo, Nb, and Ti while the MC carbide is rich in Ti and Nb, respectively

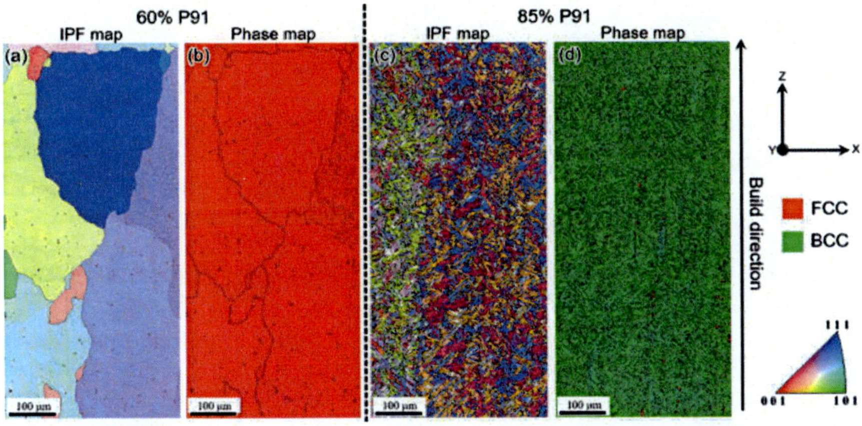

Fig. 4 IPF and phase maps obtained using EBSD for **a**, **b** 60% and **c**, **d** 85% P91 graded alloy blocks showing a coarse-grained FCC structure for the 60% P91 alloy and a fine martensitic structure for the 85% P91 alloy

had grains mostly oriented between (111) and (001) planes (blue, pink, and orange grains, refer to IPF color map) while the other side had grains oriented mostly close to (101) plane (green grains, refer to IPF color map). Hence, the partition between these two types of grains signifies the presence of a prior austenite grain boundary.

The hardness maps obtained along the build direction in the XZ plane of the graded alloys builds are shown in Fig. 5. The hardness of 60% P91 graded alloy is very low in comparison with the pure P91 steel and 740H superalloy. On the other hand, the hardness is predominantly uniform along the build direction for the 85% P91 alloy and the average hardness is higher than the corresponding value for the 60% P91 graded alloy. The striking difference between the hardness of 60 and 85% P91-graded alloys can be attributed to the microstructure of the matrix phase in these alloys. The 60% P91 alloy consists of an FCC phase structure which is generally softer than the martensitic matrix found in the 85% P91 alloy. The 60% P91 alloy has coarse grains, whereas the 85% P91 alloy has a fine martensitic structure (Fig. 4). In addition to the hardness of the graded alloy blocks, there were other weak and strong regions that could be observed from the hardness maps. In the 60% P91 graded alloy build (Fig. 5a), the P91 steel close to the interface between P91 steel and 60% P91 alloy is much higher in comparison with the same material away from the interface. Similarly, a region of low hardness was identified at the interface between 85% P91 alloy and 740H superalloy, as seen in Fig. 5b.

In order to probe into the reason for the formation of harder P91 steel close to the 60% P91 alloy and a weak region between 85% P91and 740H superalloy, further microstructure analysis was performed using EBSD and the IPF and phase maps from these regions are shown in Fig. 6. The IPF and phase maps from the interface between P91 steel and 60% P91 alloy show that the martensite in the P91 steel region is very fine (Fig. 6a and b) in comparison with the P91 steel farther away from the

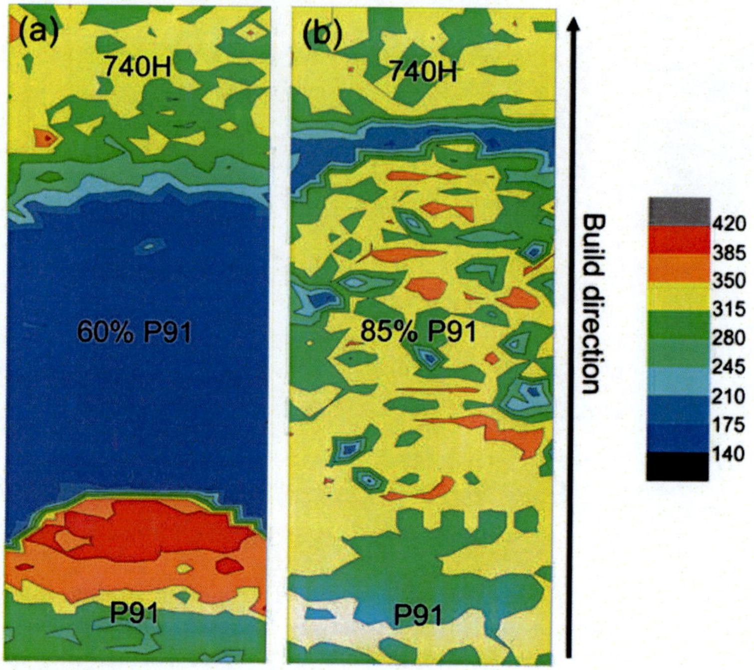

Fig. 5 Hardness maps obtained along the build direction in the XZ plane of **a** 60% and **b** 85% P91 graded alloy. The length and width of the hardness maps are 40 and 11 mm, respectively. The color bar corresponds to the hardness in $HV_{0.3}$

interface where the martensite grains are coarser as it can be seen from Fig. 6c and d. The difference in the grain structure of the P91 steel, near and away from the interface can be attributed to the cyclic re-austenitization of the martensite. When the first few layers of 60% P91 alloy are deposited over the previously deposited P91 steel, the temperature of the P91 steel close to the interface increases to the extent that the martensite can undergo re-austenitization followed by cooling to form refined martensite. It has been demonstrated in our previous work [23] that cyclic re-austenitization in steels can lead to refinement of martensite with increase in hardness. From the IPF and phase maps obtained from the interface between 85% P91 alloy and 740H superalloy (Fig. 6e and f), a band of two-phase region consisting of martensite from 85% P91 alloy and γ phase from 740H superalloy is clearly visible. Hence, the region with low hardness at the interface between 85% P91 alloy and 740H superalloy can be attributed to the difference in the matrix phase structure of the two alloys leading to the formation of a mixture of martensite and γ phases.

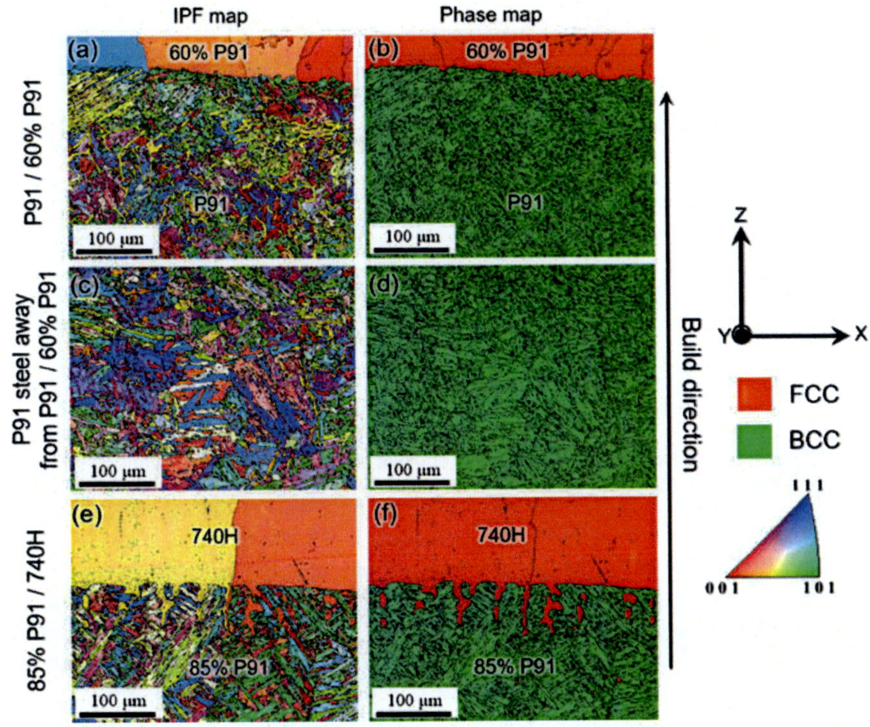

Fig. 6 IPF and phase maps obtained using EBSD at **a**, **b** interface between P91 steel and 60% P91 graded alloy, **c**, **d** P91 steel farther away from the interface between P91 steel and 60% P91 graded alloy and **e**, **f** interface between 85% P91 graded alloy and 740H superalloy

Design of Post-heat Treatment and Mechanical Testing

A two-step post-heat treatment comprising of homogenization and aging was adopted where the homogenization step will dissolve the secondary phases and relieve the residual stresses, whereas the aging step will induce the precipitation of the strengthening phases. The calculated phase fractions as a function of temperature for the 60 and 85% P91 graded alloy composition listed in Table 1 are shown in Fig. 7. The homogenization temperature was identified as 1150°C since all the secondary phases except MX will dissolve at this temperature based on the predicted phase fractions. The homogenization time was chosen to be 1 h because a longer time period can lead to grain growth at high temperatures. Hence, the homogenization for the graded alloy builds was performed at 1150°C for 1 h followed by water quenching.

The aging temperature of 760°C was found to be suitable because at that temperature the fraction of MX in 60% P91 alloy and the fractions of $M_{23}C_6$ as well as MX phases in 85% P91 alloy were found to be maximum based on the thermodynamic predictions. In addition, 760°C was identified as the optimum aging temperature for

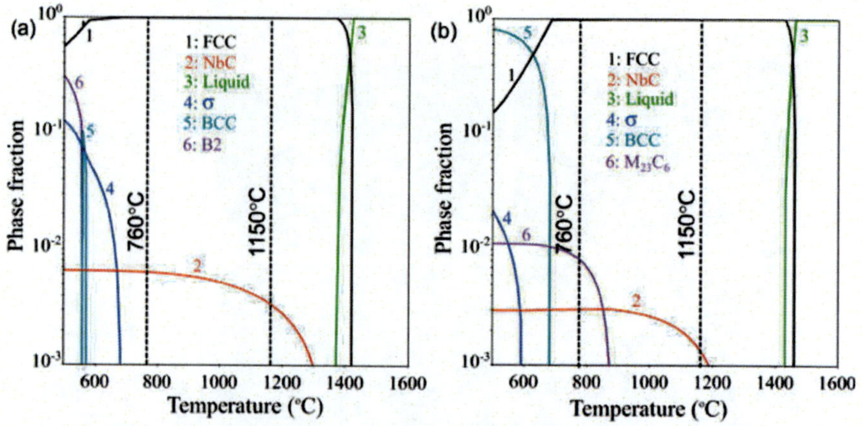

Fig. 7 Fraction of phases as a function of temperature predicted using thermodynamic calculations for **a** 60% and **b** 85% P91 graded alloys

P91 steel [21] and 740H superalloy [22] and moreover, different aging temperatures cannot be used for a multi-material build, as mentioned previously. In order to identify the optimum aging time, the samples were subjected to aging heat treatment after homogenization for different times at 760°C followed by air cooling. The time durations for aging were identified as 2, 4, 8, and 12 h because the optimum aging time for P91 steel [21] and 740H superalloy [22] are 2 and 12 h, respectively. Two time periods were chosen in between (4 and 8 h) in order to identify the optimal condition.

In order to identify the optimum aging time, the peak hardness needs to be determined. The hardness maps obtained for different aging times at 760°C for 60 and 85% P91 graded alloy builds are shown in Figs. 8 and 9, respectively. The hardness maps for aged 60% P91 builds showed no improvement in hardness after aging (Fig. 8). This observation agrees well with the CALPHAD calculations which predicted a lack of strengthening precipitates at 760°C. This proves that there is a good correlation between the CALPHAD predictions and experimental observations. On the contrary, the hardness of the 85% P91 graded alloy sample improved significantly after aging as shown in Fig. 9. The width of the soft zone, which is a mixture of martensite and γ phases (Fig. 6f), was found to be unchanged after 2 h of aging while reduced slightly after aging for 4 h. After aging for 8 and 12 h, the soft zone width reduced considerably showing that a longer aging time is required to allow the diffusion. However, as the aging time is increased, the pure P91 steel becomes weaker with low hardness since it has reached an overaged condition. Thus, the optimum aging time was identified as 8 h at 760°C to attain reduced width of the soft zone without lowering the hardness of pure P91 steel extensively.

The stress–strain curves and the tensile properties obtained after mechanical testing equipped with DIC are shown in Fig. 10 and Table 3, respectively. No significant improvement was observed in the yield strength and ultimate tensile strength of

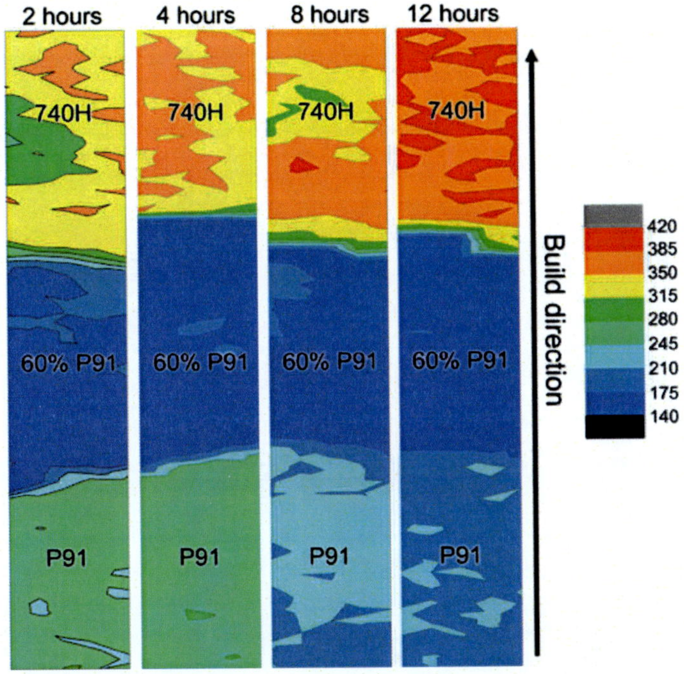

Fig. 8 Hardness maps for 60% P91 graded alloy after aging at 760°C for different durations. The length and width of the hardness maps are 50 and 4 mm, respectively. The color bar corresponds to the hardness in $HV_{0.3}$

the heat treated 60% P91 graded alloy builds. However, the ductility of the as-built alloy is higher than the heat treated 60% P91 graded alloy possibly due to the precipitation of the fine and brittle MC carbides during aging which can act as crack initiation sites. On the other hand, there is a considerable improvement in the yield strength and ductility of 85% P91 graded alloy build after post-heat treatment (Table 3). The precipitation of $M_{23}C_6$ precipitates could have contributed to the strengthening of the graded alloy and hence, enhanced tensile properties has been achieved.

In addition, DIC was used to calculate the strains during the mechanical testing to identify the region where the necking starts. The locations of failure (region enclosed by a white rectangle) identified from screenshots of the strain map are shown in Fig. 11. The 60% P91 graded alloy failed in the graded alloy block in as-built and heat treated conditions since the strength did not improve with the application of post-heat treatment (Fig. 11a and b). However, the location of failure changed with application of post-heat treatment for the 85% P91 graded alloy. In the as-built condition, the sample failed at the interface between the 85% P91 alloy and 740H superalloy as shown in Fig. 11(c) since the hardness in this region was low due to the presence of a two-phase region with martensite and γ phases. After post-heat treatment, the location of failure shifted to the pure P91 block (Fig. 11(d)). The failure in the pure

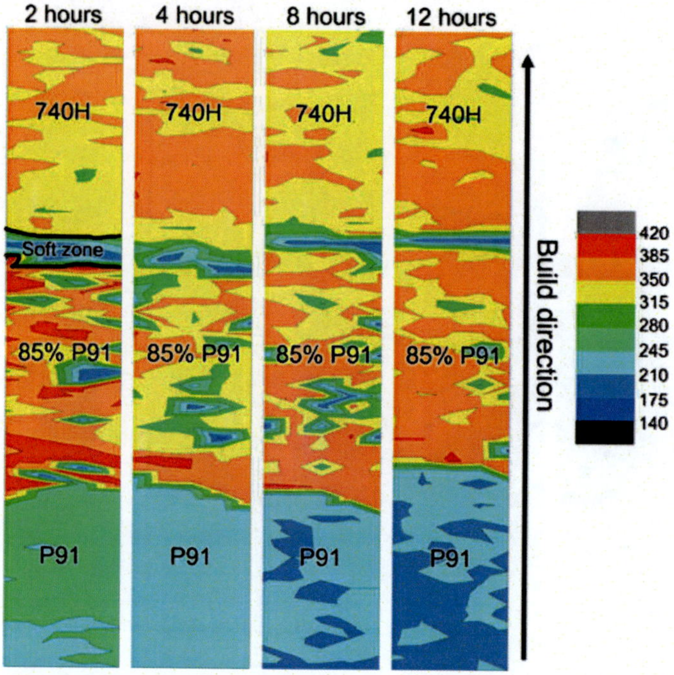

Fig. 9 Hardness maps for 85% P91 graded alloy after aging at 760°C for different durations. The length and width of the hardness maps are 50 and 4 mm, respectively. The color bar corresponds to the hardness in $HV_{0.3}$

P91 region is inevitable since it would have reached an overaged condition as the optimum aging time for this material is 2 h [21]. Therefore, the designed post-heat treatment for the 85% P91 graded alloy has strengthened the material to achieve improved mechanical performance.

Conclusions

A CALPHAD-based ICME framework has been employed to design the composition of the graded alloy that can potentially act as an interlayer between P91 steel and 740H superalloy. The designed compositions were fabricated using WAAM along with microstructure characterization and mechanical property testing. The salient outcomes from the present study can be summarized as follows.

- Guided by the high-throughput calculations, the bimetallic alloy builds with 60 and 85% P91 as interlayers were fabricated successfully with desired composition and reduced dilution using WAAM.

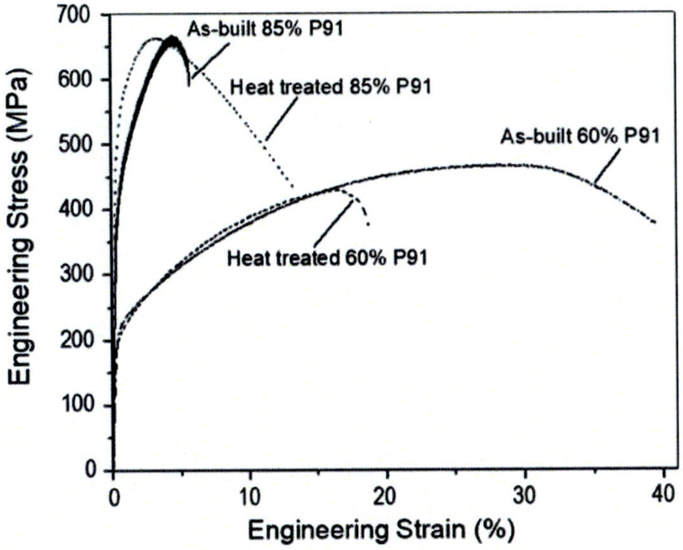

Fig. 10 Stress–Strain curves obtained using mechanical testing equipped with DIC for 60 and 85% P91 graded alloys builds in as-built and heat treated conditions

Table 3 Tensile properties and failure locations for 60 and 85% P91 builds fabricated using WAAM in different conditions (2 samples were tested for each condition to determine the standard deviation)

Material	Condition	Yield strength (MPa)	Ultimate tensile strength (MPa)	Elongation (%)	Location of failure
60% P91	As-built	219 ± 2	459 ± 11	32.5 ± 9.9	60% P91 block
	Heat treated	212 ± 2	440 ± 14	21.2 ± 3.1	60% P91 block
85% P91	As-built	419 ± 5	665 ± 18	6.5 ± 1.2	Interface between 85% P91 and 740H blocks
	Heat treated	538 ± 22	667 ± 6	13.2 ± 0.1	P91 block

- The matrix of 60% P91 alloy had a coarse-grained FCC structure, while the 85% P91 alloy comprises a fine-grained martensitic matrix and secondary phases such as the Laves phase and MC carbides. There is a good correlation between the predicted phases and the observed microstructure in the WAAM builds.
- The hardness of 60% P91 graded alloy is much lower than the 85% P91 alloy. Moreover, the interface between the 85% P91 alloy and 740H superalloy is low due to the presence of a two-phase mixture of martensite and γ phases, while the fine martensitic structure of the P91 steel close to the interface between P91 steel and 60% P91 alloy increased its hardness.

Fig. 11 Location of failure identified using the DIC strain maps for **a, b** 60% and **c, d** 85% P91 graded alloy builds in as-built and heat treated conditions. The color bar corresponds to the strain in %

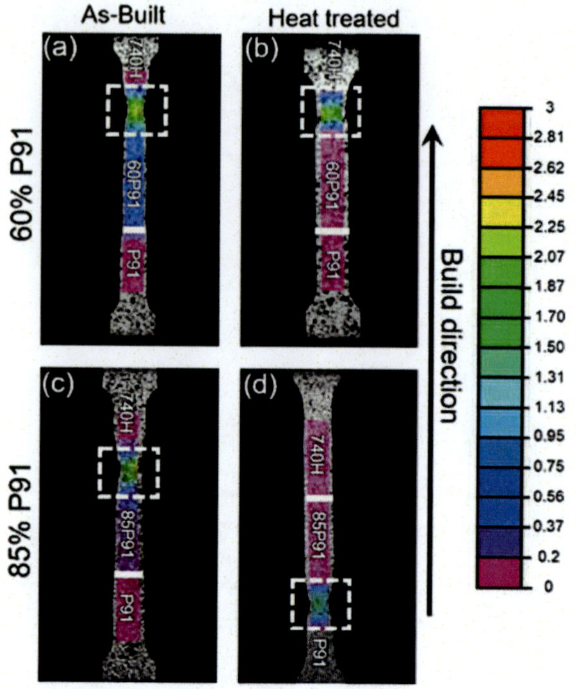

- 60% P91 graded alloy showed no improvement in the hardness even after applying the post-heat treatment due to lack of effective strengthening precipitates, while 85% P91 graded alloy showed considerable improvement in the hardness after aging with an optimum aging time of 8 h.
- The location of failure for the 60% P91 graded alloy remained within the graded alloy block in as-built and heat treated conditions whereas the as-built 85% P91 graded alloy build failed at the interface between the intermediate block and 740H superalloy interface and it shifted to the pure P91 after the post-heat treatment.
- This work demonstrated that the CALPHAD-based ICME design frame is essential for the successful bimetallic printing using the WAAM combined with post-heat treatment.

Acknowledgements The authors gratefully acknowledge the financial support from the National Energy Technology Laboratory, Department of Energy, United States under the award number DE-FE0031637.

Disclaimer This report was prepared as an account of work sponsored by an agency of the United States Government. Neither the United States Government nor any agency thereof, nor any of their employees, makes any warranty, express or implied, or assumes any legal liability or responsibility for the accuracy, completeness, or usefulness of any information, apparatus, product, or process disclosed, or represents that its use would not infringe privately owned rights. Reference herein to any specific commercial product, process, or service by trade name, trademark, manufacturer, or otherwise does not necessarily constitute or imply its endorsement, recommendation, or favoring by the United States Government or any agency thereof. The views and opinions of authors expressed herein do not necessarily state or reflect those of the United States Government or any agency thereof.

References

1. De Barbadillo JJ, Baker BA, Virginia W (2014) Microstructure stability of Alloy 740H and its effect on material properties. Proceedings of the ASME Symposium on Elevated Temperature Application of Materials for Fossil, Nuclear and Petrochemical Industries, 1–13.
2. Pandey C, Giri A, Mahapatra MM (2016) Evolution of phases in P91 steel in various heat treatment conditions and their effect on microstructure stability and mechanical properties. Mater. Sci. Eng. A. 664:58–74. https://doi.org/10.1016/j.msea.2016.03.132.
3. Sun Z, Karppi R (1996) The application of electron beam welding for the joining of dissimilar metals: An overview. J. Mater. Process. Technol. 59:257–267. https://doi.org/10.1016/0924-0136(95)02150-7.
4. Xue P, Ni DR, Wang D, Xiao BL, Ma ZY (2011) Effect of friction stir welding parameters on the microstructure and mechanical properties of the dissimilar Al-Cu joints. Mater. Sci. Eng. A. 528:4683–4689. https://doi.org/10.1016/j.msea.2011.02.067.
5. Feng J, Songbai X, Wei D (2012) Reliability studies of Cu/Al joints brazed with Zn-Al-Ce filler metals. Mater. Des. 42:156–163. https://doi.org/10.1016/j.matdes.2012.05.028.
6. Guo Y, Liu G, Jin H, Shi Z, Qiao G (2011) Intermetallic phase formation in diffusion-bonded Cu/Al laminates. J. Mater. Sci. 46:2467–2473. https://doi.org/10.1007/s10853-010-5093-0.
7. Onuike B, Heer B, Bandyopadhyay A (2018) Additive manufacturing of Inconel 718—Copper alloy bimetallic structure using laser engineered net shaping (LENS^{TM}). Addit. Manuf. 21:133–140. https://doi.org/10.1016/j.addma.2018.02.007.
8. Gorsse S, Hutchinson C, Gouné M, Banerjee R (2017) Additive manufacturing of metals: a brief review of the characteristic microstructures and properties of steels, Ti-6Al-4V and high-entropy alloys. Sci. Technol. Adv. Mater. 18:584–610. https://doi.org/10.1080/14686996.2017.1361305.
9. Herzog D, Seyda V, Wycisk E, Emmelmann C (2016) Additive manufacturing of metals. Acta Mater. 117:371–392. https://doi.org/10.1016/J.ACTAMAT.2016.07.019.
10. Imran MK, Masood SH, Brandt M, Bhattacharya S, Mazumder J (2011) Direct metal deposition (DMD) of H13 tool steel on copper alloy substrate: Evaluation of mechanical properties. Mater. Sci. Eng. A. 528:3342–3349. https://doi.org/10.1016/j.msea.2010.12.099.
11. Domack MS, Baughman JM (2005) Development of nickel-titanium graded composition components. Rapid Prototyp. J. 11:41–51. https://doi.org/10.1108/13552540510573383.
12. Sahasrabudhe H, Harrison R, Carpenter C, Bandyopadhyay A (2015) Stainless steel to titanium bimetallic structure using LENS^{TM}. Addit. Manuf. 5:1–8. https://doi.org/10.1016/j.addma.2014.10.002.
13. Carroll BE, Otis RA, Borgonia JP, Suh, JO Dillon RP, Shapiro AA, Hofmann DC, Liu ZK, Beese AM (2016) Functionally graded material of 304L stainless steel and inconel 625 fabricated by directed energy deposition: Characterization and thermodynamic modeling. Acta Mater. 108:46–54. https://doi.org/10.1016/j.actamat.2016.02.019.
14. Ali H, Ghadbeigi H, Mumtaz K (2018) Processing Parameter Effects on Residual Stress and Mechanical Properties of Selective Laser Melted Ti6Al4V. J. Mater. Eng. Perform. 27:4059–4068. https://doi.org/10.1007/S11665-018-3477-5/FIGURES/19.
15. Hauser T, Reisch RT, Seebauer S, Parasar A, Kamps T, Casati R, Volpp J, Kaplan AFH (2021) Multi-Material Wire Arc Additive Manufacturing of low and high alloyed aluminium alloys with in-situ material analysis. J. Manuf. Process. 69:378–390. https://doi.org/10.1016/J.JMAPRO.2021.08.005.

16. Chang T, Fang X, Liu G, Zhang H, Huang K (2022) Wire and arc additive manufacturing of dissimilar 2319 and 5B06 aluminum alloys. J. Mater. Sci. Technol. 124:65–75. https://doi.org/10.1016/J.JMST.2022.02.024.
17. Ahsan MRU, Tanvir ANM, Ross T, Elsawy A, Oh MS, Kim DB (2019) Fabrication of bimetallic additively manufactured structure (BAMS) of low carbon steel and 316L austenitic stainless steel with wire + arc additive manufacturing. Rapid Prototyp. J. 26:519–530. https://doi.org/10.1108/RPJ-09-2018-0235.
18. Mohan Kumar S, Rajesh Kannan A, Pravin Kumar N, Pramod R, Siva Shanmugam N, Vishnu AS, Channabasavanna SG (2021) Microstructural Features and Mechanical Integrity of Wire Arc Additive Manufactured SS321/Inconel 625 Functionally Gradient Material. J. Mater. Eng. Perform. 30:5692–5703. https://doi.org/10.1007/S11665-021-05617-3/FIGURES/13.
19. Sridar S, Klecka MA, Xiong W (2022) Interfacial characteristics of P91 steel - Inconel 740H bimetallic structure fabricated using wire-arc additive manufacturing. J. Mater. Process. Technol. 300:117396. https://doi.org/10.1016/J.JMATPROTEC.2021.117396.
20. Agrawal A, Saboo A, Xiong W, Olson G, Choudhary A (2019) Martensite start temperature predictor for steels using ensemble data mining. 2019 IEEE International Conference on Data Science and Advanced Analytics (DSAA), 521–530. https://doi.org/10.1109/DSAA.2019.00067.
21. Li K, Klecka MA, Chen S, Xiong W (2021) Wire-arc additive manufacturing and post-heat treatment optimization on microstructure and mechanical properties of Grade 91 steel. Addit. Manuf. 37:101734. https://doi.org/10.1016/J.ADDMA.2020.101734.
22. Sridar S, Klecka MA, Xiong W (2022) Effect of post-deposition heat treatment on the microstructure and mechanical properties of Inconel 740H fabricated using wire-arc additive manufacturing (Unpublished work).
23. Sridar S, Zhao Y, Xiong W (2020) Cyclic re-austenitization of copper-bearing high-strength low-alloy steels fabricated by laser powder bed fusion, Mat. Char. 166:110437. https://doi.org/10.1016/j.matchar.2020.110437.

Microstructure Evolution During Post-heat Treatment of Haynes 282 Alloy Processed by Wire-Arc Additive Manufacturing

Luis Fernando Ladinos Pizano, Soumya Sridar, Chantal Sudbrack, and Wei Xiong

Abstract Post-heat treatment optimization is imperative to improve the mechanical properties of superalloys prepared by additive manufacturing. In this work, the effect of solution treatment on the microstructural heterogeneity and γ' precipitation of Haynes 282 fabricated by wire-arc additive manufacturing (WAAM) has been investigated. The results suggest that the standard solution treatment carried out at 1150 °C for 2 h is insufficient to remove the grain texture developed during WAAM. Instead, a heat treatment at 1250 °C for 2 h facilitates homogenization and recrystallization without causing excessive coarsening. Furthermore, solution treatment temperature affects the kinetics of γ' precipitate growth. By increasing the solution heat treatment temperature from 1150 to 1250 °C, strengthening γ' precipitates grow faster, achieving the peak hardness earlier. Moreover, the increase in solution temperature favors the development of a bimodal distribution of γ' precipitates during aging. This work demonstrates the need for an effective post-heat treatment to eliminate the heterogeneities that are formed during the WAAM process and alter the γ' size distribution to improve the mechanical performance of Haynes 282 alloy.

Keywords Haynes 282 · Additive manufacturing · CALPHAD · Precipitation behavior · Strengthening mechanism

Introduction

The global energy demand has driven the development of more efficient and ecological power plants. An impactful way to increase thermal efficiency and reduce plant pollution is to increase the operating temperature and pressure of industrial gas turbines used for power generation. However, the turbine operating conditions are

L. F. L. Pizano · S. Sridar · W. Xiong (✉)
Physical Metallurgy and Materials Design Laboratory, Department of Mechanical Engineering and Materials Science, University of Pittsburgh, Pittsburgh, PA 15261, USA
e-mail: weixiong@pitt.edu

C. Sudbrack
National Energy Technology Laboratory, Albany, OR 97321, USA

E. A. Ott et al. (eds.), *Proceedings of the 10th International Symposium on Superalloy 718 and Derivatives*, The Minerals, Metals & Materials Series,
https://doi.org/10.1007/978-3-031-27447-3_46

limited by the mechanical properties of the key structural components, promoting the design of materials with excellent mechanical performance at high temperatures (>760 °C).

Haynes 282 is a precipitation-strengthened nickel-base superalloy with high creep strength along with excellent tensile and fatigue behavior at 760 °C, which has target use in advanced ultra-super-critical (A-USC) turbines [1]. Moreover, Haynes 282 shows good fabricability and weldability as the volume fraction of the reinforcing phase (γ') is relatively low (19%) compared to other commercial nickel-base superalloys. However, a low γ' content in general negatively impacts mechanical properties. To compensate for the low volume fraction of γ', Haynes 282 incorporates a generous amount of Mo, improving mechanical performance through solid solution strengthening [2]. Likewise, γ' precipitates and a distribution of micron-sized MC and $M_{23}C_6$ carbides significantly improve the mechanical behavior of Haynes 282. While γ' improves both tensile properties and creep resistance, MC and $M_{23}C_6$ carbides stabilize the grain structure at high temperatures, favoring creep resistance [3].

The microstructure and mechanical performance of Haynes 282 are optimized by applying a post-heat treatment comprising of solution heat treatment and aging steps. The standard heat treatment developed by Haynes International Inc. consists of a solution treatment above the solvus temperature of γ' and $M_{23}C_6$ carbides, carried out between 1121 °C and 1149 °C. Subsequently, two aging heat treatments at 1010 and 788 °C are recommended to precipitate $M_{23}C_6$ carbides and γ', respectively [4]. The traditional manufacturing processes for Haynes 282 are forging and casting. However, the product forms are limited by the hardness of the material and the ability to create hollow structures such as cooling ducts. Due to the geometric freedom offered by additive manufacturing (AM) and the excellent weldability of Haynes 282, the trend to process Haynes 282 by AM has increased in recent years.

AM is a computer-aided manufacturing technique capable of producing 3D objects from the addition of material layer-by-layer. The versatility of this methodology facilitates the production of customizable components with complex geometries and final finishes. Wire arc additive manufacturing (WAAM) is a direct energy deposition (DED) technique used in the fabrication and repair of metal components with wire as starting material. The solidification process along with the heating and cooling cycles associated with WAAM introduces microstructural variation compared to samples produced by conventional techniques. Therefore, studying the effect of WAAM process conditions and post-heat treatments on the microstructure and mechanical properties is imperative.

Recently, it has been shown that additively manufactured Haynes 282 can match or even exceed the mechanical properties of wrought alloy by modifying the post-heat treatment conditions. Ramakrishnan and Dinda [5] reported that Haynes 282 fabricated by powder-based direct laser metal deposition is able to achieve an ultimate tensile strength of 1200 MPa and ductility of 18% after application of single-step aging heat treatment at 788 °C for 16 h. However, most of the works reported up to now [6] are dedicated exclusively to optimizing the characteristics of γ' precipitates and mechanical properties without considering the heterogeneity associated with the columnar grain structure.

This work focuses on the design of post-heat treatments capable of minimizing grain structure heterogeneity and maximizing the mechanical strength through γ' precipitation in Haynes 282 fabricated using WAAM. The effect of solution heat treatment temperature on grain structure and γ' precipitation during subsequent aging heat treatments has been extensively studied. By elucidating the relationship between solution heat treatment and second aging heat treatment, this paper provides guidance for the design of consecutive post-heat treatments for Haynes 282.

Materials and Methods

Experimental Details

The Haynes 282 sample was fabricated on a stainless steel 304 substrate using the WAAM system GEFERTEC ARC 605 equipped with a Fronius TPS 400i PULSE power source. The diameter of the wire used as feedstock in the WAAM process was 1 mm and its chemical composition as provided by the supplier, Haynes international, USA, is listed in Table 1. The distance from the torch to the substrate was maintained at 14 mm, while the torch travel speed and wire feed rate were 950 mm/min and 6.5 m/min, respectively. An average voltage of 24 V was supplied to the system, producing a current of 140 A. To prevent oxidation of the sample during the manufacturing process, Cronigon Ni10 (mixture of 70% Ar and 30% He) was used as torch shielding gas at a rate of 20 L/min. Furthermore, the scanning strategy consisted of a single bead pattern as shown in Fig. 1. In order to improve the mechanical performance of Haynes 282 samples, the effect of different post-heat treatments on the microstructure and hardness was evaluated. The details of the conditions studied are summarized in Table 2. The heat treatments were performed in a box furnace. To avoid oxidation, the samples were encapsulated in quartz tubes under vacuum and backfilled with Argon.

Microstructure characterization was carried out in the XZ plane along the build direction of the sample as shown in Fig. 1c. The metallographic preparation consisted of a grinding process with SiC abrasive paper followed by a polishing using suspensions of diamond and silica up to a particle size of 0.05 μm. The etchant used to reveal the γ' phase was chromic acid (H_2CrO_4). The characteristics of the precipitates and the grain structure were studied with the scanning electron microscope (SEM, FEI Apreo) operated at 20 kV and equipped with an energy dispersive spectroscope (EDS, EDAX Elite 150) and electron backscatter diffraction (EBSD) detector. Subsequently, the γ' size distribution was calculated by analyzing SEM micrographs with

Table 1 Composition (wt. %) of the Haynes 282 wire used for fabrication of the wall build

Elements	Al	B	C	Co	Cr	Fe	Mn	Mo	Ni	Si	Ti
Composition (wt. %)	1.5	0.005	0.06	10	20	1.5	0.3	8.5	55.8	0.15	2.1

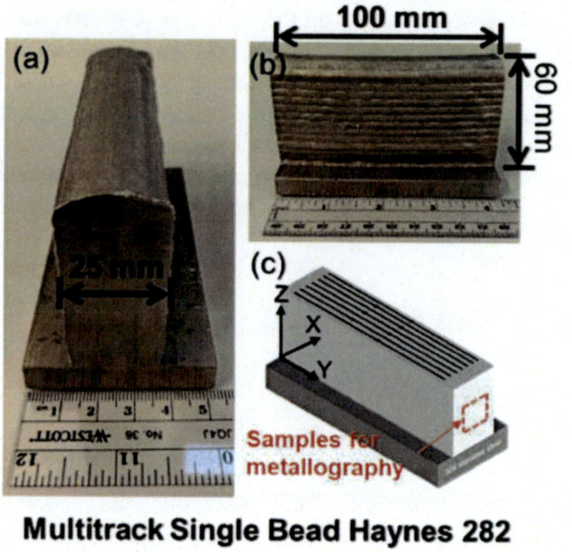

Fig. 1 The thickwall build of WAAM Haynes 282 investigated here in **a** XZ side view and **b** YZ side view with **c** a schematic of the scanning strategy followed during WAAM printing with Z indicating the build direction

Table 2 Conditions and identifications (ID) for all samples studies in this work. After every heat treatment the samples were water quenched

Sample ID	Solution heat treatment	First aging	Second aging
As-built			
ST1150	1150 °C/2 h		
1150-FA	X	1010 °C/2 h	
1150-FA-SA2h	X	X	788 °C/2 h
1150-FA-SA4h	X	X	788 °C/4 h
1150-FA-SA6h	X	X	788 °C/6 h
1150-FA-SA8h	X	X	788 °C/8 h
ST1250	1250 °C/2 h		
1250-FA	X	1010 °C/2 h	
1250-FA-SA2h	X	X	788 °C/2 h
1250-FA-SA4h	X	X	788 °C/4 h
1250-FA-SA6h	X	X	788 °C/6 h
1250-FA-SA8h	X	X	788 °C/8 h

ImageJ software. Microhardness was measured using the LM800-LECO durometer and a Vicker pyramidal indenter. The load and dwell time were 0.3 kg and 10 s, respectively. The average Vickers hardness ($HV_{0.3}$) was determined from 35 indentations separated by 150 μm using 0.3 kg load.

Computational Details

The phase fractions as a function of temperature along with the Scheil solidification diagram and the variation in chemical composition of MC and $M_{23}C_6$ carbides as a function of temperature were predicted using Thermo-Calc software. To perform these thermodynamic calculations, the TCNI11 database developed for Ni-base superalloys together with the chemical composition listed in Table 1 were used.

Results and Discussion

Effect of WAAM Process Conditions on Microstructure

The cooling rate (around 100 °C/s [7]) and the thermal cycles associated with the WAAM process determine the initial microstructures of Haynes 282. SEM images and EDS maps of the secondary phases found in the as-built condition together with the Haynes 282 Scheil solidification diagram are presented in Fig. 2. Ti- and Mo-rich MC primary carbides with irregular morphologies and a size range between 0.30 and 0.45 μm were located in interdendritic regions and Ti-rich nitrides were identified within the γ dendrite core. However, γ′ precipitates were not observed throughout the build.

According to Ojo et al. [8], the solidification process begins with the formation of γ dendrites. As the γ dendrites grow, elements with partition coefficients less than

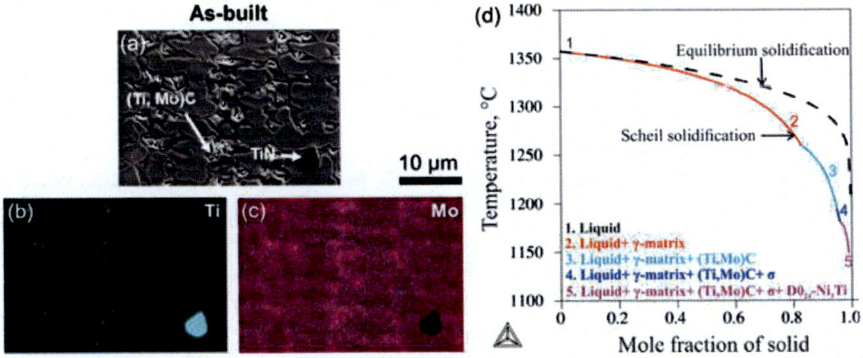

Fig. 2 **a** SEM image of the as-built sample in XZ plane. The dark gray continuous phase represents the γ matrix while the black and light gray precipitates are identified as TiN and (Ti, Mo)C carbides, respectively. Energy dispersive X-ray spectroscopy (EDS) elemental maps, revealing that MC carbides are rich in **b** Ti and **c** Mo. **d** Scheil solidification diagram of Haynes 282, showing the formation of (Ti, Mo)C carbides during the non-equilibrium solidification process

1 such as Ti, Mo, and C begin to segregate into the interdendritic liquid. The liquid continues to be enriched in MC carbides-forming elements until the composition exceeds the maximum solubility of the γ matrix, causing the formation of the non-invariant eutectic γ/MC. This reaction practically consumes the C content of the alloy and is defined as $L \leftrightarrow \gamma + MC$. Subsequently, the interdendritic liquid continues to enrich with γ'-forming alloying elements such as Ti and Al, causing the formation of the eutectic γ/γ'. These reactions explain the presence of (Ti, Mo)C carbides located in interdendritic regions. However, the solidification rates associated with WAAM process prevent the precipitation of γ'. According to the Scheil solidification diagram in Fig. 2d, γ dendrite formation starts at 1360 °C, while the precipitation of γ/MC eutectic takes place at 1260 °C. To decrease the segregation of alloying elements and, therefore, increase the degree of supersaturation in the γ matrix, solution heat treatment followed by water quenching is recommended.

Microstructural Evolution During Solution Heat Treatments

The standard solution heat treatment for wrought Haynes 282 is intended to dissolve secondary phases and reduce residual stresses. However, the solution heat treatment for Haynes 282 fabricated using WAAM should additionally recrystallize the grain structure. The distribution and morphology of the precipitates found in samples ST1150 and ST1250 can be observed from the SEM micrographs shown in Fig. 3. Figure 3a and d shows that MC carbides and their distribution along the interdendritic regions persist after solution heat treatments, which is expected since MC carbides typically have good stability at 1150 °C. However, as the temperature increases, the γ matrix is able to dissolve more alloying elements, causing the size and volume fraction of the MC carbides to decrease in the ST1250 sample.

On the other hand, Fig. 3d and e indicates that the morphology of the MC carbides changes from irregular to blocky as the solution temperature rises from 1150 to 1250 °C and that the dissolution of alloying elements causes the dendritic subgrains to disappear above 1150 °C. Likewise, the MC carbides located at the grain boundaries completely disappeared in the ST1250 sample as shown in Fig. 3f. In general, the dissolution of secondary phases is attributed to the diffusion of atoms across the precipitate/matrix interface into the matrix. The reports by Cieslak et al. [9] and Cao et al. [10] suggest that the dissolution of MC carbides in nickel-based superalloys occurs between 1230 and 1290 °C. These observations explain the presence of MC carbides in the ST1150 and ST1250 samples.

Figure 4 presents the Haynes 282 phase fraction diagram together with the variation in the composition of (Ti, Mo)C carbides as a function of temperature obtained using Thermo-Calc. The diagram in Fig. 4a shows that the fraction of MC carbides decreases from 0.0033 to 0.0013 between 1150 and 1250 °C, which agrees with the experimental observations made from Fig. 3a and d. Likewise, thermodynamic calculations indicate that MC carbides can completely dissolve into the matrix at 1270 °C as suggested by Cieslak et al. [9] and Cao et al. [10]. Figure 4b shows that Ti and Mo

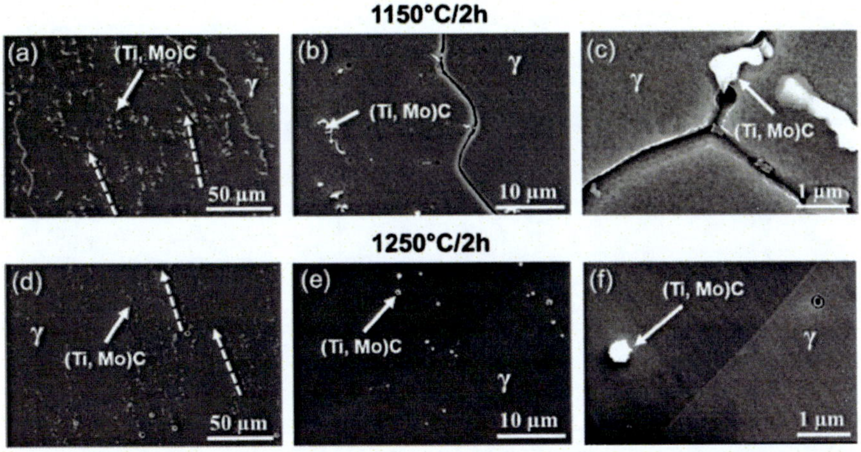

Fig. 3 Distribution and morphology of (Ti, Mo)C carbides as a function of solution temperature, dotted arrows highlight the columnar distribution of precipitates. Distribution of (Ti, Mo)C carbides of **a** ST1150 and **b** ST1250 samples. Carbides with irregular and cubic morphology after solution heat treatment at **c** 1150 °C and **d** 1250 °C, respectively, and **e**, **f** MC carbides located at grain boundaries

are the main constituents of MC carbides, which is consistent with the EDS maps presented in Fig. 2b and c. Furthermore, the solubility of Ti in the carbides reduce as the temperature increases, while the solubility of Mo increases with temperature. The dissolution of (Ti, Mo)C carbides and the reduction of Ti solubility in the remaining carbides at high temperatures increase the degree of supersaturation of the γ matrix in the ST1250 sample, which can improve the precipitation kinetics in the subsequent aging steps.

Figure 5 shows the inverse pole figure (IPF) maps of the as-built, ST1150, and ST1250 samples. According to the IPF maps, the grain structure of the as-built and ST1150 conditions consist of a columnar grain structure, while the ST1250 sample exhibits equiaxed grains and some twin boundaries. Although the directional solidification process of AM tends to introduce texture to the material, the as-built samples did not present strong texture.

The grain morphology is governed by the local direction of heat flow and competitive grain growth during solidification. Wei et al. [11] showed that the grains of nickel-base alloys with FCC matrix grow preferentially when the direction of easy growth <100> is aligned parallel to the direction of maximum heat flux. Wei et al. [12] also suggest that the direction of maximum heat flux in AM is perpendicular to the edge of the molten pool towards the substrate, promoting the formation of columnar structures as observed in the as-built sample.

On the other hand, the epitaxial nucleation and the competitive grain growth define the crystallographic orientation of the grains. According to Li and Tan [13], if the first layer is made up of columnar grains at the base and a large area of equiaxed grains at the top, the subsequent layer will be unable to completely remelt the equiaxed area,

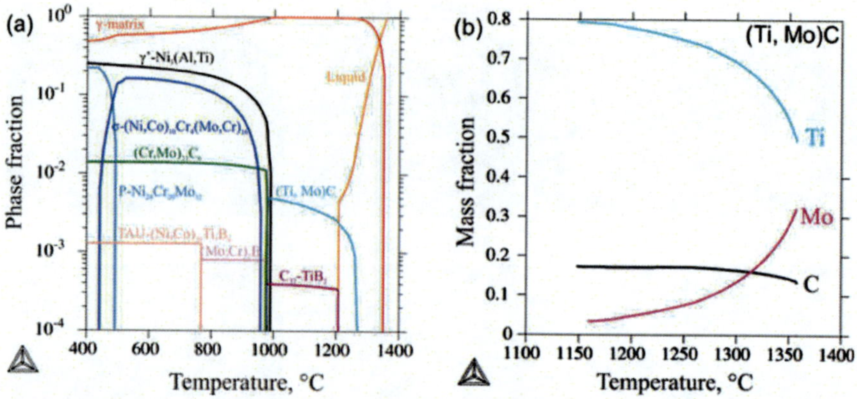

Fig. 4 **a** The phase fraction diagram for Haynes 282 calculated by Thermo-Calc software using TCNI11 database shows that (Ti, Mo)C carbides dissolve in the γ matrix as the temperature increases. **b** The temperature-dependent composition diagram of (Ti, Mo)C carbides indicates that the solubility of Ti in MC carbides decreases with temperature

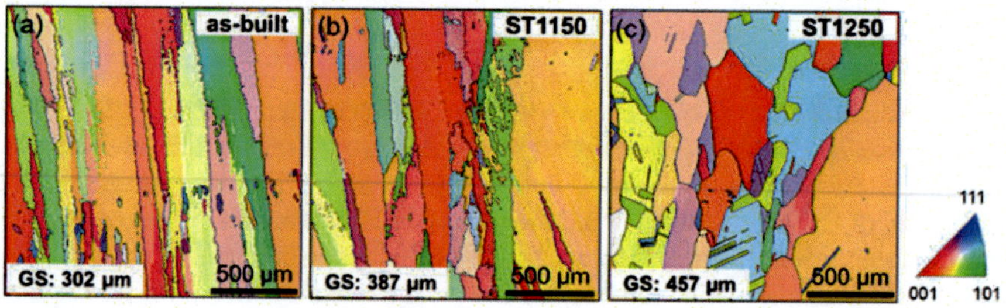

Fig. 5 Inverse pole figure (IPF) maps of the **a** as-built, **b** ST1150, and **c** ST1250 conditions in XZ plane using the FCC γ matrix for indexing

causing new columnar grains to grow epitaxially on equiaxed grains. Consequently, the new columnar grains will adopt the random crystallographic orientations of the equiaxed grains that act as the substrate. As this process introduces new 'competitors,' competitive growth is unable to filter out the most favorable orientations, resulting in a material with no strong texture such as the as-built Haynes 282 sample. As Christofidou et al. [14] and Shaikh et al. [15] reported, the temperature of standard solution heat treatment is not sufficient to recrystallize the grain structure of Haynes 282 processed by AM. However, the results suggest that the heat treatment at 1250 °C for 2 h allowed the sample to completely recrystallize, decreasing the heterogeneity of the grain structure.

Residual stresses and grain size have a significant effect on the mechanical performance of alloys. In general, grain coarsening and residual tensile stresses decrease the mechanical properties. The contribution of the microstructural characteristics on the alloy strength can be carried out through the hardness test. The microhardness

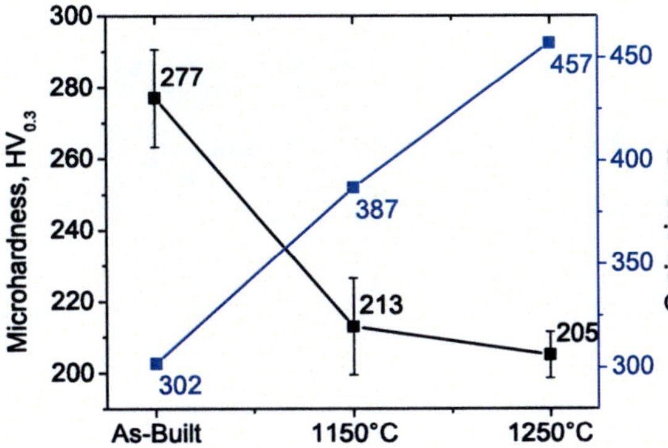

Fig. 6 The microhardness and grain size of Haynes 282 as a function of solution treatment temperature

and grain size of the as-built, ST1150, and ST1250 samples, are shown in Fig. 6. The profiles in this figure indicate that as the solution temperature increases, the microhardness drops and the grain size increases, reaching a value of 205 $HV_{0.3}$ and 457 μm, respectively.

Thermal cycles intrinsic to additive manufacturing cause the material to expand and contract at different rates, increasing the residual stresses and therefore the hardness of the as-built samples. On the other hand, the softening of samples ST1150 and ST1250 is attributed to temperature-assisted recovery and grain coarsening processes. The combined effect of grain growth and relaxation of residual stresses causes the hardness of the ST1150 sample to decrease by 64 $HV_{0.3}$ compared to the as-built sample. However, even though the grain size increases continuously, the hardness of the ST1250 sample decreases by only 8 $HV_{0.3}$ compared to the ST1150 sample. Therefore, the relief of residual stresses is the main cause of the drop in hardness. Similar results have been reported in several works available in the literature [16, 17]. These results suggest that solution heat treatments above 1150 °C are sufficient to relieve the residual stresses developed during the AM process.

Precipitation of $M_{23}C_6$ Carbides Throughout the First Aging

To increase the thermal stability of the grains at high temperatures and thus, the creep resistance, Haynes 282 is aged at 1010 °C for 2 h. This heat treatment is aimed at the formation of Cr- and Mo-rich $M_{23}C_6$ carbides at the grain boundaries. Figure 7 shows an SEM micrograph, EDS maps, and the temperature-dependent chemical composition plot calculated using Thermo-Calc for $(Cr, Mo)_{23}C_6$ carbides found in 1150-FA and 1250-FA samples. Figure 7a indicates that $(Cr, Mo)_{23}C_6$ carbides tend to distribute over grain boundaries, whereas Fig. 7b and c proves that Cr and Mo are the main constituents of this phase, which is corroborated by the computational prediction shown in Fig. 7d.

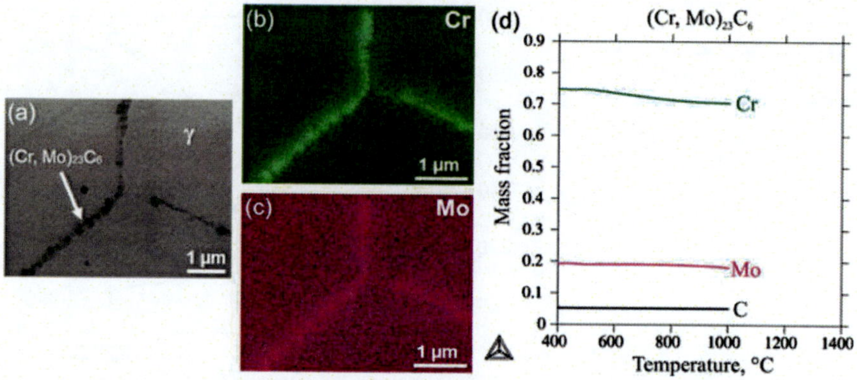

Fig. 7 **a** Representative SEM micrograph of microstructure of the 1150 + FA and 1250 + FA samples, EDS maps of **b** Mo and **c** Cr and **d** the evolution of the chemical composition of the $M_{23}C_6$ carbides as a function of temperature, indicating the presence of Cr and Mo in these precipitates

Unlike the MC-type primary carbides that precipitate during the solidification process in interdendritic regions from the liquid phase due to the segregation of alloying elements, the $M_{23}C_6$-type secondary carbides precipitate from the γ matrix. However, due to the high interfacial energy involved in the formation of the $\gamma/M_{23}C_6$ interface, the $M_{23}C_6$ carbides grow preferentially on the grain boundaries. The work carried out by El-Bagoury [18] suggests that the $M_{23}C_6$ carbides precipitate due to the decomposition of the primary MC carbides at high temperatures or the presence of remaining carbon in the matrix. The reaction between the MC carbides and the matrix results in the formation of $M_{23}C_6$ carbides and γ′ ($MC + \gamma \rightarrow M_{23}C_6 + \gamma'$), or in some cases, it results in the precipitation of M_6C carbides and γ′ ($MC + \gamma \rightarrow M_6C + \gamma'$). According to Adelajda et al. [19], taking into account the chemical composition of Haynes 282, the previous reactions are described as

$$(Ti,\ Mo)C + (Ni,\ Cr,\ Al,\ Ti) \rightarrow Cr_{21}Mo_2C_6 + Ni_3(Al,\ Ti)$$
$$(Ti,\ Mo)C + (Ni,\ Co,\ Al,\ Ti) \rightarrow Mo_3(Ni,\ Co)_3C + Ni_3(Al,\ Ti)$$

On the other hand, Lvov et al. [20] suggest that these reactions are dominated by the diffusion of carbon from the MC carbide to the matrix and the diffusion of Ni, Cr, and Co in the opposite direction. Since the first aging temperature is above the γ′ solvus temperature (997 °C), γ′ phase does not precipitate, and the hardness does not increase significantly after this heat treatment. The hardness of samples 1150-FA and 1250-FA increases around 20 $HV_{0.3}$ compared to samples ST1150 and ST1250, respectively. Therefore, a second aging heat treatment is required to induce γ′ precipitation and improve the mechanical properties.

Effect of Solution Temperature on γ′ Phase Precipitation During Second Aging

The precipitation of γ′ prevents the movement of dislocations through the material, improving the mechanical strength of Haynes 282. However, the improper selection of the aging treatment parameters negatively affects the mechanical behavior of the alloy. High aging temperatures lead to low degrees of undercooling and supersaturation of the γ matrix, hindering the nucleation process, while low temperatures prevent diffusion of the essential alloying elements and therefore γ′ growth. Likewise, long aging times cause over-aging of γ′ precipitates. Therefore, the design of second aging conditions is crucial. According to Kruger [21], the volume fraction and diameter of γ′ in the fully aged condition of Haynes 282 are about 19% and 20 nm, respectively. On the other hand, Shin et al. [3] showed that solution treatment conditions have a profound effect on γ′ precipitation, suggesting that the temperature and time of the second aging should be optimized based on the conditions of the previous solution heat treatment.

The morphology and size distribution of the γ′ phase as a function of solution temperature and second aging time are seen in the SEM micrographs and histograms shown in Figs. 8 and 9, respectively. The SEM micrographs show that the spherical morphology of γ′ remains invariant regardless of solution temperature and aging time, which can be attributed to the 0.2% lattice mismatch between γ and γ′ phases [22]. The histograms indicate that γ′ size increases with the second aging time regardless of the solution temperature. These results are expected since the growth of the γ′ phase is controlled by diffusion, therefore, the γ′ size is proportional to the aging time and temperature. However, the γ′ size increases faster in the samples that were exposed to 1250 °C during the previous solution heat treatment, which is more evident at 2 and 4 h of aging. Likewise, 1250-FA-SA6h and 1250-FA-SA8h samples show a bimodal distribution of the γ′ phase. While the smallest (d1) and largest (d2) diameters of γ′ phase in sample 1250-FA-SA6h are 25 and 53 nm, the particle sizes of sample 1250-FA-SA8h are 30 and 61 nm, respectively.

The difference in γ′ size between samples 1150-FA-SAxh and 1250-FA-SAxh can be explained by the precipitation kinetics. According to Mao [23], the volumetric fraction of γ′ increases with the degree of supersaturation of the matrix prior to the aging heat treatment, facilitating the subsequent coarsening process. As discussed above, MC carbides are partially dissolved during solution treatment at 1250 °C, increasing the degree of supersaturation of the matrix. As a result, the γ′ nucleation and growth kinetics in the 1250-FA-SAxh samples are faster compared to the 1150-FA-SAxh samples, causing the γ′ size of the 1250-FA-SAxh samples to be larger.

On the other hand, γ′ can be divided into three different types: primary γ′, cooling γ′, and aging γ′, which refer to precipitates remaining from previous sub-solvus heat treatments, precipitates formed during quenching, and γ′ particles that form from the aging heat treatment, respectively. Jian [23] showed that γ′ can precipitate during the cooling step of solution heat treatment if the cooling rate is low enough below the solvus temperature of γ′ phase (997 °C). Therefore, low cooling rates

Fig. 8 The precipitation of γ' during second aging heat treatment carried out at 788 °C for x = 2, 4, 6, and 8 h for samples **a** 1150-FA-SAxh and **b** 1250-FA-SAxh. The average γ' diameter is indicated below the metallography of each sample

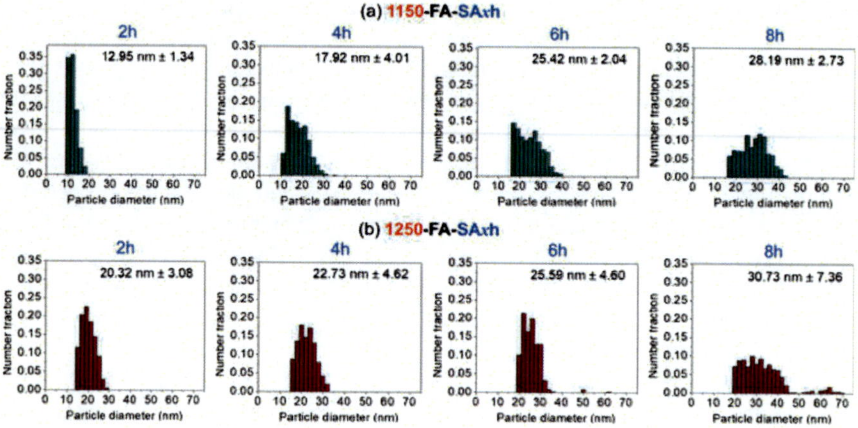

Fig. 9 The histograms of the γ' size distribution for **a** 1150-FA and **b** 1250-FA samples with different aging heat treatment conditions

and high degrees of matrix supersaturation contribute to the formation of cooling γ'. These observations suggest that although the 1150-FA-SAhx and 1250-FA-SAhx samples were quenched in the same way, the high degree of matrix supersaturation in the 1250-FA-SAhx samples contributes to the formation of cooling γ' after first-aging heat treatment, leading to a bimodal distribution of γ' precipitates after the second aging heat treatment. The bimodal distribution is clearer in samples 1250-FA-SA6h and 1250-FA-SA8h since the coarsening of γ' highlights the difference in size between cooling γ' and aging γ'.

The mechanical behavior of Haynes 282 is mainly controlled by the size distribution and volume fraction of γ'. If the precipitates are significantly smaller than the length of the dislocation line, the dislocations tend to overcome the precipitate by shearing. On the other hand, if the material is overaged, the dislocations pass the precipitates by looping, resulting in a drop in mechanical strength [24].

Hardness evaluation can be used as a rapid screening mechanism to measure the effect of γ' on the mechanical strength of materials. Figure 10 shows the hardness of 1150-FA-SAxh and 1250-FA-SAxh samples, highlighting the contribution of aging heat treatments. In general, it is observed that the hardness of 1150-FA-SAxh samples (Fig. 10a) are higher than the corresponding 1250-FA-SAxh samples (Fig. 10b). According to these results, high solution heat treatment temperature favors the softening of the material. On the other hand, the precipitation of $M_{23}C_6$ carbides during the first aging has a similar effect on hardness regardless of the previous solution treatment condition.

Furthermore, the maximum hardness of the 1150-FA-SAxh samples (390 $HV_{0.3}$) is reached after 8 h of aging at 788 °C when the γ' diameter is 28 nm, while the peak hardness of the 1250-FA-SAxh samples (374 $HV_{0.3}$) is observed at 6 h when the γ' diameter is 25 nm. The degree of supersaturation of the matrix after the solution heat treatment at 1250 °C causes the peak hardness and the overaged state of the ST1250-FA-SAxh samples to be reached faster compared to the 1150-FA-SAxh samples, resulting in the drop in hardness of 1250-FA-SA8h sample. Adelajda et al. [19] reported that the hardness of forged Haynes 282 after standard two-step aging heat treatment is 373 $HV_{0.3}$, showing that the hardness of Haynes 282 samples manufactured by WAAM are comparable. Interestingly, the contribution of γ' precipitates to the hardness of samples 1150-FA-SA2h and 1250-FA-SA2h is similar despite the difference in particle size being 7.32 nm. However, as the size of the γ' increases, the changes in hardness become more significant. Samples 1150-FA-SA4h and 1250-FA-SA4h indicate that a 4.81 nm change in γ' size produces a hardness variation of 16 $HV_{0.3}$. The results of this work demonstrate that to maximize the hardness

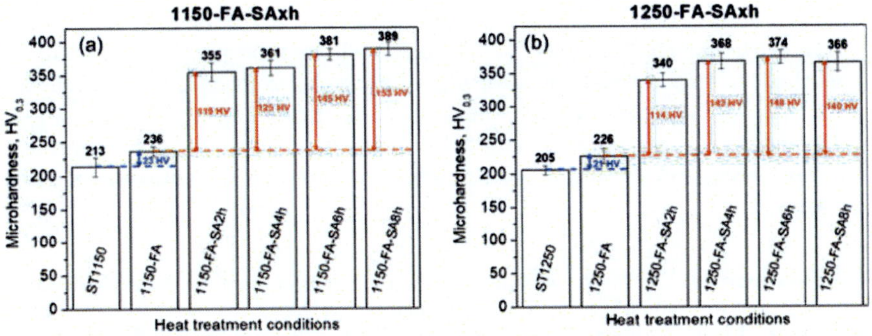

Fig. 10 The effect of heat treatments on the microhardness of Haynes 282 samples heat treated at **a** 1150 and **b** 1250 °C

of Haynes 282, the second aging heat treatment must be optimized based on the conditions of the previous solution heat treatment.

Conclusions

This work is focused on evaluating the effect of different heat treatments on the microstructure and hardness of Haynes 282 deposited using WAAM. The important outcomes of this investigation are as follows.

- The redistribution of alloying elements during the WAAM process allowed the precipitation of MC (M = Ti and Mo) carbides in interdendritic regions. However, the high cooling rate during solidification suppressed the precipitation of γ' phase and $M_{23}C_6$ carbides.
- Solution treatment at 1150 °C eliminated the segregation of alloying elements in interdendritic regions, dissolving the dendritic subgrains observed in the as-built sample. However, the temperature was insufficient to dissolve the primary (Ti, Mo)C carbides. Nevertheless, the volume fraction of (Ti, Mo)C carbides decreased, and the grain structure recrystallized only with solution heat treatment at 1250 °C.
- The first aging heat treatment carried out at 1010 °C promoted the formation of $(Cr, Mo)_{23}C_6$ secondary carbides at grain boundaries without affecting (Ti, Mo)C carbide distribution. The temperature of the solution heat treatment had no noticeable effect on the contribution of the $M_{23}C_6$ carbides to hardness.
- The solution heat treatment temperature has an important effect on the precipitation kinetics of the γ' phase. The degree of supersaturation of the matrix increases with the solution treatment temperature, enhancing the γ' growth rate. Likewise, samples exposed to 1250 °C during solution treatment show a bimodal γ' distribution.
- The hardness profiles show that the peak hardness associated with the γ' precipitation during the second aging was reached faster as the temperature of the previous solution treatment increases. The samples with solution treatment at 1150 °C reached their maximum value at 8 h while the samples exposed to 1250 °C reached the peak hardness at 6 h of the second aging heat treatment.
- The mechanical properties should be determined as future work to comprehensively understand the impact of solution treatments on the structure–property relationship.

Acknowledgements This project is supported via a subcontract from the Pennsylvania State University from the National Energy Technology Laboratory, U. S. Department of Energy, under the University Coalition for Fossil Energy Research program with grant No. DE-FE0026825. The authors would like to appreciate many valuable discussions with Drs. Ping Wang, Bingtao Li, and Jeremy Caron at the Haynes International, Inc. In addition, the authors are grateful for the technical support on printing from GEFERTEC GmbH.

Disclaimer This report was prepared as an account of work sponsored by an agency of the United States Government. Neither the United States Government nor any agency thereof, nor any of their employees, makes any warranty, express or implied, or assumes any legal liability or responsibility for the accuracy, completeness, or usefulness of any information, apparatus, product, or process disclosed, or represents that its use would not infringe privately owned rights. Reference herein to any specific commercial product, process, or service by trade name, trademark, manufacturer, or otherwise does not necessarily constitute or imply its endorsement, recommendation, or favoring by the United States Government or any agency thereof. The views and opinions of authors expressed herein do not necessarily state or reflect those of the United States Government or any agency thereof.

References

1. Shingledecker J, Purgert R, Rawls P (2013) Current status of the U.S. DOE/OCDO A-USC materials technology research and development program, Advances in Materials Technology for Fossil Power Plants: Proceedings from the Seventh International Conference, 41–52.
2. Pike LM (2008) Development of a fabricable gamma-prime (γ') strengthened superalloy. Superalloys 2008, 191–200. https://doi.org/10.7449/2008/Superalloys_2008_191_200.
3. Shin KY, Kim JH, Terner M, Kong BO, Hong HU (2019) Effects of heat treatment on the microstructure evolution and the high-temperature tensile properties of Haynes 282 superalloy. Mater. Sci. Eng. A. 751:311–322. https://doi.org/10.1016/j.msea.2019.02.054.
4. Haynes® 282® alloy Principal Features (2021) Haynes International. https://haynesintl.com/docs/default-source/pdfs/new-alloy-brochures/high-temperature-alloys/brochures/282-brochure.pdf?sfvrsn=20. Accessed 14 September 2022.
5. Ramakrishnan A, Dinda GP (2019) Microstructure and mechanical properties of direct laser metal deposited Haynes 282 superalloy. Mater. Sci. Eng. A. 748:347–356. https://doi.org/10.1016/j.msea.2019.01.101.
6. Ghiaasiaan R, Ahmad N, Gradl PR, Shao S, Shamsaei N (2022) Additively manufactured Haynes 282: effect of unimodal vs. bimodal γ'- microstructure on mechanical properties. Mater. Sci. Eng. A. 831:14234. https://doi.org/10.1016/j.msea.2021.142234.
7. S. Gudur, V. Nagallapati, S. Pawar, G. Mavvala, S. Simhambhatla (2021) A study on the effect of substrate heating and cooling on bead geometry in wire arc additive manufacturing and its correlation with cooling rate. Mater Today Procee. 41:431–436. https://doi.org/10.1016/j.matpr.2020.10.071.
8. Ojo OA, Richards NL, Chaturvedi MC (2006) Study of the Fusion Zone and Heat-Affected Zone Microstructures in Tungsten Inert Gas-Welded inconel 738LC Superalloy. Metall Mater Trans A. 37:421–433. https://doi.org/10.1007/s11661-006-0013-2.
9. Cieslak MJ, Knorovsky GA, Headley Jr. TJ, Romig AD (1989) The Solidification Metallurgy of Alloy 718 and Other Nb-Containing Superalloys, International Symposium on Alloy 718 Metallurgy and Applications, 1–6. https://doi.org/10.7449/1989/Superalloys_1989_59_68.
10. Cao WD, Kennedy RL, Willis MP (1991) Differential thermal analysis (DTA) study of the homogenization process in alloy 718. Superalloys 718, 625 and Various Derivatives, 147–160.
11. Wei HL, Mazumder J, DebRoy T (2015) Evolution of solidification texture during additive manufacturing. Sci Rep. 5: 1–7. https://doi.org/10.1038/srep16446.
12. Wei HJ, Mukherjee T, Debroy T (2016) Grain growth modeling for additive manufacturing of nickel based superalloys. Proceedings of the 6th International Conference on Recrystallization and Grain Growth (ReX&GG) 2016, 265–269. https://doi.org/10.1007/978-3-319-48770-0_39.
13. Li X, Tan W (2018) Numerical investigation of effects of nucleation mechanisms on grain structure in metal additive manufacturing. Comput Mater Sci. 153:159–169. https://doi.org/10.1016/j.commatsci.2018.06.019.

14. Christofidou KA, Pang HT, Li W, Pardhi Y, Jones CN, Jones NG, Stone HJ (2020) Microstructural Control and Optimization of Haynes 282 Manufactured Through Laser Powder Bed Fusion, Superalloys 2020, 1014–1023. https://doi.org/10.1007/978-3-030-51834-9_99.
15. Shaikh AS, Schulz F, Minet-Lallemand K, Hryha E (2021) Microstructure and mechanical properties of Haynes 282 superalloy produced by laser powder bed fusion. Mater Today Commun. 26:102038. https://doi.org/10.1016/j.mtcomm.2021.102038.
16. Huang W, Yang J, Yang H, Jing G, Wang Z, Zeng X (2019) Heat treatment of Inconel 718 produced by selective laser melting: Microstructure and mechanical properties. Mater. Sci. Eng. A. 750:98–107. https://doi.org/10.1016/j.msea.2019.02.046.
17. Fayed EM, Shahriari D, Saadati M, Brailovski V, Jahazi M, Medraj M (2020) Influence of homogenization and solution treatments time on the microstructure and hardness of Inconel 718 fabricated by laser powder bed fusion process. Materials. 13:2574. https://doi.org/10.3390/ma13112574.
18. El-Bagoury N (2016) Ni based superalloy: casting technology, metallurgy, development, properties and applications. Int J Eng Sci Res Technol. 5:108–152.
19. Polkowska A, Polkowski W, Warmuzek M, Cieśla N, Włoch G, Zasada D, Purgert RM (2019) Microstructure and Hardness Evolution in Haynes 282 Nickel-Based Superalloy During Multivariant Aging Heat Treatment. J Mater Eng Perform. 28:3844–3851. https://doi.org/10.1007/s11665-019-3886-0.
20. Lvov G, Levit VI, Kaufman MJ (2004) Mechanism of Primary MC Carbide Decomposition in Ni-Base Superalloys. Metall Mater Trans A. 35:1669–1679. https://doi.org/10.1007/s11661-004-0076-x.
21. Kruger KL (2017) Haynes 282 alloy, in: Augusto Di Gianfrancesco (ed) Materials for Ultra-Supercritical and Advanced Ultra-Supercritical Power Plants, Woodhead Publishing, p 511–545. https://doi.org/10.1016/B978-0-08-100552-1.00015-4.
22. Geddes B, Leon H, Huang X (2010) Superalloys : alloying and performance., ASM International, Ohio, USA.
23. Mao J (2002) Gamma prime precipitation modeling and strength responses in Gamma prime precipitation modeling and strength responses in powder metallurgy superalloys powder metallurgy superalloys, West Virginia University. https://researchrepository.wvu.edu/etd.
24. Adlakha I, Garg P, Solanki KN (2019) Revealing the atomistic nature of dislocation-precipitate interactions in Al-Cu alloys. J Alloys Compd. 797:325–333. https://doi.org/10.1016/j.jallcom.2019.05.110.

Characterization of the Anisotropic Behaviour of Inconel 718 Parts Manufactured by Wire Arc Additive Manufacturing

Karin Hartl, Christopher Wallis, Pier Paolo Curti, Martin Bielik, and Martin Stockinger

Abstract The usage of additive manufacturing as a process for component production is becoming increasingly important, as it offers enormous potential for material savings and therefore cost reduction. In particular, Wire Arc Additive Manufacturing (WAAM) processes are arousing a great deal of interest in several industries by its high deposition rates at low equipment acquisition costs and the low buy-to-fly ratio. This process is being specifically investigated for aerospace and space applications, as it allows the production of large structural complex near-net-shape components in small batches. However, a major drawback of this technology is the high anisotropic behaviour of the manufactured structures in the as-welded state. Since the Ni–Fe alloy Inconel 718 is an anisotropic material, in which introduced textures strongly influence the mechanical properties, the impact of the WAAM processing route on the mechanical properties as well as the underlying microstructure is specifically focused on in this study. Using a plasma arc as heat source and Inconel 718 wire as feedstock material, test walls are produced in order to characterize the created material. In addition to the identification of factors influencing the process, temperature cycles are measured at different positions during the build-up. The resulting microstructure is subsequently evaluated macroscopically as well as microscopically and examined regarding pores and precipitates. SEM/EDS analysis is carried out to investigate the underlying microstructure of the additively manufactured parts. Furthermore, mechanical properties are evaluated in the build-up direction as well as transversal to this direction in order to characterize the anisotropy of the material.

Keywords Additive manufacturing · PMD® process · Inconel 718 · Mechanical properties

K. Hartl (✉) · M. Stockinger
Montanuniversität Leoben, Franz Josef-Str. 18, 8700 Leoben, Austria
e-mail: karin.hartl@unileoben.ac.at

C. Wallis · P. P. Curti · M. Bielik
RHP Technology GmbH, Forschungs- Und Technologiezentrum, 2444 Seibersdorf, Austria

E. A. Ott et al. (eds.), *Proceedings of the 10th International Symposium on Superalloy 718 and Derivatives*, The Minerals, Metals & Materials Series,
https://doi.org/10.1007/978-3-031-27447-3_47

Introduction

Over the past few years, additive manufacturing (AM) has been an innovative and promising area of component production that is increasingly attracting the interest of industry and research. The associated possibility of producing near-net shape components in small batches with a minimum amount of scrap material caused by subtractive manufacturing processing routes is becoming progressively important, in particular for cost-intensive materials [1–4]. While powder-bed processes initially dominated the research field of AM, increasing variations of wire-based AM routes are emerging. A high material efficiency, up to 100%, can be reached for wire-based methods even though the complexity and surface quality of the produced parts decrease [1]. Especially the Wire Arc Additive Manufacturing (WAAM) method, which is characterized by high deposition rates and thus an enormous reduction of production time compared to powder-bed processes, is attracting increased interest for the production of large components [5–7]. WAAM uses focused thermal energy to fuse material by melting while it is deposited. The thermal energy for WAAM can be applied in various types like a laser beam, electron beam, or (plasma) arc. Since the material is deposited locally and only where it is necessary, a near-net shape part is created. A subsequent machining process (e.g. milling) removes just a small amount of material to receive the final part, thus achieving a low buy-to-fly ratio (BTF). The BTF is a useful indicator of the efficiency of a process in terms of material consumption. It describes the mass ratio of the raw material's weight needed for manufacturing a component and the final part's weight [2]. Especially for high cost materials like titanium and its alloys a high BTF up to 12-25:1 is common for traditional processes while AM-process routes can reduce the ratio to 3-12:1, as reported by Liu and Shin [8].

In this study a plasma transferred arc energy source in combination with metal wire feeding is used at RHP Technology GmbH, which is called Plasma Metal Deposition (PMD®). The PMD® system is built up in a M3DP facility by SBI GmbH and equipped with a building platform of 0.6 × 2.0 m and allows the manufacturing of large metallic structures made of titanium alloys, steels, aluminium alloys, Ni–Fe superalloys as well as special materials like Invar or magnesium alloys. To protect the material from oxidation, the airtight process chamber can be flushed with protective gas (e.g. argon with the purity of 99.999%). Figure 1 depicts the plasma torch and the wire feeders (on the left) where the wire is introduced into the plasma plume. Here, the wire melts and fuses to the material beneath.

In this way, components can be manufactured from high-strength materials that exhibit comparably good or even improved mechanical properties.

However, this requires precise control of the influencing parameters of the process. The most important variables influencing the process are the voltage and current, the protective gas in use, the nozzle distance, the wire feed speed (WFS) as well as the nozzle travel speed (TS). The build-up can also vary in terms of the welding strategy such as unidirectional, bidirectional, or oscillating while an external temperature supply (e.g., heat mat) can be used to reduce process-related residual stress evolution.

Fig. 1 Plasma torch and wire feeder (*Courtesy* SBI GmbH)

These factors are influencing the cooling time, which is of major importance for microstructure, residual stresses/warping, and mechanical properties. With regard to the deposited material, the main influencing factors are the wire diameter, the cast and helix, and the surface, while in addition, there is the possibility to use a filled wire. Depending on the parameters applied and the heat conduction potential of the material selected, variations in wall width and layer thickness are obtained.

Taking all these aspects into account, a large number of varying parameter sets can be found, where certain variations can have a significant influence on the properties of the resulting component. Additionally, the certification of the pre-material as well as the built component regarding microstructure and mechanical properties is difficult. Therefore, the use of such manufactured components is still limited [9–11].

Furthermore, anisotropies of the mechanical properties are likely to occur parallel and perpendicular to the build-up direction in such processes. Due to the continuous heat input in the underlying layers during the application of further welding layers, the underlying microstructure is constantly changing. In order to gain an insight into the behavior of an additively manufactured component, thin walled samples were made of the Ni–Fe alloy Inconel 718 using the PMD® process. During the build-up process of the test coupons, the temperature development in certain layers was measured at different locations. The components produced in this way were then analyzed in terms of microstructure and the mechanical properties, both horizontally and vertically according to the extraction position.

Material and Methods

For this investigation, Inconel 718 in form of a wire with a diameter of 1.2 mm by Deutsche Nickel GmbH and voestalpine High Performance Metals International GmbH was used as feedstock material. To avoid interference between the substrate and the welding layers, Inconel 718 was selected as substrate plates (20 mm × 10 mm

Table 1 Chemical composition of Inconel 718 in [wt.-%] according to AMS 5662

	Ni	Fe	Cr	Nb	Mo	Ti	Al	Co
[wt.-%]	50–55	bal	17–21	4.75–5.50	2.80–3.30	0.65–1.15	0.20–0.80	1

× 1 mm). In the current research the focus was set on the additive material, thus, no further investigations were performed on the substrate material.

Inconel 718 is one of the most well-known and widely used Ni–Fe alloys, due to its excellent mechanical and thermal stability as well as its outstanding corrosion resistance. The chemical composition of the alloy is shown in Table 1.

The main contributors to the thermal and mechanical stability are, on the one hand, the face-centered cubic γ matrix up to the melting point, and various intermetallic precipitates. The most significant precipitations are γ′ with a composition of Ni_3(Al, Ti) (L12 ordered fcc structure) and the γ″-phase with a composition of Ni_3Nb ($D0_{22}$ ordered bct structure). Those mentioned phases usually precipitate in the range of 600–900 °C. The δ-phase, which is the stable form of γ″ (Ni_3Nb with orthorhombic $D0_a$ structure) precipitates at temperatures above 700 °C with the highest precipitation rate around 900 °C and it dissolves at approximately 1000 °C, depending on the Niobium content of the alloy and material's previous history [12–14]. Since the δ and γ″ phases have the same composition, the precipitation of the δ-phase is accompanied by a decline in the γ″ fraction. As a consequence, the strength is negatively affected, since the precipitation hardening effect is dominated by γ″ [13, 15, 16]. On the other hand, the grain boundary pinning effect of the δ-phase is beneficial to limit grain growth. Especially in welding processes, a brittle intermetallic Laves phase often occurs. This phase is formed by a high Nb diffusion during solidification and preferably consists of elements with a high atomic diameter. Its occurrence can have enormously negative effects on the mechanical properties of the welded material [17, 18].

Test walls were set up in both bidirectional and oscillating strategies using the PMD® process. For the bidirectional test walls, the setup shown in Fig. 2 was used to record temperature data. For this purpose, several thermocouples were placed both in the substrate plate and in the built structure at different heights. The thermocouples were inserted into the still mushy material at the appropriate layer height.

Based on these measurements, the temperature development can be traced according to time and position. For this arrangement, walls consisting of about 60 layers were built up bidirectionally, which exhibit a wall thickness of approx. 5 mm and a height of approx. 80 mm. In addition, further walls were built up using this welding strategy to a height of approx. 130 mm, in order to create horizontal and vertical flat tensile specimens. Those walls, which were welded with an oscillating welding strategy, were used due to their larger wall thickness (~20 mm) for the extraction of round tensile and Charpy specimens in horizontal as well as in vertical direction.

The walls were manufactured with a constant power input. The voltage was kept consistent at 28 V, while the current varied depending on the part's height. In the

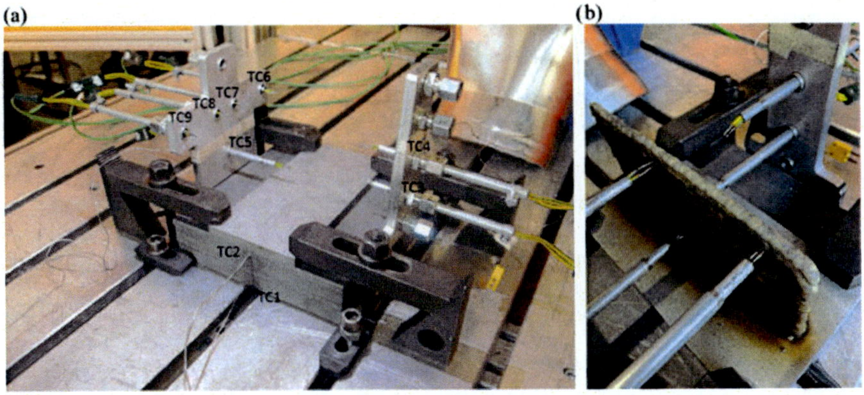

Fig. 2 **a** Setup for temperature measurements and **b** resulting test wall

first layers, a higher energy input (higher current) is required because the substrate is absorbing heat. With increasing height, the substrate's absorbing influence becomes less prominent; therefore, the current is reduced. Furthermore, the travel speed of the nozzle head is increased to reduce heat input and avoid overheating of the weld seam. The main welding parameters are listed in Table 2. In particular, the applied current (I), the travel speed (TS) are of great importance in this context. The wire feed speed (WFS) is slightly increased with increasing build-up height. Before the actual application of the first layer, the substrate plate was preheated twice (PH) to guarantee a good connection of welded material to the substrate. This was accomplished by moving the nozzle head twice over the subsequent build-up area without depositing material.

Due to the material's low tendency to oxidation, the manufacturing process was performed with local shielding gas (Argon) in an atmospheric condition.

Table 2 Parameters of the PMD® process used for the test walls

Layer	I [A]	TS [mm/min]	WFS [m/min]
PH	140	200	0
First layers	140	160	1.5
Middle layers	120–130	200–240	1.5
Last layers	100–110	280–300	2.0

Results

Temperature Development

For the temperature data generation during the build-up process, those bidirectional applied walls were used, which consist of 60 layers.

The temperature data recorded by two thermocouples are compared to provide a representative overview. The first thermocouple is placed in layer 14, at a height of about 22 mm from the underlying substrate plate. The second thermocouple is inserted at a height of 38 mm, corresponding to layer 25. The resulting temperature curves of the two thermocouples can be seen in Fig. 3. The measurement data clearly shows the process-related thermal influence by the subsequent layers, causing a cyclic thermal heat treatment. The first thermocouple T1 represents the temperature development more precisely than T2. The reason is found in the imprecise positioning of the thermocouple T2, which can occur when putting to the thermocouple in the mushy material. When it is positioned too close to the wall's lateral surface, it can get exposed to the plasma plume, causing a melting of the thermocouple (no signal) or an undetectability of the maximum temperature (cut-off of temperature peaks, as it is the case here). However, the thermal cyclic behavior still can be observed in T2, which corresponds to the data received from thermocouple T1.

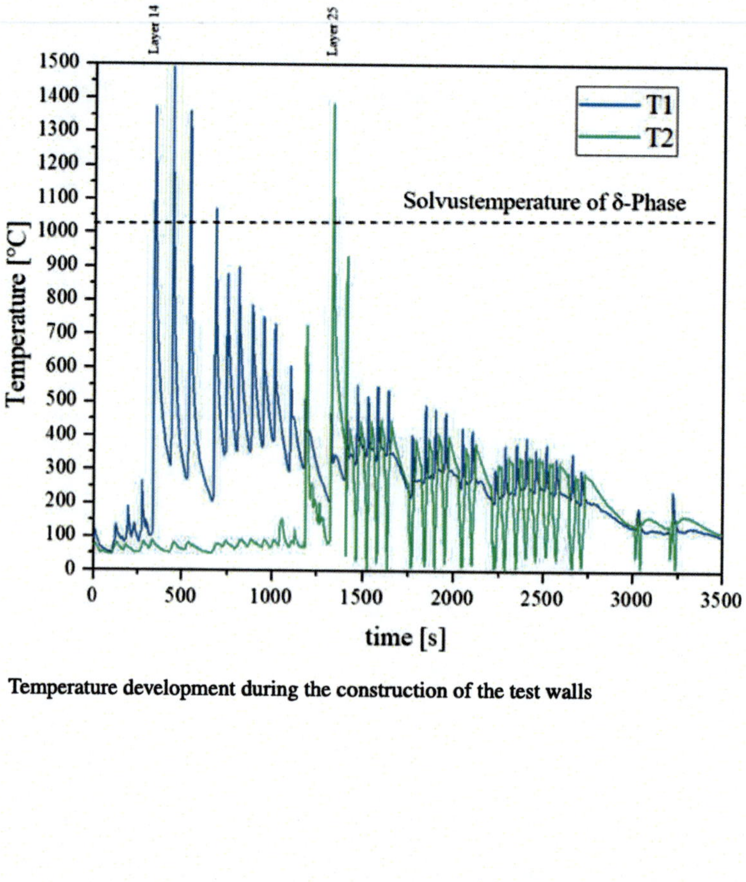

Fig. 3 Temperature development during the construction of the test walls

According to the data obtained, heating rates of up to 320 K/s and cooling rates of 20 °C/s can be observed when depositing an additional layer. This results in peak temperatures of 1500 °C. During this process, the δ-solvus temperature, here specified at 1025 °C according to Chenna Krishna et al. [19], is cyclically exceeded and undershot several times in a very short time. Whether cyclic dissolution and re-precipitation of the δ phase actually occurs in this process depends on the underlying kinetics, since the times exceeding the δ-solvus temperature are on average not longer than 10 s.

Microstructural Analysis

For microstructural analysis, the test walls used for the generation of temperature data were partitioned into several segments in order to analyze the cross section. Figure 4 shows a stereomicroscopic image of the cross section as well as optical microscopic (OM) images. The specimen was etched using the Beraha III etching method to reveal the weld structure and dendritic pattern. It is clearly visible that the structure is predominantly arranged in the build-up direction and also interconnected across the welding layers. In fact, the dendrites grow into that direction with the highest temperature gradient (heat flux), which in this case is the direction to the substrate. The OM images show the structure in the top and bottom areas of the test wall. The microstructure with its needle-like dendrites seems to become a finer appearance towards the top. This could be the result of the cyclic heat treatment of the underlying layers.

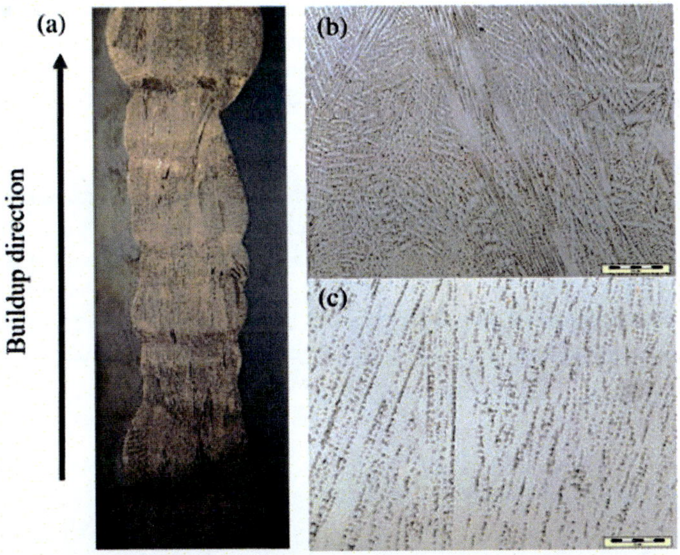

Fig. 4 **a** stereomicroscopic micrograph of the cross section of the test wall and **b** OM micrograph in the top region and **c** in the bottom region of the wall

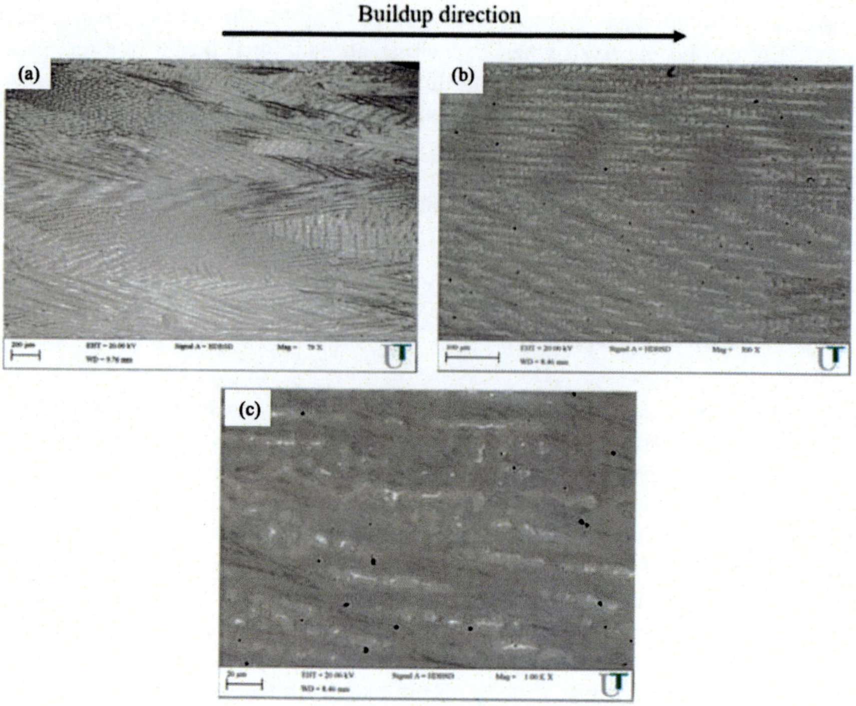

Fig. 5 SEM images of the microstructure in **a** 80x, **b** 300x, **c** 1000x magnification

When reporting the microstructural results, great attention was first paid to the pore analysis. However, even with the assistance of large-scale optical microscopy, it was not possible to perform a statistically representative distribution analysis since the pore frequency is insufficient. Therefore, Scanning Electron Microscopy (SEM) and Energy Dispersive X-Ray Analysis (EDX) were performed to detect potential micropores.

The SEM analysis again reveals a uniformly oriented structure in the build-up direction. At magnifications of 300x and 1000x, bright lines as well as small, dark areas can be detected, as can be seen in Fig. 5. The bright areas occur in elongated clusters and have an average size of 10 μm.

EDX results of the bright and dark regions show that the bright ones are likely to represent Laves phases and the dark smaller ones are titanium nitride precipitates. Figure 6 shows the spectra obtained for the two regions.

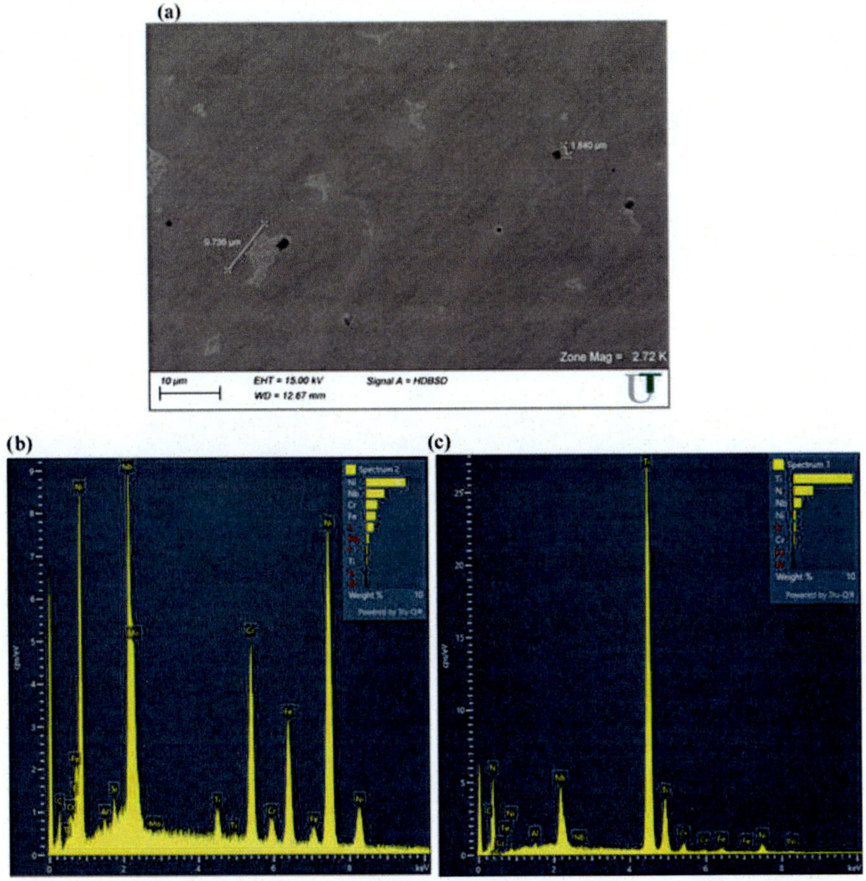

Fig. 6 Spectra obtained from EDX for **a** bright areas and **b** dark areas

In addition, an EDX mapping was created, indicating that the bright domains contain a high amount of nickel, niobium, chrome, and iron whereas the dark domains contain titanium and nitrogen. By this method, the dark areas could be excluded from representing micropores (Fig. 7).

Fig. 7 Resulting element distribution of the EDX mapping for the area shown in **a** for **b** titanium, **c** chrome, **d** niobium, and **e** nickel

SEM analysis likewise failed to indicate a significant number of pores. Indeed, some minor adhesion defects between two welding layers were detected. These are clearly resolvable, especially in the SEM, and could have significant negative effects on the mechanical stability. Figure 8 shows an adhesion defect in SEM at a magnification of 300x.

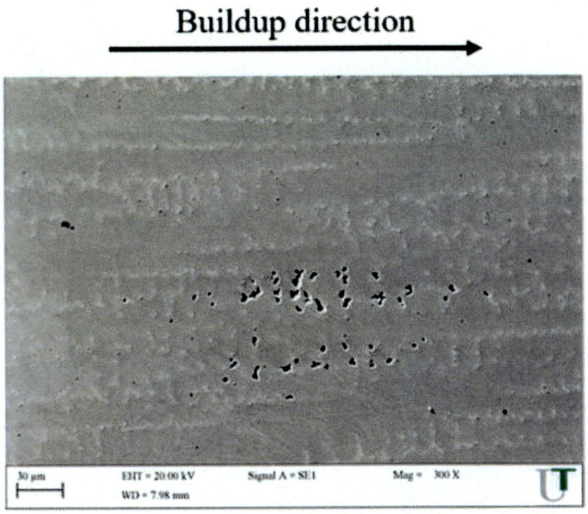

Fig. 8 Representation of an adhesion defect between two welding layers

Mechanical Properties

In order to investigate the effect of the assumed anisotropic behavior of the mechanical properties, tensile specimens were taken both horizontally and vertically from the test walls in the as-built condition. Flat tensile specimens were machined in accordance with DIN 50125 Form E. Round tensile specimens were manufactured based on Form B with an M10 thread. The notched bar specimens were prepared in accordance with DIN EN ISO 148. The measurement was conducted using standardized test conditions (DIN EN 6892 for tensile tests, DIN EN ISO 148-1). The sample positions were taken into account in each case in order to be able to attribute potential anomalies in the mechanical properties during the tensile test to the position. The specimen extraction positions of the samples can be seen in Fig. 9. Index 1 stands for the upper part of the wall for the horizontally taken specimens and for the left side for the vertical specimens, i.e., the position at which the nozzle starts depositing.

In the following section, the mechanical properties in terms of yield strength (YS), ultimate tensile strength (UTS) and uniform elongation (eu) are compared from top to bottom and left to right (Fig. 10). Furthermore, the obtained stress–strain curves of the tensile specimens are compared horizontally and vertically from top, center, and bottom, as well as from left, middle, and right (Fig. 11).

The flat tensile tests clearly indicate that the specimens taken horizontally from the top exhibit a significantly lower strength with a slightly increased uniform elongation in comparison to the samples of the bottom. The strength and fracture elongation of the horizontal specimen approach those of the vertically taken specimens at extraction locations towards the bottom. This is not exactly reflected in the behavior of the round tensile specimens. However, even with the round tensile specimens, the uniform elongation values in the top extraction region exceed those of the lower

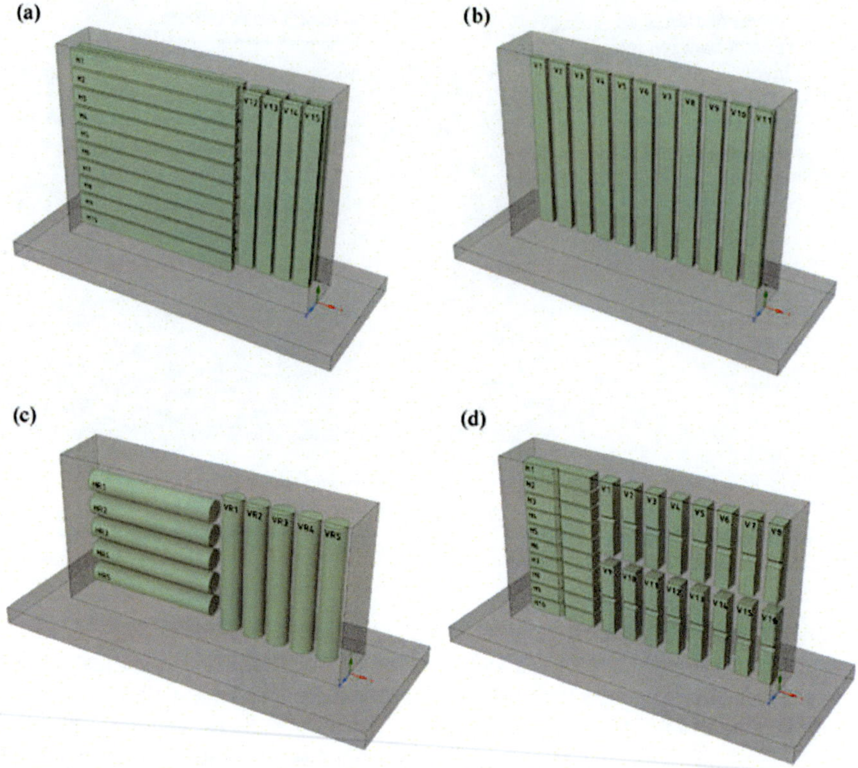

Fig. 9 Sampling locations from the test walls **a** and **b** flat tensile testing samples **c** round tensile testing samples and **d** Charpy impact testing samples

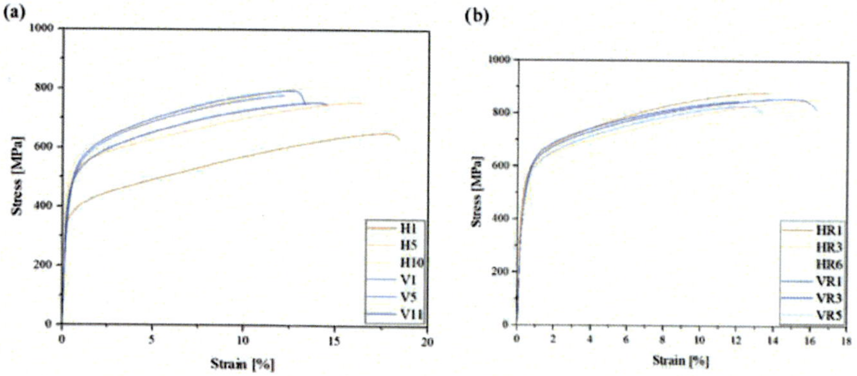

Fig. 10 Stress–strain diagrams of **a** flat tensile specimens and **b** round tensile specimens

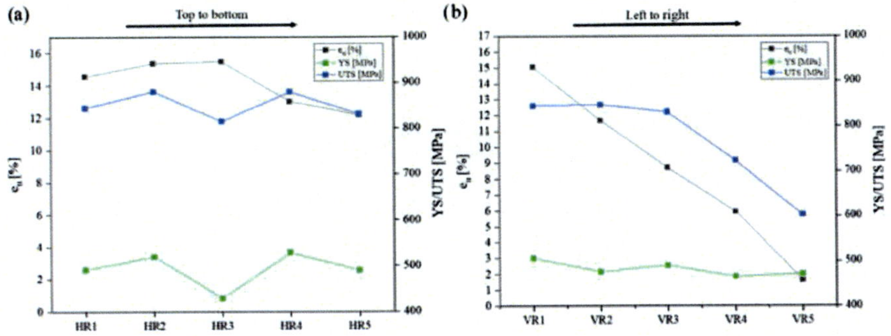

Fig. 11 Visualization of the mechanical properties of the round tensile specimens in **a** horizontal and **b** vertical direction

regions, while maintaining comparatively consistent high-strength values. The vertically taken specimens reveal a noticeable edge influence. In this case, the strength as well as the uniform elongation decrease steadily from left to right, while according to Fig. 9, the "left" sample was extracted in the middle of the test wall.

The Charpy impact tests can be seen in Fig. 12 demonstrating a higher impact strength of the vertical specimens compared to the horizontal samples. Here, the specimens taken horizontally from the top can again be attributed to a higher impact strength than those at the bottom.

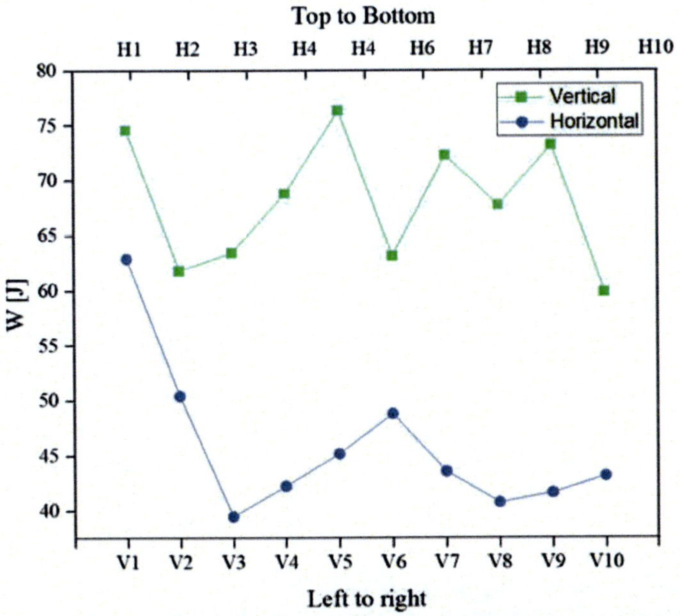

Fig. 12 Results of the impact strength in vertical and horizontal direction

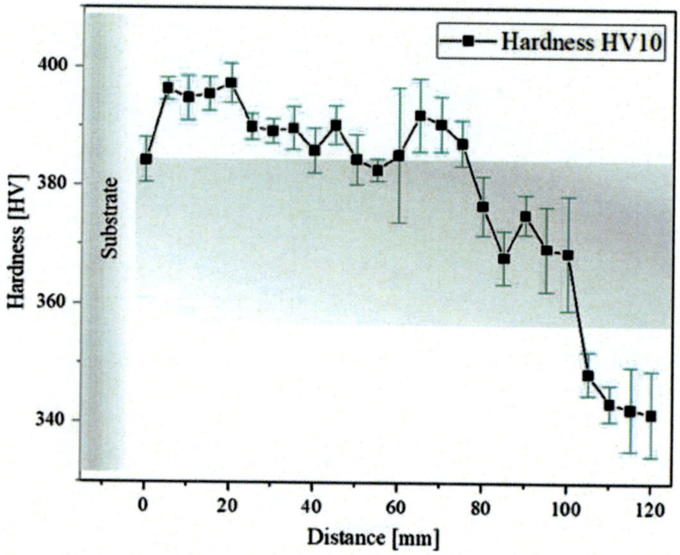

Fig. 13 Representation of the obtained hardness values along the wall build-up

Due to the discrepancy between the resulting mechanical properties of the upper and lower layers, the hardness along the test wall was additionally measured. The resulting hardness values shown in Fig. 13 exhibit a noticeable decrease in hardness, especially in the final layers which is in accordance with the results from flat horizontal tensile test specimens extracted from the top (Fig. 11a).

Discussion

In this investigation, the main emphasis lies in the as-built condition with no subsequent heat treatment or additional processing. In this condition, the specimens partially exhibit anisotropic behavior. However, due to the highly anisotropic microstructural appearance, a more pronounced anisotropy of the mechanical properties was expected. The low anisotropy is an aspect to be considered positive, even though the values in this condition are not comparable to values of Inconel 718 produced conventionally via casting and forming. As-forged Inconel 718 is probably best comparable to the as-built condition. In this state, the material exhibits a YS of about 900–1000 MPa and UTS of 1200–1300 MPa, while the values achieved for the PMD® processed Inconel 718 with a YS of around 500 and 800 MPa UTS are considerably lower.

Unlike a large number of observations in the field of wire-based AM in Inconel 718, in this study the top region of the wall tends to have lower strengths than the bottom region [20–22]. Due to the cyclic heat treatment the lower layers are subjected to higher temperatures frequently, thus, it could be expected that they exhibit a coarser grain structure than the top region, resulting in lower strength and hardness with increased elongation and toughness values. In the study considered here, a finer dendrite structure is perceived in the upper layers, suggesting similar results. Nevertheless, a diametrical result has been observed for the mechanical properties as well as in the hardness testing along the wall cross section. In this case, it is suggested that the lower regions are experiencing a strength enhancing effect which is considered due to the cyclic heat input during wall manufacturing. Since the precipitation kinetics of the strength-increasing γ' and γ'' are faster than those of the δ-phase, more frequent precipitation of these could occur in this particular heat input pattern. An investigation of the distribution and growth behavior of the γ' and γ'' precipitates using atomic probe tomography by Theska et al. [23] showed that these phases grew in response to even short heat treatment times, enhancing the strength-increasing effect. In addition, it was observed that with prolonged heat treatment, the former clusters of precipitates dissolve and the precipitates become more homogeneously distributed. Possibly, the cyclic temperature profile of the PMD® process applied here results in a similar behavior of the γ' and γ'' precipitates and could explain the higher strength and hardness values in the bottom region. Since these phases are only detectable by high-resolution examination methods, they were not analyzed in the present study.

Furthermore, the samples taken horizontally and vertically to the build-up direction likewise behave differently from the majority of published observations. They postulate higher strength values for the specimens taken horizontally than for the vertical samples [21, 24, 25]. In the context of the present investigations, this cannot be confirmed, especially for the top areas of the walls from which the flat tensile specimens were taken. However, the horizontal strength values increasingly approach those of the vertical ones the further to the bottom they were extracted. In the case of the round tensile specimens, no significant discrepancy between the horizontal and the vertical specimens can be detected. On the other hand, the results of the impact strength obtained, where the vertical specimens perform better, are again in agreement with the effect observed in other studies. Here, however, the welding strategy may influence the results. While the flat tensile specimens were taken from bidirectionally welded walls, an oscillating strategy was chosen for the geometry of the round tensile and Charpy specimens. The resulting microstructure of such a wall structure has not been considered so far.

However, these as-built investigations are sparsely published, since in most cases a precipitation-hardening heat treatment is performed prior to sampling. In addition, the manufacturing parameters vary enormously in some cases, which complicates a concrete comparison with other investigations.

Conclusion and Outlook

The utilization of WAAM technologies such as the PMD® process opens up a multitude of positive aspects for usage in the industry. The selective application of layers for the production of a component instead of using subtractive manufacturing routes has an enormous potential to save costs and to optimize the production-development cycle. This process is particularly suitable for large components, such as turbine casings, which are made of superalloys or titanium alloys. In view of these possible applications, a thruster nozzle was produced in advance using this process from Inconel 718, the partially machined state of which is shown in Fig. 14.

For this reason, the process will be investigated specifically for superalloys and the manufactured structures will be studied in detail in terms of both microstructural investigations and mechanical properties, which have been introduced in the course of this study. The characterization of the as-built state should thus serve as the basis for further investigations.

In order to achieve a more isotropic behavior of components and to further reduce the Laves phase produced by means of the PMD® process, it is recommended to conduct a heat treatment of the component before finishing. In order to increase the strength, a heat treatment with age hardening would be suitable. Whether classical heat treatment routes of conventionally produced Inconel 718 also reflect the optimum conditions for PMD® manufactured components or whether the time–temperature relations need to be adjusted will be investigated in the future.

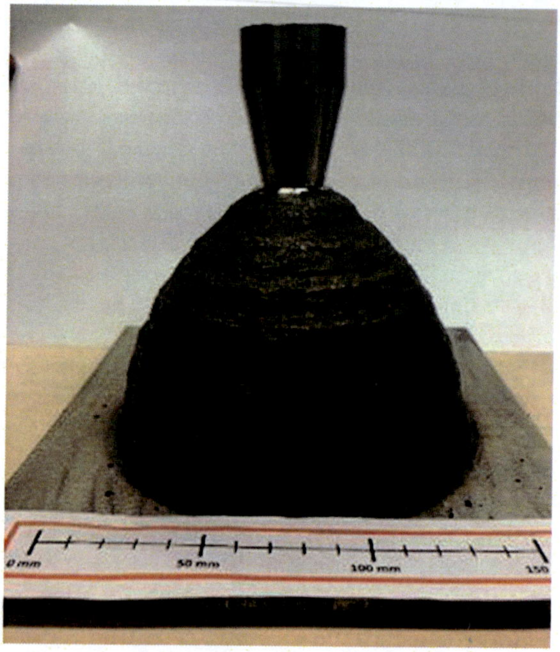

Fig. 14 Partly machined Inconel 718 thruster nozzle manufactured by the PMD® process

Another method of modifying the mechanical properties is provided by work-hardening treatment processes integrated into the WAAM process. One promising method is inter-pass rolling. In this process, a solid rolling treatment of the top layer is applied after the deposition of a layer. This increases the dislocation density in the material and enforces recrystallization when heat is introduced during the deposition of further layers. The resulting grain refinement improves the mechanical properties of the material and reduces anisotropic behavior [26, 27].

Furthermore, it is intended to create a model of the PMD process in the finite element software DEFORM™, which offers an AM module. The simulation can then be verified by the generated temperature–time curves during the wall build-up process. On the one hand, the virtual representation of the process allows process parameters to be analyzed in a time- and resource-saving manner. In addition, by integrating a suitable microstructure model, the effect of cyclic heat input on the layers at different heights can also be analyzed.

References

1. Ding D, Pan Z, Cuiuri D, Li H. Wire-feed additive manufacturing of metal components: technologies, developments and future interests. Int J Adv Manuf Technol 2015;81(1–4):465–81. https://doi.org/10.1007/s00170-015-7077-3.
2. Ford S, Despeisse M. Additive manufacturing and sustainability: an exploratory study of the advantages and challenges. Journal of Cleaner Production 2016;137(4):1573–87. https://doi.org/10.1016/j.jclepro.2016.04.150.
3. Frazier WE. Metal Additive Manufacturing: A Review. J. of Materi Eng and Perform 2014;23(6):1917–28. https://doi.org/10.1007/s11665-014-0958-z.
4. Saboori A, Aversa A, Marchese G, Biamino S, Lombardi M, Fino P. Application of Directed Energy Deposition-Based Additive Manufacturing in Repair. Applied Sciences 2019;9(16):3316. https://doi.org/10.3390/app9163316.
5. Ahn D-G. Directed Energy Deposition (DED) Process: State of the Art. Int. J. of Precis. Eng. and Manuf.-Green Tech. 2021;8(2):703–42. https://doi.org/10.1007/s40684-020-00302-7.
6. Martina F, Mehnen J, Williams SW, Colegrove P, Wang F. Investigation of the benefits of plasma deposition for the additive layer manufacture of Ti–6Al–4V. Journal of Materials Processing Technology 2012;212(6):1377–86. https://doi.org/10.1016/j.jmatprotec.2012.02.002.
7. Russell R, Wells D, Waller J, Poorganji B, Ott E, Nakagawa T et al. Qualification and certification of metal additive manufactured hardware for aerospace applications. In: Additive Manufacturing for the Aerospace Industry. Elsevier; 2019, p. 33–66.
8. Liu S, Shin YC. Additive manufacturing of Ti6Al4V alloy: A review. Materials & Design 2019;164(3):107552. https://doi.org/10.1016/j.matdes.2018.107552.
9. Najmon JC, Raeisi S, Tovar A. Review of additive manufacturing technologies and applications in the aerospace industry. Additive Manufacturing for the Aerospace Industry 2019:7–31. https://doi.org/10.1016/B978-0-12-814062-8.00002-9.
10. Katz-Demyanetz A, Popov, Vladimir V., Jr., Kovalevsky A, Safranchik D, Koptioug A. Powder-bed additive manufacturing for aerospace application Techniques, metallic and metal/ceramic composite materials and trends. MANUFACTURING REVIEW 2019;6.
11. Seifi M, Salem A, Beuth J, Harrysson O, Lewandowski JJ. Overview of Materials Qualification Needs for Metal Additive Manufacturing. JOM 2016;68(3):747–64. https://doi.org/10.1007/s11837-015-1810-0.

12. Frank R, Roberts C, Zhang J. Effect of Nickel Content on Delta Solvus Temperature and Mechanical Properties of Alloy 718. In: 7th International Symposium on Superalloy 718 and Derivatives (2010). John Wiley & Sons, Inc; 2010.
13. Azadian S, Wei L-Y, Warren R. Delta phase precipitation in Inconel 718. Materials Characterization 2004;53(1):7–16. https://doi.org/10.1016/j.matchar.2004.07.004.
14. Beaubois V, Huez J, Coste S, Brucelle O, Lacaze J. Short term precipitation kinetics of delta phase in strain free Inconel* 718 alloy. Materials Science and Technology 2004;20(8):1019–26. https://doi.org/10.1179/026708304225019830.
15. Brooks JW, Bridges PJ. Metallurgical Stability of Inconel Alloy 718. In: Superalloys 1988 (Sixth International Symposium). TMS; 1988.
16. Oblak JM, Paulonis DF, Duvall DS. Coherency strengthening in Ni base alloys hardened by DO22 γ' precipitates. Metall Mater Trans B 1974;5(1):143–53. https://doi.org/10.1007/BF02642938.
17. Xiao H, Li S, Han X, Mazumder J, Song L. Laves phase control of Inconel 718 alloy using quasi-continuous-wave laser additive manufacturing. Materials & Design 2017;122:330–9. https://doi.org/10.1016/j.matdes.2017.03.004.
18. Sui S, Tan H, Chen J, Zhong C, Li Z, Fan W et al. The influence of Laves phases on the room temperature tensile properties of Inconel 718 fabricated by powder feeding laser additive manufacturing. Acta Materialia 2019;164:413–27. https://doi.org/10.1016/j.actamat.2018.10.032.
19. Chenna Krishna S, Rao GS, Singh SK, Narayana Murty SVS, Venkatanarayana G, Jha AK et al. Processing and Characterization of Sub-delta Solvus Forged Hemispherical Forgings of Inconel 718. J. of Materi Eng and Perform 2016;25(12):5477–85. https://doi.org/10.1007/s11665-016-2377-9.
20. Kirka MM, Unocic KA, Raghavan N, Medina F, Dehoff RR, Babu SS. Microstructure Development in Electron Beam-Melted Inconel 718 and Associated Tensile Properties. JOM 2016;68(3):1012–20. https://doi.org/10.1007/s11837-016-1812-6.
21. Deng D, Moverare J, Peng RL, Söderberg H. Microstructure and anisotropic mechanical properties of EBM manufactured Inconel 718 and effects of post heat treatments. Materials Science and Engineering: A 2017;693:151–63. https://doi.org/10.1016/j.msea.2017.03.085.
22. Deng D. Additively Manufactured Inconel 718: Microstructures and Mechanical Properties. Linköping: Department of Management and Engineering, Linköping University; 2018.
23. Theska F, Nomoto K, Godor F, Oberwinkler B, Stanojevic A, Ringer SP et al. On the early stages of precipitation during direct ageing of Alloy 718. Acta Materialia 2020;188:492–503. https://doi.org/10.1016/j.actamat.2020.02.034.
24. Ni M, Chen C, Wang X, Wang P, Li R, Zhang X et al. Anisotropic tensile behavior of in situ precipitation strengthened Inconel 718 fabricated by additive manufacturing. Materials Science and Engineering: A 2017;701:344–51. https://doi.org/10.1016/j.msea.2017.06.098.
25. Popovich VA, Borisov EV, Popovich AA, Sufiiarov V, Masaylo DV, Alzina L. Impact of heat treatment on mechanical behaviour of Inconel 718 processed with tailored microstructure by selective laser melting. Materials & Design 2017;131:12–22. https://doi.org/10.1016/j.matdes.2017.05.065.
26. Donoghue J, Antonysamy AA, Martina F, Colegrove PA, Williams SW, Prangnell PB. The effectiveness of combining rolling deformation with Wire–Arc Additive Manufacture on β-grain refinement and texture modification in Ti–6Al–4V. Materials Characterization 2016;114:103–14. https://doi.org/10.1016/j.matchar.2016.02.001.
27. Colegrove PA, Coules HE, Fairman J, Martina F, Kashoob T, Mamash H et al. Microstructure and residual stress improvement in wire and arc additively manufactured parts through high-pressure rolling. Journal of Materials Processing Technology 2013;213(10):1782–91. https://doi.org/10.1016/j.jmatprotec.2013.04.012.

Keyhole TIG Welding of New Co-Lean Nickel-Based Superalloy G27

Achmad Ariaseta, Dario Pick, Joel Andersson, and Olanrewaju Ojo

Abstract The influence of keyhole TIG (K-TIG) welding parameters on the weld geometry and defects of a new Co-lean nickel-based superalloy G27 was studied, and the microstructures of the heat-affected zone (HAZ) and fusion zone (FZ) of the K-TIG-welded Alloy G27 were characterized. No cracks are found in the FZ and HAZ. Minimum weld width and face underfill statistically were significantly influenced by travel speed and interaction current*travel speed. Root excess weld metal was only significantly influenced by travel speed. Face excess weld metal was significantly influenced by all the factors, including their interaction. On the other hand, all the factors, including their interaction, did not significantly influence the average pore diameter. In the FZ microstructure, interdendritic microconstituents are identified as (Nb, Ti)C particles and γ/Laves eutectic constituents. In addition, a plate-like phase is observed surrounding the Laves phase, and γ' precipitates are found to be inhomogeneously precipitated in the FZ. In the partially melted zone (PMZ), (Nb, Ti)C, Laves phase, γ' precipitates, and plate-like particles are found in the liquated and resolidified regions, suggesting the solidification behavior in PMZ is likely to follow a similar pattern to the one observed in FZ.

Keywords K-TIG welding · Weld geometry · Weld defects · Microstructure · G27 superalloy

A. Ariaseta (✉) · D. Pick · J. Andersson
Department of Engineering Science, University West, 46186 Trollhättan, Sweden
e-mail: achmad.ariaseta@hv.se

A. Ariaseta
Faculty of Mining and Petroleum Engineering, Department of Metallurgical Engineering, Institut Teknologi Bandung, Jl. Ganesha 10, Bandung 40132, Indonesia

O. Ojo
Department of Mechanical Engineering, University of Manitoba, Winnipeg, MB R3T 5V6, Canada

E. A. Ott et al. (eds.), *Proceedings of the 10th International Symposium on Superalloy 718 and Derivatives*, The Minerals, Metals & Materials Series,
https://doi.org/10.1007/978-3-031-27447-3_48

Introduction

G27 is a new patented wrought γ'-strengthened Ni-based superalloy that Carpenter Technology Corporation has recently developed to obtain properties comparable to those of Waspaloy in terms of high-temperature strength, stress rupture, notch ductility, stability, and dwell fatigue crack growth resistance [1]. This new alloy is intended for high-temperature applications such as gas turbine components in aircraft engines with service temperature requirements of up to about 760 °C. The alloy G27 is formulated to be Co-free, therefore, producible through conventional metallurgical processes, i.e., Vacuum Induction Melting (VIM)/Vacuum Arc Remelting (VAR) and VIM/Electroslag Remelting (ESR), with a significantly reduced raw material cost than for Waspaloy. However, Co can be added to the alloy to achieve enhanced stress rupture strength at the cost of dwell fatigue crack growth resistance [1].

The new alloy G27 is a promising material candidate to be utilized in the hot sections of aero-engines that have been preferably fabricated by joining small pieces of superalloys by the welding process instead of casting the single large parts. The fabrication strategy has several advantages, such as enhancing the design flexibilities, e.g., it becomes feasible to join cast (complex geometry) alloys with wrought (high strength) alloys and even join completely different superalloys [2]. Another benefit is decreasing the total weight of the component, increasing the engine's fuel efficiency, which eventually leads to reduced CO_2 emission to the environment.

In the fabrication of hot structural parts of the aero-engine in the aerospace industry, one of the most standard techniques used in the fabrication is still tungsten inert gas (TIG) welding [3]. However, TIG welding processes have undergone significant improvements that paved the way for the fabrication of hot structural parts of the aero-engine, and now feasible to use faster travel speed and lower heat input. One of the most recent TIG welding processes, so-called keyhole TIG (K-TIG) welding, enables one to weld with a faster travel speed and lower heat input to achieve a fully penetrating weld in a single pass in comparison to the traditional TIG owing to a highly concentrated arc [4–6]. This process excellence is critical when welding sophisticated materials such as superalloys in the aero-engine. In addition, K-TIG welding is more cost-effective and much easier to operate than other high-energy density welding processes, e.g., electron and laser beam welding [7], making it highly suitable for on-site fabrication [8].

In welding superalloys, it is of high importance to produce a sound weld that satisfies the tight weld quality requirement in the aerospace industry in terms of weld geometry and defects. In addition, understanding the microstructures in the HAZ and FZ is essential since they influence the properties and integrity of the weldment and can become the foundation for developing a suitable post-weld heat treatment (PWHT) of the alloy. Since Alloy G27 is a new alloy, there are no reports from the literature regarding the influence of K-TIG welding parameters on the weld geometry and defects, and no knowledge available with regards to the resulting microstructures in HAZ and FZ of the welded alloy. Therefore, the article aims to study the influence of K-TIG welding parameters on the weld geometry and defects

Table 1 Chemical compositions of Alloy G27 in wt.% used in the present study

Element	Ni	Al	Ti	Nb	S	Cr	Ta	V	Zr	Mo	Co
wt.%	Bal	1.90	1.88	3.70	0.001	15.11	<0.02	0.1	0.03	4.02	0.18
Element (cont.)	Si	Mn	C	P	Cu	Ca	Mg	B	N	O	Fe
wt.% (cont.)	0.06	0.06	0.028	0.007	0.04	<0.0005	0.0034	0.005	0.006	0.001	15.00

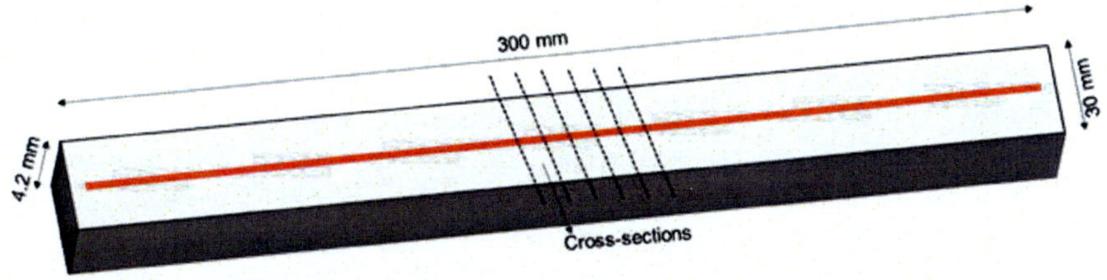

Fig. 1 The schematic of bead on plate weld run on the Alloy G27 plates and the corresponding location where the cross-sections were excised

when welding new Co-lean Alloy G27 and to characterize the microstructures of welded alloy in the HAZ and FZ.

Experimental Procedure

Material and Welding

As-received Alloy G27 plates in as-hot-rolled condition with the chemical composition (in wt.%) shown in Table 1 and thickness of 4.2 mm were used for the bead on plate weld runs using the robotic K-TIG welding process. High-purity argon gas (99.99%) was utilized as the shielding gas, with a 10 l/min flow rate. The schematic of the bead on plate weld run on the Alloy G27 plates is shown in Fig. 1.

Statistical Design of Experiment (DOE)

The DOE used in this work to study the influence of K-TIG welding parameters on the weld geometry and defects was based on a two-level full factorial design with the two factors, i.e., welding current and travel speed. Three middle points, i.e., Samples 5, 6, and 7, were added, resulting in seven experiments. The levels defined for welding parameters with their notations, units, and levels for the K-TIG welding are shown

Table 2 K-TIG welding parameters and their level used in this study

No.	K-TIG welding parameters	Units	Process parameter levels		
			−1	0	1
1	Welding current	A	425	430	435
2	Travel speed	mm/s	11	11.5	12

in Table 2. The welding current and travel speed ranges were selected to produce a fully penetrating weld without burning through the plate. The distance between the K-TIG electrode tip and the Alloy G27 plate was kept at 1.5 mm throughout the DOE runs. After welding the samples, the typical features of weld geometry and defects specified in AWS D17.1M:2010 [9] as the responses were measured in the as-welded alloy, i.e., face and root excess weld metal, underfill, and average pore diameter. In addition, measurements on minimum weld width (Wm) and cracks were also carried out. However, no single cracks are found in the FZ and HAZ of any sample. Then, the results were input into Modde software (MODDE 12, Umetrics, Umeå, Sweden) for the DOE analysis. Linear regression was used to model each response according to Eq. (1):

$$y = \beta_0 + \beta_1 \,\text{current} + \beta_2 \,\text{speed} + \beta_{12} \,\text{current.speed} + \varepsilon \qquad (1)$$

where y is the response and dependent variable; the β terms are constants, of which all except β_0 are called regression coefficients; ε is the residual error. The factor levels were represented on a coded scale. The welding current levels of 425, 430, and 435 A are coded −1, 0, and 1, respectively. The travel speed levels of 11, 11.5, and 12 mm/s are coded −1, 0, and 1, respectively. The calculation of regression coefficients was carried out in Modde using the multiple linear regression method, and a 95% confidence interval of these coefficients was also calculated. In addition, a square test was automatically carried out in Modde to detect possible non-linearity or square terms in the regression models. The reproducibility test was also performed in Modde and set a value of 0–1 in Modde, wherein a value higher than 0.5 suggests good reproducibility [10]. Other diagnostic parameters, R^2 and *p*-value, were also calculated by Modde.

Weld Geometry and Weld Defects Quantification

Five cross-sections transverse to the welding direction were excised from all the welded plates at the center of the plates (Fig. 1). The mounted cross-sections were prepared by multistep grinding with the grit sizes down to 52 μm, followed by three polishing steps with 9 and 3 μm diamond suspensions and 0.02 μm colloidal SiO_2. Electrolytic etching using 10% oxalic acid at 2 V was performed to reveal the bead geometry. Macrographs of the sections were captured from etched samples

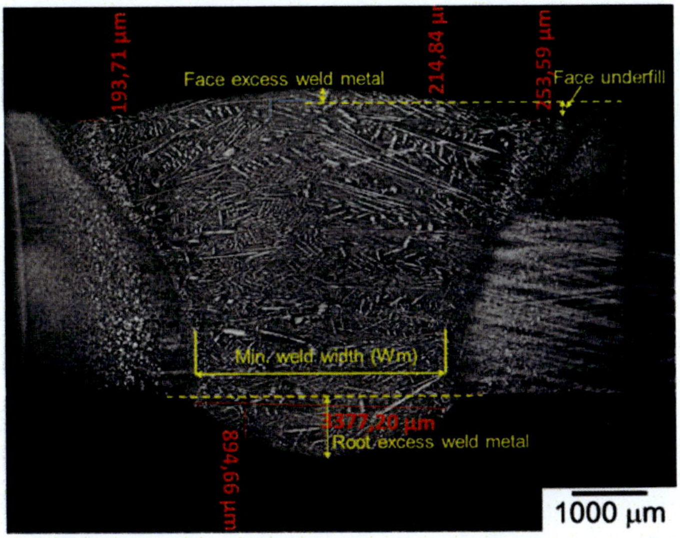

Fig. 2 LOM image showing the example of weld bead geometry of K-TIG welded Alloy G27 and the corresponding bead geometry measurements

using a light optical microscope (LOM) to get precise measurements of Wm, excess weld metal, and underfill, where the average values were obtained from five cross-sections. An example of a LOM image of weld bead geometry and the corresponding measurements are shown in Fig. 2.

The average pore diameter was measured quantitatively using a LOM on an as-polished surface. ImageJ software was used to quantify the pore diameter of a welded plate in the entire FZ area of five cross-sections. The average pore diameter was obtained from the images captured from five cross-sections of one welded plate.

The measured face and root excess weld metal, face underfill, and average pore diameter in the DOE runs were compared with the maximum acceptable limit specified in AWS D17.1M:2010 [9]. Note that Wm is not specified in the AWS D17.1M:2010. Thus, 1.5 mm was used as the minimum acceptable value for Wm in aerospace applications as Ref. [11].

Scanning Electron Microscopy (SEM)

High-resolution SEM microstructural characterization was performed on as-polished and electrolytically etched samples using a field emission gun scanning electron microscope (FEG-SEM) operated in the in-lens and backscatter electron (BSE) modes at an accelerating voltage of 10 kV. The FEG-SEM was also equipped with Oxford energy-dispersive X-ray spectroscopy (EDS). The SEM characterization was carried out on samples welded using the DOE's lowest (lowest current and fastest travel speed) and highest (highest current and lowest travel speed) heat inputs. It is revealed that there is no substantial microstructural difference in both samples. In this paper, only the sample welded using the lowest heat input is presented.

Results and Discussion

Influence of K-TIG Welding Parameters on Weld Geometry and Weld Defects

All measured DOE responses of Samples 1–7 are shown in Fig. 3. These data were taken directly from Table 3. The middle points in all DOE responses, i.e., Samples 5–7, showed similar values. This was reflected in the calculated reproducibility value of 0.999, 0.914, 0.989, 0.904, and 1.000 for Wm, face and root excess weld metal, face underfill, and pore diameter, respectively. Furthermore, all the DOE runs produced welds with Wm larger than 1.5 mm. All samples also had excess weld metals and pore diameters lower than the maximum allowable limit stipulated in AWS D17.1M:2010 [9] for aerospace applications. However, only Sample 3 (lowest current and highest travel speed) met the AWS D17.1M:2010 criteria in terms of face underfill.

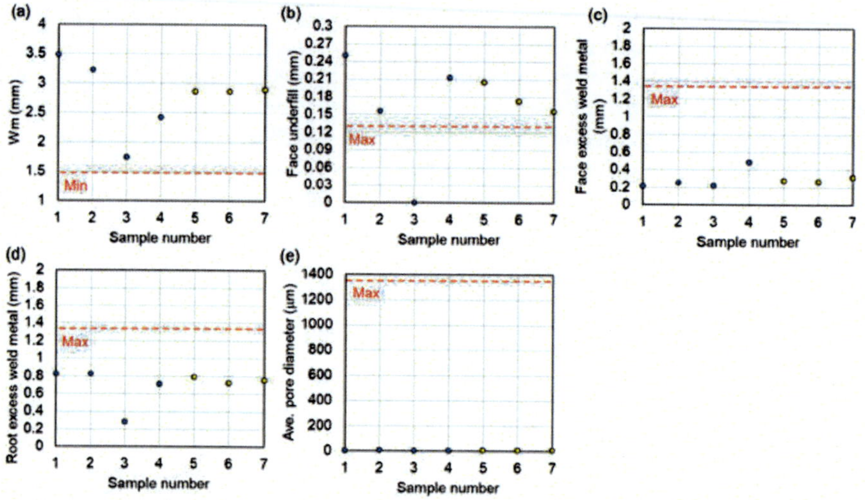

Fig. 3 Measured responses of K-TIG welded G27 samples (samples with varied parameters are in blue, while replicates are in yellow); measured face and root excess weld metal, face underfill, and average pore diameter are compared with the maximum acceptable limit stipulated in AWS D17.1M:2010, while 1.5 mm is used as the minimum acceptable limit for Wm as a reference, shown in red dotted lines [11]

Table 3 Full factorial design of K-TIG welding and the measured responses

Experiment no.	Current (A)	Travel speed (mm/s)	Wm (mm)	Face underfill (mm)	Face excess weld metal (mm)	Root excess weld metal (mm)	Ave. pore diameter (μm)
1	425	11	3.49	0.25	0.22	0.83	6
2	435	11	3.23	0.16	0.25	0.83	8
3	425	12	1.75	0	0.22	0.28	6
4	435	12	2.43	0.21	0.49	0.71	6
5	430	11.5	2.86	0.21	0.27	0.79	7
6	430	11.5	2.87	0.17	0.27	0.72	7
7	430	11.5	2.90	0.16	0.32	0.76	7

The analysis carried out in Modde modeled the responses of the Wm, face and root excess weld metal, face underfill, and average pore diameter of the seven samples to know which parameters had the most substantial influence on the responses and whether there are any significant interactions among these parameters. Therefore, regression coefficients and their 95% confidence interval are plotted in Fig. 4. The statistically significant parameters (p-value <0.05) are marked in green. In contrast, those that were not statistically significant (p-value $\geq$0.05) that included zero in their confidence intervals are marked in red. All the calculated p-values are presented in Table 4, where values lower than 0.05 are marked in bold. It is shown that Wm and face underfill statistically were significantly influenced by travel speed and interaction current*travel speed, where the travel speed and interaction current*travel speed have the strongest influence on the Wm and face underfill, respectively. Root excess weld metal was only significantly influenced by travel speed. Face excess weld metal was significantly influenced by all the factors, including their interaction, where the current was the most substantial influencing factor. On the other hand, all the factors, including their interaction, did not significantly influence the average pore diameter.

The next step is to know the comparison between the calculated results of the regression model and all the experimentally measured responses. In this step, the average pore diameter is excluded because all factors and their interaction statistically did not significantly influence the response. The regression models of Wm, face excess weld metal, root excess weld metal, and face underfill are presented in Eqs. 2–5 consecutively. No non-linearity or square terms were detected in all regression models of those responses based on the square test automatically performed in Modde. Plots comparing the calculated and measured responses are presented in Fig. 5. The dotted line crosses the points where the measured values are equal to the calculated values. The R^2 values of all measured responses are shown in Table 5.

Wm, face excess weld metal, root excess weld metal, and face underfill exhibit high R^2 values of 0.968, 0.979, 0.920, and 0.944, respectively, suggesting that the calculated results were close to the experimental results. For those responses shown

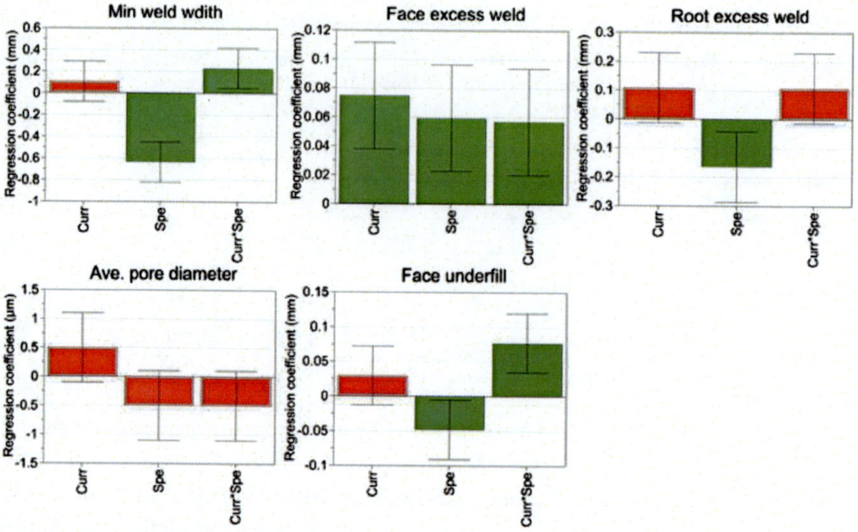

Fig. 4 Regression coefficients for different factors and interaction of factors affecting weld geometry and defects (green bars show significant factors, while red bars show non-significant factors)

Table 4 P-values calculated using Modde software for the regression constants that had significant (significant values are marked in bold) and non-significant influence on the responses (Wm, face and root excess weld metal, face underfill, and pore diameter)

Factor	*p*-value of Wm	*p*-value of face excess weld metal	*p*-value of root excess weld metal	*p*-value of face underfill	*p*-value of pore diameter
Current	**0.167**	0.008	**0.215**	**0.114**	**0.077**
Travel speed	0.002	0.014	0.036	0.037	**0.077**
Current*travel speed	0.029	0.016	**0.268**	0.011	**0.077**

in Fig. 5, the points are reasonably close around the 1:1 dotted line, suggesting a good fit. The *p*-values of the regression models of Wm, face excess weld metal, root excess weld metal, and face underfill generated in the ANOVA table were 0.005, 0.009, 0.038, and 0.022, respectively (Table 5). This suggests that the regression models of Wm, face excess weld metal, and face underfill as presented in Eqs. 2, 3, and 5, respectively, were statistically good or significant.

$$\text{Wm (mm)} = 4.113 + 0.106\,\text{current} - 0.635\,\text{speed} + 0.232\,\text{current.speed} \quad (2)$$

$$\text{Face excess weld metal (mm)} = 0.314 + 0.075\text{current} + 0.059\text{speed} + 0.057\text{current.speed} \quad (3)$$

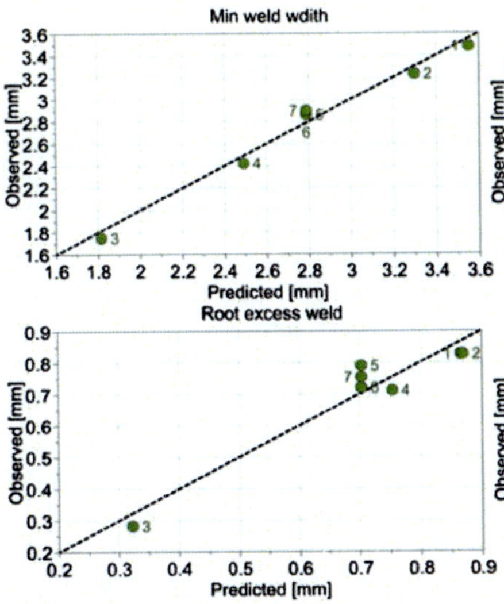

Fig. 5 Comparison of the experimentally measured responses with the predicted values from the regression model

Table 5 R^2 and *p*-values of regression models calculated using Modde software for the responses (Wm, face and root excess weld metal, and face underfill)

Response	R^2	*p*-value of the regression model (from ANOVA table)
Wm	0.979	0.005
Face excess weld metal	0.968	0.009
Root excess weld metal	0.853	0.038
Face underfill	0.944	0.022

$$\text{Root excess weld metal (mm)} = 0.779 + 0.108\,\text{current} - 0.165\,\text{speed} + 0.107\,\text{current.speed} \quad (4)$$

$$\text{Face underfill (mm)} = 0.192 + 0.030\,\text{current} - 0.048\,\text{speed} + 0.077\,\text{current.speed} \quad (5)$$

SEM Microstructural Characterization

Base Material Microstructure

The primary microstructural features of the Alloy G27 base materials are depicted in the low magnification backscatter electron (BSE) image in Fig. 6a. The microstructure consists of unrecrystallized grains and bright particles, identified as Nb-rich carbides containing Ti and C, i.e., (Nb, Ti)C by SEM–EDS analysis. One example of the typical EDS profiles and corresponding chemical compositions of (Nb, Ti)C found in the base material microstructure are displayed in Fig. 7 and Table 6, respectively. Note that it is not possible to measure the concentrations C by SEM–EDS analysis quantitatively; therefore, C is excluded from the analysis. Apart from (Nb, Ti)C, fine γ' with the size of ~10 nm or less are observed to be present uniformly throughout the microstructure on grain boundary and within grain interior (Fig. 6b).

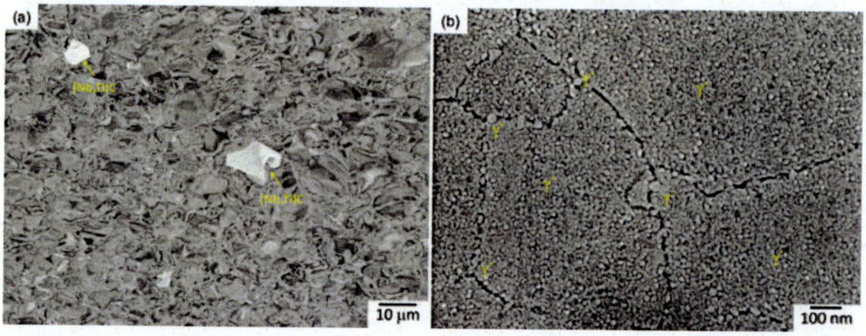

Fig. 6. **a** Low magnification SEM-BSE image of base material microstructure of as-received Alloy G27 showing the presence of unrecrystallized grains and (Nb, Ti)C particles; **b** high magnification SEM in-lens magnification (BSE images)

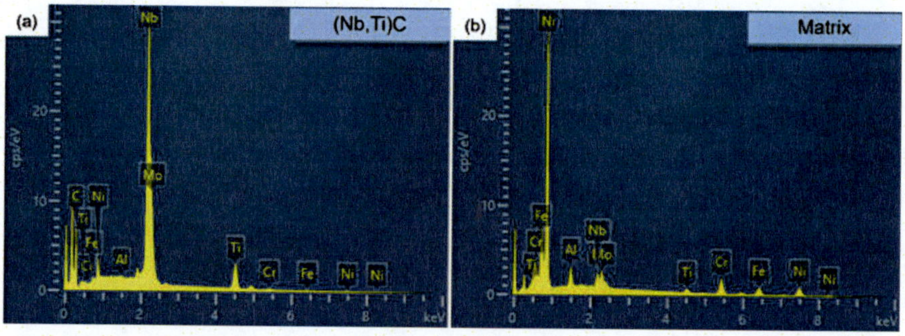

Fig. 7 Typical EDS profile of **a** (Nb, Ti)C and **b** matrix in the base metal microstructure of Alloy G27

Table 6 Chemical composition of (Nb, Ti)C and matrix in the base material of Alloy G27 obtained from semi-quantitative SEM–EDS point analysis

Element (wt.%)	Al	Ti	Cr	Fe	Ni	Nb	Mo
(Nb, Ti)C	0.14	15.28	1.40	1.03	7.06	75.10	0.00
Matrix	1.55	1.60	13.60	14.75	61.81	3.27	3.43

FZ Microstructure

Figure 8a displays the low magnification SEM-BSE image of the as-polished FZ microstructure consisting of a dendritic microstructure with a high number density of microconstituents present in the interdendritic region imaged in bright contrast. SEM–EDS analysis in higher magnification (Fig. 8b) reveals two types of microconstituents, i.e., particle A, which exhibits blocky and rod-like morphologies, and particle B, which exhibits an irregular shape distributed semi-continuously along the interdendritic region. Examples of high-resolution SEM-BSE images showing the blocky particle A and irregular-shaped particle B are depicted in Fig. 8c and d, respectively. Examples of particles A and B's chemical compositions based on SEM–EDS analysis are depicted in Table 7. The SEM–EDS analysis suggests that particle A appears to be an MC-type carbide that contains Nb as the main metallic element and Ti as a secondary metallic element, i.e., (Nb, Ti)C. The Nb content of (Nb, Ti)C particles is found to be varied based on SEM–EDS analysis, which also can be seen from the varied elemental contrast in some (Nb, Ti)C particles (Fig. 9b). The blocky and rod-like morphologies are also typical for MC carbide in precipitation-hardening superalloys [12–15]. It is observed in Table 7 that the chemical composition of particle B has a substantial similarity with γ/Laves eutectics in Nb-bearing Ni-based superalloys such as Alloy 718 [16] and ATI 718Plus [17] in terms of Nb, Mo, Cr, Fe, and Ni enrichments. It is worth noting that Alloy G27 is also considered an Nb-bearing superalloy with an Nb content of 3.7 wt.%. Therefore, particle B appears to be γ/Laves eutectics type constituent based on the SEM–EDS analysis. However, the γ/Laves eutectics does not exhibit the eutectic-type morphology, which is the typical morphology of γ/Laves eutectics constituent found in other Nb-carrying superalloys such as alloy 718 and ATI 718Plus [16, 18, 19].

Figure 9a displays a high-resolution SEM in-lens image of γ/Laves eutectic constituent in the interdendritic region of the electrolytically etched welded sample. Interestingly, plate-like phase is revealed around the Laves particle after etching. Due to the extremely fine size of the plate-like phase, it is impossible to analyze the plate-like phase's chemical composition using SEM–EDS. The existence of a plate-like phase around the interdendritic Laves eutectic particles is also found in the FZ of TIG-welded [20] and the deposit of WAAM-built [21] ATI 718Plus as reported by Asala et al. The plate-like phase is identified as η-phase using TEM selected area diffraction pattern by the same authors and found to be enriched in Ni, Nb, Al, and Ti from the TEM-EDS analysis. More in-depth characterization using TEM is needed

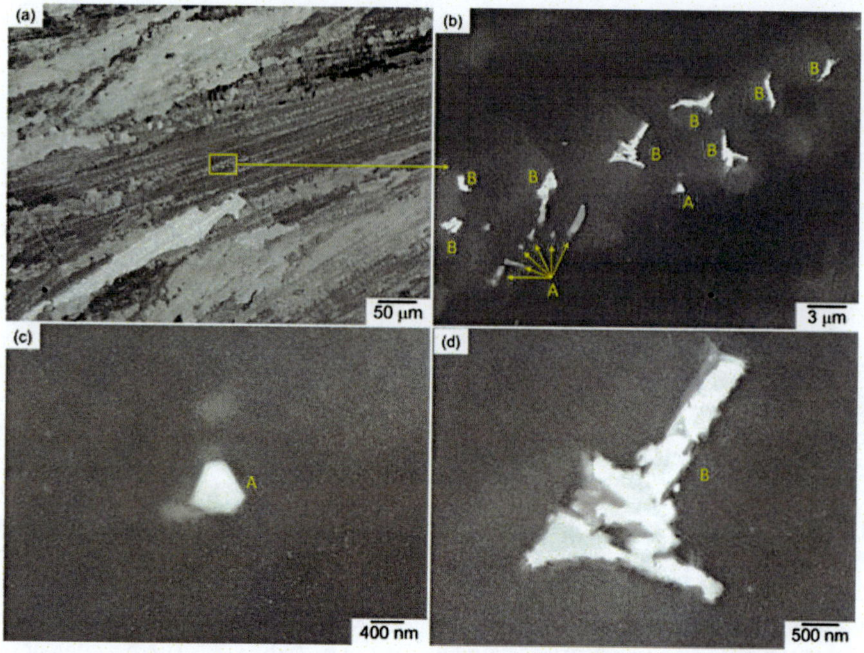

Fig. 8 SEM-BSE images of fusion zone in as-polished K-TIG welded Alloy G27 sample; **a** low magnification image showing dendritic microstructure with the presence of microconstituents in the interdendritic regions; **b** higher magnification image marked in **a** showing the different types of interdendritic microconstituents, i.e., particle A is (Nb, Ti)C, while particle B is γ/Laves eutectics; **c** high-resolution image of the blocky (Nb, Ti)C; **d** high-resolution image of irregular-shaped of γ/Laves eutectics

Table 7 Chemical composition of interdendritic microconstituents (particles A and B) in the FZ of K-TIG welded Alloy G27 obtained from semi-quantitative SEM–EDS point analysis

Element (wt.%)	Al	Ti	Cr	Fe	Ni	Nb	Mo
Particle A	0.24	14.16	2.20	1.13	4.44	72.74	5.09
Particle B	0.57	1.88	12.63	11.36	45.25	19.64	8.66
Interdendritic γ	1.62	2.53	12.87	12.22	60.13	6.27	4.37
γ/Laves eutectic in FZ of alloy 718 [16]	–	2.0	13.2	11.4	48.6	19.1	4.4
γ/Laves eutectic in FZ of ATI 718Plus [17]	1.02	1.35	13.2	6.03	42.07	20.04	3.65

to positively identify the plate-like phase in the FZ of K-TIG welded Alloy G27, which is the future work of this study.

A high magnification image (Fig. 9b) of the region marked red in Fig. 9(a) reveals the presence of γ′ precipitates with a few tens of nm size in the γ phase immediately adjacent to the Laves eutectic constituent. The spherical morphology of γ′ precipitates

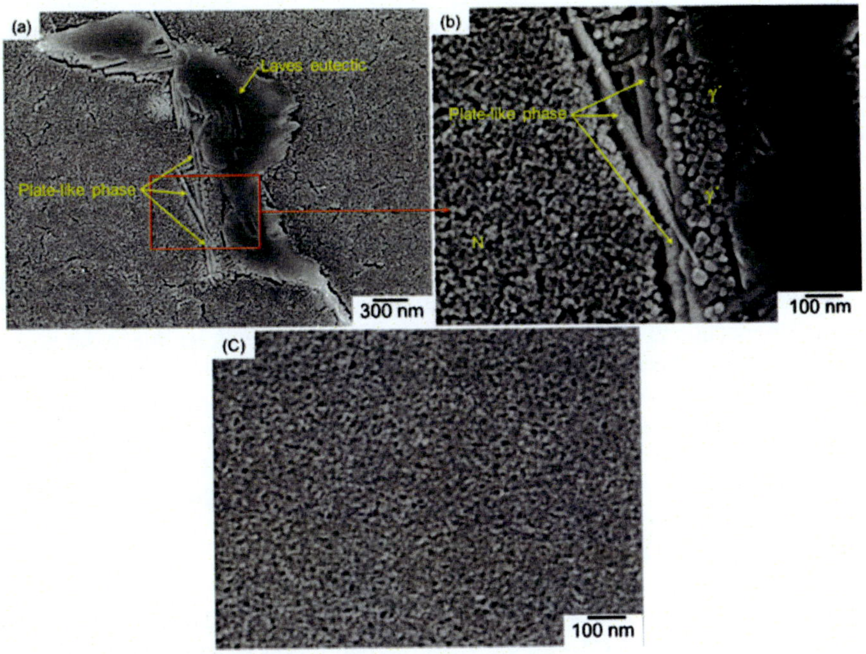

Fig. 9. **a** SEM in-lens image showing the presence of plate-like phase around the γ/Laves eutectic in the interdendritic region of the etched K-TIG welded Alloy G27 sample, **b** higher magnification image of an area marked red in **a** showing the presence of γ′ in the γ phase immediately adjacent to the Laves particle; **c** high magnification image of primary dendrite core

indicates that γ′ was likely to form via solid-state precipitation during the weld cooling, not through a solidification reaction. The γ′ phase that precipitates as a solidification product, i.e., γ-γ′ eutectics, commonly exhibits an irregular shape as reported in other precipitation-hardening superalloys [12, 22, 23]. The existence of γ′ precipitates is not clearly resolvable from the SEM image in the interdendritic region away from Laves particle, as shown in an area marked N in Fig. 9b and the dendrite core (Fig. 9c). This observation indicates that γ′ particles did not form uniformly in the FZ.

In other Nb-bearing Ni-based superalloys such as Alloy 718 and ATI 718Plus, Nb-rich phases such as MC-type carbide and Laves phase in the interdendritic region are in general well-known to form through eutectic reactions during the weld solidification process owing to elemental microsegregation [16, 18, 20, 24, 25]. The formation of both microconstituents in Alloy G27 might follow a similar solidification reaction sequence as those reported in Alloy 718 and ATI 718Plus. DuPont et al. [25] reported that the general solidification reaction sequence of Nb-bearing Ni-based superalloys with a high Nb concentration (>2 wt.%) and low C concentration (<0.075 wt.%) could be described by a three-step process. Recall that Nb and C concentrations in Alloy G27 are 3.7 and 0.028 wt.%, respectively. Firstly, the solidification is started by forming primary γ dendrites through a reaction of L → γ, and during the growth of the dendrite, solute elements with partition coefficients less than unity are rejected to the interdendritic liquid. Once the solubility limit of these solute elements in the interdendritic liquid is surpassed, the secondary solidification microconstituents start to form. During the primary dendrite growth, strong MC-type carbide forming elements, i.e., Nb and Ti, enrich the interdendritic liquid continuously, which may eventually lead to the formation of MC-type carbides. This is most likely to occur through a univariant eutectic-type reaction of L → γ + MC over a range of temperatures [26]. During further cooling, the residual interdendritic liquid would continue to get enriched by Nb, which may ultimately lead to the Laves eutectics formation through a reaction of L → γ + Laves over a range of temperatures due to the supersaturation of Nb. ThermoCalc simulation using the TTNI8 database (Fig. 10) shows that the liquid is enriched with Nb and Ti atoms as the solidification progressed in Alloy G27. This suggests that Nb and Ti segregate into the liquid during the weld solidification, which ultimately may result in the formation of Nb-rich MC carbide and Laves phase via eutectic reactions as previously described. Meanwhile, it is still unclear whether the plate-like phase around the Laves particles in FZ of Alloy G27 was formed through solidification or precipitation reaction.

The existence of intermetallic constituent, i.e., Laves phase, in the FZ of Alloy G27 needs careful consideration. In Alloy 718, Schirra et al. [27] found that Laves phase has a weak interface with the γ matrix, and it acts as a preferential site for easy crack initiation and propagation due to its inherent brittle nature. In addition, the formation of Laves phase would also cause the matrix a depletion of principal alloying elements required for the strengthening. It was reported in as-welded ATI 718Plus by Vishwakarma et al. [24] that precipitation of the main strengthening phase γ′, which contains a substantial amount of Nb (approximately 14 wt.%), in FZ was retarded due to the Laves phase formation. Regarding the plate-like phase, their existence in the interdendritic region may also potentially degrade the mechanical properties

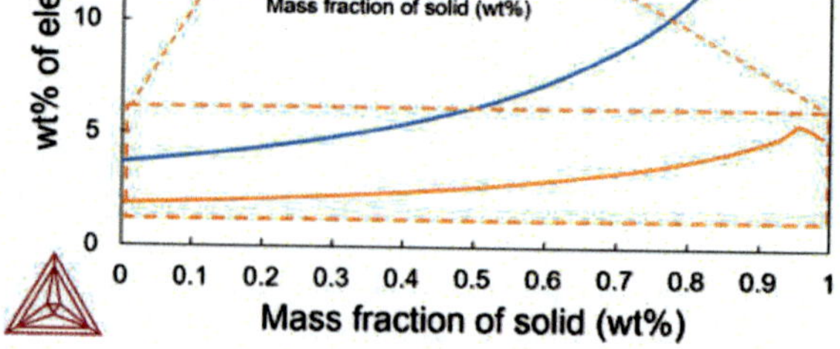

Fig. 10 ThermoCalc simulation showing the enrichment of Nb and Ti in the liquid during the weld solidification of Alloy G27

of the weldment, especially on the notch sensitivity and intergranular fatigue crack resistance [21]. Recall that γ′ phase does not form homogeneously in the FZ of K-TIG welded Alloy G27. Inhomogeneous distribution of γ′ may negatively influence the mechanical properties at high temperatures [28]. Thus, a suitable PWHT is needed to eliminate Laves and plate-like phases and produce uniform precipitation of γ′ particles.

HAZ Microstructure

The HAZ microstructure of welded Alloy G27 immediately adjacent to the fusion boundary, as shown in Fig. 11a, consists of a partially melted zone (PMZ) with numerous liquated and resolidified regions in the matrix and grain boundary. PMZ has been described as the region outside the FZ where liquation happens during welding due to the heating that exceeds the alloy's equilibrium solidus temperature.

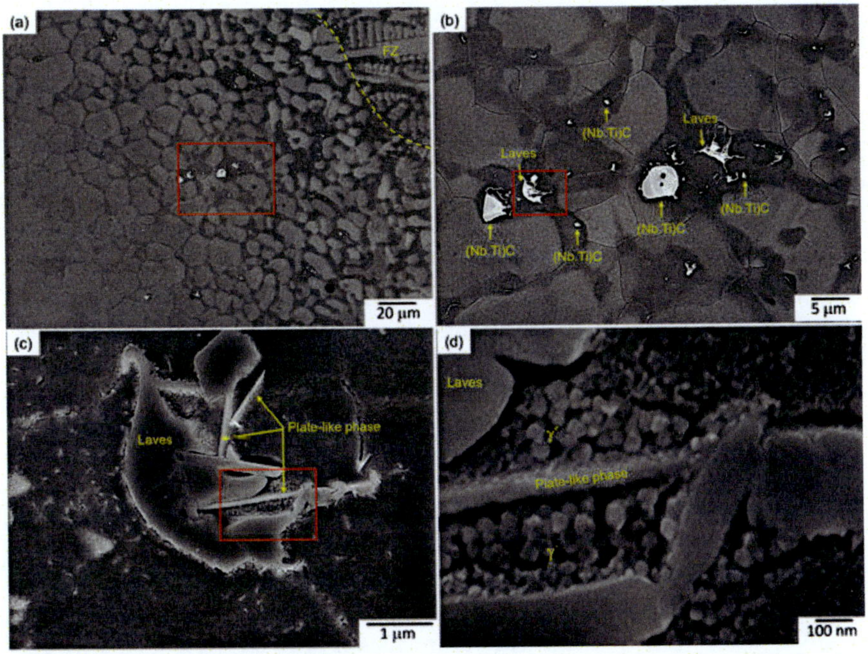

Fig. 11. **a** Low magnification SEM-BSE image of HAZ showing PMZ region and **b** (Nb, Ti)C particles and γ/Laves eutectic in the resolidified areas in PMZ, **c** higher magnification SEM in-lens image of a region marked in **b** depicting Laves phase particle and plate-like phase surrounding it; **d** higher magnification SEM in-lens image of a region marked in **c** showing the presence of γ′ in the γ phase immediately adjacent to the Laves particle

Microconstituents imaged in bright contrast (Fig. 11b) are also present in the liquated and resolidified regions. Based on the SEM–EDS analysis, the first microconstituent is found to be rich in Nb and contains a small amount of Ti, while the other one is found to be enriched in Nb, Mo, Ti, Ni, Fe, and Cr, suggesting that they appear to be (Nb, Ti)C and Laves phase, respectively, that are formed as resolidified products from the former PMZ liquid. Moreover, high magnification images (Figs. 11c and d) reveal that plate-like phase and γ' precipitates are present around the Laves phase. Nano-sized fine particles surrounding the Laves phase are also found. These features are similar to the ones observed in FZ (cf. Figure 9a and b). Therefore, it is possible that the solidification behavior of PMZ liquid, to some extent, is likely to follow a similar pattern to the one observed in FZ. The initial solid to form from the PMZ liquid would be γ solid solution followed by γ/(Nb, Ti)C eutectic and γ/Laves eutectic reactions, as observed in the HAZ/PMZ microstructure. As in the FZ, the formation of a plate-like phase around the Laves particle in HAZ is also still unclear, whether through solidification or precipitation reaction.

Conclusions

The influence of K-TIG welding parameters on the weld geometry and defects of a new Co-lean nickel-based superalloy G27 was studied using statistical DOE, and the microstructures of the HAZ and FZ of the K-TIG-welded Alloy G27 were characterized. The conclusions are

1. All the K-TIG welding DOE runs produced welds with Wm larger than 1.5 mm, excess weld metals, and pore diameter lower than the maximum allowable limit stipulated in AWS D17.1M:2010 for aerospace applications. However, only the sample welded using the lowest current and highest travel speed met the AWS D17.1M:2010 criteria in terms of face underfill. No cracks are found in the FZ and HAZ.
2. Wm and face underfill statistically were significantly influenced by travel speed and interaction current*travel speed. Root excess weld metal was only significantly influenced by travel speed. Face excess weld metal was significantly influenced by all the factors, including their interaction. On the other hand, all the factors, including their interaction, did not significantly influence the average pore diameter.
3. Microconstituents in the interdendritic regions of FZ of K-TIG welded Alloy G27 are identified as (Nb, Ti)C particles and γ/Laves eutectic constituents, suggesting microsegregation during the weld solidification. In addition, plate-like phase is also revealed around the Laves phase, and γ' precipitates are found to be inhomogeneously precipitated in the FZ.
4. Suitable PWHT is needed to eliminate Laves and plate-like phases and produce uniform γ' precipitation in the FZ.

5. In the HAZ immediately adjacent to the fusion boundary where equilibrium supersolidus melting takes place, i.e., PMZ, (Nb, Ti)C, and Laves phase with plate-like phase and γ' around it are found in the liquated and resolidified regions, suggesting the solidification behavior in PMZ is likely to follow a similar pattern to the one observed in FZ.

Acknowledgements This work is based on the research within the FEAST (Weld Feasibility Heat Treatment Studies of New Superalloys, registration number: 2019-02787) project funded by the Swedish funding agency VINNOVA. We highly appreciate the support and input from Dr. Ceena Joseph, Mr. Bengt Pettersson, Mr. Jimmy Johansson at GKN Aerospace Sweden AB, and Mr. Fredrik Olofsson at Brogren Industries AB. We also appreciate the support from Carpenter Technology Corporation in providing the material. Special acknowledgment to Mattias Igerstand from University West for performing the K-TIG welding operations.

References

1. Heck KA, Zhou N, Kernion SJ, Rickert D, Van Weereld F (2020) A New Co-free Ni-Based Alloy for Gas Turbine and Exhaust Valve Applications. In: Tin S, Clews J, Feng Qi, O'Brien C, Hardy M, Cormier J, Marcin J, Suzuki A (eds) Superalloys 2020 (TMS), p 142–152
2. Andersson J (2011) Weldability of precipitation hardening superalloys–influence of microstructure. Ph.D. thesis, Chalmers University of Technology
3. Andersson J (2018) Review of Weldability of Precipitation Hardening Ni-and Fe-Ni-Based Superalloys. In: Ott E, Liu X, Andersson J, Bi Z, Bockenstedt K, Dempster I, Groh J, Heck K, Jablonski P, Kaplan M, Nagahama D, Sudbrack C (eds) Superalloy 718 and Derivatives: Energy, Aerospace, and Industrial Applications. The Minerals, Metals & Materials Society, Pittsburgh; Springer, New York, pp 899–916
4. Cui S, Liu Z, Fang Y, Luo Z, Manladan SM, Yi S (2017) Keyhole process in K-TIG welding on 4 mm thick 304 stainless steel. J. Mater. Process. Technol. 243:217–228
5. Liu Z, Fang Y, Cui S, Luo Z, Liu W, Liu Z, Jiang Q, Yi S (2016) Stable keyhole welding process with K-TIG. J. Mater. Process. Technol. 238:65–72
6. Fei Z, Pan Z, Cuiuri D, Li H, Wu B, Ding D, Su L, Gazder AA (2018) Investigation into the viability of K-TIG for joining armour grade quenched and tempered steel. J. Manuf. Process. 32:482–93
7. Kuzmikova L, Larkin N, Pan Z, Callaghan M, Li H, Norrish J (2012) Investigation into feasibility of hybrid laser-GMAW process for welding high strength quenched and tempered steel. Australas. Weld. J. 57:1–9
8. Fei Z, Pan Z, Cuiuri D, Li H, Van Duin S, Yu Z (2019) Microstructural characterization and mechanical properties of K-TIG welded SAF2205/AISI316L dissimilar joint. J. Manuf. Process. 45:340–355
9. Khan A., Hilton P., Blackburn J., Allen C (2012) Meeting weld quality criteria when laser welding Ni-based alloy 718. International Congress on Applications of Lasers & Electro-Optics, Laser Institute of America, pp 549–557
10. Eriksson L, Johansson E, Kettaneh-Wold N, Wikström C, Wold S (2000) Design of experiments. Principles and Applications. Learn ways AB
11. Alvarez P, Vázquez L, Ruiz N, Rodríguez P, Magaña A, Niklas A, Santos F (2019) Comparison of hot cracking susceptibility of TIG and laser beam welded alloy 718 by varestraint testing. Metals. 9:985
12. Abedi H, Ojo (2022) Microstructural Analysis of Fusion Zone in Gas Tungsten Arc-Welded Newly Developed Co-Based Superalloy. Metall Mater Trans A. 53:377–381

13. Hanning F, Khan AK, Andersson J, Ojo O (2020) Advanced microstructural characterization of cast ATI 718Plus®—effect of homogenization heat treatments on secondary phases and repair welding behaviour. WELD WORLD. 64:523–533
14. Ariaseta A, Kobayashi S, Takeyama M, Wang Y, Imano S (2020) Characterization of Recrystallization and Second-Phase Particles in Solution-Treated Additively Manufactured Alloy 718. Metall Mater Trans A. 51:973–981
15. Otsuka T, Wakabayashi H, Igarashi D, Ariaseta A, Imano S, Wang Y, Kobayashi S, Takeyama M (2018) Effect of Homogenization on Creep of Additive Manufactured Alloy 718. In: Ott E, Liu X, Andersson J, Bi Z, Bockenstedt K, Dempster I, Groh J, Heck K, Jablonski P, Kaplan M, Nagahama D, Sudbrack C (eds) Superalloy 718 and Derivatives: Energy, Aerospace, and Industrial Applications. The Minerals, Metals & Materials Society, Pittsburgh; Springer, New York, pp 501–513
16. Cieslak M, Knorovsky G, Headley T, Romig Jr A (1989) The solidification metallurgy of alloy 718 and other Nb-containing superalloys. In: Loria EA (ed) Superalloy 718-Metallurgy and Applications. The Minerals, Metals & Materials Society, Pittsburgh, pp 59–68
17. Andersson J, Sjöberg G, Viskari L, Brederholm A, Hänninen H, Knee C (2008) Hot cracking of Allvac 718Plus, alloy 718 and Waspaloy at varestraint testing. In: Jahazi M, Elboujdaini M, Patnaik P (eds) 4th International Symposium on Aerospace Materials and Manufacturing Processes: Advances in Processing and Repair of Aerospace Materials. Canadian Institute of Mining, Metallurgy and Petroleum, Montreal, pp 401–413
18. Knorovsky G, Cieslak M, Headley T, Romig A, Hammetter W (1989) Inconel 718: a solidification diagram. Metall Trans A. 20:2149–2158
19. Singh S, Andersson J (2019) Varestraint weldability testing of cast ATI® 718Plus™—a comparison to cast Alloy 718. WELD WORLD. 63:389–939
20. Asala G, Andersson J, Ojo OA (2017) Precipitation behavior of γ' precipitates in the fusion zone of TIG welded ATI 718Plus®. J. Adv. Manuf. Technol. 87:2721–2729
21. Asala G, Khan A, Andersson J, Ojo O (2017= Microstructural analyses of ATI 718Plus® produced by wire-ARC additive manufacturing process. Metall Mater Trans A. 48:4211–4228
22. Ojo O, Richards N, Chaturvedi M (2006) Study of the fusion zone and heat-affected zone microstructures in tungsten inert gas-welded INCONEL 738LC superalloy. Metall Mater Trans A. 37:421–433
23. Ojo O, Richards N, Chaturvedi M (2004) Microstructural study of weld fusion zone of TIG welded IN 738LC nickel-based superalloy. Scr. Mater. 51:683–688
24. Vishwakarma K, Richards N, Chaturvedi M (2008) Microstructural analysis of fusion and heat affected zones in electron beam welded ALLVAC® 718PLUS™ superalloy. Mater. Sci. Eng. A. 480:517–528
25. DuPont JN, Notis M, Marder A, Robino C, Michael J (1998) Solidification of Nb-bearing superalloys: Part I. Reaction sequences. Metall Mater Trans A. 29:2785–2796
26. Radhakrishnan B, Thompson R (1989) Solidification of the nickel-base superalloy 718: a phase diagram approach. Metall Trans A. 20:2866–2868
27. Schirra JJ, Caless RH, Hatala RW (1991) The effect of Laves phase on the mechanical properties of wrought and cast+ HIP Inconel 718. Superalloys. In: Loria EA (ed) Superalloys 718, 625 and Various Derivatives. The Minerals, Metals & Materials Society, Pittsburgh, pp 375–388
28. Strum M, Summers L, Morris Jr J (1983) The aging response of a welded iron-based superalloy. Weld. J. 62:235–242

Author Index

E. A. Ott et al. (eds.), *Proceedings of the 10th International Symposium on Superalloy 718 and Derivatives*, The Minerals, Metals & Materials Series,
https://doi.org/10.1007/978-3-031-27447-3

Subject Index

E. A. Ott et al. (eds.), *Proceedings of the 10th International Symposium on Superalloy 718 and Derivatives*, The Minerals, Metals & Materials Series,
https://doi.org/10.1007/978-3-031-27447-3

S

T